Oxidative Stress in Applied Basic Research and Clinical Practice

Series Editor
Donald Armstrong

For further volumes:
http://www.springer.com/series/8145

Ashok Agarwal · R. John Aitken
Juan G. Alvarez
Editors

Studies on Men's Health and Fertility

Humana Press

Editors
Ashok Agarwal, PhD
Center for Reproductive Medicine
Cleveland Clinic
Lerner College of Medicine
Cleveland, OH, USA

R. John Aitken, PhD, ScD
Department of Biological Sciences
University of Newcastle
Callaghan, NSW, Australia

Juan G. Alvarez, MD, PhD
Department of Male Infertility
Centro Androgen, La Coruña
La Coruña, Spain

ISBN 978-1-61779-775-0 e-ISBN 978-1-61779-776-7
DOI 10.1007/978-1-61779-776-7
Springer New York Dordrecht Heidelberg London

Library of Congress Control Number: 2012933783

Printed on acid-free paper

Humana Press is part of Springer Science+Business Media (www.springer.com)

We dedicate this book to the late
Professor Thaddeus Mann FRS,
University of Cambridge, and
Professor Bayard Storey,
Emeritus Professor of Obstetrics and
Gynecology, University of Pennsylvania,
who pioneered our understanding
of reactive oxygen species and
oxidative stress in the control
of mammalian sperm function.

Foreword

> Oxidative stress is a universal phenomenon of aerobic life, you cannot escape it, nor should you wish to [1].

In the early days of research in the field, oxygen radicals and other "reactive oxygen/nitrogen species" (RONS) were universally thought of as deleterious molecules that must be eliminated at all costs by high levels of endogenous or exogenous antioxidants. Indeed, at high levels they *are* deleterious, e.g. to spermatozoa, other parts of the reproductive system and indeed to all cells and tissues. Sperm must be protected by their own antioxidants and by those in the bodily secretions surrounding them. Yet we now realise that RONS play key physiological roles, helping us to adapt to stress, defending us against infection and regulating physiological/pathological processes such as signal transduction and the intensity of inflammation [2–5].

The reproductive system is a beautiful example of all these principles. RONS at the correct level help to modulate uterine function, ovulation, the progress (or failure) of pregnancy and the behaviour of sperm, e.g. in response to inflammation in the surrounding tissues or even to electromagnetic radiation. Sperm generate reactive oxygen species (ROS) in mitochondria and by NADPH oxidase enzymes and these ROS may regulate sperm function (e.g. capacitation). Yet ROS can also damage sperm, e.g. during storage or handling procedures for in vitro fertilisation or during thermal stress. The highly polyunsaturated sperm lipids are a particular target.

This book "Studies on Men's Health and Fertility" is therefore extremely timely. Edited by three experts who have contributed enormously to the field, Ashok Agarwal, Juan Alvarez and Robert John Aitken, it examines all aspects of the roles of RONS in male fertility/infertility and semen quality; as well as their role in other conditions such as testicular torsion, variococele and erectile dysfunction. Each chapter is well written, carefully edited and appropriately referenced. I learned a great deal from reading this book, and I am sure that you the reader will do so as well.

I recommend it strongly.

Singapore — Barry Halliwell

References

1. Halliwell B. Reactive species and antioxidants. Redox biology is a fundamental theme of aerobic life. Plant Physiol. 2006;141:312–22.
2. Woo Ha, Yim SH, Shin DH, Kang D, Yu DY, Rhee SG. Inactivation of peroxiredoxin I by phosphorylation allows localized H_2O_2 accumulation for cell signaling. Cell. 2010; 140:517–28.
3. Halliwell B. Free radicals and antioxidants: quo vadis? Trends Pharmacol Sci. 2011; 32:125–30.
4. Hultqvist M, Olsson LM, Gelderman KA, Holmdahl R. The protective role of ROS in autoimmune disease. Trends Immunol. 2009;30:201–8.
5. Lee K, Won HY, Bae MA, Hong JH, Hwang ES. Spontaneous and aging-dependent development of arthritis in NADPH oxidase 2 deficiency through altered differentiation of CD11b+ and Th/Treg cells. Proc Natl Acad Sci USA. 2011;108:9548–53.

Contents

Contributors

Ashok Agarwal, PhD Center for Reproductive Medicine, Cleveland Clinic, Lerner College of Medicine, Cleveland, OH, USA

R. John Aitken, PhD, ScD Department of Biological Sciences, University of Newcastle, Callaghan, NSW, Australia

Naif Al-hathal, MD Department of Surgery, St. Mary's Hospital, Montréal, QC, Canada

Juan G. Alvarez, MD, PhD Department of Male Infertility, Centro Androgen, La Coruña, La Coruña, Spain

Mark A. Baker, PhD Department of Environmental and Life Science, University of Newcastle, Callaghan, NSW, Australia

David Benjamin, BA Center for Reproductive Medicine, Cleveland Clinic, Cleveland, OH, USA

Arthur L. Burnett, MD Department of Urology, Johns Hopkins Hospital, Baltimore, MD, USA

Aldo E. Calogero, MD Department of Internal Medicine and Systemic Diseases, Policlinico "G. Rodolico", Catania, Italy

Anna Czernikiewicz, MSc Institute of Human Genetics, Polish Academy of Sciences, Department of Reproductive Biology and Stem Cells, Poznan, Poland

Geoffry N. De Iuliis, PhD, BSc (Hons) Department of Biological Sciences, University of Newcastle, Life Sciences, University Drive, Callaghan, NSW, Australia

Dikmen Dokmeci, MD, PhD Department of Pharmacology, Faculty of Medicine, Trakya University, Edirne, Turkey

Joël R. Drevet, PhD GReD Laboratory, CNRS UMR 6293, INSERM U1013, Clermont Université, Aubiere, France

Stefan S. du Plessis, PhD, MBA Division of Medical Physiology, Stellenbosch University, Tygerberg, South Africa

Constanze Fischer-Hammadeh, MD Department of Obstetrics and Gynecology, University of Saarland, Homburg, Saarland, Germany

Margot Flint, MSc Division of Medical Physiology, Stellenbosch University, Tygerberg, South Africa

Monika Fraczek, PhD Institute of Human Genetics, Polish Academy of Sciences, Department of Reproductive Biology and Stem Cells, Poznan, Poland

Sheila R. Gokul Center for Reproductive Medicine, Cleveland Clinic, Cleveland, OH, USA

Jaime Gosalvez, BSc, PhD Department of Biology, University Autónoma Madrid, Madrid, Spain

Mohammed Hamad, PhD Department of Pharmacology and Bioscience, Petra University, Amman, Jordan

Mohamad Eid Hammadeh, DVM, BSc, PhD Department of Obstetrics and Gynecology, University of Saarland, Homburg, Saarland, Germany

Ralf Henkel, PhD Department of Medical Bioscience, University of the Western Cape, Bellville, Western Cape, South Africa

Anthony H. Kashou, BS Center For Reproductive Medicine, Cleveland Clinic, Cleveland, OH, USA

Fariba Khosravi, MSc ReproMed, Department of Andrology, Toronto, ON, Canada

Bruce V. King, PhD, Elect Eng (Hons), BSc Department of Physics, School of Mathematical and Physical Sciences, University of Newcastle, Callaghan, NSW, Australia

Adam John Koppers, PhD Department of Anatomy and Developmental Biology, Monash University, Clayton, VIC, Australia

Maciej Kurpisz, MD, PhD Institute of Human Genetics, Polish Academy of Sciences, Department of Reproductive Biology and Stem Cells, Poznan, Poland

Eve de Lamirande, PhD Urology Research Laboratory, McGill University Health Center, Royal Victoria Hospital, Montréal, QC, Canada

Sandro La Vignera, MD Department of Internal Medicine and Systemic Diseases, Policlinico "G. Rodolico", Catania, Italy

Stuart A. Meyers, DVM, PhD School of Veterinary Medicine, Anatomy, Physiology, and Cell Biology, University of California, Davis, CA, USA

Arozia Moazzam, MD Center for Reproductive Medicine, Cleveland Clinic, Cleveland, OH, USA

Biljana Musicki, PhD Department of Urology, Johns Hopkins Hospital, Baltimore, MD, USA

Cristian O'Flaherty, DVM, PhD Urology Research Laboratory, McGill University Health Center, Royal Victoria Hospital, Montréal, QC, Canada

Eleonora Bedin Pasqualotto, MD, PhD Department of Gynecology, University of Caxias do Sul, Caxias do Sul, RS, Brazil

Fabio Firmbach Pasqualotto, MD, PhD Department of Urology, University of Caxias do Sul, Caxias do Sul, RS, Brazil

Catriona Paul, PhD Departments of Pharmacology and Therapeutics and of Obstetrics and Gynecology, McGill University and the MUHC-RI, Montréal, QC, Canada

Khaled Refaat, MD Department of Obstetrics and Gynecology, Al Azhar University, Assuit, Egypt

Bernard Robaire, PhD Departments of Pharmacology and Therapeutics and of Obstetrics and Gynecology, McGill University and the MUHC-RI, Montréal, QC, Canada

Tamer M. Said, MD, PhD, HCLD, (ABB) Andrology Laboratory and Reproductive Tissue Bank, The Toronto Institute for Reproductive Medicine, Toronto, ON, Canada

Murat Savas, MD Department of Urology, Harran University Medical School, Arastirma Hospital, Sanliurfa, Turkey

Lucky H. Sekhon, MD, BSc OBGYN, Mount Sinai School of Medicine, Medical Center, New York, NY, USA

Johanna Selvaratnam, BSc, MSc Departments of Pharmacology and Therapeutics and of Obstetrics and Gynecology, McGill University and the MUHC-RI, Montréal, QC, Canada

Rakesh K. Sharma, PhD Center For Reproductive Medicine, Cleveland Clinic, Cleveland, OH, USA

Koji Shiraishi, MD, PhD Department of Urology, Yamaguchi University, Ube, Yamaguchi, Japan

Kelton Tremellen, MBBS, (Hons), PhD, FRANZCOG, CREI
School of Pharmacy and Medical Science, University of South Australia, Adelaide, SA, Australia

Repromed, Dulwich, SA, Australia

Enzo Vicari, MD Department of Internal Medicine and Systemic Diseases, Policlinico "G. Rodolico", Catania, Italy

Armand Zini, MD Department of Surgery, St. Mary's Hospital, Montréal, QC, Canada

Part I
Basic Research

Chapter 1
Electromagnetic Radiation and Oxidative Stress in the Male Germ Line

Geoffry N. De Iuliis, Bruce V. King, and R. John Aitken

Abstract The beneficial impacts of mobile-based communications on society are considerable. Health concerns over the broadcast of radio frequency electromagnetic waves, which carry the information for this medium, are now gaining momentum but are not without its controversies. Studies in the past that aim to determine whether concerns are warranted are sometimes lacking in impact because of poor understanding of radiation science. Nevertheless, the studies completed to date are important in developing the field toward the goal of confirming or disproving claims that radio frequency electromagnetic radiation (RF-EMR) is a serious health issue. We focus on what has been achieved to date, toward determining the effects of RF-EMR on the male reproductive system and information presented which may underpin the potential mechanisms at play. We suggest that oxidative stress may have a key role in the detrimental effects observed in the human spermatozoon and that this cell type may be a unique model to determine the potential mechanism of action given its sensitivities to such stressors.

Keywords Electromagnetic radiation • Oxidative stress • Male germ line • Sperm oxidative stress • DNA damage

---

G.N. De Iuliis, PhD, Bsc (Hons)(✉)
Department of Biological Sciences, University of Newcastle, Life Sciences, University Drive, Callaghan, NSW 2308, Australia
e-mail: geoffry.deiuliis@newcastle.edu.au

B.V. King, PhD, Elect Eng (Hons), BSc
Department of Physics, School of Mathematical and Physical Sciences, University of Newcastle, Callaghan, NSW 2308, Australia

R.J. Aitken, PhD, ScD
Department of Biological Sciences, University of Newcastle, Callaghan, NSW 2308, Australia

A. Agarwal et al. (eds.), *Studies on Men's Health and Fertility*, Oxidative Stress in Applied Basic Research and Clinical Practice, DOI 10.1007/978-1-61779-776-7_1,

1.1 Introduction

This chapter will aim to outline the basic principles of electromagnetic radiation (EMR) and what is known about their interactions with biological systems, and then will discuss some early and the most recent findings of EMR and mobile phone exposure and male fertility. The main intention will be to work towards shedding light on the potential mechanisms of action of this radiation on spermatozoa focusing on oxidative stress as the mediator.

In recent times, there has been some controversy over the impact of the physical factor, radio frequency electromagnetic radiation (RF-EMR) broadly on human health. Several studies have found an association between human health and exposure to RF-EMR, with emphasis on a range of clinical conditions including childhood leukaemia, brain tumours, genotoxicity and neurodegenerative disease [1, 2]. One such controversial area surrounds studies that indicate elevated risk of brain tumours after 10 years of mobile phone use [3]. However, for every one of these studies, there seems to be another refuting the claims [4]. Nevertheless, these studies are important in developing the field toward the goal of confirming or disproving claims that RF-EMR is a serious health issue. Although work which focuses on the rest of the body aids our understanding of the purported phenomena, this chapter will only focus on work relating to reproduction. To date, the "real" clinical effects of RF-EMR on human health and reproduction are not proven and still controversial; however, if we obtain a basic understanding of the physics of EMR and experimental design together with our knowledge of male reproduction and sperm cell biology, we are well placed to take this field forward markedly.

1.2 EMR Defined

EMR is a form of radiation that ranges from extremely high-energy cosmic and gamma rays at frequencies above 10^{18} Hz down through the visible spectrum (frequencies near 10^{15} Hz) to the relatively low-energy microwave (10^{10} Hz or 10 GHz) and radio frequencies (10^{8} Hz or 100 MHz) (Fig. 1.1). The part of the spectrum used for mobile phone communications is in the frequency range from 800 MHz to 2.5 GHz, labelled Global System for Mobile Communications (GSM) in Fig. 1.1. EMR may be considered to comprise alternating electric, E, and magnetic, B, fields. The E and B fields both generate forces on charged particles in materials, but the forces due to the electric fields are normally much larger, except in magnetic materials. However, in the context of mobile phone exposure, the magnetic component of the radiation may be more significant due to its considerable penetrative ability inside not only in human body, but also in buildings.

As will be seen in the discussion below on mobile phone communications, the EM wave may contain components oscillating at a range of different frequencies. This is the case with modulated EM waves where the amplitude or frequency of

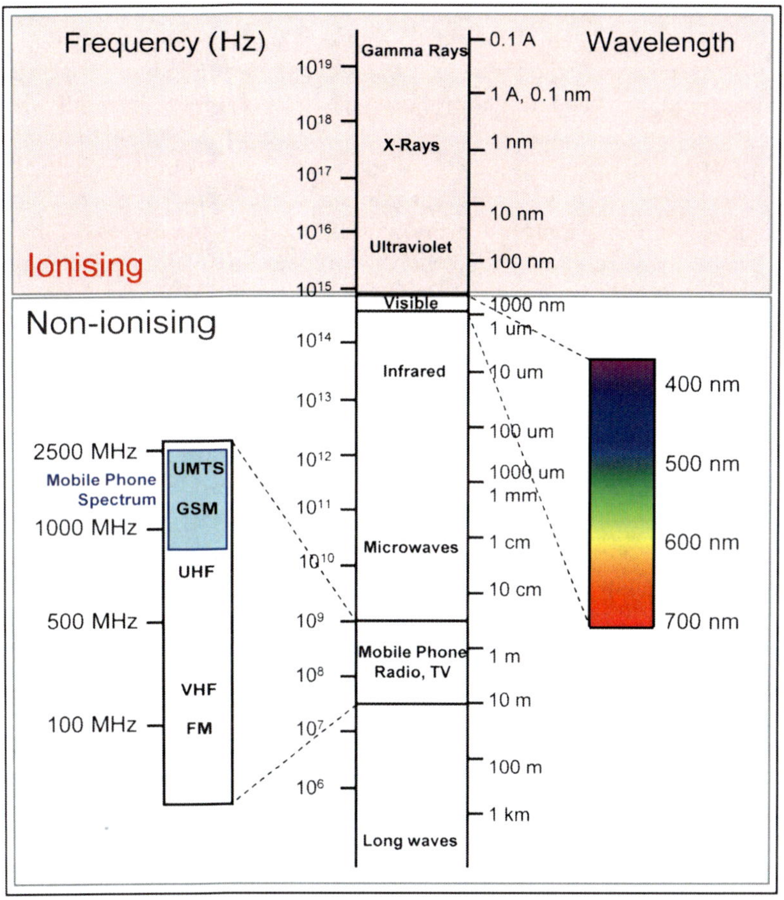

Fig. 1.1 Electromagnetic spectrum. The *blue box* indicates the mobile phone frequency spectrum which begins to enter the microwave spectrum. The energy of mobile phone frequency radiation is far lower than that of ionising radiation (adapted from Physics Central (http://www.physicscentral.com/experiment/askaphysicist/physics-answer.cfm?uid=20110119110703))

the so-called carrier wave is varied in time, in order to carry information. The modulated EM wave contains a component oscillating with the carrier frequency f as well as sideband frequencies $f \pm \Delta f$. The challenge for communications engineers is to maximise the information carried while minimising the frequency spectrum, $2\Delta f$, used.

Most of the radiation studies related to mobile phone communications are done at frequencies used by the GSM phone network. GSM is a digital standard first offered commercially in 1991 and is currently the world's most popular standard for mobile telephony systems, with over 80% of the global mobile market using the standard. GSM networks operate in a number of different carrier frequency ranges with most 2G GSM networks operating in the 900 or 1,800 MHz bands. Where these bands were already allocated, the 850 and 1,900 MHz bands were used instead (e.g. in Canada and the United States). The more recent 3G networks operate in the 2,100 MHz frequency band in Europe. Although the technology is rapidly evolving, with the incorporation of data transmission and the progression to 4G protocols, the basics of EM radiation emission by GSM mobile phones are relatively unchanged. During a GSM call, speech is converted from analogue sound waves to digital data by the phone itself and transmitted through the mobile phone network by digital means. The EM wave emitted by the phone comprises a range of frequencies. As an example, for a GSM-900 phone, the frequency band 890–915 MHz is used for transmission from the mobile phone to the base station and the band 935–960 MHz from the base station to the mobile phone. In each band, there are 124 separate carrier frequencies spaced 200 kHz apart, starting in the above example at 890.2 MHz. Each 200 kHz frequency is segmented in time, so eight separate channels of information can be sent on each carrier. The digitally encoded information from the codec for all channels together is then used to modulate the frequency of the carrier at a digital rate of 270 kbit s^{-1}. An individual GSM-900 mobile phone will then generate an EM wave with a time-varying frequency within a 200 kHz band on a carrier frequency between 890 and 915 MHz. The intensity of the EMR will also vary in time, since the encoding of the eight separate channels occurs within a 4.615 ms period. So, to properly measure the effect of all mobile phone irradiation on biological systems, experiments should be conducted using pulsed radiation for a range of frequencies within the 850–2,100 MHz band.

The alternating electric, E, and magnetic, B, fields in the EMR interact with materials by exerting forces on charged particles, changing charge distributions in the material. In nonmagnetic materials, the E field causes polarisation (or separation) of bound charges, orientation of permanent dipoles (pairs of opposite charges) and movement of electrons and ions. The first two effects are taken into account by the permittivity, ε, which is a measure of how easily the polarisation of the material changes due to an electric field. Materials primarily affected by the first two processes are called dielectrics. The third effect, the movement of both electrons and positively and negatively charged ions, is accounted for by the conductivity, σ, and materials affected by the third process are known as conductors. The permittivity is typically expressed as a complex quantity

$$\varepsilon = \varepsilon_0(\varepsilon' - j\varepsilon'') = \varepsilon_0\varepsilon' - j(\sigma / \omega) \qquad (1.1)$$

where ε_0 is the permittivity of free space (8.85×10^{12} F m^{-1}), $\varepsilon_0\varepsilon'$ is the real part of the complex permittivity (termed the dielectric constant), $j = \sqrt{-1}$ and $\omega = 2\pi f$ is the angular frequency in radians per second. Both ε' and σ increase with increasing water content in the tissue being low for fat and high for blood. The variation of these parameters over the communication frequency range is not large. For testes, ε' is 58 and conductivity is 1.34 S m^{-1} at 900 MHz, whereas at 2.5 GHz, ε' and σ are 57.5 and 2.21 S m^{-1}, respectively [5]. Since ε' is high, tissue such as testicles may then be considered to be lossy dielectrics. With a lossy dielectric, the transmitted wave is attenuated as it travels into the material. Energy is transferred from the wave to the dielectric as kinetic energy of the charged particles in the dielectric. The loss is related to the average permittivity for biological tissue and depends on frequency, generally decreasing as the frequency increases since the charges cannot respond to rapid changes in high frequency fields. ε represents the conduction of ions as well as friction associated with the alignment of dipoles and vibrational and rotational motion in molecules. The depth of penetration of EM radiation, defined as the distance at which power absorption is approximately 14% of the surface value, is about 4 cm at 1 GHz and 2.5 cm at 2 GHz in tissue. Real world irradiations are, however, more complicated because scattering and refraction of EM waves at interfaces means that energy is deposited in a non-uniform manner into tissue. The energy absorbed from the wave is directly related to the internal E field at the point of absorption. But the incident and internal fields can be quite different depending on the size and shape of the body, its electrical properties, its orientation with respect to the field and its frequency of the EM radiation.

The power absorbed by the sample is related to the specific absorption rate (SAR) where "specific" indicates that the parameter is normalised with respect to mass. The SAR (in W kg^{-1}) is then the power absorbed per unit mass or

$$\text{SAR} = \int_{\text{sample}} \frac{\sigma(r)\,|E(r)|^2}{\rho(r)}\,\mathrm{d}r \tag{1.2}$$

where σ is the sample electrical conductivity (in S m^{-1}), ρ is the sample density (kg m^{-3}) and $|E(r)|^2$ is the square of the magnitude of the electric field, $E(r)$, at point r in the sample. The actual SAR delivered to a region of the body will depend heavily on the depth of the region below the skin, the electrical characteristics of the tissue between the skin and irradiated region and on the exact location and geometry of the RF source.

The US standard is that phones have a SAR level at or below 1.6 W kg^{-1} taken over a volume containing 1 g of tissue, whereas European standards require a SAR maximum of 2 W kg^{-1} averaged over 10 g of tissue. For tissue of density 10^3 kg m^{-3} and 1 Ωm, a SAR of 10 W kg^{-1} corresponds to a field of 100 V m^{-1} and B field of 0.3 μT. However, the actual SAR absorbed by tissue depends on its depth below the surface, the electrical characteristics of the tissue between the source and the target and the presence of external factors which may influence the EMR delivered to the skin. For example, most men put a mobile phone in a front trouser pocket [6]. A 1 W phone placed in the position of the front trouser pocket [7] generates SAR levels of 2 W kg^{-1} in the testes over the frequency range 0.9–4 GHz. This SAR rose to 4 W kg^{-1}, if the effect of metal objects, such as keys, in the pocket was included. Given that GSM-850–900 handsets can have a peak power level of 2 W, then peak SARs in the testes could reach to more than 10 W kg^{-1} under worst case scenarios. However, at typical phone power levels of 0.5 W, SARs would be a more realistic 2 W kg^{-1}.

EMR energy absorption increases the average energy level of random molecular excitation, resulting in a temperature rise. The power absorbed into a region of tissue will cause an initial increase in temperature ΔT in the time interval Δt given by the bioheat equation [8]

$$k\nabla^2 T - \rho^2 C m_b T + \rho \mathrm{SAR} = c\rho \frac{\partial T}{\partial t}$$

where T is the temperature above the mean arterial temperature, k is the thermal conductivity of the tissue (typically 0.5 W m^{-1} K^{-1}), C is the heat capacity of the tissue (typically 3,700 J K^{-1} kg^{-1}), ρ is the density of tissue and blood (typically 1.06 kg m^{-3}) and m_b is the volumetric perfusion rate of blood (typically 0.5–10×10^{-6} m^3 kg^{-1} s^{-1}). The first two terms in the above equation represent heat loss from the irradiated area by conduction through the tissue and by blood flow, respectively. The third term is the heat gain due to the irradiation and the fourth term is the temperature rise of the irradiated region with time. On commencement of an irradiation, there is minimal heat loss so the temperature increases linearly with time. For typical tissue, a SAR of 1 W kg^{-1} will cause a 1°C temperature increase per second. As the temperature of the irradiated region rises above the surroundings, heat energy will be transferred away from the irradiated region by thermal conduction through tissue and convective heat flow through the blood, reducing the rate of temperature increase. The sum effect of these channels for heat loss results in an effective thermal conductivity of approximately 10 W m^{-2} K^{-1}. Finally, the system will come to thermal equilibrium with the EM energy delivered to the irradiated region balanced by the heat energy leaving the region in any time interval. An irradiated spherical region of tissue of mass 10 g would then typically show an equilibrium temperature rise of about 1°C at a SAR of 2 W kg^{-1}. It is very well known that the heating of tissue will induce a stress response that is invariably damaging for the tissue involved. To study non-thermal effects of RF irradiation, the subject of this chapter, the equilibrium temperature increase should be kept below 0.1°C, requiring typical SARs of less than 0.2 W kg^{-1}.

1.3 Physical Models of the Interaction of Mobile Phone Radiation with Cells

At low power levels where thermal effects are unimportant, EMR still maintains the ability to affect cells. Radiation in the very high-energy gamma-ray frequency range, for example, can directly induce ionisation and lead to radical formation. While this has clear implications for biology, the energy associated with the visible region and down to the radio frequency is not sufficient to remove electrons from atomic or molecular orbitals, i.e. they are not ionising radiations. For example, radiofrequency EMR, at the gigahertz frequencies used in mobile phone communications, can be considered to be a stream of particles, or photons, with energies one million times less than the energies required to directly alter the chemistry of molecules.

Several hypotheses exist which may explain the interaction of RF-EMR with biology, including the male reproductive system and spermatozoa. The difficulty comes in identifying a physical mechanism by which RF radiation at low levels, 10^6 times lower than required to directly ionise molecules, can cause biochemical effects. Indeed, as summarised by Phillips et al. [9] "we are at the stage of having inconsistent results and no proven mechanism to explain RF-induced effects on DNA damage". This is in spite of the fact that there are structures in the body that are extremely sensitive to EMR. For example, a single photon of visible light can cause a change in rhodopsin causing a polarisation in a rod cell in the retina [1]. Different authors have surveyed the range of physical mechanisms which may result in RF-induced biochemical effects. Sheppard et al. [2] argue that possible mechanisms including coherence, resonance, signal averaging, field non-uniformity in inhomogeneous dielectric structures and nonlinear effects all produce effects far below the levels of the electric fields associated with normal bodily processes of wound healing and excitation of muscles and the nervous system. Challis [10] identified one possible non-thermal process—free radical generation in biomolecules with large hyperfine splittings and fast relaxation. In this process, if the EM frequency is resonant with the difference between energy levels in a molecule, then the energy from the external signal can be concentrated, leading to an amplified response at the driving frequency. However, the degree of amplification decreases with the losses in the system. Since most ions are associated with water, the energy dissipation by collisions of water molecules increases the loss of the system at RF frequencies and limits the degree of amplification which can be achieved in resonant systems.

Two other processes commonly used to justify RF effects on cells are the development of additional potentials across membranes which lead to a change in ion transport [11–15] and alteration of normal vibrations of molecular bonds, perhaps affecting proteins and the activities and interactions [16, 17] through alterations in protein conformation. An alteration of ion transport across cell membranes is possible, but only for fields of several hundreds of millivolts, much higher than the resting voltages across membranes, even of organelles such as the mitochondria. Modelling of transient voltages across organelle structures show that if the organelle membrane is thicker than the cell membrane and the organelle contains a high ion concentration, it is possible at RF frequencies for the voltage across the organelle membrane to be larger than that across the cell membrane [15]. Indeed, the change in voltage is of the order of ER [18], where E is external field and R is cell radius in the direction parallel to the field—changes may give rise to 100 mV changes in membrane potential. However, if a voltage is applied across a tissue, most of the voltage drop appears across the membrane at low frequencies. At gigahertz frequencies, the capacitance of the membrane effectively shorts out the membrane resistance at gigahertz frequencies, so the above effects are unlikely in that frequency range.

Cotgreave [19] argues that cellular proteins have different structures and would be expected to behave differently when exposed to RF. In addition, many proteins are in electrostatic contact, so RF-EMR may affect organisation of proteins within the cell. Studies have shown denaturation, aggregation and stability of proteins are affected by RF exposure [20, 21]. Indeed, the efficiency of a protein as an enzyme depends on its conformation. Several side chains of amino acids in proteins are polar and so will behave differently in EM fields. Experiments at high irradiation levels (3 h of 1.95 GHz irradiation at 51 $W kg^{-1}$) have been shown to affect protein folding [22]. So a possible physical model suggests that resonances with charges and dipoles in each protein conformation change the barrier heights for intermediate processes in the refolding pathways [22]. However, dissipative effects discussed previously would reduce the effect of resonant excitation of dynamic excitations, making it difficult to imagine a mechanism at gigahertz frequencies.

In summary, even though several hypotheses exist for effects of non-ionising, mobile phone range EMR on biology, there is no clear proven mechanism. This is a key point to which much of the debate about the reported detrimental effects of EMR are based. Nonetheless, because there is no known mechanism at this point does not prove there is no effect. With this unknown and the many conflicting reports in the literature over the past decade, this field remains controversial.

1.4 Sperm Oxidative Stress and DNA Damage

From what is currently known about sperm cell damage and dysfunction, there is certainly scope for RF-EMR to contribute to this affliction. The three main types of cellular damage which can account for the adverse observations made are membrane, protein and DNA damage. The factors responsible for contributing to this damage in the male germ line can be grouped into several categories. Quality of spermatogenesis defines the susceptibility of cells to damage [23]. Biological factors including diet and stress as well as chemical and physical factors may all have direct and/or indirect effects on the "health" of spermatozoa. These categories are not necessarily mutually exclusive and combination between them is most probable. The complex, multifactorial nature of male infertility makes it a challenging area of research; however, cellular damage originating from environmental factors including the impacts of EMR on the male germ line must be understood if this issue is to be managed. The continued effort to determine the clinical significance of environmental-born EMR exposure on reproduction (as well as human health) and to gain an understanding of the mechanisms involved is urgently required as the exposure of RF-EMR to humans will only escalate.

There is growing evidence that environmental factors may be a key factor in male infertility [24]. Sperm concentration or microscopic analysis of sperm quality has dominated focus in the past, when studying xenobiotic or other environmental exposures; however, recently, more attention has been centred on the effects of sperm DNA integrity. There is a wealth of reports that link environmental exposures to sperm DNA damage and reduced fertility [25–27], with RF-EMR recently included. To put the risks in some perspective, evidence suggests the ability of DNA damaged cells to initiate fertilisation is somewhat compromised; however, it is not necessarily precluded from fertilising an oocyte. Indeed, DNA damage in the male germ line has been linked with a range of adverse clinical outcomes, including poor fertility rates in vitro, but also subsequent disruptions of embryonic development, increased rates of miscarriage and poor childhood health [27–29] including cancer. Therefore, it is crucial that we understand the clinical implications and mechanisms of RF-EMR if indeed it plays any part in elevating DNA damage in the male germ line.

Regardless of the recent attention sperm DNA damage has enjoyed, the aetiology of this damage remains unknown. While the cellular mechanisms underpinning these effects have not been completely resolved, it has been suggested that oxidative stress derived from numerous possible pathways could be a key factor [17, 30]. There are several reports that also strongly link this mode of action to the adverse effects observed after EMR exposure [31–34], strengthening the potential role of this environmental factor in this affliction. Oxidative stress has also been implicated in a range of other infertility pathologies, including loss of sperm motility and vitality, which is also a common observation after RF-EMR exposure. Failure of sperm–oocyte fusion is also another result of oxidative stress [35]. We now know that human spermatozoa are capable of generating significant amounts of ROS [36, 37], both spontaneously and when exposed to xenobiotic or physical environmental factors [38]. Furthermore, these highly specialised cells are intrinsically sensitive to ROS and may enter a state of oxidative stress [39] with little hindrance. Therefore, the induction of ROS by environmental factors may account for the majority of cellular damage and dysfunction observed in human spermatozoa. Mobile phone radiation has the potential to elevate ROS leading to a state of oxidative stress which in turn impacts sperm motility vitality and DNA integrity; however, the fundamental mechanism by which ROS is generated is unknown. Two main ideas that lead to a state of oxidative stress in spermatozoa by RF-EMR are the disruption of the sperm mitochondria [40] and the activation of plasma membrane NADH oxidases [41].

Many studies have reported the presence of reactive oxygen species within human spermatozoa and its consequences for the gamete [42–44]. From our understanding of the cell biology of human spermatozoa, it is perhaps no surprise that these cells are then susceptible to oxidative stress and DNA damage. The minute volume of cytoplasm in these highly specialised cells limits the antioxidant capacity usually afforded to other cell types. Once oxidative stress is initiated, fertilisation is compromised through the loss of motility and ability to fuse to the oocyte. These outcomes arise due to lipid peroxidation of the abundant redox-sensitive polyunsaturated fatty acids in the plasma membrane. These peroxides also have the capacity to further propagate oxidative stress by a lipid peroxidation cascade [45]. Oxidative stress also leads to a range of protein damage, including alkylation (by lipid peroxides products) and oxidation [46]. The sperm chromatin does not escape the negative effects of this stress where DNA damage, ranging from oxidation, adduct formation and strand breaks result [26, 39, 47]. Further strengthening the central role of oxidative stress in sperm damage, several studies have shown that supplementary antioxidants have had some protective role against damage induced by EMR [48–50]. The area of antioxidant treatment for male infertility is a rapidly growing one; nevertheless, it is largely driven by empirical data. The work in this area confirms the major importance of ROS and oxidative stress in spermatozoa exposed to RF-EMR;

however, in a broader sense more understanding of oxidant–antioxidant interactions are needed here before rational clinical applications can be established. Similarly, care must be taken when aligning the adverse observations made after RF-EMR or mobile phone exposure to clinical relevance. The biophysics of EMR is exceedingly complex, evident by some examples of poor experimental design and lack of insight offered to date. There have been few experiments done on spermatozoa and the evidence for low-level genotoxic effects is weak [51] at this current time. More appropriately designed studies need to be conducted if any headway is to be made in this field.

1.5 Studies on EMR and Male Infertility

Within the last decade, there have been several well-designed and executed studies on the effects of mobile phone radiation on spermatozoa, covering three main areas, clinical/epidemiological, human (in vitro) and in vivo animal models. The association between RF-EMR and male infertility was initially suggested by an epidemiological study in 2001 where it was speculated that differences in fertility parameters from Chinese males in various professions may be due to EMR exposures [52]. More pertinent epidemiological data from a study in 2005 found negative correlations between mobile phone usage and various attributes of semen quality, particularly motility [53]. Studies on male reproduction in mouse or rat models showing that mobile phone radiation affected testicular histology, including decreased seminiferous tubule diameter, appeared in 1999 [54], but wasn't investigated further until 2003 [55] and 2004 [56] with the effects of low frequency EMR on murine models. There were some conflicting data within these two latter studies; however, both indicated that testis histology was altered in the exposed group. The former study also showed a decrease in testis size, while the latter did not. The latter study did, however, show one of the first instances of DNA damage (fragmentation within spermatogonia only) in the male germ line after mobile phone exposure.

This work was immediately followed by an experimental study involving exposure of male mice to RF-EMR via a wave guide. Exposures were at a frequency of 900 MHz at 90 mW kg^{-1} for 12 h day^{-1} for 7 days. This study revealed a significant impact on the integrity of the sperm mitochondrial genome, but no effect on the nuclear DNA or microscopic parameters [57], somewhat confirming the detrimental effects of RF-EMR on DNA integrity. Then further negative impacts of mobile phone usage on semen quality in human males were observed in a study that found significant reductions in sperm motility after exposure to a mobile phone after only 5 min exposure (talk mode) at a 10 cm range in vitro [58]; shortly after, a study also reported losses of motility and vitality after mobile phone exposure for 6 h day^{-1} for 18 weeks on rats held 1 cm from mobile phones measured at an SAR ranging from 0.9 to 1.8 W kg^{-1} in standby and talk modes, respectively. The spermatozoa of animals exposed to mobile phones also exhibited an up-regulation of CAD-1 and ICAM-1 RNA levels [59] (proteins associated with cell adhesion).

In three elegant studies, some concerning links with mobile phone use and male infertility were presented. Fejes et al. [53] found over 371 men that the duration of possession and use of mobile phones negatively correlated with semen quality. Wdowiak et al. [60] similarly showed that lower motility, vitality and poor morphology correlated with the frequency of mobile phone use over the 304 men studied. Agarwal et al. [61] confirmed these findings showing that sperm cell motility, vitality, normal morphology as well as sperm counts were defective in men who used mobile phones more frequently [61]. Also within this earlier time frame of research, various researchers proposed that the RF-EMR also exerts a range of negative genotoxic effects on different cell types including mature sperm cells [57, 62, 63]. These effects include chromatid exchange, aneuploidy and defective chromosome recombination. The "real world" clinical significance of this body of work was not confirmed; however, these studies were greatly important in providing a platform for continuing high-quality research and focusing effort into the potential harmful effects of mobile phone use. The work in this field up until 2007 is reviewed by Deepinder et al. [64].

From 2008 to the present (2011), several additional studies have shown adverse effects; in a clinical setting, in model systems and studies on human spermatozoa in vitro. Importantly, some studies have begun to shed light on how RF-EMR may drive the adverse effects observed by some researchers in the past. Nevertheless, this latest period of research has also generated several studies showing no detrimental effects of RF-EMR or mobile phone use.

Despite several groups reporting no effects of RF-EMR on male reproduction, other groups with very similar experimental design have reported a common finding of lower motility [65]. Some groups also showed that signs of oxidative stress in the exposed cohort, where increases in markers such as 8-OH-dG [31] and lipid peroxidation and the reduction of antioxidant levels [65], were present in human sperm in vitro, as well as in murine models. One study reported an increase in testicular sperm count in the rat after 1.95 GHz cellular phone radiation with a SAR or 0.08–0.4 $W kg^{-1}$ after 5 $h day^{-1}$ for 5 weeks [66]. A further study in the rat using exposures of 90 min day^{-1}, 5 days $week^{-1}$ for 12 weeks and using 848.5 MHz frequency at an SAR of 2.0 $W kg^{-1}$ found no changes compared to controls in testis histology or several other markers including lipid peroxidation, expression of p53, bcl2 or caspase [67], again exemplifying the range of conflicting data.

A detailed human (in vitro) study was completed by our research group in 2009 [31]. Our aims were to uncover a chain of cause and effect from RF-EMR radiation to the resulting motility and vitality loss and increased levels of DNA damage. We completed this by exposing purified human spermatozoa in a waveguide in a power-dependant fashion (0.4–27.5 $W kg^{-1}$ at 1.8 GHz). In step with increasing SAR, motility and vitality were significantly reduced, while the mitochondrial generation of reactive oxygen species and DNA fragmentation were significantly elevated ($P<0.001$). Furthermore, we also observed highly significant relationships between SAR, mitochondrial ROS levels, the oxidative DNA damage bio-marker, 8-OH-dG, and DNA fragmentation after RF-EMR exposure. This study has identified that RF-EMR may interact with the mitochondria in the sperm mid-piece, which then

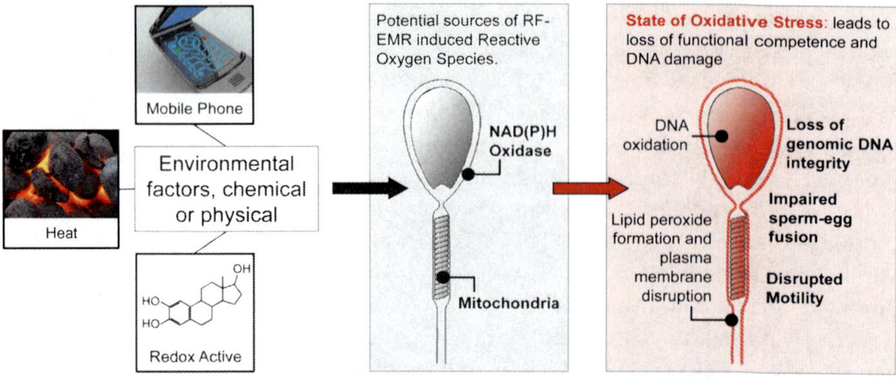

Fig. 1.2 Proposed oxidative stress model of the effects of mobile phone frequency radiation on the human spermatozoon. Evidence suggests that, like several other factors, radio frequency electromagnetic radiation (RF-EMR) can induce a non-thermal oxidative stress response in the gamete, possibly through interactions with NAD(P)H oxidases in the plasma membrane or by perturbation of mitochondria. This stress then leads to the range of adverse effects commonly observed under experimental conditions

leads to ROS generation and a state of oxidative stress. This stress manifests in a loss of motility and vitality (through lipid peroxidation) and the presence of oxidative DNA damage and DNA strand breaks in the nucleus. Our conclusion from this study was that RF-EMR in both the power density and frequency range of mobile phones enhances mitochondrial reactive oxygen species generation by human spermatozoa, decreasing the motility and vitality of these cells while stimulating DNA base adduct formation and, ultimately DNA fragmentation. This study shed light on a potential mechanism by which "real-life" mobile phone radiation may affect biology. These findings confirm other published data that RF-EMR can indeed impact the male germ line and further that extensive mobile phone use may have negative impacts on males of reproductive age, potentially affecting both their fertility and the health and wellbeing of their offspring. This work was then supported by Agarwal et al. [68], showing that human sperm in vitro suffered the same oxidative stress by and increase in ROS levels and a decrease in the total antioxidant capacity of the cells after exposure to a mobile phone for only 1 h. Whereas we hypothesis that the source of ROS is the sperm mitochondria, Agarwal et al. suggest that NADH oxidase on the plasma membrane is responsible. Nonetheless, there is growing confidence that ROS has a key role to play in the potential mechanism of adverse effects of RF-EMR on the male germ line (Fig. 1.2). The relationship of oxidative stress to the detrimental effects of mobile phone use is also reviewed in this recent article [69].

Certainly in the last 6 years, several papers have implicated ROS as the mediator for RF-EMR-based cellular damage [65]. This hypothesis is not confined to the germ line, where similar studies in other cell types also seem to conform to this proposed pathway [70, 71]. Several studies have demonstrated that antioxidant molecules such as melatonin, caffeic acid, phenyl ester, vitamin C and E have some protective or preventive effect against oxidative stress caused by RF-EMR [49, 50, 72]. More recent reports have also found similar results with other types of antioxidants including catechins (from green tea), *N*-acetylcysteine [73], again strengthening the hypothesised role of ROS in the effects of RF-EMR on the body.

Besides some conformity and progress of our knowledge in this field, we cannot ignore the controversy in many findings. The roots of this disparity may arise from two main factors, firstly, the intention of the study, and secondly, the experimental setup. There are a fraction of studies within the field which attempt to obtain mechanistic information and the molecular consequences of RF-EMR on spermatozoa and male reproduction; this focus, in part, removes itself from clinical relevance, however is key to establishing a causative effect and may therefore resolve some of these inconsistencies. Understanding its aetiology and resolving any potential adverse health effects of this radiation is a major priority. The second aspect, the experimental design, is particularly difficult to address at this point in time as the interactions between RF-EMR and biology are complex and generally not well understood. Further to this, as with various other human studies, clinical investigation into RF-EMR is extremely difficult to control.

1.6 Conclusions

Historically, the human spermatozoon has been commonly defined as being of poor quality and declining [74], in contrast to the majority of spermatozoa from other mammals. Together with the hypothesis of testicular dysgenesis syndrome (TDS), and its increasing prevalence in the recent past [75–77], the impacts of environmental factors, including those, whose effects are capable of inducing ROS in the male gamete, must be identified. Mobile phone usage is increasing worldwide at an astounding rate. The impressive development and use of mobile telecommunication services in the last decade have drastically increased the amount of RF-EMR exposure in our daily lives. With the ever-increasing uptake of this technology come concerns regarding the harmful effects of mobile phone exposure on human health, including reproductive systems. As part of its charter to protect public health, the World Health Organization in 1996 put into place the International EMF Project to assess the scientific evidence of possible health effects of EMR in the range of 30 Hz to 300 GHz [78]. Despite more than a decade of research in this field, the potential harmful effects of mobile phone radiation remain controversial and in a clinical sense inconclusive.

The challenges in uncovering the aetiology of male infertility are many. Nevertheless, in recent times, some progression to this question is being made. Recent evidence has indicated that RF-EMR can induce a state of oxidative stress in spermatozoa and that supplemental antioxidants can prevent RF-EMR-induced cellular damage. The ROS generated in these experiments is likely to originate from sperm cell mitochondria [79], or oxidases on the sperm plasma membrane.

It is also conceivable that the limited intrinsic antioxidants may be suppressed within spermatozoa, further escalating the potential for oxidative stress. Other hypotheses for the non-thermal mechanism of action such as protein mis-folding [80], activation and changes in expression [81] have also been implicated. Many high-quality studies from several groups around the world have brought us to close to a potential mechanism of action; however, the question remains, whether mobile phone use (or other environmental-born EMR) has a significant reproductive toxic effect. To date, there is no clear mechanism by which the non-ionising, mobile phone range EMR may affect biology. This is a key point to which much of the debate about the reported detrimental effects of EMR are based. With this unknown and the many conflicting reports in the literature over the past decade, this field remains controversial. It will be a challenge to answer this question because of the complex nature of the interaction of RF-EMR and the human body and the relatively subtle effects this type of radiation has on biological systems. This is an important time for the field, as biologist gain an understanding of the basic physics of mobile phone radiation, more appropriate and well-defined studies will be conducted which will eventually address the question, does mobile phone use pose a risk to reproduction and human health in general.

References

1. Darnell J, Lodish H, Baltimore D, editors. Nerve cells and the electrical properties of cell membranes. New York: Scientific American Books; 1986.
2. Sheppard AR, Swicord ML, Balzano Q. Quantitative evaluations of mechanisms of radiofrequency interactions with biological molecules and processes. Health Phys. 2008;95(4):365–96.
3. Dubey RB, Hanmandlu M, Gupta SK. Risk of brain tumors from wireless phone use. J Comput Assist Tomogr. 2010;34(6):799–807.
4. de Vocht F, Burstyn I, Cherrie JW. Time trends (1998-2007) in brain cancer incidence rates in relation to mobile phone use in England. Bioelectromagnetics. 2011;32(5):334–9.
5. Joines WT, Zhang Y, Li CX, Jirtle RL. The measured electrical-properties of normal and malignant human tissues from 50 to 900 Mhz. Med Phys. 1994;21(4):547–50.
6. Ichikawa F, Chipchase J, Grignani R. Where's the phone? A study of mobile phone location in public spaces. In: Paper presented at mobile technology, applications and systems, 2005 2nd international conference, Guangzhou, China; 2005.
7. Whittow WG, Panagamuwa CJ, Edwards RM, Vardaxoglou JC. On the effects of straight metallic jewellery on the specific absorption rates resulting from face-illuminating radio communication devices at popular cellular frequencies. Phys Med Biol. 2008;53(5):1167–82.
8. Foster KR, Glaser R. Thermal mechanisms of interaction of radiofrequency energy with biological systems with relevance to exposure guidelines. Health Phys. 2007;92(6):609–20.
9. Phillips JL, Singh NP, Lai H. Electromagnetic fields and DNA damage. Pathophysiology. 2009;16(2–3):79–88.
10. Challis LJ. Mechanisms for interaction between RF fields and biological tissue. Bioelectromagnetics. 2005;Suppl 7:S98–106.
11. Johnson RD, Navratil M, Poe BG, et al. Analysis of mitochondria isolated from single cells. Anal Bioanal Chem. 2007;387(1):107–18.
12. Blank M. Do electromagnetic fields interact with electrons in the Na, K-ATPase? Bioelectromagnetics. 2005;26(8):677–83.

13. Blank M. Electric stimulation of protein synthesis in muscle. Adv Chem Ser. 1995;250: 143–53.
14. Luben RA. Membrane signal-transduction mechanisms and biological effects of low-energy electromagnetic fields. Adv Chem Ser. 1995;250:437–50.
15. Kotnik T, Miklavcic D. Theoretical evaluation of voltage inducement on internal membranes of biological cells exposed to electric fields. Biophys J. 2006;90(2):480–91.
16. Leszczynski D, Joenvaara S, Reivinen J, Kuokka R. Non-thermal activation of the hsp27/p38MAPK stress pathway by mobile phone radiation in human endothelial cells: molecular mechanism for cancer- and blood-brain barrier-related effects. Differentiation. 2002;70(2–3): 120–9.
17. Seger R, Friedman J, Kraus S, Hauptman Y, Schiff Y. Mechanism of short-term ERK activation by electromagnetic fields at mobile phone frequencies. Biochem J. 2007;405:559–68.
18. Kotnik T, Pucihar G, Miklavcic D. Induced transmembrane voltage and its correlation with electroporation-mediated molecular transport. J Membr Biol. 2010;236(1):3–13.
19. Cotgreave IA. Biological stress responses to radio frequency electromagnetic radiation: are mobile phones really so (heat) shocking? Arch Biochem Biophys. 2005;435(1):227–40.
20. Porcelli M, Cacciapuoti G, Fusco S, et al. Non-thermal effects of microwaves on proteins: thermophilic enzymes as model system. FEBS Lett. 1997;402(2–3):102–6.
21. de Pomerai DI, Smith B, Dawe A, et al. Microwave radiation can alter protein conformation without bulk heating. FEBS Lett. 2003;543(1–3):93–7.
22. Mancinelli F, Caraglia M, Abbruzzese A, d'Ambrosio G, Massa R, Bismuto E. Non-thermal effects of electromagnetic fields at mobile phone frequency on the refolding of an intracellular protein: myoglobin. J Cell Biochem. 2004;93(1):188–96.
23. De Iuliis GN, Thomson LK, Mitchell LA, et al. DNA damage in human spermatozoa is highly correlated with the efficiency of chromatin remodeling and the formation of 8-hydroxy-2′-deoxyguanosine, a marker of oxidative stress. Biol Reprod. 2009;81(3):517–24.
24. Pacey AA. Environmental and lifestyle factors associated with sperm DNA damage. Hum Fertil. 2010;13(4):189–93.
25. Sakkas D, Alvarez JG. Sperm DNA fragmentation: mechanisms of origin, impact on reproductive outcome, and analysis. Fertil Steril. 2010;93(4):1027–36.
26. Robaire B, Delbes G, Hales BF. Toxicants and human sperm chromatin integrity. Mol Hum Reprod. 2010;16(1):14–22.
27. Aitken RJ, De Iuliis GN. On the possible origins of DNA damage in human spermatozoa. Mol Hum Reprod. 2010;16(1):3–13.
28. Aitken RJ, De Iuliis GN, McLachlan RI. Biological and clinical significance of DNA damage in the male germ line. Int J Androl. 2009;32(1):46–56.
29. Evenson D, Wixon R. Meta-analysis of sperm DNA fragmentation using the sperm chromatin structure assay. Reprod Biomed Online. 2006;12(4):466–72.
30. Grundler W, Kaiser F, Keilmann F, Walleczek J. Mechanisms of electromagnetic-interaction with cellular-systems. Naturwissenschaften. 1992;79(12):551–9.
31. De Iuliis GN, Newey RJ, King BV, Aitken RJ. Mobile phone radiation induces reactive oxygen species production and DNA damage in human spermatozoa in vitro. PLoS One. 2009; 4(7):e6446.
32. Naziroglu M, Gumral N. Modulator effects of L-carnitine and selenium on wireless devices (2.45 GHz)-induced oxidative stress and electroencephalography records in brain of rat. Int J Radiat Biol. 2009;85(8):680–9.
33. Guney M, Ozguner F, Oral B, Karahan N, Mungan T. 900 MHz Radiofrequency-induced histopathologic changes and oxidative stress in rat endometrium: protection by vitamins E and C. Toxicol Ind Health. 2007;23(7):411–20.
34. Ozguner F, Altinbas A, Ozaydin M, et al. Mobile phone-induced myocardial oxidative stress: protection by a novel antioxidant agent caffeic acid phenethyl ester. Toxicol Ind Health. 2005;21(9):223–30.
35. Aitken RJ, Sawyer D. The human spermatozoon—not waving but drowning. Adv Exp Med Biol. 2002;518:85–98.

36. Delamirande E, Gagnon C. Impact of reactive oxygen species on spermatozoa—a balancing act between beneficial and detrimental effects. Hum Reprod. 1995;10:15–21.
37. Aitken J, Fisher H. Reactive oxygen species generation and human spermatozoa—the balance of benefit and risk. Bioessays. 1994;16(4):259–67.
38. Aitken RJ, Krausz C. Oxidative stress, DNA damage and the Y chromosome. Reproduction. 2001;122(4):497–506.
39. Aitken RJ, De Iuliis GN. On the possible origins of DNA damage in human spermatozoa. Mol Hum Reprod. 2010;16(1):3–13.
40. Koppers AJ, De Iuliis GN, Finnie JM, McLaughlin EA, Aitken RJ. Significance of mitochondrial reactive oxygen species in the generation of oxidative stress in spermatozoa. J Clin Endocrinol Metab. 2008;93(8):3199–207.
41. Agarwal A, Desai NR, Kesari KK. Pathophysiology of cell phone radiation: oxidative stress and carcinogenesis with focus on male reproductive system. Reprod Biol Endocrinol. 2009;7:114.
42. Aitken RJ, De Iuliis GN, Finnie JM, Hedges A, McLachlan RI. Analysis of the relationships between oxidative stress, DNA damage and sperm vitality in a patient population: development of diagnostic criteria. Hum Reprod. 2010;25(10):2415–26.
43. Zini A, Delamirande E, Gagnon C. Reactive oxygen species in semen of infertile patients—levels of superoxide dismutase-like and catalase-like activities in seminal plasma and spermatozoa. Int J Androl. 1993;16(3):183–8.
44. Irvine DS, Twigg JP, Gordon EL, Fulton N, Milne PA, Aitken RJ. DNA integrity in human spermatozoa: relationships with semen quality. J Androl. 2000;21(1):33–44.
45. Twigg J, Fulton N, Gomez E, Irvine DS, Aitken RJ. Analysis of the impact of intracellular reactive oxygen species generation on the structural and functional integrity of human spermatozoa: lipid peroxidation, DNA fragmentation and effectiveness of antioxidants. Hum Reprod. 1998;13(6):1429–36.
46. Kato Y, Osawa T. Detection of lipid-lysine amide-type adduct as a marker of PUFA oxidation and its applications. Arch Biochem Biophys. 2010;501(2):182–7.
47. Makker K, Agarwal A, Sharma R. Oxidative stress & male infertility. Indian J Med Res. 2009;129(4):357–67.
48. Oktem F, Ozguner F, Mollaoglu H, Koyu A, Uz E. Oxidative damage in the kidney induced by 900-MHz-emitted mobile phone: protection by melatonin. Arch Med Res. 2005;36(4):350–5.
49. Oral B, Guney M, Ozguner F, et al. Endometrial apoptosis induced by a 900-MHz mobile phone: preventive effects of vitamins E and C. Adv Ther. 2006;23(6):957–73.
50. Ozguner F, Bardak Y, Comlekci S. Protective effects of melatonin and caffeic acid phenethyl ester against retinal oxidative stress in long-term use of mobile phone: a comparative study. Mol Cell Biochem. 2006;282(1–2):83–8.
51. Verschaeve L, Juutilainen J, Lagroye I, Miyakoshi J, Saunders R, de Seze R, Tenforde T, van Rongen E, Veyret B, Xu Z. In vitro and in vivo genotoxicity of radiofrequency fields. Muta Res. 2010;705(3):252–68.
52. Chia SE, Tay SK. Occupational risk for male infertility: a case-control study of 218 infertile and 227 fertile men. J Occup Environ Med. 2001;43(11):946–51.
53. Fejes I, Zavaczki Z, Szollosi J, et al. Is there a relationship between cell phone use and semen quality? Arch Androl. 2005;51(5):385–93.
54. Dasdag S, Ketani MA, Akdag Z, et al. Whole-body microwave exposure emitted by cellular phones and testicular function of rats. Urol Res. 1999;27(3):219–23.
55. Hong R, Liu Y, Yu YM, Hu K, Weng EQ. [Effects of extremely low frequency electromagnetic fields on male reproduction in mice]. Zhonghua Lao Dong Wei Sheng Zhi Ye Bing Za Zhi. 2003;21(5):342–5.
56. Lee JS, Ahn SS, Jung KC, Kim YW, Lee SK. Effects of 60 Hz electromagnetic field exposure on testicular germ cell apoptosis in mice. Asian J Androl. 2004;6(1):29–34.
57. Aitken RJ, Bennetts LE, Sawyer D, Wiklendt AM, King BV. Impact of radio frequency electromagnetic radiation on DNA integrity in the male germline. Int J Androl. 2005;28(3): 171–9.

58. Erogul O, Oztas E, Yildirim I, et al. Effects of electromagnetic radiation from a cellular phone on human sperm motility: an in vitro study. Arch Med Res. 2006;37(7):840–3.
59. Yan JG, Agresti M, Bruce T, Yan YH, Granlund A, Matloub HS. Effects of cellular phone emissions on sperm motility in rats. Fertil Steril. 2007;88(4):957–64.
60. Wdowiak A, Wdowiak L, Wiktor H. Evaluation of the effect of using mobile phones on male fertility. Ann Agric Environ Med. 2007;14(1):169–72.
61. Agarwal A, Deepinder F, Sharma RK, Ranga G, Li J. Effect of cell phone usage on semen analysis in men attending infertility clinic: an observational study. Fertil Steril. 2008;89(1): 124–8.
62. Tice RR, Hook GG, Donner M, McRee DI, Guy AW. Genotoxicity of radiofrequency signals. I. Investigation of DNA damage and micronuclei induction in cultured human blood cells. Bioelectromagnectics. 2002;23(2):113–26.
63. d'Ambrosio G, Massa R, Scarfi MR & Zeni O. Cytogenetic damage in human lymphocytes following GMSK phase modulated microwave exposure. Bioelectromagnetics. 2002;23(1): 7–13.
64. Deepinder F, Makker K, Agarwal A. Cell phones and male infertility: dissecting the relationship. Reprod Biomed Online. 2007;15(3):266–70.
65. Mailankot M, Kunnath AP, Jayalekshmi H, Koduru B, Valsalan R. Radio frequency electromagnetic radiation (RF-EMR) from GSM (0.9/1.8 GHz) mobile phones induces oxidative stress and reduces sperm motility in rats. Clinics (Sao Paulo). 2009;64(6):561–5.
66. Imai N, Kawabe M, Hikage T, Nojima T, Takahashi S, Shirai T. Effects on rat testis of 1.95-GHz W-CDMA for IMT-2000 cellular phones. Syst Biol Reprod Med. 2011;57(4):204–9.
67. Lee HJ, Pack JK, Kim TH, et al. The lack of histological changes of CDMA cellular phone-based radio frequency on rat testis. Bioelectromagnetics. 2010;31(7):528–34.
68. Agarwal A, Desai NR, Makker K, et al. Effects of radiofrequency electromagnetic waves (RF-EMW) from cellular phones on human ejaculated semen: an in vitro pilot study. Fertil Steril. 2009;92(4):1318–25.
69. Desai NR, Kesari KK, Agarwal A. Pathophysiology of cell phone radiation: oxidative stress and carcinogenesis with focus on male reproductive system. Reprod Biol Endocrinol. 2009;7:114.
70. Tomruk A, Guler G, Dincel AS. The influence of 1800 MHz GSM-like signals on hepatic oxidative DNA and lipid damage in nonpregnant, pregnant, and newly born rabbits. Cell Biochem Biophys. 2010;56(1):39–47.
71. Sokolovic D, Djindjic B, Nikolic J, et al. Melatonin reduces oxidative stress induced by chronic exposure of microwave radiation from mobile phones in rat brain. J Radiat Res (Tokyo). 2008;49(6):579–86.
72. Oktem F, Ozguner F, Mollaoglu H, Koyu A, Uz E. Oxidative damage in the kidney induced by 900-MHz-emitted mobile phone: protection by melatonin. Arch Med Res. 2005;36(4): 350–5.
73. Seyhan N, Ozgur E, Guler G. Mobile phone radiation-induced free radical damage in the liver is inhibited by the antioxidants n-acetyl cysteine and epigallocatechin-gallate. Int J Radiat Biol. 2010;86(11):935–45.
74. Phillips KP, Tanphaichitr N. Human exposure to endocrine disrupters and semen quality. J Toxicol Environ Health B Crit Rev. 2008;11(3–4):188–220.
75. Toppari J, Larsen JC, Christiansen P, et al. Male reproductive health and environmental xenoestrogens. Environ Health Perspect. 1996;104 Suppl 4:741–803.
76. Agarwal A, Burns WR, Sabanegh E, Dada R, Rein B. Is male infertility a forerunner to cancer? Int Braz J Urol. 2010;36(5):527–36.
77. Sharpe RM, Skakkebaek NE. Testicular dysgenesis syndrome: mechanistic insights and potential new downstream effects. Fertil Steril. 2008;89(2 Suppl):e33–8.
78. Repacholi MH. Low-level exposure to radiofrequency electromagnetic fields: health effects and research needs. Bioelectromagnetics. 1998;19(1):1–19.

79. Iorio R, Delle Monache S, Delle Monache S, Bennato F, et al. Involvement of mitochondrial activity in mediating ELF-EMF stimulatory effect on human sperm motility. Bioelectromagnetics. 2011;32(1):15–27.
80. Mousavy SJ, Riazi GH, Kamarei M, et al. Effects of mobile phone radiofrequency on the structure and function of the normal human hemoglobin. Int J Biol Macromol. 2009;44(3): 278–85.
81. Gerner C, Haudek V, Schandl U, et al. Increased protein synthesis by cells exposed to a 1,800-MHz radio-frequency mobile phone electromagnetic field, detected by proteome profiling. Int Arch Occup Environ Health. 2010;83(6):691–702.

Chapter 2
Mitochondria as a Source of ROS in Mammalian Spermatozoa

Adam John Koppers

Abstract The influence of reactive oxygen species (ROS) on sperm function and male fertility is well documented; in contrast, the role of the mitochondria in the generation of aberrant oxidative stress is a recent development. The mitochondria, comprised of complex machinery for energy production, are also equally complex in term of oxidative stress with multiple sites of ROS generation and multiple neutralizing enzymatic and nonenzymatic antioxidants. Knockout mouse models of enzymatic antioxidants do not lead to drastic changes in male fertility; however, increases susceptibility to external toxins and aging. Taking into account the numerous intrinsic and external factors that have been related to mitochondria ROS generation including fatty acids, apoptosis, cigarette smoking, and paternal age, it is likely that multiple risk factors will increase the likelihood of excessive mitochondria ROS generation in mammalian spermatozoa. A clinical focus on mitochondria-targeted antioxidant therapies and research may provide greater insight into oxidative stress-related male infertility and potential treatments.

Keywords Mitochondria • Reactive oxygen species • Mammalian spermatozoa • Mitochondrial ROS production • Lipid peroxidation • Smoking

2.1 Introduction

The clinical significance of oxidative stress in the etiology of defective sperm function was first indicated by Thaddeus Mann and colleagues at the University of Cambridge, recognizing the ability of the antioxidant catalase to prevent motility loss

A.J. Koppers, PhD (✉)
Department of Anatomy and Developmental Biology, Monash University, Wellington Road, Clayton, VIC 3800, Australia
e-mail: adam.koppers@monash.edu

A. Agarwal et al. (eds.), *Studies on Men's Health and Fertility*, Oxidative Stress in Applied Basic Research and Clinical Practice, DOI 10.1007/978-1-61779-776-7_2,

in human sperm [1]. The role of oxidative stress in male infertility has since been well documented over the last few decades [2–5]. This phenomenon is not restricted to humans, but is also evident in a number of other species including boar [6], mouse [7], rabbit [8], rat [9], and horse [10].

Unlike somatic cells, spermatozoa are unique in that they are no longer capable of gene transcription [11]. This is compounded by the lack of cytoplasmic space to house antioxidant enzymes, leaving the spermatozoon in a precarious position. However, the male reproductive tract compensates for the lack of internal antioxidant capacity by saturating spermatozoa in a complex mixture of reactive oxygen species (ROS)-scavenging enzymes and small molecular mass antioxidants. This is evident when spermatozoa are traversing the epididymis, a process that may take 12 days [12], and after they are ejaculated into seminal plasma. Therefore, a significant increase in ROS generation and/or severe decrease in antioxidants over a chronic time frame are needed to shift the balance and cause a state of oxidative stress within spermatozoa.

The principal sources of endogenous ROS in semen are spermatozoa themselves (intrinsic) [13] and leukocytes (extrinsic) [14–16]. Every human semen sample contains differential levels of contaminating leukocytes, the majority being neutrophils and macrophages. Upon stimulation, neutrophils are extremely efficient generators of ROS particularly superoxide anion (O_2^-) and hydrogen peroxide (H_2O_2). The importance of oxidative stress due to leukocyte contamination has been shown to be critical during chronic infection [17] or when of epididymal origin [18], as both have been associated with the induction of significant sperm DNA damage [19]. Leukocytospermia and oxidative stress will be discussed in greater detail in Chap. 26.

Often considered a likely source of ROS in spermatozoa, the presence of NADPH oxidase in spermatozoa has been inconclusive. There is data indicating that the addition of NADPH can elicit a ROS response when measured by chemiluminescence [20]. However, Richer and Ford [21] were unable to confirm NADP-dependent ROS generation using electron paramagnetic resonance spectroscopy. Since then, a number of studies have shown the presence of a calcium-dependant NADPH oxidase, NOX 5, to be present within sperm [22–24]. The identified NOX 5 appears to be a unique form, as it is not controlled by protein kinase C as occurs in the leukocyte form of NADPH oxidase [24]. Subsequently, evidence for a NOX 2 was also reported by Shukla et al. [25] within the head region of mouse spermatozoa. No definitive conclusion on the role of NADPH oxidase in mammalian spermatozoa has yet been reached, and as such, highlights the importance of research into other sources of ROS.

Novel understandings on the subcellular source of aberrant ROS generation in spermatozoa have been developed in recent years. Although the generation of mitochondrial ROS by spermatozoa was already known in rabbit [26] and rat [9], the study by Koppers et al. [27] highlighted its importance in human sperm by showing a correlation between low sperm motility and increased mitochondrial ROS production.

This chapter will present evidence for the role the mitochondria play in the generation of excessive ROS production within mammalian spermatozoa and the importance of mitochondrial antioxidants. Links to intrinsic and external factors and subsequent consequences will then be discussed.

2.2 Mitochondria and ROS Generation

One of the first reports of mitochondrial ROS production came from Boveris and Chance [28], who used antimycin A, an inhibitor of the mitochondrial electron transport chain (ETC) complex III, to induce O_2^- and H_2O_2 generation. Since then, it has not only become evident that mitochondria are capable of low levels of ROS production during normal physiology [29, 30], but ROS are critical factors in a number of different signaling pathways involved in apoptosis [31].

Mitochondrial ROS generation is an unavoidable by-product of oxidative phosphorylation due to the nature of reduction–oxidation reactions that occur during generation of the proton gradient across the inner mitochondrial membrane. It is recognized that the two major sites of ROS production arise from the autooxidation of intermediate semiquinones at either complexes I or III [32]. The semiquinones at these sites are nonenzymatically oxidized by molecular O_2 to yield O_2^-.

The majority of mitochondrial O_2^- is released into the matrix and the rest into the intermembrane space where it is rapidly converted to H_2O_2 by superoxide dismutase (SOD).

$$2O_2^- + 2H^+ \Rightarrow O_2 + H_2O_2 \qquad (2.1)$$

Three different nitric oxide synthases (NOS) have been identified in mammals: neuronal NOS (nNOS), inducible or macrophage NOS (iNOS), and endothelial NOS (eNOS). The mitochondrial NOS (mtNOS) has been identified as the splice variant of the nNOS. While there is still some conjecture about the existence of mtNOS, both mitochondria and submitochondrial preparations have been shown to yield rates of 0.25–0.90 nmol NO min^{-1} mg $protein^{-1}$ [33]. The importance of this is critical as the reaction between O_2^- and nitric oxide (NO) yields peroxynitrite ($ONOO^-$).

$$O_2^- + NO = ONOO^- \qquad (2.2)$$

The intramitochondrial metabolites, O_2^-, H_2O_2, NO, and $ONOO^-$, are prooxidants, potentially leading to oxidative stress and damage. Two of them, O_2^- and NO, are free radicals; however, are considered less reactive and do not participate in destructive propagation reactions, although they do participate in termination reactions yielding H_2O_2 and $ONOO^-$. The latter two species are potentially harmful due to the potential to generate the reactive hydroxyl radical.

While the mitochondrial ETC is capable of ROS production under normal physiological conditions, mitochondrial dysfunction can result in an increase in ROS production and has been implicated in numerous pathological conditions including Alzheimer's disease [34] and ischemia [35]. In most cases of pathological mitochondrial dysfunction, elevated oxidative damage is thought to play a role [36–40]. This is because the mitochondrion is a major source of the ROS within the cell that leads to oxidative damage [41, 42]. As mitochondria are particularly susceptible to oxidative damage, this contributes to mitochondrial dysfunction and cell death in a range of diseases. Due to the susceptible nature of mitochondrial DNA (mtDNA) in comparison to nuclear DNA, a negative loop can develop wherein excessive mitochondrial ROS generation leads to oxidative damage to the mtDNA. This is turn has been shown to cause mitochondrial dysfunction which further exacerbates or initiates mitochondrial ROS production, thus creating an increasing cycle of mitochondrial dysfunction and oxidative stress.

2.2.1 *Mitochondrial ROS and Spermatozoa*

2.2.1.1 Mitochondrial ROS Production

The first reports of mitochondrial ROS generation in mammalian spermatozoa were established in the rabbit using cytochrome c peroxidase assays [26] and an in-depth study in the rat via chemiluminescence [9]. The first analysis of human spermatozoa using the novel MitoSOX Red superoxide fluorescent probe by Koppers et al. [27] indicated that the mitochondria was indeed a major source of ROS and oxidative stress observed in defective human spermatozoa. The level of mitochondrial ROS generation within the dysfunctional sperm population highly correlated with decreased sperm motility. Confirming these results, stimulation of mitochondrial ROS via inhibition of different ETC complexes resulted in decreased motility and lipid peroxidation, a hallmark of oxidative stress. Although only a recent advance regarding the understanding of the subcellular sources of ROS in mammalian spermatozoa, a number of studies provide further support for this hypothesis.

2.2.1.2 Mitochondrial Membrane Potential, ROS, and Oxidative DNA Damage

The rate of ROS formation from the ETC is found to be increased when the electron flow slows down (resulting in a reduced state) and/or when the concentration of oxygen increases [30]. However, the generation of ROS from mitochondria is not straight forward and is becoming an increasingly complex process. The contribution of ROS between different sites of ROS formation, e.g., Complex I vs. III, can greatly vary between different tissues and can also change depending on the mitochondrial membrane potential ($\Delta\Psi_m$) [43]. For example, complex III is responsible for the majority of O_2^- produced in both heart and lung mitochondria [44, 45]; in contrast, complex I is the major ROS site in the brain [43].

In human spermatozoa, a number of studies have now shown that the $\Delta\Psi_m$ can be used as an indicator of sperm function and therefore quality. Spermatozoa that exhibit high $\Delta\Psi_m$ generally have been shown to have high fertilizing capacity as represented by increased ability to acrosome react, higher motility, and normal morphology [46]. In contrast, spermatozoa with low MMP showed correlations with decreased DNA stability [47] and motility [48–50]. Indicative of all these results is that low $\Delta\Psi_m$ spermatozoa cells are associated with reduced IVF rates [49].

Given the relationship between $\Delta\Psi_m$ and ROS generation, the most critical study to date was performed by Wang et al. [51], whom in accordance with other studies showed that patients exhibiting abnormal semen parameters had a significantly lower $\Delta\Psi_m$ than control subjects. More importantly, they also showed patients had higher ROS levels and the two parameters negatively correlated with the ROS produced ($r=-0.45$). The $\Delta\Psi_m$ also was positively correlated with sperm concentration ($r=0.62$). As such, the $\Delta\Psi_m$ has been linked as a strong indicator of sperm function, which may be due to its inverse relationship with ROS generation.

While these studies do not provide causative evidence for a link between $\Delta\Psi_m$ and ROS generation, it is further supported in a subsequent study by De Iuliis et al. [52], which found that the 8-hydroxy-2′-deoxyguanosine (8OHdG) formation in human spermatozoa is negatively associated with $\Delta\Psi_m$. To further emphasize such a relationship, exposure of purified human spermatozoa to electromagnetic radiation (EMR) results in significant increases in mitochondrial ROS and 8OHdG formation, which are highly correlative with each other ($R^2=0.727$) [53]. While no study to date has analyzed the correlation between all three factors, mitochondrial ROS, $\Delta\Psi_m$, and oxidative DNA damage, these data combined does suggest all three have significant relationship.

2.2.1.3 Lipid Peroxidation in Midpiece

Along with decreased motility, one of the best markers of oxidative stress in spermatozoa is increased levels of lipid peroxidation. Validated in a number of subsequent studies, it has also now been shown that lipid peroxidation directly affects sperm functions including motility and sperm–oocyte fusion [54, 55].

The direct stimulation of mitochondrial ROS by inhibition of the ETC results in the subsequent increase in lipid peroxidation in human spermatozoa [27]. Interestingly, the lipid peroxidation observed in this study is distinctly localized to only the midpiece. A similar result was also observed following stimulation with fatty acids (FAs) (discussed below). This is indicative that lipid peroxidation within the midpiece is also a marker of mitochondrial ROS generation due to the unique localization of mitochondria to the midpiece. In light of these results, it is important to note that a number of external stimuli have been used to generate oxidative stress in human spermatozoa such as iron (II) [56].

While still a novel concept, there is now an increasing evidence that the mitochondria is a major source of ROS in mammalian spermatozoa, and as such, is likely to be a key focus in future research of oxidative stress-related male infertility.

2.2.2 Mitochondrial Antioxidants and Fertility

The deleterious effects resulting from the formation of ROS in the mitochondrion are, to a large extent, prevented by various antioxidant systems, and as such, mammalian mitochondria possess a relative abundance of enzymatic and nonenzymatic antioxidants designed to manage and counteract the ROS generated. An in-depth presentation of the literature is well beyond the scope of this chapter; however, a brief description is provided to represent the well-characterized ROS scavenging systems that act within the mitochondria and their known relationships with male fertility.

2.2.2.1 Superoxide Dismutase

O_2^- is one of the major ROS generated in the mitochondria and is critically maintained at the lowest possible concentration. O_2^- can spontaneously dismutate or more commonly is enzymatically converted to H_2O_2 by a group of metalloenzymes called SOD [57]. O_2^- is capable of propagating the generation of other ROS, either through the spontaneous reaction with NO creating peroxynitrite or via reducing transition metals within the mitochondrial ETC, which subsequently react with H_2O_2 producing hydroxyl radical. The mitochondrial matrix contains a specific form of SOD with manganese in the active site (MnSOD) [57], which eliminates the O_2^- formed in the matrix or on the inner side of the inner membrane. The inner membrane space contains a different SOD which contains copper and zinc instead of manganese (CuZnSOD) [58].

Not surprisingly, due to the important role of the enzyme, homozygous MnSOD knockout mice do not survive more than a few days following birth, in comparison to the heterozygous mutant mice which are viable and fertile [59, 60]. The life span and rate of aging are similar between heterozygous and wild-type animals, despite more accumulated DNA damage and increased prevalence of cancer during their life span [61]. In contrast, the overexpression of MnSOD to six to ten times above the normal level results in developmental abnormalities and decreased fertility of mice [62]. This decrease in fertility may be related to the importance of low-level ROS signals that have been shown to be critical during capacitative events, which will be discussed in Chap. 4.

2.2.2.2 Cytochrome c

Found in the intermembrane space of mitochondria, cytochrome c is an integral link in the ETC and has a dual role in O_2^- removal. In both functions, cytochrome c is reduced by either an electron from the ETC or by O_2^- and subsequently is regenerated (oxidized) by cytochrome c oxidase. A testis-specific isoform of cytochrome c has been identified; the knockout mouse model for the testis-specific isoform is fertile despite lower motility percentage in sperm from both vas deferens and epididymis. These knockout mice do exhibit loss of germ cells following aging (>4 months), due to increased levels of atrophy not apoptosis [63].

2.2.2.3 Glutathione Peroxidase

Glutathione peroxidases (GPxs) are a group of selenoenzymes that utilize reduced glutathione (GSH) for the reduction of H_2O_2 to H_2O. GPxs participate in the recycling of ROS and some peroxidized compounds that are products from the attacks of organic molecules by ROS. The classical GPx (GPx1) is known to be ubiquitously expressed in all mammalian tissues [64] and can be detected in various cellular compartments including the cytosol, mitochondrial matrix, and intermembrane space. Homozygous knockout mice possessing no GPx1 activity are healthy, fertile, develop normally, and do not show any signs of tissue damage and oxidative stress [65]. Interestingly, exposure of GPx1 knockout mice to toxins known to induce oxidative stress resulted in hypertension [66], suggesting that GPx1 is more likely to be involved in protecting tissues and mitochondria against acute oxidative stress rather than the basal levels of mitochondrial ROS production. Its lack of importance in fertility is supported by Chabory et al. [67], which showed that Gpx1 is expressed in very low levels compared to other GPx family members within the epididymis.

Phospholipid hydroperoxide GPx (GPx4) has broad selectivity, which allows it to reduce phospholipid hydroperoxides, H_2O_2, and cholesterol peroxides [68]. Three forms of GPx4 exist; nuclear, short, and the long form found exclusively in the mitochondria. The long form RNA transcript is predominately expressed in testis in mice [69]. Localized to the sperm midpiece, mitochondrial GPx4 has been hypothesized to play a role as a structural protein rather than an enzymatic antioxidant since it has been shown to have completely lost its solubility and its scavenging enzymatic properties [70]. Two independent knockout models have also confirmed that when mGPx4 is not expressed, it leads to male infertility due to structural malformations of the sperm midpiece and not due to a loss in antioxidant capability.

2.2.2.4 Glutathione Reductase

GSH can either scavenge O_2^- and hydroxyl radicals nonenzymatically or possibly more importantly by acting as an electron donor to several enzymes already discussed above. Once GSH is oxidized to GSSG (oxidized glutathione), it cannot be exported to cytosol and has to be reduced back to GSH in the mitochondrial matrix. The reduction is catalyzed by glutathione reductase (GRD), which is present in the matrix of mitochondria [71]. A strain of hypomorphic mice ($Gr1^{a1Neu}$) that exhibited GRD activities that were just above basal levels (less than 10% in liver) compared to control mice were fertile and showed no phenotypic changes [72].

2.2.2.5 Peroxiredoxins

Peroxiredoxins are a recently discovered group of peroxidases that reduce H_2O_2 and lipid hydroperoxides, with at least six members currently identified in mammals [73]. Two isoforms of peroxiredoxins (PRDX3 and PRDX5) are found in mammalian mitochondria. The PRDX5 gene is also ubiquitously expressed in bovine tissues, with the highest expression found in testis [74]. In human spermatozoa, PRDX5 is localized to the acrosome, postacrosomal region, and midpiece. Therefore, PRDX5 may play a role in regulating ROS actions in the mitochondria as well as other functions. Although not all mitochondrial related, a number of peroxiredoxin knockout mice have been generated. *PRDX1* knockout mice are fertile yet have a significant reduction in lifespan due to increased levels of oxidative protein and DNA damage and higher frequency of anemia and malignant cancer [75]. A similar anemic phenotype was also observed in the *PRDX2* knockout [76]. *PRDX3* knockout mice are fertile, but were shown to be more sensitive to lipopolysaccharide-induced oxidative stress [77]. *PRDX6* knockout mice are also viable yet are shown to be more sensitive to hyperoxia and paraquat-induced oxidative stress [78, 79]. No knockout for *PRDX5* is reported in the literature.

2.2.2.6 Catalase

Catalase is another enzyme which catalyzes H_2O_2 into O_2 and H_2O. The role of catalase in H_2O_2 removal is thought to be insignificant compared to that of GPx [80]. Shown to be present at low levels in both human and rat spermatozoa [4, 81, 82], it is known to be absent from rabbit spermatozoa [26], mouse spermatozoa [7], and bull spermatozoa [83]. Unsurprisingly, due to its absence from mouse sperm, the knockout mouse models are fertile [84], although this does not preclude it from a role in spermatozoa in humans and the rat.

In general, knockout mouse models of mitochondrial antioxidants have not resulted in infertility, indicating that these enzymes are not critical for fertility. In contrast, they have indicated greater susceptibility to oxidative stress due to toxins or aging. The oxidative stress "challenged" mice in these experiments may better represent a more realistic model in humans considering the myriad environmental and external stimuli encountered in day-to-day activities.

2.2.3 *Intrinsic Causes of Mitochondrial ROS Generation*

There is evidence for the role of mitochondrial ROS in the etiology of male infertility in human spermatozoa. Although the factors responsible for stimulating increased free radical release from the mitochondria of spermatozoa are still unresolved, recent research suggests the involvement for a number of intrinsic factors including FAs and apoptosis.

2.2.3.1 Fatty Acids

The fatty acid (FA) composition within spermatozoa plays a crucial role in the ability of these cells to achieve fertilization. The high unsaturated FA content within the sperm plasma membrane is critical for function, although also results in increased susceptibility as targets for lipid peroxidation. Greater detail on this topic will be covered in Chap. 6.

A number of studies have described differing influences of FAs on mitochondrial function. Research by Cocco et al. [85] showed that the addition of endogenous fatty acids to isolated bovine heart mitochondria caused 81% and 32% inhibition of the ETC complexes I and III, respectively. The inhibition of mitochondrial ETC complexes was specific to only unsaturated fatty acids, as saturated fatty acids did not elicit an effect [85]. Interestingly, in line with these results, inhibition of complex I, not complex III, in human spermatozoa results in peroxidative damage and reduction in fertility potential [27].

It has been previously indicated by Aitken et al. [86] that addition of polyunsatutared fatty acids (PUFAs) to human spermatozoa results in increased ROS generation, lipid peroxidation, and motility loss. However, at the time the source of the ROS was not determined, although lipoxygenase and cycloxygenase pathways were excluded as mediators of this response. Subsequent data have now been shown by Koppers et al. [87], recognizing that the mitochondria are the source of ROS and damaging oxidative stress following addition of endogenous PUFA. The addition of any unsaturated FA analyzed in the study (omega-3, -6, -9) to human spermatozoa results in increased mitochondrial ROS generation, albeit at differing levels. The oxidative stress created by FA-stimulated mitochondria resulted in a significant decrease in sperm motility and the associated induction of oxidative DNA damage. The same study demonstrated that dysfunctional human spermatozoa, separated via density gradient centrifugation, contain significantly more FA than their functional counterparts. This FA excess applied to all classes of FA (saturated, monounsaturated, and polyunsaturated) and was also observed regardless of whether analyzed as the total FA content of these cells or only the unesterified (free) FA component. This confirms an earlier study by Ollero et al. [88], who observed an increase in the PUFA (particularly DHA) content of defective spermatozoa.

Koppers et al. [87] were the first to reveal that highly significant correlations exist between spontaneous mitochondrial ROS production by human spermatozoa and their free (unesterified) unsaturated FA content for both omega-6 ($R^2=0.605$) and omega-3 ($R^2=0.615$) PUFAs. The findings are also consistent with previous analysis of the FA content of human spermatozoa, which observed an increase in the cellular content of both saturated and unsaturated FAs in spermatozoa from men seeking infertility treatment compared to a cohort of normozoospermic controls [89]. Overall these results have established a causal link between the levels of unsaturated FAs in human spermatozoa, mitochondrial ROS generation, and its adverse effect on function.

Recent evidence suggests that systemic deficiencies in lipid metabolism seem an unlikely cause of increased FA content in spermatozoa given the apparent discrepancy between the FA profiles of blood serum and spermatozoa [90]. Alternatively, the high FA content of defective spermatozoa likely reflects a fundamental error in the remodeling of human sperm cells during spermiogenesis, which would be associated with the increased retention of residual cytoplasm and an enhanced cytoplasmic volume.

2.2.3.2 Apoptosis

Apoptosis due to oxidative stress is usually only observed in pathological conditions [91, 92]. Increased mitochondrial ROS is one known trigger of the intrinsic apoptotic pathway, which is achieved via an increase in the permeability of the outer mitochondrial membrane through the opening of transition pores. The opening of the permeability transition pore is favored by oxidative stress through oxidation of intracellular GSH and other critical sulfhydryl groups [93]. Under normal conditions, various antiapoptotic factors (including Bcl-xL) prevent increases in mitochondrial permeability as long as they remain bound to the outer membrane; the translocation of Bax to the mitochondria removes these antiapoptotic factors initiating apoptosis [94]. The opening of permeability transition pores results in the gradual loss of cytochrome c from the intermembrane space during apoptosis. As cytochrome c is released, the respiratory chain becomes more reduced as electron flow between complexes III and IV slows down resulting in increased semiquinone formation and thus ROS [95].

Mature human spermatozoa have been recognized to exhibit many but not all features of apoptotic signal transduction pathways including the externalization of phosphatidylserine (PS) (annexin-V binding), activation of caspases 1, 3, 8, and 9 [96–98], mitochondrial dysfunction, and ROS generation [99–101]. One pathology that has been associated with apoptosis in spermatozoa is varicocele. Spermatozoa from men with this condition show higher levels of externalization of PS mitochondrial dysfunction and nuclear DNA damage [102]. In vitro exposure of human spermatozoa to oxidative stress mediator H_2O_2 can trigger the activation of caspases and externalization of PS [103]. In contrast, addition of antioxidants (catalase) to spermatozoa prior to H_2O_2 will prevent this apoptotic response [104]. Activation of the apoptotic cascade by PI3 kinase inhibitor, wortmannin, has also shown to result in increased mitochondrial ROS generation, caspase activation, and PS externalization [99].

2.2.4 Environmental and External Factors Related to Mitochondrial ROS

2.2.4.1 Environmental

The generation of ROS in all cell types can be exacerbated by a multitude of external and environmental factors. Industrial development is exposing men and women to an increasing number of different chemicals and by-products of manufacturing in their environment. Many of these toxic compounds are found to induce oxidative stress and have been suggested to pose a serious threat towards the normal and reproductive health of humans around the world and especially in developing nations [105]. Some environmental compounds that are known to induce mitochondrial ROS generation include phthalates [106], estrogens [107], and EMR [53].

One of the most heavily investigated compounds for effects on reproductive function are phthalates. Phthalates are most commonly found in plastic food packaging and can be released into foods upon contact, increasing human exposure levels of these compounds [108]. Phthalate exposure in rats has shown to induce significant levels of DNA damage in spermatozoa as well as impairing normal spermatogenesis [109]. Additional studies have shown that exposure of germ cells to phthalates in vitro results in oxidative stress, mitochondrial dysfunction, cytochrome c release, and apoptosis [110, 111].

An in-depth mouse study investigating the in vivo effects of exposure to various phthalates found treatment resulted in higher concentration in semen and this negatively correlated with semen quality. Increased ROS generation was also observed in spermatozoa as well as other markers of oxidative stress including lipid peroxidation and DNA damage. Although the source of ROS was not investigated, bis(2-ethylhexyl) phthalate DEHP treatment also resulted in mitochondrial dysfunction, a major cause of mitochondrial ROS as discussed [112].

2.2.4.2 Paternal Age

The consequences of paternal age on human fertility have come to the fore due to changes in human reproductive patterns, a combination of social changes, prolonged life expectancy, and a reliance on Assisted reproductive technology (ART) [113]. The association between increased paternal age and decreased sperm parameters has been well documented. Jung et al. [114] reported that older subjects (>50 years) showed a 27% decrease in progressive motility compared to younger men (21–25 years). Another study also found that older subjects (>55 years) exhibited approximately 25% lower total sperm count, semen volume, and sperm concentration compared to the younger age group (30–35 years) [115]. Furthermore, recent studies have also shown that ROS production and oxidative stress are increased in human spermatozoa during aging, suggesting a possible role for the decline in fertility [116].

Although the source of ROS in the study by Cocuzza et al. [116] was not investigated, other evidence related to aging suggests that it may be of mitochondrial origin. Many hallmarks of reduced fertility including oxidative DNA damage and lipid peroxidation have been shown to be increased in various tissues during aging [117–119].

Increased age in humans has also shown to be linked to increased levels of mutation in mtDNA [120]; this is a direct result of the constant mitochondria ROS generation in combination with the lack of protection to mtDNA in comparison to that provided to nuclear DNA by histones or protamines [121]. Subsequently, mutations or deletions in mtDNA lead to defects in oxidative phosphorylation, calcium homeostasis, and other related mtDNA diseases [122]. With increased paternal age associated with decreased fertility, it is likely that the two are linked by the associating factor of mitochondrial ROS.

2.2.4.3 Smoking

The carcinogenic and toxic effects of cigarette smoking on the health of an individual are well documented. Smoking has also been shown to result in an increased likelihood of male infertility, with significantly higher levels of seminal ROS generation [123] and DNA damage [124] found in these men. Previous research has shown a significant elevation in the activity of MnSOD (mitochondrial isozyme) in cigarette smokers [125], indicating that cigarette smoking may cause increased mitochondrial ROS generation, rather than via a cytosolic or membrane oxidase system.

In smokers, the overall mitochondrial ETC function is significantly decreased which also correlates with increased peroxidative damage to lymphocyte membranes [126]. An extremely large Danish study of 2,542 healthy men, without bias towards reproductive history, discovered that compared with nonsmokers, smokers exhibited significantly decreased sperm concentration, semen volume, motility, and total sperm count [127]. Although the changes observed were only minor (20–30% decrease), such effects may further compound an already-existing minor reproductive condition or combine with other xenobiotics to further reduce fertility. Overall, the smokers presented in the study exhibited increased seminal ROS production and decreased antioxidant levels. A number of independent studies have also shown that smoking results in higher levels of DNA damage in the human spermatozoa in comparison to nonsmokers [124, 128, 129]. Also, a higher rate of smoking observed in men is associated with corresponding increases in the rate of aneuploidy in spermatozoa [130, 131]. Overall, the results on smoking in men provide further evidence for the role that oxidative stress and DNA damage play in male infertility. Although only suggestive evidence exists, these data also provide a rationale for further investigation into the role of mitochondria in smoking-related oxidative stress in the male germ line.

2.2.5 Clinical Implications

2.2.5.1 Mitochondrial ROS and DNA Damage

A strong marker for oxidative DNA damage is the major oxidized base adduct formed when DNA is subjected to attack by ROS, 8OHdG. Clinically, this marker has been shown to be higher in spermatozoa of subfertile patients [132]. The importance and clinical implications of oxidative DNA damage will be covered in Chap. 11, although it is important to further emphasize the ability of mitochondrial ROS to contribute to nuclear DNA damage. Admittedly mtDNA damage is a common factor and contributor of excessive ROS generation; it is also critical to note that paternal mtDNA is not passed onto the offspring during fertilization and development. However, despite the physical barriers that separate the sperm head from the midpiece, mitochondrial ROS has been shown to be capable of inducing oxidative damage to nuclear in a number of recent studies via different stimuli including activation of apoptosis though inhibition of PI3-kinase inhibitor, wortmannin [99], or through direct stimulation of mitochondrial ROS through exposing spermatozoa to PUFAs or EMR [53, 87].

2.2.5.2 Importance in IVF and Targeted Antioxidant Therapy

Clinically, the importance of oxidative stress-mediated male infertility has given rise to the trial of antioxidant therapies. While a number of research studies have focused on a single antioxidant's effects, current commercial treatments such as Menevit© use a broad range of antioxidant compounds in order to address oxidative stress-related male infertility. In light of evidence presented in this chapter, it may prove beneficial to develop antioxidant trials and therapies that specifically target mitochondria. From this perspective, the most logical candidate is coenzyme Q10 (CoQ10), a well-known mitochondrial antioxidant component.

Although CoQ10 has a number of physiological roles including redox carrier, activator of uncoupling proteins, and in mitochondrial pore formation [133], CoQ10 in its reduced form ($CoQH_2$) is a powerful antioxidant through its ability to inhibit destructive lipid peroxidation chains. The effectiveness of $CoQH_2$ as an inhibitor of lipid peroxidation is through its ability to break the complex chain reactions produced during lipid peroxidation cascades. In this process, CoQ10 prevents the formation of lipid peroxyl radicals ($LOO^•$) production during initiation phase of lipid peroxidation. The reduced form of CoQ10 reduces the initiating peroxyl radical via the formation of a semiquinone and H_2O_2. Alternatively, $CoQH_2$ can eliminate $LOO^•$ directly. This is achieved again via the formation of a semiquinone, the same mechanism by which α-tocopherol prevents lipid peroxidation [134].

There are a number of pathologies where oxidative stress is a well-documented contributing factor that also exhibits increases in the synthesis of CoQ10 including Alzheimer's, prion, other neurodegenerative disease, and diabetes [135]. During both aging and in heart disease, there is a significant lowering of CoQ10 content in the target organ [136], creating a state of vulnerability to oxidative stress. Since mitochondrial ROS generation is a continual and unavoidable by-product of cellular respiration, conditions associated with a decreased availability of CoQ10 in the male reproductive tract may well be associated with oxidative stress. Given the preliminary evidence for CoQ10 as a major antioxidant in the male reproductive tract [137–139], further studies should be undertaken to examine the potential inclusion of this coenzyme as a candidate for the antioxidant therapy of male infertility such as the in vivo study by Balercia et al. [140] that showed increased sperm motility in infertile men with asthenozoospermia treated with CoQ10 for 6 months.

However, while it is an effective and a viable option for those undergoing ART, antioxidant therapy is an approach that only addresses the symptoms of this condition; it does not address the underlying causative mechanisms.

2.3 Conclusions

- Greater understanding of the source(s) of aberrant ROS generation in mammalian spermatozoa is critical to the study of oxidative stress and male infertility.
- Increasing evidence suggests that the mitochondria are a major source of intracellular ROS in spermatozoa.
- Due to the high ROS generation in mitochondria, specific antioxidants in this organelle are critical. Due to the breadth and variety found, knockout mouse models have not generated significant data in the reproductive field.
- A number of intrinsic and external factors including FAs, apoptosis, environmental toxicants, cigarette smoking, and paternal age have all been related to mitochondria ROS generation. Exposure to multiple risk factors may further increase the likelihood of excessive mitochondria ROS generation.
- A clinical focus on mitochondria-targeted antioxidant therapies may provide a greater and more efficient alleviation of oxidative stress-related male infertility.

References

1. Jones R, Mann T, Sherins R. Peroxidative breakdown of phospholipids in human spermatozoa, spermicidal properties of fatty acid peroxides, and protective action of seminal plasma. Fertil Steril. 1979;31(5):531–7.
2. Aitken RJ, Clarkson JS. Significance of reactive oxygen species and antioxidants in defining the efficacy of sperm preparation techniques. J Androl. 1988;9(6):367–76.
3. Aitken J, Fisher H. Reactive oxygen species generation and human spermatozoa: the balance of benefit and risk. Bioessays. 1994;16(4):259–67.
4. Alvarez JG, Touchstone JC, Blasco L, Storey BT. Spontaneous lipid peroxidation and production of hydrogen peroxide and superoxide in human spermatozoa. Superoxide dismutase as major enzyme protectant against oxygen toxicity. J Androl. 1987;8(5):338–48.
5. Sharma RK, Agarwal A. Role of reactive oxygen species in male infertility. Urology. 1996;48(6):835–50.
6. Guthrie HD, Welch GR. Determination of intracellular reactive oxygen species and high mitochondrial membrane potential in Percoll-treated viable boar sperm using fluorescence-activated flow cytometry. J Anim Sci. 2006;84(8):2089–100.
7. Alvarez JG, Storey BT. Assessment of cell damage caused by spontaneous lipid peroxidation in rabbit spermatozoa. Biol Reprod. 1984;30(2):323–31.
8. Holland MK, Alvarez JG, Storey BT. Production of superoxide and activity of superoxide dismutase in rabbit epididymal spermatozoa. Biol Reprod. 1982;27(5):1109–18.
9. Vernet P, Fulton N, Wallace C, Aitken RJ. Analysis of reactive oxygen species generating systems in rat epididymal spermatozoa. Biol Reprod. 2001;65(4):1102–13.
10. Ball BA, Vo A. Osmotic tolerance of equine spermatozoa and the effects of soluble cryoprotectants on equine sperm motility, viability, and mitochondrial membrane potential. J Androl. 2001;22(6):1061–9.
11. Dadoune JP. The nuclear status of human sperm cells. Micron. 1995;26(4):323–45.
12. Amann RP, Johnson L, Thompson Jr DL, Pickett BW. Daily spermatozoal production, epididymal spermatozoal reserves and transit time of spermatozoa through the epididymis of the rhesus monkey. Biol Reprod. 1976;15(5):586–92.

13. Gomez E, Buckingham DW, Brindle J, Lanzafame F, Irvine DS, Aitken RJ. Development of an image analysis system to monitor the retention of residual cytoplasm by human spermatozoa: correlation with biochemical markers of the cytoplasmic space, oxidative stress, and sperm function. J Androl. 1996;17(3):276–87.
14. Aitken RJ, West K, Buckingham D. Leukocytic infiltration into the human ejaculate and its association with semen quality, oxidative stress, and sperm function. J Androl. 1994;15(4): 343–52.
15. Aitken RJ, Baker HW. Seminal leukocytes: passengers, terrorists or good samaritans? Hum Reprod. 1995;10(7):1736–9.
16. Aitken RJ, Buckingham DW, Brindle J, Gomez E, Baker HW, Irvine DS. Analysis of sperm movement in relation to the oxidative stress created by leukocytes in washed sperm preparations and seminal plasma. Hum Reprod. 1995;10(8):2061–71.
17. Kullisaar T, Turk S, Punab M, et al. Oxidative stress in leucocytospermic prostatitis patients: preliminary results. Andrologia. 2008;40(3):161–72.
18. Haidl G, Allam JP, Schuppe HC. Chronic epididymitis: impact on semen parameters and therapeutic options. Andrologia. 2008;40(2):92–6.
19. Alvarez JG, Sharma RK, Ollero M, et al. Increased DNA damage in sperm from leukocytospermic semen samples as determined by the sperm chromatin structure assay. Fertil Steril. 2002;78(2):319–29.
20. Aitken RJ, Fisher HM, Fulton N, et al. Reactive oxygen species generation by human spermatozoa is induced by exogenous NADPH and inhibited by the flavoprotein inhibitors diphenylene iodonium and quinacrine. Mol Reprod Dev. 1997;47(4):468–82.
21. Richer SC, Ford WC. A critical investigation of NADPH oxidase activity in human spermatozoa. Mol Hum Reprod. 2001;7(3):237–44.
22. Sabeur K, Ball BA. Characterization of NADPH oxidase 5 in equine testis and spermatozoa. Reproduction. 2007;134(2):263–70.
23. Banfi B, Molnar G, Maturana A, et al. A Ca(2+)-activated NADPH oxidase in testis, spleen, and lymph nodes. J Biol Chem. 2001;276(40):37594–601.
24. Armstrong JS, Bivalacqua TJ, Chamulitrat W, Sikka S, Hellstrom WJ. A comparison of the NADPH oxidase in human sperm and white blood cells. Int J Androl. 2002;25(4):223–9.
25. Shukla S, Jha RK, Laloraya M, Kumar PG. Identification of non-mitochondrial NADPH oxidase and the spatio-temporal organization of its components in mouse spermatozoa. Biochem Biophys Res Commun. 2005;331(2):476–83.
26. Holland MK, Storey BT. Oxygen metabolism of mammalian spermatozoa. Generation of hydrogen peroxide by rabbit epididymal spermatozoa. Biochem J. 1981;198(2):273–80.
27. Koppers AJ, De Iuliis GN, Finnie JM, McLaughlin EA, Aitken RJ. Significance of mitochondrial reactive oxygen species in the generation of oxidative stress in spermatozoa. J Clin Endocrinol Metab. 2008;93(8):3199–207.
28. Boveris A, Chance B. The mitochondrial generation of hydrogen peroxide. General properties and effect of hyperbaric oxygen. Biochem J. 1973;134(3):707–16.
29. Inoue M, Sato EF, Nishikawa M, et al. Mitochondrial generation of reactive oxygen species and its role in aerobic life. Curr Med Chem. 2003;10(23):2495–505.
30. Turrens JF. Mitochondrial formation of reactive oxygen species. J Physiol. 2003;552(Pt 2): 335–44.
31. Brookes PS, Levonen AL, Shiva S, Sarti P, Darley-Usmar VM. Mitochondria: regulators of signal transduction by reactive oxygen and nitrogen species. Free Radic Biol Med. 2002;33(6):755–64.
32. St-Pierre J, Buckingham JA, Roebuck SJ, Brand MD. Topology of superoxide production from different sites in the mitochondrial electron transport chain. J Biol Chem. 2002;277(47): 44784–90.
33. Brookes PS. Mitochondrial nitric oxide synthase. Mitochondrion. 2004;3(4):187–204.
34. Beal MF. Aging, energy, and oxidative stress in neurodegenerative diseases. Ann Neurol. 1995;38(3):357–66.

35. Stone D, Darley-Usmar V, Martin JF. Calcium fluxes and reperfusion damage: the role of mitochondria. In: Parratt JR, editor. Myocardial response to acute injury. Basingstoke: Macmillan; 1992.
36. Ames BN, Shigenaga MK, Hagen TM. Oxidants, antioxidants, and the degenerative diseases of aging. Proc Natl Acad Sci USA. 1993;90(17):7915–22.
37. Beckman KB, Ames BN. Mitochondrial aging: open questions. Ann N Y Acad Sci. 1998;854: 118–27.
38. Beckman KB, Ames BN. The free radical theory of aging matures. Physiol Rev. 1998;78(2): 547–81.
39. Finkel T. Radical medicine: treating ageing to cure disease. Nat Rev Mol Cell Biol. 2005;6(12): 971–6.
40. Green K, Brand MD, Murphy MP. Prevention of mitochondrial oxidative damage as a therapeutic strategy in diabetes. Diabetes. 2004;53 Suppl 1:S110–8.
41. Balaban RS, Nemoto S, Finkel T. Mitochondria, oxidants, and aging. Cell. 2005;120(4): 483–95.
42. Murphy MP. How mitochondria produce reactive oxygen species. Biochem J. 2009;417(1): 1–13.
43. Barja G. Mitochondrial oxygen radical generation and leak: sites of production in states 4 and 3, organ specificity, and relation to aging and longevity. J Bioenerg Biomembr. 1999;31(4): 347–66.
44. Turrens JF, Freeman BA, Levitt JG, Crapo JD. The effect of hyperoxia on superoxide production by lung submitochondrial particles. Arch Biochem Biophys. 1982;217(2):401–10.
45. Turrens JF, Boveris A. Generation of superoxide anion by the NADH dehydrogenase of bovine heart mitochondria. Biochem J. 1980;191(2):421–7.
46. Gallon F, Marchetti C, Jouy N, Marchetti P. The functionality of mitochondria differentiates human spermatozoa with high and low fertilizing capability. Fertil Steril. 2006;86(5): 1526–30.
47. Troiano L, Granata AR, Cossarizza A, et al. Mitochondrial membrane potential and DNA stainability in human sperm cells: a flow cytometry analysis with implications for male infertility. Exp Cell Res. 1998;241(2):384–93.
48. Marchetti C, Obert G, Deffosez A, Formstecher P, Marchetti P. Study of mitochondrial membrane potential, reactive oxygen species, DNA fragmentation and cell viability by flow cytometry in human sperm. Hum Reprod. 2002;17(5):1257–65.
49. Marchetti C, Jouy N, Leroy-Martin B, Defossez A, Formstecher P, Marchetti P. Comparison of four fluorochromes for the detection of the inner mitochondrial membrane potential in human spermatozoa and their correlation with sperm motility. Hum Reprod. 2004;19(10):2267–76.
50. Donnelly GB, Glassman J, Long C, et al. Granulocyte-colony stimulating factor (G-CSF) may improve disease outcome in elderly patients with diffuse large cell lymphoma (DLCL) treated with CHOP chemotherapy. Leuk Lymphoma. 2000;39(1–2):67–75.
51. Wang X, Sharma RK, Gupta A, et al. Alterations in mitochondria membrane potential and oxidative stress in infertile men: a prospective observational study. Fertil Steril. 2003;80 Suppl 2:844–50.
52. De Iuliis GN, Thomson LK, Mitchell LA, et al. DNA damage in human spermatozoa is highly correlated with the efficiency of chromatin remodeling and the formation of 8-hydroxy-2′-deoxyguanosine, a marker of oxidative stress. Biol Reprod. 2009;81(3):517–24.
53. De Iuliis GN, Newey RJ, King BV, Aitken RJ. Mobile phone radiation induces reactive oxygen species production and DNA damage in human spermatozoa in vitro. PLoS One. 2009;4(7):e6446.
54. Twigg J, Fulton N, Gomez E, Irvine DS, Aitken RJ. Analysis of the impact of intracellular reactive oxygen species generation on the structural and functional integrity of human spermatozoa: lipid peroxidation, DNA fragmentation and effectiveness of antioxidants. Hum Reprod. 1998;13(6):1429–36.
55. Aitken RJ, Clarkson JS, Fishel S. Generation of reactive oxygen species, lipid peroxidation, and human sperm function. Biol Reprod. 1989;41(1):183–97.

56. Aitken RJ, Wingate JK, De Iuliis GN, McLaughlin EA. Analysis of lipid peroxidation in human spermatozoa using BODIPY C11. Mol Hum Reprod. 2007;13(4):203–11.
57. Fridovich I. Superoxide radical and superoxide dismutases. Annu Rev Biochem. 1995;64: 97–112.
58. Okado-Matsumoto A, Fridovich I. Subcellular distribution of superoxide dismutases (SOD) in rat liver: Cu, Zn-SOD in mitochondria. J Biol Chem. 2001;276(42):38388–93.
59. Lebovitz RM, Zhang H, Vogel H, et al. Neurodegeneration, myocardial injury, and perinatal death in mitochondrial superoxide dismutase-deficient mice. Proc Natl Acad Sci USA. 1996;93(18):9782–7.
60. Li Y, Huang TT, Carlson EJ, et al. Dilated cardiomyopathy and neonatal lethality in mutant mice lacking manganese superoxide dismutase. Nat Genet. 1995;11(4):376–81.
61. Van Remmen H, Ikeno Y, Hamilton M, et al. Life-long reduction in MnSOD activity results in increased DNA damage and higher incidence of cancer but does not accelerate aging. Physiol Genomics. 2003;16(1):29–37.
62. Raineri I, Carlson EJ, Gacayan R, et al. Strain-dependent high-level expression of a transgene for manganese superoxide dismutase is associated with growth retardation and decreased fertility. Free Radic Biol Med. 2001;31(8):1018–30.
63. Narisawa S, Hecht NB, Goldberg E, Boatright KM, Reed JC, Millan JL. Testis-specific cytochrome c-null mice produce functional sperm but undergo early testicular atrophy. Mol Cell Biol. 2002;22(15):5554–62.
64. Lenzen S, Drinkgern J, Tiedge M. Low antioxidant enzyme gene expression in pancreatic islets compared with various other mouse tissues. Free Radic Biol Med. 1996;20(3):463–6.
65. Ho YS, Magnenat JL, Bronson RT, et al. Mice deficient in cellular glutathione peroxidase develop normally and show no increased sensitivity to hyperoxia. J Biol Chem. 1997;272(26): 16644–51.
66. de Haan JB, Bladier C, Griffiths P, et al. Mice with a homozygous null mutation for the most abundant glutathione peroxidase, Gpx1, show increased susceptibility to the oxidative stress-inducing agents paraquat and hydrogen peroxide. J Biol Chem. 1998;273(35):22528–36.
67. Chabory E, Damon C, Lenoir A, et al. Epididymis seleno-independent glutathione peroxidase 5 maintains sperm DNA integrity in mice. J Clin Invest. 2009;119(7):2074–85.
68. Thomas JP, Maiorino M, Ursini F, Girotti AW. Protective action of phospholipid hydroperoxide glutathione peroxidase against membrane-damaging lipid peroxidation. In situ reduction of phospholipid and cholesterol hydroperoxides. J Biol Chem. 1990;265(1):454–61.
69. Knopp EA, Arndt TL, Eng KL, et al. Murine phospholipid hydroperoxide glutathione peroxidase: cDNA sequence, tissue expression, and mapping. Mamm Genome. 1999;10(6):601–5.
70. Ursini F, Heim S, Kiess M, et al. Dual function of the selenoprotein PHGPx during sperm maturation. Science. 1999;285(5432):1393–6.
71. Kelner MJ, Bagnell RD, Montoya MA, Lanham KA. Structural organization of the human gastrointestinal glutathione peroxidase (GPX2) promoter and 3′-nontranscribed region: transcriptional response to exogenous redox agents. Gene. 2000;248(1–2):109–16.
72. Pretsch W. Glutathione reductase activity deficiency in homozygous Gr1a1Neu mice does not cause haemolytic anaemia. Genet Res. 1999;73(1):1–5.
73. Wood ZA, Poole LB, Karplus PA. Peroxiredoxin evolution and the regulation of hydrogen peroxide signaling. Science. 2003;300(5619):650–3.
74. Leyens G, Donnay I, Knoops B. Cloning of bovine peroxiredoxins-gene expression in bovine tissues and amino acid sequence comparison with rat, mouse and primate peroxiredoxins. Comp Biochem Physiol B Biochem Mol Biol. 2003;136(4):943–55.
75. Neumann CA, Krause DS, Carman CV, et al. Essential role for the peroxiredoxin Prdx1 in erythrocyte antioxidant defence and tumour suppression. Nature. 2003;424(6948):561–5.
76. Lee TH, Kim SU, Yu SL, et al. Peroxiredoxin II is essential for sustaining life span of erythrocytes in mice. Blood. 2003;101(12):5033–8.
77. Li L, Shoji W, Takano H, et al. Increased susceptibility of MER5 (peroxiredoxin III) knockout mice to LPS-induced oxidative stress. Biochem Biophys Res Commun. 2007;355(3): 715–21.

78. Wang Y, Feinstein SI, Manevich Y, Ho YS, Fisher AB. Lung injury and mortality with hyperoxia are increased in peroxiredoxin 6 gene-targeted mice. Free Radic Biol Med. 2004;37(11):1736–43.
79. Wang X, Phelan SA, Forsman-Semb K, et al. Mice with targeted mutation of peroxiredoxin 6 develop normally but are susceptible to oxidative stress. J Biol Chem. 2003;278(27): 25179–90.
80. Antunes F, Han D, Cadenas E. Relative contributions of heart mitochondria glutathione peroxidase and catalase to H(2)O(2) detoxification in in vivo conditions. Free Radic Biol Med. 2002;33(9):1260–7.
81. Jeulin C, Soufir JC, Weber P, Laval-Martin D, Calvayrac R. Catalase activity in human spermatozoa and seminal plasma. Gamete Res. 1989;24(2):185–96.
82. Tramer F, Rocco F, Micali F, Sandri G, Panfili E. Antioxidant systems in rat epididymal spermatozoa. Biol Reprod. 1998;59(4):753–8.
83. Bilodeau JF, Chatterjee S, Sirard MA, Gagnon C. Levels of antioxidant defenses are decreased in bovine spermatozoa after a cycle of freezing and thawing. Mol Reprod Dev. 2000;55(3): 282–8.
84. Ho YS, Xiong Y, Ma W, Spector A, Ho DS. Mice lacking catalase develop normally but show differential sensitivity to oxidant tissue injury. J Biol Chem. 2004;279(31):32804–12.
85. Cocco T, Di Paola M, Papa S, Lorusso M. Arachidonic acid interaction with the mitochondrial electron transport chain promotes reactive oxygen species generation. Free Radic Biol Med. 1999;27(1–2):51–9.
86. Aitken RJ, Wingate JK, De Iuliis GN, Koppers AJ, McLaughlin EA. Cis-unsaturated fatty acids stimulate reactive oxygen species generation and lipid peroxidation in human spermatozoa. J Clin Endocrinol Metab. 2006;91(10):4154–63.
87. Koppers AJ, Garg ML, Aitken RJ. Stimulation of mitochondrial reactive oxygen species production by unesterified, unsaturated fatty acids in defective human spermatozoa. Free Radic Biol Med. 2010;48(1):112–9.
88. Ollero M, Gil-Guzman E, Lopez MC, et al. Characterization of subsets of human spermatozoa at different stages of maturation: implications in the diagnosis and treatment of male infertility. Hum Reprod. 2001;16(9):1912–21.
89. Khosrowbeygi A, Zarghami N. Fatty acid composition of human spermatozoa and seminal plasma levels of oxidative stress biomarkers in subfertile males. Prostaglandins Leukot Essent Fatty Acids. 2007;77(2):117–21.
90. Conquer JA, Martin JB, Tummon I, Watson L, Tekpetey F. Fatty acid analysis of blood serum, seminal plasma, and spermatozoa of normozoospermic vs. asthenozoospermic males. Lipids. 1999;34(8):793–9.
91. Kroemer G, Dallaporta B, Resche-Rigon M. The mitochondrial death/life regulator in apoptosis and necrosis. Annu Rev Physiol. 1998;60:619–42.
92. Hoeberichts FA, Woltering EJ. Multiple mediators of plant programmed cell death: interplay of conserved cell death mechanisms and plant-specific regulators. Bioessays. 2003;25(1): 47–57.
93. Chernyak BV. Redox regulation of the mitochondrial permeability transition pore. Biosci Rep. 1997;17(3):293–302.
94. Finucane DM, Bossy-Wetzel E, Waterhouse NJ, Cotter TG, Green DR. Bax-induced caspase activation and apoptosis via cytochrome c release from mitochondria is inhibitable by Bcl-xL. J Biol Chem. 1999;274(4):2225–33.
95. Cai J, Jones DP. Superoxide in apoptosis. Mitochondrial generation triggered by cytochrome c loss. J Biol Chem. 1998;273(19):11401–4.
96. Oehninger S, Morshedi M, Weng SL, Taylor S, Duran H, Beebe S. Presence and significance of somatic cell apoptosis markers in human ejaculated spermatozoa. Reprod Biomed Online. 2003;7(4):469–76.
97. Glander HJ, Schaller J. Binding of annexin V to plasma membranes of human spermatozoa: a rapid assay for detection of membrane changes after cryostorage. Mol Hum Reprod. 1999; 5(2):109–15.

98. Paasch U, Grunewald S, Agarwal A, Glandera HJ. Activation pattern of caspases in human spermatozoa. Fertil Steril. 2004;81 Suppl 1:802–9.
99. Koppers AJ, Mitchell LA, Wang P, Lin M, Aitken RJ. Phosphoinositide 3-kinase signalling pathway involvement in a truncated apoptotic cascade associated with motility loss and oxidative DNA damage in human spermatozoa. Biochem J. 2011;436(3):687–98.
100. Grunewald S, Sharma R, Paasch U, Glander HJ, Agarwal A. Impact of caspase activation in human spermatozoa. Microsc Res Tech. 2009;72(11):878–88.
101. Barroso G, Taylor S, Morshedi M, Manzur F, Gavino F, Oehninger S. Mitochondrial membrane potential integrity and plasma membrane translocation of phosphatidylserine as early apoptotic markers: a comparison of two different sperm subpopulations. Fertil Steril. 2006;85(1):149–54.
102. Wu GJ, Chang FW, Lee SS, Cheng YY, Chen CH, Chen IC. Apoptosis-related phenotype of ejaculated spermatozoa in patients with varicocele. Fertil Steril. 2009;91(3):831–7.
103. Lozano GM, Bejarano I, Espino J, et al. Relationship between caspase activity and apoptotic markers in human sperm in response to hydrogen peroxide and progesterone. J Reprod Dev. 2009;55(6):615–21.
104. Libman J, Gabriel MS, Sairam MR, Zini A. Catalase can protect spermatozoa of FSH receptor knock-out mice against oxidant-induced DNA damage in vitro. Int J Androl. 2010;33(6): 818–22.
105. Aitken RJ, Skakkebaek NE, Roman SD. Male reproductive health and the environment. Med J Aust. 2006;185(8):414–5.
106. Ghosh J, Das J, Manna P, Sil PC. Hepatotoxicity of di-(2-ethylhexyl)phthalate is attributed to calcium aggravation, ROS-mediated mitochondrial depolarization, and ERK/NF-kappaB pathway activation. Free Radic Biol Med. 2010;49(11):1779–91.
107. Felty Q, Xiong WC, Sun D, et al. Estrogen-induced mitochondrial reactive oxygen species as signal-transducing messengers. Biochemistry. 2005;44(18):6900–9.
108. Rock G, Labow RS, Tocchi M. Distribution of di(2-ethylhexyl) phthalate and products in blood and blood components. Environ Health Perspect. 1986;65:309–16.
109. Agarwal DK, Maronpot RR, Lamb JC, Kluwe WM. Adverse effects of butyl benzyl phthalate on the reproductive and hematopoietic systems of male rats. Toxicology. 1985;35(3):189–206.
110. Richburg JH, Boekelheide K. Mono-(2-ethylhexyl) phthalate rapidly alters both Sertoli cell vimentin filaments and germ cell apoptosis in young rat testes. Toxicol Appl Pharmacol. 1996;137(1):42–50.
111. Kasahara E, Sato EF, Miyoshi M, et al. Role of oxidative stress in germ cell apoptosis induced by di(2-ethylhexyl)phthalate. Biochem J. 2002;365(Pt 3):849–56.
112. Pant N, Shukla M, Kumar Patel D, et al. Correlation of phthalate exposures with semen quality. Toxicol Appl Pharmacol. 2008;231(1):112–6.
113. Heck KE, Schoendorf KC, Ventura SJ, Kiely JL. Delayed childbearing by education level in the United States, 1969–1994. Matern Child Health J. 1997;1(2):81–8.
114. Jung A, Schuppe HC, Schill WB. Comparison of semen quality in older and younger men attending an andrology clinic. Andrologia. 2002;34(2):116–22.
115. Levitas E, Lunenfeld E, Weisz N, Friger M, Potashnik G. Relationship between age and semen parameters in men with normal sperm concentration: analysis of 6022 semen samples. Andrologia. 2007;39(2):45–50.
116. Cocuzza M, Athayde KS, Agarwal A, et al. Age-related increase of reactive oxygen species in neat semen in healthy fertile men. Urology. 2008;71(3):490–4.
117. Lee HC, Lim ML, Lu CY, et al. Concurrent increase of oxidative DNA damage and lipid peroxidation together with mitochondrial DNA mutation in human lung tissues during aging—smoking enhances oxidative stress on the aged tissues. Arch Biochem Biophys. 1999;362(2):309–16.
118. Wei YH, Kao SH, Lee HC. Simultaneous increase of mitochondrial DNA deletions and lipid peroxidation in human aging. Ann N Y Acad Sci. 1996;786:24–43.
119. Lu CY, Lee HC, Fahn HJ, Wei YH. Oxidative damage elicited by imbalance of free radical scavenging enzymes is associated with large-scale mtDNA deletions in aging human skin. Mutat Res. 1999;423(1–2):11–21.

120. Wallace DC. Mitochondrial DNA mutations in disease and aging. Environ Mol Mutagen. 2010;51(5):440–50.
121. Wei YH, Lee CF, Lee HC, et al. Increases of mitochondrial mass and mitochondrial genome in association with enhanced oxidative stress in human cells harboring 4,977 BP-deleted mitochondrial DNA. Ann N Y Acad Sci. 2001;928:97–112.
122. James AM, Murphy MP. How mitochondrial damage affects cell function. J Biomed Sci. 2002;9(6 Pt 1):475–87.
123. Saleh RA, Agarwal A, Sharma RK, Nelson DR, Thomas Jr AJ. Effect of cigarette smoking on levels of seminal oxidative stress in infertile men: a prospective study. Fertil Steril. 2002;78(3):491–9.
124. Fraga CG, Motchnik PA, Wyrobek AJ, Rempel DM, Ames BN. Smoking and low antioxidant levels increase oxidative damage to sperm DNA. Mutat Res. 1996;351(2):199–203.
125. St Clair DK, Jordan JA, Wan XS, Gairola CG. Protective role of manganese superoxide dismutase against cigarette smoke-induced cytotoxicity. J Toxicol Environ Health. 1994;43(2):239–49.
126. Miro O, Alonso JR, Jarreta D, Casademont J, Urbano-Marquez A, Cardellach F. Smoking disturbs mitochondrial respiratory chain function and enhances lipid peroxidation on human circulating lymphocytes. Carcinogenesis. 1999;20(7):1331–6.
127. Ramlau-Hansen CH, Thulstrup AM, Aggerholm AS, Jensen MS, Toft G, Bonde JP. Is smoking a risk factor for decreased semen quality? A cross-sectional analysis. Hum Reprod. 2007;22(1):188–96.
128. Viloria T, Garrido N, Fernandez JL, Remohi J, Pellicer A, Meseguer M. Sperm selection by swim-up in terms of deoxyribonucleic acid fragmentation as measured by the sperm chromatin dispersion test is altered in heavy smokers. Fertil Steril. 2007;88(2):523–5.
129. Sepaniak S, Forges T, Gerard H, Foliguet B, Bene MC, Monnier-Barbarino P. The influence of cigarette smoking on human sperm quality and DNA fragmentation. Toxicology. 2006;223(1–2):54–60.
130. Rubes J, Lowe X, Moore II D, et al. Smoking cigarettes is associated with increased sperm disomy in teenage men. Fertil Steril. 1998;70(4):715–23.
131. Shi Q, Ko E, Barclay L, Hoang T, Rademaker A, Martin R. Cigarette smoking and aneuploidy in human sperm. Mol Reprod Dev. 2001;59(4):417–21.
132. Kodama H, Yamaguchi R, Fukuda J, Kasai H, Tanaka T. Increased oxidative deoxyribonucleic acid damage in the spermatozoa of infertile male patients. Fertil Steril. 1997;68(3):519–24.
133. Bentinger M, Brismar K, Dallner G. The antioxidant role of coenzyme Q. Mitochondrion. 2007;7(Suppl):S41–50.
134. Mukai K, Kikuchi S, Urano S. Stopped-flow kinetic study of the regeneration reaction of tocopheroxyl radical by reduced ubiquinone-10 in solution. Biochim Biophys Acta. 1990;1035(1):77–82.
135. Turunen M, Olsson J, Dallner G. Metabolism and function of coenzyme Q. Biochim Biophys Acta. 2004;1660(1–2):171–99.
136. Littarru GP, Tiano L. Clinical aspects of coenzyme Q10: an update. Curr Opin Clin Nutr Metab Care. 2005;8(6):641–6.
137. Mancini A, Milardi D, Conte G, Festa R, De Marinis L, Littarru GP. Seminal antioxidants in humans: preoperative and postoperative evaluation of coenzyme Q10 in varicocele patients. Horm Metab Res. 2005;37(7):428–32.
138. Mancini A, De Marinis L, Oradei A, et al. Coenzyme Q10 concentrations in normal and pathological human seminal fluid. J Androl. 1994;15(6):591–4.
139. Mancini A, Conte G, Milardi D, De Marinis L, Littarru GP. Relationship between sperm cell ubiquinone and seminal parameters in subjects with and without varicocele. Andrologia. 1998;30(1):1–4.
140. Balercia G, Mancini A, Paggi F, et al. Coenzyme Q10 and male infertility. J Endocrinol Invest. 2009;32(7):626–32.

Chapter 3
Cryostorage and Oxidative Stress in Mammalian Spermatozoa

Stuart A. Meyers

Abstract Although cryopreservation of ejaculated sperm has been in clinical and agricultural use for decades, it is not completely clear how the damage that sperm incur as a result of cryopreservation contributes to fertilization failure, or embryonic or fetal loss. Oxygen is required for life, but oxidative metabolism, particularly during low temperature storage, of biological molecules can be potentially toxic due to the formation of reactive oxygen species (ROS) that can modify cell functions or viability. A limited ability to store antioxidant enzymes combined with a membrane rich in unsaturated fatty acids makes spermatozoa particularly susceptible to oxidative stress and peroxidative attack by ROS, specifically superoxide anion and hydrogen peroxide. This chapter outlines the primary mechanisms of sperm damage during cryopreservation and loss of subsequent fertility and discusses the potential mechanisms of DNA/chromosomal fragmentation and damage, lipid peroxidation, and intracellular ice formation and associated cell damage. The origin of ROS in sperm is discussed as well as how ROS are processed and ultimately scavenged by sperm.

Keywords Cryostorage • Oxidative stress • Mammalian spermatozoa • Cryopreservation effects • Sperm fertility • Origin of reactive oxygen species • Semen • Sperm

3.1 Introduction: Cryopreservation Effects on Sperm Fertility

The spermatozoon, like all cells living under aerobic conditions, constantly faces the oxygen paradox; oxygen is required for life, but oxidative metabolism of biological molecules can be potentially toxic due to the formation of reactive oxygen

S.A. Meyers, DVM, PhD (✉)
School of Veterinary Medicine, Anatomy, Physiology, and Cell Biology, University of California, Davis, CA, USA
e-mail: smeyers@ucdavis.edu

A. Agarwal et al. (eds.), *Studies on Men's Health and Fertility*, Oxidative Stress in Applied Basic Research and Clinical Practice, DOI 10.1007/978-1-61779-776-7_3,

species (ROS) that can modify cell functions or viability. Although oxidative stress was suggested as an important factor in disruption of sperm function over 60 years ago [1, 2], it is only within the past 10–15 years that the importance of oxidative stress has gained a wider understanding. Recent studies indicate that ROS play an important role in normal sperm function; however, an imbalance in the production or degradation of ROS may have adverse consequences for sperm function. Excessive production of ROS could be responsible in part for loss of viable, fertilization-competent sperm either directly or indirectly by intensifying the cellular response to other stressors, such as osmotic imbalance.

Long-term storage of numerous mammalian somatic and germ cell types has been accomplished using cryopreservation, typically in liquid nitrogen (cryogenic tanks) or in other subzero environments (−80°C ultra-low-type freezer). Conventional cryopreservation for sperm is termed equilibrium, or slow, freezing of the bulk of water, followed by storage at extremely low temperatures, usually −196°C, although other types of low-temperature freezing such as vitrification and lyophilization have been reported [3–7]. The process of cryopreservation has profound effects on cells, many of which result in sublethal damage and subsequent reduction of function. Many cell types do not tolerate frozen storage above −80°C and undergo severe deterioration with subsequent lethal damage. Some hematopoietic and human embryonic stem cell lines and gametes undergo sublethal and lethal changes associated with the complex interaction of low-temperature cryoprotective agent (CPA) reagents, osmotic and oxidative balance, solute and electrolyte balance, and ice crystallization [8, 9]. However, in the presence of CPAs, which can be cytotoxic, many cell types can survive low-temperature storage. As a routine consequence of cryogenic storage, approximately 25–75% of cells stored this way are lost due to necrotic and apoptotic cell death and this is dependent on cell type, freezing rate, and CPA. Cryoprotectants have been broadly classified as penetrating and nonpenetrating substances. Penetrating CPAs are small nonionic molecules (glycerol, dimethylsulfoxide, propylene glycol, ethylene glycol, methylformamide) with high water solubility while the nonpenetrating CPAs are long-chain polymers or sugars (methylcellulose, sucrose, raffinose, trehalose). One of the most significant origins of cell damage and death during freezing is that of intracellular and extracellular ice formation. Cryoprotectants act primarily by reducing the speed at which ice forms and the size of ice crystals [10, 11]. With adequate CPA, freezing damage can be minimized. However, high concentrations of CPA can cause osmotic and toxic cell damage. Thus, CPA addition and removal can have long-lasting cellular effects.

Although cryopreservation of ejaculated sperm has been in clinical and agricultural use for decades, it is not completely clear how the damage that sperm incur as a result of cryopreservation contributes to fertilization failure, or embryonic or fetal loss. This problem is multifactorial and has species-specific contributing factors [12]. Frozen ejaculated semen has enabled the cattle industry to make significant genetic progress since the 1950s when artificial insemination (AI) was embraced commercially. However, due to species variability and individual male variability, the use of AI with frozen semen has been less successful in the swine, sheep, and horse breeding industries. A considerable issue has been that of male variation, but in swine and horses the relatively poor sperm survivability has magnified the high variability among males intended to be breeding males [10, 13]. In pigs, the use of frozen semen for AI has resulted consistently in decreased farrowing rates and litter sizes [14, 15], thus limiting the usefulness of frozen semen for this economically important industry. In sheep, frozen semen has also resulted in decreased pregnancy rates [10], although insemination practices have recently increased pregnancy rates. In human medicine, sperm cryopreservation is widely used for insemination both in vivo and in vitro, but successful pregnancy rates are highly variable and male dependent. The use of assisted reproductive technologies, such as intracytoplasmic sperm injection, has obviated a portion of the need, in large part, to greatly improve cryopreservation success in humans. The bovine artificial insemination industry has had great success for more than 50 years owing to active selection for bulls with high fertility with frozen sperm. It is no surprise that a considerable amount of sublethal and lethal damage occurs as a result of exposure to the extreme temperature and osmolality effects associated with cryopreservation [16–19]. Cryopreservation poses a severe osmotic insult to the cell that leads to generation of ROS; this process can be minimized but not eliminated by using CPA agents. Additionally, there is substantial evidence that cryopreservation results in increased DNA damage, aneuploidy, and chromosome fragmentation [5, 20–22]. Lipid peroxidation can also contribute directly to specific sublethal effects, like chromatin cross-linking, base changes, and DNA strand breaks [23–26]. A limited ability to store antioxidant enzymes combined with a membrane rich in unsaturated fatty acids makes spermatozoa particularly susceptible to oxidative stress and peroxidative attack by ROS, specifically superoxide anion and hydrogen peroxide [27].

There is clinical evidence that damage to sperm DNA results in downstream impaired embryo development and pregnancy in mice and humans [28–31]. High levels of ROS have been associated with sperm DNA damage in the semen of 25% of infertile men [32]. Recent evidence indicates that men with high DNA fragmentation indices (DFIs) have significantly higher rates of spontaneous abortion in their partners [33]. Further, apoptotic processes may contribute to DNA damage in sperm [34]. It has been proposed that some sperm with DNA damage may have initiation and possibly escape of apoptosis [35]. There is also evidence for a number of other inducers of DNA damage that are associated with embryonic loss and infertility, such as tobacco use, environmental toxins, chronic orchitis/genital tract inflammation, varicoceles, radiation, chemotherapeutic drugs, and testicular hyperthermia [34].

3.2 Cryopreservation and Oxidative Stress

Spermatozoa and seminal plasma have several mechanisms that generate ROS and this is discussed below. However, sperm and seminal plasma possess a number of enzymes and low-molecular-weight antioxidants that scavenge ROS in order to prevent cellular damage. During spermiogenesis, when sperm shape is modified and streamlined by losing most of the cytoplasm and, thus, intrinsic cytoplasmic enzyme scavenger systems, sperm attain vulnerability to oxidative attack [36]. Spermatozoa subsequently contain a limited volume of cytoplasm and are predominantly dependent upon the antioxidant support of seminal plasma. Further, sperm preparation for cryopreservation often involves the removal of seminal plasma by centrifugation, thereby increasing the susceptibility of spermatozoa to oxidative stress. In addition, research suggests that the antioxidant activity of the spermatozoa themselves may be decreased by cryopreservation [37, 38]. Freeze thawing of equine, human, and bovine spermatozoa was associated with an increase in ROS generation [39–43]. Recent studies have demonstrated that rhesus macaque sperm also suffer from excessive ROS generation when frozen and subsequently thawed [44, 45] in addition to severe DNA and chromosomal damage [5].

Oxidative stress denotes a condition associated with an increased rate of cellular damage induced by ROS [46]. The cause of the pro-oxidant–antioxidant shift may be due to an increase in ROS production, a decrease in antioxidant capacity, or possibly a combination of the two. Oxidative stress induced by the generation of ROS in vitro results in a reduction in sperm motility, viability, ionophore-induced acrosome reaction and sperm–oocyte fusion. Hydrogen peroxide appears to be the primary ROS responsible for these changes, and membrane lipid peroxidation is believed to be an important mechanism of action [36, 47–50].

The lipid peroxidation cascade is initiated when ROS attack polyunsaturated fatty acids (PUFAs) in the sperm cell membrane [50–52]. Spermatozoa are particularly susceptible to oxidative attack because they contain high concentrations of unsaturated fatty acids and, as terminally differentiated cells, have limited repair mechanisms. As a consequence of lipid peroxidation, the plasma membrane loses the fluidity and integrity it requires for participation in the membrane fusion events associated with fertilization. In addition to membrane effects, lipid peroxidation can also damage DNA. Peroxidation of DNA can lead to chromatin cross-linking, base changes, and DNA strand breaks. Several researchers have reported DNA damage in spermatozoa associated with membrane lipid peroxidation and oxidative stress [36, 53–55].

3.3 The Nature of Cellular Damage During Freezing

Cryopreservation damage, or cryoinjury, to any cell is, thus, due to three primary factors: *osmotic stress/dehydration, oxidative stress, and intracellular ice formation.* During cryopreservation and subsequent thawing, sperm are exposed to a variety of changing osmotic and oxidative conditions to which the cell must respond in order to maintain or recover function. Freezing and thawing sperm result in exposure to osmotic change, extracellular and intracellular ice crystals, production of ROS, and resultant downstream effects on motility, lipid phase change, membrane integrity, mitochondrial function, DNA integrity, cell signaling, metabolism, and apoptotic and necrotic cell death. Many of these negative effects occur mainly during the thawing process owing to the relative stability of the frozen aqueous state. In general, it is not likely that cells surviving low-temperature freezing sustain damage by intracellular ice since this phenomenon is known to cause cell rupture and ensuing lethal effects that are not compatible with cell survival. Both osmotic and oxidative changes have been shown to induce ROS in sperm and contribute to cellular damage that may be lethal or sublethal [19, 21, 37, 44, 45, 49, 50, 53, 54, 56–59].

3.3.1 Origin of ROS in Semen and Sperm

Reduced forms of oxygen, collectively termed ROS, interact with various types of molecules and include free radicals that are charged molecules with unpaired electrons. Examples of free radicals are superoxide anion (O_2^-•) and hydroxyl radical (•OH). Not all ROS are free radicals, however, and molecules such as hydrogen peroxide (H_2O_2), ozone, and singlet oxygen are also ROS and contribute to oxidative physiology and pathology [60, 61].

Within an ejaculate, there are two potential sources of ROS: spermatozoa [62, 63] and leukocytes [64]. It is believed that spermatozoa and leukocytes possess similar mechanisms for ROS generation: an NADPH oxidase (NOX) located in or near the cell membrane [60, 65–68].

In mammalian semen, leukocytes are not normally part of an ejaculate and are not present in significant numbers in normal ejaculates in most species. Human semen, however, commonly contains leukocytes and, hence, sperm have a consistent production of ROS and exposure to the subsequent oxidative effects. Leukocytes utilize the generation of ROS in the oxidative burst of microbial-killing phagocytosis and are capable of generating a significant amount of ROS in semen [69].

All NOX family members are transmembrane proteins with an NAD(P)H binding site at the carboxylic end, an FAD-binding region, six conserved transmembrane domains, and four highly conserved heme-binding histidines. It has been reported that human spermatozoa contain a NOX similar to that reported in phagocytic leukocytes [66]. However, NAD(P)H failed to stimulate extracellular (O_2^-•) production as detected by MCLA (2-methyl-6-(*p*-methoxyphenyl)-3,7-dihydroimidazo [1,2-a] pyrazin-3-one), whereas addition of progesterone or ultrafiltrates of fetal cord serum, follicular fluid, or seminal plasma (stimulators of sperm) to human spermatozoa did stimulate (O_2^-•) generation [70]. Another group using a highly sensitive spin trapping technique has also reported the absence of NOX activity in human spermatozoa [71]. Richer and Ford [72] proposed that the majority of the chemiluminescence promoted by NAD(P)H is derived from redox cycling independent of O_2^- • generation by spermatozoa [72]. Recent research has determined that lucigenin chemiluminescence in response to NAD(P)H in rat epididymal sperm preparations is due to the presence of an NAD(P)H-dependent cytochrome P450 reductase that originates largely from contaminating epididymal cells [73]. This enzyme directly reduces lucigenin and tetrazolium salts, thus initiating their redox cycling with molecular oxygen and production of (O_2^-•). Consequently, not only is the existence of an NAD(P)H oxidase in spermatozoa subject to scrutiny, but the

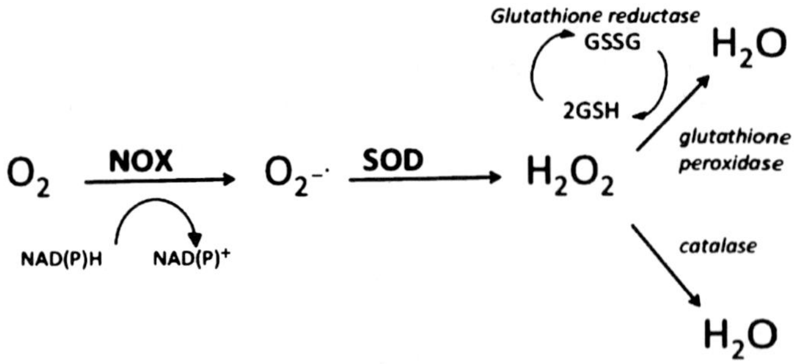

Fig. 3.1 Conversion of diatomic oxygen to ROS in sperm. NAD(P)H oxidase shown as NOX; superoxide dismutase shown as SOD

mechanism of (O_2^-•) detection is susceptible to interference. In addition to driving ROS production in sperm by exogenous NAD(P)H, the reaction can be inhibited using the classic inhibitor diphenylene iodonium (DPI). Although DPI has been shown to be a less specific inhibitor than earlier proposed, it has been shown to inhibit ROS production in sperm from a variety of species (reviewed by Aitken and Curry) [36]. As these authors point out, the NOX family influence in sperm ROS has not been clearly elucidated. Mouse sperm [74] and stallion [68] sperm have been shown to express NOX2 and NOX5, respectively. Further, dual oxidase (DUOX) has been recently identified in human sperm using a proteomic method [75]. These studies suggest that NOX systems are capable of generating ROS in sperm, but it is not known whether this mechanism predominates during normal sperm function or during cryopreservation as a stress response.

The major ROS that are generated by sperm and play significant biological roles are the superoxide anion (O_2^-•) which is the most common ROS generated by spermatozoa, and rapidly dismutates either spontaneously or catalyzed by superoxide dismutase (SOD) to H_2O_2, which is the second major ROS in sperm (Fig. 3.1). Hydrogen peroxide is not a free radical, and in contrast to (O_2^-•) it is more stable and can readily cross the plasma membrane [76]. An increase in ROS generation has been attributed to abnormal or damaged spermatozoa [77–79]. Generation of ROS by sperm has been the subject of a number of studies in recent years to elucidate the subcellular origin of reduced oxygen forms. In different cell types, ROS can be produced variably by intracellular oxidases and peroxidases but also by cytochrome p450, nitric oxide synthase, and leakage of electrons from the electron transport chain (reviewed in [61, 67]). ROS generation continues as a cascade of reactions as discussed for Fig. 3.1 culminating in H_2O_2 production. In many cell types, reactive nitrogen species (RNS) are concurrently produced that are also agents of oxidative stress to cells [60]. As ROS are processed by enzyme systems, they may also react with RNS products, such as nitric oxide (NO), to generate peroxinitrite (superoxide + NO), hypochlorous acid (H_2O_2 + NO), and the iron-catalyzed Fenton reaction to result in hydroxyl radical (•HO). RNS have been shown to affect sperm function and fertility [80, 81].

Sperm mitochondria also contribute to ROS production through active leakage of electrons and have been shown to be a significant source of ROS leading to cell damage [82, 83]. Mitochondrial ROS production has also been associated with cell regulation and initiation of apoptosis, or programmed cell death [84]. As apoptotic cell processes have been associated with sperm cryoinjury and death, it seems reasonable that mitochondria, the source of energy for sperm motility, could be specifically damaged during cryopreservation leading to a major source of oxidative change in sperm.

When generated by sperm or leukocytes in semen, ROS interact with numerous molecules, including lipids, proteins, carbohydrates, and nucleic acids, and cause often irreversible effects. Sperm lipid membranes contain an abundance of highly unsaturated fatty acids and are, therefore, very susceptible to oxidative damage [52, 67].

3.3.2 Mechanisms to Scavenge ROS in Sperm

Production of ROS by sperm must be balanced by scavenging systems for minimizing cell damage. Sperm-associated glutathione peroxidase is the primary scavenger for H_2O_2 and converts H_2O_2 to water. Catalase has been described in human semen to similarly break down H_2O_2 to water [85]. Consequently, H_2O_2 appears to be the primary ROS responsible for oxidative damage to spermatozoa in vitro, although superoxide anion is also a significant but short-lived stressor [36, 78, 86–88]. The NOX family of NOXs have been shown to be crucial components in a wide range of cells, including sperm that generate ROS (see excellent review by Bedard and Krause [60]). These proteins function by transferring electrons across lipid membranes using oxygen as electron acceptor. The product of this transfer is superoxide.

3.3.3 Influence of Low and Ultra-Low Temperatures on ROS and Free Radicals

Since ROS affects living cells as a result of freezing, cryogenic storage, and subsequent thawing, we should consider some of the effects from low-temperature storage. The enzyme-driven chemical reactions described in Fig. 3.1 and the generation of ROS occur in nature at body temperature. Sperm function has been studied at body temperatures (37°C) as well as at ambient environmental temperatures since it has been shown that maintaining sperm at body temperature after ejaculation or collection is highly detrimental to maintenance of sperm motility and viability. Ejaculated sperm, depending on species, can function close to optimally at these temperatures. It has not been studied how low temperature influences spermatozoal enzyme systems, although it is likely that in mammalian systems they are optimized for mammalian body temperatures and any ROS that are generated due to osmotic and oxidative stress are scavenged suboptimally at subzero temperatures. As alluded to earlier, low temperatures can predispose cells to oxidative attack by affecting primary metabolic processes, membrane integrity of the plasma, and mitochondrial membranes, and compromise the activity of enzymes that may protect cells from ROS and RNS. Much of our knowledge of low-temperature cell biology derives from studies in plants, yeast, and cell-free frozen solutions [89]. Plants are highly vulnerable to oxidative stress since O_2 is a terminal electron acceptor in respiration and is produced by normal photosynthetic mechanisms. The technique of electron paramagnetic resonance (EPR) spectroscopy which monitors unpaired electrons has been used to determine that superoxide anion trapped in ice at −196°C remains unaltered for a minimum of 7 days [90]. This suggests that there are several main considerations for sperm stored at subzero temperatures: (1) production and stability of cooling-induced ROS, particularly free radicals, at low temperature; (2) functionality of SOD, catalase, and glutathione peroxidase at low temperatures; and (3) presence of unfrozen water in the intracellular and extracellular compartments. Regarding the first, any ROS formed prior to attainment of cellular freezing could remain stable during the cryopreservation and subzero period. It is not known as yet for mammalian cells whether new ROS can be formed at subzero temperatures. Secondly, it is not likely that enzyme scavenging systems would be functional at temperatures associated with cooling to subzero temperatures nor warming during the thaw process until ambient temperature is reached to allow normal enzyme conformation and subsequent function. For scavenging to be at least partially optimal, temperatures would likely need to be in the 15–37°C range to prevent additional oxidative damage from surviving and newly generated ROS. A paucity of information exists for these subjects currently. In plants, experimental overexpression of SOD genes in transgenic alfalfa crops demonstrated enhanced winter survival indicating that SOD expression minimizes ROS damage when applied to this crop [91]. Plants have a number of other adaptive responses to surviving winter extreme temperatures that mammalian cells do not have. However, animal cells have some similar capabilities and can be potentially modified or therapeutically managed to better survive cryopreservation.

Very small amounts of water are known to be associated with proteins, carbohydrates, and membranes associated with cells and there is controversy surrounding the role of this water when cells are cryopreserved [92]. It has been estimated that intact cells retain up to 5% of total intracellular water that may be excluded from freezing due to very tight association with protein and membrane structure. Some researchers consider this water to be nonfreezable and that it may play a significant role in tissue damage. An association of ROS with this unfreezable cellular water has not been made, but it is tempting to speculate that unfrozen water could potentially serve as a reservoir for ROS produced during cryopreservation and its associated dehydration process.

3.3.4 ROS Are Generated by Osmotic Stress During Cryopreservation

Several steps in a typical cryopreservation protocol are responsible for the osmotic stress that cells experience. First, during the loading of cells with cryopreservative agent (CPA), they typically experience a hyperosmotic stress and shrink on initial exposure to 1 M glycerol. This initial and immediate response is linear (an ideal osmometer) and occurs by passive membrane diffusion in which water travels down its concentration gradient and requires no cellular control. This adjustment in cell volume follows the Boyle van't Hoff (BVH) relationship,

$$\frac{V}{V_{is}} = \frac{M_{iso}}{M}\left[1 - \frac{V_b}{V_{is}}\right] + \frac{V_b}{V_{is}}.$$

The BVH equation for determination of cell volume: V_b is the inactive cell volume that is excluded from osmotic response [11].

and is known to occur in sperm in usually less than 1–2 s. Subsequently, the cells re-expand as they equilibrate to the CPA. Then, if volume regulation occurs, there will be a slow change in volume over a time period of minutes to tens of minutes. This final change would be evidence of regulated volume increase or decrease (RVI or RVD). They return to near-isotonic volume (but slightly expanded because of the CPA loading). Next, the cells are exposed to a slow cooling protocol. During this phase of cryopreservation, the cells are dehydrated and typically must lose about 90% of their cell water to survive [11, 93] which means substantial shrinkage of the cell: the exact amount depending on the inactive cell volume, V_b. Finally, after thawing, cryoprotectant must be washed out of the cell. If done in a single step, the cell, which is now internally in a hyperosmotic state because of the presence of CPA, is placed in an isotonic solution. The cell responds by rapidly swelling in volume followed by a slower return to isosmotic volume as the CPA leaves the cell. Although most spermatozoa have narrowly defined osmolal conditions in which cell function is lost, a mechanism for this loss in cell function has not been determined. Since sperm cells emerge from freezing with primarily loss of function (motility) rather than compromised structure, it is likely that events other than formation of intracellular ice contribute to cryodamage of these cells.

Most cells, including spermatozoa, are capable of regulating their intracellular osmotic balance and, hence, cell volume, when exposed to osmotic challenge. This is a highly conserved cell function and is extremely important for homeostatic balance in mammals for essentially all cell functions. This volume regulation is linked to function of the cell's cytoskeleton and plasma membrane system. Spermatozoa are exposed to various environments during their life in vivo that impart osmotic challenges. The distal cauda epididymis, for example, is known to be significantly hypertonic in comparison to testicular and caput epididymal fluids. Consequently, when sperm are ejaculated, they encounter a rapid exposure to the relative hypotonicity of the seminal fluid and female reproductive tract, and swell by osmosis. However, excessive cell volume change, as when exposed to cryopreservation, could impair cellular regulatory and metabolic components, resulting in detrimental changes including cell lysing. Volume regulation is important to sperm function and may be damaged by cryopreservation as phospholipid membranes are physically stretched during the process resulting in plasma membrane and mitochondrial injury. When sperm are exposed to anisosmotic conditions, the volume change has been associated with increases in phosphorylation of protein tyrosine residues, a signaling event coincidentally required for sperm capacitation and fertilization [16, 36, 88, 94–96]. It has been hypothesized that when sperm undergo slow freezing, the ability to volume regulate, recover motility, and respond to subsequent post-insemination environments may be compromised. Therefore, the result could be decreased or compromised fertility.

3.3.5 Interactions Between Osmotic and Oxidative Stress

Signal transduction events in spermatozoa are generally believed to be highly regulated by cross talk among intracellular pathways. ROS have been implicated in the promotion of tyrosine phosphorylation associated with sperm capacitation [36, 97]. In addition, recent evidence indicates that components of the extracellular signal-regulated kinase (ERK) family of mitogen-activated protein kinases (MAPKs) are also present in spermatozoa and might be involved in motility and capacitation [98–100]. It was suggested that $O_2^- \bullet$ could modulate the ERK pathway in human sperm [101] and that H_2O_2 could have a similar role in the modulation of MAPK. In somatic cells, the involvement of ROS in cellular signaling is extensive, including activation of MAPK cascades [102]. As described above, osmotic shock can also activate p38 MAPK and Jak/STAT pathways in mammalian cells [103, 104]. Interestingly, Qin and coworkers reported that in B cells ROS were involved in mediating osmotic stress-induced activation of Syk, a protein-tyrosine kinase [105]. Further, the rise in intracellular calcium as seen during freezing [106] will drive ROS generation by the enzyme NOX 5 because the latter possesses EF-hand motifs (calcium-binding regions) that give this oxidase a sensitivity to calcium [66]. The increase in temperature-induced cell calcium is due to a sudden decline in Ca^{2+}/Mg^{2+}-ATPase activity that occurs as sperm pass through their lipid-phase transition temperatures, thus resulting in elevated calcium.

3.3.5.1 Protection of Cryopreserved Sperm from Oxidative Damage

The addition of enzyme scavengers or antioxidants to sperm preparations in vitro has been successful at counteracting some of the effects of oxidative stress on sperm motility, viability, lipid peroxidation, sperm–oocyte fusion, and DNA fragmentation [36, 41, 49, 50, 52, 78, 95]. Accordingly, antioxidant addition to semen extenders has been implicated for cryopreservation of sperm to minimize the potential effects of oxidative stress during the freeze–thaw process. Not surprisingly, there is a wide variation in the outcome of these studies with considerable species differences. Nonetheless, researchers have demonstrated positive effects with the addition of enzyme scavengers and antioxidants to bull, boar, human, and rabbit spermatozoa [48, 53, 56, 86, 107–110].

3.4 Conclusions

The loss of significant numbers of viable cells during cryopreservation clearly leads to a decrease in clinical usefulness of any stored cell type. Consequently, an understanding of the mechanisms by which cryoinjury to sperm occurs is exceedingly important. In this terminally differentiated cell, the plasma membrane serves as the main physical barrier to the extracellular environment, and is therefore a primary site of freeze–thaw damage. Such damage includes membrane destabilization due to lateral lipid rearrangement, loss of lipids from the membrane, and peroxidation of membrane lipids as a result of ROS formation. These events can affect sperm motility and response to osmotic stress and signaling pathways, and thus the ability to reach, bind to, and react with the zona pellucida during fertilization could become compromised [58].

Sperm, although seemingly simple, have a number of cellular compartments, where each is susceptible to various types of environmental stresses. Thus, it is no surprise that a considerable amount of sublethal and lethal damage occurs as a result of exposure to the extreme temperature and osmolality effects from cryopreservation [16–18, 19, 44, 45, 50, 88]. As the cell is exposed to plunging temperatures, extracellular ice crystallization begins which results in the concentration of the surrounding solutes in the unfrozen aqueous channels between ice crystals [11]. This poses a severe osmotic insult to the cell that can be minimized but not eliminated by using CPA agents. This osmotic change has been shown to result in induction of oxidative change that is, in itself, damaging to sperm.

From the information presented in this chapter, it seems that most if not all cell damage incurred from cryopreservation occurs during the cooling and thawing processes, as extracellular space alternates between hypertonic and hypotonic conditions. However, we have seen that at least some oxidative cell damage may occur at steady state during the cryostorage phase of the process of cryopreservation. Clearly, much research remains to be performed with regard to the role of oxidative stress during the phases of cryopreservation.

3.5 Key Points

- Cryopreservation damage to sperm is caused by osmotic/dehydration effects, induction of ROS, and intracellular ice formation at subzero temperatures.
- Superoxide anion and hydrogen peroxide are the two most important ROS in spermatozoal response to cryopreservation.
- Intracellular water remains an unknown source or reservoir of ROS that are produced during various stages of the cryopreservation process.
- Antioxidant therapy may alleviate some, but not necessarily all, generation and scavenging of ROS.

References

1. Tosic J, Walton A. Formation of hydrogen peroxide by spermatozoa and its inhibitory effect of respiration. Nature. 1946;158:485.
2. Macleod J. The role of oxygen in the metabolism and motility of human spermatozoa. Am J Physiol. 1943;138:512–8.
3. Klooster KL, Burruel VR, Meyers SA. Loss of fertilization potential of desiccated rhesus macaque spermatozoa following prolonged storage. Cryobiology. 2011;62(3):161–6.
4. Li MW, Biggers JD, Elmoazzen HY, Toner M, McGinnis L, Lloyd KC. Long-term storage of mouse spermatozoa after evaporative drying. Reproduction. 2007;133(5):919–29.
5. Li MW, Meyers S, Tollner TL, Overstreet JW. Damage to chromosomes and DNA of rhesus monkey sperm following cryopreservation. J Androl. 2007;28(4):493–501.
6. Sanchez R, Risopatron J, Schulz M, et al. Canine sperm vitrification with sucrose: effect on sperm function. Andrologia. 2011;43(4):233–41.
7. Katkov II, Isachenko V, Isachenko E, et al. Low- and high-temperature vitrification as a new approach to biostabilization of reproductive and progenitor cells. Int J Refrig. 2006; 29:346–57.
8. Buchanan SS, Gross SA, Acker JP, Toner M, Carpenter JF, Pyatt DW. Cryopreservation of stem cells using trehalose: evaluation of the method using a human hematopoietic cell line. Stem Cells Dev. 2004;13(3):295–305.
9. Reubinoff BE, Pera MF, Vajta G, Trounson AO. Effective cryopreservation of human embryonic stem cells by the open pulled straw vitrification method. Hum Reprod. 2001;16(10): 2187–94.
10. Massip A, Blesbois E. Cryobiology of gametes and the breeding of domestic animals. In: Lane N, Fuller J, Benson E, editors. Life in the frozen state. Boca Raton: CRC Press LLC; 2004. p. 347–70.
11. Mazur P. Principles of cryobiology. In: Fuller BJ, Lane N, Benson EE, editors. Life in the frozen state. Boca Raton: CRC; 2004.
12. Watson PF. The causes of reduced fertility with cryopreserved semen. Anim Reprod Sci. 2000;60:481–92.
13. Massip A. Cryopreservation of bovine oocytes: current status and recent developments. Reprod Nutr Dev. 2003;43(4):325–30.
14. Graham EF, Schmehl MKL, Evensen BK, Nelson DS. Viability assays for frozen semen. Cryobiology. 1978;15(2):242–4.
15. Johnson LA, Weitze KF, Fiser P, Maxwell WMC. Storage of boar semen. Anim Reprod Sci. 2000;62(1–3):143–72.
16. Pommer AC, Rutllant J, Meyers SA. The role of osmotic resistance on equine spermatozoal function. Theriogenology. 2002;58(7):1373–84.
17. Pommer AC, Rutllant J, Meyers SA. Phosphorylation of protein tyrosine residues in fresh and cryopreserved stallion spermatozoa under capacitating conditions. Biol Reprod. 2003;68(4): 1208–14.
18. Rutllant J, Pommer AC, Meyers SA. Osmotic tolerance limits and properties of rhesus monkey (Macaca mulatta) spermatozoa. J Androl. 2003;24(4):534–41.
19. Correa LM, Thomas A, Meyers SA. The macaque sperm actin cytoskeleton reorganizes in response to osmotic stress and contributes to morphological defects and decreased motility. Biol Reprod. 2007;77(6):942–53.
20. Baumber J, Ball BA, Linfor JJ, Meyers SA. Reactive oxygen species and cryopreservation promote DNA fragmentation in equine spermatozoa. J Androl. 2003;24(4):621–8.
21. Gandini L, Lombardo F, Lenzi A, Spano M, Dondero F. Cryopreservation and sperm DNA integrity. Cell Tissue Bank. 2006;7(2):91–8.
22. Toro E, Fernandez S, Lopez-Teijon M, Alvarez J, Velillal E. Cryopreservation of human sperm results in an increase in sperm DNA fragmentation. J Androl. 2009;30:90–1.

23. Hughes CM, Lewis SE, McKelvey-Martin VJ, Thompson W. A comparison of baseline and induced DNA damage in human spermatozoa from fertile and infertile men, using a modified comet assay. Mol Hum Reprod. 1996;2(8):613–9.
24. Kodama H, Yamaguchi R, Fukuda J, Kasai H, Tanaka T. Increased oxidative deoxyribonucleic acid damage in the spermatozoa of infertile male patients. Fertil Steril. 1997;68(3):519–24.
25. Twigg JP, Irvine DS, Aitken RJ. Oxidative damage to DNA in human spermatozoa does not preclude pronucleus formation at intracytoplasmic sperm injection. Hum Reprod. 1998; 13(7):1864–71.
26. Barratt CL, Aitken RJ, Bjorndahl L, et al. Sperm DNA: organization, protection and vulnerability: from basic science to clinical applications—a position report. Hum Reprod. 2010; 25(4):824–38.
27. Baker MA, Aitken RJ. Reactive oxygen species in spermatozoa: methods for monitoring and significance for the origins of genetic disease and infertility. Reprod Biol Endocrinol. 2005;3:67.
28. Ahmadi A, Ng SC. Fertilizing ability of DNA-damaged spermatozoa. J Exp Zool. 1999; 284(6):696–704.
29. Evenson DP, Jost LK, Marshall D, et al. Utility of the sperm chromatin structure assay as a diagnostic and prognostic tool in the human fertility clinic. Hum Reprod. 1999;14(4):1039–49.
30. Zini A, Bielecki R, Phang D, Zenzes MT. Correlations between two markers of sperm DNA integrity, DNA denaturation and DNA fragmentation, in fertile and infertile men. Fertil Steril. 2001;75(4):674–7.
31. Cho C, Jung-Ha H, Willis WD, et al. Protamine 2 deficiency leads to sperm DNA damage and embryo death in mice. Biol Reprod. 2003;69(1):211–7.
32. Irvine DS, Twigg JP, Gordon EL, Fulton N, Milne PA, Aitken RJ. DNA integrity in human spermatozoa: relationships with semen quality. J Androl. 2000;21(1):33–44.
33. Wehbi E, Meriano J, Laskin C, Jarvi KA. Adverse Ivf/Icsi outcomes associated with higher levels of sperm DNA fragmentation. J Urol. 2009;181(4):688.
34. Zini A, Libman J. Sperm DNA damage: clinical significance in the era of assisted reproduction. CMAJ. 2006;175(5):495–500.
35. Sakkas D, Seli E, Bizzaro D, Tarozzi N, Manicardi GC. Abnormal spermatozoa in the ejaculate: abortive apoptosis and faulty nuclear remodelling during spermatogenesis. Reprod Biomed Online. 2003;7(4):428–32.
36. Aitken RJ, Curry BJ. Redox regulation of human sperm function: from the physiological control of sperm capacitation to the etiology of infertility and DNA damage in the germ line. Antioxid Redox Signal. 2011;14(3):367–81.
37. Bilodeau JF, Chatterjee S, Sirard MA, Gagnon C. Levels of antioxidant defenses are decreased in bovine spermatozoa after a cycle of freezing and thawing. Mol Reprod Dev. 2000;55(3): 282–8.
38. Gadea J, Selles E, Marco MA, et al. Decrease in glutathione content in boar sperm after cryopreservation. Effect of the addition of reduced glutathione to the freezing and thawing extenders. Theriogenology. 2004;62(3–4):690–701.
39. Chatterjee S, de Lamirande E, Gagnon C. Cryopreservation alters membrane sulfhydryl status of bull spermatozoa: protection by oxidized glutathione. Mol Reprod Dev. 2001;60(4): 498–506.
40. Chatterjee S, Gagnon C. Production of reactive oxygen species by spermatozoa undergoing cooling, freezing, and thawing. Mol Reprod Dev. 2001;59(4):451–8.
41. Ball BA, Medina V, Gravance CG, Baumbe J. Effect of antioxidants on preservation of motility, viability and acrosomal integrity of equine spermatozoa during storage at 5 degrees C. Theriogenology. 2001;56(4):577–89.
42. Ball BA, Vo AT, Baumber J. Generation of reactive oxygen species by equine spermatozoa. Am J Vet Res. 2001;62(4):508–15.
43. Wang Y, Sharma RK, Agarwal A. Effect of cryopreservation and sperm concentration on lipid peroxidation in human semen. Urology. 1997;50(3):409–13.

44. McCarthy MJ, Baumber J, Kass PH, Meyers SA. Osmotic stress induces oxidative cell damage to rhesus macaque spermatozoa. Biol Reprod. 2010;82(3):644–51.
45. McCarthy MJ, Meyers SA. Antioxidant treatment in the absence of exogenous lipids and proteins protects rhesus macaque sperm from cryopreservation-induced cell membrane damage. Theriogenology. 2011;76(1):168–76.
46. Sofikitis N, Miyagawa I, Dimitriadis D, Zavos P, Sikka S, Hellstrom W. Effects of smoking on testicular function, semen quality and sperm fertilizing capacity. J Urol. 1995;154(3):1030–4.
47. Aitken RJ, De Iuliis GN, Finnie JM, Hedges A, McLachlan RI. Analysis of the relationships between oxidative stress, DNA damage and sperm vitality in a patient population: development of diagnostic criteria. Hum Reprod. 2010;25(10):2415–26.
48. Alvarez JG, Storey BT. Role of superoxide-dismutase in protecting rabbit spermatozoa from O-2 toxicity due to lipid-peroxidation. Biol Reprod. 1983;28(5):1129–36.
49. Barbas JP, Mascarenhas RD. Cryopreservation of domestic animal sperm cells. Cell Tissue Bank. 2009;10(1):49–62.
50. Tatone C, Di Emidio G, Vento M, Ciriminna R, Artini PG. Cryopreservation and oxidative stress in reproductive cells. Gynecol Endocrinol. 2010;26(8):563–7.
51. Aitken RJ, Paterson M, Fisher H, Buckingham DW, van Duin M. Redox regulation of tyrosine phosphorylation in human spermatozoa and its role in the control of human sperm function. J Cell Sci. 1995;108(Pt 5):2017–25.
52. Aitken RJ, Roman SD. Antioxidant systems and oxidative stress in the testes. Adv Exp Med Biol. 2008;636:154–71.
53. Bailey JL, Bilodeau JF, Cormier N. Semen cryopreservation in domestic animals: a damaging and capacitating phenomenon. J Androl. 2000;21(1):1–7.
54. Baumber J, Ball BA, Linfor JJ, Meyers SA. Reactive oxygen species and cryopreservation promote DNA fragmentation in equine spermatozoa. J Androl. 2003;24(4):621–8.
55. Linfor JJ, Meyers SA. Detection of DNA damage in response to cooling injury in equine spermatozoa using single-cell gel electrophoresis. J Androl. 2002;23(1):107–13.
56. Bilodeau JF, Blanchette S, Cormier N, Sirard MA. Reactive oxygen species-mediated loss of bovine sperm motility in egg yolk Tris extender: protection by pyruvate, metal chelators and bovine liver or oviductal fluid catalase. Theriogenology. 2002;57(3):1105–22.
57. Fraser L, Strzezek J. Effects of freezing-thawing on DNA integrity of boar spermatozoa assessed by the neutral comet assay. Reprod Domest Anim. 2005;40(6):530–6.
58. Holt WV. Basic aspects of frozen storage of semen. Anim Reprod Sci. 2000;62(1–3):3–22.
59. Ricker JV, Linfor JJ, Delfino WJ, et al. Equine sperm membrane phase behavior: the effects of lipid-based cryoprotectants. Biol Reprod. 2006;74(2):359–65.
60. Bedard K, Krause KH. The NOX family of ROS-generating NADPH oxidases: physiology and pathophysiology. Physiol Rev. 2007;87(1):245–313.
61. Pourova J, Kottova M, Voprsalova M, Pour M. Reactive oxygen and nitrogen species in normal physiological processes. Acta Physiol (Oxf). 2010;198(1):15–35.
62. Aitken RJ, Clarkson JS. Significance of reactive oxygen species and antioxidants in defining the efficacy of sperm preparation techniques. J Androl. 1988;9(6):367–76.
63. Aitken RJ, Clarkson JS, Fishel S. Generation of reactive oxygen species, lipid peroxidation, and human sperm function. Biol Reprod. 1989;41(1):183–97.
64. Aitken RJ, West KM. Analysis of the relationship between reactive oxygen species production and leucocyte infiltration in fractions of human semen separated on Percoll gradients. Int J Androl. 1990;13(6):433–51.
65. Aitken RJ. Molecular mechanisms regulating human sperm function. Mol Hum Reprod. 1997;3(3):169–73.
66. Banfi B, Molnar G, Maturana A, et al. A Ca(2+)-activated NADPH oxidase in testis, spleen, and lymph nodes. J Biol Chem. 2001;276(40):37594–601.
67. Ford WC. Regulation of sperm function by reactive oxygen species. Hum Reprod Update. 2004;10(5):387–99.
68. Sabeur K, Ball BA. Characterization of NADPH oxidase 5 in equine testis and spermatozoa. Reproduction. 2007;134(2):263–70.

69. Ford WC, Whittington K, Williams AC. Reactive oxygen species in human sperm suspensions: production by leukocytes and the generation of NADPH to protect sperm against their effects. Int J Androl. 1997;20 Suppl 3:44–9.
70. de Lamirande E, Harakat A, Gagnon C. Human sperm capacitation induced by biological fluids and progesterone, but not by NADH or NADPH, is associated with the production of superoxide anion. J Androl. 1998;19:215–25.
71. Armstrong JS, Bivalacqua TJ, Chamulitrat W, Sikka S, Hellstrom WJ. A comparison of the NADPH oxidase in human sperm and white blood cells. Int J Androl. 2002;25(4):223–9.
72. Richer SC, Ford WC. A critical investigation of NADPH oxidase activity in human spermatozoa. Mol Hum Reprod. 2001;7(3):237–44.
73. Baker MA, Krutskikh A, Curry BJ, McLaughlin EA, Aitken RJ. Identification of cytochrome P450-reductase as the enzyme responsible for NADPH-dependent lucigenin and tetrazolium salt reduction in rat epididymal sperm preparations. Biol Reprod. 2004;71(1):307–18.
74. Shukla S, Jha RK, Laloraya M, Kumar PG. Identification of non-mitochondrial NADPH oxidase and the spatio-temporal organization of its components in mouse spermatozoa. Biochem Biophys Res Commun. 2005;331(2):476–83.
75. Baker MA, Reeves G, Hetherington L, Muller J, Baur I, Aitken RJ. Identification of gene products present in Triton X-100 soluble and insoluble fractions of human spermatozoa lysates using LC-MS/MS analysis. Proteomics Clin Appl. 2007;1(5):524–32.
76. Halliwell B, Cross CE. Reactive oxygen species, antioxidants, and acquired immunodeficiency syndrome. Sense or speculation? Arch Intern Med. 1991;151(1):29–31.
77. Aitken RJ, West K, Buckingham D. Leukocytic infiltration into the human ejaculate and its association with semen quality, oxidative stress, and sperm function. J Androl. 1994;15(4): 343–52.
78. Ball BA. Oxidative stress, osmotic stress and apoptosis: impacts on sperm function and preservation in the horse. Anim Reprod Sci. 2008;107(3–4):257–67.
79. Ball BA, Vo A. Detection of lipid peroxidation in equine spermatozoa based upon the lipophilic fluorescent dye C11-BODIPY581/591. J Androl. 2002;23(2):259–69.
80. Jang HY, Kim YH, Kim BW, et al. Ameliorative effects of melatonin against hydrogen peroxide-induced oxidative stress on boar sperm characteristics and subsequent in vitro embryo development. Reprod Domest Anim. 2010;45(6):943–50.
81. Ramya T, Misro MM, Sinha D, Nandan D, Mithal S. Altered levels of seminal nitric oxide, nitric oxide synthase, and enzymatic antioxidants and their association with sperm function in infertile subjects. Fertil Steril. 2011;95(1):135–40.
82. Koppers AJ, De Iuliis GN, Finnie JM, McLaughlin EA, Aitken RJ. Significance of mitochondrial reactive oxygen species in the generation of oxidative stress in spermatozoa. J Clin Endocrinol Metab. 2008;93(8):3199–207.
83. Koppers AJ, Garg ML, Aitken RJ. Stimulation of mitochondrial reactive oxygen species production by unesterified, unsaturated fatty acids in defective human spermatozoa. Free Radic Biol Med. 2010;48(1):112–9.
84. Brookes PS, Levonen AL, Shiva S, Sarti P, Darley-Usmar VM. Mitochondria: regulators of signal transduction by reactive oxygen and nitrogen species. Free Radic Biol Med. 2002; 33(6):755–64.
85. Sanocka D, Miesel R, Jedrzejczak P, Kurpisz MK. Oxidative stress and male infertility. J Androl. 1996;17(4):449–54.
86. Aitken RJ, Baker MA. Oxidative stress, sperm survival and fertility control. Mol Cell Endocrinol. 2006;250(1–2):66–9.
87. Aitken RJ, Buckingham DW, Carreras A, Irvine DS. Superoxide dismutase in human sperm suspensions: relationship with cellular composition, oxidative stress, and sperm function. Free Radic Biol Med. 1996;21(4):495–504.
88. Burnaugh L, Ball BA, Sabeur K, Thomas AD, Meyers SA. Osmotic stress stimulates generation of superoxide anion by spermatozoa in horses. Anim Reprod Sci. 2010;117(3–4):249–60.
89. Benson E, Bremner DH. Oxidative stress in the frozen plant: a free radical point of view. In: Fuller BJ, Lane N, Benson EE, editors. Life in the frozen state. Boca Raton: CRC; 2004. p. 205–41.

90. Dambrova M, Baumane L, Kalvinsh I, Wikberg JE. Improved method for EPR detection of DEPMPO-superoxide radicals by liquid nitrogen freezing. Biochem Biophys Res Commun. 2000;275(3):895–8.
91. McKersie BD, Bowley SR, Jones KS. Winter survival of transgenic alfalfa overexpressing superoxide dismutase. Plant Physiol. 1999;119(3):839–48.
92. Crowe JH, Crowe LM, Tablin F, Wolkers W, Oliver AE, Tsvetkova N. Stabilization of cells during freeze-drying: the trehalose myth. In: Fuller BJ, Lane N, Benson EE, editors. Life in the frozen state. Boca Raton: CRC; 2004. p. 581–601.
93. Mazur P, Leibo SP, Seidel Jr GE. Cryopreservation of the germplasm of animals used in biological and medical research: importance, impact, status, and future directions. Biol Reprod. 2008;78(1):2–12.
94. Aitken RJ, Baker MA, De Iuliis GN, Nixon B. New insights into sperm physiology and pathology. Handb Exp Pharmacol. 2010;198:99–115.
95. Baumber J, Sabeur K, Vo A, Ball BA. Reactive oxygen species promote tyrosine phosphorylation and capacitation in equine spermatozoa. Theriogenology. 2003;60(7):1239–47.
96. Sakkas D, Leppens-Luisier G, Lucas H, et al. Localization of tyrosine phosphorylated proteins in human sperm and relation to capacitation and zona pellucida binding. Biol Reprod. 2003;68(4):1463–9.
97. Aitken RJ, Henkel RR. Sperm cell biology: current perspectives and future prospects. Asian J Androl. 2011;13(1):3–5.
98. Ashizawa K, Hashimoto K, Higashio M, Tsuzuki Y. The addition of mitogen-activated protein kinase and p34(cdc2) kinase substrate peptides inhibits the flagellar motility of demembranated fowl spermatozoa. Biochem Biophys Res Commun. 1997;240(1):116–21.
99. Lu Q, Sun QY, Breitbart H, Chen DY. Expression and phosphorylation of mitogen-activated protein kinases during spermatogenesis and epididymal sperm maturation in mice. Arch Androl. 1999;43(1):55–66.
100. Luconi M, Barni T, Vannelli GB, et al. Extracellular signal-regulated kinases modulate capacitation of human spermatozoa. Biol Reprod. 1998;58(6):1476–89.
101. de Lamirande E, Gagnon C. The extracellular signal-regulated kinase (ERK) pathway is involved in human sperm function and modulated by the superoxide anion. Mol Hum Reprod. 2002;8(2):124–35.
102. Droge W. Free radicals in the physiological control of cell function. Physiol Rev. 2002; 82(1):47–95.
103. Bode JG, Gatsios P, Ludwig S, et al. The mitogen-activated protein (MAP) kinase p38 and its upstream activator MAP kinase kinase 6 are involved in the activation of signal transducer and activator of transcription by hyperosmolarity. J Biol Chem. 1999;274(42):30222–7.
104. Gatsios P, Terstegen L, Schliess F, et al. Activation of the Janus kinase signal transducer and activator of transcription pathway by osmotic shock. J Biol Chem. 1998;273(36):22962–8.
105. Qin SF, Ding JY, Takano T, Yamamura H. Involvement of receptor aggregation and reactive oxygen species in osmotic stress-induced Syk activation in B cells. Biochem Biophys Res Commun. 1999;262(1):231–6.
106. Irvine DS, Aitken RJ. Measurement of intracellular calcium in human-spermatozoa. Gamete Res. 1986;15(1):57–71.
107. Askari HA, Check JH, Peymer N, Bollendorf A. Effect of natural antioxidants tocopherol and ascorbic acids in maintenance of sperm activity during freeze-thaw process. Arch Androl. 1994;33(1):11–5.
108. Breininger E, Beorlegui NB, O'Flaherty CM, Beconi MT. Alpha-tocopherol improves biochemical and dynamic parameters in cryopreserved boar semen. Theriogenology. 2005;63(8):2126–35.
109. Killian G, Honadel T, Mcnutt T, Henault M, Wegner C, Dunlap D. Evaluation of butylated hydroxytoluene as a cryopreservative added to whole or skim milk diluent for bull semen. J Dairy Sci. 1989;72(5):1291–5.
110. Pena FJ, Johannisson A, Wallgren M, Martinez HR. Antioxidant supplementation in vitro improves boar sperm motility and mitochondrial membrane potential after cryopreservation of different fractions of the ejaculate. Anim Reprod Sci. 2003;78(1–2):85–98.

Chapter 4
Sperm Capacitation as an Oxidative Event

Eve de Lamirande and Cristian O'Flaherty

Abstract Capacitation is associated with a mild oxidative stress. Spermatozoa are the only cell type in which a reciprocal reactive oxygen species (ROS)-induced ROS formation is demonstrated, the superoxide anion ($O_2^{\bullet-}$) promoting nitric oxide ($NO^{\bullet}$) synthesis and vice versa. Both $O_2^{\bullet-}$ and $NO^{\bullet}$ are essential and are synthesized by sperm enzymes, an oxidase, and nitric oxide synthase (NOS), and they have specific targets. Semenogelin, zinc ions, and the sulfhydryl/disulfide couple are natural regulator for both the oxidase and NOS and are also expected to directly regulate some of the enzymes involved in transduction cascades triggered by $O_2^{\bullet-}$ and $NO^{\bullet}$. ROS promote all the signal transduction pathways involved during capacitation leading to the late protein Tyr phosphorylation. Sperm hyperactivation and acrosome reaction are also modulated by ROS. Therefore, ROS play major role in sperm activation and related stimulation of several processes, multiplicity of enzymatic pathways and types of regulation, cross talks, apparent redundancy of mechanisms, etc. This is expected to insure that spermatozoa acquire their fertility potential.

Keywords Sperm capacitation • Mild oxidative event • Reactive oxygen species • Cell physiology • Sperm activation • Targets for ROS

4.1 Introduction

Spermatozoa are complex and compartmentalized cells subjected to sequential maturational steps in the testis, the epididymis, and finally the female genital tract [1]. This latter step is associated with a series of timely and finely regulated changes,

---

E. de Lamirande, PhD (✉) • C. O'Flaherty, DVM, PhD
Urology Research Laboratory, McGill University Health Center, Royal Victoria Hospital, Montréal, QC, Canada H3A 1A1
e-mail: edelamirande@yahoo.com

A. Agarwal et al. (eds.), *Studies on Men's Health and Fertility*, Oxidative Stress in Applied Basic Research and Clinical Practice, DOI 10.1007/978-1-61779-776-7_4,

collectively called capacitation, that are essential for the acquisition of fertilizing ability [1–6]. Various events occur during capacitation, such as increases in intracellular pH and calcium (Ca^{2+}), production of low and controlled levels of reactive oxygen species (ROS), increase in membrane fluidity due to loss of cholesterol, and activation of several signal transduction cascades and related protein kinases resulting in the subsequent phosphorylation of numerous proteins on serine (Ser), threonine (Thr), and tyrosine (Tyr) residues [1–10]. In parallel, spermatozoa develop a very specific type of movement, called hyperactivation, which is characterized by high speed, low linearity, and high amplitude of lateral head displacement [1, 3, 6]. Hyperactivation involves higher force, which allows spermatozoa to detach from the oviductal epithelium after capacitation is completed, swim through media of higher viscosity and elasticity as can be found in the female genital tract and around the oocytes, and finally penetrate through the zona pellucida [1, 3, 6].

Once capacitated spermatozoa can attach to the zona pellucida that surrounds the eggs; as a result, they undergo the acrosome reaction, which is an exocytotic event involving the release of enzymes (acrosin, hyaluronidase, etc.) that hydrolyze proteins of the zona pellucida, thus helping their progression toward the egg [1, 3, 7, 11, 12]. The acrosome reaction is a complex and irreversible process that occurs over a short period of time (within minutes); it is also associated with ROS generation as well as specific ion fluxes, activation of signal transduction cascades, and phosphorylation of proteins [1–3, 7, 9, 13, 14].

Sperm activation normally occurs in the female genital tract and its physiological inducers are presently not all known; we can expect that they are many and also that they may interact with each other. Sperm capacitation, hyperactivation, and acrosome reaction can be studied in vitro using defined media and various inducers, but it should always be remembered that all these are models and that their use has to be established with more than one inducer or conditions.

Numerous cellular processes, including acquisition of fertilizing ability by spermatozoa, are regulated by ROS [1, 3, 13, 15–18]. Some of the best-studied ROS for their positive role in cell biology are the superoxide anion ($O_2^{\bullet -}$), its dismutation product, hydrogen peroxide (H_2O_2), nitric oxide ($NO^{\bullet}$), and the peroxynitrite anion ($ONOO^-$) (Table 4.1). The effect of these ROS is limited not only by the action of enzymes, such as superoxide dismutase (SOD), catalase, glutathione peroxidase (GPX), thioredoxin (TRX), and peroxiredoxin (PRDX), but also by small molecules, such as vitamins (E, C, etc.) and sulfhydryl (SH)-containing substances (glutathione, protein cysteine residues, Cys) that dispose of ROS [19–23]. We are aware that the ROS and ROS scavengers listed above represent only few players in a very complex scheme of ROS reactions, but they appear as the most important for their actions in cell physiology. The reader is referred to very-well-written books for more extensive data on ROS chemistry and biology [19, 24].

In this chapter, we summarize the present knowledge on the positive role of ROS during sperm activation. We center our attention mainly on studies performed on human spermatozoa and use data reported on other species as supplement, or complement, when needed for the benefit of the discussion. The deleterious effects of serious oxidative stresses in which ROS production far exceeds cell defenses and

Table 4.1 Main ROS involved in cell physiology.

ROS	Origin (best recognized)	Half-life	Reactivity	Cell permeant	Specific scavenger (best recognized)
$O_2^{\bullet-}$	One electron reduction of O_2 (metabolism, specific oxidases)	1 ms	Low	No	Superoxide dismutase
H_2O_2	$O_2^{\bullet-}$ dismutation	Minutes to hours	Low to medium	Yes	Catalase, glutathione peroxidase, thioredoxin, peroxiredoxin
$NO^{\bullet}$	L-Arg conversion to L-citrulline by nitric oxide synthase	1–7 s	Low	Yes	None
$ONOO^-$	Reaction of $O_2^{\bullet-}$ with $NO^{\bullet}$	Minutes	High	No	Glutathione peroxidase, peroxiredoxin

causes loss of motility and viability, DNA damage, lipid peroxidation, etc. are emphasized in other chapters of this book.

We first introduce the physiology of ROS and present the first evidences for a positive action of ROS in human spermatozoa as well as relevance to clinical situations. We then report on the sperm production of ROS and more specifically on the methods of measurement, time course during capacitation, and natural modulators (zinc and semenogelin) and generators (oxidases, NOX; nitric oxide synthase, NOS). The study of ROS effects on signal transduction cascades, protein phosphorylation events, and the sulfhydryl/disulfide (SH/SS) couple on sperm proteins follows. We then briefly report on what is known of ROS involvement in other sperm activation events, namely, hyperactivation and acrosome reaction, and finally offer some conclusions, main points to remember, as well as a general outlook on what future research could aim for.

4.2 ROS in Cell Physiology and Sperm Activation: First Data and Clinical Relevance

It was first counterintuitive that ROS may be essential for cell physiology because the general idea is that ROS are toxic. However, the effects ROS produce on cells depend on several factors, such as the reactivity of each ROS and the cell compartment in which it is produced, but even more on the amount formed over a definite period of time and how much this overcomes the scavenging capacity of the system [3, 8, 13, 15–18].

Cell activation is often related to a mild oxidative stress. It is worth stressing at this point that ROS possess the main characteristics of second messengers: minute amounts are locally synthesized by the cells via specialized enzymes (e.g., oxidases, NOS) or normal metabolism; ROS are small and diffusible and have a short half-life and precise targets; and finally their effects are easily reversible. As such, ROS appear as simple, economical, and efficient molecules to trigger signal transduction cascades, affect ion pumps, etc. [3, 8, 15–18].

The number of reports on the essential role of ROS in cell activation increased strikingly over the last 10 years. It is now recognized that activating stimuli, whether physiological (e.g., hormone, such as insulin) or pharmacological (e.g., medication, enzyme activators, or inhibitors), often promote ROS formation [3, 8, 15–18] via oxidases, such as those of the NOX family for $O_2^{\bullet-}$ [25–28] and NOS (endothelial, neuronal, and epithelial isoforms) for $NO^{\bullet}$ [29–34].

It is during our studies on the dose-dependent toxicity of ROS ($O_2^{\bullet-}$ and H_2O_2) on sperm motility and viability [35, 36] that we realized that the lowest concentrations promote hyperactivation [37, 38]. This result was definitely surprising and suggested that capacitation, as a closely related event, might also be driven by ROS. This hypothesis was corroborated as subsequent experiments clearly showed that exogenous addition of ROS ($O_2^{\bullet-}$ from xanthine + xanthine oxidase; H_2O_2 by direct addition or from glucose + glucose oxidase; $NO^{\bullet}$ from spontaneous degradation of chemicals, such as sodium nitroprusside or NONOates) promotes capacitation [8, 37–45]. Conversely, ROS scavengers, such as SOD and catalase, as well as NOS inhibitors (L-NAME, L-NMMA, 7-nitroindazole), prevented capacitation triggered by agents as diverse as bovine serum albumin (BSA), progesterone, L-arginine (L-Arg), ultrafiltrate from fetal cord serum (FCSu), etc. [8, 37–46], further strengthening the concept that ROS are essential for capacitation. The importance of ROS is further evidenced as they appear to control several events of the capacitation process, including the late Tyr phosphorylation of two fibrous sheath proteins of 80 and 105 kDa (p80 and p105) [8, 39, 40, 42–44, 46] which is often considered as a hallmark of capacitation.

Clinical data also support a role for ROS in sperm activation. First, we observed over the years that about 16% of men presenting at the infertility clinic and having normal sperm parameters according to the World Health Organization [47] have recurrent spontaneous sperm hyperactivation in whole semen; this appears linked to a lower SOD-like activity ($O_2^{\bullet-}$ scavenging capacity) in seminal plasma (by 37%) and spermatozoa (by 45%) as compared to what is found in samples from normospermic men [48, 49]. On the other side, we have seen two cases in which an excessively high (2–2.5-fold) $O_2^{\bullet-}$ scavenging capacity in seminal plasma was associated with complete failure at capacitation, hyperactivation, and in vitro fertilization (unpublished data). Therefore, imbalance in the scavenging of $O_2^{\bullet-}$ seems to lead to fertility problems. Furthermore, the concentration of ROS in the medium after sperm washing correlates positively with success in in vitro fertilization and there is then even a positive trend between sperm ROS formation and the four cell stage formation [50]. The sharp increase in oxygen (O_2) pressure, from 5 ± 1 to 41 ± 4 mmHg, in fluids from the female genital tract (golden hamster) around the time of ovulation [51] may be a promoting factor for the increased synthesis of ROS in vivo.

4.3 Generation of ROS

It is very important first to verify the triggering effects of exogenously added ROS and inhibition due to ROS scavengers and then to make sure that more than one capacitation agent is tested (see above) to ascertain that the involvement of ROS is a general phenomenon and not related to only one condition [46]. As stated above, any of the ROS mentioned induces capacitation when added exogenously and this seems to suggest a nonspecific effect of ROS. On the other hand, specific scavengers (SOD, catalase) or NOS inhibitors all block capacitation, which rather indicates that each of the ROS acts on defined and specific targets and this is much more in line with the known high level of organization cells possess. This led us to the hypothesis that capacitating spermatozoa generate ROS at extremely low levels and in the close proximity to their targets so to achieve a defined effect before these ROS spontaneously degrade or react with other cellular components. Therefore, the next challenge was the measurement of ROS in capacitating spermatozoa.

4.3.1 *Measurements*

The assessment of very low levels of ROS needs special caution because the probes themselves and any of the products under study may influence cells or cause interference in the measurement techniques. First of all, most of the probes can affect capacitation, even to the point of inhibition, simply because they react with ROS to produce a signal and, at the same time, they remove too much of these same ROS that are essential to trigger capacitation. An example is that of luminol and lucigenin, two common luminescence probes for the determination of high ROS levels in the semen from infertile men [52–54]. Both probes are then used at concentrations of 200–400 μM, which is very high as compared to the amount of ROS produced by capacitating spermatozoa. As a consequence, the probes themselves block capacitation (at least in our experimental conditions) because they scavenge all $O_2^{\bullet-}$ present (SOD-like effect) [55].

We chose a modified Cypridina luciferin analog (MCLA) as luminescence amplifier and opted for a suboptimal concentration (20 μM) because it allowed $O_2^{\bullet-}$ measurements without affecting capacitation [55]. We first tested MCLA at a lower concentration (10 μM), but the signal was too short due to depletion of the probe; on the other hand, MCLA at higher level (50 μM) blocked capacitation because of extensive $O_2^{\bullet-}$ scavenging. Therefore, the assay performed with MCLA at 20 μM allows comparison between sperm samples (control, treated, capacitated, etc.) and a gross, rather than exact, determination of the amounts of $O_2^{\bullet-}$ formed [46, 55]. To our approximation, 1×10^6 capacitating human spermatozoa would generate about 1 nmole of $O_2^{\bullet-}$ per hour [46].

We confirmed that MCLA-amplified luminescence is specific for $O_2^{\bullet-}$ and not modified by H_2O_2, $NO^\bullet$, or $ONOO^-$ [46, 55]. Then, as for most probes for $O_2^{\bullet-}$

(cytochrome C, nitroblue tetrazolium, etc.), every sample should be tested in the absence and presence of SOD so that only the SOD-inhibitable signal, actually related to $O_2^{\bullet-}$, is considered [19, 55]. Moreover, because MCLA is highly sensitive, even the incubation media (without or with capacitation inducer or enzymatic modulators) should be tested so that their contribution is subtracted [46, 55].

The levels of $O_2^{\bullet-}$ synthesized by spermatozoa is of the same order of magnitude whether capacitation is promoted by FCSu, progesterone, L-Arg, etc. [40, 46, 56], confirming the validity of the data. Then, when comes a situation as for BSA, another known capacitation inducer but that causes a major interference (autoluminescence) in MCLA assay [46, 57], the role of $O_2^{\bullet-}$ has to be deduced from the inhibitory effect of SOD on capacitation [5, 46] and the fact that all the other inducers tested promote $O_2^{\bullet-}$ formation.

Compounds that appear to modify sperm $O_2^{\bullet-}$ production may actually affect the luminescence assay itself rather than cells. For example, yellow compounds, such as 7-nitroindazole (NOS inhibitor), chelerythrine (inhibitor of protein kinase C, PKC), or tyrphostin A47 (inhibitor of protein tyrosine kinase, PTK), often cause interferences in luminescence assays (e.g., striking increase of baseline in the absence of cells) (unpublished observations) and therefore need special care, more controls, and corroboration using other chemicals with similar effects on cells and enzymes. Another example is that of uric acid (end product of xanthine + xanthine oxidase incubation) that inhibits the reaction between $O_2^{\bullet-}$ and luminol, and therefore blocks luminescence, even at 10^{-4} M [35, 36, 58, 59]. These are only examples to stress the importance to assess experimental conditions, always to verify for possible interferences and whenever possible use more than one compound of every class of chemical.

MCLA is cell impermeant and, therefore, useful for evaluation of extracellular $O_2^{\bullet-}$ [53]. Alternatively, dihydrorhodamine 123 concentrates mainly to the sperm mitochondria allowing intracellular $O_2^{\bullet-}$ determination; then, fluorescence increases more in FCSu-treated than in control spermatozoa but addition of SOD, and therefore inhibition of capacitation, does not modify this effect indicating that the raise in $O_2^{\bullet-}$ synthesis probably relates simply to a higher metabolic activity [60, 61]. Dihydroethidium, another $O_2^{\bullet-}$-specific probe, localizes mostly to the sperm head, but the increased fluorescence appears to correlate with a loss of sperm motility and/or altered membrane permeability [62] rather than capacitating conditions [46].

The fluorescent probe dichlorofluorescin diacetate (DCFH-DA) was used to detect H_2O_2 in spermatozoa from infertile men [63], but this chemical is rather nonspecific and reacts with most ROS. Heparin-induced capacitation in bull spermatozoa is associated with increased formation of H_2O_2 as determined by fluorimetry using the *p*-hydroxyphenylacetic acid-horseradish peroxidase system; the fact that diphenyliodonium, an inhibitor of the sperm oxidase, prevents the rise in H_2O_2 indicates that H_2O_2 probably originates from $O_2^{\bullet-}$ dismutation [64]. This is an interesting example of how the use of probes and inhibitors can help to determine which of the ROS plays an initiating role.

The classical methods to detect $NO^{\bullet}$ by measuring L-citrulline (end product of $NO^{\bullet}$ formation) or nitrates and nitrites ($NO^{\bullet}$ metabolites) [65, 66] are not sensitive enough for the very low levels of NO$^{\bullet}$ formed during capacitation [45]. Electrodes sensitive to NO$^{\bullet}$ have been used but only with spermatozoa treated with L-Arg (substrate for NOS) and/or A23187 (calcium ionophore) [29]. Electron spin resonance, using the spin trap sodium *N*-methyl-D-glucamine dithiocarbamate, allows the direct, real-time detection of NO$^{\bullet}$ and evidences that capacitating spermatozoa (BSA as inducer) produce sevenfold higher amounts of NO$^{\bullet}$ than control cells [43]. However, the need of expensive equipment and not common expertise limits the use of this technique. The fluorescence probes diaminofluorescein-2 (Daf2) and its cell-permeant analog Daf2-diacetate (Daf2-DA) are recognized for their specificity and sensitivity and now extensively used, but not often on capacitating spermatozoa [67, 68].

ROS other than NO$^{\bullet}$ and related to capacitation, $O_2^{\bullet-}$, H_2O_2, and $ONOO^-$ do not significantly affect Daf2 fluorescence, and Daf2 (up to 25 μM) and Daf2-DA (10 μM) do not alter sperm capacitation nor $O_2^{\bullet-}$ production [46]. The extracellular NO$^{\bullet}$ formation of spermatozoa (10×10^6 cells; Daf2 25 μM) incubated with FCSu as inducer appears lower than 1 nmole NO$^{\bullet}$ per hour [46]. It may be that NO$^{\bullet}$ synthesis is really that low, but it may be that the NO$^{\bullet}$ formed aims faster and better at its cellular targets than at Daf2. Sperm intracellular NO$^{\bullet}$, as detected after loading with Daf2-DA, is localized mostly in mitochondria and related to metabolic activity but is also present in the sperm head (Fig. 4.1a) [46]. The proportion of spermatozoa with a brighter fluorescence in the head (white arrows) increases following a challenge with FCSu as early as 15 min after the beginning of the incubation and peaks at about 1 h (Fig. 4.1a, b). Of interest also is that this proportion of cells with more fluorescent head similarly increases whether capacitation is induced by FCSu, BSA, L-Arg, etc. (Fig. 4.1c); as expected, the NOS inhibitor L-NMMA (1 mM) prevents this effect and capacitation, confirming the role of NO$^{\bullet}$ and its detection with Daf2-DA in spermatozoa [46].

Again, any product added to the sperm incubation medium needs to be tested for possible interferences. For example, the need for PKC in NO$^{\bullet}$ production cannot be assessed with chelerythrine because this chemical strikingly increases (threefold) the fluorescence of Daf2 incubated with NO$^{\bullet}$-releasing agent (diethylamine-NONOate, DA-NONOate) in cell-free assay and causes an artificial increase in the fluorescence of sperm heads so that they are all very bright; other PKC inhibitors (calphostin and GÖ6976) are, therefore, better choices for studies on spermatozoa [56].

The need for, and production of, both $O_2^{\bullet-}$ and NO$^{\bullet}$ in spermatozoa may imply the formation of the peroxynitrite anion ($ONOO^-$), an ROS with longer half-life and higher reactivity but for which there is no specific probe or scavenger for a test in vitro [19]. Indirect evidence for the formation of $ONOO^-$ may come from an increase in protein Tyr nitration [19]. Exogenous $ONOO^-$ (50 μM) promotes sperm capacitation, the associated Tyr phosphorylation of p80 and p105, as well as Tyr nitration of several sperm proteins [69]. Furthermore, the increase in Tyr nitration of three proteins (105, 116, and 200 kDa) in capacitating spermatozoa (FCSu as inducer) is blocked by SOD and L-NMMA [46], therefore providing an indirect evidence for the endogenous formation of $ONOO^-$ from $O_2^{\bullet-}$ and NO$^{\bullet}$ during this process (Fig. 4.2). The surprise at this point rather comes from the inhibitory effect of catalase on Tyr nitration (Fig. 4.2); this result is, however, rather coherent with our previous one showing that catalase also blocks capacitation triggered by

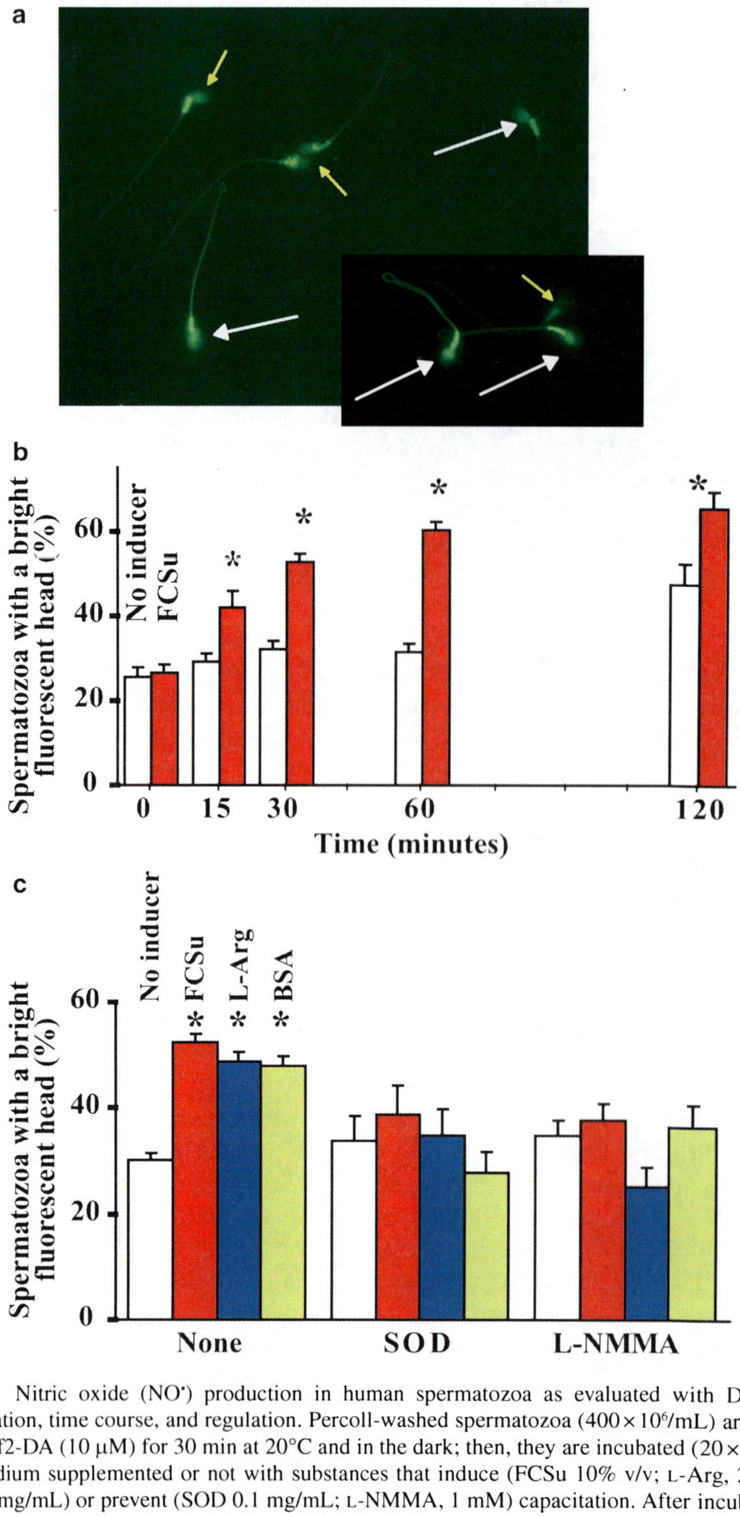

Fig. 4.1 Nitric oxide (NO$^{\bullet}$) production in human spermatozoa as evaluated with Daf2-DA. Localization, time course, and regulation. Percoll-washed spermatozoa (400×10^6/mL) are loaded with Daf2-DA (10 μM) for 30 min at 20°C and in the dark; then, they are incubated (20×10^6/mL) in a medium supplemented or not with substances that induce (FCSu 10% v/v; L-Arg, 2.5 mM; BSA, 3 mg/mL) or prevent (SOD 0.1 mg/mL; L-NMMA, 1 mM) capacitation. After incubation at 37°C, spermatozoa are fixed with formaldehyde (2%) and mounted with a mix of glycerol and water containing DABCO (1.5%, w/v) as antifading agent [46]. (**a**) Fluorescence is concentrated

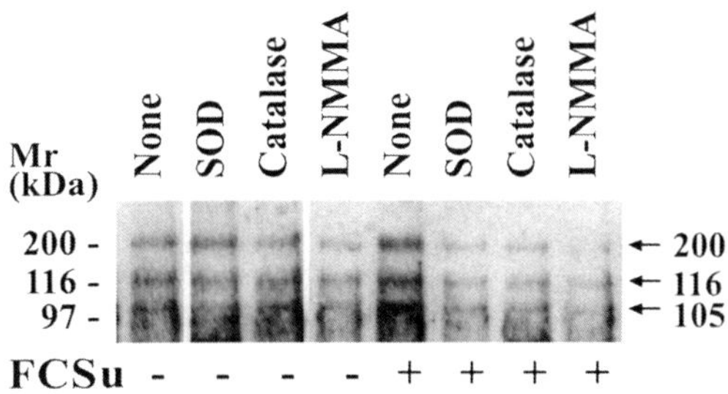

Fig. 4.2 Protein Tyr nitration associated with sperm capacitation is prevented by SOD, catalase, and L-NMMA. Percoll-washed spermatozoa are incubated in BWW medium supplemented (+, capacitating) or not (–, control) with FCSu (10% v/v) and in the absence (none) or presence of SOD (0.1 mg/mL), catalase (0.03 mg/mL), or L-NMMA (1 mM) for 3 h at 37°C. Then, sperm proteins (equivalent to 0.2×10^6 cells/well) are immunoblotted with an anti-nitro-Tyr antibody [46, 69]

exogenous $NO^•$ (NONOates) and supports a role for H_2O_2 in $ONOO^-$ formation [45]. A plausible explanation involves a subsequent oxidation of $NO^•$ to the nitrosonium cation (NO^+) which can then react with H_2O_2 (from $O_2^{•-}$ dismutation) to give $ONOO^-$ [70, 71]. The nature of the Tyr-nitrated proteins and whether this modification is essential for, or only an effect of, capacitation is still under debate.

Therefore, as we attempted to stress above, it is possible to measure ROS produced during sperm capacitation even if levels are extremely low. These tests should not impair cell function and need real care and caution, not only for the choice of the probe and the technique (specificity, interferences, etc.) and tools (activators, inhibitors, chelators, etc.) but also to ascertain that the ROS measured effectively play a role in sperm capacitation (hyperactivation, acrosome reaction). The possibility of ROS reactions with cell components and between themselves should always be kept in mind as well as the need to preserve cell physiology.

4.3.2 Time Course of ROS Formation

Spermatozoa start to produce $O_2^{•-}$ and $NO^•$ as soon as they are incubated with a capacitation inducer [43, 46, 55]. ROS production reaches a maximum at about 30 min for $O_2^{•-}$ and 1 h for $NO^•$ and then decreases slowly over the next hours

Fig. 4.1 (continued) in sperm mitochondria and head. The heads have a low (*short yellow arrows*) or a high (*long white arrows*) fluorescence. (**b**) The percentage of spermatozoa with a bright fluorescent head, indicative of a higher $NO^•$ synthesis, increases in FCSu-treated spermatozoa (*red bars*) as early as 15 min after the beginning of incubation and is higher (*) than that of untreated spermatozoa (*white bars*) at all times ($p<0.05$, $n=5$). (**c**) The percentage of spermatozoa with a bright fluorescent head increases with the three capacitation inducers (FCSu, L-Arg, BSA) tested here but not when conditions (SOD or L-NMMA) prevent this process. These tests are done with 1 h of incubation since it allows the best difference between control and capacitating spermatozoa

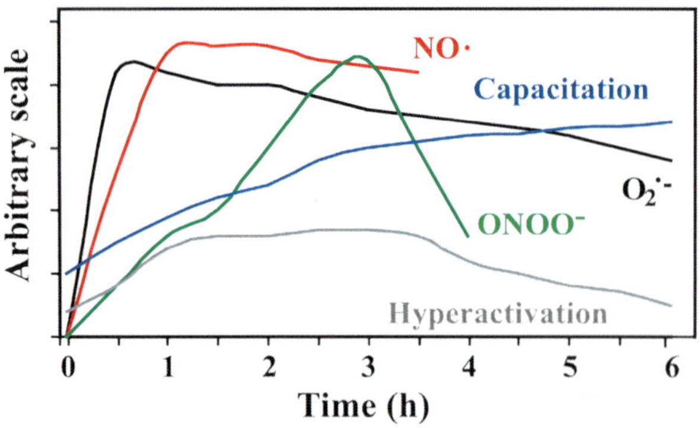

Fig. 4.3 Time course of ROS production in relation to in vitro capacitation and hyperactivation. Tendency curves are proposed rather than numbers in order to give a global idea on how each of the processes varies over time. Percoll-washed spermatozoa are treated with FCSu as capacitation inducer. ROS ($O_2^{\bullet -}$, $NO^{\bullet}$, and $ONOO^-$) are determined as described in the text. Capacitation is evaluated by the lysophosphatidylcholine-induced acrosome reaction followed by labeling with *Pisum sativum* agglutinin conjugated with fluorescein isothiocyanate [2, 55]. Sperm hyperactivation is measured with the CellSoft™ (Cryo Resources, Montgomery, NY) computer-assisted digital image analysis system [2, 55]. Production of $O_2^{\bullet -}$ (*black line*) and $NO^{\bullet}$ (*red line*) starts immediately at the beginning of the incubation and reaches a maximum at 30 min [2, 55] and 1 h [46], respectively, and then slowly decreases over the next hours. Protein Tyr nitration (*green line*), as a measure of $ONOO^-$ formation, is maximal at 3 h and then decreases [46]. Sperm hyperactivation (*gray line*) in vitro is observed between 1 and 3 h and the level of capacitation (*blue line*) is increasing over time [2, 55]

(Fig. 4.3). This time course for $O_2^{\bullet -}$ and $NO^{\bullet}$ appears independent of the triggering agent. Differences between $O_2^{\bullet -}$ and $NO^{\bullet}$ arise when we look at the period of time for which these ROS are essential. Addition of SOD to spermatozoa 30 min after the beginning of incubation does not anymore prevent capacitation [40, 46], indicating that extracellular $O_2^{\bullet -}$ plays a role only at early steps in this process. On the other hand, NOS inhibitors (L-NAME or L-NMMA) block capacitation even if added after 1–4 h of incubation [42, 46], attesting the need for $NO^{\bullet}$ for the whole process. Data obtained with dibutyryl-cAMP (dbcAMP, a cell-permeant analog of cAMP), a model compound that promotes capacitation but bypasses the initial steps of this event, confirm these assumptions. Then, NOS inhibitors (L-NAME or L-NMMA), but not SOD, prevent capacitation due to dbcAMP [43, 46].

At this point, it is important to realize that ROS generation involved in capacitation, at least for what we can detect, is extracellular for $O_2^{\bullet -}$, but in the sperm head for $NO^{\bullet}$ [46]. Therefore, we expect $O_2^{\bullet -}$ to aim at a target on the cell membrane since this ROS, being ionized, should not easily diffuse through plasma membrane. On the other hand, $NO^{\bullet}$ is not charged and a part of what spermatozoa synthesize could pass across cell membranes and act both at intra- and extracellular levels. We can also hypothesize that spermatozoa possess more than one NOS, that one could be on the membrane and the other intracellular, and that each of the NOS acts at different times and on specific targets.

Protein Tyr nitration reaches a maximum 3 h after the beginning of capacitation (Fig. 4.3) [46], a time at which both $O_2^{\bullet-}$ and $NO^{\bullet}$ synthesis are slowly decreasing. A plausible hypothesis is that this reflects an accumulation of protein modifications that are reversible but on a slow rate.

4.3.3 *ROS-Induced ROS*

The close relation between $O_2^{\bullet-}$ and $NO^{\bullet}$ synthesis rises the possibility of an ROS-induced ROS formation as occurs in other cell types. As examples, $NO^{\bullet}$ generation increases in cultured glial cells treated with xanthine + xanthine oxidase, an effect abolished by SOD [72] and, conversely, H_2O_2 stimulates NOS in porcine aortic endothelial cells via a complex signaling pathway [57].

Several observations confirm the hypothesis of an ROS-induced ROS system operating during sperm capacitation [46]. The first indication is that SOD blocks capacitation due to L-Arg (NOS substrate) and $NO^{\bullet}$ itself (from DA-NONOate) and that the NOS inhibitor L-NMMA prevents that due to $O_2^{\bullet-}$ (from xanthine + xanthine oxidase) (Fig. 4.4a) [46]. Second is that endogenous $NO^{\bullet}$ synthesis stops in the presence of SOD (Figs. 4.1c and 4.4b) and conversely that $O_2^{\bullet-}$ formation is arrested by NOS inhibitors (Fig. 4.4c) [46]. Additionally, spermatozoa treated with DA-NONOate initiate the synthesis $O_2^{\bullet-}$ (Fig. 4.4c) and, conversely, spermatozoa treated with $O_2^{\bullet-}$ (from xanthine + xanthine oxidase) produce more $NO^{\bullet}$, an effect blocked in the presence of SOD (Fig. 4.4b) [46]; it should be stressed here that the amounts of ROS ($O_2^{\bullet-}$ and $NO^{\bullet}$) then generated are of the same magnitude as those measured in spermatozoa incubated with FCSu or L-Arg [46]. Moreover, H_2O_2 promotes synthesis of $NO^{\bullet}$ but not of $O_2^{\bullet-}$ suggesting that there is no self-induction of the oxidase [46]. Finally, $ONOO^-$ has no effect on either $O_2^{\bullet-}$ or $NO^{\bullet}$ generation pointing out that this ROS is a final product rather than initiating agent [46].

Pooling of all these data points out to spermatozoa as the first known cells with a reciprocal (two sided) ROS-induced ROS system, $O_2^{\bullet-}$ inducing $NO^{\bullet}$ production and vice versa.

4.3.4 *ROS-Generating Enzymes in Human Sperm Capacitation: NOS and Oxidase, One or Two Enzymes?*

The three isoforms of NOS (epithelial, neuronal, and inducible) are found in spermatozoa both in the head and flagellum (immunoblotting and immunocytochemistry data) [30, 31, 66]. The oxidase responsible for sperm $O_2^{\bullet-}$ generation during capacitation is, however, still elusive. It should be at the membrane level because $O_2^{\bullet-}$ is released in the extracellular milieu and SOD (cell impermeant) prevents capacitation [55]. The first consideration given to the possibility that the sperm enzyme is of the same type as the NADPH oxidase of neutrophils (NOX1) is now

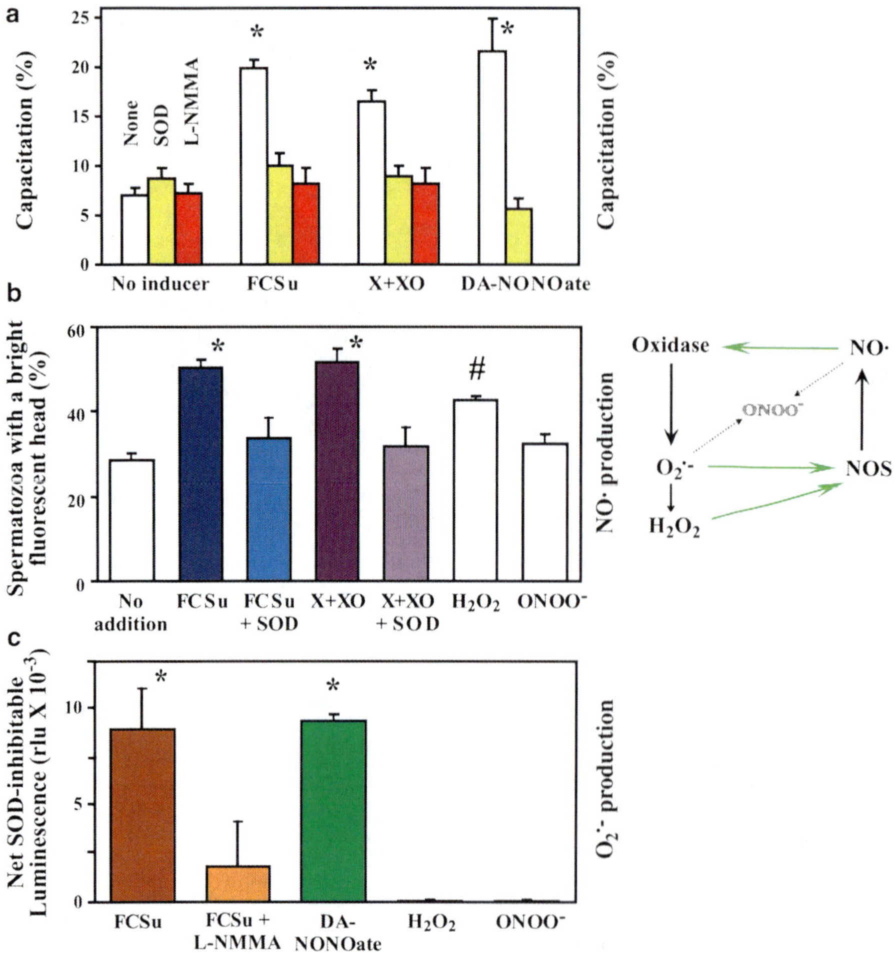

Fig. 4.4 ROS-induced ROS production in human spermatozoa. Percoll-washed spermatozoa are incubated without or with FCSu (10%, v/v), the combination of xanthine (0.1 mM) and xanthine oxidase (0.1 U/mL) (X+XO), DA-NONOate, SOD (0.1 mg/mL), L-NMMA (1 mM), H_2O_2 (25 μM), or $ONOO^-$ (25 μM). Capacitation is evaluated after 3.5 h of incubation by induction of the acrosome reaction [2]. $NO^\bullet$ production is estimated by the percentage of spermatozoa with a bright head (see Fig. 4.1) [46]. Net SOD-inhibitable luminescence (MCLA as a probe) [55] is the measure of $O_2^{\bullet-}$ synthesis. (**a**) Capacitation due to X+XO is blocked by the NOS inhibitor L-NMMA and that induced by DA-NONOate (0.1 mM) is prevented by SOD. (**b**) Spermatozoa treated with FCSu or X+XO produce higher levels of $NO^\bullet$ (*$p<0.05$). SOD blocks this effect confirming the involvement of $O_2^{\bullet-}$ in $NO^\bullet$ synthesis. H_2O_2, but not $ONOO^-$, also promotes $NO^\bullet$ synthesis but to a lower extent (#$p<0.05$) than FCSu or X+XO. (**c**) L-NMMA blocks sperm $O_2^{\bullet-}$ generation due to FCSu and DA-NONOate (25 μM) promotes $O_2^{\bullet-}$ formation, substantiating the role of $NO^\bullet$ in $O_2^{\bullet-}$ synthesis. Spermatozoa treated with DA-NONOate or FCSu produce similar amounts of $O_2^{\bullet-}$. H_2O_2 and $ONOO^-$ do not have such an effect

Table 4.2 Oxidases from spermatozoa and neutrophils (NOX1) are very different

Sperm oxidase	Neutrophil NADPH oxidase, NOX1
Very low levels of $O_2^{\bullet-}$ produced	High levels of $O_2^{\bullet-}$ produced
Synthesis of $O_2^{\bullet-}$ over hours	Burst of $O_2^{\bullet-}$ synthesis for 30–60 min
Aims at cell signaling and transduction	Aims at defense against pathogens
Diphenyliodonium inhibits at 100 μM	Diphenyliodonium inhibits at 10 μM
No need for PKC, ERK, PTK, etc.	PKC, ERK, PTK, etc. needed for activation
$p67^{phox}$, $p47^{phox}$, $p21^{rac}$ not found	$p67^{phox}$, $p47^{phox}$, $p21^{rac}$ are parts of the oxidase
PMA does not activate $O_2^{\bullet-}$ formation	PMA stimulates the oxidative burst
NADPH does not activate	NADPH is essential
FCSu, FFu, progesterone stimulate	FCSu, FFu, progesterone have very little effect
Zn^{2+}, 50 μM, totally inhibits $O_2^{\bullet-}$ production	Zn^{2+}, 150 μM, causes 15% drop in $O_2^{\bullet-}$ formed
Sg, 1 mg/mL, blocks 80% of $O_2^{\bullet-}$ production	Sg, 1 mg/mL, blocks 44% of $O_2^{\bullet-}$ production

invalidated because of several important differences (Table 4.2). Components of the neutrophil oxidase are not found in human spermatozoa neither by PCR analysis [73] nor by immunoblotting (antibodies raised against $p67^{phox}$, $p47^{phox}$, and $p21^{rac}$; unpublished data). Mouse spermatozoa possess $p67^{phox}$, $p47^{phox}$, and $p40^{phox}$ but not $p22^{phox}$ and the oxidase activity decreases as epididymal maturation proceeds [27], suggesting that this oxidase may not be that related to capacitation. Also, the time course of the two oxidases is very different [55]. Moreover, contrarily to what is seen with the neutrophil oxidase [19, 25], activation of the sperm enzyme is independent of signal transduction cascades involving PTK, PKC, extracellular signal-regulated kinase (ERK) pathway, etc. [55, 56, 74]. Several NOX (NADH/NADPH oxidase family of enzymes) isoforms are found in reproductive (male and female) tissues, but NOX activation processes are so highly variable between cell types and tissues [25, 26] that they cannot be considered as plausible candidates as the oxidase involved in sperm capacitation. As a supplementary example, PKC activation drives NOX5 of pachytene spermatocytes and mature equine spermatozoa [75] but not the sperm oxidase involved in capacitation [56].

Over the years, we evidenced several other differences between sperm and neutrophil oxidases (Table 4.2). When evaluated with MCLA-amplified luminescence, neutrophils produce during the first 5 min after stimulation with PMA (phorbol 12-myristate 13-acetate, 100 nM) about 250-fold higher amounts of $O_2^{\bullet-}$ than spermatozoa during their first 30 min of capacitation [76]. Also, progesterone, FCSu, and FFu (follicular fluid ultrafiltrate) stimulate sperm $O_2^{\bullet-}$ synthesis [40], but have almost no effect on neutrophils [76]. Furthermore, natural inhibitors of sperm capacitation and $O_2^{\bullet-}$ generation, such as zinc ions (Zn^{2+}) and semenogelin (Sg) (section below), inhibit the sperm oxidase much more efficiently (higher extent at lower concentration) than that of neutrophils [76].

Standing at this point, we tried to put together all our observations on the oxidase and NOS involved in sperm capacitation, on the close interaction between these enzymes and/or the ROS produced (see above), as well as on the confirmed association among ROS ($O_2^{\bullet-}$ and $NO^{\bullet}$), capacitation, and its related events (e.g., Tyr phosphorylation of p80/p105). We realized that this could imply that a single enzyme is responsible for the production of the two ROS. NOS are versatile enzymes and, even if they are best known to generate NO• from the conversion of L-Arg to L-citrulline, they can also reduce nitrite to NO• under anoxia or even become oxidases. NOS possess two functionally distinct domains, one being an oxygenase (N-terminal) and responsible for NO• formation and the other being a reductase (C-terminal) that synthesizes $O_2^{\bullet-}$ [32–34, 77, 78]. NOS produce NO•, $O_2^{\bullet-}$, or both ROS simultaneously depending on the isoform present as well as various conditions, such as availability of substrate (L-Arg, oxygen) and cofactor (NADPH, tetrahydrobiopterin) and uncoupling due to oxidation of Cys residues at the active center and consequent release of Zn^{2+} [32–34, 77, 78].

Whether this double activity of NOS prevails during capacitation is not known. On one side, discrepancies between the formation and action of $O_2^{\bullet-}$ and NO• would not favor a model in which NOS generates both ROS. As stated above, the time courses for the formation and need of these ROS diverge significantly [46]. Also, $O_2^{\bullet-}$ acts upstream, and NO• downstream and/or upstream, of the cAMP/PKA couple [46]. Finally, the double phosphorylation of the Thr-glutamic acid-Tyr (Thr-Glu-Tyr) motif that increases in sperm proteins from 1 h of capacitation is under the control of NO• but not of $O_2^{\bullet-}$ and H_2O_2 (see section below) [79, 80]. However, even with these differences, we cannot completely rule out the possibility that sperm NOS may generate both ROS. Capacitation may involve more than one NOS each having specific regulators, locations, and targets; it is also possible that $O_2^{\bullet-}$ and NO• synthesis is switched on/off sequentially and/or that NOS produces simultaneously $O_2^{\bullet-}$ and NO• at some time points of capacitation.

Our hypothesis model is rather that the sperm oxidase at the cell membrane is initiator of ROS production and that the ROS synthesized activate an intracellular NOS to provide NO• for long-term effect. The localization of NOS in all sperm compartments [30] would support this idea. Whether $O_2^{\bullet-}$ and/or NO• are synthesized sequentially or simultaneously at the initiation of capacitation may not be of paramount importance considering the obvious cross talk between ROS and their generators. This would rather represent another situation in which more than one mechanism exists to give flexibility to the process and to guarantee that capacitation and the subsequent fertilization occur.

4.3.5 Modulators of ROS Generation

Up to date, the formation of both $O_2^{\bullet-}$ and NO• increases with capacitation whether induced with BSA, progesterone, ultrafiltrates from biological fluids (FCSu, FFu), etc. [40, 46, 55, 56]. However, this does not point out to a specific substance as initiator of ROS synthesis. The cell-permeant analog of cAMP, dbcAMP, stimulates NO•, but not $O_2^{\bullet-}$, synthesis [46], presenting cAMP as a possible intracellular modulator of NO• synthesis [46, 56].

A major factor for sperm ROS generation is naturally the presence and amount of cofactors acting as reducing equivalents, mostly under the form of NAD(P)H. Dehydrogenases involved in cellular metabolism are potential candidates to provide

NAD(P)H. Interestingly, inhibitors of NADH formation by lactate dehydrogenase C4 (sodium oxamate) and of NADPH synthesis by isocitrate dehydrogenase (sodium oxalomalate) block bull sperm capacitation induced by heparin but not that due to exogenous $O_2^{\bullet-}$ (xanthine + xanthine oxidase) [81, 82]. This finding strongly indicates a role for the two enzymes as suppliers of NAD(P)H, for sperm ROS synthesis related to capacitation [81, 82]. On the other hand, exogenous NADPH, which is also a cofactor for NOS, stimulates the generation of $NO^{\bullet}$ but not that of $O_2^{\bullet-}$ [56], which may solve the previous controversy [40, 83] as for the mechanism by which NADPH promotes human sperm capacitation. There is no real opposition between these sets of data because intracellular NAD(P)H formed by dehydrogenases as those mentioned above [81, 82] may act on an NOS in the cytosol, or a transmembrane oxidase with a binding site for NADPH oriented toward the inside of the cell. The actual localization of the ROS generators is still unknown which makes several hypotheses plausible.

The regulation of intracellular levels of ROS in spermatozoa is also under the control of antioxidant enzymes (Table 4.1) and, in this line of research, PRDX now appear as very attractive players because of their dual action, being both targets for ROS (which allows ROS action) and scavengers at higher ROS levels (then protecting cells from oxidative stress). PRDX are small (~20–31 kDa) acidic proteins with one or two essential Cys residues at the active site that reduce both organic and inorganic hydroperoxides and $ONOO^-$ [21–23]. The SH groups in PRDX (Cys residues) are reactive enough to be direct targets for H_2O_2, which explains that they are readily oxidized and inactivated in cells exposed to low levels of H_2O_2 [21–23]. This allows H_2O_2-dependent signaling [3, 15–18]; Cys oxidation in PRDX is reversible and the enzymes are reactivated by the thioredoxin/thioredoxin reductase system [21–23]. The various PRDX isoforms are present and differentially localized in subcellular compartment of human spermatozoa [20]. Their involvement in capacitation is not yet established, but it is suggested because they naturally react with H_2O_2 not only at high levels but also at the low levels that induce sperm capacitation [20].

All together, observations noted above and in the preceding sections raise the possibility that the oxidase and/or the NOS at the sperm membrane level are always in a "switched-on" state and that their action depends on the presence of adsorbed seminal plasma inhibitors. The removal or degradation of these substances would allow the synthesis of ROS and initiation of capacitation. This mechanism is plausible since spermatozoa lose their surface-adsorbed material during their transit in the female genital tract [1, 2, 4, 9].

Seminal plasma contains several factors, including cholesterol [2, 10], phosphatidylcholine-binding proteins [84], glycodelin-S [85], to regulate capacitation but, as for now, the main physiological inhibitors for $O_2^{\bullet-}$ and $NO^{\bullet}$ generation are probably Zn^{2+} and semenogelin (SgI, 52 kDa, and SgII, 71 and 76 kDa, collectively called Sg in this chapter), these factors being present in seminal plasma at concentrations (Zn^{2+}: 2–5 mM, Sg: 10–20 mg/mL) [86–88] 100-fold higher than those needed to block ROS formation and capacitation [56, 86, 87, 89, 90]. Therefore, Zn^{2+} and Sg may act as regulators of ROS synthesis and prevent premature development of capacitation and consequent decreased fertility [87, 91].

Zn^{2+} is of paramount importance for sperm function and chromatin integrity and is found in all sperm compartments [92]. At mM levels, Zn^{2+} blocks the high ROS formation in spermatozoa from infertile men [93] and at lower levels (50 μM) Zn^{2+} inhibits capacitation and associated synthesis of both $O_2^{\bullet-}$ and $NO^{\bullet}$ [56, 87]. Conversely, Zn^{2+} chelators (TPEN, 200 μM; EDTA, 100 μM) promote these events, but not in the presence of SOD or NOS inhibitors adding again to the idea that Zn^{2+} plays a central role in ROS synthesis [56, 87].

It is important at this point to stress the interaction between Zn^{2+} and Sg. Each Sg molecule can bind up to ten Zn^{2+} [86–88] and Zn^{2+} promotes the binding of Sg to sperm proteins in cell-free assays [94]. However, the inhibitory effect of Sg in sperm capacitation and ROS synthesis is not due to its adsorbed Zn^{2+} since it occurs even in the presence of a Zn^{2+} chelator (TPEN) [90]. The binding of Zn^{2+} to Sg reduces the amount of nonprotein-bound Zn^{2+} in seminal plasma to levels low enough that the prostate-specific antigen (PSA) is activated and degrades Sg [86–88].

Regulation of capacitation onset appears as a major role for Sg. In vitro, Sg, as well as its degradation peptides, blocks sperm capacitation triggered by various stimuli (FCSu, L-Arg, BSA, etc.), an effect that is reversed by the exogenous addition of both $O_2^{\bullet-}$ (xanthine + xanthine oxidase) and $NO^{\bullet}$ (DA-NONOate) [86, 89, 90]. As expected, Sg completely blocks $O_2^{\bullet-}$ and $NO^{\bullet}$ formation [86, 89, 90]. Sg and its degradation peptides may also have other effects on spermatozoa (activation of hyaluronidase, inhibition of movement, antibacterial activity, hyperpolarization of plasma membranes, binding to heparin, eppin, and fibronectin, etc.) as reviewed before [86, 88, 95] without relation to capacitation.

Surprisingly, spermatozoa not only bind [86, 88, 95], but also internalize [90] Sg. Sg and its degradation peptides are present in Triton-soluble and -insoluble sperm fractions, as well as on the sperm head (without or with permeabilization) [90]. In all cases, Sg levels decrease with incubation with FCSu as capacitation inducer (Fig. 4.5). Other conditions that promote capacitation (exogenous $O_2^{\bullet-}$ + $NO^{\bullet}$ or Zn^{2+} chelator) reproduce these effects and, conversely, the combination of SOD and NOS inhibitors, or Zn^{2+}, prevents the drop in Sg and its degradation peptides present in sperm fractions [90]. Furthermore, Sg degradation in sperm fractions is Zn^{2+} inhibitable [90]. The very basic isoelectric point of Sg (pI 9.5) [86, 88] and the high content of acidic (often sialylated) proteins on sperm surface [96] may be compared to the interaction between cationic peptides and highly negatively charged membrane of bacteria which leads to internalization of these peptides [97]. However, the mechanism by which Sg passes sperm membrane still needs to be proven.

At this point, we are again confronted to a situation, where we cannot determine the initial event (see above, which of the ROS is first synthesized). Sg blocks ROS synthesis and ROS are needed to dispose of Sg in spermatozoa [90]. We can hypothesize that Sg adsorbs on, and enters into, spermatozoa at ejaculation and, as a result, prevents premature capacitation. Sperm transit in the female genital tract, or incubation in vitro with inducers, promotes the release of Sg and Zn^{2+} from the cell surface, which allows the initiation of ROS synthesis. ROS would favor further Sg processing, and the reduction of sperm Sg then would amplify ROS formation. Whether this cycle starts with degradation and/or release of Sg or with the formation of

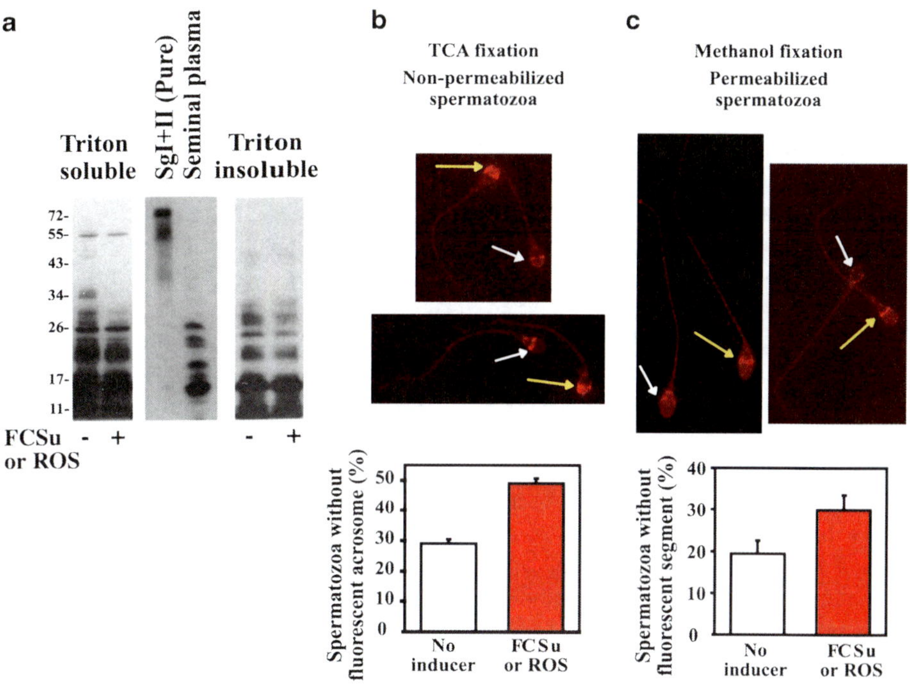

Fig. 4.5 The decrease of Sg levels on and in spermatozoa during capacitation is dependent on ROS. Percoll-washed spermatozoa are incubated for 30 min at 37°C in the absence (−) or presence (+) of FCSu (10%, v/v) or ROS (xanthine, 0.1 mM + xanthine oxidase, 0.1 U/mL + DA-NONOate 0.1 mM). (**a**) Triton-soluble and -insoluble sperm proteins are immunoblotted with an anti-Sg polyclonal antibody [90]. Pure Sg and seminal plasma are given as standards so that it is clear that the bands observed in spermatozoa correspond to those of Sg and its degradation peptides. (**b**) Spermatozoa are smeared, fixed with trichloroacetic acid (TCA; non-permeabilized), and labeled with the anti-Sg antibody [90]. Sg labeling is on the acrosome (*yellow arrow*) or only on the equatorial segment (*white arrow*). (**c**) Spermatozoa are smeared, fixed, and permeabilized with methanol and then Triton, and labeled with the anti-Sg antibody [90]. The labeling is on the equatorial segment (*yellow arrow*) or only at the base of the head (*white arrow*). In (**b**) and (**c**), treatment with FCSu or ROS (*red bars*) is associated with an increased proportion of spermatozoa with lower Sg on the acrosome (TCA) or on the equatorial segment (methanol and Triton)

minute amounts of ROS is not known. The two options are plausible [90] and could prevail in spermatozoa to warrant a sure initiation and progression of capacitation.

A similar cycle may exist not only as we look at ROS production that is switched on by the release of Zn^{2+}, but also as we consider that the release of Zn^{2+} can be accelerated by ROS [56]. Zn^{2+} often binds to proteins via coordination complexes with SH groups of Cys residues and this is one of the mechanisms by which Zn^{2+} regulates enzymes, such as NOS and PKC [98–100]. Oxidation of SH groups leading to the formation of disulfide bonds (SS) or reaction of these SH groups with alkylating agents (e.g., *N*-ethylmaleimide and diamide) promotes capacitation and

ROS synthesis [56, 101] and is also expected to cause the removal of Zn^{2+} from proteins [98–100]. Whether this occurs in spermatozoa is still not proven. The low levels of Zn^{2+} measured in incubation media are not significantly different whether spermatozoa are incubated in control or capacitating conditions (unpublished data). This may indicate that our hypothesis is not valid or that Zn^{2+} is immediately shuttled to other proteins [98, 100]. Nevertheless, as in the preceding discussion, whether Zn^{2+} is released before ROS are formed or ROS cause the release of Zn^{2+} is not known and the two possibilities are valid and may be concrete and needed to assure that capacitation occurs.

As said above, the oxidase involved during capacitation appears to act upstream of the transduction cascades known to be involved in this process. None of the inhibitors tested for PKA, PKC, PI3K (phosphoinositide 3-kinase), Akt (protein kinase B), ERK pathway, and PTK had any effect on $O_2^{\bullet-}$ generation during capacitation [2, 17, 56]. At the present time, Sg [89, 90], Zn^{2+} [56, 90], $NO^{\bullet}$ [46], and reactives related to the SH/SS couple [56, 101] are the known modulators of $O_2^{\bullet-}$ generation.

On the other hand, $NO^{\bullet}$ formation is not only regulated by Sg [90], Zn^{2+} [56, 90], $O_2^{\bullet-}$ [46], and reactives related to the SH/SS couple [56] but also blocked by inhibitors of PKC, PI3K, Akt, ERK pathway, and PTK [56]. These data, in conjunction with the need of $NO^{\bullet}$ for the whole capacitation period [46], could indicate that there is more than one NOS involved. At the membrane level, one would be regulated as the oxidase, by Sg, Zn^{2+}, and reactives directed to SH/SS couple and activated at the initiation of capacitation. The second NOS could be intracellular and modulated by transduction cascades related to capacitation. In agreement with this, one of the phospho-Akt substrate in capacitating spermatozoa is of 140 kDa [3, 102], which is suggestive of NOS, this enzyme being a downstream effector for Akt (see section below). This hypothesis will need further confirmation. Of interest is that the two PKA inhibitors tested (H89 and KT5720) do not affect $NO^{\bullet}$ generation [56]. In line with this data is that in the signal transduction cascades involved during capacitation the cAMP/PKA axis and the PI3K/Akt pathway seem to act in parallel (see section below) [5, 17].

4.4 Targets for ROS

4.4.1 Signal Transduction Cascades and Phosphorylation Events

Protein Tyr phosphorylation increases during sperm capacitation, from about 2 h of incubation, and is often considered as a marker for this process [2, 17, 39, 44, 103, 104]. However, over the years, we described several transduction cascades forming a tightly regulated network of phosphorylation events, all regulated by ROS, that occur as earlier steps and are prerequisites for capacitation.

Soon after the onset of capacitation, there is a rise in cAMP levels [1, 2, 39, 43, 105, 106]; both exogenous $O_2^{\bullet-}$ and $NO^{\bullet}$ can trigger this event [3, 17, 39, 43,

105, 107] maybe via activation of adenylyl cyclase [108] since phosphodiesterase activity does not seem to be modified (unpublished data) (Fig. 4.6). cAMP then stimulates PKA and a maximum activity is reached at 30 min [105]. This rise in PKA is translated to a rapid increase in the phosphorylation of two proteins (80 and 105 kDa) containing Ser and Thr residues within the Arg-X-X-(Ser/Thr) motif and collectively called phospho-PKA substrates [3, 17, 106]. In the boar, the equivalent two phospho-PKA substrates (59 and 96 kDa) obtained after differential extraction of the midpiece/large tail fraction appear to be members of the Odf2 (outer dense fiber) family of proteins [109]. This rise in phospho-PKA substrates also reaches a maximum at 30 min, occurs with various capacitation inducers (FCSu, progesterone, albumin, and FFu), including exogenous ROS ($O_2^{\bullet-}$, H_2O_2, and NO$^\bullet$), and is prevented by SOD, catalase, NOS inhibitor (L-NAME), and PKA inhibitors (H89 and Rp-cAMP-S) [3, 17, 106]. Of interest, the rise in phospho-PKA substrates is also blocked by inhibitors of nonreceptor-type PTK (herbimycin A and PP2) but not when spermatozoa are stimulated with dbcAMP + isobutylmethylxanthine (IBMX, phosphodiesterase inhibitor), thus suggesting that a PTK acts upstream of PKA and maybe needed for adenylyl cyclase activation [3, 17, 106]. It is important to mention that this early-acting PTK is probably different of that responsible for the late Tyr phosphorylation of fibrous sheath proteins (p85 and p105). PKC, receptor-type PTK, or MEK does not appear to be involved in the regulation of the cAMP/PKA/phospho-PKA substrates axis [3, 17, 106].

The ERK pathway then appears, in time, involved in human sperm capacitation (Fig. 4.6) [3, 5, 17, 74, 79, 80, 110–112]. Most of the elements of the cascade, from Shc to Ras, as well as the ERK module common to all MAPK (mitogen-activated protein kinase) pathways [113, 114], and made of Raf (MAP kinase kinase kinase; phosphorylates Ser/Thr), MEK (MAP kinase kinase; dual specificity for Ser/Thr and Tyr), and ERK1 and ERK2 (MAP kinase; both for Ser/Thr), are present in human spermatozoa [74, 111, 112, 115, 116].

Sperm capacitation is associated not only with the phosphorylation of MEK (45 kDa; low amounts in human spermatozoa), but surprisingly also of MEK-like proteins (55, 94, and 115 kDa; higher levels) that reaches a maximum level at 60 min [112] (Fig. 4.6). The anti-phospho-MEK antibody recognizes these four proteins and PD98059 (inhibits MEK phosphorylation and activation) blocks their phosphorylation, as well as capacitation, confirming these as MEK-like proteins [112].

Capacitation inducers, such as FCSu and BSA, as well as exogenous ROS ($O_2^{\bullet-}$, H_2O_2, and NO$^\bullet$), promote the phosphorylation of MEK-like proteins and, conversely, SOD, catalase, and NOS inhibitor prevent this process emphasizing again the major role of ROS [112]. PKA inhibitors (H89 and Rp-cAMP-S) prevent the phosphorylation of MEK-like proteins [112] suggesting that PKA or phospho-PKA substrates are intermediate players involved in the cross talk between these phosphorylation pathway [3, 17, 112]. PKC, PTK (both nonreceptor and receptor type) and the ERK pathway also regulate the phosphorylation of MEK-like proteins [112].

Then, MEK (and MEK-like proteins) as dual-specificity kinase phosphorylates both Thr and Tyr residues within the Thr-Glu-Tyr motif present not only in ERK1 and ERK2 but also in several proteins, such as ERK5, ERK7, and MOK [117, 118].

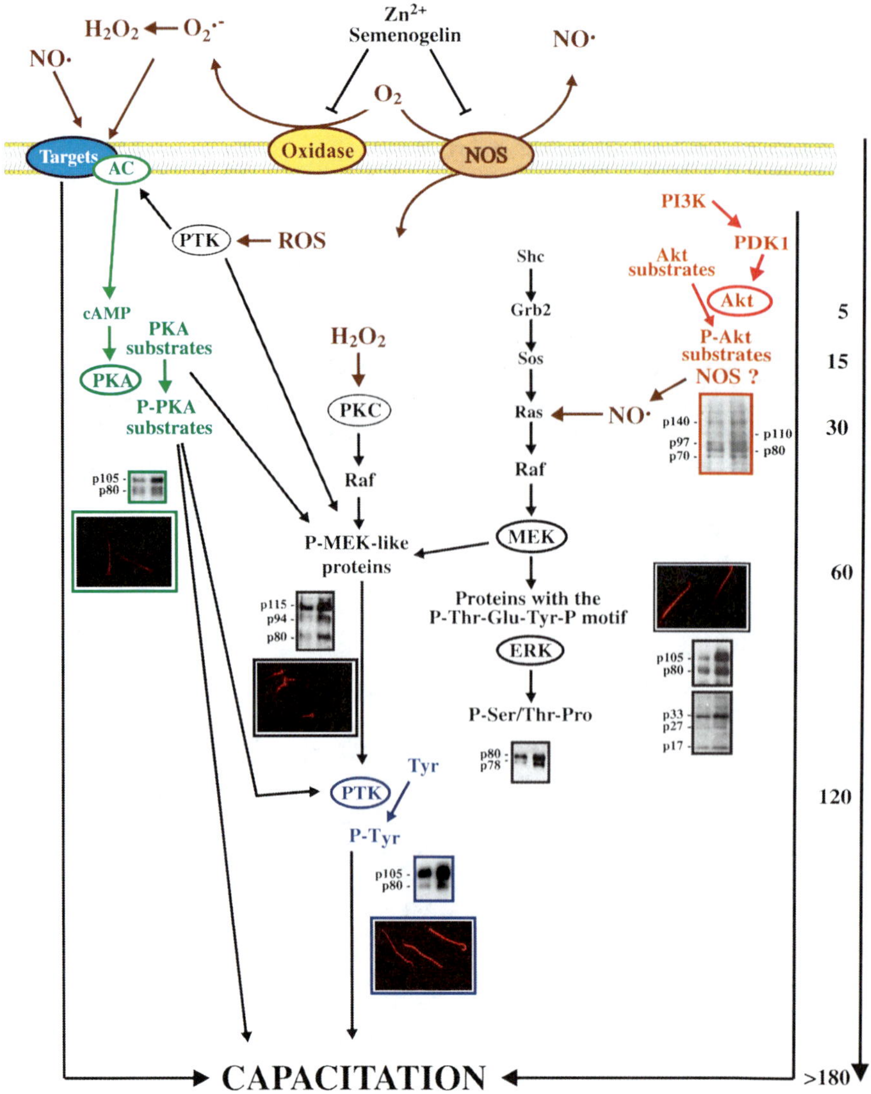

Fig. 4.6 Proposed schema for phosphorylation events known to happen and be regulated by ROS during sperm capacitation. The main axes are presented: cAMP/PKA/P-PKA substrates (*green*), P-MEK and ERK cascade (*black*), PI3K/Akt (*red*), and late P-Tyr (*blue*). Blots (control and capacitating spermatozoa, left and right lines, respectively) and characteristic labeling pattern obtained with the various anti-phospho-antibodies are also presented. The time course of events is given on the *right*. ROS generated from the beginning of the capacitation period appear to act on adenylyl cyclase (AC) as one of the first targets and stimulate the production of cAMP. This cAMP activates PKA to phosphorylate its specific phospho-PKA (P-PKA) substrates. PKA and/or the P-PKA substrates might be involved in the phosphorylation of MEK-like (P-MEK-like) proteins and subsequently of Tyr residues (P-Tyr) of fibrous sheath proteins. PKA and the P-PKA substrates may also participate in other pathways associated with capacitation. A nonreceptor-type PTK, possibly activated by ROS, seems to modulate the increase in P-PKA substrates, maybe via activation of AC.

Sperm capacitation is associated with a transient increase in the Thr-Glu-Tyr phosphorylation of ERK1 and ERK2 (p42 and p44) that peaks at 5 min [3, 74]. However, the Thr-Glu-Tyr phosphorylation of 27–33 kDa proteins (high for up to 2 h), as well as that of 80 and 105 kDa proteins (from 1 h and on) (Fig. 4.6), appears even more important because it is sustained and rises as capacitation proceeds [3, 74]. Inhibitors of MEK (PD98059 and U126) prevent capacitation and this phosphorylation confirming the ERK-like nature of these proteins [3, 74]. The double phosphorylation of the 27–33 kDa proteins depends on $O_2^{\bullet-}$ [74], but the role of $NO^{\bullet}$ is still unknown. The Thr-Glu-Tyr phosphorylation of 80 and 105 kDa proteins is also controlled by receptor-type PTK and the ERK pathway, but not by PKC or PKA [79, 80]. Of interest is that ROS regulation is this time (for 80 and 105 kDa proteins) limited to $NO^{\bullet}$, and not $O_2^{\bullet-}$ and H_2O_2, which, in other words, means that NOS inhibitor prevents and exogenous $NO^{\bullet}$ promotes this event [80]. This is, therefore, a very good example that ROS may act specifically on their targets.

Downstream, ERK1, ERK2, and ERK-like proteins phosphorylate substrates containing Ser or Thr within the motif proline (Pro)-X-X-Ser/Thr-Pro [114]. A monoclonal anti-phospho-Ser/Thr-Pro antibody (MPM2) [119] recognizes a doublet (78 and 80 kDa) of Triton-insoluble sperm proteins, the intensity of which increases with capacitation (Fig. 4.6) [74]. SOD and PD98059 block this phosphorylation, but the involvement of $NO^{\bullet}$ is still unknown [3, 17, 74]. In the mouse, MPM2 antibody recognizes several sperm proteins (70–250 kDa), but the intensity of bands appears independent of MEK activity [110]. Therefore, the increased levels of various phosphorylations noted during capacitation are not only modulated by different ROS and kinases, but they may also vary from one species to the other.

The PI3K/Akt axis also participates in capacitation (Fig. 4.6). PI3K and Akt are involved in cell growth, proliferation, differentiation, etc. [120, 121] and also in sperm capacitation [3, 17, 102, 122]. PI3K phosphorylates the phosphoinositide 3,4,5 (PIP_3), an intermediate that activates, directly or via its downstream target Akt, different enzymes, such as PKC, PKA, and NOS [123]. The PI3K/Akt axis regulates

Fig. 4.6 (continued) The rise in P-MEK-like proteins that occurs 1 h after the beginning of capacitation is triggered by ROS. PKA, all the ERK pathway components, and PKC also modulate the rise of P-MEK-like proteins. We can hypothesize that H_2O_2 activates PKC, which in turn phosphorylates Raf, the kinase normally responsible for the phosphorylation of MEK and MEK-like proteins. Inhibitors of PKA and of all elements of the ERK pathway prevent the increase in P-Tyr which allows suggesting that P-MEK-like proteins and P-PKA substrates are intermediates between the early events and the late increase in P-Tyr during capacitation. The phosphorylation of the Thr-Glu-Tyr motif in sperm proteins increases progressively from 1 h after the beginning of capacitation and is controlled by the entire ERK pathway and the PI3K/Akt axis. $NO^{\bullet}$, but not $O_2^{\bullet-}$ or H_2O_2, triggers and modulates this phosphorylation. The PI3K/Akt axis is present in human spermatozoa and regulates the phosphorylation of the Thr-Glu-Tyr motif. PI3K, through its downstream effectors PDK1 and Akt, could activate NOS and the $NO^{\bullet}$ formed could stimulate Ras and ERK pathway and later cause the increase in P-Tyr. Besides their effects on most of the kinases mentioned above, ROS may also inactivate several protein phosphatases (for Ser/Thr and for Tyr) and consequently prevent protein dephosphorylation at any step of capacitation (not shown on the schema to keep clarity)

capacitation and associated phosphorylation of Thr-Glu-Tyr and of Tyr on sperm proteins (FCSu or albumin as inducer) [102]. Furthermore, the phosphorylation of Arg-X-Arg-X-X-phospho-Ser/Thr, characteristic of Akt substrates, of several proteins (70, 80, 97, 110, and 140 kDa) increases during capacitation [102], but the modulation by ROS is presently not known. On the other hand, the Akt inhibitor prevents the rise of intracellular NO• formation in spermatozoa [56], suggesting that one of the effectors of Akt during capacitation could be NOS, specially considering that one of the phospho-Atk substrate is of 140 kDa [102], as NOS is [19, 43]. Wortmannin (PI3K inhibitor) or the Akt inhibitor blocks Tyr and Thr-Glu-Tyr phosphorylation events [102] but not when capacitation is induced by exogenous H_2O_2 or NO• [5]; they do not affect the increases of either phospho-PKA substrates (unpublished data) or phospho-MEK-like proteins [5]. Therefore, the PI3K/Akt axis seems to modulate capacitation independently of the cAMP/PKA pathway but rather via an action on the ERK cascade. We could hypothesize that this is mediated by NO• since this ROS is known to activate Ras [124, 125].

Protein Tyr phosphorylation appears as a late event of capacitation occurring downstream of all other phosphorylation events mentioned above (Fig. 4.6) and is found in all species where it was studied, from rodents to domestic animals [65, 126–130], as well as in men [2, 17, 43, 130–132]. Tyr phosphorylation of the two Triton-insoluble proteins (81 and 105 kDa, antigenically related to the A-kinase-anchoring proteins, AKAPs) [44, 104, 130] is increased with all capacitation inducers tested (FCSu, FFu, BSA, progesterone, cell-permeant analogs of cAMP, etc.) as well as exogenous ROS ($O_2^{\bullet-}$, H_2O_2, NO•) and is prevented by SOD, catalase [2, 3, 17, 44, 130, 131], and inhibitors of NOS [42, 130].

Several proteins are Tyr phosphorylated during capacitation beside the two well-known fibrous sheath proteins of 80 and 115 kDa (Fig. 4.6) [3, 17, 31, 111, 122, 130, 133, 134]. We can mention, between others, Triton-soluble proteins of 37, 42, and 47 kDa [14], AKAP 3, and valosin [135]. There are also several PTK, both of receptor [136] and nonreceptor type (Src, c-yes, Lyk, cAbl, etc.) [133, 134, 137–139], participating in these phosphorylation events. Although the association and/or role of both these PTK and their substrates in capacitation are indisputable, their regulation by ROS is not yet ascertained. We could expect that Src and Lyk are regulated by ROS as they are in other types of cells, but no data is actually available.

Therefore, ROS modulate all signal transduction cascades presently known to be related to capacitation, including cAMP/PKA, ERK pathway (MEK-like, ERK and ERK-like, and also ERK substrates), and PI3K/Akt axis [3, 17]. All these are needed and converge to the late downstream Tyr phosphorylation of two proteins of 80 and 105 kDa that was first thought to be modulated only by cAMP/PKA [131, 132].

The regulation of the pathways described above has to be looked at with a wide point of view since cross talks occur and the participation of other players is also needed. For example, PKC is one of those enzymes known to be directly activated by ROS, in part via the release of Zn^{2+} from this kinase [99, 140]. PKC is needed for capacitation and participates to the phosphorylation of the Thr-Glu-Tyr motif [79, 80], MEK-like proteins [112], and Tyr residues of fibrous sheath proteins (unpublished observations). Inhibition of PKC blocks the phosphorylation of MEK-like proteins triggered by FCSu or H_2O_2, suggesting that ROS generated during capacitation may activate PKC and one of its downstream effectors, Raf [3, 5, 17, 112, 141]. ROS appear to activate PKC and then Raf in bovine tracheal smooth muscle cells [141], supporting the possibility that this ROS-dependent stimulation of Raf through PKC may occur as well during human sperm capacitation.

PTKs, of both receptor and nonreceptor types, are susceptible of direct regulation through ROS [142] and are of paramount importance for the late Tyr phosphorylation of sperm proteins during capacitation [39, 42–44, 55]. It is conceivable that different PTKs must act considering the different time courses for the phosphorylation of proteins, such as PKA substrates [106], MEK-like proteins [112], and those with the Thr-Glu-Tyr motif [79, 80], and of the late Tyr residues [39, 42–44, 55].

The specificity of ROS in their action should not be disputed even if in most cases $O_2^{\bullet-}$, H_2O_2, and $NO^{\bullet}$ seem to have similar effects [2, 14, 40, 106, 112] since phosphorylation of the Thr-Glu-Tyr motif depends only on $NO^{\bullet}$ [80]. Furthermore, ROS act, over time, at all steps of capacitation with activation of PKA [106] and Akt [102] at early steps but then also on phosphorylation of MEK-like proteins [112], of the Thr-Glu-Tyr motif [3, 5, 17, 74, 79, 80], and finally of Tyr residues. We have to remember also that $NO^{\bullet}$ synthesis is essential for the whole course of capacitation [46], maybe because one of the downstream effectors for Akt is NOS.

In this section, we looked at how ROS modulate signal transduction cascades and pathways related to sperm capacitation and some of the cross talks that are involved. The next section reviews some of the mechanisms by which ROS can directly affect enzyme activity.

4.4.2 Targets at the Molecular Level

ROS are known to interact with all cellular components [19] and we are used to think of lipid and DNA as first targets for ROS in spermatozoa [143, 144]. However, $O_2^{\bullet-}$ and $NO^{\bullet}$ have relatively low reactivity and, especially at the low levels they are generated, do not promote lipid peroxidation or oxidation of DNA bases as the toxic hydroxyl radical [19, 20]. As noted above, $ONOO^-$, which is also very reactive, is endogenously formed during capacitation and causes Tyr nitration of sperm proteins [46]. This modification is associated to capacitation but the cause–effect relationship still needs to be ascertained. Tyr nitration can be toxic when it happens at high levels, but this is not the case during capacitation as sperm motility and viability are not decreased and capacitation proceeds [46].

The relatively weak reactivity and low amounts of $O_2^{\bullet-}$ and $NO^{\bullet}$ formed by spermatozoa point to SH groups (e.g., Cys on proteins) as some of the main targets for ROS during capacitation [56, 101, 145]. We have to realize that a limited number of Cys on proteins are prone to oxidation at low ROS levels. The acidic Cys residues, those with a pK_a (acidity constant) lower than 7.0, are present as Cys thiolate anion (Cys-S^-) at physiologic pH, which markedly increases their vulnerability to oxidation as compared to most other Cys (Cys-SH, pK_a around 8.5) [146–150].

Worth mentioning is that many enzymes related to signal transduction cascades, as those involved in sperm capacitation, have these acidic Cys residues that render them susceptible to ROS modulation. For example, ROS activate adenylyl cyclase, PKA, PKC, and NOS but inhibit protein Tyr and Ser/Thr phosphatases via oxidation of Cys [146–150]. One major advantage for cells to use this oxidation of Cys on proteins is that the resulting products, Cys sulfenic acid, SS bridges, nitrosylated Cys (Cys-NO), etc., are readily reduced back to Cys by various cellular reductants [146–150].

The SH/SS couple on sperm proteins is subject to extensive and complex redox modifications during capacitation [101, 145]. First, there is a major time-dependent increase in SH content of Triton-soluble proteins during the first 30–60 min of capacitation [101, 145]. This is quite surprising because capacitation, as an oxidative process, is expected to be rather associated with a decrease in SH content of proteins. However, this could suggest an important rearrangement of SH-carrying proteins on the sperm membrane during the initiation of capacitation [101, 151].

A deeper analysis using two-dimension (2D) gel electrophoresis points out that alterations in SH levels of at least ten proteins during capacitation are both decreases (by 45–95%, five proteins, 20–60 kDa) and increases (by 200–400%, five proteins, 20–60 kDa) (Fig. 4.7) [145]. The time course of these modifications parallels that of $O_2^{\bullet-}$ formation by capacitating spermatozoa and, as could be expected, SOD and/or catalase prevent all these changes [145]. The identity of these proteins is still unknown but, as mentioned above, they could be elements of signal transduction pathways.

Sulfhydryl groups such as those of glutathione and Cys of many sperm proteins are relatively labile and expected to be involved in exchange mechanisms, leading to glutathionylation of proteins [152, 153]. This protein modification usually switches off or stabilizes enzyme, and its involvement in sperm capacitation still needs to be ascertained. However, glutathione, reduced (GSH) and even better oxidized (GSSG), prevents the striking increases and redistribution of SH groups on bull sperm surface that occurs during cryopreservation as well as the associated oxidative stress and premature capacitation [151]. Therefore, protein glutathionylation could also play a role in the regulation of sperm capacitation.

Enzymes, such as PKC, Ras, adenylyl cyclase, are all susceptible to activation by $O_2^{\bullet-}$ and H_2O_2 [99, 108, 140, 154, 155]. On the other hand, the oxidation of the vicinal Cys residues at the active center of protein phosphatases causes a selective and reversible inactivation of the enzyme [142, 148, 156, 157]. Protein Ser/Thr phosphatases, such as PP1, PP2A, and both Ser/Thr- and Tyr-directed classes of ERK protein phosphatases, are frequent targets for ROS [156–161]. This mechanism could, at least in part, explain the increases in phosphorylation levels that occur during capacitation, but this hypothesis still needs to be proven.

Cys residues are also targets for $NO^{\bullet}$, which results in Cys nitrosylation [162–165]. Exogenous $NO^{\bullet}$ (from *S*-nitrosocysteine and DA-NONOate) promotes extensive Cys nitrosylation in sperm proteins, but the high number of proteins then modified [166] may be due to exposure of cells to relatively high levels of $NO^{\bullet}$ as compared to what they endogenously produce. In our hands, spermatozoa have only four protein bands indicative of Cys nitrosylation (Fig. 4.8). One of these (21 kDa)

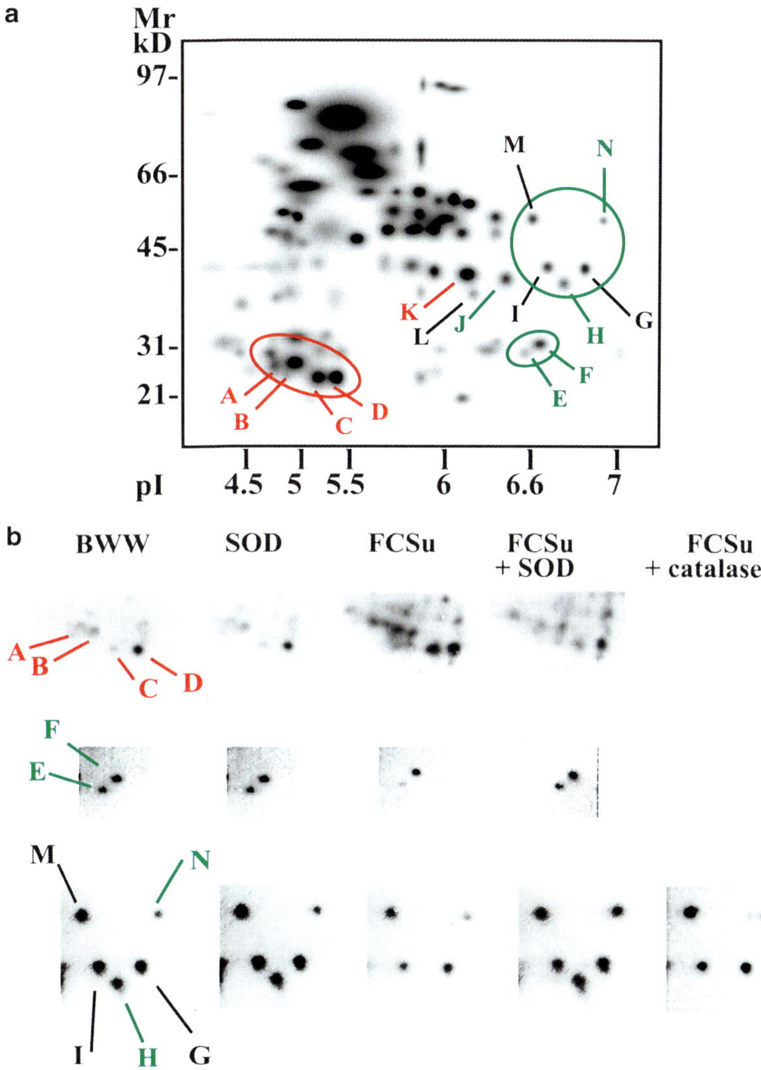

Fig. 4.7 Sulfhydryl groups change during capacitation. Percoll-washed spermatozoa are treated without (BWW medium alone) or with FCSu (10%, v/v) and in the absence or presence of SOD (0.1 mg/mL) or catalase (0.1 mg/mL) for 30 min at 37°C. Then, Triton-soluble sperm proteins are labeled with 3-(*N*-maleimidylpropionyl) biocytin, separated by 2D gel electrophoresis, blotted, and probed with streptavidin conjugated to horseradish peroxidase [145]. (**a**) Digitized image of a blot after scanning. Some spots chosen for their most evident differences in capacitating spermatozoa are identified with a letter, *red* for those that present an increased intensity with FCSu treatment, *green* for those with a decrease, and *black* for those with no change. (**b**) SOD reverses the effects due to FCSu in all cases; catalase also reverses effects due to FCSu but not those seen on spot H

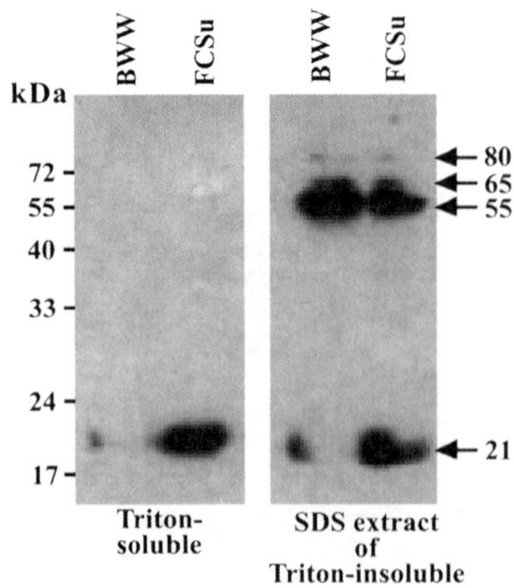

Fig. 4.8 Capacitation induced by FCSu is associated with an early increase in nitrosylated Cys level of sperm proteins. Percoll-washed spermatozoa are incubated for 20 min in the absence or presence of FCSu (10%, v/v). Cells are then extracted with Triton (0.2%) and the residual Triton-insoluble pellet with SDS (sodium dodecylsulfate, 0.1%, w/v). The biotin switch assay is then performed [162] to assess Cys nitrosylation. Both sperm extracts show an increased Cys nitrosylation of a 21 kDa protein in capacitating spermatozoa. Three other proteins (55, 65, and 80 kDa) are also detected in the sperm SDS extract, but there seem to be no differences between control and treated spermatozoa

is more intense during the beginning of capacitation and we could hypothesize that this is Ras^{p21}, a protein involved in sperm capacitation [5, 17, 74, 166] and known to be subject to Cys nitrosylation [124, 167].

4.5 Associated Events

4.5.1 Hyperactivation

Sperm hyperactivation that occurs during capacitation is blocked by SOD and catalase and, as counterpart, $O_2^{\bullet-}$ (xanthine + xanthine oxidase) and H_2O_2 (as such is or from glucose + glucose oxidase) trigger this specific type of motility [2, 37, 38, 41]. Furthermore, hyperactivation induced by FCSu is blocked by Zn^{2+} and Sg [2, 89] that are seminal plasma inhibitors of ROS production.

The role of $NO^{\bullet}$ in sperm hyperactivation is still not clear. Some $NO^{\bullet}$ donors, such as sodium nitroprusside, trigger hyperactivation [31] but others (DA-NONOate and spermine-NONOate) do not [45]. This may be due to differences in experimental design or in kinetics for the release of NO• from the chemicals tested. However, proper controls, such as the effects of nitrite and nitrate (decomposition products of NO•) and of preincubated NO• donor (to test what is left after NO• release) are sometimes missing. For example, spermine-NONOate, whether fresh or after a 24-h preincubation, promotes hyperactivation to a same extent, indicating that the effect observed is not related to NO• [45].

The sharp peak in O_2 tension in the fluids from the female genital tract (golden hamster) at the time of ovulation [51] is proposed to induce ROS formation and as a result hyperactivation, thus providing enough force to spermatozoa to detach from the oviduct and to progress further toward the eggs [6]. The scarcity of data on the mechanisms of ROS action during sperm hyperactivation allows only hypothesizing that ROS may act via the stimulation of cAMP synthesis (activation of adenylyl cyclase) and PKA activity both of these being key players in sperm motility [1, 6, 35, 168].

4.5.2 Acrosome Reaction

ROS also modulate the acrosome reaction [14, 31, 66, 131, 169–171]. Exogenous ROS, $O_2^{\bullet-}$ (xanthine + xanthine oxidase) or H_2O_2, at the same level as those used to stimulate capacitation, promote the acrosome reaction in previously capacitated spermatozoa [14, 169]. Furthermore, SOD and catalase block the acrosome reaction whether the triggering agent is the calcium ionophore (A23187) or lysophosphatidylcholine (LPC), and independently of the inducer previously used for capacitation [14]. Interestingly, the level of $O_2^{\bullet-}$ formation reached by acrosome reacting spermatozoa is twice that seen during capacitation indicating a stepwise increase in ROS production [14].

The acrosome reaction induced by LPC or A23187 is associated with an increase in protein Tyr phosphorylation that adds over that seen during capacitation, some of the proteins affected during the two events having similar molecular masses [14, 103, 172]. This increase in Tyr phosphorylation related to the acrosome reaction is partly reversed with SOD or catalase but completely abolished in the presence of both antioxidants [14].

The current indications for a role of NO• in the acrosome reaction are based mostly on data showing that NO• donors or L-Arg promote, and NOS inhibitors prevent, this process [31, 66, 103, 170, 173]. However, there is very little evidence for actual NO• synthesis, except for that noted in spermatozoa in which the acrosome reaction is induced with follicular fluid (by formation of L-citrulline from L-Arg and by a 24-h accumulation of nitrite) [66].

The targets for ROS and signal transduction events modulated by ROS during the acrosome reaction may be different from those reported for capacitation. This is expected since these two events proceed according to specific pathways [1, 174] and ROS, as in other cellular systems, enter the normal flow of events (e.g., activation/inhibition of enzymes).

Therefore, even though experimental evidences point out for an important role for ROS in sperm hyperactivation and acrosome reaction, the relative scarcity of data as compared to those reporting on capacitation leaves open an interesting area for studies on the mechanisms of ROS action in these events associated to capacitation.

4.6 General Discussion

We reviewed what is presently known on the positive effects of a mild oxidative stress in sperm activation. All these reports demonstrate that the controlled generation of low amounts of ROS by spermatozoa themselves plays an essential role in the modulation of signal transduction events related to the acquisition of fertilizing ability.

We identified main actors involved at the initiation of capacitation: (1) the sperm oxidase and NOS as ROS generators; (2) Zn^{2+} and Sg as physiological inhibitors originating from the seminal plasma; and (3) SH/SS redox couple as target for ROS, Zn^{2+} chelators, and also regulator for ROS generators. One of the very interesting observations at this point is that there seem to be several cross talks and interactions involving ROS, Zn^{2+}, Sg, and the protein SH/SS couple. First, $O_2^{\bullet-}$ induces $NO^{\bullet}$ synthesis and vice versa, and this type of double-sided ROS-induced ROS formation is, at the present time, found only in spermatozoa. Second, Sg and Zn^{2+} inhibit ROS generation and capacitation, but ROS also help to the disposal (degradation and release) of Sg so that capacitation can proceed. Third, modification of the SH/SS couple promotes capacitation and ROS formation, maybe via the release of Zn^{2+}, but ROS could also promote the release of Zn^{2+} by oxidation of protein SH. Finally, even if Zn^{2+} binds to Sg and promotes Sg binding to spermatozoa, Zn^{2+} is not needed for Sg action but anyhow prevents Sg degradation. It is also probable that other interactions also play a role in the initiation of capacitation. We cannot at this time know what is really the first event, whether it is the removal of Sg or Zn^{2+}, induction of $O_2^{\bullet-}$ or $NO^{\bullet}$ production, modification of the SH/SS couple, or something else. What is more important is to realize that several mechanisms are possible, that they potentially all interact together, and that this apparent redundancy provides superior reliability and resilience to the system and gives better guarantee that capacitation can proceed.

We also described several points at which ROS, $O_2^{\bullet-}$ and/or $NO^{\bullet}$, modulate the signal transduction pathways known to be activated during capacitation and ultimately leading to the well-known Tyr phosphorylation of fibrous sheath proteins. Pathways involving cAMP/PKA, PI3K/Akt, and ERK cascade and ultimately leading to protein Tyr phosphorylation are all regulated by ROS and these ROS could act at various levels and on several kinases (directly or via other proteins). The phosphorylation events related to these enzymatic systems are needed for the early, intermediate, and late steps of capacitation. Some of these pathways seem to be independent and act in parallel, but others are rather characterized by cross talks and interdependency. Taken together, all these cascades and their elements demonstrate the complexity of signal transduction during capacitation and its regulation by ROS and, again, evidence redundancy of mechanisms to assure that spermatozoa acquire their fertility potential.

We have to keep an opened mind on the possible, and even probable, involvement of other ROS (H_2O_2, $ONOO^-$, etc.) at minute amounts during capacitation. Also, ROS surely act on other targets than those mentioned here that could include not only other kinases and phosphatases but also ion channels, actin cytoskeleton, phospholipases (A2, C, D), etc. [24]. Future research may prove that they link those that we already evidenced.

It is important to go on with these studies because unraveling the positive actions of ROS may suggest new diagnosis tools as well as improvement of artificial reproduction techniques. Also, they may very well indicate that it is really time to reconsider protocols involving the use of high levels of antioxidants in vitro (media for assisted reproduction) and in vivo (treatments with vitamins, Zn^{2+}, etc.) as they may prove to be inefficient or even decrease rather than improve fertility.

4.7 Key Points

- Capacitation is associated with a mild oxidative stress. The generation of ROS ($O_2^{\bullet-}$ and $NO^{\bullet}$), by spermatozoa themselves, is essential and occurs as an early step of capacitation.
- $O_2^{\bullet-}$ and $NO^{\bullet}$ are synthesized by sperm enzymes, oxidase and NOS, and these ROS appear to have specific targets. Exogenous addition of ROS, combined with the use of inhibitors and ROS scavengers, gives important indications on cellular processes involved but does not replace actual measurements of endogenously formed ROS.
- The oxidase is likely at the membrane level since $O_2^{\bullet-}$ is released extracellularly. There may be more than one NOS, one being stimulated at the beginning of capacitation and another being part of the signal transduction cascades related to this process.
- Spermatozoa are at the present time the only cell type in which a reciprocal (double sided) ROS-induced ROS formation is demonstrated, $O_2^{\bullet-}$ promoting $NO^{\bullet}$ synthesis and vice versa.
- Sg is a natural regulator for ROS generators, both the oxidase and NOS.
- Zn^{2+} is also a major physiological regulator of ROS generation, but is also expected to directly regulate some of the enzymes (e.g., PKC) involved in signal transduction cascades.
- The SH/SS couple is involved both as regulator of ROS formation and target for ROS action.
- Both $O_2^{\bullet-}$ and $NO^{\bullet}$ promote the signal transduction pathways known to be involved during capacitation and as a consequence also the late protein Tyr phosphorylation that occurs downstream of all these. ROS may act via oxidation of

protein SH to activate or inhibit several enzymes (kinases, phosphatases, etc.) of these cascades.

- Sperm hyperactivation and acrosome reaction are also modulated by ROS. ROS may then be acting on other targets since these events are differently controlled.
- Sperm activation is characterized by stimulation of several processes, multiplicity of enzymatic pathways and types of regulation, cross talks, apparent redundancy of mechanisms, etc. This is expected to insure that spermatozoa acquire their fertility potential.

References

1. Yanagimachi R. Mammalian fertilization. In: Knobil E, Neill J, editors. The physiology of reproduction. New York: Raven Press Ltd; 1994. p. 189–317.
2. de Lamirande E, Leclerc P, Gagnon C. Capacitation as a regulatory event that primes spermatozoa for the acrosome reaction and fertilization. Mol Hum Reprod. 1997;3(3):175–94.
3. de Lamirande E, O'Flaherty C. Sperm activation: role of reactive oxygen species and kinases. Biochim Biophys Acta. 2008;1784(1):106–15.
4. DeJonge C. Biological basis for human capacitation. Hum Reprod Update. 2005;11(3): 205–14.
5. O'Flaherty C, de Lamirande E, Gagnon C. Reactive oxygen species modulate independent protein phosphorylation pathways during human sperm capacitation. Free Radic Biol Med. 2006;40(6):1045–55.
6. Suarez SS. Control of hyperactivation in sperm. Hum Reprod Update. 2008;14(6):647–57.
7. Baldi E, Luconi M, Bonaccorsi L, Muratori M, Forti G. Intracellular events and signaling pathways involved in sperm acquisition of fertilizing capacity and acrosome reaction. Front Biosci. 2000;5:E110–23.
8. Herrero MB, Gagnon C. Nitric oxide: a novel mediator of sperm function. J Androl. 2001;22(3):349–56.
9. Olds-Clarke P. Unresolved issues in mammalian fertilization. Int Rev Cytol. 2003;232: 129–84.
10. Saez F, Ouvrier A, Drevet JR. Epididymis cholesterol homeostasis and sperm fertilizing ability. Asian J Androl. 2011;13(1):11–7.
11. Baker HW, Liu DY, Garrett C, Martic M. The human acrosome reaction. Asian J Androl. 2000;2(3):172–8.
12. Kirkman-Brown JC, Punt EL, Barratt CL, Publicover SJ. Zona pellucida and progesterone-induced Ca^{2+} signaling and acrosome reaction in human spermatozoa. J Androl. 2002;23(3): 306–15.
13. Aitken RJ, Baker MA, De Iuliis GN, Nixon B. New insights into sperm physiology and pathology. Handb Exp Pharmacol. 2010;198:99–115.
14. de Lamirande E, Tsai C, Harakat A, Gagnon C. Involvement of reactive oxygen species in human sperm arcosome reaction induced by A23187, lysophosphatidylcholine, and biological fluid ultrafiltrates. J Androl. 1998;19(5):585–94.
15. Droge W. Free radicals in the physiological control of cell function. Physiol Rev. 2002;82(1): 47–95.
16. Filomeni G, Rotilio G, Ciriolo MR. Disulfide relays and phosphorylative cascades: partners in redox-mediated signaling pathways. Cell Death Differ. 2005;12(12):1555–63.
17. O'Flaherty C, de Lamirande E, Gagnon C. Positive role of reactive oxygen species in mammalian sperm capacitation: triggering and modulation of phosphorylation events. Free Radic Biol Med. 2006;41(4):528–40.

18. Suzuki YJ, Forman HJ, Sevanian A. Oxidants as stimulators of signal transduction. Free Radic Biol Med. 1997;22(1–2):269–85.
19. Halliwell B, Gutteridge JMC. Free radicals in biology and medicine. New York: Oxford University Press; 2007.
20. O'Flaherty C, Rico de Souza A. Hydrogen peroxide modifies human sperm peroxiredoxins in a dose-dependent manner. Biol Reprod. 2011;84(1):238–47.
21. Peshenko IV, Shichi H. Oxidation of active center cysteine of bovine 1-Cys peroxiredoxin to the cysteine sulfenic acid form by peroxide and peroxynitrite. Free Radic Biol Med. 2001;31(3):292–303.
22. Rhee SG, Kang SW, Jeong W, Chang TS, Yang KS, Woo HA. Intracellular messenger function of hydrogen peroxide and its regulation by peroxiredoxins. Curr Opin Cell Biol. 2005; 17(2):183–9.
23. Rhee SG, Chae HZ, Kim K. Peroxiredoxins: a historical overview and speculative preview of novel mechanisms and emerging concepts in cell signaling. Free Radic Biol Med. 2005;38(12): 1543–52.
24. Forman HJ, Fukuto J, Torres M. Signal transduction by reactive oxygen and nitrogen species: pathways and chemical principles. Dordrecht: Kluwer Academic Publishers; 2003.
25. Bedard K, Krause KH. The NOX family of ROS-generating NADPH oxidases: physiology and pathophysiology. Physiol Rev. 2007;87(1):245–313.
26. Sabeur K, Ball BA. Characterization of NADPH oxidase 5 in equine testis and spermatozoa. Reproduction. 2007;134(2):263–70.
27. Shukla S, Jha RK, Laloraya M, Kumar PG. Identification of non-mitochondrial NADPH oxidase and the spatio-temporal organization of its components in mouse spermatozoa. Biochem Biophys Res Commun. 2005;331(2):476–83.
28. Sumimoto H, Miyano K, Takeya R. Molecular composition and regulation of the Nox family NAD(P)H oxidases. Biochem Biophys Res Commun. 2005;338(1):677–86.
29. Donnelly ET, Lewis SE, Thompson W, Chakravarthy U. Sperm nitric oxide and motility: the effects of nitric oxide synthase stimulation and inhibition. Mol Hum Reprod. 1997;3(9): 755–62.
30. Herrero MB, Perez MS, Viggiano JM, Polak JM, de Gimeno MF. Localization by indirect immunofluorescence of nitric oxide synthase in mouse and human spermatozoa. Reprod Fertil Dev. 1996;8(5):931–4.
31. Herrero MB, de Lamirande E, Gagnon C. Nitric oxide is a signaling molecule in spermatozoa. Curr Pharm Des. 2003;9(5):419–25.
32. Kawashima S. The two faces of endothelial nitric oxide synthase in the pathophysiology of atherosclerosis. Endothelium. 2004;11(2):99–107.
33. Vasquez-Vivar J, Kalyanaraman B. Generation of superoxide from nitric oxide synthase. FEBS Lett. 2000;481(3):305–6.
34. Weaver J, Porasuphatana S, Tsai P, Pou S, Roman LJ, Rosen GM. A comparative study of neuronal and inducible nitric oxide synthases: generation of nitric oxide, superoxide, and hydrogen peroxide. Biochim Biophys Acta. 2005;1726(3):302–8.
35. de Lamirande E, Gagnon C. Reactive oxygen species and human spermatozoa. II. Depletion of adenosine triphosphate plays an important role in the inhibition of sperm motility. J Androl. 1992;13(5):379–86.
36. de Lamirande E, Gagnon C. Reactive oxygen species and human spermatozoa. I. Effects on the motility of intact spermatozoa and on sperm axonemes. J Androl. 1992;13(5):368–78.
37. de Lamirande E, Gagnon C. A positive role for the superoxide anion in triggering hyperactivation and capacitation of human spermatozoa. Int J Androl. 1993;16(1):21–5.
38. de Lamirande E, Gagnon C. Human sperm hyperactivation and capacitation as parts of an oxidative process. Free Radic Biol Med. 1993;14(2):157–66.
39. Aitken RJ, Harkiss D, Knox W, Paterson M, Irvine DS. A novel signal transduction cascade in capacitating human spermatozoa characterised by a redox-regulated, cAMP-mediated induction of tyrosine phosphorylation. J Cell Sci. 1998;111(5):645–56.

40. de Lamirande E, Harakat A, Gagnon C. Human sperm capacitation induced by biological fluids and progesterone, but not by NADH or NADPH, is associated with the production of superoxide anion. J Androl. 1998;19(2):215–25.
41. Griveau JF, Renard P, Le Lannou D. An in vitro promoting role for hydrogen peroxide in human sperm capacitation. Int J Androl. 1994;17(6):300–7.
42. Herrero MB, de Lamirande E, Gagnon C. Nitric oxide regulates human sperm capacitation and protein-tyrosine phosphorylation in vitro. Biol Reprod. 1999;61(3):575–81.
43. Herrero MB, Chatterjee S, Lefièvre L, de Lamirande E, Gagnon C. Nitric oxide interacts with the cAMP pathway to modulate capacitation of human spermatozoa. Free Radic Biol Med. 2000;29(6):522–36.
44. Leclerc P, de Lamirande E, Gagnon C. Regulation of protein-tyrosine phosphorylation and human sperm capacitation by reactive oxygen derivatives. Free Radic Biol Med. 1997;22(4): 643–56.
45. Zini A, de Lamirande E, Gagnon C. Low levels of nitric oxide promote human sperm capacitation in vitro. J Androl. 1995;16(5):424–31.
46. de Lamirande E, Lamothe G. Reactive oxygen-induced reactive oxygen formation during human sperm capacitation. Free Radic Biol Med. 2009;46(4):502–10.
47. World Health Organization. WHO laboratory manual for the examination of human semen and sperm-cervical mucus interaction. Cambridge: Cambridge University Press; 1999.
48. de Lamirande E, Gagnon C. Human sperm hyperactivation in whole semen and its association with low superoxide scavenging capacity in seminal plasma. Fertil Steril. 1993;59(6): 1291–5.
49. de Lamirande E, Gagnon C. Human sperm hyperactivation in whole semen and its association with low superoxide scavenging capacity in seminal plasma and spermatozoa. Fertil Steril. 1993;59(Suppl):1291–5.
50. Henkel RR. Leukocytes and oxidative stress: dilemma for sperm function and male fertility. Asian J Androl. 2011;13(1):43–52.
51. Kaufman DL, Mitchell JA. Intrauterine oxygen tension during the oestrous cycle in the hamster: patterns of change. Comp Biochem Physiol Comp Physiol. 1994;107(4):673–8.
52. Aitken RJ, Buckingham DW, West KM. Reactive oxygen species and human spermatozoa: analysis of the cellular mechanisms involved in luminol- and lucigenin-dependent chemiluminescence. J Cell Physiol. 1992;151(3):466–77.
53. de Lamirande E, Leduc BE, Iwasaki A, Hassouna M, Gagnon C. Increased reactive oxygen species formation in semen of patients with spinal cord injury. Fertil Steril. 1995;63(3): 637–42.
54. Plante M, de Lamirande E, Gagnon C. Reactive oxygen species released by activated neutrophils, but not by deficient spermatozoa, are sufficient to affect normal sperm motility. Fertil Steril. 1994;62(2):387–93.
55. de Lamirande E, Gagnon C. Capacitation-associated production of superoxide anion by human spermatozoa. Free Radic Biol Med. 1995;18(3):487–95.
56. de Lamirande E, Lamothe G, Villemure M. Control of superoxide and nitric oxide formation during human sperm capacitation. Free Radic Biol Med. 2009;46(10):1420–7.
57. Thomas S, Kotamraju S, Zielonka J, Harder DR, Kalyanaraman B. Hydrogen peroxide induces nitric oxide and proteosome activity in endothelial cells: a bell-shaped signaling response. Free Radic Biol Med. 2007;42(7):1049–61.
58. Lundin A, Richardsson A, Thore A. Continuous monitoring of ATP-converting reactions by purified firefly luciferase. Anal Biochem. 1976;75(2):611–20.
59. Radi R, Rubbo H, Thomson L, Prodanov E. Luminol chemiluminescence using xanthine and hypoxanthine as xanthine oxidase substrates. Free Radic Biol Med. 1990;8(2):121–6.
60. de Lamirande E, Gagnon C. Increased production of intra- and extracellular superoxide anion by capacitating human spermatozoa. J Androl. 1995;16(Suppl):P54.
61. de Lamirande E, Jiang H, Zini A, Kodama H, Gagnon C. Reactive oxygen species and sperm physiology. Rev Reprod. 1997;2(1):48–54.

62. De Iuliis GN, Wingate JK, Koppers AJ, McLaughlin EA, Aitken RJ. Definitive evidence for the nonmitochondrial production of superoxide anion by human spermatozoa. J Clin Endocrinol Metab. 2006;91(5):1968–75.
63. Mahfouz R, Sharma R, Lackner J, Aziz N, Agarwal A. Evaluation of chemiluminescence and flow cytometry as tools in assessing production of hydrogen peroxide and superoxide anion in human spermatozoa. Fertil Steril. 2009;92(2):819–27.
64. O'Flaherty C, Beorlegui N, Beconi MT. Participation of superoxide anion in the capacitation of cryopreserved bovine sperm. Int J Androl. 2003;26(2):109–14.
65. Lewis B, Aitken RJ. A redox-regulated tyrosine phosphorylation cascade in rat spermatozoa. J Androl. 2001;22(4):611–22.
66. Revelli A, Soldati G, Costamagna C, et al. Follicular fluid proteins stimulate nitric oxide (NO) synthesis in human sperm: a possible role for NO in acrosomal reaction. J Cell Physiol. 1999;178(1):85–92.
67. Funahashi H. Induction of capacitation and the acrosome reaction of boar spermatozoa by L-arginine and nitric oxide synthesis associated with the anion transport system. Reproduction. 2002;124(6):857–64.
68. Lampiao F, Strijdom H, du Plessis SS. Direct nitric oxide measurement in human spermatozoa: flow cytometric analysis using the fluorescent probe, diaminofluorescein. Int J Androl. 2006;29(5):564–7.
69. Herrero MB, de Lamirande E, Gagnon C. Tyrosine nitration in human spermatozoa: a physiological function of peroxynitrite, the reaction product of nitric oxide and superoxide. Mol Hum Reprod. 2001;7(10):913–21.
70. Cristol JP, Guerin MC, Torreilles J. Measurement of nitric oxide and biological systems. C R Acad Sci III. 1994;317(6):549–60.
71. Hughes MN. Relationships between nitric oxide, nitroxyl ion, nitrosonium cation and peroxynitrite. Biochim Biophys Acta. 1999;1411(2–3):263–72.
72. Manning P, Cookson MR, McNeil CJ, Figlewicz D, Shaw PJ. Superoxide-induced nitric oxide release from cultured glial cells. Brain Res. 2001;911(2):203–10.
73. Baker MA, Krutskikh A, Aitken RJ. Biochemical entities involved in reactive oxygen species generation by human spermatozoa. Protoplasma. 2003;221(1–2):145–51.
74. de Lamirande E, Gagnon C. The extracellular signal-regulated kinase (ERK) pathway is involved in human sperm function and modulated by the superoxide anion. Mol Hum Reprod. 2002;8(2):124–35.
75. Serrander L, Jaquet V, Bedard K, et al. NOX5 is expressed at the plasma membrane and generates superoxide in response to protein kinase C activation. Biochimie. 2007;89(9):1159–67.
76. de Lamirande E. The capacitation inducers progesterone and biological fluid ultrafiltrates, and the capacitation inhibitors zinc and semenogelin, modulate differently the extracellular superoxide production in human spermatozoa and in neutrophils. J Androl. 2011;32(65).
77. Forstermann U. Janus-faced role of endothelial NO synthase in vascular disease: uncoupling of oxygen reduction from NO synthesis and its pharmacological reversal. Biol Chem. 2006;387(12):1521–33.
78. Xia Y. Superoxide generation from nitric oxide synthases. Antioxid Redox Signal. 2007;9(10):1773–8.
79. Thundathil J, de Lamirande E, Gagnon C. Different signal transduction pathways are involved during human sperm capacitation induced by biological and pharmacological agents. Mol Hum Reprod. 2002;8(9):811–6.
80. Thundathil J, de Lamirande E, Gagnon C. Nitric oxide regulates the phosphorylation of the threonine-glutamine-tyrosine motif in proteins of human spermatozoa during capacitation. Biol Reprod. 2003;68(4):1291–8.
81. O'Flaherty C, Beorlegui N, Beconi MT. Heparin- and superoxide anion-dependent capacitation of cryopreserved bovine spermatozoa: requirement of dehydrogenases and protein kinases. Free Radic Res. 2006;40(4):427–32.

82. O'Flaherty CM, Beorlegui NB, Beconi MT. Lactate dehydrogenase-C4 is involved in heparin- and NADH-dependent bovine sperm capacitation. Andrologia. 2002;34(2):91–7.
83. Aitken RJ, Fisher HM, Fulton N, et al. Reactive oxygen species generation by human spermatozoa is induced by exogenous NADPH and inhibited by the flavoprotein inhibitors diphenylene iodonium and quinacrine. Mol Reprod Dev. 1997;47(4):468–82.
84. Manjunath P, Bergeron A, Lefebvre J, Fan J. Seminal plasma proteins: functions and interaction with protective agents during semen preservation. Soc Reprod Fertil Suppl. 2007;65: 217–28.
85. Chiu PC, Chung MK, Tsang HY, et al. Glycodelin-S in human seminal plasma reduces cholesterol efflux and inhibits capacitation of spermatozoa. J Biol Chem. 2005;280(27): 25580–9.
86. de Lamirande E. Semenogelin, the main protein of the human semen coagulum, regulates sperm function. Semin Thromb Hemost. 2007;33(1):60–8.
87. Jonsson M, Linse S, Frohm B, Lundwall A, Malm J. Semenogelins I and II bind zinc and regulate the activity of prostate-specific antigen. Biochem J. 2005;387(Pt 2):447–53.
88. Robert M, Gagnon C. Semenogelin I: a coagulum forming, multifunctional seminal vesicle protein. Cell Mol Life Sci. 1999;55(6–7):944–60.
89. de Lamirande E, Yoshida K, Yoshiike TM, Iwamoto T, Gagnon C. Semenogelin, the main protein of semen coagulum, inhibits human sperm capacitation by interfering with the superoxide anion generated during this process. J Androl. 2001;22(4):672–9.
90. de Lamirande E, Lamothe G. Levels of semenogelin in human spermatozoa decrease during capacitation: involvement of reactive oxygen species and zinc. Hum Reprod. 2010;25(7): 1619–30.
91. Cormier N, Sirard MA, Bailey JL. Premature capacitation of bovine spermatozoa is initiated by cryopreservation. J Androl. 1997;18(4):461–8.
92. Stoltenberg M, Sorensen MB, Danscher G. Histochemical demonstration of zinc ions in ejaculated human semen. Int J Androl. 1997;20(4):229–36.
93. Gavella M, Lipovac V, Vucic M, Sverko V. In vitro inhibition of superoxide anion production and superoxide dismutase activity by zinc in human spermatozoa. Int J Androl. 1999;22(4): 266–74.
94. Yoshida K, Kawano N, Yoshiike M, Yoshida M, Iwamoto T, Morisawa M. Physiological roles of semenogelin I and zinc in sperm motility and semen coagulation on ejaculation in humans. Mol Hum Reprod. 2008;14(3):151–6.
95. Yoshida K, Krasznai ZT, Krasznai Z, et al. Functional implications of membrane modification with semenogelins for inhibition of sperm motility in humans. Cell Motil Cytoskeleton. 2009;66(2):99–108.
96. Oshio S, Kaneko S, Iizuka R, Mohri H. Sialic acid in purified human sperm. Arch Androl. 1987;18(3):225–30.
97. Yeaman MR, Yount NY. Mechanisms of antimicrobial peptide action and resistance. Pharmacol Rev. 2003;55(1):27–55.
98. Giles NM, Watts AB, Giles GI, Fry FH, Littlechild JA, Jacob C. Metal and redox modulation of cysteine protein function. Chem Biol. 2003;10(8):677–93.
99. Knapp LT, Klann E. Superoxide-induced stimulation of protein kinase C via thiol modification and modulation of zinc content. J Biol Chem. 2000;275(31):24136–45.
100. Maret W. Zinc coordination environments in proteins determine zinc functions. J Trace Elem Med Biol. 2005;19(1):7–12.
101. de Lamirande E, Gagnon C. Paradoxical effect of reagents for sulfhydryl and disulfide groups on human sperm capacitation and superoxide production. Free Radic Biol Med. 1998;25(7): 803–17.
102. Nauc V, de Lamirande E, Leclerc P, Gagnon C. Inhibitors of phosphoinositide 3-kinase, LY294002 and wortmannin, affect sperm capacitation and associated phosphorylation of proteins differently: Ca^{2+}-dependent divergences. J Androl. 2004;25(4):573–85.
103. Aitken RJ, Paterson M, Fisher H, Buckingham DW, van Duin M. Redox regulation of tyrosine phosphorylation in human spermatozoa and its role in the control of human sperm function. J Cell Sci. 1995;108(5):2017–25.

104. Carrera A, Moos J, Ning XP, et al. Regulation of protein tyrosine phosphorylation in human sperm by a calcium/calmodulin-dependent mechanism: identification of A kinase anchor proteins as major substrates for tyrosine phosphorylation. Dev Biol. 1996;180(1):284–96.
105. Lefievre L, Jha KN, de Lamirande E, Visconti PE, Gagnon C. Activation of protein kinase A during human sperm capacitation and acrosome reaction. J Androl. 2002;23(5):709–16.
106. O'Flaherty C, de Lamirande E, Gagnon C. Phosphorylation of the arginine-X-X-(serine/threonine) motif in human sperm proteins during capacitation: modulation and protein kinase A dependency. Mol Hum Reprod. 2004;10(5):355–63.
107. Zhang H, Zheng RL. Promotion of human sperm capacitation by superoxide anion. Free Radic Res. 1996;24(4):261–8.
108. Tan CM, Xenoyannis S, Feldman RD. Oxidant stress enhances adenylyl cyclase activation. Circ Res. 1995;77(4):710–7.
109. Harrison RAP. Rapid PKA-catalysed phosphorylation of boar sperm proteins induced by the capacitating agent bicarbonate. Mol Reprod Dev. 2004;67(3):337–52.
110. Jha KN, Salicioni AM, Arcelay E, et al. Evidence for the involvement of proline-directed serine/threonine phosphorylation in sperm capacitation. Mol Hum Reprod. 2006;12(12):781–9.
111. Luconi M, Krausz C, Barni T, Vannelli GB, Forti G, Baldi E. Progesterone stimulates p42 extracellular signal-regulated kinase (p42erk) in human spermatozoa. Mol Hum Reprod. 1998;4(3):251–8.
112. O'Flaherty C, de Lamirande E, Gagnon C. Reactive oxygen species and protein kinases modulate the level of phospho-MEK-like proteins during human sperm capacitation. Biol Reprod. 2005;73(1):94–105.
113. Kolch W. Meaningful relationships: the regulation of the Ras/Raf/MEK/ERK pathway by protein interactions. Biochem J. 2000;351(Pt 2):289–305.
114. Widmann C, Gibson S, Jarpe MB, Johnson GL. Mitogen-activated protein kinase: conservation of a three-kinase module from yeast to human. Physiol Rev. 1999;79(1):143–80.
115. Morte C, Iborra A, Martinez P. Phosphorylation of Shc proteins in human sperm in response to capacitation and progesterone treatment. Mol Reprod Dev. 1998;50(1):113–20.
116. Naz RK, Ahmad K, Kaplan P. Expression and function of ras proto-oncogene proteins in human sperm cells. J Cell Sci. 1992;102(Pt 3):487–94.
117. Miyata Y, Nishida E. Distantly related cousins of MAP kinase: biochemical properties and possible physiological functions. Biochem Biophys Res Commun. 1999;266(2):291–5.
118. Yan C, Luo H, Lee JD, Abe J, Berk BC. Molecular cloning of mouse ERK5/BMK1 splice variants and characterization of ERK5 functional domains. J Biol Chem. 2001;276(14):10870–8.
119. Westendorf JM, Rao PN, Gerace L. Cloning of cDNAs for M-phase phosphoproteins recognized by the MPM2 monoclonal antibody and determination of the phosphorylated epitope. Proc Natl Acad Sci USA. 1994;91(2):714–8.
120. Kumar A, Carrera AC. New functions for PI3K in the control of cell division. Cell Cycle. 2007;6(14):1696–8.
121. Marone R, Cmiljanovic V, Giese B, Wymann MP. Targeting phosphoinositide 3-kinase: moving towards therapy. Biochim Biophys Acta. 2008;1784(1):159–85.
122. Luconi M, Carloni V, Marra F, Ferruzzi P, Forti G, Baldi E. Increased phosphorylation of AKAP by inhibition of phosphatidylinositol 3-kinase enhances human sperm motility through tail recruitment of protein kinase A. J Cell Sci. 2004;117(Pt 7):1235–46.
123. Wymann MP, Marone R. Phosphoinositide 3-kinase in disease: timing, location, and scaffolding. Curr Opin Cell Biol. 2005;17(2):141–9.
124. Lander HM, Milbank AJ, Tauras JM, et al. Redox regulation of cell signalling. Nature. 1996;381(6581):380–1.
125. Lander HM, Hajjar DP, Hempstead BL, et al. A molecular redox switch on p21(ras). Structural basis for the nitric oxide-p21(ras) interaction. J Biol Chem. 1997;272(7):4323–6.
126. Galantino-Homer HL, Visconti PE, Kopf GS. Regulation of protein tyrosine phosphorylation during bovine sperm capacitation by a cyclic adenosine 3′5′-monophosphate-dependent pathway. Biol Reprod. 1997;56(3):707–19.

127. Petrunkina AM, Simon K, Gunzel-Apel AR, Topfer-Petersen E. Specific order in the appearance of protein tyrosine phosphorylation patterns is functionally coordinated with dog sperm hyperactivation and capacitation. J Androl. 2003;24(3):423–37.
128. Pommer AC, Rutllant J, Meyers SA. Phosphorylation of protein tyrosine residues in fresh and cryopreserved stallion spermatozoa under capacitating conditions. Biol Reprod. 2003;68(4):1208–14.
129. Tardif S, Dubé C, Chevalier S, Bailey JL. Capacitation is associated with tyrosine phosphorylation and tyrosine kinase-like activity of pig sperm proteins. Biol Reprod. 2001;65(3): 784–92.
130. Visconti PE, Stewart-Savage J, Blasco A, et al. Roles of bicarbonate, cAMP, and protein tyrosine phosphorylation on capacitation and the spontaneous acrosome reaction of hamster sperm. Biol Reprod. 1999;61(1):76–84.
131. Aitken RJ, Buckingham DW, Harkiss D, Paterson M, Fisher H, Irvine DS. The extragenomic action of progesterone on human spermatozoa is influenced by redox regulated changes in tyrosine phosphorylation during capacitation. Mol Cell Endocrinol. 1996;117(1):83–93.
132. Leclerc P, de Lamirande E, Gagnon C. Cyclic adenosine 3′,5′ monophosphate-dependent regulation of protein tyrosine phosphorylation in relation to human sperm capacitation and motility. Biol Reprod. 1996;55(3):684–92.
133. Baker MA, Hetherington L, Aitken RJ. Identification of SRC as a key PKA-stimulated tyrosine kinase involved in the capacitation-associated hyperactivation of murine spermatozoa. J Cell Sci. 2006;119(15):3182–92.
134. Leclerc P, Goupil S. Regulation of the human sperm tyrosine kinase c-yes. Activation by cyclic adenosine 3′,5′-monophosphate and inhibition by $Ca^{(2+)}$. Biol Reprod. 2002;67(1): 301–7.
135. Ficarro S, Chertihin O, Westbrook VA, et al. Phosphoproteome analysis of capacitated human sperm. Evidence of tyrosine phosphorylation of a kinase-anchoring protein 3 and valosin-containing protein/p97 during capacitation. J Biol Chem. 2003;278(13):11579–89.
136. Naz RK, Rajesh PB. Role of tyrosine phosphorylation in sperm capacitation/acrosome reaction. Reprod Biol Endocrinol. 2004;2:75.
137. Bailey JL. Factors regulating sperm capacitation. Syst Biol Reprod Med. 2010;56(5): 334–48.
138. Baker MA, Hetherington L, Curry B, Aitken RJ. Phosphorylation and consequent stimulation of the tyrosine kinase c-Abl by PKA in mouse spermatozoa; its implications during capacitation. Dev Biol. 2009;333(1):57–66.
139. Lawson C, Goupil S, Leclerc P. Increased activity of the human sperm tyrosine kinase SRC by the cAMP-dependent pathway in the presence of calcium. Biol Reprod. 2008;79(4): 657–66.
140. Gopalakrishna R, Anderson WB. Ca^{2+}- and phospholipid-independent activation of protein kinase C by selective oxidative modification of the regulatory domain. Proc Natl Acad Sci USA. 1989;86(17):6758–62.
141. Abe MK, Kartha S, Karpova AY, et al. Hydrogen peroxide activates extracellular signal-regulated kinase via protein kinase C, Raf-1, and MEK1. Am J Respir Cell Mol Biol. 1998; 18(4):562–9.
142. Chiarugi P, Cirri P. Redox regulation of protein tyrosine phosphatases during receptor tyrosine kinase signal transduction. Trends Biochem Sci. 2003;28(9):509–14.
143. Aitken RJ, De Iuliis GN, McLachlan RI. Biological and clinical significance of DNA damage in the male germ line. Int J Androl. 2009;32(1):46–56.
144. Kodama H, Kuribayashi Y, Gagnon C. Effect of sperm lipid peroxidation on fertilization. J Androl. 1996;17(2):151–7.
145. de Lamirande E, Gagnon C. Redox control of changes in protein sulfhydryl levels during human sperm capacitation. Free Radic Biol Med. 2003;35(10):1271–85.
146. Barford D. The role of cysteine residues as redox-sensitive regulatory switches. Curr Opin Struct Biol. 2004;14(6):679–86.

147. Kim JR, Yoon HW, Kwon KS, Lee SR, Rhee SG. Identification of proteins containing cysteine residues that are sensitive to oxidation by hydrogen peroxide at neutral pH. Anal Biochem. 2000;283(2):214–21.
148. Salmeen A, Barford D. Functions and mechanisms of redox regulation of cysteine-based phosphatases. Antioxid Redox Signal. 2005;7(5–6):560–77.
149. Sen CK. Cellular thiols and redox-regulated signal transduction. Curr Top Cell Regul. 2000;36:1–30.
150. Wu Y, Kwon KS, Rhee SG. Probing cellular protein targets of H2O2 with fluorescein-conjugated iodoacetamide and antibodies to fluorescein. FEBS Lett. 1998;440(1–2):111–5.
151. Chatterjee S, de Lamirande E, Gagnon C. Cryopreservation alters membrane sulfhydryl status of bull spermatozoa: protection by oxidized glutathione. Mol Reprod Dev. 2001; 60(4):498–506.
152. Dalle-Donne I, Milzani A, Gagliano N, Colombo R, Giustarini D, Rossi R. Molecular mechanisms and potential clinical significance of S-glutathionylation. Antioxid Redox Signal. 2008;10(3):445–73.
153. Ghezzi P. Regulation of protein function by glutathionylation. Free Radic Res. 2005;39(6): 573–80.
154. Adler V, Yin Z, Tew KD, Ronai Z. Role of redox potential and reactive oxygen species in stress signaling. Oncogene. 1999;18(45):6104–11.
155. Dimon-Gadal S, Gerbaud P, Keryer G, Anderson W, Evain-Brion D, Raynaud F. In vitro effects of oxygen-derived free radicals on type I and type II cAMP-dependent protein kinases. J Biol Chem. 1998;273(35):22833–40.
156. Hecht D, Zick Y. Selective inhibition of protein tyrosine phosphatase activities by H_2O_2 and vanadate in vitro. Biochem Biophys Res Commun. 1992;188(2):773–9.
157. Sommer D, Coleman S, Swanson SA, Stemmer PM. Differential susceptibilities of serine/threonine phosphatases to oxidative and nitrosative stress. Arch Biochem Biophys. 2002;404(2):271–8.
158. Brondello JM, Pouyssegur J, McKenzie FR. Reduced MAP kinase phosphatase-1 degradation after p42/p44MAPK-dependent phosphorylation. Science. 1999;286(5449):2514–7.
159. Camps M, Nichols A, Arkinstall S. Dual specificity phosphatases: a gene family for control of MAP kinase function. FASEB J. 2000;14(1):6–16.
160. Levinthal DJ, Defranco DB. Reversible oxidation of ERK-directed protein phosphatases drives oxidative toxicity in neurons. J Biol Chem. 2005;280(7):5875–83.
161. Muda M, Boschert U, Dickinson R, et al. MKP-3, a novel cytosolic protein-tyrosine phosphatase that exemplifies a new class of mitogen-activated protein kinase phosphatase. J Biol Chem. 1996;271(8):4319–26.
162. Jaffrey SR, Erdjument-Bromage H, Ferris CD, Tempst P, Snyder SH. Protein S-nitrosylation: a physiological signal for neuronal nitric oxide. Nat Cell Biol. 2001;3(2):193–7.
163. Miersch S, Mutus B. Protein S-nitrosation: biochemistry and characterization of protein thiol-NO interactions as cellular signals. Clin Biochem. 2005;38(9):777–91.
164. Moon KH, Kim BJ, Song BJ. Inhibition of mitochondrial aldehyde dehydrogenase by nitric oxide-mediated S-nitrosylation. FEBS Lett. 2005;579(27):6115–20.
165. Yang Y, Loscalzo J. S-nitrosoprotein formation and localization in endothelial cells. Proc Natl Acad Sci USA. 2005;102(1):117–22.
166. Lefièvre L, Chen Y, Conner SJ, et al. Human spermatozoa contain multiple targets for protein S-nitrosylation: an alternative mechanism of the modulation of sperm function by nitric oxide? Proteomics. 2007;7(17):3066–84.
167. Oliveira CJ, Schindler F, Ventura AM, et al. Nitric oxide and cGMP activate the Ras-MAP kinase pathway-stimulating protein tyrosine phosphorylation in rabbit aortic endothelial cells. Free Radic Biol Med. 2003;35(4):381–96.
168. Gagnon C, de Lamirande E. Controls of sperm motility. In: De Jonge C, Barratt CL, editors. The sperm cell: production, maturation, fertilization and regeneration. Cambridge: Cambridge University Press; 2006. p. 108–33.

169. Griveau JF, Renard P, Le Lannou D. Superoxide anion production by human spermatozoa as a part of the ionophore-induced acrosome reaction process. Int J Androl. 1995;18(2):67–74.
170. O'Flaherty C, Rodriguez P, Srivastava S. L-arginine promotes capacitation and acrosome reaction in cryopreserved bovine spermatozoa. Biochim Biophys Acta. 2004;1674(2): 215–21.
171. O'Flaherty CM, Beorlegui NB, Beconi MT. Reactive oxygen species requirements for bovine sperm capacitation and acrosome reaction. Theriogenology. 1999;52(2):289–301.
172. Aitken RJ, Buckingham DW, Carreras A, Irvine DS. Superoxide dismutase in human sperm suspensions: relationship with cellular composition, oxidative stress, and sperm function. Free Radic Biol Med. 1996;21(4):495–504.
173. Revelli A, Costamagna C, Moffa F, et al. Signaling pathway of nitric oxide-induced acrosome reaction in human spermatozoa. Biol Reprod. 2001;64(6):1708–12.
174. Revelli A, Ghigo D, Moffa F, Massobrio M, Tur-Kaspa I. Guanylate cyclase activity and sperm function. Endocr Rev. 2002;23(4):484–94.

Chapter 5
Protection of Epididymal Spermatozoa from Oxidative Stress

Joël R. Drevet

Abstract Mammalian spermatozoa leaving the testis via the *rete* testis and efferent ducts start a long journey through the epididymis. During their trip along this single tubule, fertilizing potential will gradually be acquired by these spermatozoa. *In fine*, for most mammals, spermatozoa are stored in the distal epididymal compartment for undetermined periods of time, obviously depending on the male sexual activity. During this phase of post-testicular spermatozoa maturation and storage, the silent nature of these highly differentiated cells renders them particularly fragile and susceptible to attacks, one of which being oxidative stress. It is one of the tasks of the epididymis to provide efficient protection to the male gametes against the deleterious effects of oxidative damage that, if not counteracted, could hamper its structures and function. This is done through the concerted actions of both nonenzymatic and enzymatic primary antioxidants. This chapter intends to give the reader an updated view of the means by which the mammalian epididymis protects transiting spermatozoa from oxidative injuries.

Keywords Primary antioxidants • Nonenzymatic antioxidant • Enzymatic antioxidant • Disulfide bridging • Epididymal spermatozoa • Oxidative stress

5.1 Introduction

Inherent to the mammalian spermatogenetic process, spermatozoa exiting the male gonad have features that make them unique in terms of structure and physiology. Through the meiotic process, the male germ cell becomes haploid and during late

J.R. Drevet, PhD (✉)
GReD Laboratory, CNRS UMR 6293, INSERM U1013, Clermont Université,
24, avenue des Landais, Aubiere 63171, France
e-mail: joel.drevet@univ-bpclermont.fr

A. Agarwal et al. (eds.), *Studies on Men's Health and Fertility*, Oxidative Stress in Applied Basic Research and Clinical Practice, DOI 10.1007/978-1-61779-776-7_5,

spermatogenesis undergoes a spectacular program of cell differentiation in which among many changes the size of its nucleus is reduced to about one-tenth that in a somatic cell. This is permitted by the gradual replacement of testicular histones, first by transition proteins, and then by smaller positively charged molecules called protamines. This compacted state of the sperm chromatin signs the end of transcriptional activity in the male germ cell. In addition, supporting Sertoli cells engulf most of the residual cytoplasm of the male gamete as well as the subcellular organelles that normally sustain cell metabolic activity, especially protein translation. Finally, the late events of spermatogenesis also give this cell an engine and propeller in the form of a flagellar structure connected to the mitochondria-containing sperm midpiece. Although these dramatic structural modifications elegantly and efficiently serve the purpose of fertilization, they have direct consequences on the subsequent physiology of spermatozoa. The absence of transcription/translation events in this now silent cell makes it unable to elicit any kind of response when challenged by external and/or internal stressors. This nonresponsive situation is aggravated by the fact that being devoid of most of its cytoplasm the sperm cell is consequently devoid of the cytosolic tools (molecules and/or enzymes) that usually serve to protect somatic cells. Thus, at the end of spermatogenesis, spermatozoa are particularly fragile and have to rely on their environment both to mature and survive. One of the multiple tasks of the epididymis luminal environment is to provide efficient protection for transiting and stored spermatozoa.

5.2 Oxidative Stress: A Major Challenge for Epididymal Spermatozoa

Aerobic organisms use oxygen to support cell metabolism but, doing so, they generate highly reactive oxygen derivatives. Oxidative stress results from an imbalanced situation, where generation of reactive oxygen species (ROS) exceeds recycling and/or when there is failure of all the systems eukaryotic cells have evolved to fight the inherent dangerous by-products of oxygen consumption. This equilibrium between ROS generation and removal is precisely regulated not only within the cells but also in the extracellular compartments through a large variety of pathways. Although generally presented as negative bystanders, it should not be forgotten that ROS are normal products of cell metabolism and physiologically important in regulating cell functions, such as innate immunity processes, signal transduction, and cell fate, through for example apoptosis. However, when present in excess, their high level of reactivity and possible penetration of most cell compartments render them dangerous for cell metabolism and cell structural integrity. This is particularly true in the epididymis, and with no other cell than the spermatozoa the "friend and foe" roles of ROS are highlighted. The silent nature of epididymal maturing spermatozoa mentioned above makes them particularly vulnerable to any oxidative stress situation in this compartment. In addition, during epididymal maturation of spermatozoa, the lipid fraction of the male gamete plasma membrane is subjected to modifications that increase the representation in polyunsaturated fatty acids (PUFAs) bearing double bonds that greatly facilitate ROS-mediated sperm membrane oxidative damage [1, 2]. The level of unsaturated fatty acids in sperm plasma membrane reaches a figure that is nowhere else seen being above 50% of the total FA of the membrane with docosahexanoic acid (DHA) alone, representing more than 20% of the sperm plasma membrane PUFA [3]. Thus, a post-testicular structural modification of spermatozoa designed to facilitate sperm/oocyte plasma fusion at fertilization puts the spermatozoa in a risky situation of oxidative injury. It has been argued that, paradoxically, it could be seen as a means to protect spermatozoa from peroxidative attacks since such a large amount of double bond-containing FA would probably exert a quenching effect by blocking a large quantity of free radicals. However, it is rather unlikely since, when attacked by free radicals, PUFAs in membranes readily generate new intermediates propagating peroxidative injury throughout in the so-called radical chain reaction [4].

The "oxygen paradox" is very evident in the epididymal compartment because while post-testicular spermatozoa are very susceptible to oxidative injury they evolve in a luminal environment that is surprisingly rather pro-oxidant. This is particularly the case of the proximal part of the epididymis tubule, where an oxidative environment serves at least two major functions. It is first used to complete the structural maturation of the male gamete through spontaneous or enzyme-mediated disulfide-bridging events that both required hydrogen peroxide (H_2O_2) or/and lipid peroxides (LOOH). Secondly, it is part of the epididymal immune strategy. As it is the case in the testis, the epididymal luminal compartment must establish a tolerogenic environment toward spermatozoa while preserving itself from infectious situations. Recent data suggest that a peculiar state of inflammation is triggered in the proximal epididymis by the spermatozoa themselves [5]. ROS are normal products of inflammatory situations and part of the ROS generated in the caput epididymis is most likely attributable to the inflammatory status of the tissue. In addition to the caput epididymidis, the terminal portion of the epididymis tubule, the so-called cauda, is also a compartment where spermatozoa risk oxidative injury. However, here, the reasons for oxidative damage are somehow different, since the risk of oxidative insult comes essentially from spermatozoa themselves. The millions of cauda-stored spermatozoa ready to fulfill their function although maintained in a nonpermissive environment (low oxygen tension, low energy supply, presence of proteins that immobilize them) are likely to leak free radicals from now operational mitochondria. Taken together, these situations, where ROS are both required and feared by epididymal spermatozoa, imply that very efficient regulation of the luminal concentration of these potentially dangerous intermediates is essential whatever their origin.

5.3 Mammalian Epididymis Antioxidant Equipment

Before going into the specific tools with which the epididymis protects spermatozoa from oxidative insults, it should be noted that the characteristic anatomical features of this tissue favor low oxidative stress. In particular, the scrotal position of the

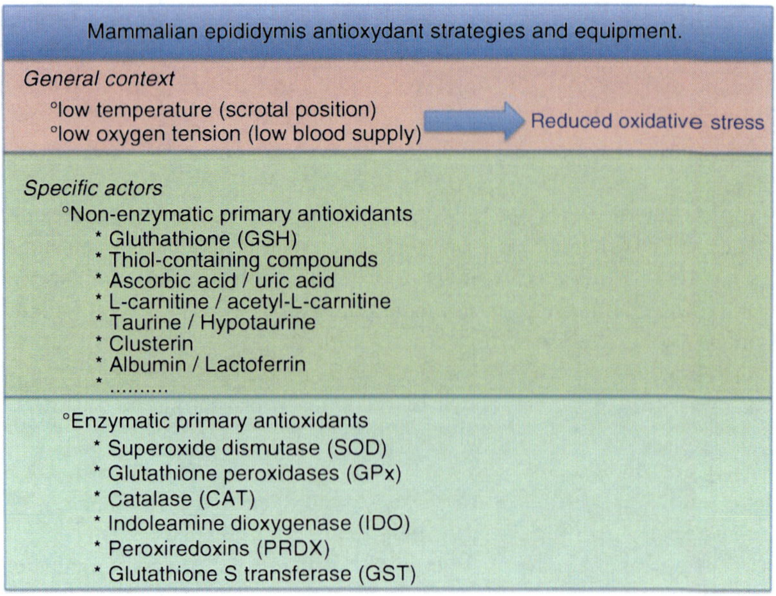

Fig. 5.1 Diagrammatic representation of the major antioxidant strategies used by the mammalian epididymis to cope with oxidative stress

epididymis helps maintain a temperature about 2–3°C below body temperature. In addition, blood supply to the tail of the epididymis corresponding to the storage part of the organ is reduced, giving a low oxygen tension (PO_2). Both low epididymal temperature and low PO_2 contribute to reduced oxidative stress on the spermatozoa and consequently enhanced sperm survival (Fig. 5.1).

To maintain a proper balance between ROS' beneficial and detrimental actions around epididymal transiting spermatozoa, the luminal compartment of the mammalian epididymis possesses several tools acting in combination (Fig. 5.1). As in all tissues, cells, and/or biological fluids, ROS are disposed of in the mammalian epididymis by both nonenzymatic and enzymatic players grouped under the name of primary antioxidants. It appears that the mammalian epididymis relies more on enzymatic antioxidant scavengers to protect transiting spermatozoa against ROS-mediated insults efficiently than on nonenzymatic scavengers. This is supported by two observations: first, the mammalian epididymis was shown to express a large variety of antioxidant enzymes, some of them being unique to this tissue [6–8]. Second, it was recently shown with a transgenic mouse model that, when the major epididymal luminal scavenging enzyme is absent, spermatozoa readily suffer oxidative injury [9]. These observations support the idea, at least in mouse, that nonenzymatic scavengers present in the luminal fluid are not sufficient to compensate efficiently for the loss of enzymatic antioxidant protection.

5.3.1 *Nonenzymatic Epididymal Scavengers*

Many molecules with intrinsic radical-scavenging activity exist in the epididymis compartment. I do not intend here to present them in an exhaustive fashion. I only focus on those that are quite classical and those that have been highlighted because they are present in unusual concentrations in this tissue.

5.3.1.1 Glutathione and Thiol-Containing Compounds

The tripeptide glutathione (GSH) is generally the most important/abundant nonenzymatic scavenger in and outside cells. Its cysteine residue provides a reactive thiol group (SH) that interacts readily with free radicals. Reduction of oxidized GSH is then ensured by GSH reductase and nicotinamide adenine dinucleotide (NADPH). It was repeatedly reported in many mammal models that GSH concentrations are rather low in the epididymis compartment, suggesting that GSH is not a major antioxidant actor in this tissue. However, it is important to note that γGT (gamma glutamyl transpeptidase), the enzyme that regenerates GSH, is quite abundantly expressed in the caput epididymidis [10]. In addition, data have been produced pleading in favor of a significant role of GSH in protecting epididymal sperm from oxidative insult since cauda epididymidis and stored spermatozoa have been shown to be particularly vulnerable to GSH depletion [11]. Thus, this common and powerful scavenger is most probably one important player in the global antioxidant defense strategy of the epididymis. Observations indicating that thiol content in normozoospermic semen was found to be significantly higher than in samples originating from vasectomized men [12] also support this idea and strongly suggest that the epididymis contributes significantly to the accumulation of thiol-containing antioxidants in the semen.

Thioredoxins (Trx) are other thiol-containing molecules that could indirectly act as antioxidants within the epididymis. They are 12-kDa oxidoreductase enzymes that facilitate the reduction of other proteins by cysteine thiol-disulfide exchange, found in nearly all known organisms, and have been shown to be essential for life in mammals [13]. Although Trx is an enzyme and therefore I should have evoked it in the next section of this review devoted to epididymal enzymatic scavengers, it also acts as an electron donor to peroxidases and other enzymes [14]. In that sense, Trx functions more as a disulfide intermediate rather than an enzymatic antioxidant. There is very little information regarding the level of Trx in the mammalian epididymis, so it is quite difficult to ascribe to Trx an important role in protecting epididymal spermatozoa from oxidative insult. However, it was reported that human spermatozoa do possess specific Trx called SPTRX for spermatozoa-specific Trx [15]. SPTRX expression is restricted to postmeiotic sperm cells and are considered to play roles in the extensive reorganization of sperm protein disulfide bonds that occur during either spermiogenesis or/and epididymal maturation to stabilize cytoskeletal sperm structures [15, 16].

5.3.1.2 Ascorbic Acid and Uric Acid

In addition to thiols, ascorbic acid (AA) and uric acid (UA) are the major individual antioxidants present in mammalian semen, humans included [17]. However, unlike thiol-containing molecules, semen ascorbate and urate contents of vasectomized men do not differ from those of normozoospermic men suggesting that AA and UA are unlikely to be prominent epididymal antioxidants [12]. Besides the scavenging effects AA and UA have on H_2O_2 and on the hydroxyl radical, respectively, they both have been proposed to be good $ONOO^-$ scavengers [18]. It is suggested that it is through this action that both molecules exert their antioxidant potential [19].

5.3.1.3 L-Carnitine/Acetyl-L-Carnitine

The primary role of L-carnitine is in transferring long-chain fatty acids across mitochondrial membranes, thus facilitating oxidation within mitochondria during energy production. It is found concentrated in tissues, such as muscles, in which energy demand is high. Intriguingly, L-carnitine was shown to be present at very high concentrations in the mammalian epididymis and spermatozoa (in the mM range) far above circulating levels (in the μM range). Epididymal intraluminal carnitine concentration increases gradually along the epididymis, reaching its maximum in the cauda compartment [20]. The prevalent view is that carnitine is not synthesized by the epididymis epithelia but rather is transported from the systemic compartment. Although, it is quite conceivable that, as the energy substrate for spermatozoa carnitine might support sperm respiration and motility, it is however difficult to understand why this should be the case in a compartment where spermatozoa are immotile. Therefore, the exact role of the high epididymal carnitine content remains an enigma. In addition to its contribution to cellular energy metabolism, L-carnitine and acetyl-L-carnitine were shown to protect cells from apoptosis in various ways [21], one being through their antioxidant properties [22]. It was suggested recently that the antioxidant properties of L-carnitine and/or acetyl-L-carnitine are not direct but secondary to their anti-inflammatory potential [23]. It is furthermore believed to be mediated at least partly by the downregulation of the proinflammatory sphingomyelin/ceramide pathway.

5.3.1.4 Taurine/Hypotaurine

Taurine (Tau) is one of the most abundant free amino acids in several tissues and it is known to have beneficial physiological effects, including an antioxidant potential [24]. The presence of taurine and hypotaurine in significant quantities in the male genital tract and their actions on spermatozoa have long been reported [25, 26]. More precisely, taurine was reported to be particularly abundant in the cauda epididymidis with a concentration largely exceeding that found in the caput [27]. Although Tau is incapable of directly scavenging the classical ROS such as superoxide anion, hydroxyl radical, and H_2O_2 [28], there are numerous studies suggesting that it is an effective inhibitor of ROS generation (reviewed in [29]). One of the mechanisms put forward is that taurine may upregulate/restore antioxidant enzymes, including superoxide dismutase (SOD) and glutathione peroxidase (GPx) [30]. Another potential indirect antioxidant action of taurine could be the prevention of mitochondria-associated ROS generation as a result of calcium accumulation. Taurine interferes with calcium overload [31] by its osmolyte effect on the stimulation of Na^{2+}/Ca^{2+} exchanger [29]. Recently, several papers have underlined the mitochondrial action of taurine and its effect in attenuating mitochondrial ROS production [29, 32]. Briefly, taurine was shown to form conjugates with mitochondrial tRNAs, thus facilitating the translation of mitochondrial proteins and *in fine* energy production. If not present in adequate quantity, this could lead to impaired electron transport flux, lower rate of ATP generation, superoxide anion generation, and mitochondria damage. Therefore, in the cauda epididymis, taurine could be essential for maintaining an adequate energy balance and limiting the rate of superoxide generation [29].

5.3.1.5 Clusterin

Clusterin (Clu) is one of the major secretions of the mammalian epididymis epithelium accounting, for example, for 30% of the human epididymis secretome [33]. Clusterin is also named apolipoprotein J (Apo-J). It is a 75–80 kDa disulfide-linked heterodimeric glycoprotein quite conserved between many species and quite ubiquitously expressed in most mammalian tissues and body fluids, including semen. It contains a cysteine-rich domain stabilized by five disulfide bridges. Because clusterin can partner with a broad range of molecules, it is difficult to assign it a particular function. It is now considered that clusterin functions as an extracellular chaperone protecting cells from various stresses, including oxidation. For some authors, clusterin is best viewed as a marker of cell response to pro-oxidant situations. This would be in agreement with the idea that the epididymis luminal environment is in a pro-oxidant context, where clusterin is present to exert its chaperone function.

5.3.1.6 Albumin/Lactoferrin

As was the case with clusterin, albumins are major components of the mammalian epididymis fluid. Albumin possesses two types of antioxidant properties [34]. First, it is involved in free radical trapping and, second, it specifically binds metals ions, such as copper and iron, thus preventing the production of free radicals by these transition metals [35]. Lactoferrin, which is also present in the mammalian epididymal fluid, also provides an indirect antioxidant protection by blocking iron, thus avoiding its immediate reaction with H_2O_2 along the Fenton/Haber–Weiss biochemical pathway. Albumin-free radical-trapping property is mediated by a sulfhydryl group (Cys34) that can form disulfide with several compounds. Through this reduced Cys34, albumin may scavenge hydroxyl radicals as well as reactive nitrogen

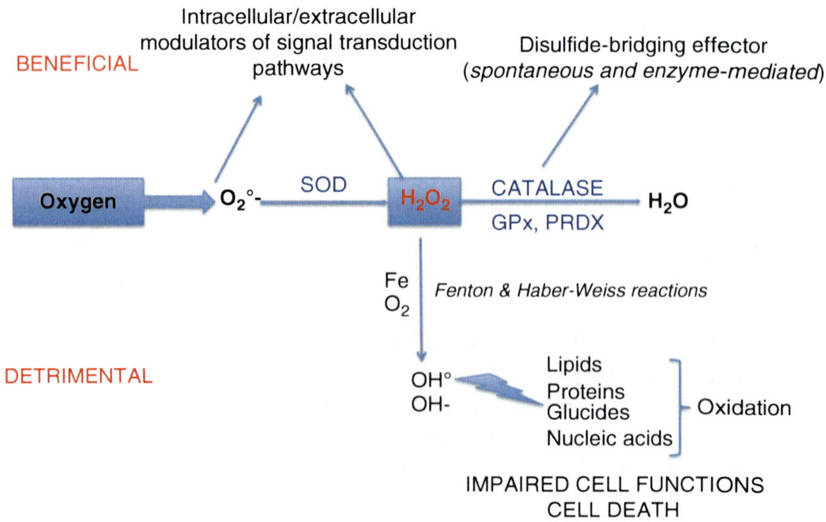

Fig. 5.2 Hydrogen peroxide (H_2O_2) generation and recycling by the classical SOD/GPx/catalase enzymatic triad. H_2O_2 arises from the activity of superoxide dismutase (SOD) that recycles superoxide anion ($O_2^{\bullet-}$) coming from oxygen catabolism. Glutathione peroxidases catalase and peroxiredoxins recycle H_2O_2 into something neutral, H_2O. H_2O_2 concentrations are precisely kept under control in and out the cell because H_2O_2 has both beneficial H_2O_2 and detrimental effects on cell physiology. On the one hand, excessive accumulation of H_2O_2 due to increased generation or/and defective recycling will lead (in the presence of iron (Fe) and oxygen *via* the classical Fenton and Haber–Weiss biochemical reactions) to the production of very aggressive free radicals. Eukaryotic cells have no enzymatic equipment to deal with these free radicals that will damage every cell constituents starting with lipids in membranes. Excessive free radical-mediated damages will first impair cell functions and could, if not properly counteracted, lead to cell death. On the other hand, H_2O_2 is necessary for some physiological processes. First, it acts as a secondary messenger in signal transduction pathways and it also modulates signal transduction cascades via its effect on cellular proteins that are sensible to the redox state of the cell. Second, H_2O_2 transforms free thiol groups carried by cysteine-containing proteins into disulfide bonds, either spontaneously or via the action of enzymes, such as disulfide isomerases, thiol peroxidases, and glutathione peroxidases

species, such as peroxynitrite ($ONOO^-$), that are powerful oxidants [34]. In addition, albumins bind with high affinity to long-chain fatty acids and PUFAs, perhaps protecting them from lipid peroxidation [36]. With all these antioxidant effects, albumin in the epididymal fluid should contribute to the protection of spermatozoa against oxidative insults.

5.3.2 Enzymatic Scavengers

Primary antioxidant enzymes work in a tightly associated manner constituting the commonly called antioxidant catalytic triad (Fig. 5.2). SOD (EC 1.15.1.1) recycles the spontaneous formation of superoxide anion and doing so generates H_2O_2.

To adapt H_2O_2 concentrations precisely, cells rely on three enzymes, GPx (EC 1.11.1.9), catalase (CAT; EC 1.11.1.6), and peroxiredoxins (PRDX: EC 1.11.1.5). GPx and PRDX activities cover both the intracellular and the extracellular compartments while CAT located in peroxisomes confers an additional protection to the cytoplasmic compartment of most cells. These H_2O_2-recycling enzymes do not work in the same way. Catalase is switched on in situations of acute stress during the so-called oxidative burst when enormous amounts of H_2O_2 are produced. For more physiological adjustments of H_2O_2 concentrations, cells rather rely on GPx and PRDX. In addition, while catalase only recycles H_2O_2, GPx are more versatile in their substrate preferences. This extends the protecting role of GPx since, unlike CAT, in addition to H_2O_2 GPx can also recycle organic peroxidized molecules, such as phospholipid hydroperoxides. GPx therefore act both as scavengers and repairing enzymes. Thus, although catalase is a powerful H_2O_2-recycling enzyme, GPx and PRDX are viewed as the key regulators of H_2O_2 concentration and consequently of H_2O_2-mediated attacks in and around most cells. This is a particularly important role considering the range of actions devoted to H_2O_2 in cell physiology. Roughly, H_2O_2 is a "Dr. Jekyll and Mr. Hyde" molecule. When present in above-physiological concentrations, it gives rise to very aggressive free radicals ($OH^•$, OH^-) *via* the classical Fenton and Haber–Weiss biochemical reactions against which eukaryotic cells are devoid of efficient protection. Excessive generation of such free radicals will affect all organic cellular components ranging from lipids in membranes to nuclear DNA material, ultimately leading to cell death (see Fig. 5.2). However, certain amounts of H_2O_2 and lipid peroxides (LOOH) are necessary for normal cell physiology since these molecules also act as secondary messengers modulating intracellular signal transduction pathways [37–39]. In addition, H_2O_2 and LOOH are necessary substrates for numerous enzymes that use them to mediate disulfide-bridging events in thiol-containing proteins. This is the case of disulfide isomerases/thiol peroxidases. Disulfide-bridging events are one type of posttranslational modification important for protein maturation. When occurring between sulfydryl (SH–) groups on one protein, they participate in proper folding while when disulfide bonds affect different proteins they are involved in protein–protein interactions. Both phenomena greatly contribute to the activity of the two proteins. During the last decade, it has been reported that some GPx can work as *bona fide* disulfide isomerases, provided they contain, in their primary amino acid sequence (outside of their scavenger catalytic site), a cysteine residue that will be involved in disulfide bridging of thiol-containing protein targets [40]. To mediate disulfide bonds in thiol-carrying proteins, GPx require H_2O_2 or LOOH as cosubstrate. Thus, GPx can either catabolize/neutralize H_2O_2 using GSH as a cofactor or/and mediate disulfide-bridging events using H_2O_2 or LOOH and thiol-containing proteins. The common point between these reactions is the presence of H_2O_2 or other organic hydroperoxides in the environment. Because GPx are bifunctional enzymes that can either work as antioxidant scavengers or as thiol peroxidases, they are with H_2O_2 at the center of the "oxygen paradox" in the mammalian epididymis. This is detailed below in the context of the epididymis and spermatozoa.

5.3.2.1 Superoxide Dismutase

The mammalian epididymis expresses both the copper/zinc form of SOD (Cu-Zn SOD1) as well as a secreted form of the SOD enzyme (eSOD or SOD3). Both catalyze the dismutation of superoxide anion to generate H_2O_2. Although commonly seen as an antioxidant enzyme, to my view SOD is more a pro-oxidant enzyme, since it transforms a short-lived and low-permeable free radical into an activated form of oxygen, H_2O_2, that can readily pass any cell membrane. SOD1 is highly expressed throughout the epididymis epithelium from the caput down to the cauda [41]. That is not very surprising since this epithelium has an active metabolism requiring sustained mitochondrial activity, the major cytosolic source of superoxide anion. Thus, SOD1 may be responsible for epithelial production of H_2O_2 part of which may reach the luminal compartment, where it could potentially harm spermatozoa. However, to counteract this, epithelial epididymal cells express, at quite high levels, various GPx (see below) that deal with the production and the circulation of H_2O_2 [6–8]. In addition to SOD1, the cauda epididymis epithelium also expresses SOD3, a secreted SOD that is closely associated with spermatozoa [42]. The antioxidant relevance of this sperm-associated SOD is difficult to apprehend. It has been suggested that this SOD could perhaps participate in H_2O_2 signaling during redox-regulated tyrosine phosphorylation events associated with the induction of motility and the initiation of capacitation [43–45].

5.3.2.2 Glutathione Peroxidases

As said above, the mammalian GPx gene family encodes bifunctional enzymes that can work either as classical ROS scavengers or as disulfide isomerases, thus introducing disulfide bridges in thiol-containing proteins. These dual effects are nowhere else better illustrated than in epididymal maturing spermatozoa, where the concomitant actions of several GPx ensure the achievement of structural maturation of sperm cells as well as their protection against ROS-induced damage. I review here the roles played by the sperm-associated forms of GPx4 (mitochondrial GPx4 and nuclear GPx4), the secreted GPx5 protein as well as the epithelial proteins, GPx1, GPx3, and cGPx4 (cellular GPx4), all functioning in the mammalian epididymis at different stages of spermatozoa epididymal journey and in different compartments. Alvarez and Storey [46] were among the first to point out the role played by GPx in protecting mammalian spermatozoa from loss of motility caused by spontaneous lipid peroxidation. Many years later, it was reported that failure of the expression of a GPx in the spermatozoa was correlated with infertility in human [47, 48]. These last 5 years, the development of mouse GPx knockout models [9, 49–52] associated with infertility or subfertility has demonstrated that GPx do play important roles in epididymal sperm protection and more largely in mammalian sperm physiology.

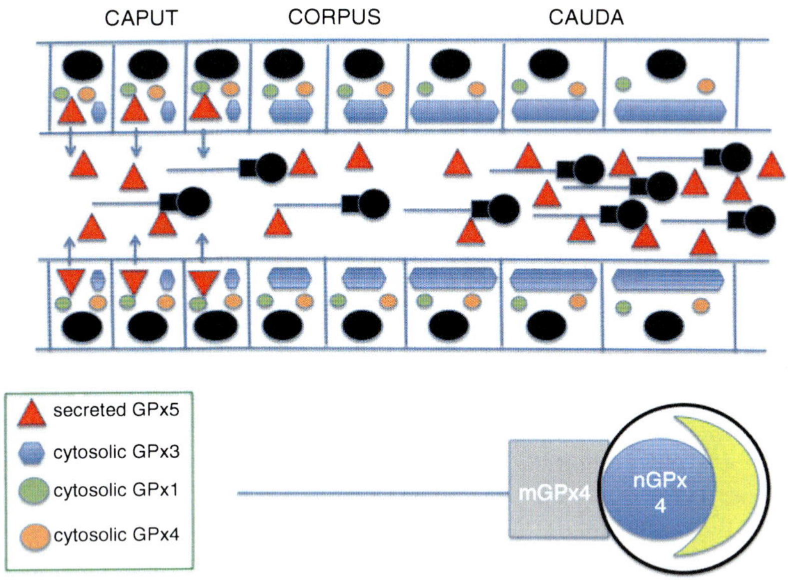

Fig. 5.3 Glutathione peroxidase localizations in the mammalian epididymis. Diagrammatic representation of GPx expressed by the epididymis epithelium. GPx5 is abundantly expressed by the caput epithelium and the protein is secreted in the epididymal duct. The luminal GPx5 protein accompanies transiting spermatozoa and is stored with them in the cauda luminal compartment. GPx3 is cytosolic and increasingly expressed by the epididymis epithelium from the caput to the cauda. Besides these two major GPx, the epididymis epithelium (proximal to distal) expresses the cytosolic GPx (GPx1 and cGPx4) at lower levels. In addition, epididymal transiting spermatozoa carry two sperm-specific isoforms of GPx4, the mitochondria-associated mGPx4 (in the sperm midpiece) and the nucleus-associated nGPx4

5.3.3 *Mammalian GPx Epididymal Coverage*

It has been shown that the mammalian epididymis expresses several GPx and, to date, it is the organ in which, although at different levels and in different subterritories, most of the known GPx are expressed, from GPx1 to GPx8 [6, 8]. Within the GPx multigenic family, four members (GPx1, GPx3, GPx4, and GPx5) are particularly well represented and characterized and are located in the epididymal epithelium, luminal compartment, and spermatozoa. Figure 5.3 presents a diagram of the localization of these GPx in the mouse epididymis, the picture being approximately the same in all the mammals that have been tested so far (reviewed in [7]). Briefly, GPx1, GPx3, and cGPx4 are cytosolic enzymes expressed by the epididymal epithelial cells (essentially principal cells) while GPx5 is a secreted protein. GPx5 and GPx3 are quantitatively the most abundant GPx in the whole epididymis, representing altogether more than 95% of the epididymal GPx both at the mRNA and protein levels.

Their differences essentially reside in the fact that GPx3 is a cytosolic enzyme increasingly expressed from the caput to the cauda epithelia, whereas GPx5 is a secreted enzyme whose expression and secretion are restricted to the caput epithelium. Another difference between GPx3 and GPx5 is the fact that GPx3 is a classical selenium-dependent enzyme, whereas GPx5 belongs to the noncanonical selenium-independent group (together with GPx6, and the predicted GPx7 and GPx8). Although long suspected to be an inefficient GPx, it has been demonstrated that GPx5 and other selenium-independent forms can act as true H_2O_2 scavengers [9, 53–55]. GPx1 and cGPx4 are expressed all along the epididymis epithelium at low levels compared with the other epididymal GPx. Both are cytosolic enzymes. Thus, the epididymis epithelium is mainly protected by GPx3 whereas the luminal compartment of the epididymis is protected by GPx5. In many mammals, GPx5 is a major secretion of the proximal epididymis duct [56]. It moves along the epididymal duct with maturing spermatozoa and accumulates with them in the caudal storage compartment. The strong GPx5 caudal luminal content together with the high cauda epithelial cytosolic expression of GPx3 suggests that this territory is involved in protecting sperm cells and the epididymal tissue from peroxidative injuries. The importance of GPx5 as a luminal scavenger in human has been questioned because here the GPx5 gene is subjected to differential expression giving birth to two distinct transcripts. The most abundant transcript was shown to be devoid of residues important for its GPx-catalytic function [57], and thus the GPx5 variant from this transcript should therefore not code for an active ROS scavenger. However, this GPx5 isoform does contain appropriate cysteine residues that could allow it to work as a disulfide intermediate/exchanger like the other GPx. The second human GPx5 transcript does encode the full-length and presumably active enzyme. In addition, GPx6, a gene closely related to GPx5, probably coming from a quite recent duplication event, is also significantly expressed in the human epididymis, where it could back up GPx5 [58].

To complete the picture of the epididymal localization of mammalian GPx, it should be noted that spermatozoa themselves carry GPx proteins which are added to the spermatozoa during testicular spermatogenesis. This is the case for the sperm nucleus-associated isoform of GPx4 (snGPx4 or nuclear GPx4, nGPx4) and the mitochondria-associated isoform of GPx4 (mGPx4). Along with the cytosolic or cellular GPx4 variant (cGPx4), mGPx4 and snGPx4 arise from differential expression of the single-copy GPx4 gene [59–63]. Both the sperm-associated mGPx4 and nGPx4 are precisely localized in the sperm midpiece compartment and in the nucleus (Fig. 5.3), respectively, during the final cytodifferentiation step of spermatogenesis. The sperm-associated and sperm-restricted GPx4 variants (mGPx4 and nGPx4) have been shown to be associated with intracellular proteins and to function as disulfide isomerases rather than as classical ROS-scavenging GPx.

Concerning the sperm midpiece-located mGPx4, it has been estimated that this isoform of GPx4 constitutes up to 50% of the sperm midpiece protein content embedding the helix of mitochondria [64]. For this reason, it was proposed that mGPx4 was the selenoprotein of the sperm midpiece, a role given earlier to a protein called SMCP for Sperm Mitochondria-associated Cysteine-rich Protein [65]. In the sperm midpiece, the mGPx4 protein is suggested to be more a structural protein than an active enzyme since it has been shown to have completely lost its solubility and scavenging enzymatic properties [64]. It is, however, probable that the sperm midpiece-located mGPx4 is involved in local structural reorganization based on protein disulfide-bridging events. It has been shown that disulfide bonds in the late stages of spermatogenesis and during epididymal transit are important for several sperm structures (besides the nucleus), such as the plasma membrane, midpiece, and acrosome [66–70]. In the sperm midpiece, during late spermatogenesis, it has been shown that mitochondria attach to outer dense fiber proteins of the axoneme and that disulfide bonds in several proteins are involved in this process. As a result, the spermatid cytoplasm is reduced and the sperm plasma membrane is connected to the sperm midpiece. In addition, it has been shown that the acrosome contains the greatest relative amount of disulfides (SS), compared with the head and the tail in guinea-pig spermatozoa [71], suggesting that there are regionalized disulfide-bridging events during sperm maturation. Regarding the sperm nucleus-specific isoform of GPx4 (nGPx4), it was shown to result from differential expression of its gene owing to the use of an alternative promoter located in the first intron [62]. This results in the expression of a GPx4 isoform having an N-terminus sequence rich in arginine residues allowing its nuclear localization and binding to chromatin [61]. In sperm nuclei, the GPx4 variant has been proposed to act as a protamine disulfide isomerase responsible for stabilizing the condensed chromatin by cross-linking protamine disulfides [61]. Condensation of sperm chromatin is an essential process in sperm differentiation, which starts during postmeiotic spermatogenesis with the replacement of somatic histones by transition proteins and finally by protamines. It appears that the sperm DNA-packaging process is not totally completed when spermatozoa leave the testis and that it continues in the early stages of epididymal maturation. During epididymal transit, oxidation of protamine thiols plays an important role in compacting sperm DNA further and also locking it in a highly condensed state. The cross-linking of protamine disulfides induced by ROS is comparable to GSH oxidation and peroxide reduction catalyzed by GPx. Therefore, it has been proposed that the sperm-nucleus GPx4 variant could use protamine cysteine residues as reducing partners and could act as a protamine disulfide isomerase [61]. For its activity, the nGPx4 isoform would thus not depend on GSH availability, which decreases significantly in late spermatogenesis and early maturation within the epididymis transit. In agreement with this hypothesis is the observation that in selenium-deficient animals, in which the concentrations of selenium-dependent GPx such as nGPx4 are greatly reduced, nearly all sperm cells recovered from the vas deferens possess incompletely compacted nuclei. In addition, in vitro experiments have shown that dithiothreitol provokes rat sperm DNA decondensation, an effect that is restored by adding H_2O_2 [61]. Finally, it has been shown that the use of an nGPx4 inhibitor blocks the condensation of sperm DNA. Altogether, these data strongly support the idea that the sperm nucleus-located GPx4 variant is responsible for protamine disulfide bridging within the sperm nucleus.

Thus, neither mGPx4 nor nGPx4 appears to have direct scavenging functions in epididymal spermatozoa, whereas GPx5 is thought to work as a true epididymal luminal scavenger.

5.3.4 Lessons from the Mouse GPx Knockout Models

In 2009, the group of Marcus Conrad (Germany) generated a transgenic mouse model in which the mGPx4 was disrupted [52] by introducing an in-frame translational stop into the mitochondrial leader sequence of mGPx4. Analysis of this model revealed that *mGPx4*$^{-/-}$ mice are viable, in contrast to early embryonic lethality recorded in the *GPx4*$^{-/-}$ mice in which the somatic isoform (cGPx4) as well as the two sperm-specific variants, mGPx4 and snGPx4, were absent [72, 73]. Interestingly, the mouse *mGPx4*$^{-/-}$ model showed male infertility associated with impaired sperm integrity. Essentially and quite logically, *mGPx4*$^{-/-}$ spermatozoa showed important structural abnormalities in the midpiece region leading to an increase in bent flagella, sperm heads detached from the flagellum, abnormal distribution of mitochondria along the midpiece, and abnormal organization of the sperm axoneme [52]. In addition and confirming the disulfide-bridging function of the protein mGPx4, deficient spermatozoa exhibit a higher protein thiol content and their phenotype resembles what occurs in severe selenodeficiency situations [74, 75]. Not surprisingly also, sperm motility was significantly reduced in *mGPx4*$^{-/-}$ males. The authors showed that male infertility could be bypassed by intra-cytoplasmic sperm injection (ICSI), suggesting that the male gametes were inefficient because they were unable to move properly as a consequence of sperm midpiece structural abnormalities and not because of their incapacity to initiate fertilization. Confirmation of these findings with regard to mGPx4 function was reported by Liang et al. [51] and Imai et al. [50]. Using a different strategy, Liang and collaborators generated transgenic mouse strains that carried mutations inhibiting either the expression of cytosolic or mitochondrial GPx4 and, consequently, overexpressing the other isoform. Their data confirmed that the mitochondrial GPx4 variant is testis- and male germ cell-specific. They also confirmed that when mGPx4 is not expressed, male infertility results, essentially because of structural malformations of the sperm midpiece. The strategy used by Imai's group was to establish a spermatocyte-specific GPx4 knockout mouse with the Cre-loxP system. Again, this new transgenic mouse model showed oligoasthenozoospermia resulting in male infertility, confirming that a decrease in GPx4 activity in spermatozoa results in male infertility in mice [52].

In 2005, Conrad et al. generated a transgenic mouse model in which they specifically abolished the expression of the sperm nucleus GPx4 isoform. In contrast to the full GPx4 knockout, *nGPx4*$^{-/-}$ animals are viable and fully fertile suggesting that the nGPx4 isoform is not responsible for the developmental defects observed when all the GPx4 isoforms are deleted [72, 73]. When spermatozoa from these *nGPx4*$^{-/-}$ animals were investigated more closely, they did not show any obvious phenotype. When spermatozoa from *nGPx4*$^{-/-}$ animals were compared with those of wild-type (WT) animals, they were found to show a delay in completion of post-testicular sperm nucleus compaction. In the caput epididymis of *nGPx4*$^{-/-}$ animals, sperm nuclei were less compacted than in spermatozoa from the caput epididymis of WT animals. This delayed compaction was resumed later on since there was no difference in the state of sperm nuclei compaction for spermatozoa collected from the cauda compartment of *nGPx4*$^{-/-}$ and WT animals [49]. These data support the idea that nGPx4 acts as a thiol peroxidase on thiol-containing sperm nuclear protamines in the caput compartment of the epididymis. The fact that normal sperm DNA compaction is recovered in spermatozoa stored in the cauda compartment of the nGPx4$^{-/-}$ animals suggests that one or more disulfide isomerases probably compensate for the lack of nGPx4 expression when the sperm cells travel along the epididymal tubule. Another possibility is that the cytosolic isoform (cGPx4), which is still expressed in testis of the nGPx4$^{-/-}$ mice, may partly back up nGPx4 deficiency since it is small enough to enter the nuclear pore [49, 76]. Finally, although H_2O_2 is commonly believed to be rather inefficient in mediating –S–S– bridging directly, one cannot exclude the idea that spontaneous disulfide bridging occurs during epididymal migration of sperm cells providing there is sufficient H_2O_2 in the epididymal lumen.

The epididymal secreted GPx5 knockout model has brought some clear evidence that the epididymal lumen contains significant amounts of H_2O_2 that could be available for spontaneous or enzyme-mediated sulfoxidation. It has also been shown that epididymal GPx5 is a true ROS scavenger protecting epididymis-transiting sperm cells from ROS-mediated loss of integrity. The epididymis-specific GPx (GPx5) occupies a special position in the GPx family and was initially suspected not to behave as a true GPx. The peculiarity of GPx5 is the absence of the selenocysteine (SeCys) residue in its catalytic site [6, 8] contrary to the other well-studied members of the mammalian GPx family (GPx1 to GPx4). In GPx5, the SeCys residue is replaced by a cysteine residue. Because of this particularity, the scavenger activity of GPx5 was questioned. In cannonical GPx, if the SeCys residue was replaced by a cysteine amino acid, there was a dramatic drop in the enzyme activity [77]. However, it was shown in vitro that GPx5-transfected mammalian cells survive much better in oxidative conditions (increasing H_2O_2 concentrations in the cell medium) than control cells, suggesting that GPx5, at least in vitro, is efficient in recycling H_2O_2 [53]. It was also demonstrated that mice subjected to a selenium-free diet, depleting their Se-dependent GPx activities, show an overall increase in peroxidative injury in all tissues except in the epididymis, where GPx5 mRNA and protein levels were increased, backing up the failing Se-dependent activities [54]. This strongly suggests that in vivo as well, the Se-independent GPx5 protein acts as a true scavenger. Final clues proving the real scavenging role of GPx5 in the epididymal environment came from the generation and analysis of a mouse strain that does not express GPx5 [9]. Lack of GPx5 expression in the epididymial lumen of the *GPx5*$^{-/-}$ animals established an oxidative stress in the cauda epididymidis. GPx5 deficiency was not followed by any change in the ratio of free thiols to sulfoxide in spermatozoa, suggesting that GPx5 has nothing to do with disulfide-bridging events and therefore behaves as a conventional ROS-scavenging GPx. To cope with the pro-oxidative situation in the cauda compartment, the cauda epididymidal epithelium of the *GPx5*$^{-/-}$ animals transcriptionally upregulated the three cytosolic GPx normally expressed there (GPx1, GPx3, and cGPx4). Upregulation of these epididymal GPx was sufficient to maintain the total GPx activity of the tissue at a normal value. Transcription of epithelial catalase was also increased in the cauda epididymidis of *GPx5*$^{-/-}$ animals, reinforcing the idea that the tissue was facing an increase in H_2O_2 [9] since catalase metabolizes only this substrate. These observations suggest that luminal ROS and especially H_2O_2 accumulate in the cauda compartment when GPx5 is no longer present. Despite the antioxidant response of the tissue, we have shown that the cauda epithelium of the *GPx5*$^{-/-}$ animals suffers oxidative injuries. This was also the case for the cauda-stored spermatozoa. In particular, cauda-stored spermatozoa in *GPx5*$^{-/-}$ animals showed a higher level of DNA oxidation, indicated by the increase in 8-oxo-deoxyguanosine residues (8-oxodG) associated with increased fragmentation and a slight nuclear decompaction state when compared to WT cauda-stored spermatozoa [9]. Although PRDX were not investigated in that study, it is possible that they also contributed to protect the cauda epididymidal epithelium and spermatozoa against the pro-oxidant situation generated in the GPx5-deficient context since several PRDX were very recently shown to be present on spermatozoa [78, 79].

Interestingly, in the caput epididymidis, sperm nuclei of the *GPx5*$^{-/-}$ animals were significantly more condensed than those of WT animals, suggesting that absence of H_2O_2 recycling via GPx5 in the caput luminal compartment left more H_2O_2 available for the disulfide-bridging activity of the nGPx4 protein or favored spontaneous disulfide-bridging events in sperm nucleus protamines. If this is the correct hypothesis, then GPx5 that is secreted early in the caput lumen indirectly participates in sperm DNA compaction by regulating the luminal epididymal concentration in H_2O_2. In the cauda compartment of the *GPx5*$^{-/-}$ animals, we hypothesize that what we see are the results of prolonged exposure to the damaging effect of H_2O_2 on spermatozoa that leads to DNA oxidation, increasing fragmentation, nucleus decompaction, and lipid peroxidation [9]. Spermatozoa themselves may contribute to this situation since it has been reported that sperm mitochondria are inactive and devoid of membrane potential in the caput, whereas they are completely mature and functional, showing a membrane potential, in the cauda epididymidis [80]. Thus, it is possible that cauda-stored spermatozoa, although not in optimal conditions of oxygen tension, pH, and energy substrate to sustain full mitochondrial activity, may contribute to the generation of free radicals *via* a leakage of the electron transport chain.

Oxidative damage of cauda-stored spermatozoa has been shown to increase in aging *GPx5*$^{-/-}$ animals [9]. The oxidative insults on sperm DNA recorded in over 12-month-old *GPx5*$^{-/-}$ males provoked a phenotype of subfertility when these males were mated with WT female mice of proven fertility. We have observed a significant decline in male fertility that was not due to impaired fertilization but to a clear rise in developmental defects, miscarriages, and perinatal mortality [9]. In the absence of impact on fertilization rate and because female mice were perfectly normal, the type of defects observed in embryos generated from aging *GPx5*$^{-/-}$ males indicate that loss of sperm DNA integrity is responsible. It has, thus, been assumed that oxidation of sperm DNA explains the effects recorded in the offspring of aging *GPx5*$^{-/-}$ males, as has been suggested elsewhere [81–83]. Such developmental defects due to alterations of paternal chromosomal material have already been reported in humans [84, 85].

Taken together, these data clearly demonstrate that GPx5 is an important luminal scavenger that protects cauda sperm cells from the damaging effects of H_2O_2. The physiological importance of GPx5 during aging has been highlighted, in agreement with the well-known "free radical theory of aging" which maintains that a decline in ROS-scavenging activities with age allows free radicals to affect cell constituents and cell physiology in many ways. GPx5, therefore, appears as quite an important enzyme that ultimately contributes to the maintenance of sperm DNA integrity and consequently to embryo viability. When it is absent, sperm DNA oxidation is too extensive for the reparative capacities of the oocyte leading to abnormal development. Without this protective protein, male mice run a higher risk of siring offspring with developmental defects, including some severe enough to lead to miscarriage. This could be particularly relevant clinically for the fertility of the aging male and also have an important effect on assisted reproductive technologies [9, 68, 82, 86–88] in which cryopreservation of the male gametes and micromanipulation in different media can be the sources of oxidative insults on the paternal chromosomal set.

5.3.4.1 Catalase

Among the primary antioxidant enzymes that have been studied in the epididymis, catalase is the least represented both at the mRNA and protein levels [89–91]. Since catalase is a cytosolic and not a secreted antioxidant, its contribution in protecting epididymal spermatozoa is only indirect. It is interesting to note that the cauda epididymis of GPx5-deficient animals responds to increased oxidative stress by transcriptional upregulation of cauda epithelial cell catalase. This demonstrates that, in a situation of oxidative burst, catalase is called on to protect the epididymis epithelium from H_2O_2-mediated damage. It, thus, limits the detrimental impacts of permeable H_2O_2 on the epididymis cell wall and has only a secondary role in protecting luminal spermatozoa from oxidative insults.

5.3.4.2 Indoleamine Dioxygenase

Indoleamine 2,3-dioxygenase (IDO) was once proposed to be a putative antioxidant in the mammalian epididymis because it uses as a cofactor the superoxide anion to catalyze the oxidative degradation of tryptophan into kynurenines and because, uniquely, this tissue shows constitutive expression of this enzyme. However, IDO is a heme oxygenase which, in the presence of oxygen, will readily regenerate as many superoxide anions as it consumes so that the intrinsic antioxidant impact of IDO is negligible. Some of the by-products of IDO activity are known to have antioxidant and pro-oxidant properties. We have recently shown that IDO activity in the epididymis leads essentially to the formation of kynurenic acid and 3-OH kynurenine [5]. Several reports have shown that 3-OH kynurenine induces ROS generation, mainly H_2O_2 and hydroxyl radicals, *in vitro* in primary striatal cultures as well as in various neuronal cell lines [92–96]. However, other reports demonstrated antioxidant properties of 3-OH kynurenine in the vascular system, the eye lens, and the brain [97–100]. Whether the epididymal kynurenines have antioxidant properties will have to wait further investigations.

5.3.4.3 Peroxiredoxins

PRDX are enzymes with a dual role as scavengers of ROS and modulators of ROS signaling. In mammals, six PRDX genes have been characterized, four of which, PRDX1 and PRDX4 to 6, have been found on spermatozoa at different subcellular localizations, such as head, acrosome, mitochondrial sheat, and flagellum [79]. They essentially originate from the expression of the corresponding genes in the testis during spermatogenesis. Although they may act as antioxidants in these discrete sperm locations, it is likely that they act essentially as modulators of ROS-signaling events leading to the ultimate maturation of these cells upon capacitation and acrosome reaction [45]. Spermatozoa PRDX are themselves susceptible to oxidative stress since it was shown that they are modified when spermatozoa are challenged with H_2O_2 impairing their signaling function [79].

5.3.4.4 Glutathione *S*-transferase

Glutathione *S*-transferases (GSTs) constitute a family of enzymes that catalyze the conjugation of GSH to various compounds and doing so protect cellular constituents from oxidative attacks. They are not considered as primary antioxidants although in some situations GST harbors a GSH-dependent peroxidase activity and therefore may assist primary scavengers in recycling H_2O_2 [101]. The mammalian epididymis has been shown to express some GST isoforms, such as Yo and Yb1 of the mu subfamily [102, 103]. Spermatozoa have also been shown to carry GST of the mu and pi subclasses [104] at a significant level. Whether they are active players in protecting the epididymis epithelium and spermatozoa from peroxidative injuries is difficult to say with our present knowledge.

5.4 Conclusions

The data presented in this review clearly illustrate the ambiguous situation existing in maturing epididymal spermatozoa. Firstly, during the epididymal journey, spermatozoa use ROS to mediate disulfide-bridging events that are necessary for the completion of their structural modifications. Sperm DNA compaction is one of these structural changes that are not completed when spermatozoa enter the epididymis tubule. The increased sperm DNA compaction ensured by protamine oxidation in the epididymis is a crucial phenomenon that serves both to protect paternal DNA from mutational effects and to reduce the volume of the sperm head allowing an optimum velocity of mature spermatozoa, both being critical for the success of fertilization. Secondly, however, these ROS-mediated oxidation events have to be particularly well balanced, since spermatozoa are very susceptible to oxidative insults that may have dramatic effects on their integrity and consequently on their fertilizing potential. Although the epididymal lumen contains several non-enzymatic antioxidants, GPx proteins appear to be the masters in controlling this fine equilibrium, acting either as disulfide isomerases or true GPx H_2O_2 scavenger. Whereas nGPx4 uses H_2O_2 or other organic hydroperoxides to perform oxidation of protamines which further compacts the sperm nucleus and locks it in this condensed state, the luminal GPx5 protein controls the amount of luminal H_2O_2 available for optimal oxidation and also protects maturing spermatozoa against H_2O_2-mediated damage. PRDX that have recently been shown to be present on spermatozoa may also participate in this fine regulation of beneficial vs. detrimental effects of ROS on spermatozoa. The precise interplay between GPx proteins and H_2O_2 during the last steps in the generation of fully competent spermatozoa in the male genital tract may explain why oxidative stress is a parameter so frequently associated with male infertility whether it comes from infectious situations and infiltrating leucocytes, environmental toxicants, metabolic syndromes, or aging, all situations that are known to lead to excessive generation of ROS or alteration of ROS scavengers.

References

1. Jones R, Mann T, Sherins RJ. Adverse effects of peroxidized lipid on human spermatozoa. Proc R Soc Lond Biol Sci. 1978;201(1145):413–7.
2. Jones R, Mann T, Sherins RJ. Peroxidative breakdown of phospholipids in human spermatozoa, spermicidal properties of fatty acid peroxides, and protective action of seminal plasma. Fertil Steril. 1979;31(5):531–7.
3. Zalata AA, Chrostophe AB, Depuydt CE, et al. The fatty acid composition of phospholipids of spermatozoa from infertile patients. Mol Hum Reprod. 1998;4(2):11–118.
4. Halliwell B, Gutteridge JMC. Free radical in biology and medicine. Oxford, UK: Charendon Press; 1999.
5. Jrad-Lamine A, Henry-Berger J, Gourbeyre P, et al. Deficient-tryptophan catabolism along the kynurenine pathway reveals that the epididymis is in a unique tolerogenic state. J Biol Chem. 2011;286(10):8030–42.
6. Drevet JR. Glutathione peroxidases expression in the mammalian epididymis and vas deferens. In: Francavilla F, Francavilla S, Forti G, editors. Andrology 2000. L'Aquila, Italy: Collana di "Study Abruzzesi"; 2000. p. 427–61.
7. Vernet P, Aitken RJ, Drevet JR. Antioxidant strategy in the epididymis. Mol Cell Endocrinol. 2004;216:31–9.
8. Drevet JR. The antioxidant glutathione peroxidase family and spermatozoa: a complex story. Mol Cell Endocrinol. 2006;250:70–9.
9. Chabory E, Damon C, Lenoir A, et al. Epididymis seleno-independent glutathione peroxidase 5 (GPx5) contributes to the maintenance of sperm DNA integrity. J Clin Invest. 2009;119:2074–85.
10. Hinton BT, Palladino MA, Mattmueller DR, et al. Expression and activity of gamma-glutamyl transpeptidase in the rat epididymis. Mol Reprod Dev. 1991;28(1):40–6.

11. Zubkova EV, Robaire B. Effect of glutathione depletion on antioxidant enzymes in the epididymis, seminal vesicles, and liver and on spermatozoa motility in the aging brown Norway rat. Biol Reprod. 2004;71(3):1002–8.
12. Potts RJ, Jefferies TM, Notarianni LJ. Antioxidant capacity of the epididymis. Hum Reprod. 1999;14(10):2513–6.
13. Nordberg J, Arnér ES. Reactive oxygen species, antioxidants, and the mammalian thioredoxin system. Free Radic Biol Med. 2001;31(11):1287–312.
14. Arnér ES, Holmgren A. Physiological functions of thioredoxin and thioredoxin reductase. Eur J Biochem. 2000;267(20):6102–9.
15. Miranda-Vizuete A, Ljung J, Damdimopoulos AE, et al. Characterization of Sptrx, a novel member of the thioredoxin family specifically expressed in human spermatozoa. J Biol Chem. 2001;276(34):31567–74.
16. Sadek CM, Damdimopoulos AE, Pelto-Huikko M, et al. Sptrx-2, a fusion protein composed of one thioredoxin and three tandemly repeated NDP-kinase domains is expressed in human testis germ cells. Genes Cells. 2001;6(12):1077–90.
17. Lewis SE, Sterling ES, Young IS, et al. Comparison of individual antioxidants of sperm and seminal plasma in fertile and infertile men. Fertil Steril. 1997;67(1):142–7.
18. Regoli F, Winston GW. Quantification of total oxidant scavenging capacity of antioxidants for peroxynitrite, peroxyl radicals, and hydroxyl radicals. Toxicol Appl Pharmacol. 1999;156(2):96–105.
19. Aitken RJ, Ryan AL, Baker MA, et al. Redox activity associated with the maturation and capacitation of mammalian spermatozoa. Free Radic Biol Med. 2004;36(8):994–1010.
20. Pruneda A, Yeung CH, Bonet S, et al. Concentrations of carnitine, glutamate and myo-inositol in epididymal fluid and spermatozoa from boars. Anim Reprod Sci. 2007;97:344–55.
21. Chiu Ming NG, Blackman MR, Wang C, et al. The role of carnitine in the male reproductive system. Ann N Y Acad Sci. 2004;1033:177–88.
22. Vicari E, Calogero AE. Effects of treatment with carnitines in infertile patients with prostato-vesiculo-epididymitis. Hum Reprod. 2001;16(11):2338–42.
23. Abd-Allah AR, Helal GK, Al-Yahya AA, et al. Pro-inflammatory and oxidative stress pathways which compromise sperm motility and survival may be altered by L-carnitine. Oxid Med Cell Longev. 2009;2(2):73–81.
24. Wright CE, Tallan HH, Lin YY, et al. Taurine: biological update. Annu Rev Biochem. 1986;55:427–53.
25. Mrsny RJ, Waxman L, Meizel S. Taurine maintains and stimulates motility of hamster sperm during capacitation in vitro. J Exp Zool. 1979;210(1):123–8.
26. Alvarez JG, Storey BT. Taurine, hypotaurine, epinephrine and albumin inhibit lipid peroxidation in rabbit spermatozoa and protect against loss of motility. Biol Reprod. 1983;29(3):548–55.
27. Hinton BT. The testicular and epididymal luminal amino acid microenvironment in the rat. J Androl. 1990;11(6):498–505.
28. Aruoma OI, Halliwell B, Hoey BM, et al. The antioxidant action of taurine, hypotaurine and their metabolic precursors. Biochem J. 1988;256(1):251–5.
29. Schaffer SW, Azuma J, Mozaffari M. Role of antioxidant activity of taurine in diabetes. Can J Physiol Pharmacol. 2009;87:91–9.
30. Vohra BP, Hui X. Taurine protects against carbon tetrachloride toxicity in the cultured neurons and in vivo. Arch Physiol Biochem. 2001;109(1):90–4.
31. Wu H, Jin Y, Wei J, Jin H, et al. Mode of action of taurine as a neuroprotector. Brain Res. 2005;1038(2):123–31.
32. Suzuki T, Suzuki T, Wada T, et al. Taurine as a constituent of mitochondrial tRNAs: new insights into the functions of taurine and human mitochondrial diseases. EMBO J. 2002;21(23):6581–9.
33. Dacheux JL, Belghazi M, Lanson Y, et al. Human epididymal secretome and proteome. Mol Cell Endocrinol. 2006;250(1–2):36–42.

34. Roche M, Rondeau P, Singh NR, et al. The antioxidant properties of serum albumin. FEBS Lett. 2008;582:1783–7.
35. Halliwell B. Albumin—an important extracellular antioxidant? Biochem Pharmacol. 1988; 37(4):569–71.
36. Quinlan GJ, Martin GS, Evans TW. Albumin: biochemical properties and therapeutic potential. Hepatology. 2005;41(6):1211–9.
37. Forman HJ, Maiorino M, Ursini F. Signaling functions of reactive oxygen species. Biochemistry. 2010;49(5):835–42.
38. Bartz RR, Piantadosi CA. Clinical review: oxygen as a signaling molecule. Crit Care. 2010;14(5):234.
39. Conrad M, Sandin A, Förster H, et al. 12/15-lipoxygenase-derived lipid peroxides control receptor tyrosine kinase signaling through oxidation of protein tyrosine phosphatases. Proc Natl Acad Sci USA. 2010;107(36):15774–9.
40. Delaunay A, Pflieger D, Barrault MB, et al. A thiol peroxidase is an H_2O_2 receptor and redox-transducer in gene activation. Cell. 2002;111:471–81.
41. Jervis KM, Robaire B. Dynamic changes in gene expression along the rat epididymis. Biol Reprod. 2001;65(3):696–703.
42. Perry AC, Jones R, Hall L. Isolation and characterization of a rat cDNA clone encoding a secreted superoxide dismutase reveals the epididymis to be a major site of its expression. Biochem J. 1993;293(Pt 1):21–5.
43. Lewis B, Aitken RJ. Impact of epididymal maturation on the tyrosine phosphorylation patterns exhibited by rat spermatozoa. Biol Reprod. 2001;64(5):1545–56.
44. O'Flaherty C, de Lamirande E, Gagnon C. Positive role of reactive oxygen species in mammalian sperm capacitation: triggering and modulation of phosphorylation events. Free Radic Biol Med. 2006;41(4):528–40.
45. de Lamirande E, O'Flaherty C. Sperm activation: role of reactive oxygen species and kinases. Biochim Biophys Acta. 2008;1784(1):106–15.
46. Alvarez JG, Storey BT. Role of glutathione peroxidase in protecting mammalian spermatozoa from the loss of motility caused by spontaneous lipid peroxidation. Gamete Res. 1989;2: 77–90.
47. Imai H, Suzuki K, Ishizaka K, et al. Failure of the expression of phospholipid hydroperoxide glutathione peroxidase in the spermatozoa of human infertile males. Biol Reprod. 2001;64: 674–83.
48. Foresta C, Flohé L, Garolla A, et al. Male infertility is linked to the selenoprotein phospholipid hydroperoxide glutathione peroxidase. Biol Reprod. 2002;67:967–71.
49. Conrad M, Moreno SG, Sinowatz F, et al. The nuclear form of phospholipids hydroperoxide glutathione is a protein thiol peroxidase contributing to sperm chromatin stability. Mol Cell Biol. 2005;25:7637–44.
50. Imai HN, Hakkaku R, Iwamoto J, et al. Depletion of Selenoprotein GPx4 in spermatocytes causes male infertility in mice. J Biol Chem. 2009;284(47):32009–522.
51. Liang H, Yoo SH, Na R, et al. Short form glutathione peroxidase 4 is the essential isoform required for survival and somatic mitochondrial functions. J Biol Chem. 2009;284(45): 30836–44.
52. Schneider M, Förster H, Boersma A, et al. Mitochondrial glutathione peroxidase 4 disruption causes male infertility. FASEB J. 2009;23(9):3233–42.
53. Vernet P, Rigaudiere N, Ghyselinck NB, et al. In vitro expression of a mouse tissue specific glutathione-peroxidase-like protein lacking the selenocysteine can protect stably transfected mammalian cells against oxidative damage. Biochem Cell Biol. 1996;74:125–31.
54. Vernet P, Rock E, Mazur A, et al. Selenium-independent epididymis-restricted glutathione peroxidase 5 protein (GPX5) can back up failing Se-dependent GPXs in mice subjected to selenium deficiency. Mol Reprod Dev. 1999;54:362–70.
55. Herbette S, Roeckel-Drevet P, Drevet JR. Seleno-independent glutathione peroxidases. More than simple antioxidant scavengers. FEBS J. 2007;274:2163–80.

56. Belleannée C, Labas V, Teixeira-Gomes AP, et al. Identification of luminal and secreted proteins in bull epididymis. J Proteomics. 2011;74(1):59–78.
57. Hall L, Williams K, Perry AC, et al. The majority of human glutathione peroxidase type 5 (GPX5) transcripts are incorrectly spliced: implications for the role of GPX5 in the male reproductive tract. Biochem J. 1998;333(Pt 1):5–9.
58. Pontius JU, Wagner L, Schuller GD, Pontius JU, Wagner L, Schuller GD. UniGene a unified view of the transcriptome. In: McEntyre J, Ostell J, editors. The NCBI Handbook. Bethesda, MD: National Center for Biotechnology Information; 2003.
59. Pushpa-Rebka TR, Bursdall AI, Oleksa M, et al. Rat phospholipids hydroxyperoxide glutathione peroxidase. cDNA cloning and identification of multiple transcription and translation start sites. J Biol Chem. 1995;270:26993–9.
60. Godeas C, Tramer F, Micali F, et al. Distribution and possible novel role of phospholipid hydroperoxide glutathione peroxidase in rat epididymal spermatozoa. Biol Reprod. 1997;57: 1502–8.
61. Pfeifer HM, Conrad M, Roethlein D, et al. Identification of a specific sperm nuclei selenoenzyme necessary for protamine thiol cross-linking during sperm maturation. FASEB J. 2001; 15:1236–8.
62. Moreno SG, Laux G, Brielmeier M, et al. Testis-specific expression of the nuclear form of phospholipid hydroperoxide glutathione peroxidase (PHGPx). J Biol Chem. 2003;384: 635–43.
63. Maiorino M, Scapin M, Ursini F, et al. Distinct promoters determine alternative transcription of GPx4 into phospholipids-hydroperoxide glutathione peroxides variants. J Biol Chem. 2003;278:34286–90.
64. Ursini F, Heim S, Kiess M, et al. Dual function of the selenoprotein PHGPx during sperm maturation. Science. 1999;285:1393–6.
65. Kleene KC. The mitochondrial capsule selenoprotein-a structural protein in the mitochondrial capsule of mammalian sperm. In: Burk RF, editor. Selenium in biology and human health. New York, NY: Springer; 1994. p. 135–48.
66. Sivashanmugam P, Rajalakshmi M. Sperm mauration in rhesus monkey: changes in ultrastructure, chromatin condensation, and organization of lipid bilayer. Anat Rec. 1997;247: 25–32.
67. Mammoto A, Matsumoto N, Tahara M, et al. Involvement of s sperm protein sensitive to sulfhydryl-depleting reagents in mouse sperm-egg fusion. J Exp Zool. 1997;278:178–88.
68. Francavilla S, Cordeschi G, Gabriele A, et al. Chromatin defects in normal and malformed human ejaculated and epididymal spermatozoa: a cytochemical ultrastructural study. J Reprod Fertil. 1996;106(2):259–68.
69. Mate KE, Kosower NS, White IG, et al. Fluorescent localization of thiols and disulfides in marsupial spermatozoa by brombimane labelling. Mol Reprod Dev. 1994;37:318–25.
70. Cummins JM, Fleming AD, Crozet N, et al. Labelling of living mammalian spermatozoa with the fluorescent thiol alkylating agent, monobromobimane (MB): immobilization upon exposure to ultraviolet light and analysis of acrosomal status. J Exp Zool. 1986;237: 375–82.
71. Huang TT, Kosower NK, Yanagimachi R. Localization of thiol and disulfide groups in guinea pig spermatozoa during maturation and capacitation using bimane fluorescent labels. Biol Reprod. 1984;31:797–809.
72. Imai H, Hirao F, Sakamoto T, et al. Early embryonic lethality caused by targeted disruption of the mouse PHGPx gene. Biochem Biophys Res Commun. 2003;305(2):278–86.
73. Yant LJ, Ran Q, Rao L, et al. The selenoprotein GPx4 is essential for mouse development and protects from radiation and oxidative damage insults. Free Radic Biol Med. 2003;34: 496–502.
74. Flohé L. Selenium in mammalian spermiogenesis. Biol Chem. 2007;388(10):987–95.
75. Shalini S, Bansal MP. Dietary selenium deficiency as well as excess supplementation induces multiple defects in mouse epididymal spermatozoa: understanding the role of selenium in male fertility. Int J Androl. 2008;31(4):438–49.

76. Weiss K. Importins and exportins: how to get in and out of the nucleus. Trends Biochem Sci. 1998;23(5):185–9.
77. Maiorino M, Aumann KD, Brigelius-Flohe R, et al. Probing the presumed catalytic triad of selenium-containing peroxidases by mutational analysis of phospholipid hydroperoxide glutathione peroxidase (PHGPx). Biol Chem Hoppe Seyler. 1995;376:651–60.
78. Manadhar G, Miranda-Vizuete A, Pedrajas JR, et al. Peroxiredoxin 2 and peroxidase enzymatic activity of mammalian spermatozoa. Biol Reprod. 2009;80:1168–77.
79. O'Flaherty C, de Souza AR. Hydrogen peroxide modifies human sperm peroxiredoxins in a dose-dependent manner. Biol Reprod. 2011;84(2):238–47.
80. Koppers AJ, De Iuliis GN, Nixon B, et al. Significance of mitochondrial reactive oxygen species in the generation of oxidative stress in spermatozoa. J Clin Endocrinol Metab. 2008;93(8):3199–207.
81. Aitken RJ, De Iuliis GN. On the possible origins of DNA damage in human spermatozoa. Mol Hum Reprod. 2009;6(1):3–13.
82. Aitken RJ. GPx5 protects the family jewels. J Clin Invest. 2009;119:1849–51.
83. De Iullis GN, Thompson LK, Mitchell LA, et al. DNA damage in human spermatozoa is highly correlated with the efficiency of chromatin remodeling and the formation of 8-hydroxy-2′-deoxyguanosine, a marker of oxidative stress. Biol Reprod. 2009;81:517–24.
84. Tesarik J, Greco E, Mendoza C. Late, but not early, paternal effect on human embryo development is related to sperm DNA fragmentation. Hum Reprod. 2004;19:611–5.
85. Tesarik J, Mendoza-Tesarik R, Mendoza C. Sperm nuclear DNA damage: update on the mechanism, diagnosis and treatment. Reprod Biomed Online. 2006;12:715–21.
86. Baker MA, Aitken RJ. Reactive oxygen species in spermatozoa: methods for monitoring and significance for the origins of genetic disease and infertility. Reprod Biol Endocrinol. 2005;29:3–67.
87. Aitken RJ, De Iuliis GN, McLachlan RI. Biological and clinical significance of DNA damage in the male germ line. Int J Androl. 2008;32:46–56.
88. Thompson LK, Fleming SD, Aitken RJ, et al. Cryopreservation-induced human sperm DNA damage is predominantly mediated by oxidative stress rather than apoptosis. Hum Reprod. 2009;24:1061–2070.
89. Gu W, Hecht NB. Developmental expression of glutathione peroxidase, catalase, and manganese superoxide dismutase mRNAs during spermatogenesis in the mouse. J Androl. 1996;17(3):256–62.
90. Zini A, Schlegel PN. Catalase mRNA expression in the male rat reproductive tract. J Androl. 1996;17(5):473–80.
91. Zini A, Schlegel PN. Identification and characterization of antioxidant enzyme mRNAs in the rat epididymis. Int J Androl. 1997;20(2):86–91.
92. Okuda S, Nishiyama N, Saito H, et al. Hydrogen peroxide-mediated neuronal cell death induced by an endogenous neurotoxin, 3-hydroxykynurenine. Proc Natl Acad Sci USA. 1996;93(22):12553–8.
93. Okuda S, Nishiyama N, Saito H, et al. 3-Hydroxykynurenine, an endogenous oxidative stress generator, causes neuronal cell death with apoptotic features and region selectivity. J Neurochem. 1998;70(1):299–307.
94. Wei H, Leeds P, Chen RW, et al. Neuronal apoptosis induced by pharmacological concentrations of 3-hydroxykynurenine: characterization and protection by dantrolene and Bcl-2 overexpression. J Neurochem. 2000;75(1):81–90.
95. Goldstein LE, Leopold MC, Huang X, et al. 3-Hydroxykynurenine and 3-hydroxyanthranilic acid generate hydrogen peroxide and promote alpha-crystallin cross-linking by metal ion reduction. Biochemistry. 2000;39(24):7266–75.
96. Lee HJ, Bach JH, Chae HS, et al. Mitogen-activated protein kinase/extracellular signal-regulated kinase attenuates 3-hydroxykynurenine-induced neuronal cell death. J Neurochem. 2004;88(3):647–56.
97. Christen S, Peterhans E, Stocker R. Antioxidant activities of some tryptophan metabolites: possible implication for inflammatory diseases. Proc Natl Acad Sci USA. 1990;87(7):2506–10.

98. Christen S, Thomas SR, Garner B, et al. Inhibition by interferon-gamma of human mononuclear cell-mediated low density lipoprotein oxidation. Participation of tryptophan metabolism along the kynurenine pathway. J Clin Invest. 1994;93(5):2149–58.
99. Luthra M, Balasubramanian D. 3-Hydroxykynurenine and 3-hydroxyanthranilic acid may act as endogenous antioxidants in the eye lens. Exp Eye Res. 1992;55(4):641–3.
100. Leipnitz G, Schumacher C, Dalcin KB, et al. In vitro evidence for an antioxidant role of 3-hydroxykynurenine and 3-hydroxyanthranilic acid in the brain. Neurochem Int. 2007;50(1): 83–94.
101. Kanai T, Takahashi K, Inoue H. Three distinct-type glutathione *S*-transferases from Escherichia coli important for defense against oxidative stress. J Biochem. 2006;140(5): 703–11.
102. Mueller A, Hermo L, Robaire B. The effects of aging on the expression of glutathione *S*-transferases in the testis and epididymis of the Brown Norway rat. J Androl. 1998;19(4): 450–65.
103. Andonian S, Hermo L. Immunocytochemical localization of the Ya, Yb1, Yc, Yf, and Yo subunits of glutathione S-transferases in the cauda epididymidis and vas deferens of adult rats. J Androl. 1999;20(1):145–57.
104. Hemachand T, Gopalakrishnan B, Salunke DM, et al. Sperm plasma-membrane-associated glutathione *S*-transferases as gamete recognition molecules. J Cell Sci. 2002;115(Pt 10): 2053–65.

Chapter 6
Lipid Peroxidation in Human Spermatozoa

Juan G. Alvarez and R. John Aitken

Abstract Oxidation of lipids can be a blessing or a curse as far as spermatozoa are concerned. Beatitudes are conferred via the oxidative generation of oxysterols, which then drive sperm capacitation by promoting the removal of cholesterol from the sperm plasma membrane. Conversely, the anathema involves peroxidation of polyunsaturated fatty acids (PUFA) to generate lipid peroxides that have a detrimental effect on spermatozoa, disrupting DNA integrity and limiting their competence for fertilization. Spermatozoa actively detoxify and remove toxic lipid peroxides from the sperm plasma membrane, but once these defense mechanisms have been overwhelmed, lipid peroxidation spreads rapidly through the cell leading to membrane damage, leakage of ATP, and a rapid loss of sperm motility and viability. The excessive presence of unesterified PUFA may be instrumental in the genesis of oxidative stress through the ability of these amphiphiles to interfere with the mitochondrial electron transport chain and promote cellular generation of superoxide anion.

Keywords Lipid peroxidation • Human spermatozoa • Induction of sperm capacitation • Unsaturated fatty acids • Propagation of peroxidative damage

J.G. Alvarez, MD, PhD
Department of Male Infertility, Centro Androgen, La Coruña 15004,
La Coruña, Spain

R.J. Aitken, PhD, ScD (✉)
Department of Biological Sciences, University of Newcastle, Callaghan,
NSW 2308, Australia
e-mail: John.Aitken@newcastle.edu.au

A. Agarwal et al. (eds.), *Studies on Men's Health and Fertility*, Oxidative Stress in Applied Basic Research and Clinical Practice, DOI 10.1007/978-1-61779-776-7_6,

6.1 Introduction

The vulnerability of mammalian spermatozoa to oxidative stress has been appreciated since 1943 when John MacLeod [1] demonstrated that human spermatozoa lost motility in oxygenated media and that this effect could be rescued by catalase. The major implication of these findings, that mammalian spermatozoa must be capable of reducing oxygen to hydrogen peroxide, was confirmed 3 years later when Tosic and Walton [2] revealed that bull spermatozoa could manufacture hydrogen peroxide. As a consequence of these pioneering studies, spermatozoa were shown to be professional generators of reactive oxygen species (ROS) decades before this activity was discovered in phagocytic leukocytes. Since these original reports, the cellular production of ROS has been confirmed in the spermatozoa of a wide variety of mammals including man, mouse, hamster, rat, rabbit, and horse [3]. Furthermore, the toxic impact of hydrogen peroxide on sperm physiology has also been confirmed many times, influencing a range of structures and activities including sperm motility [4], sperm–oocyte fusion, and DNA integrity [5]. Systematic analysis of these effects demonstrated that at low hydrogen peroxide concentrations the spermatozoa capacitated rapidly, levels of sperm–oocyte fusion were enhanced, and DNA damage was reduced. At higher levels of hydrogen peroxide exposure, DNA damage increased, while the levels of motility and sperm–oocyte fusion remained high, while at the highest levels of oxidant treatment all aspects of sperm function were adversely affected [5]. These findings effectively capture the full spectrum of effects that ROS are capable of exerting on human spermatozoa, encompassing the positive redox-regulated changes that drive sperm capacitation and chromatin compaction on the one hand [6] and the destructive effects of oxidative stress on the other [7]. They also suggest an intermediate situation wherein the fertilizing potential of the spermatozoa remains unimpaired, while the integrity of DNA in the sperm nucleus is significantly disrupted. This must be the case when adverse effects, such as childhood cancer or neurological disease, are seen in the offspring of men who have suffered oxidative damage to their sperm chromatin as a consequence of such factors as age or heavy smoking [3, 8]. In the following chapter, we consider this Janus-like impact of ROS on sperm function from the perspective of the unusual lipid composition of these cells.

6.2 The Positive Impact of Lipid Peroxidation: The Induction of Sperm Capacitation

We have known since the excellent studies published by Claude Gagnon and Eve de Lamirande in the early 1990s that sperm capacitation is an oxidative process that is heavily dependent on the active generation of ROS [9]. While several independent laboratories have confirmed the oxidative drive to capacitation [10–12], the way in which changes in the redox status of human spermatozoa allows these cells to attain a capacitated state has remained largely unresolved. To date, there is evidence that both primary cAMP production and the resultant stimulation of tyrosine phosphorylation are oxidatively induced by as a consequence of ROS-mediated changes in soluble adenylyl cyclase and tyrosine phosphatase activities respectively [6, 13–15]. In addition, exciting data has recently come to light, suggesting that oxidative changes to cholesterol in the sperm plasma membrane may also facilitate capacitation [16].

It has been known for some time that one of the key changes driving sperm capacitation is a loss of cholesterol from the sperm plasma membrane under the absorptive influence of albumin [17]. Such a change is thought to enhance the fluidity of the plasma membrane, promoting critical intermolecular interactions that promote the development of a capacitated state. However, the mechanisms by which cholesterol is induced to leave the plasmalemma and bind to albumin have never been satisfactorily explained. Brouwers et al. [16] have provided a possible answer to this question by revealing that cholesterol becomes oxidized in response to the ROS generated during capacitation. The oxysterols generated as a consequence of this process are more hydrophilic than the parent molecule and can therefore move more freely out of the plasma membrane to bind acceptor proteins, such as albumin.

Paradoxically, the redox mechanisms that drive capacitation through the oxidation and subsequent depletion of cholesterol from the sperm plasma membrane may also ultimately damage these cells. Thus oxysterol accumulation has been widely linked with the activation of apoptosis and cell death in other cell types [18]. Of particular interest is the fact that mitochondria appear to be key mediators of oxysterol-induced pathology via mechanisms involving stimulation of the intrinsic apoptotic cascade and activation of excessive ROS generation [19–21]. If this is the case, then we might postulate a dual role for cholesterol oxidation during the life history of a sperm cell [22]. Thus, during capacitation ROS-induced cholesterol oxidation can be considered biologically useful because it facilitates the removal of cholesterol from the sperm plasma membrane and the attainment of a capacitated state. However, if the spermatozoa do not manage to find and fertilize an egg, then the continued generation of ROS will result in an accumulation of oxysterols in the mitochondrial as well as the plasma membrane. The presence of oxysterols in the mitochondrial membranes might then precipitate enhanced mitochondrial ROS generation and induction of the intrinsic apoptotic cascade. This sedate progression from a state of capacitation to apoptotic cell death is probably an adaptive response that ensures the efficient removal of moribund, postmature spermatozoa from the female reproductive tract. Under these circumstances, the presence of apoptotic markers such as phosphatidylserine on the sperm surface would be expected to inform the infiltrating phagocyte population that the consumption of these cells should be "silent," i.e., should not be accompanied by an oxidative burst or the production of proinflammatory cytokines [22]. These beautifully orchestrated biochemical changes illustrate the important, complex role that oxidized lipid products have in regulating the physiological function of spermatozoa in terms of their competence for fertilization and efficiency with which these cells can be removed from the female tract once they are beyond their "sell-by" date. However, in cases of male infertility, this carefully choreographed sequence of events becomes disrupted and the focus of interest shifts from the oxysterol content of these cells to the oxidation status of the fatty acids in the sperm plasma membrane [23–25].

6.3 The Negative Impact of Unsaturated Fatty Acids and Lipid Peroxidation

The association between male infertility and the peroxidation of polyunsaturated fatty acids (PUFA) has been appreciated since the landmark studies of Thaddeus Mann and colleagues at the University of Cambridge. Mann et al. drew attention to the vulnerability of human spermatozoa to lipoperoxidative damage because of their particularly high PUFA content [23]. They also demonstrated that when washed human spermatozoa were treated with as little as 30 nmol of lipid peroxide/mL, they become irreversibly immotile within a few minutes. In addition, this seminal paper [23] highlighted the significance of seminal plasma antioxidants in protecting spermatozoa from oxidative damage.

6.3.1 Extracellular Antioxidants Protect Against Lipid Peroxidation

In order to prevent intracellular oxygen radical-induced damage, spermatozoa mostly rely on superoxide dismutase (SOD) and glutathione peroxidase (GPX) as the primary antioxidant enzymes, since mammalian spermatozoa have been shown to lack catalase activity. Spermatozoa are largely bereft of cytoplasm and cytoplasmic space, and therefore, these enzymes must exert their antioxidant activity while bound to the cytoskeleton of the axoneme, as it has been suggested for glycolytic enzymes [26]. The role of intracellular GPX and SOD in preventing lipid peroxidation and motility loss has been previously reported [4, 27]. Inhibition of intracellular SOD results in the release of very high levels of superoxide anion to the extracellular medium thus providing support for the important role of SOD in preventing intracellular oxygen radical-induced damage. On the other hand, in order to prevent extracellular oxygen radical-induced damage, these cells rely heavily, but not exclusively, on the presence of extracellular antioxidants in reproductive tract fluids that accompany the spermatozoa on their journey through the male and into the female reproductive tracts.

The presence of such antioxidants in seminal plasma is important because it has an impact on the methods that are used to prepare spermatozoa for assisted conception purposes. Thus, if spermatozoa are washed free of seminal plasma and then pelleted in a simple defined culture medium in preparation for swim-up, then ROS-generating germ cells and leukocytes can attack normal spermatozoa in the same cell suspension in the absence of antioxidant interference. Such attack can result in extensive DNA damage [28] loss of motility and impairment of sperm–oocyte fusion [29]. However, if the culture medium is supplemented with antioxidants such as vitamin E, then this kind of collateral damage can be significantly reduced [29].

The biochemical composition of the antioxidants present in epididymal and seminal plasma is known to include a variety of antioxidant enzymes (a secreted form of SOD, specific isoforms of GPX, and a variety of peroxiredoxins) as well as a wide range of small molecular mass free radical scavengers including vitamin C, vitamin E, uric acid, acetylcarnitine, tryptophan, tyrosine, and taurine [30–32]. Another form of antioxidant, which is rarely discussed, is lactoferrin. The latter was identified as the sperm-coating antigen many years ago [33] and serves an extremely important role in limiting the ability of transition metals, particularly iron and copper, to access the fatty acids in the sperm plasma membrane. Limiting the local availability of free transition metals is significant as far as spermatozoa are concerned because these metals they catalyze both the first chain initiation and the propagation of lipid peroxidation chain reactions.

6.3.2 Propagation of Peroxidative Damage with Transition Metals

The presence of redox-active metals such as iron and copper in sperm suspensions leads to the breakdown of lipid peroxides to form peroxyl (ROO$^{\bullet}$) and alkoxyl (RO$^{\bullet}$) radicals, according to the following equations:

$$\underset{\text{lipid hydroperoxide}}{\text{ROOH}} + Fe^{2+} \rightarrow \underset{\text{alkoxyl radical}}{\text{RO}^{\bullet}} + OH^{-} + Fe^{3+}$$

$$\underset{\text{lipid hydroperoxide}}{\text{ROOH}} + Fe^{3+} \rightarrow \underset{\text{peroxyl radical}}{\text{ROO}^{\bullet}} + H^{+} + Fe^{2+}$$

In order to stabilize these lipid radicals, abstraction of a hydrogen atom from an adjacent carbon takes place, generating the corresponding acid (ROOH) or alcohol (ROH). Unfortunately, the abstraction of a hydrogen atom from an adjacent lipid creates another carbon-centered radical that combines with molecular oxygen to recreate another lipid peroxide. In order to stabilize, the latter must again abstract a hydrogen atom from a nearby lipid, creating yet another carbon radical, which combines with molecular oxygen to create yet more lipid peroxides. In this manner, a chain reaction is created that, if unchecked, can propagate the peroxidative damage throughout the plasma membrane, leading to a rapid loss of membrane dependent functions. It is because of these catalytic properties that transition metals such as iron and copper are so damaging to sperm function [23, 25, 34].

The ability of transition metals to stimulate lipid peroxidation has been widely used as a diagnostic tool in assessing human sperm quality. Thus, when a ferrous ion promoter (ferrous sulfate and ascorbate) is added to suspensions of human spermatozoa, a sudden increase in lipid peroxide generation is observed [23]. This increase is not due to the de novo first-chain initiation of lipid peroxidation but rather features the catalytic decomposition of preexisting lipid peroxides [35]. It has been proposed that the existence of lipid peroxides in these cells might be due to the ability of chain breaking antioxidants such as vitamin E to impede the propagation of lipid peroxidation cascades initiated during the lifespan of spermatozoa. The resulting accumulation of lipid peroxides can then be induced to decompose during Fe(2+)-promoted lipid peroxidation assays, stimulating lipoperoxidative chain reactions and production of lipid decomposition products such as malondialdehyde and 4-hydroxyalkenals that can be readily measured [35].

Iron-promoted lipid peroxidation assays therefore provides the andrologist with a picture of the amount of lipid peroxidation the spermatozoa have been exposed to during their lifetime. Such measurements of lipid peroxidation potential in highly purified, leucocyte-free sperm suspensions have revealed the existence of powerful inverse correlations with the motility of the spermatozoa, their viability, their competence for sperm–oocyte fusion and, most significantly, the quality of sperm movement in the original semen samples [36, 37]. Similar negative correlations have been observed between sperm function and ROS generation but, unlike the lipid peroxidation measurements, these relationships were obfuscated by the presence of leucocytes [36]. Measurements of lipid peroxidation levels in the presence of transition metal promoters have therefore secured an important place in the diagnosis of human sperm function [37, 38].

An important aspect of the lipid peroxidation process is that it is very cell-dependent. Thus, a suspension of spermatozoa progressively motile spermatozoa will exhibit very little lipid peroxide formation, whereas dead or moribund cells in the same suspension will be highly loaded with the by-products of this process [38]. Lipid peroxidation may therefore be induced at a relatively late stage in the life history of a sperm cell and only occurs once the antioxidant defense mechanisms have been overwhelmed.

6.3.3 Protective Mechanisms Operative in Spermatozoa

If the antioxidants present in epididymal and seminal plasma do not successfully intercept ROS, or if excessive ROS are generated by the spermatozoa themselves, then oxygen radical-induced damage and the initiation and propagation of the lipid peroxidation chain reaction may take place. If this happens, then the spermatozoon has to respond by removing and neutralizing the lipid peroxide. The removal is effected by phospholipase A2 (PLA2) which responds to the presence of a peroxidized fatty acid by moving into the plasma/mitochondrial membrane and cleaving out the offending peroxide. The fatty acids esterified to the sn-2 position of the glycerol backbone of sperm phospholipids are highly unsaturated (largely arachidonic and docosohexaenoic acid) and therefore very vulnerable to free radical attack because the carbon hydrogen dissociation energies are lowest at the bis-allylic methylene group [39]. As a result, the hydrogen abstraction event that initiates lipid peroxidation is promoted. Once the lipid peroxide has been cleaved by PLA2, two downstream events must now occur. First, the parent phospholipid (now a lysophospholipid) has to be repaired by replacing the lost fatty acid through the action of an acyl transferase. The plasmalogen fraction of the phospholipid pool appears to be particularly susceptible to peroxidative attack [23] and is the preferential target for acyl transferase activity [36]. Second, once the peroxidized fatty acid has been cleaved by PLA2, it can then be processed by GPX and/or thioredoxin reductase [40] in order to convert the toxic lipid peroxide to a harmless fatty alcohol. In addition, toxic lipid peroxides can be removed from the cell following PLA2 cleavage and transfer to albumin in the extracellular medium, as a consequence of the lipid binding characteristics of albumin. The latter is highly effective at protecting spermatozoa from oxidative stress by virtue of its ability to bind and neutralize cytotoxic lipid hydroperoxides generated as a consequence of sperm oxidative metabolism [39, 41]. Nevertheless, since PUFA content and oxygen radical production in mature spermatozoa are relatively low and homogenously distributed among the sperm population, at constant temperature and pO2, the rate-limiting factor that governs the rate of lipid peroxidation and sperm motile lifespan and function is intracellular antioxidant enzymatic activity.

6.3.4 Fatty Acids and the Induction of Lipid Peroxidation

So far in this review, the emphasis has been placed on the susceptibility of unsaturated fatty acids to oxidative attack. However, free PUFA may also be a significant cause of oxidative stress in these cells. This suggestion hinges on the ability of free, unesterified PUFA to stimulate ROS generation by sperm mitochondria [42]. In this study, a wide range of free PUFA (arachidonic, linoleic, and docosahexaenoic acids [DHAs]) were found to trigger mitochondrial ROS generation by human spermatozoa, while saturated fatty acids and methyl esters of PUFA were ineffective [42]. On the basis of these results, it was concluded that the amphiphilic properties of these molecules were central to their ROS generating capacity.

While such data indicate that amphiphilic PUFA *can* stimulate ROS generation by human spermatozoa, they do not tell us whether a superabundance of free PUFA is critically involved in the pathophysiology of defective sperm function. The possible involvement of excess unsaturated fatty acids in the pathological generation of ROS found in the infertile patient population is suggested by the finding that dysfunctional defective spermatozoa recovered from the low-density region of Percoll gradients are characterized by significantly higher levels of PUFA compared with their normal functional counterparts isolated from the high-density region of such gradients [43]. These studies have recently been extended by Koppers et al. [44] who demonstrated that exposure to unesterified PUFA stimulated a time- and dose-dependent increase in *mitochondrial* ROS generation in populations in human spermatozoa. Moreover, the result of such exposure was to bring about the reduced motility and high levels of DNA damage, typically found in the patient population. Furthermore, the free PUFA content of human spermatozoa was found to be positively correlated with the spontaneous levels of mitochondrial superoxide generation in these cells [44].

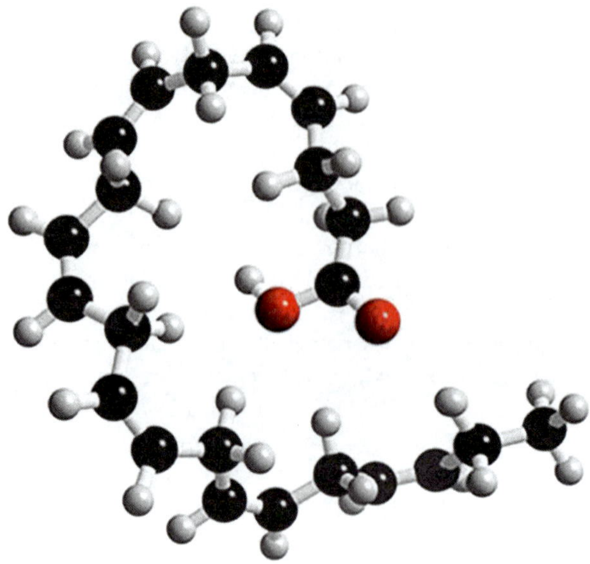

Fig. 6.1 Molecular structure of DHA

6.3.5 Loss of Polyunsaturated Fatty Acids During Sperm Maturation

It has been previously reported that ejaculated spermatozoa show high cell-to-cell variability in life span and, consequently, in their susceptibility toward lipid peroxidation [45]. The concentration of DHA significantly differs in subsets of human spermatozoa at different stages of maturation. Immature germ cells and immature sperm have the highest DHA content, while mature sperm have the lowest. DHA content in mouse sperm obtained from the seminiferous tubules was threefold higher than that found in mouse sperm obtained from the epididymis, consistent with the findings observed in ejaculated human sperm [45]. The results of this study indicated that there is a net loss in DHA content during the process of sperm maturation and that sperm retain relatively low levels of membrane-bound DHA in order to minimize the occurrence of lipid peroxidation and, at the same time, optimize membrane fluidity, leading to successful completion of the membrane fusion reactions that take place during the fertilization process.

DHA is a PUFA that contains 22 carbons and 6 double bonds along its aliphatic chain of which the last one is located three carbons away from the terminal methyl group. That is why DHA is designated 22:6n-3 or also omega-3. Given the presence of these six nonconjugated double bonds with interposed bis-allylic methylene groups, the DHA molecule adopts a circular configuration (Fig. 6.1).

This molecular configuration explains, to a great extent, the high entropy and mobility of membrane phospholipids. This high mobility of DHA confers upon the plasma membrane the required membrane fluidity to support, at least in part, sperm motility and sperm membrane fusion properties.

6.4 Conclusions

As it has been previously indicated, five major factors determine the rate of lipid peroxidation in mammalian sperm: (1) membrane-bound PUFA content, (2) oxygen radical production, (3) partial pressure of oxygen (pO2) in the extracellular medium, (4) temperature in the extracellular medium, and (5) antioxidant defense mechanisms.

PUFA are the actual substrate for lipid peroxidation reactions. In a way, they are like a "flammable jacket" that covers the sperm cell. However, as previously indicated, PUFA play a key role in the regulation of membrane fluidity, which is of paramount importance in the maintenance sperm motility and the membrane fusion reactions that take place during the process of fertilization. Therefore, PUFA cell content must be carefully regulated. That is perhaps why during the process of sperm maturation DHA content in immature sperm is significantly reduced, with mature sperm just retaining a critical mass of DHA that will allow sperm to reach the site of fertilization and complete the process of fertilization.

Oxygen radical production, on the other hand, is like the spark that ignites this flammable jacket and initiates lipid peroxidation by abstraction of a hydrogen from the bis-allylic methylene group of sperm membrane-bound PUFA. Oxygen radical production is determined, for the most part, by the rate of oxidative phosphorylation in the mitochondria and by the pO2 in the extracellular medium. The higher the flow of electrons through the inner mitochondrial membrane and the higher the pO2 in the extracellular medium, the higher oxygen radical production. While immature sperm produce high levels of oxygen radicals, mature sperm isolated from the gradient pellet produce very low levels consistent with their physiological role in the process of sperm capacitation [43].

Temperature, on the other hand, determines both oxygen radical production and the rate of reaction of oxygen radicals with membrane-bound PUFA. The activation energy of these reactions is lowered at higher temperatures. In fact, it has been shown that the rate of lipid peroxidation in mammalian spermatozoa increases exponentially above 37°C and decreases significantly at temperatures below 36°C [46]. This would favor storage of sperm in the epididymis, where the temperature is below 36°C.

Concerning antioxidant defense mechanisms, spermatozoa have developed through evolution a strategy to prevent and counteract the occurrence of lipid peroxidation. This strategy includes (1) the presence of antioxidant enzymes, including SOD and GPX which eliminate superoxide anion and hydrogen peroxide and hydroperoxides, respectively, (2) PLA2 which removes lipid peroxides from the sn-2 position of phospholipids, thus preventing and reducing the propagation reactions of lipid peroxidation in the sperm membrane, and (3) sperm's metabolic strategy. Concerning the latter, the preferential conversion of glucose to lactate under aerobic conditions through the Embden–Meyerhof pathway may be an important evolutionary feature of sperm, perhaps intended to minimize the accumulation of reducing equivalents in the mitochondria, thus reducing oxygen radical production by the mitochondria.

It is well known that sperm produce oxygen radicals, that the bulk of these radicals are produced by the mitochondria, and that oxygen radical-induced damage could result in loss of sperm motility, loss of acrosomal contents, and oxidative DNA damage [47–52]. Conversion of glucose to lactate under aerobic conditions would (1) decrease production of mitochondrial NADH and FADH-reducing equivalents by the Krebs cycle and (2) decrease electron flow in the inner mitochondrial membrane, thereby downregulating oxygen radical formation. This metabolic feature of sperm is especially important outside the protective environment of the epididymis where oxygen radical-induced damage is minimized by a lower temperature and lower pO_2.

References

1. MacLeod J. The role of oxygen in the metabolism and motility of human spermatozoa. Am J Physiol. 1943;138:512–8.
2. Tosic J, Walton A. Formation of hydrogen peroxide by spermatozoa and its inhibitory effect on respiration. Nature. 1946;158:485.
3. Aitken RJ, Curry BJ. Redox regulation of human sperm function: from the physiological control of sperm capacitation to the etiology of infertility and DNA damage in the germ line. Antioxid Redox Signal. 2011;14:367–81.
4. Alvarez JG, Storey BT. Role of glutathione peroxidase in protecting mammalian spermatozoa from loss of motility caused by spontaneous lipid peroxidation. Gamete Res. 1989;23:77–90.
5. Aitken RJ, Gordon E, Harkiss D, Twigg JP, Milne P, Jennings Z, Irvine DS. Relative impact of oxidative stress on the functional competence and genomic integrity of human spermatozoa. Biol Reprod. 1998;59:1037–46.
6. Aitken RJ, Harkiss D, Knox W, Paterson M, Irvine DS. A novel signal transduction cascade in capacitating human spermatozoa characterised by a redox-regulated, cAMP-mediated induction of tyrosine phosphorylation. J Cell Sci. 1998;111:645–56.
7. Gharagozloo P, Aitken RJ. The role of sperm oxidative stress in male infertility and the significance of oral antioxidant therapy. Hum Reprod. 2011;26:1628–40.
8. Fraga CG, Motchnik PA, Wyrobek AJ, Rempel DM, Ames BN. Smoking and low antioxidant levels increase oxidative damage to DNA. Mutat Res. 1996;351:199–203.
9. de Lamirande E, Gagnon C, de Lamirande E, Gagnon C. Human sperm hyperactivation and capacitation as parts of an oxidative process. Free Radic Biol Med. 1993;14:157–66.
10. Griveau JF, Renard P, Le Lannou D. An in vitro promoting role for hydrogen peroxide in human sperm capacitation. Int J Androl. 1994;17:300–7.
11. Bize I, Santander G, Cabello P, Driscoll D, Sharpe C. Hydrogen peroxide is involved in hamster sperm capacitation in vitro. Biol Reprod. 1991;44:389–403.
12. Aitken RJ, Ryan AL, Baker MA, McLaughlin EA. Redox activity associated with the maturation and capacitation of mammalian spermatozoa. Free Radic Biol Med. 2004;36:994–1010.
13. Rivlin J, Mendel J, Rubinstein S, Etkovitz N, Breitbart H. Role of hydrogen peroxide in sperm capacitation and acrosome reaction. Biol Reprod. 2004;70:518–22.
14. Ecroyd HW, Jones RC, Aitken RJ. Endogenous redox activity in mouse spermatozoa and its role in regulating the tyrosine phosphorylation events associated with sperm capacitation. Biol Reprod. 2003;69:347–54.
15. Lewis B, Aitken RJ. A redox-regulated tyrosine phosphorylation cascade in rat spermatozoa. J Androl. 2001;22:611–22.
16. Brouwers JF, Boerke A, Silva PFN, Garcia-Gil N, van Gestel RA, Helms JB, van der Lest CHA, Gadella BM. Mass spectrometric detection of cholesterol oxidation in bovine sperm. Biol Reprod. 2011;85:128–36.

17. Davis BK. Influence of serum albumin on the fertilizing ability in vitro of rat spermatozoa. Proc Soc Exp Biol Med. 1976;151:240–3.
18. Ryan L, O'Callaghan YC, O'Brien NM. The role of the mitochondria in apoptosis induced by 7beta-hydroxycholesterol and cholesterol-5beta, 6beta-epoxide. Br J Nutr. 2005;94:519–25.
19. Liu H, Wang T, Huang K. Cholestane-3beta,5alpha,6 beta-triol-induced reactive oxygen species production promotes mitochondrial dysfunction in isolated mice liver mitochondria. Chem Biol Interact. 2009;179:81–7.
20. Laskar A, Yuan XM, Li W. Dimethyl sulfoxide prevents 7beta-hydroxycholesterol-induced apoptosis by preserving lysosomes and mitochondria. J Cardiovasc Pharmacol. 2010;56: 263–7.
21. Kim DE, Youn YC, Kim YK, Hong KM, Lee CS. Glycyrrhizin prevents 7-ketocholesterol toxicity against differentiated PC12 cells by suppressing mitochondrial membrane permeability change. Neurochem Res. 2009;34:1433–42.
22. Aitken RJ. The capacitation-apoptosis highway: oxysterols and mammalian sperm function. Biol Reprod. 2011;85:9–12.
23. Jones R, Mann T, Sherins R. Peroxidative breakdown of phospholipids in human spermatozoa, spermicidal properties of fatty acid peroxides, and protective action of seminal plasma. Fertil Steril. 1979;31:531–753.
24. Alvarez JG, Touchstone JC, Blasco L, Storey BT. Spontaneous lipid peroxidation and production of hydrogen peroxide and superoxide in human spermatozoa. Superoxide dismutase as major enzyme protectant against oxygen toxicity. J Androl. 1987;8:338–48.
25. Aitken RJ, Clarkson JS, Fishel S. Generation of reactive oxygen species, lipid peroxidation, and human sperm function. Biol Reprod. 1989;41:183–97.
26. Kim YH, Haidl G, Schaefer M, Egner U, Herr JC. Compartmentalization of a unique ADP/ATP carrier protein SFEC (sperm flagellar energy carrier, AAC4) with glycolytic enzymes in the fibrous sheath of the human sperm flagellar principal piece. Dev Biol. 2007;302:463–76.
27. Alvarez JG, Storey BT. Role of superoxide dismutase in protecting rabbit spermatozoa from O_2 toxicity due to lipid peroxidation. Biol Reprod. 1983;28:1129–36.
28. Twigg J, Irvine DS, Houston P, Fulton N, Michael L, Aitken RJ. Iatrogenic DNA damage induced in human spermatozoa during sperm preparation: protective significance of seminal plasma. Mol Hum Reprod. 1998;4:439–45.
29. Aitken RJ, Clarkson JS. Significance of reactive oxygen species and antioxidants in defining the efficacy of sperm preparation techniques. J Androl. 1988;9:367–76.
30. O'Flaherty C, de Souza AR. Hydrogen peroxide modifies human sperm peroxiredoxins in a dose-dependent manner. Biol Reprod. 2011;84:238–47.
31. van Overveld FW, Haenen GR, Rhemrev J, Vermeiden JP, Bast A. Tyrosine as important contributor to the antioxidant capacity of seminal plasma. Chem Biol Interact. 2000;127:151–61.
32. Rhemrev JP, van Overveld FW, Haenen GR, Teerlink T, Bast A, Vermeiden JP. Quantification of the nonenzymatic fast and slow TRAP in a post-addition assay in human seminal plasma and the antioxidant contributions of various seminal compounds. J Androl. 2000;21:913–20.
33. Hekman A, Rümke P. The antigens of human seminal plasma. With special reference to lactoferrin as a spermatozoa-coating antigen. Fertil Steril. 1969;20:312–23.
34. Holland MK, White IG. Heavy metals and human spermatozoa. III. The toxicity of copper ions for spermatozoa. Contraception. 1988;38:685–95.
35. Aitken RJ, Harkiss D, Buckingham DW. Analysis of lipid peroxidation mechanisms in human spermatozoa. Mol Reprod Dev. 1993;35:302–15.
36. Gomez E, Irvine DS, Aitken RJ. Evaluation of a spectrophotometric assay for the measurement of malondialdehyde and 4-hydroxyalkenals in human spermatozoa: relationships with semen quality and sperm function. Int J Androl. 1998;21:81–94.
37. Aitken RJ, Harkiss D, Buckingham D. Relationship between iron-catalysed lipid peroxidation potential and human sperm function. J Reprod Fertil. 1993;98:257–65.
38. Rhemrev JP, Vermeiden JP, Haenen GR, De Bruijne JJ, Rekers-Mombarg LT, Bast A. Progressively motile human spermatozoa are well protected against in vitro lipid peroxidation imposed by induced oxidative stress. Andrologia. 2001;33:151–8.

39. Alvarez JG, Storey BT. Differential incorporation of fatty acids into and peroxidative loss of fatty acids from phospholipids of human spermatozoa. Mol Reprod Dev. 1995;42:334–46.
40. Björnstedt M, Hamberg M, Kumar S, Xue J, Holmgren A. Human thioredoxin reductase directly reduces lipid hydroperoxides by NADPH and selenocystine strongly stimulates the reaction via catalytically generated selenols. J Biol Chem. 1995;270:11761–4.
41. Twigg J, Fulton N, Gomez E, Irvine DS, Aitken RJ. Analysis of the impact of intracellular reactive oxygen species generation on the structural and functional integrity of human spermatozoa: lipid peroxidation, DNA fragmentation and effectiveness of antioxidants. Hum Reprod. 1998;13:1429–36.
42. Aitken RJ, Wingate JK, De Iuliis GN, Koppers AJ, McLaughlin EA. Cis-unsaturated fatty acids stimulate reactive oxygen species generation and lipid peroxidation in human spermatozoa. J Clin Endocrinol Metab. 2006;91:4154–63.
43. Ollero M, Gil-Guzman E, Lopez MC, Sharma RK, Agarwal A, Larson K, Evenson D, Thomas Jr AJ, Alvarez JG. Characterization of subsets of human spermatozoa at different stages of maturation: implications in the diagnosis and treatment of male infertility. Hum Reprod. 2001;16:1912–21.
44. Koppers AJ, Garg ML, Aitken RJ. Stimulation of mitochondrial reactive oxygen species production by unesterified, unsaturated fatty acids in defective human spermatozoa. Free Radic Biol Med. 2010;48:112–9.
45. Ollero M, Powers D, Alvarez JG. Variation of docosahexaenoic acid content in subsets of human spermatozoa at different stages of maturation: implications for sperm lipoperoxidative damage. Mol Reprod Dev. 2000;55:326–34.
46. Alvarez JG, Storey BT. Spontaneous lipid peroxidation in rabbit and mouse epididymal spermatozoa: dependence of rate on temperature and oxygen concentration. Biol Reprod. 1985;32:342–51.
47. Alvarez JG, Storey BT. Spontaneous lipid peroxidation in rabbit epididymal spermatozoa. Biol Reprod. 1982;27:1102–8.
48. Holland MK, Alvarez JG, Storey BT. Production of superoxide dismutase and activity of superoxide dismutase in rabbit epididymal spermatozoa. Biol Reprod. 1982;27:1109–18.
49. Alvarez JG, Holland MK, Storey BT. Spontaneous lipid peroxidation in rabbit spermatozoa: a useful model for the reaction of O_2 metabolites with cells. In: Lubbers DW, Acker H, Leninger-Follert E, Goldstick TK, editors. Oxygen transport to tissue—V. New York: Plenum; 1984. p. 433–43.
50. Alvarez JG, Storey BT. Assessment of cell damage caused by spontaneuous lipid peroxidation in rabbit spermatozoa. Biol Reprod. 1984;30:323–32.
51. Alvarez JG, Touchstone JC, Blasco L, Storey BT. Spontaneous lipid peroxidation and production of hydrogen peroxide and superoxide in human spermatozoa. Superoxide dismutase as major enzyme protectant against oxygen toxicity. J Androl. 1987;8:338–48.
52. Fraga CG, Motchnik PA, Shigenaga MK, Helbock HJ, Jacob RA, Ames BN. Ascorbic acid protects against endogenous oxidative DNA damage in human sperm. Proc Natl Acad Sci. 1991;88:11003–6.

Chapter 7
Age and Oxidative Stress in the Germ Line

Bernard Robaire, Catriona Paul, and Johanna Selvaratnam

Abstract There is increasing evidence that the male reproductive system declines with advancing age Studies examining germ cells of older men have raised concerns regarding several aspects of germ cell quality. Increasing paternal age has been linked to genetic diseases (achondroplasia, Apert syndrome, and Marfan syndrome) in the offspring of these fathers; the incidence of autism and schizophrenia is also associated with increasing paternal age. Men above the age of 35 have increased incidences of anneuploidy in their sperm, decreased sperm motility, and increased chromatin aberrations in sperm associated with further problems such as decreased pregnancy rate in the partners of older males. In several tissues, aging is associated with oxidative stress. Rodent studies show that aging male germ cells display an increase in reactive oxygen species (ROS) and a reduction in the antioxidant enzymes normally present to neutralize ROS and protect the cellular structures against ROS-induced damage. This oxidative stress may be the cause of the DNA damage seen in the germ cells and could also have an effect at the stem cell level resulting in reduced germ cell quality with age.

Keywords Immune theory of aging • Telomere theory of aging • Oxidative stress theory of aging • Aging and male fertility • Germ stem cells • DNA damage and repair

B. Robaire, PhD (✉) • C. Paul, PhD • J. Selvaratnam, MSc
Departments of Pharmacology and Therapeutics and of Obstetrics and Gynecology, McGill University and the MUHC-RI, 3655 Promenade Sir William Osler, Montréal, QC, Canada H3G 1Y6
e-mail: bernard.robaire@mcgill.ca

A. Agarwal et al. (eds.), *Studies on Men's Health and Fertility*, Oxidative Stress in Applied Basic Research and Clinical Practice, DOI 10.1007/978-1-61779-776-7_7,

7.1 Introduction

Aging is the deterioration in efficiency of the body's physiological processes that is observed after the reproductive phase of life [1]. Although this phase of life is well defined in women, it has not been clearly established in men. In females, fertility reaches its biological limits as oocytes are depleted from the ovary, i.e., in woman this is associated with the attainment of menopause. However, for centuries men of old age have been considered just as fertile and capable of producing healthy offspring as their young counterparts. The ability of the male germ line to continuously renew itself has meant that males as old as 80–90 years of age can father children. Although these aged men have the ability to become fathers, are they really as fertile as they were in their youth? As a man ages, his spermatogonial stem cells (SSCs) continuously produce mature germ cells, but is the quality of these spermatozoa, produced from aged stem cells, equivalent to that of those produced by young men?

As couples delay childbearing, there is increasing concern over the quality of male germ cells and the health of the children born to older fathers. Investigation into germ cells from aging men has increased public concern for the issue of paternal age. The spermatozoa of men greater than 35 years of age have a higher incidence of aneuploidy and partners of these men have a higher rate of miscarriage and trophoblast diseases [2–5].

In this chapter, we will first discuss some of the most prevalent theories of aging. We will then provide an overview of how aging affects male fertility and discuss models of germ cell aging. Finally, we will address stem cell DNA damage and repair and the role of oxidative stress in aging germ cells.

7.2 Theories of Aging

Aging is a complex process that is not yet fully understood, but that affects every tissue in the body and leads to deterioration of bodily function over time. Current theories of aging were developed from the idea of Weismann who proposed the theory of programmed death to explain why aging evolved as a price paid for the propagation of the species at the expense of the individual, and more importantly, suggested that there was a limitation to the number of somatic cell divisions [6]. This idea was later experimentally confirmed by Hayflick and Moorhead [87], and over the years, has been developed along with numerous theories of aging. These theories can be divided into two categories: programmed and damage theories of aging. Programmed theories of aging include: (1) programmed longevity—this is associated with increasing programmed genetic instability [7]; (2) endocrine theory—biological clocks regulate aging through the action of hormones [8]; (3) immunological theory—the immune system is programmed to decline as the organism ages, thereby increasing vulnerability to pathogens and gradually leading to death [9]. Damage theories include: (1) wear and tear theory—the idea that parts of the body gradually wear out after many years of use which results in aging [6]; (2) rate of living theory—the greater an organism's basal rate of oxygen metabolism, the shorter is its life span [10]; (3) cross-linking theory—the accumulation of cross-linking proteins damages cells and tissues, resulting in slowing down of bodily processes and aging [6]; (4) DNA damage theory—DNA damage accumulation occurs over time, thus resulting in aging; (5) telomere theory—shortening of the telomere ends occurs following multiple cellular replications, resulting in loss of DNA until a critical level is reached, leading to cell death [11]; and (6) free radical theory or reactive oxygen species (ROS)—highly reactive free radicals cause damage to macromolecular components (such as nucleic acids, lipids, sugars, and proteins) of cells and accumulation of such damage leads to gradual dysfunction and aging [12]. We will discuss below three of the most prevalent theories of aging, i.e., the telomere theory, the immune theory, and the free radical/ROS theory.

7.2.1 Telomere Theory of Aging

The telomere theory of aging was formulated in the 1990s after scientists discovered that telomeres shorten during the aging of human fibroblasts [11]. Telomeres represent the terminal region of the DNA helix, made up of repetitive DNA sequences that aid in preventing incomplete replication and instability [13]. They are essentially "caps" on the ends of linear chromosomes, composed of 10–20 kb of hexameric repeats, TTAGGG [14]. In somatic cells, telomeres shorten with each cell division, eventually reducing the number of repeat sequences [15]. This reduction ultimately renders the chromosomes unstable and the cell is no longer able to replicate [15]. Hence, progressive somatic cell division occurs over a finite period of time, which represents the often-mentioned "biological clock." Eventually, somatic cells will enter a growth-arrested yet viable state called senescence [16].

A key element of this theory is that life span is determined by cell division and not time; it proposes that there is a direct correlation between telomere shortening and aging [17]. Further work has determined that telomerase, the enzyme capable of restoring telomeric ends lost during cell division, is absent in many somatic cells, rendering them incapable of maintaining proper telomeric structure [15]. However, telomerase is present in germ cells [18], and hence the theory is limited in that it cannot account for aging of these cells, given that telomere length is maintained. Further, many believe the theory to be oversimplified and not fully predictive [14], particularly because much of the work has been done in the murine model, in which telomere biology differs greatly from that of the human [19]. Indeed, it has recently been found that, in sperm, telomere length increases as men age [20].

7.2.2 Immune Theory of Aging

Aging is associated with a decline in adaptive immunity [21]. Adaptive immunity is the component of the immune system comprising the B cells, cytotoxic T cells, and helper T cells and utilizes specific antigen recognition and memory as a means to fight infectious disease. It is distinguished from the innate system, which is able to mount a much more immediate response to invading pathogens. In the elderly, increased morbidity and mortality are directly associated with an increased susceptibility to infection [22]. This is coincident with a decline in adaptive immunity [21]. Indeed, aging has been associated with a decreased proportion of naïve T cells [23], a reduced proliferative response of adaptive immune cells, diminished antigen recognition, and deficits in signal transduction [24–26]. Advanced age is associated with a greater incidence of immune cells in the epididymis of the Brown Norway rat [27], but little is known of the interaction of immune cells and germ cells during aging,

7.2.3 Oxidative Stress Theory of Aging

The free radical theory of aging was first proposed by Harman [12]. This theory essentially implies that biological aging is a result of the production of increased levels of ROS. The theory maintains that the normal respiratory chain function of the mitochondria results in sustained ROS production throughout an organism's life span, resulting in an accumulation of ROS-induced damage [12]. It has been proposed that ROS of mitochondrial origin, when produced consistently over time, can cause accumulation of mitochondrial DNA (mtDNA) mutations, potentially leading to deterioration of the respiratory chain function. This can lead to even greater production of ROS because of disturbed electron flow [28] and, hence, a vicious cycle of ROS production can be established. Strong genetic evidence points to a role for the respiratory chain of the mitochondrion in controlling how quickly an organism ages. Studies in *Caenorhabditis elegans* (*C. elegans*) have shown that mutations in the CLK1 protein that is responsible for the final step in the synthesis of a critical transporter of electrons between complexes I/II and III lead to a 50% increase in life span [29]. In fact, disruption of any essential step of the biosynthesis pathway of ubiquinone increases life span and decreases ROS production [30]. Using RNAi technology, it has been shown that *C. elegans* displays a tenfold overrepresentation of mitochondrial genes, particularly involving complexes I, III, and IV subunits, among genes whose inhibition extends life span [31].

The oxidative stress theory has been further substantiated in a number of model organisms where the supplementation or depletion of ROS-scavengers has either increased or decreased life span, respectively [28]. A simple test of the involvement of ROS in aging organisms involves administration of antioxidants. A large number of studies have examined the efficacy of administration of antioxidants to aging organisms and how this affects longevity. Supplementation with vitamin E significantly increases the life span in nematodes [32], *Paramecia* [33], and diphtheria *Zapronius paravittiger* [34]. It has also been shown in *Podospora anserine* that glutathione (GSH) addition increases lifespan by 13% and also strongly reduces the formation of end products of ROS-induced lipid damage [35].

In *Drosophila,* overexpression of the antioxidant enzymes superoxide dismutase (SOD) and catalase (CAT) results in lower accumulation of carbonylated proteins

[36] and approximately a 34% increase in life expectancy [37]. The overexpression of Cu, Zn SOD in motorneurons of *Drosophila* also resulted in a 40% increase in life span [38]. Caloric restriction can substantially increase life span and cause decreased ROS production along with increasing the expression of antioxidant defenses [39].

Aging is thus characterized by an increase in oxidative stress concordant with a decrease in antioxidant capacity. Higher levels of ROS are found in aged tissue [40] and these ROS can irreparably damage cellular macromolecules, such as DNA, polyunsaturated fatty acids, and proteins [41–43]. Indeed, aging has been associated with increased levels of oxidative modifications of proteins [36, 44, 45], as well as increased lipid peroxidation products [46], and 8-hydroxydeoxyguanosine DNA-adduct formation [47]. Although the free radical theory of aging still remains the most widely accepted theory of aging, it has recently come under scrutiny because a series of studies using mouse transgenic and null mutations models for proteins such as MnSOD, thioredoxin 2, and CAT have not substantiated the role of ROS modulation in altering life span expectancy [48, 49].

7.3 Aging and Male Fertility

In 1912, Weinberg noted that the skeletal disorder achondroplasia was prominent in younger rather than older siblings; he was the first to suggest the possible link between paternal age and genetic diseases in offspring [4]. Thus, despite the ability of men to father children well into old age, there appears to be a positive relationship between paternal age and genetic disease in their offspring [50]. Older men have greater difficulties in achieving full-term pregnancy with their partners; increasing paternal age is associated with a reduction in the quality of spermatozoa, i.e., decline in sperm motility and morphology [51] and a higher incidence of DNA fragmentation [52]; there are inconsistent reports as to whether, in man, there is a decline in sperm production rate with advancing age [53]. However, fathers over 40 have both a greater difficulty to conceive [54–56] and their partners have a higher miscarriage rate [57]. With respect to genetic diseases, increased incidences of disorders such as Apert syndrome [58], Marfan syndrome, and achondroplasia [50, 59] have been linked to paternal aging. In addition, several other diseases of complex etiology have been associated with a paternal age effect; these include schizophrenia [60], Alzheimer's disease [61], autism [62], and cardiac defects [63], which all have increased relative risk in children born to older fathers. Spontaneous mutation frequency in pachytene spermatocytes, round spermatids, and caudal spermatozoa is elevated in aged mice, and 80% of these mutations are characterized as single base substitutions [64]. There is an increased rate of basal sperm chromatin damage and a decreased ability to respond to external oxidative stress as rats age [65]. Thus, there appears to be an accumulation of DNA damage and/or mutations in the germ line as men age and this affects the health and wellbeing of their offspring. Studies using mouse and rat animal models have further substantiated the link between paternal age and adverse effects on progeny outcome.

7.4 Models of Germ Cell Aging

7.4.1 Mouse Models

To examine the mechanisms of aging, several inbred strains of mice, such as the senescence-accelerated mouse (SAM) [66], and transgenic mouse models, such as Klotho [67], or mice having null mutations or overexpresssing several enzymes associated with oxidative stress [48] have been developed. There are at least fourteen strains of senescence-prone inbred strains (SAMP) and four senescence-resistance strains (SAMR); the SAMP life span is 40% shorter than SAMR. Interestingly, SAMP1 mice have an early maturation and rapid decline in their spermatozoa production with advancing age [68]. Studies comparing testicular histology between SAMP1 and SAMR1 mice reveal that regressive changes in the testes of SAMP1 occur by 6–7 months, with degradation of seminiferous tubules and spermatogenic epithelium defects appearing much earlier than expected [66]. All SAMP mice have a shorter reproductive period, terminating by the age of approximately 6-months; by 9–10 months of age, SAMP1 mice have a depopulation of the seminiferous tubules, morphologically abnormal germ cells, and atrophy in SAMP1 testes [66]. However, surprisingly, by the age of 14–16 months the spermatogenic epithelium is restored, indicating that SAMP mice possibly have mechanisms/factors that can reverse the abnormalities associated with aging. These findings have been explained in additional studies that show four genes (ankyrin repeat and SOCS box-containing 8 (Asb-8), germ cell-specific gene 1, T-complex polypeptide 1b, and activator of cAMP responsive element modulator in testis) implicated in the regulation of late-stage spermatogenesis that are downregulated in aged SAMP1 mice [68]. This study highlights the difficulties in discriminating whether accelerated aging in animal models is due to the normal aging processes or is a possible manifestation of pathologies.

The klotho mutant mouse model has been proposed as one that resembles human aging [69]. Disruption of a single gene, klotho, that encodes a secreted protein results in a mutant mouse that manifests phenotypes resembling premature aging [67]. Homozygous klotho mice have shortened life spans to 60.7 days, decreased activity, and displayed infertility and atrophy of the genital organs. The klotho mouse model has been used to investigate the molecular mechanisms of normal human aging as well as human diseases in which there is premature aging or accelerated aging, such as progeria and Werner syndrome, but has not yet been used to investigate mechanisms of male germ cell aging.

Studies using mouse models that lack any known or induced mutations have demonstrated reduced quality in aged spermatozoa; histological observations reveal changes in testicular architecture, such as an increased number of vacuoles in germ cells, thinning of the seminiferous epithelia, and reduction in the number of spermatocytes [70]. Additionally, an increase in the frequency of germ cell mutations was identified [71]. Mouse models have also been used to assess both stem cell aging and the stem cell niche; these studies are described below.

7.4.2 *Rat Models*

The Brown Norway (BN) rat is a highly robust model for the study of male reproductive aging [72–75]. This strain of rat has a long life span (up to 40 months), does not exhibit many of the age-related pathologies found in other rat strains, such as pituitary and Leydig cell tumors, and does not become obese. There are striking changes in the testes and epididymides of these animals, even though no systemic disease is apparent. Aging of the testis is marked by a gradual decrease in the percentage of normal seminiferous tubules and total sperm count [72] as well as the ability of Leydig cells to produce testosterone [76]; the expression of several genes and the activities of enzymes associated with testosterone production are also affected as a function of aging [77, 78]. At the cellular level, regressed testes display anomalies in the structure of the endoplasmic reticulum and nuclei of Sertoli cells (the niche-forming "nurse" cells that surround the germ cells and ensure their normal development), between which large intracellular spaces are observed rather than the normally embedded germ cells [79]. Importantly, the decreases seen in testis function in the BN rat with age have been reported also in aging men [80, 81]. There are also dramatic changes in the epididymal epithelial architecture [82] and epididymal gene expression [83] in aging BN rats.

Mating of male BN rats of increasing age (3–24 months) to young females resulted in an increase in preimplantation loss, a decrease in the average fetal weight, and a greater than threefold increase in neonatal deaths [84]. Together, these results clearly indicate that the quality of spermatozoa decreases as males age in this model. The actual basis for these observations remains unclear; however, it has been noted that there was a large increase with age in the number of sperm with an abnormal flagellar midpiece, suggesting that the formation of spermatozoa in the testes of older males was defective [85]. The percentage of motile spermatozoa was significantly decreased in the cauda epididymidis of old rats and the proportion of spermatozoa that retained their cytoplasmic droplet (a sign of lack of sperm maturation) was markedly elevated. Some of these effects are likely to be due to changes taking place in spermatozoa during the process of spermatogenesis (e.g., formation of the flagellum), while others could occur during sperm maturation in the epididymis (e.g., acquisition of motility).

7.5 Germ Stem Cells and Aging

Inevitably, if aging germ stem cells (GSCs) are unable to efficiently prevent, intercept, and repair oxidative damage caused by oxidative stress, they will accumulate cellular damage over the years. The prevention of ROS-mediated damage is mainly dealt with by antioxidants; these are reduced in aging spermatozoa [86]. The combined effects of increased ROS production, reduced antioxidant activity, and reduced efficiency of the aging cell's repair mechanisms will result in aging germ cells with

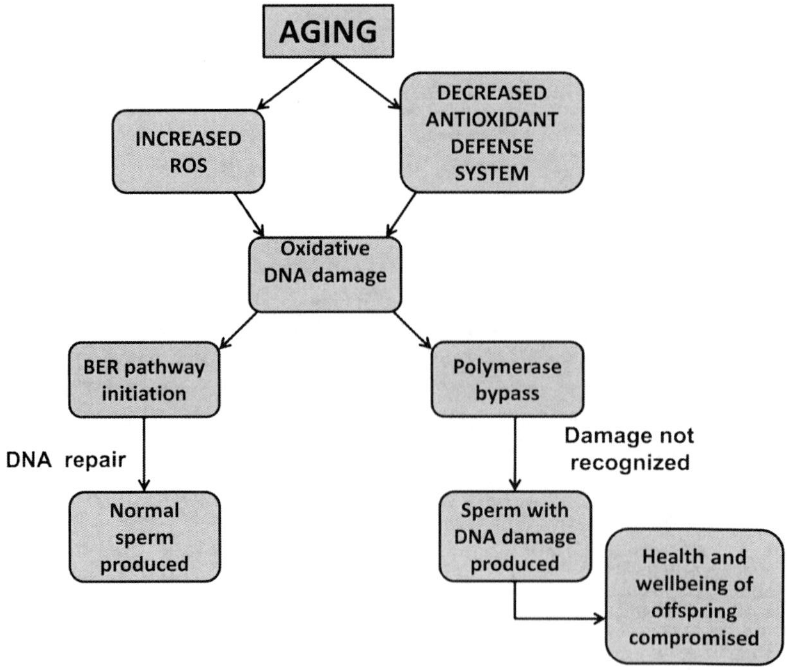

Fig. 7.1 Schematic representation of the impact of aging on male germ cells and of potential consequences for progeny outcome

a reduced capacity to manage oxidative stress. The interactions and consequences of increased ROS production and decreased antioxidant defenses with advancing age are depicted in Fig. 7.1.

Although there have been many studies looking at the role of stem cells in aging, only a handful have examined the role of stem cells specifically in terms of testis aging. The regenerative potential of any organ within the body, including the testis, is dependent on its stem cells. However, stem cells in some tissues appear to have a finite number of divisions for repopulation and regeneration, i.e., the Hayflick limit (around fifty cell divisions) discussed above [87]. This hypothesis was put forth based on lung fibroblast divisions in vitro; since then, it has been demonstrated that cells in a number of other tissues can undergo many more divisions in a lifetime without changing morphology and function; for example, intestinal stem cells undergo around 1,000 divisions in the lifetime of a mouse and probably many more in the lifetime of a human [88]. The testis also does not appear to conform to the Hayflick limit as the production of sperm requires many more divisions of SSCs to continually produce mature spermatozoa. If SSCs were to conform to the Hayflick limit, it would only be possible to produce sperm for a maximum of approximately 10 years. Since men produce sperm into old age, clearly the Hayflick hypothesis does not hold for the testis. Indeed, SSCs are estimated to have undergone 380 mitotic divisions by the time a man reaches the age of 30 years (840 mitotic divisions by 50 years) [4]; however, with each additional replication, the possibility of causing errors in the germ line increases, making paternal age a major potential source of new mutations [89].

As one of the main morphological features of the aging testis is regression due to germ cell loss, the question as to what is responsible for causing this loss is central. There are a small number of theories suggesting that it is either an accumulation of DNA damage in the SSCs or that there is an age-associated decrease in the support of the Sertoli cells and thus the nurturing niche of the SSCs is no longer able to support the maintenance and division of these cells. While some studies suggest that it is the failure of the SSC niche that causes diminished fertility in older males, and not the ability of SSCs to self-renew [90, 91], others demonstrate, using SSC transplantation in mice, that not only the ability of the niche to sustain spermatogenesis decreases with age, but also that the number and the activity of SSCs decline with advancing age [92]. Using transplantation of SSCs to determine age-related alterations, another group concluded that there was no change with age; however, this group did not study advanced age, as had the other investigators, since they used only animals that were no older than 1 year of age [93].

7.6 Aging and DNA Damage and Repair: Relationship to Oxidative Stress

Several studies have been done on the effects of oxidative stress and antioxidants on male germ cells during aging; however, there are relatively few studies designed to understand whether/how oxidative stress is related to DNA damage in male germ cells in the testes. Under normal conditions, sperm maturing in the epididymis generate low levels of ROS, a process that is thought to be related to capacitation [94]. One of the main sources of ROS is electron leakage from the mitochondria of spermatozoa [95]. Although ROS are required for normal processes during spermatogenesis and fertilization, any imbalance between the generation of ROS and the antioxidant defense system, as occurs in aging, can compromise DNA integrity and thus fertilization potential [96]. DNA damage has been linked routinely to aging in many somatic tissues, including the brain and the liver [97, 98]; the testis appears to be no exception. As discussed above, the DNA damage theory of aging is now widely accepted and states that an accumulation of DNA damage leads to the dysfunction of the cell and thereby aging. Once the DNA is damaged, a variety of responses can ensue, including cessation of transcription, cell cycle arrest, mutagenesis, and cell death [99]. Any one of these responses could result in the disruption of spermatogenesis, mutation in the germ line, epigenetic modifications, and ultimately passing on an error in the genetic/epigenetic information to offspring.

Spermatogenesis is a long and complex process that includes numerous cell divisions, DNA replication, chromatid exchange during homologous recombination, and extensive repackaging of chromatin during spermiogenesis [100]; therefore, there are numerous points at which DNA damage and mutations could arise. Many different types of damage can occur in male germ cells, including single- and double-strand breaks; however, one of the main types of damage seen in male germ cells in the aging male is caused by oxidative stress. Oxidative stress is a common pathology seen in men with fertility problems [101, 102]. This type of stress can not only result in damage to proteins and lipids, but also cause DNA (including mtDNA) damage. DNA bases are particularly susceptible to oxidation mediated by ROS. Due to its low redox potential, guanine is highly susceptible to oxidative lesions. The most common DNA lesions resulting from oxidative stress are the oxidized base 8-oxoguanine (8-oxoG) and the oxidized nucleoside 8-oxo-deoxyguanosine (8-oxodG). The 8-oxoG lesion has the ability to mimic the base thymine (T) in the *syn* conformation and, accordingly, can pair with adenine (A), forming an 8-oxoG:A base pair that can allow escape from proof reading and bypass of DNA polymerases [103]. Failure to remove 8-oxoG prior to replication can result in G- to T-transversion mutations [104]. Animal studies have shown that such mutations can contribute to around 20% of the mutations found in young mice; however, this is increased to 40% in aged mice. Given that the mutation frequency is sevenfold higher in pachytene spermatocytes, tenfold higher in round spermatids, and sixfold higher in spermatozoa in aged mice, this translates to a rather large mutational load in male germ cells [64, 71].

In addition to nuclear DNA, the mtDNA can also be affected by oxidative stress. In fact, mtDNA is highly sensitive to oxidative stress because the mitochondrial genome is present in multiple copies and lacks histones and protamines or any DNA-associated proteins that would protect it from attack; hence, its structure is more open and easily accessible to ROS. Furthermore, the mtDNA is located near the electron transport chain where free radicals are continuously being leaked [105]. Aging has been associated with an accumulation of damage to somatic mtDNA [106]. In the case of the germ line, however, the passing on of mtDNA damage to the offspring may not be of great concern as it is the maternal mtDNA which is inherited and the paternal mtDNA is eliminated after fertilization [107, 108].

DNA base excision repair (BER) is likely to be one of the mechanisms involved in maintaining genomic stability in the male germ line as it is one of the main pathways responsible for repairing spontaneous base damage caused by oxidative stress. Several BER genes are highly expressed in the testis and more specifically in the germ cells [109]. There are two BER pathways; the involvement of each one depends on the number of bases to be repaired. The short patch BER pathway is involved when only one base is excised by the glycosylase Ogg1 and Ape endonuclease followed with DNA synthesis by β-polymerase and ligation by DNA ligase III [78]. In the long patch BER pathway, several nucleotides are repaired and the new bases are placed by the polymerase δ and ε enzymes in conjunction with proliferating cell nuclear antigen (PCNA); the gap is ultimately resected by DNA ligase I [110]. Mitochondrial oxidative DNA damage is also repaired by the BER pathway. If the damage is not repaired, however, this can result in mutations and ultimately may be passed on to the offspring.

Spermatozoa are highly susceptible to oxidative DNA damage caused by ROS as they have limited levels of antioxidants under normal conditions and have many substrates for free radical attack, such as unsaturated fatty acids. It has been shown that ROS can cause DNA damage in the sperm nucleus and in their mitochondria and that these damaged sperm can retain their motility and fertilizing capability [111]. This can result in the sperm carrying the oxidative DNA damage on to the next generation and cause severe abnormalities [112]. For example, men who smoke exhibit higher DNA fragmentation indices and their offspring have an increased incidence of childhood cancer [113], further enforcing the importance of the impact of oxidative stress on the male germ line.

Using techniques such as the sperm chromatin structure assay (SCSA®), chromomycin A3 assay (CMA3), and Comet analysis, chromatin integrity in sperm has been found to be altered in aging males [52, 65, 114]. Additional studies on sperm chromosome 1 have illustrated increased (twofold) frequencies of segmental duplications and deletions as well as breaks in the 1q12 site of this chromosome with advancing age [115].

Despite the number of studies investigating DNA damage in sperm from the aging male, there are relatively few studies on DNA damage in earlier germ cells such as spermatogonia, spermatocytes, and round spermatids. This may be due to the difficulty in isolating pure populations of these specific cell types from the testis. However, the studies that have been done suggest that there are alterations in the DNA of aging males as early as the spermatocyte stage, where a number of chromosomal aberrations have been detected [116]. Spermatids from the aging male have an increase in aneuploidies [117, 118]. There is also an increase in micronuclei with age in hamster spermatids, indicating chromosome loss with age [119].

A higher index of arrested germ cell divisions has been observed in testes from older men [120]. This study showed the presence of abnormal cells with structural disorganization in the area of the seminiferous epithelium where spermatocytes would normally be located. The authors suggested a dysfunction in the Sertoli cell barrier; however, meiotic arrest could be due to accumulation of DNA damage. There is also an increased sensitivity of round spermatids and elongating spermatids to DNA damage in general (not only in association with aging) due to the alterations seen in transcription and translation in these cells as they undergo many morphological changes to become mature spermatozoa [121]. This results in lowered transcription and/or translation of DNA repair genes. A decrease in repair proteins, accompanied by an increase in ROS production in addition to decreased levels of ROS-scavengers and antioxidants, could be detrimental to these cells in the aging male (Fig. 7.1).

The increase seen in DNA damage in many tissues with aging may not only be the result of increased oxidative stress, but also a lowered capacity to repair increased levels of oxidative stress-induced DNA damage. A study using aging mice demonstrated that there was a decline in the BER pathway with age and that this corresponded to declines in levels of β-polymerase activity at the RNA and protein levels [122]. The BER pathway is reduced by more than 50% in nuclear extracts prepared from isolated mixed germ cells from aged (28-month) mice and repair using this pathway is limited at different steps in the old compared to the young; BER activity in the young animals is limited by uracil-DNA glycosylase (Ogg1), and in the old animals it is limited by AP endonuclease [109]. This reduction in the BER pathway could account in part (in conjunction with the decreased antioxidants and increased ROS) to the increase in DNA damage observed in sperm from older men. On the other hand, a study using immunohistochemical labeling of germ cells with different markers of the BER pathway reported that many of the proteins showed higher expression in the germ cells of older males [123]. It is worth mentioning, though, that it is extremely difficult to quantify protein expression using immunohistochemistry. The efficiency of the BER pathway has also been shown to decline in mtDNA in other tissues such as the brain [124] and liver [125] in mice.

There are a number of DNA repair pathways other than BER at play in the testis; however, very few studies have been undertaken regarding the role of these pathways in this tissue during aging. It is known that double-strand break repair is compromised in older males in tissues other than the testis [126, 127]. In the testis, however, one of the key components of this repair pathway, Ku80, is significantly decreased in older males; this conclusion was reached using homogenized testis samples, thus making it impossible to determine in which specific germ cell type this occurs [128].

7.7 Aging Germ Cells and Epigenetic Marks

In tissues other than the testis, it has been postulated that oxidative stress can cause changes in DNA methylation. Oxidative DNA damage (e.g., 8-oxodG) may inhibit methylation at adjacent residues, thereby causing a decrease in methylation [129]. Age-dependent modifications of the germ cell epigenome remain relatively uninvestigated at present; however, many of these modifications may influence gene expression during embryogenesis and thus explain some of the paternal age effects that have been demonstrated previously. It has been shown that the pattern of gene-specific DNA methylation is altered in the aged somatic tissues of many mammalian species, including humans [130]. It is evident that with increasing age there is an increase in methylation in a number of systems, including the testis, lung, liver, blood, and stem cells [131–134]. If there is hypermethylation occurring in aged SSCs, it is understandable that aging is accompanied by a loss of developmental capacity, i.e., reduced numbers of maturing germ cells. This has recently been demonstrated in mice where hematopoietic stem cells (HSCs) from older animals display increased stem cell self-renewal, but decreased efficiency to differentiate [135]. In contrast, it has been shown in the *Drosophila* testis that there are increasing numbers of GSCs in aging males containing misoriented centrosomes (normal orientation of which ensures correct asymmetric division) and that these GSCs have entered cell cycle arrest which may contribute to the decline seen in spermatogenic activity [136]. It appears that there are alterations in either the stem cells or their niche with age; however, it remains to be shown whether this is due to the increased oxidative stress seen in aged males.

7.8 Summary Statements

- In humans, epidemiological studies indicate that there is a reduction in germ cell quality past the age of 35 years.
- Aging is associated with increased oxidative stress and decreased antioxidants.
- Rodent studies show that aging male germ cells display an increase in ROS and reduction in the antioxidant enzymes normally present that neutralize ROS and protect the cellular structures against ROS-induced damage.
- Studies using animal models demonstrate that advanced paternal age is associated with adverse progeny outcome.
- DNA damage increases and DNA repair decreases with age.

References

1. Hayflick L. Biological aging is no longer an unsolved problem. Ann N Y Acad Sci. 2007;1100:1–13.
2. Sartorius GA, Nieschlag E. Paternal age and reproduction. Hum Reprod Update. 2010; 16:65–79.
3. Singh NP, Muller CH, Berger RE. Effects of age on DNA double-strand breaks and apoptosis in human sperm. Fertil Steril. 2003;80:1420–30.
4. Thacker PD. Biological clock ticks for men, too: genetic defects linked to sperm of older fathers. JAMA. 2004;291:1683–5.
5. Perrin MC, Brown AS, Malaspina D. Aberrant epigenetic regulation could explain the relationship of paternal age to schizophrenia. Schizophr Bull. 2007;33:1270–3.
6. Jin K. Modern biological theories of aging. Aging Dis. 2010;1:72–4.
7. Davidovic M, Sevo G, Svorcan P, Milosevic DP, Despotovic N, Erceg P. Old age as a privilege of the "selfish ones". Aging Dis. 2010;1:139–46.
8. van Heemst D. Insulin, IGF-1 and longevity. Aging Dis. 2010;1:147–57.
9. Rozemuller AJ, van Gool WA, Eikelenboom P. The neuroinflammatory response in plaques and amyloidangiopathy in Alzheimer's disease: therapeutic implications. Curr Drug Targets CNS Neurol Disord. 2005;4:223–33.
10. Brys K, Vanfleteren JR, Braeckman BP. Testing the rate-of-living/oxidative damage theory of aging in the nematode model *Caenorhabditis elegans*. Exp Gerontol. 2007;42:845–51.
11. Harley CB, Futcher AB, Greider CW. Telomeres shorten during ageing of human fibroblasts. Nature. 1990;345:458–60.
12. Harman D. Aging: a theory based on free radical and radiation chemistry. J Gerontol. 1956;11:298–300.
13. Shampay J, Szostak JW, Blackburn EH. DNA sequences of telomeres maintained in yeast. Nature. 1984;310:154–7.
14. Holt SE, Shay JW, Wright WE. Refining the telomere-telomerase hypothesis of aging and cancer. Nat Biotechnol. 1996;14:836–9.
15. Mera SL. The role of telomeres in ageing and cancer. Br J Biomed Sci. 1998;55:221–5.
16. Counter CM. The roles of telomeres and telomerase in cell life span. Mutat Res. 1996; 366:45–63.
17. Ishikawa F. Aging clock: the watchmaker's masterpiece. Cell Mol Life Sci. 2000;57:698–704.
18. Achi MV, Ravindranath N, Dym M. Telomere length in male germ cells is inversely correlated with telomerase activity. Biol Reprod. 2000;63:591–8.

19. Rangarajan A, Weinberg RA. Opinion: comparative biology of mouse versus human cells: modelling human cancer in mice. Nat Rev Cancer. 2003;3:952–9.
20. Kimura M, Cherkas LF, Kato BS, Demissie S, Hjelmborg JB, Brimacombe M, Cupples A, Hunkin JL, Gardner JP, Lu X, Cao X, Sastrasinh M, Province MA, Hunt SC, Christensen K, Levy D, Spector TD, Aviv A. Offspring's leukocyte telomere length, paternal age, and telomere elongation in sperm. PLoS Genet. 2008;4:e37.
21. DeVeale B, Brummel T, Seroude L. Immunity and aging: the enemy within? Aging Cell. 2004;3:195–208.
22. Plackett TP, Boehmer ED, Faunce DE, Kovacs EJ. Aging and innate immune cells. J Leukoc Biol. 2004;76:291–9.
23. Miller RA. The aging immune system: primer and prospectus. Science. 1996;273:70–4.
24. Chakravarti B, Abraham GN. Aging and T-cell-mediated immunity. Mech Ageing Dev. 1999;108:183–206.
25. Weksler ME. Changes in the B-cell repertoire with age. Vaccine. 2000;18:1624–8.
26. Szakal AK, Aydar Y, Balogh P, Tew JG. Molecular interactions of FDCs with B cells in aging. Semin Immunol. 2002;14:267–74.
27. Serre V, Robaire B. Distribution of immune cells in the epididymis of the aging Brown Norway rat is segment-specific and related to the luminal content. Biol Reprod. 1999;61:705–14.
28. Dufour E, Larsson NG. Understanding aging: revealing order out of chaos. Biochim Biophys Acta. 2004;1658:122–32.
29. Felkai S, Ewbank JJ, Lemieux J, Labbe JC, Brown GG, Hekimi S. CLK-1 controls respiration, behavior and aging in the nematode *Caenorhabditis elegans*. EMBO J. 1999;18:1783–92.
30. Asencio C, Rodriguez-Aguilera JC, Ruiz-Ferrer M, Vela J, Navas P. Silencing of ubiquinone biosynthesis genes extends life span in *Caenorhabditis elegans*. FASEB J. 2003;17:1135–7.
31. Lee SS, Lee RY, Fraser AG, Kamath RS, Ahringer J, Ruvkun G. A systematic RNAi screen identifies a critical role for mitochondria in *C. elegans* longevity. Nat Genet. 2003;33:40–8.
32. Balin AK. Testing the free radical theory of aging. In: Adelman RC, Roth GS, editors. Testing the theories of aging. Boca Raton: CRC Press; 1983. p. 137–82.
33. Thomas J, Nyberg D. Vitamin E supplementation and intense selection increase clonal life span in *Paramecium tetraurelia*. Exp Gerontol. 1988;23:501–12.
34. Kakkar R, Bains JS, Sharma SP. Effect of vitamin E on life span, malondialdehyde content and antioxidant enzymes in aging *Zaprionus paravittiger*. Gerontology. 1996;42:312–21.
35. Munkres K, Rana RS. Antioxidants prolong life span and inhibit the senescence-dependent accumulation of fluorescent pigment (lipofuscin) in clones, of *Podospora anserina*. Mech Ageing Dev. 1978;7:407–15.
36. Sohal RS, Sohal BH, Orr WC. Mitochondrial superoxide and hydrogen peroxide generation, protein oxidative damage, and longevity in different species of flies. Free Radic Biol Med. 1995;19:499–504.
37. Orr WC, Sohal RS. Extension of life-span by overexpression of superoxide dismutase and catalase in *Drosophila melanogaster*. Science. 1994;263:1128–30.
38. Parkes TL, Elia AJ, Dickinson D, Hilliker AJ, Phillips JP, Boulianne GL. Extension of *Drosophila* lifespan by overexpression of human SOD1 in motorneurons. Nat Genet. 1998;19:171–4.
39. Barouki R. Ageing free radicals and oxidative stress. Med Sci (Paris). 2006;22:266–72.
40. Schoneich C. Reactive oxygen species and biological aging: a mechanistic approach. Exp Gerontol. 1999;34:19–34.
41. Lucas DT, Szweda LI. Cardiac reperfusion injury: aging, lipid peroxidation, and mitochondrial dysfunction. Proc Natl Acad Sci USA. 1998;95:510–4.
42. Inal ME, Kanbak G, Sunal E. Antioxidant enzyme activities and malondialdehyde levels related to aging. Clin Chim Acta. 2001;305:75–80.
43. van der Loo B, Bachschmid M, Spitzer V, Brey L, Ullrich V, Luscher TF. Decreased plasma and tissue levels of vitamin C in a rat model of aging: implications for antioxidative defense. Biochem Biophys Res Commun. 2003;303:483–7.

44. Starke-Reed PE, Oliver CN. Protein oxidation and proteolysis during aging and oxidative stress. Arch Biochem Biophys. 1989;275:559–67.
45. Stadtman ER. Protein oxidation and aging. Science. 1992;257:1220–4.
46. Choi JH, Yu BP. Brain synaptosomal aging: free radicals and membrane fluidity. Free Radic Biol Med. 1995;18:133–9.
47. Ames BN, Shigenaga MK. Oxidants are major contributors to cancer and aging. In: Halliwell B, Aruoma OI, editors. DNA and free radicals. New York: Ellis Horwood; 1993. p. 1–15.
48. Pérez VI, Bokov A, Van Remmen H, Mele J, Ran Q, Ikeno Y, Richardson A. Is the oxidative stress theory of aging dead? Biochim Biophys Acta. 2009;1790:1005–14.
49. Jang YC, Remmen VH. The mitochondrial theory of aging: insight from transgenic and knockout mouse models. Exp Gerontol. 2009;44:256–60.
50. Crow JF. The origins, patterns and implications of human spontaneous mutation. Nat Rev Genet. 2000;1:40–7.
51. Kidd SA, Eskenazi B, Wyrobek AJ. Effects of male age on semen quality and fertility: a review of the literature. Fertil Steril. 2001;75:237–48.
52. Wyrobek AJ, Eskenazi B, Young S, Arnheim N, Tiemann-Boege I, Jabs EW, Glaser RL, Pearson FS, Evenson D. Advancing age has differential effects on DNA damage, chromatin integrity, gene mutations, and aneuploidies in sperm. Proc Natl Acad Sci USA. 2006;103:9601–6.
53. Robaire B, Oakes C, Zubkova EV. Effects of aging on spermatogenesis and sperm function. In: Kandeel FR, Swerdloff RS, Pryor JL, editors. Male reproductive dysfunction, pathophysiology and treatment. New York: Informa Healthcare USA, Inc.; 2007. p. 101–11.
54. de La Rochebrochard E, de Mouzon J, Thepot F, Thonneau P. Fathers over 40 and increased failure to conceive: the lessons of in vitro fertilization in France. Fertil Steril. 2006;85:1420–4.
55. de La Rochebrochard E, McElreavey K, Thonneau P. Paternal age over 40 years: the "amber light" in the reproductive life of men? J Androl. 2003;24:459–65.
56. de La Rochebrochard E, Thonneau P. Paternal age >or=40 years: an important risk factor for infertility. Am J Obstet Gynecol. 2003;189:901–5.
57. Kleinhaus K, Perrin M, Friedlander Y, Paltiel O, Malaspina D, Harlap S. Paternal age and spontaneous abortion. Obstet Gynecol. 2006;108:369–77.
58. Yoon SR, Qin J, Glaser RL, Jabs EW, Wexler NS, Sokol R, Arnheim N, Calabrese P. The ups and downs of mutation frequencies during aging can account for the Apert syndrome paternal age effect. PLoS Genet. 2009;5:e1000558.
59. Rolf C, Nieschlag E. Reproductive functions, fertility and genetic risks of ageing men. Exp Clin Endocrinol Diabetes. 2001;109:68–74.
60. Dalman C, Allebeck P. Paternal age and schizophrenia: further support for an association. Am J Psychiatry. 2002;159:1591–2.
61. Bertram L, Busch R, Spiegl M, Lautenschlager NT, Muller U, Kurz A. Paternal age is a risk factor for Alzheimer disease in the absence of a major gene. Neurogenetics. 1998;1:277–80.
62. Tsuchiya KJ, Matsumoto K, Miyachi T, Tsujii M, Nakamura K, Takagai S, Kawai M, Yagi A, Iwaki K, Suda S, Sugihara G, Iwata Y, Matsuzaki H, Sekine Y, Suzuki K, Sugiyama T, Mori N, Takei N. Paternal age at birth and high-functioning autistic-spectrum disorder in offspring. Br J Psychiatry. 2008;193:316–21.
63. Yang Q, Wen SW, Leader A, Chen XK, Lipson J, Walker M. Paternal age and birth defects: how strong is the association? Hum Reprod. 2007;22:696–701.
64. Walter CA, Intano GW, McMahan CA, Kelner K, McCarrey JR, Walter RB. Mutation spectral changes in spermatogenic cells obtained from old mice. DNA Repair (Amst). 2004;3:495–504.
65. Zubkova EV, Robaire B. Effects of ageing on spermatozoal chromatin and its sensitivity to in vivo and in vitro oxidative challenge in the Brown Norway rat. Hum Reprod. 2006;21:2901–10.
66. Zakhidov ST, Gopko AV, Marshak TL, Kulibin AY, Zelenina IA. Analysis of spermatogenesis in senescence-accelerated mice. Biol Bull. 2007;34:551–7.
67. Takahashi Y, Kuro-O M, Ishikawa F. Aging mechanisms. Proc Natl Acad Sci USA. 2000; 97:12407–8.

68. Chiba T, Yao J, Higami Y, Shimokawa I, Hosokawa M, Higuchi K. Identification of differentially expressed genes in senescence-accelerated mouse testes by suppression subtractive hybridization analysis. Mamm Genome. 2007;18:105–12.
69. Nagai R, Saito Y, Ohyama Y, Aizawa H, Suga T, Nakamura T, Kurabayashi M, Kuroo M. Endothelial dysfunction in the klotho mouse and downregulation of klotho gene expression in various animal models of vascular and metabolic diseases. Cell Mol Life Sci. 2000;57:738–46.
70. Tanemura K, Kurohmaru M, Kuramoto K, Hayashi Y. Age-related morphological changes in the testis of the BDF1 mouse. J Vet Med Sci. 1993;55:703–10.
71. Walter CA, Intano GW, McCarrey JR, McMahan CA, Walter RB. Mutation frequency declines during spermatogenesis in young mice but increases in old mice. Proc Natl Acad Sci USA. 1998;95:10015–9.
72. Wright WW, Fiore C, Zirkin BR. The effects of aging in the seminiferous epithelium of the Brown Norway rat. J Androl. 1993;14:110–7.
73. Wang C, Leung A, Sinha-Hikim A. Reproductive aging in the male Brown-Norway rat: a model for the human. Endocrinology. 1993;133:2773–81.
74. Chen H, Hardy MP, Huhtaniemi I, Zirkin BR. Age-related decreased Leydig cell testosterone production in the Brown Norway rat. J Androl. 1994;15:551–7.
75. Robaire B, Syntin P, Jervis K. The coming of age of the epididymis. In: Jegou B et al., editors. Testis, epididymis and technologies in the year 2000. New York: Springer; 2000. p. 229–62.
76. Zirkin BR, Santulli R, Strandberg JD, Wright WW, Ewing LL. Testicular steroidogenesis in the aging Brown Norway rat. J Androl. 1993;14:118–23.
77. Zirkin BR, Chen H. Regulation of Leydig cell steroidogenic function during aging. Biol Reprod. 2000;63:977–81.
78. Paul C, Nagano M, Robaire B. Aging results in differential regulation of DNA repair pathways in pachytene spermatocytes in the Brown Norway rat. Biol Reprod. 2011;85(6):1269–78.
79. Levy S, Serre V, Hermo L, Robaire B. The effects of aging on the seminiferous epithelium and the blood-testis barrier of the Brown Norway rat. J Androl. 1999;20:356–65.
80. Neaves WB, Johnson L, Petty CS. Seminiferous tubules and daily sperm production in older adult with varied numbers of Leydig cells. Biol Reprod. 1987;36:301–8.
81. Vermeulen A. Andropause. Maturitas. 2000;34:5–15.
82. Serre V, Robaire B. Segment specific morphological changes in the aging Brown Norway rat epididymis. Biol Reprod. 1998;58:497–513.
83. Jervis KM, Robaire B. Changes in gene expression during aging in the Brown Norway rat epididymis. Exp Gerontol. 2002;37:897–906.
84. Serre V, Robaire B. Paternal age affects fertility and progeny outcome in the Brown Norway rat. Fertil Steril. 1998;70:625–31.
85. Syntin B, Robaire B. Sperm motility and structural changes during aging in the Brown Norway rat. J Androl. 2001;22:235–44.
86. Weir CP, Robaire B. Spermatozoa have decreased antioxidant enzymatic capacity and increased reactive oxygen species production during aging in the Brown Norway rat. J Androl. 2007;28:229–40.
87. Hayflick L, Moorhead PS. The serial cultivation of human diploid cell strains. Exp Cell Res. 1961;25:585–621.
88. Rubin H. The disparity between human cell senescence in vitro and lifelong replication in vivo. Nat Biotechnol. 2002;20:675–81.
89. Crow JF, Johnson TE. Research on environmental effects in genetic studies of aging: comments. J Gerontol B Psychol Sci Soc Sci. 2005;60(1):7–11.
90. Schmidt JA, Abramowitz LK, Kubota H, Wu X, Niu Z, Avarbock MR, Tobias JW, Bartolomei MS, Brinster RL. In vivo and in vitro aging is detrimental to mouse spermatogonial stem cell function. Biol Reprod. 2011;84:698–706.
91. Ryu BY, Orwig KE, Oatley JM, Avarbock MR, Brinster RL. Effects of aging and niche microenvironment on spermatogonial stem cell self-renewal. Stem Cells. 2006;24:1505–11.
92. Zhang X, Ebata KT, Robaire B, Nagano MC. Aging of male germ line stem cells in mice. Biol Reprod. 2006;74:119–24.

93. Ehmcke J, Joshi B, Hergenrother SD, Schlatt S. Aging does not affect spermatogenic recovery after experimentally induced injury in mice. Reproduction. 2007;133:75–83.
94. Aitken RJ, Baker MA, De Iuliis GN, Nixon B. New insights into sperm physiology and pathology. Handb Exp Pharmacol. 2010;198:99–115.
95. Koppers AJ, De Iuliis GN, Finnie JM, McLaughlin EA, Aitken RJ. Significance of mitochondrial reactive oxygen species in the generation of oxidative stress in spermatozoa. J Clin Endocrinol Metab. 2008;93:3199–207.
96. O'Flaherty C, de Lamirande E, Gagnon C. Reactive oxygen species modulate independent protein phosphorylation pathways during human sperm capacitation. Free Radic Biol Med. 2006;40:1045–55.
97. Brasnjevic I, Hof PR, Steinbusch HW, Schmitz C. Accumulation of nuclear DNA damage or neuron loss: molecular basis for a new approach to understanding selective neuronal vulnerability in neurodegenerative diseases. DNA Repair (Amst). 2008;7:1087–97.
98. Moller P, Lohr M, Folkmann JK, Mikkelsen L, Loft S. Aging and oxidatively damaged nuclear DNA in animal organs. Free Radic Biol Med. 2010;48:1275–85.
99. Essers J, Vermeulen W, Houtsmuller AB. DNA damage repair: anytime, anywhere? Curr Opin Cell Biol. 2006;18:240–6.
100. Clermont Y. Kinetics of spermatogenesis in mammals: seminiferous epithelium cycle and spermatogonial renewal. Physiol Rev. 1972;52:198–236.
101. Agarwal A, Makker K, Sharma R. Clinical relevance of oxidative stress in male factor infertility: an update. Am J Reprod Immunol. 2008;59:2–11.
102. Gharagozloo P, Aitken RJ. The role of sperm oxidative stress in male infertility and the significance of oral antioxidant therapy. Hum Reprod. 2011;26:1628–40.
103. Hsu GW, Ober M, Carell T, Beese LS. Error-prone replication of oxidatively damaged DNA by a high-fidelity DNA polymerase. Nature. 2004;431:217–21.
104. Neeley WL, Essigmann JM. Mechanisms of formation, genotoxicity, and mutation of guanine oxidation products. Chem Res Toxicol. 2006;19:491–505.
105. Muller FL, Liu Y, Van Remmen H. Complex III releases superoxide to both sides of the inner mitochondrial membrane. J Biol Chem. 2004;279:49064–73.
106. Larsson NG. Somatic mitochondrial DNA mutations in mammalian aging. Annu Rev Biochem. 2010;79:683–706.
107. Sutovsky P, Moreno RD, Ramalho-Santos J, Dominko T, Simerly C, Schatten G. Ubiquitin tag for sperm mitochondria. Nature. 1999;402:371–2.
108. Wai T, Ao A, Zhang X, Cyr D, Dufort D, Shoubridge EA. The role of mitochondrial DNA copy number in mammalian fertility. Biol Reprod. 2010;83:52–62.
109. Intano GW, McMahan CA, Walter RB, McCarrey JR, Walter CA. Mixed spermatogenic germ cell nuclear extracts exhibit high base excision repair activity. Nucleic Acids Res. 2001;29:1366–72.
110. Christmann M, Tomicic MT, Roos WP, Kaina B. Mechanisms of human DNA repair: an update. Toxicology. 2003;193:3–34.
111. Aitken RJ, Gordon E, Harkiss D, Twigg JP, Milne P, Jennings Z, Irvine DS. Relative impact of oxidative stress on the functional competence and genomic integrity of human spermatozoa. Biol Reprod. 1998;59:1037–46.
112. Meseguer M, Martinez-Conejero JA, O'Connor JE, Pellicer A, Remohi J, Garrido N. The significance of sperm DNA oxidation in embryo development and reproductive outcome in an oocyte donation program: a new model to study a male infertility prognostic factor. Fertil Steril. 2008;89:1191–9.
113. Zenzes MT. Smoking and reproduction: gene damage to human gametes and embryos. Hum Reprod Update. 2000;6:122–31.
114. Schmid TE, Eskenazi B, Baumgartner A, Marchetti F, Young S, Weldon R, Anderson D, Wyrobek AJ. The effects of male age on sperm DNA damage in healthy non-smokers. Hum Reprod. 2007;22:180–7.
115. Sloter ED, Marchetti F, Eskenazi B, Weldon RH, Nath J, Cabreros D, Wyrobek AJ. Frequency of human sperm carrying structural aberrations of chromosome 1 increases with advancing age. Fertil Steril. 2007;87:1077–86.

116. Sloter E, Nath J, Eskenazi B, Wyrobek AJ. Effects of male age on the frequencies of germinal and heritable chromosomal abnormalities in humans and rodents. Fertil Steril. 2004;81:925–43.
117. Lowe X, Collins B, Allen J, Titenko-Holland N, Breneman J, van Beek M, Bishop J, Wyrobek AJ. Aneuploidies and micronuclei in the germ cells of male mice of advanced age. Mutat Res. 1995;338:59–76.
118. Xiao Y, Tates AD, Boei J, Natarajan AT. Aging and diethylstilbestrol-induced aneuploidy in male germ cells: a transgenic mouse model. Chromosoma. 1998;107:507–13.
119. Allen JW, Collins BW, Setzer RW. Spermatid micronucleus analysis of aging effects in hamsters. Mutat Res. 1996;316:261–6.
120. Miething A. Arrested germ cell divisions in the ageing human testis. Andrologia. 2005;37:10–6.
121. Hecht NB. Molecular mechanisms of male germ cell differentiation. Bioessays. 1998;20:555–61.
122. Cabelof DC, Raffoul JJ, Yanamadala S, Ganir C, Guo Z, Heydari AR. Attenuation of DNA polymerase beta-dependent base excision repair and increased DMS-induced mutagenicity in aged mice. Mutat Res. 2002;500:135–45.
123. El-Domyati MM, Al-Din AB, Barakat MT, El-Fakahany HM, Xu J, Sakkas D. Deoxyribonucleic acid repair and apoptosis in testicular germ cells of aging fertile men: the role of the poly(adenosine diphosphate-ribosyl)ation pathway. Fertil Steril. 2009;91:2221–9.
124. Gredilla R, Garm C, Holm R, Bohr VA, Stevnsner T. Differential age-related changes in mitochondrial DNA repair activities in mouse brain regions. Neurobiol Aging. 2008;31:993–1002.
125. de Souza-Pinto NC, Hogue BA, Bohr VA. DNA repair and aging in mouse liver: 8-oxodG glycosylase activity increase in mitochondrial but not in nuclear extracts. Free Radic Biol Med. 2001;30:916–23.
126. Ren K, Pena de Ortiz S. Non-homologous DNA end joining in the mature rat brain. J Neurochem. 2002;80:949–59.
127. Vyjayanti VN, Rao KS. DNA double strand break repair in brain: reduced NHEJ activity in aging rat neurons. Neurosci Lett. 2006;393:18–22.
128. Um JH, Kim SJ, Kim DW, Ha MY, Jang JH, Chung BS, Kang CD, Kim SH. Tissue-specific changes of DNA repair protein Ku and mtHSP70 in aging rats and their retardation by caloric restriction. Mech Ageing Dev. 2003;124:967–75.
129. Zawia NH, Lahiri DK, Cardozo-Pelaez F. Epigenetics, oxidative stress, and Alzheimer disease. Free Radic Biol Med. 2009;46:1241–9.
130. Richardson B. Impact of aging on DNA methylation. Ageing Res Rev. 2003;2:245–61.
131. Oakes CC, Smiraglia DJ, Plass C, Trasler JM, Robaire B. Aging results in hypermethylation of ribosomal DNA in sperm and liver of male rats. Proc Natl Acad Sci USA. 2003;100:1775–80.
132. Maegawa S, Hinkal G, Kim HS, Shen L, Zhang L, Zhang J, Zhang N, Liang S, Donehower LA, Issa JP. Widespread and tissue specific age-related DNA methylation changes in mice. Genome Res. 2010;20:332–40.
133. Rakyan VK, Down TA, Maslau S, Andrew T, Yang TP, Beyan H, Whittaker P, McCann OT, Finer S, Valdes AM, Leslie RD, Deloukas P, Spector TD. Human aging-associated DNA hypermethylation occurs preferentially at bivalent chromatin domains. Genome Res. 2010;20:434–9.
134. Teschendorff AE, Menon U, Gentry-Maharaj A, Ramus SJ, Weisenberger DJ, Shen H, Campan M, Noushmehr H, Bell CG, Maxwell AP, Savage DA, Mueller-Holzner E, Marth C, Kocjan G, Gayther SA, Jones A, Beck S, Wagner W, Laird PW, Jacobs IJ, Widschwendter M. Age-dependent DNA methylation of genes that are suppressed in stem cells is a hallmark of cancer. Genome Res. 2010;20:440–6.
135. Rossi DJ, Bryder D, Zahn JM, Ahlenius H, Sonu R, Wagers AJ, Weissman IL. Cell intrinsic alterations underlie hematopoietic stem cell aging. Proc Natl Acad Sci USA. 2005;102:9194–9.
136. Cheng J, Türkel N, Hemati N, Fuller MT, Hunt AJ, Yamashita YM. Centrosome misorientation reduces stem cell division during ageing. Nature. 2008;456:599–604.

Chapter 8
Heat and Oxidative Stress in the Germ Line

Koji Shiraishi

Abstract Spermatogenesis is highly dependent on scrotal temperature. In the testis, germ cells but not somatic cells are vulnerable to heat stress. In response to heat stress, germ cells undergo apoptosis, autophagy, necrosis, and cell cycle arrest; these behaviors are different in each testicular component. Heat induces oxidative stress in the testicles in a variety of ways, mainly by lipid peroxidation of the cellular membrane and mitochondria-derived reactive oxygen species (ROS), and heat-induced oxidative stress is involved in all of these cellular behaviors. Heat-shock factor 1 (HSF1) protects the cells by regulating the expression of heat-shock proteins (HSPs), promoting cell survival. Paradoxically, HSF1 promotes apoptosis of germ cells by heat stress, indicating that injured germ cells actively undergo apoptosis to maintain the quality of gametes. The pattern of heat stress (degree, duration, and interval of the elevated temperature) in humans (e.g., cryptorchidism, varicocele, and environmental heat exposure) is completely different compared to in vitro and in vivo animal experiments.

Keywords Heat stress • Oxidative stress • Reactive oxygen species • Spermatogenesis • Germ cells • Apoptosis • Thermotolerance • Heat-shock proteins

8.1 Introduction

In general, the ability of cells to tolerate low body temperatures is better than their ability to tolerate high body temperatures, and there is less resistance to body temperatures above the set-point temperature. In particular, an optimum testicular temperature, which is lower than the core body temperature, is crucial

K. Shiraishi, MD, PhD (✉)
Department of Urology, Yamaguchi University, Ube, Yamaguchi, Japan
e-mail: sirakkkay@aol.com

A. Agarwal et al. (eds.), *Studies on Men's Health and Fertility*, Oxidative Stress in Applied Basic Research and Clinical Practice, DOI 10.1007/978-1-61779-776-7_8,

for spermatogenesis. Normal testicular function is temperature-dependent, and in most mammals, the testes are maintained between 2 to 8°C below the core body temperature by the exteriorization of the testes in special hairless sacs called the scrotum [1, 2]. In fact, slight increases in testicular temperature can disturb spermatogenesis and ultimately inhibit spermatogenesis. For example, temporal exposure of rodent testes to abdominal temperatures leads to marked disruption of spermatogenesis and infertility [1, 3]. Moreover, ejaculates from men with scrotal temperatures above the normal range show increased incidence of abnormal and immature spermatozoa [4]. On the other hand, elevation of testicular temperature has been suggested as a possible contraceptive treatment for men. Furthermore, elevation of testicular temperature impairs progressive sperm motility and viability, and sperm from heat-shocked testes results in a low in vitro fertilization rate [1]. Precoital testicular heating leads to a transient retardation in embryo growth and an increase in the rate of embryonic degeneration [1, 5–7]. Taken together, these observations indicate that elevation of testicular temperature impairs not only spermatogenesis, but also sperm function and sperm DNA status. Temperatures lower than the normal temperature would lead to reduced metabolic rate and oxidative DNA damage in the testis because of active mitosis and meiosis of the germ cells.

Because of high cellular turnover, spermatogenesis per se requires extensive tissue restructuring in the seminiferous epithelium, resulting in the production of reactive oxygen species (ROS) and reactive nitrogen species, such as superoxide, hydroxyl, peroxyl, hydroperoxyl, nitric oxide, and nitrogen dioxide, which are produced by the peroxidation and oxidation of many cellular lipids, proteins, carbohydrates, and nucleic acids. Thus, germ cells are exposed constantly to thousands of free radicals and oxidative stress. Oxidative stress has been implicated in a variety of pathophysiological states, which cause male infertility [8]. Oxidative stress in heat-induced testicular injury is thought to be involved in clinical situations and may cause all testicular disorders. Research on heat-induced injury in the testes has a long history and is well studied compared with other organs, but the involvement of oxidative stress in heat-induced testicular injury has been a recent focus of research. In addition, many studies showed that ROS-scavenging mechanisms involving antioxidants in the testis play an important role in protecting germ cells against heat stress. Many studies have investigated oxidative stress in germ cells by using semen and ejaculated sperm; however, it should be mentioned that oxidative stress in semen is different from that in the testes. Difficulties in obtaining human testicular samples to examine gaseous molecules prevent the investigation of oxidative stress in the human testis. This chapter focuses on molecular events in the testis, mainly by using animal studies, and clinical situations caused by heat-induced testicular disorders, with special attention to oxidative stress. In addition, effects of oxidative stress on Sertoli and Leydig cells are discussed, because spermatogenesis is closely regulated by the cross-talk between germ and somatic cells.

8.2 Heat Induces Oxidative Stress in the Testis

In vitro and in vivo experiments suggest that heat stress usually produces ROS, such as superoxide radicals, hydroxyl radicals, and hydrogen peroxide. The generation of ROS occurs constantly, even under physiological conditions, in all living cells, and the rate of free radical generation in the testis or spermatozoa appears to be temperature-dependent [9]. For example, the level of spontaneous lipid peroxidation by cultured mouse spermatozoa, as measured by the generation of malondialdehyde (MDA), increases with temperature elevation [10]. In addition, activities of several scavenging enzymes in the testes of rats with experimentally induced cryptorchidism were impaired and were accompanied by increased peroxidation of cellular lipids [11–13], indicating that testicular oxidative stress is determined by the balance between the generation of ROS and their scavenging systems. Effects of heat stress on cellular components are listed as follows: *Cell membrane*—changes in fluidity/stability, alteration in structure, impairment of ion transport, and modulation of the transmembrane efflux pump; *cytoplasm*—impairment of protein synthesis, denaturation of protein structure and function, aggregation of proteins, and induction of heat-shock protein (HSP) synthesis; *mitochondria*—depolarization of mitochondrial membrane potential, depletion of ATP production, production of ROS, and disruption of Ca^{2+} transport across the mitochondrial membrane; *endoplasmic reticulum (ER)*—ER stress from excessive accumulation of misfolded proteins; *nucleus*—impairment of DNA synthesis, inhibition of DNA repair enzymes, alteration of DNA conformation, and changes in gene expression and signal transduction. In particular, plasma membranes are known to be extremely sensitive to heat stress because of their complex molecular composition of lipids and proteins. Upon temperature elevation, the physical state of lipids changes from a tightly packed gel to a less tightly packed crystalline structure, and the permeability of the cell membrane increases, followed by alteration of the cellular content of several ions (Na^+, Mg^{2+}, K^+, and Ca^{2+}). For example, influxes of extracellular Ca^{2+} stimulate the activity of calmodulin-dependent protein kinases and inositol triphosphate production [14], resulting in the alteration of intracellular signal transduction cascades. These changes are universal effects observed in all mammalian cells and likely occur in the germ cells. Among the alterations described above, the plasma membrane and mitochondria are considered the major sites of ROS production. Neutrophils mainly produce ROS, leading to deterioration of spermatogenesis in testicular torsion [8, 15]; however, the involvement of neutrophils in the production of heat-induced oxidative stress is unknown. The plasma membrane of testicular cells is rich in polyunsaturated fatty acids and is therefore vulnerable to oxidation by H_2O_2 and other ROS [12]. The generation of ROS will be augmented in response to elevated metabolism accompanied by oxygen consumption. The content of the redox enzymes, glutathione reductase (GSR) and aldo-keto reductase, which function in the reduction and oxidation of ROS and the resulting carbonyl compounds, is much lower in the germ line cells than in Sertoli and Leydig cells [16]. The antioxidant mechanism against heat-induced oxidative stress is described in the following section. Direct measurements of these gaseous molecules are very difficult to obtain, especially in human testis, and direct and indirect methods to detect seminal ROS are described in other chapters. Stable products of molecules (e.g., nitrite as stable product of nitric oxide (NO) and MDA as a stable product of lipid peroxidation) and modified DNA and proteins (e.g., 8-hydroxy-2′-deoxyguanosine (8-OHdG) as a marker of the oxidation of guanine residues and 4-hydroxy-2-nonenal (4-HNE)-modified proteins as a marker of protein modification by lipid peroxidation products) have been used as surrogate markers of testicular oxidative stress in vivo.

Nitric oxide is a gaseous molecule that is difficult to examine in vivo but is widely examined, because nitric oxide synthase (NOS) expression is easy to investigate at the RNA and protein levels. At higher concentrations, NO promotes germ cell apoptosis through production of peroxynitrite by a reaction between NO and superoxide [17]. The tissue concentration of NO is determined by the type of NOS involved: inducible NOS (iNOS), endothelial NOS (eNOS), and neuronal NOS (nNOS). Basically, tissue concentrations of NO produced by iNOS are 10- to 100-fold higher than those from eNOS and nNOS and can cause necrosis of germ cells [15]. Increased NO synthesis through up-regulation of iNOS by heat stress has been implicated in cellular injury and apoptosis in various cell systems. Heat treatment (43°C for 30 min for 2 consecutive days) markedly induced iNOS expression in testicular germ cells of Cynomolgus monkeys after 3–8 days [18]. Increased iNOS immunoreactivity was noted in heat-susceptible germ cells (pachytene spermatocytes and round spermatids) 3 days after treatment. On day 28, iNOS expression in germ cells is similar to that of controls, whereas very high iNOS expression was noted in Sertoli cells, because the majority of apoptotic germ cells within this time frame had been lost through phagocytosis [19, 20]. eNOS was observed mainly in Sertoli cells and spermatogonia in rats. No obvious alteration in eNOS levels was detected in any of the heat-treatment groups [18]. In humans, NO, which is basically a vasodilator, is shown to be produced by eNOS in endothelial cells in patients with varicocele [21]. In eNOS transgenic mice, NO produced through eNOS is reported to play a functional role in spermatogenesis in cryptorchid-induced apoptosis [22], whereas a several-fold increase in the levels of NO produced by eNOS is less likely to cause oxidative stress.

8.3 Biological Effects of Heat-Induced Oxidative Stress in Germ Cells

Cellular responses against heat depend on the temperature as well as the duration and frequency of heat stress. In humans, continuous monitoring of scrotal temperature shows that scrotal temperature changes every minute, which makes evaluation of the effect of heat in human testis difficult [23, 24]. This notion is important for understanding of the pathophysiology of disorders of spermatogenesis induced by heat stress (e.g., varicocele, cryptorchidism, and environmental exposure). Testes of patients with varicocele are presumably exposed to heat stress all day or, at least,

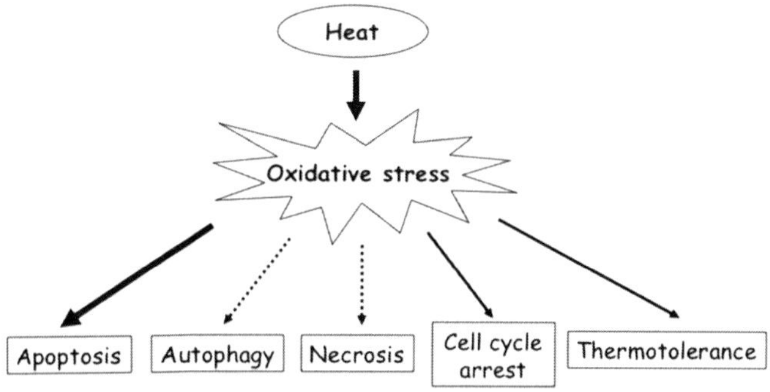

Fig. 8.1 Behavioral patterns of testicular cells, mainly germ cells, exposed to heat-induced oxidative stress. *Dashed line* indicates a possible pathway. These phenomena usually occur concurrently but not alone

several hours, but intermittently. A number of animal models designed to study the effect of heat stress on the testis have been developed, and most studies include transient exposure of the testes to elevated temperatures (typically 43°C) or placing of the testes and epididymis within the body cavity (surgical inducing cryptorchidism) for long-term exposure of the testes to the core body temperature (37°C). Both methods have been reported to cause a variety of disturbances in testicular functions, including decreased testis weight, increased apoptosis, germ cell loss, and altered fertilization capability of sperm. These models are very useful for studying germ cell death and related molecular events, but it is necessary to pay attention when adopting the phenomena obtained from animal studies to human pathophysiology.

Generally, cells with high mitotic activity (e.g., cancer cells, germ cells) are more susceptible to heat stress than are somatic cells. Major cellular responses against heat-induced oxidative stress are listed in Fig. 8.1. Direct evidence suggesting that heat stress increases the concentration of intracellular ROS in cells [25, 26] and that ROS play a role in heat-induced cell death [27, 28] is limited, but many studies have shown that ROS in heat-induced germ cell injury cause apoptosis. The association of cell survival rate and temperature sensitivity is known as an Arrhenius plot [29], and there is a "breaking point" around 42.5°C where the cell survival rate rapidly falls (Fig. 8.2). In the case of hyperthermia in cancer, most of the cell death at temperatures below 43°C occurs by apoptosis, whereas necrosis is evident above 43°C [30]. In vivo, these biological effects vary by cell type. For example, spermatocytes, especially pachytene spermatocytes, and early spermatids readily undergo apoptosis instead of showing thermotolerance. On the other hand, spermatogonia often undergo cell cycle arrest, and Sertoli and Leydig cells often express antioxidants and show thermotolerance. The nature of heat exposure is critical because 43°C for 15 min induces specific damage limited to spermatocytes and early spermatids, whereas 45°C for 15 min results in generalized, nonspecific damage to many different germ cell types. The mechanism of cell death after mild heat stress appears to be

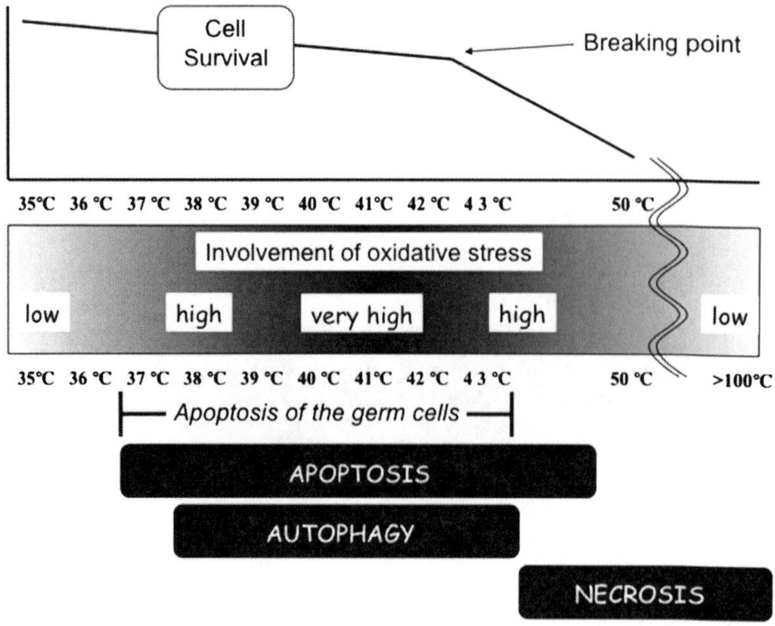

Fig. 8.2 Relationships among temperature, extent of oxidative stress, and cell death. Ranges of apoptosis, autophagy, and necrosis indicated in the *black square* are based on published literature about nongerm cells. Of note, a prolonged duration of heat stress, even at 43°C, causes necrosis rather than apoptosis

active (apoptosis), not passive (necrosis) cell death [31–33]. On the other hand, temperatures above 45°C or burns cause necrosis (Fig. 8.2), but it is likely that apoptosis and necrosis will concurrently occur at varying levels. From a clinical point of view, the mode of heat stress in vivo is quite different from experimental conditions. Most of the experimental conditions are specific for investigation of the effect of heat stress on each cell, including spermatogonia and mature sperm. In contrast, the deterioration of spermatogenesis in patients with cryptorchidism or varicocele is characterized by spermatogenic arrest at different levels, indicating that cell cycle arrest will mainly exist. Topics regarding germ cell apoptosis and cell cycle arrest are described in an independent section in this chapter.

Autophagic cell death is a completely different form of cell death compared to apoptosis and necrosis. Autophagy is a process that degrades long-lived proteins and cytoplasmic components within vesicles, which then deliver the contents to the lysosome/vacuole for degradation. The involvement of autophagy in health and disease is being increasingly reported. In the field of reproductive biology, autophagy is induced mainly in granulosa cells during folliculogenesis and shows good correlation with apoptosis [34]. Light chain 3 (LC3) serves as a marker for autophagy. Upon activation of autophagy, the 18-kDa cytosolic LC3 (LC3B-I)

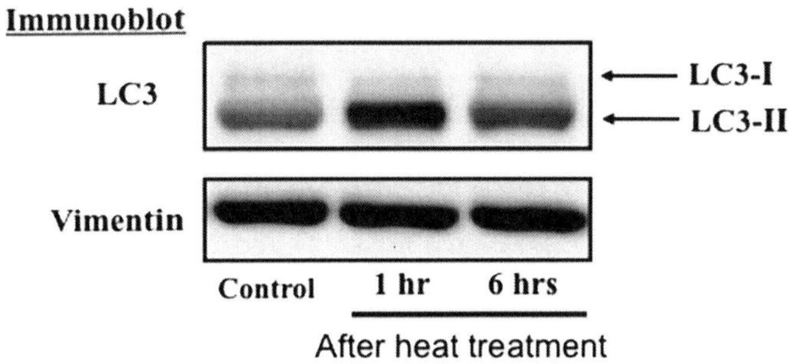

Fig. 8.3 The expression of LC3-II proteins in 8-week-old Wistar rat testis exposed to heat stress (43°C, 15 min). Representative immunoblots from total testicular lysates are shown. LC3-I (18 kDa) is converted to LC3-II (16 kDa) during autophagy induction. The apparent intensity of the LC3-II band indicates that the testis undergoes apparent basal autophagy under normal conditions. Maximum induction of autophagy is observed 1 h after heat stress. All animal experiments followed a protocol approved by the Ethics Committee on Animal Experiments from the Yamaguchi University School of Medicine and were performed according to the Guidelines for Animal Experiments of the Committee

undergoes proteolytic cleavage followed by lipid modification and is converted to the 16-kDa membrane-bound form (LC3B-II), which is specifically localized to the autophagosomal membranes. The conversion from LC3B-I to LC3B-II is used as a sensitive marker for autophagy in cells. The temperature spectrum for autophagy shown in Fig. 8.2 is based on the results of studies in nongerm cells. Figure 8.3 illustrates the presence of heat-induced autophagy in rat testis (unpublished data). Testes of 8-week-old Wistar rats were immersed in water at 43°C for 15 min under sodium pentobarbital anesthesia. The cleavage of LC3B-I to 16-kDa fragments was examined. Whole testicular lysate was electrophoresed and immunoblotted with anti-LC3 antibodies. There is basal autophagy in the normal testis and increased expression of 16-kDa LC3 fragments, indicating that heat stress induces autophagy in the testis, but localization of autophagy has not been identified. Autophagic cell death is reported to play an important role in the preimplantation embryo [35]. The involvement of autophagy has not been defined in further detail, but is considered to play important roles in normal spermatogenesis and heat-induced conditions.

8.4 Which Testicular Cells Are Vulnerable to Heat Stress?

Basically, all testicular components are the target of heat stress. Quantitative analysis of changes in the seminiferous epithelium of rats exposed to local testicular heating indicates that primary spermatocytes are the most sensitive to damage by heat and that late spermatids are relatively resistant [36]. Oxidative stress is a major cause of heat stress in germ cells and leads to apoptosis [37–39]. Many recent studies in animals with experimental cryptorchidism or local testicular heating show that the process of spermatogenesis is exquisitely sensitive to temperature, with increased germ cell apoptosis (mostly spermatocytes and early spermatids) as a consequence of only relatively short periods of mild testicular heating [1, 40, 41]. Cataldo et al. found that the initiation of translation in pachytene spermatocytes and Sertoli cells is inhibited by exposure to abdominal temperature, and that elongated spermatids are much more resistant to heat stress [42]. Needless to say, fetal testicular temperature is 37°C, which should be the same as the mother's temperature, and the cells which exist at birth, such as spermatogonia, Sertoli cells, and Leydig cells, should be considered to be resistant to heat stress. Actually, spermatogonia have a lower susceptibility to elevated temperatures [43], but abdominal temperature can markedly influence the number of spermatogonial stem cells, especially type B spermatogonia, which will generate all the sperm in later life [44]. Of course, long-term exposure or higher temperatures may result in generalized damage to many different germ cell types [45]. Leydig cells and androgen secretion do not appear to be directly affected [1]. There are few experiments on the effects of elevated environmental temperature on the secretion of hormones that control reproductive functions. In vitro and in vivo experiments about the effect of heat and oxidative stress suggest harmful effects on hormonal secretion [46, 47].

A controversial issue is whether ejaculated spermatozoa can be damaged by heat stress when deposited in the reproductive tract, because the temperature of the female reproductive tract is basically 37°C. When cultured at 40°C, the fertilizing capability of bull spermatozoa and the competence of the resultant embryos to develop to the blastocyst stage were not altered [48]. On the other hand, reduced mouse litter sizes were reported after males were subjected to a variety of heat stress regimens. For example, in vitro fertilization using sperm from testes heated 7 days before sperm harvest (an epididymal or late spermatid effect of heating) showed reduced numbers of embryos developing from the 4-cell stage onwards, and those from testes heated 21 days earlier (a spermatocyte effect) showed reduced numbers of embryos developing from the 2-cell stage onwards [49]. A further study focusing on the effects of increased whole body temperature by exposing male mice twice to 36°C for 12 h on each occasion showed reduced sperm number, pregnancy rate, and litter size with maximum effects observed 10 or 14 days after exposure to heat stress [3]. In the study by Rockett et al., control females that were mated with males subjected to transient, acute (20 min), scrotal heat stress 23–28 days previously resulted in reduced litter sizes, consistent with an effect on spermatocytes [6]. Other studies showed that heat stress affects fertility in mice and reduces pregnancy rate and embryo weight as well as fertilization rate in vitro, when sperm from males subjected to heat stress were used [5]. Even if cell death is not observed, sperm cell membrane and DNA are susceptible to heat-induced oxidative stress, and the testicular components undergo cell cycle arrest and thermotolerance, which are described in the following sections.

8.5 Heat-Induced Apoptosis of the Germ Cells

Germ cell apoptosis plays a pivotal role in normal spermatogenesis and is a highly complex process involving genes for various factors such as the Bcl-2 family, Fas, Fas ligand, and p53. Germ cell apoptosis has been considered to guarantee the quality of gametes [50–52]. Many studies with heat stress models have analyzed germ cell apoptosis with local heating at 43°C. Exposure at 39°C had no significant effect detectable by terminal deoxynucleotidyl transferase (TdT)-mediated dNTP nick-end labeling (TUNEL) [6]. This is a rapid process—spermatocyte apoptosis occurs 12–24 h after exposure to heat, continues for several days, and the number of spermatocytes returns to nearly the pretreatment level in 1–2 months. In humans, the cell cycle arrest in spermatogonia but not spermatocyte apoptosis may play a major role in the deterioration of spermatogenesis (as observed in conditions such as varicocele and cryptorchidism). In fact, it is still debated whether varicocele induces apoptosis of germ cells [53]. This is because TUNEL, which is commonly used to detect apoptotic cells, has low sensitivity. Another possible reason is that externalization of phosphatidylserine occurs at an early phase of apoptosis, and externalization at the apoptotic germ cells precedes the activation of caspase 3 [19], resulting in prompt phagocytosis by Sertoli cells in vivo before the cells can be detected by TUNEL. A previous study in patients with varicocele employed the p86 fragment of poly(ADP-ribose) polymerase (PARP) as an early marker of apoptosis and demonstrated that elevation of testicular temperature was associated with testicular oxidative stress and germ cell apoptosis [54].

Several previous studies have shown that primary spermatocyte and early spermatids readily cause apoptosis; however, the specific type of spermatocytes that cause apoptosis and the involvement of stage specificities remain controversial. In rodents, spermatocytes are sensitive to heat, independent of the spermatogenic stage of the seminiferous tubule [6, 52]. Detailed morphological studies have shown that heat stress (43°C for 15 min) induces germ cell apoptosis specifically at the early (I–IV) and late (XII–XIV) stages after 1 or 2 days. The effect of heat on spermatogenesis is not only stage-specific, but also cell-specific. Pachytene spermatocytes and early spermatids (steps 1–4) at stages I–IV and diplotene and dividing spermatocytes at stages XII–XIV were the most susceptible to heat. Nine days after heat stress, majority of the tubules were severely damaged and displayed only a few remaining apoptotic germ cells, as most of the dead cells had presumably been eliminated through phagocytosis by the Sertoli cells [19]. By day 56, spermatogenesis had recovered to the normal level and the incidence of germ cell apoptosis was comparable to that in the control. These results suggest that the adverse effects of heat stress on spermatogenesis are mediated by a stage-related activation of apoptosis involving specific germ cells, namely, secondary, early (I–IV), and late (XII–XIV) stages, but not the androgen-sensitive stages (VII–VIII) [31–33, 55]. This stage-specific susceptibility to heat is caused at least partly by the effect of supporting Sertoli cells [55] and different expressions of antioxidants in the germ cells. Similar to ischemia/reperfusion injury, it is difficult to classify the germinal stage in

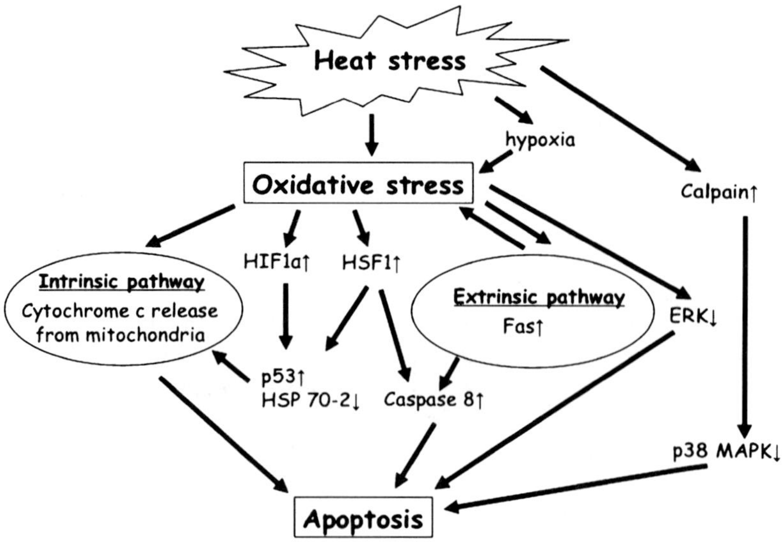

Fig. 8.4 Pathways to apoptosis caused by heat-induced oxidative stress in germ cells, mainly in primary spermatocytes. The *upward arrow* indicates increased expression and/or activity of the molecule. The *downward arrow* indicates decreased expression and/or activity of the molecule. Common apoptotic pathways (e.g., around caspases 2, 9, and 3 activation) are not shown in this figure

testicles that are severely damaged due to heat stress; this may also be one of the reasons why there are controversies regarding the stage-specific heat susceptibility of germ cells.

Intracellular signal transduction pathways activate caspase 3, which cleaves the caspase-activated deoxyribonuclease (ICAD) inhibitor and inactivates its CAD-inhibitory effect; this is a common effector of various stimuli to cause apoptosis [56] (Fig. 8.4). Activation of caspase 3, in which the p32 fragment is proteolysed to an 18-kDa fragment, is observed in testes of men with varicocele and is closely associated with scrotal temperature [54]. Heat-induced apoptosis is believed to be mediated by the intrinsic mitochondrial pathway rather than the extrinsic death receptor pathway, namely, the Fas pathway [31–33]. Studies using the *gld* and *lprcg* (lymphoproliferation complementing *gld*) mice, which harbor loss-of-function mutations in the Fas ligand and Fas, respectively, have revealed that heat-induced germ cell apoptosis is not blocked in these mice [31]. Therefore, the Fas-signaling system is not required for heat-induced germ cell apoptosis. On the other hand, the Fas system has also been involved in many aspects of germ cell apoptosis [57]. A semiquantitative RT-PCR technique revealed that Fas-mediated germ cell apoptosis in rats was activated in response to a variety of proapoptotic stimuli, including heat stress [58]. Involvement of the Fas pathway in heat-induced apoptosis of germ cells still remains controversial. These intrinsic and extrinsic pathways will occur at different phases after exposure to heat. In the intrinsic pathway, the release of cytochrome *c* from the mitochondria into the cytoplasm is a potent stimulus for caspase 9 activation. The opening of the mitochondrial permeability transition pore has been implicated as a key event in the disruption of mitochondrial membrane integrity, which is disturbed by oxidative stress, and cytochrome *c* is then released into the cytosol [59]. Mitochondria produce basal levels of ROS in the form of single-electron leakage to oxygen during normal metabolism [60].

The Bcl-2 family is a widely recognized group of apoptotic regulators. This family consists of both pro- (Bax, Bad, Bak, Bid) and antiapoptotic (Bcl-2, Bcl-xL) proteins that modulate the execution phase of the cell death pathway, and their expressions are altered in testicular germ cell apoptosis. The proapoptotic Bid/Bax/Bad moves from the cytosol to the outer mitochondrial membrane following an apoptosis-inducing signal, and the interaction between these Bcl-2 family proteins induces a conformational change, resulting in the release of cytochrome *c* from the mitochondria. Cytochrome *c* sequesters apoptotic protease-activating factor-1 (Apaf-1), and the complex activates initiator caspase 9, which in turn activates executioner caspases, leading to apoptosis [61–63]. The expressions of the initiator, caspase 9, and the effector, caspase 3, were detected within 2 h of heat treatment in heat-susceptible late pachytenes by immunofluorescence staining of active caspases 9 and 3 [32]. All the proapoptotic proteins, Bax, Bid, Bak, and Bad, are upregulated along with caspase 9 mRNA and protein, indicating the involvement of the intrinsic pathway in the H_2O_2-induced testicular germ cell apoptosis [64]. Other intrinsic factors such as the cellular redox status, cytosolic Ca^{2+} levels, ceramide, and amphipathic peptides can affect mitochondrial megachannel function [59]. In another study, increased levels of the serine-phosphorylated form of the inactive Bcl-2 in heat-susceptible germ cells after heat treatment compared with those in the controls and costaining for TUNEL and phospho-BCL-2 confirmed that Bcl-2 phosphorylation occurs only in those germ cells undergoing apoptosis [65]. Blockage of caspase 2 activation prevents heat-induced germ cell apoptosis in rats by suppressing MAPK14 [66]. Serine phosphorylation of *Bcl-2* and activation of the MAPK14-mediated mitochondria-dependent pathway are critical for heat-induced male germ cell death in monkeys [65]. Heat stress can also alter the expression of the Bax and Bcl-2 genes, where such changes are dependent on the sensitivity of heat stress [67]. On the other hand, a redistribution of Bax from a cytoplasmic to paranuclear localization occurs only in those selective germ cells before their eventual apoptosis. The combined confocal and two-photon imaging of testicular sections costained for Bax and endoplasmic reticulum (ER) revealed colocalization of Bax with ER in the paranuclear areas of late pachytene spermatocytes 2 h after heat treatment, suggesting that Bax translocates to the ER early during apoptosis of heat-induced testicular germ cells [31]. Despite this striking redistribution that was observed by immunocytochemical analysis, western blot analysis revealed that the Bax protein levels in total testis lysates remained unchanged [68]. Heat-induced Bax up-regulation and Bcl-2 down-regulation have been demonstrated in germ cells [69]; however, it is unknown whether or not these phenomena are generated in response to the heat stress.

The extrinsic pathway involves the activation of death receptors, namely Fas/tumor necrosis factor receptor, leading to the activation of initiator caspase 8, followed by activation of executioner caspases 3, 6, and 7 and, subsequently, apoptosis [61]. H_2O_2 was observed to mediate the up-regulation of Fas protein and its transcripts in spermatocytes and spermatids [64]. Vydra et al. suggested that both mitochondria-dependent intrinsic and death receptor-dependent extrinsic pathways are involved in heat-shock factor 1 (HSF1)-induced apoptosis. This suggestion was based on their findings of increased levels of Bcl-2 family proteins, p53 protein accumulation, and elevated expression levels of caspase 8 and death receptor-interacting proteins (including Fas-associated death domain protein and TNF receptor-associated death domain protein) [70].

The tumor suppressor p53 determines the cell fate: apoptosis or cell cycle arrest. Its function differs in each testicular component and across different conditions of stress. In particular, its role in p53-dependent apoptosis in primary spermatocytes is well known [51, 69, 71–73]. In a study investigating a 3-day delay of apoptosis in $p53^{-/-}$ mice with experimental cryptorchidism, p53-dependent apoptosis was found to be responsible for the initial phase of germ cell loss in experimental cryptorchidism [73]. The effects of heat stress on p53 expression and function in germ cells have not been fully investigated. Experimental vasectomy of rat induces oxidative stress in the testis [74] and the colocalization of p53-, Bax-, and TUNEL-positive cells in primary spermatocytes after experimental vasectomy of rat [69]. Furthermore, another study has reported the association between testicular oxidative stress, as measured by the expression of 4-HNE-modified proteins, and p53 expression [75]. Taken together, these results suggest that oxidative stress regulates the expression and function of p53 in germ cells. Fas is involved in heat-induced testicular germ cell apoptosis, and Fas-dependent apoptosis has been shown to be responsible for the p53-independent phase of germ cell loss in cryptorchid testes [73]. p53 is expressed in male germ cells [76], and cell death is delayed in germ cells lacking p53 in experimental cryptorchidism [71], suggesting that the p53 pathway is involved in heat stress-induced germ cell apoptosis, especially in primary spermatocytes. In addition, expression of Mta1, which was originally identified in rat metastatic breast tumors, is decreased after heat stress, following by p53 expression and apoptosis of pachytene spermatocytes [77]. Ohta et al. examined the differentiation of transplanted eGFP-labeled $p53^{+/+}$, $p53^{+/-}$, and $p53^{-/-}$ spermatogonia in cryptorchid conditions [51]. Transplanted $p53^{+/+}$ germ cells did not differentiate in the seminiferous tubules of cryptorchid testes, whereas donor germ cell differentiation was observed when $p53^{+/-}$ or $p53^{-/-}$ germ cells were transplanted into the seminiferous tubules of cryptorchid testes. Semiquantitative analyses of these histological observations indicated that the degree of differentiation was higher following the transplantation of homozygous $p53^{-/-}$ donor cells than of heterozygous $p53^{+/-}$ cells, suggesting that the p53 expression level is important for heat stress-induced germ cell apoptosis. These results indicate that heat stress-induced germ cell loss is the result of p53-dependent apoptosis [51]. It is likely that under heat stress, p53 serves to control germ cell apoptosis in cells that differentiate from type A spermatogonia to spermatocytes, but not in haploid cells or in cells undergoing meiotic division from spermatocytes, suggesting that the apoptotic mechanisms of spermatocytes and haploid spermatids differ considerably.

Calpain belongs to the family of Ca^{2+}-dependent proteases and are ubiquitously expressed in mammalian cells and involved in several physiological processes, including membrane-associated events such as cytoskeletal reorganization, enzyme mobilization, and receptor activation [78]. Further, calpain has been shown to play an important role in apoptosis, including germ cells in ischemia/reperfusion model [79]. Lipid peroxidation due to ROS generation is also reported to alter Ca^{2+} distribution and activate a Ca^{2+}-dependent apoptotic pathway [80] and probably activate calpain. Calpain activation has been demonstrated as an early event in heat stress-induced male germ cell apoptosis through p38 MAPK activation [81]. Calpastatin is a naturally occurring inhibitor of calpain that regulates its proteolytic activity. The interaction of calpastatin with calpain has been shown to prevent apoptosis [82]. The number of apoptotic cells in the testis-specific isoform of calpastatin transgenic animals has been reported to significantly decrease in response to heat [83], indicating that calpain is also involved in heat stress-induced germ cell apoptosis. The MAPK1/3 signaling pathway, though essential for controlling cell proliferation and differentiation, could also play a role in cell death. However, its role in apoptosis remains controversial, with several studies suggesting that it may play either a pro- or antiapoptotic role in this process. Available evidence also suggests that heat stress can activate MAPK1/3 [84]. Furthermore, heat stress causes testicular hypoxia, leading to the activation of hypoxia-inducible factor 1 α (HIF1-alpha), thus causing apoptosis [39]. The precise association of HIF1-alpha and oxidative stress in germ cells has not yet been fully understood; however, they are known to be closely associated [85].

Upon heat stress, antiapoptotic signal transductions and apoptotic machinery are concomitantly activated. Uncoupling proteins (UCP) are mitochondrial inner membrane proteins that mediate proton leak and reduce ATP production [86]. UCP2 is capable of protecting germ cells from apoptosis, and its presence in elongated spermatids might explain why these cells are more resistant to spontaneous apoptosis under normal conditions. In another study, when testes were exposed to heat for a short time (43°C for 5 min), UCP2 was detected in almost all cell types, and its total abundance increased by approximately sixfold [87]. Heat stress also simultaneously activates signal transduction pathways leading to antiapoptosis activity and/or cellular proliferation. Key signaling factors such as Akt, p38, extracellular signal-regulated kinase (ERK), and the heat-shock protein (HSP) HNRNPH1 protein [88] play important roles in antiapoptosis and cellular proliferation pathways.

8.6 Effects of Heat Stress on the Cell Kinetics of the Germ Cells

Transitions during spermatogenesis from mitotic divisions are accompanied by dynamic transitions in the expression of different networks of cell cycle genes, and they suggest that at least some of this differential gene expression may play key roles in regulating the progression of spermatogenic cells through the process of spermatogenesis [89]. Decreased expression of proliferating nuclear cell antigen (PCNA), which is a marker of DNA synthesis, especially in spermatogonia, has been reported in testes of men with varicocele [75, 90, 91]. Heat stress is reported to induce double-strand breaks in the DNA due to denaturation and dysfunction of heat-labile repair proteins such as DNA polymerases. Testicular DNA polymerases α, β, and γ were also decreased and the activities of DNA polymerase β and γ were related to the deleterious effect of heat stress in a cryptorchid model [92] and tissue cultures [93]. Rather than germ cell apoptosis, impairment of germ cell proliferation would be more crucial in the pathophysiology of spermatogenic disorders in men.

A typical cell cycle progresses through mitosis (M phase) alternating with a longer interphase, which is divided into three phases: G1, S, and G2 [94]. A cell grows during the G_1 phase, continues to grow and duplicates its chromosomes during the S phase, prepares for mitosis during the G_2 phase, and divides during the M phase. Spermatogonia are mitotic cells that can be expected to share mechanisms of cell cycle regulation with other mitotic cells, including most somatic cell types. During meiosis in primary spermatocytes, the DNA is replicated as in a mitotic cell cycle, but homologous chromosomes pair and chiasmata form between the homologues to facilitate genetic recombination and the subsequent proper segregation of chromosomal homologues to each secondary spermatocyte. This is followed by a second meiotic division without an intervening round of DNA replication that separates sister chromatids into haploid spermatids [95]. Because meiosis occurs uniquely in germ cells, spermatocytes can be expected to utilize at least some cell cycle control mechanisms that are not used in mitotic somatic cells or in premeiotic spermatogonia. M-phase promoting factor (MPF) is a key enzyme required for the G2/M-phase transition in both mitosis and meiosis. MPF is composed of a regulatory subunit, cyclin B, and a catalytic subunit, Cdc2. Activated MPF phosphorylates cellular machinery components involved in nuclear envelope breakdown, chromosome condensation, spindle assembly, and cyclin degradation, thus controlling both mitosis and meiosis. Basically, Cdc2 expression remains relatively constant, while the cyclin B concentration fluctuates periodically throughout the cell cycle. Although the mRNA expression of *cdc2* did not differ between spermatocytes of testes from heat-shocked mice and those from control animals [96], Cdc2 protein is fragile and susceptible to heat. Transient heat stress was reported to induce a swift decline in the Cdc2 protein level before the germ cell apoptosis was detected by the TUNEL assay [96]. Cyclin B accumulates during the interphase and peaks at the G2/M-phase transition. The association of Cdc2 with newly synthesized cyclin B initiates a series of alterations in the phosphorylation status of Cdc2, which is required for the activation of MPF. Activated MPF triggers the entry of the cell into the M phase, and cyclin B is degraded. This degradation leads to Cdc2 kinase inactivation, finally causing the cell to exit M phase [97] (Fig. 8.5).

In another study, heat stress was reported to cause altered functions of all cell cycle-related molecules. Small ubiquitin-like modifier (SUMO) proteins have been implicated in the cellular response. Heat or oxidative stress resulted in significant changes in the levels of testicular SUMO1 and SUMO2/3 conjugates, and topoisomerase 2α is one of the targets of sumoylation [98]. HSP70-2, a testes-specific gene, is expressed in spermatogenic cells during the prophase of meiosis I in male germ cells [99]. It has an essential chaperone function during the G2/M-phase transition of the meiotic phase of spermatogenesis where it is required for cyclin-dependent kinase (Cdc2) activation

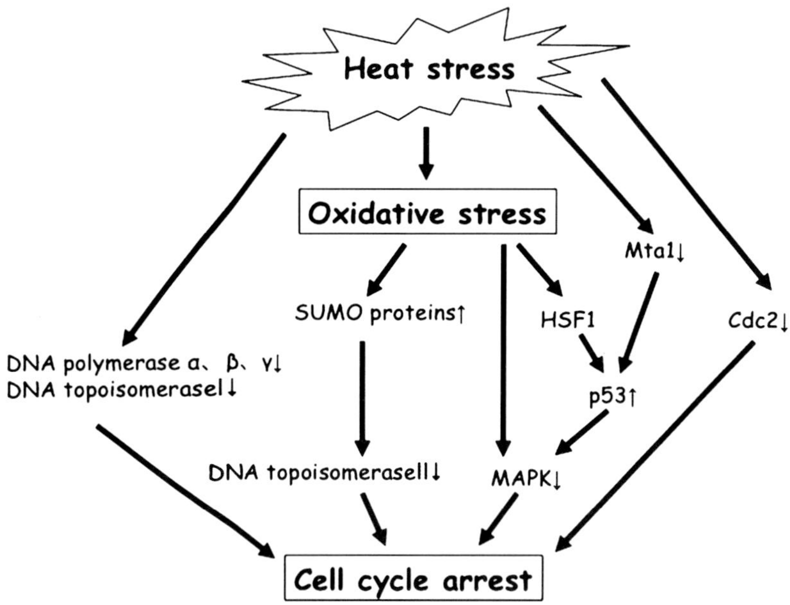

Fig. 8.5 Pathways to cell cycle arrest resulting from heat-induced oxidative stress in germ cells. These pathways are mainly reported in spermatogonia and primary spermatocytes. The *upward arrow* indicates increased expression and/or activity of the molecule. The *downward arrow* indicates decreased expression and/or activity of the molecule. Pathways indicated using *arrows* are proven in germ cells. If no association with oxidative stress (e.g., Cdc, DNA polymerases) is shown, their function and expression are regulated by oxidative stress in other cells

and Cdc2-cyclin B1 assembly [100]. Decrease in the HSP70-2-mediated CDC kinase activity, which occurs due to heat stress, has also been demonstrated in Se-induced oxidative stress. Involvement of p53 in germ cell apoptosis has already been described [101]. The presence of p53 mRNA and protein in primary spermatocytes [76] suggests that p53 plays a role in the meiotic prophase. The induced expression of p53 might be due to its increased stabilization and half-life, as suggested earlier, in response to DNA damage and oxidative stress [102]. This increased expression explains the possibility of apoptosis as well as cell cycle arrest at the G2/M-phase checkpoint following oxidative stress-mediated DNA damage (Fig. 8.5).

8.7 Heat-Induced Antioxidative Machinery and Thermotolerance in the Testis

The testis has high rates of metabolism and cell proliferation; therefore, oxidative stress can be especially damaging. Thus, the antioxidant capacity of testicular tissue is very important. Generally, cells become equipped with heat-tolerance machinery after nonlethal heat stress and are able to survive against subsequent heat stresses. This phenomenon is termed thermotolerance [103]. If the second heat attack is substituted with oxidative stress, the cells are capable of surviving even after the second attack, indicating that the initial heat stress induces a range of mechanisms to protect cells against stress. A similar phenomenon has been reported in the case of ischemia/reperfusion injury of the testis. In this case, oxidative stress plays a major role to induce ischemic preconditioning, which protects the cells from the second ischemic attack [104]. The mechanisms of ischemic preconditioning and induced thermotolerance after heat stress may be common because oxidative stress also plays a major role in ischemia/reperfusion. However, the detailed molecular mechanisms involved in acquiring this thermotolerance machinery after heat stress have not been fully investigated. Furthermore, elevation of scrotal temperature with no increase in the oxidative stress in fertile men with varicocele indicates disturbances in the oxidative stress scavenging system in infertile men with varicocele [105].

The testis contains several antioxidants that serve to protect germ cells from oxidative damage. There are two types of mechanisms by which germ cells are protected from oxidative stress: enzymatic and nonenzymatic. The primary antioxidant enzymes in mammals are superoxide dismutase (SOD), glutathione peroxidase (GPX), glutathione S-transferase (GST), and heme oxygenase 1 (HMOX1) [106, 107]. The heme oxygenase system plays an important role in protecting cells from the deleterious effects of oxidative stress and consists of the heme oxygenase proteins-inducible HMOX1 and constitutive HMOX2 [108]. There is a distinct possibility that SOD plays a role in the male reproductive system [109, 110]. Germ cells appeared to be highly susceptible to ROS and died without expressing SOD [111]. Furthermore, polymorphism in GST T1 and M1 predicts the response to varicocelectomy, indicating the importance of antioxidants in the process of human spermatogenesis [112]. Proteomic studies have demonstrated that heat stress generally decreases gene expression. Taken together, the results of the four studies, which employed a proteomics approach, indicate gene alterations in the testes after heat stress [6, 88, 113, 114]. Further, two studies have demonstrated alterations in gene expressions in testes after oxidative stress [6, 88]. Upregulated genes include heme oxygenase 1 (accession no.: AA213167), oxidative stress-induced protein (accession no.: U40930), and GST P1 (accession no.: P19157). Downregulated genes include extracellular SOD Cu-Zn (SOD3) (accession no.: U38261), GSR 1 (accession no.: X76341), GST α2 (accession no.: J03958), GST Pi1 (accession no.: D30687), microsomal GST 12 (accession no.: J03752), and thioredoxin-like protein 2 (accession no.: Q9CQM9). Generally, these proteins reduce oxidative stress. The expression of antioxidants is observed in somatic cells and, sometimes, in spermatogonia [109]. Germ cells, in general, did not appear to use the antioxidant enzymatic system as the major equipment to reduce ROS damage and, consequently, were considerably more susceptible to oxidative stress than somatic cells [106]. In other words, germ cells except for spermatogonia were not equipped to induce thermotolerance. On the other hand, spermatogonia constitute a heterogeneous population containing type A, intermediate, and type B spermatogonia, and spermatogonial stem cells [45]. Taken together, since the number or biological activity of spermatogonial stem cells can be effectively obtained by heat stress [115], spermatogonia are considered to be heat-resistant cells. Although the constitutive expression of antioxidants is present, higher antioxidant expression is usually inducible. Generally, effective antioxidant expression in the testes has been shown after 6 h of heat stress [108], which is apparently longer than the period of induction of germ cell apoptosis.

The heme oxygenase (HO) system is recognized as a member of the small HSP family and comprises two HO isozymes [116]. These two isozymes are distinct gene products; the HO-1 isozyme is known as HSP32, as it is heat and stress-inducible, whereas the HO-2 isozyme is the constitutively expressed cognate gene. In the testis, HO-2 is expressed in maturing germ cells, spermatocytes, round spermatids, occasionally in Leydig cells, as well as in the residual bodies [117]. HO-1 mRNA is present in the nucleus of the germ cell progenitor and in Sertoli and Leydig cells under both normal and stress conditions; the protein is present only in nonspermatogenic cell types. The intensity of HO-1 immunostaining induced by oxidative stress is markedly increased in Sertoli and Leydig cells in rats [108] and humans [118]. The robust expression of HO-1 in Leydig cells following heat stress suggests its role in maintaining steroidogenesis by generating antioxidants. Although these antioxidants are expressed in interstitial tissue or Sertoli cells, gaseous molecules such as ROS are homogenously distributed within the testis, and the antioxidants, irrespective of their localization, may exert protective effects on germ cells from oxidative stress.

Nonenzymatic antioxidant molecules include α-tocopherol (vitamin E), β-carotene (vitamin A), ascorbate (vitamin C), glutathione, estrogens, creatine (related to carotene); flavonoids (aromatic oxygen heterocyclic compounds widely distributed in higher plants), resveratrol (a botanical antioxidant), metallothionein (cadmium-binding protein involved in heavy metal detoxification), taurine (an aminosulfonic acid) and its precursors, and other thiols such as nonstructural polyunsaturated lipids and melatonin. Oral administration of vitamin E reduced the level of testicular-free radicals in a rat experimental model of varicocele [119], and the effectiveness of antioxidants to treat infertile men has been previously demonstrated in randomized trials [120]. Therefore, antioxidants can be effectively applied to treat heat-inducible oxidative stress in humans.

Ghrelin, an endogenous ligand for the growth hormone (GH) secretagogue receptor, has been primarily linked to the central neuroendocrine regulation of GH secretion and food intake. Further, its actions on reproductive organs have also been reported, for example, steroidogenic dysfunction is associated with increased ghrelin expression in human testes [121]. Recently, ghrelin has been suggested to be an endogenous antioxidant that functions as a free radical scavenger [122]. Ghrelin inhibits apoptosis by increasing the Bcl-2/Bax ratio, preventing release of cytochrome *c*, inhibiting ROS generation, stabilizing mitochondrial transmembrane potential, and inhibiting the activation of caspase 3 [123]. Moreover, ghrelin was able to promote antioxidant enzyme activity (particularly that of GPx) and reduce lipid peroxidation in the testis. Taken together with the finding that luteinizing hormone/human choriogonadotropic hormone (LH/hCG) induces an enzymatic system with known antioxidant properties in Leydig cells [124], it can be said that the testicular antioxidant system is also regulated by hormonal stimulation.

8.8 Heat-Shock Factors and Heat-Shock Proteins in the Germ Line Against Heat Stress

Heat-shock response is characterized by the induction of a set of HSPs and is a fundamental response in all organisms to protect themselves from heat attack [103]. The HSP family comprises molecular chaperones that regulate protein folding and degradation under stress conditions to maintain cellular homeostasis and survival. These HSPs are expressed at euthermic body temperatures (approximately 37°C) and are found to have distinct locations and functional properties under conditions of stress. In addition to heat-induced oxidative stress, a number of other stimuli are known to induce HSPs, including energy depletion, hypoxia, acidosis, ischemia/reperfusion, and viral infection. The HSP70 protein was first reported to be induced in response to heat shock in isolated germ cells [125].

Studies of oxidative stress suggest that HSP promoter activity and protein accumulation may be uncoupled [126]. Data from previous studies demonstrate that both transcriptional and posttranscriptional regulatory steps are required for HSP production [126, 127]. Increased temperature and stress can activate HSPs, which are regulated by HSFs [128]. The expression of most HSPs, including testis-specific isoforms, is generally high, but controversy exists regarding whether heat-shock genes are activated in germ cells in response to heat shock [129]. Numerous studies have demonstrated that the cytoprotective functions of HSPs can be largely attributed to the suppression of apoptosis [130, 131], whereas several HSPs have the opposite functions in the germ cells.

A unique pattern of HSP70 expression has been revealed during spermatogenesis in mice and rats. HSP70 and HSP90 were expressed in germ cells, Sertoli cells, and Leydig cells in the testes during neonatal and early developing stages and in spermatocytes and round spermatids after puberty. Further, the HSP90 protein was faintly expressed in spermatogonia during this period. In the degenerative condition, all HSP proteins were markedly expressed in germ cells after heat stress [132]. A previous study has demonstrated high HSP70-2 expression during meiosis in mice [133]. An HSP related to HSP70, Hsc70t, is expressed in late spermatids during spermiogenesis [134]. Hsc70t has been shown to be essential for spermatogenesis [135]. In contrast, HSP60 was expressed in Leydig cells [132, 136] during the neonatal and prepuberty stages, and in spermatogonia and primary spermatocytes [132, 137], in the mitochondria of spermatogonia and primary spermatocytes in stages I–V and IX–XIV [136], and mature spermatozoa [137]. HSP70-1, HSP110, and HSP27 mRNAs were highly induced in interstitial Leydig cells after heat shock, whereas these mRNAs were hardly expressed in cells within seminiferous tubules containing germ cells and somatic Sertoli cells [52]. In the rat testis, HSP105 was found to form complexes with p53 in the cytoplasm of germ cells at scrotal temperatures, while it dissociated from such complexes under heat-shock conditions such as spermatocyte apoptosis [138]. Germ cell death appeared to include p53-dependent mechanisms, and p53 is essential to maintain cellular integrity and appropriate number of germ cells during spermatogenesis [72]. Thus, one might speculate that the increased expression of HSP105 can affect the functional status of p53 in the testes of transgenic males and induce spermatocyte apoptosis. Among the small HSPs, HSP32 (HO-1) is mainly expressed in Leydig cells in response to heat stress [107, 117]. In human testes, HSP27 expression was strong in the cytoplasm of Sertoli cells, spermatogonia, and Leydig cells; moderate in the spermatocytes; weak in the spermatids; and absent in the spermatozoa [139]. HSP27 and HSP90 are mainly present in the cytoplasm of Sertoli cells, spermatogonia, spermatocytes, and spermatids, and their expressions in the nucleus were found to increase under heat stress [140]. HSP25 was neither expressed in germ cells nor somatic cells on all the days examined. HSP20 was strongly expressed in the heart and slightly in the testis of a 9-week-old rat, and the expression was localized in spermatocytes and round spermatids [141]. However, the precise role and regulation of HSPs in the testis is unknown.

In mammals, the expression of classical HSPs is regulated by heat-shock factor-1 [142]. Four HSF genes (*HSF1*, *HSF2*, HSF3, and *HSF4*) have been identified in mammals [142–144]. Apparently heat stress is not the only stimulus to activate HSF. HSF1 is ubiquitously expressed and is the most effective transactivator of stress-induced HSP expression. HSF1 associates with multiple proteins during its activation and inactivation processes, suggesting that HSF activity is regulated at multiple levels [142]. H_2O_2 targets HSF1-mediated transcription, but not through inhibition of HSF1-binding ability [145]. HSF1 remains as a monomer in both the cytoplasm and nucleus in unstressed cells, but is converted to a trimer that can bind to a DNA sequence motif, the heat-shock element (HSE), which is translocated into the nucleus in response to heat shock [129] and induces a robust activation of heat-shock genes [146, 147]. This HSF1-mediated induction of HSP expression is required for protection of cells from various pathophysiological conditions such as neurodegenerative and other degenerative diseases and confers lifespan extension and thermotolerance [146, 147]. During the past few years, comprehensive analysis has revealed numerous HSFs, particularly HSF1, whose targets include HSP and non-HSP genes, the expressions of which are regulated differentially by HSF family members. HSFs are expressed in germ cells and studies using animal models have revealed their roles in spermatogenesis [50, 148]. HSF2 is found mainly as a dimer in unstressed cells, and for the most part, neither forms a trimer nor is translocated into the nucleus during heat shock. HSF2 is considered to function in development. In fact, HSF2 is associated with development of the brain and reproductive organs [148]. Recently, HSF2 was shown to positively regulate proteasome activity, leading to the decreased expression of p53 [149]. HSF3 represents a unique HSF that has the potential to activate only nonclassical heat-shock genes to protect the cells from detrimental stress [150]. HSF4 is found as a trimer in the nucleus in unstressed cells as it uniquely lacks the HR-C domain that suppresses trimer formation [143]. HSF4 expression is ubiquitous in all cell types. The roles of HSF2, HSF3, and HSF4 on spermatogenesis are currently under investigation. As a germ cell-specific HSF, the HSFY gene is localized in the AZFb region of the Y chromosome, whose deletion causes a severe alteration in spermatogenesis [151], but the regulation of this gene by heat and oxidative stress is unknown. Ferlin et al. reported that testicular

HSP90, HSPA4, HSF1, HSF2, and HSFY were transcriptionally upregulated in the presence of varicocele [152]. Taken together with the fact that varicocele increases testicular ROS [8, 54, 75, 91, 118, 153], these expressions are perhaps regulated, at least in part, by heat-induced oxidative stress.

Generally, HSF1 overexpression protects cells from heat stress-induced cell death, whereas germ cells actively die through apoptosis [50, 52]. However, in a previous study, it was observed that neither heat-induced HSF1 activation [52] nor active HSF1 overexpression results in expression of HSP70 in spermatogenic cells [50, 70]. Although HSF1 acts as a cell-survival factor in premeiotic germ cells [52], expression of a constitutively active form of HSF1 in the spermatocytes of transgenic mice promotes germ cell apoptosis [50, 52], but it is logical to think about the character of the germ cells. In addition, Widlak et al. reported that heat-activated HSF1 paradoxically downregulated the expression of HSP 70.2 and promotes apoptosis (Fig. 8.4) [154]. HSF1-induced apoptosis in spermatocytes might be a result of the lack of induction of the antiapoptotic protein, inducing HSP70i, which has previously been shown to confer resistance to apoptosis in somatic cells [70]. In the testis, apoptosis is common and is believed to play an important role in controlling the germ cell population and eliminating defective germ cells to produce functional spermatozoa. Thus, apoptosis protects the testis against harmful stress and functions to preserve future spermatogenesis, and a study using an ischemia/reperfusion model reported that inhibition of apoptosis by caspase inhibitors does not ameliorate spermatogenesis [79].

8.9 Biological Effects of Heat-Induced Oxidative on Sertoli and Leydig Cells and Blood–Testis Barrier

Sertoli and Leydig cells do not obviously undergo apoptosis after mild heat treatment (e.g., 43°C, 15 min), although heat stress reduces the activity of Sertoli and Leydig cells followed by changes in testicular endocrine function [46, 47]. In somatic cells, heat treatment induces the accumulation of the HSPs and antioxidants, and thus leads to cell survival. Data from both bulls and boars indicate that heat stress causes an initial decline in circulating concentrations of testosterone lasting 2 weeks, but the concentrations recover even in the presence of continued heat stress [155, 156]. If left untreated through adulthood, bilateral cryptorchidism significantly lowers serum testosterone levels in patients as compared to unilateral cryptorchidism, indicating that testosterone production from Leydig cells is also impaired, at least in part, by heat stress [157]. An in vitro experiment has shown that ROS disrupts mitochondria in Leydig cells and inhibits steroidogenic acute regulatory (StAR) protein and steroidogenesis [46]. Apoptosis of Sertoli cells is not observed after exposure to heat stress [158], whereas heat-induced germ cell death is usually accompanied by alterations in Sertoli cell morphology and function [159]. Adult Sertoli cells have reverted to a dedifferentiated state in the cryptorchid testis, and thus lost their supportive role in normal spermatogenesis, leading to a cessation of spermatogenic activity [84].

After heat stress, increased expression of cytokeratin 18, liver receptor homologue-1 [84, 160], and intermediate filaments [159] and decreased expression of androgen receptor (AR), and junction-associated proteins such as zonula occludens-1 (ZO-1) and occludin [161] are observed in the testis of adult monkeys and rat Sertoli cells. Sertoli cell-specific androgen receptor ablation increases permeability of the blood–testis barrier, implying that androgens regulate the permeability of the blood–testis barrier and that scrotal heat stress induces Sertoli–Sertoli cell junction disruption [162]. A study investigating the effect of heat stress on the blood–testis barrier has shown that the biotin tracer is detected in the adluminal compartment of the testis 2 days after heat treatment; however, in the control testis, the tracer is limited to the blood–testis barrier, and no biotin is detected in the luminal compartment [163]. Heat-induced alterations in tight-junction proteins and damage to the blood–testis barrier are transient, and they recover approximately 10 days after heat treatment. Oxidative stress is involved in heat-induced interstitial damage. For example, interstitial fibrosis and tight-junction disruption are mediated by increased expression of transforming growth factor-beta generated by oxidative stress and are ameliorated by antioxidants [74, 164].

These findings suggest that not only germ cells but also Sertoli and Leydig cells are affected by heat treatment, and that the affected Sertoli cells lose their supportive roles in spermatogenesis.

8.10 Measurement Human Testicular Temperature and Its Interpretation

Scrotal structure is well adapted to serve a local thermoregulatory function. Scrotal skin is thin and is supplied with a large number of sweat glands. There is also little or no subcutaneous fat or connective tissue. The tunica dartos and cremasteric muscles vary in degree of contraction in response to changes in environmental temperature, altering the characteristics of the scrotum itself. In addition, the temperature within the testis is regulated by a countercurrent heat-exchange system between the pampiniform plexus and the testicular artery.

Scrotal temperature measured at the skin surface [23, 165, 166] can change with posture or clothing. Normal scrotal surface temperature (the external surface of the scrotum) is approximately 34°C in a normally clothed man walking about or maintaining a loose stance. Scrotal surface temperature differs from testicular temperature, and testicular temperature within the scrotum is estimated to be between 0.1 and 0.6°C higher than the scrotal surface temperature [24, 167]. Controversies exist regarding the absolute value of the scrotal temperature, because of the different modalities of its measurement. Contact thermometers measure the temperature at a depth of 5–10 mm from the skin [168, 169]. Measurement of deep-body temperature

(CoreTemp CTM204®, Terumo, Tokyo) reflects testicular temperature, but not scrotal surface temperature [169]. Yamaguchi et al. reported the usefulness of deep-body temperature to evaluate patients with varicocele and have shown that deep-body temperature increased in a standing position in patients with varicocele, but decreased in a standing position in men without varicocele [168]. In addition, mean testicular temperature of men without varicocele is reported to be 35.7°C, which is 1–2°C higher than the scrotal surface temperature, but still lower than the core abdominal temperature. Involvement of heat stress in the pathophysiology of varicocele, cryptorchidism, or environmental heat exposure is determined by this narrow 2–3°C difference. This means that completely different mechanisms will be observed in animal studies in which transient heat exposure is applied. Furthermore, the impact of other factors such as the pattern of heat exposure (cryptorchidism is continuous and varicocele is intermittent) may also be significant.

Actually, 43°C for 15 min is a good experimental condition to assess the effect of heat on spermatogenesis: heat-induced molecular events occur from minutes to hours after heat exposure, deterioration of spermatogenesis continues for several days to weeks, and then spermatogenesis recovers after 1–2 months. Experimental animals are bred under strict temperature control, which is supposed to minimize the effects of heat stress. On the other hand, humans are usually exposed to environmental stress, including changes in scrotal temperature, all day and all night, and these heat stresses are quite different in each man. Scrotal temperature was measured using a continuous recording system [166, 170, 171] and has since been applied in several studies. Lerch et al. found no difference in mean scrotal temperature between the right and left scrotum in six normal, fertile men, irrespective of whether the temperature was measured during the day or night. In contrast, for scrotal temperature measured over 24 h in 10 fertile men, Jung et al. reported a median value of 34.99°C on the right and 35.29°C on the left side [24]. In a recent study, the same authors found that mean scrotal temperature was significantly higher on the left (35.56°C) than on the right (35.37°C) in 50 clothed, male volunteers, who did not have a history of infertility and had normal results on clinical examination [166].

In humans, testicular temperature may increase as a result of occupational exposure, lifestyle, or a clinical disorder [4]. For example, occupational exposure can occur in men who work in high-temperature environments, such as bakers and welders [172], and also occupations that involve long periods in a sedentary position, such as professional drivers. Recent studies have also reported that posture and clothing can increase the scrotal temperature [172]. Clinical disorders, including cryptorchidism, where one or both testes fail to descend into the scrotum and remain in the abdominal cavity, can also result in exposure of the testes to higher than normal temperatures. Oxidative stress, which is caused by numerous factors, including elevated temperature, is shown to be widely involved in the pathophysiologies of male infertility [8]. Increased oxidative stress in human testes with varicocele has been reported, as shown by elevations of MDA [152], 8-OHdG [173], and 4-HNE-modified proteins [54, 75, 91]. One of the enigmatic issues in human studies is that elevation of testicular temperature does not necessarily result in harmful effects on spermatogenesis. Actually, when elevation of testicular temperature is observed in fertile men with varicocele, testicular oxidative stress is low [105], suggesting that the generation of oxidative stress and expression of antioxidants differ among men. In humans, a more comprehensive understanding is needed to evaluate heat stress, which includes the degree of temperature, its duration, its interval, and basal body temperature.

8.11 Conclusion

The effect of heat stress on spermatogenesis has been investigated for several decades. As described in this chapter, the mode of heat stress is very different in in vitro experiments, animal experiments, and human studies. Therefore, close attention must be paid to recognize the cells exposed and the type of heat stress. Conventionally, in vitro and animal experiments do not fit the pathophysiology of human beings. Further investigation in this area will reveal more insights regarding the mechanisms that cause male infertility, new therapies, and germ cell culture for in vitro fertilization and in vitro spermatogenesis. Oxidative stress is involved in a variety of diseases that cause male infertility. There is increasing evidence supporting the effectiveness of antioxidants in treating male infertility. Further research in this area can help develop specific treatment for heat-induced oxidative stress as opposed to nonspecific antioxidant therapy.

References

1. Setchell BP. The Parkes Lecture. Heat and the testis. J Reprod Fertil. 1998;114:179–94.
2. Ivell R. Lifestyle impact and the biology of the human scrotum. Reprod Biol Endocrinol. 2007;5:e15.
3. Yaeram J, Setchell BP, Maddocks S. Effect of heat stress on the fertility of male mice in vivo and in vitro. Reprod Fertil Dev. 2006;18:647–53.
4. Mieusset R, Bujan L, Mondinat C, et al. Association of scrotal hyperthermia with impaired spermatogenesis in infertile men. Fertil Steril. 1987;48:1006–11.
5. Jannes P, Spiessens C, van der Auwera I, et al. Male subfertility induced by acute scrotal heating affects embryo quality in normal female mice. Hum Reprod. 1998;13:372–5.
6. Rockett JC, Mapp FL, Garges JB, et al. Effects of hyperthermia on spermatogenesis, apoptosis, gene expression, and fertility in adult male mice. Biol Reprod. 2001;65:229–39.
7. Zhu BK, Setchell BP. Effects of paternal heat stress on the in vivo development of preimplantation embryos in the mouse. Reprod Nutr Dev. 2004;44:617–29.
8. Turner TT, Lysiak JJ. Oxidative stress: a common factor in testicular dysfunction. J Androl. 2008;29:488–98.
9. Yu BP. Cellular defense against damage from reactive oxygen species. Physiol Rev. 1994;74:139–62.
10. Alvarez JG, Storey BT. Spontaneous lipid peroxidation in rabbit and mouse epididymal spermatozoa: dependence of rate on temperature and oxygen concentration. Biol Reprod. 1985;32:342–51.
11. Ahutopa M, Huhtaniemi I. Impaired detoxification of reactive oxygen and consequent oxidative stress in experimentally cryptorchid rat testis. Biol Reprod. 1992;46:1114–8.

12. Peltola V, Huhtaniemi I, Ahotupa M. Abdominal position of the rat testis is associated with high level of lipid peroxidation. Biol Reprod. 1995;53:1146–50.
13. Ikeda M, Kodama H, Fukuda J, et al. Role of radical oxygen species in rat testicular germ cell apoptosis induced by heat stress. Biol Reprod. 1999;61:393–9.
14. Cividalli A, Cruciani G, Livdi E, et al. Hyperthermia enhances the response of paclitaxel and radiation in a mouse adenocarcinoma. Int J Radiat Oncol Biol Phys. 1999;44:407–12.
15. Shiraishi K, Naito K, Yoshida K. Nitric oxide promotes germ cell necrosis in the delayed phase after experimental testicular torsion of rat. Biol Reprod. 2001;65:514–21.
16. Iuchi Y, Kaneko T, Matsuki S, et al. Carbonyl stress and detoxification ability in the male genital tract and testis of rats. Histochem Cell Biol. 2004;121:123–30.
17. Bonfoco E, Krainc D, Ankarcrona M, et al. Apoptosis and necrosis: two distinct events induced, respectively, by mild and intense insults with *N*-methyl-D-aspartate or nitric oxide/superoxide in cortical cell cultures. Proc Natl Acad Sci USA. 1995;92:7162–6.
18. Guo J, Jia Y, Tao S-X, et al. Expression of nitric oxide synthase during germ cell apoptosis in testis of Cynomolgus monkey after testosterone and heat treatment. J Androl. 2009;30:190–9.
19. Shiratsuchi A, Umeda M, Ohba Y, et al. Recognition of phosphatidylserine on the surface of apoptotic spermatogenic cells and subsequent phagocytosis by Sertoli cells of the rat. J Biol Chem. 1997;272:2354–8.
20. Lue Y, Wang C, Liu YX, et al. Transient testicular warming enhances the suppressive effect of testosterone on spermatogenesis in adult cynomolgus monkeys (Macaca fascicularis). J Clin Endocrinol Metab. 2006;91:539–45.
21. Shiraishi K, Naito K. Nitric oxide produced in the testis is involved in dilatation of the internal spermatic vein that compromises spermatogenesis in infertile men with varicocele. BJU Int. 2007;99:1086–90.
22. Ishikawa T, Kondo Y, Goda K, et al. Overexpression of endothelial nitric oxide synthase in transgenic mice accelerates testicular germ cell apoptosis induced by experimental cryptorchidism. J Androl. 2005;26:281–8.
23. Mieusset R, Bujan L. Testicular heating and its possible contribution to male infertility: a review. Int J Androl. 1995;18:169–84.
24. Jung A, Eberl M, Schill WB. Improvement of semen quality by nocturnal scrotal cooling and moderate behavioural change to reduce genital heat stress in men with oligoasthenoteratozoospermia. Reproduction. 2001;121:433–8.
25. Lin P, Quamo S, Ho K, et al. Hyperthermia enhances the cytotoxic effects of reactive oxygen species to Chinese hamster and bovine endothelial cells in vitro. Radiat Res. 1991;126:43–51.
26. Flanagan SW, Moseley PL, Buettner GR. Increased flux of free radicals in cells subjected to hyperthermia: detection by electron paramagnetic resonance spin trapping. FEBS Lett. 1998;431:285–6.
27. Loven DP. A role for reduced oxygen species in heat-induced cell killing and the induction of thermotolerance. Med Hypotheses. 1988;26:39–50.
28. Lord-Fontaine S, Averil-Bates DA. Enhancement of cytotoxicity of hydrogen peroxide by hyperthermia in Chinese hamster ovary cells: role of antioxidant defenses. Arch Biochem Biophys. 1999;363:283–95.
29. Lindegaard JC. Thermosensitization induced by step-down heating. A review on heat-induced sensitization to hyperthermia alone or hyperthermia combined with radiation. Int J Hyperthermia. 1992;8:561–86.
30. Dewey WC. Arrhenius relationships from the molecule and cell to the clinic. Int J Hyperthermia. 2009;25:3–20.
31. Sinha Hikim AP, Lue Y, Yamamoto CM, et al. Key apoptotic pathways for heat-induced programmed cell death in the testis. Endocrinology. 2003;144:3167–75.
32. Sinha-Hikim AP, Lue Y, Diaz-Romero M, et al. Deciphering the pathways of germ cell apoptosis in the testis. J Steroid Biochem Mol Biol. 2003;85:175–82.
33. Vera Y, Rodriguez S, Castanares M, et al. Functional role of caspases in heat-induced testicular germ cell apoptosis. Biol Reprod. 2005;72:516–22.

34. Choi JY, Jo MW, Lee EY, et al. The role of autophagy in follicular development and atresia in rat granulosa cells. Fertil Steril. 2010;93:2532–7.
35. Tsukamoto S, Kuma A, Murakami M, et al. Autophagy is essential for preimplantation development of mouse embryos. Science. 2008;321:117–20.
36. Chowdhury A, Steinberger E. A quantitative study of the effect of heat on germinal epithelium of rat testes. Am J Anat. 1964;115:509–24.
37. Pérez-Crespo M, Pintado B, Gutiérrez-Adán A. Scrotal heat stress effects on sperm viability, sperm DNA integrity, and the offspring sex ratio in mice. Mol Reprod Dev. 2008;75:40–7.
38. Paul C, Murray AA, Spears N, et al. A single, mild, transient scrotal heat stress causes DNA damage, subfertility and impairs formation of blastocysts in mice. Reproduction. 2008; 136:73–84.
39. Paul C, Teng S, Saunders PT. A single, mild, transient scrotal heat stress causes hypoxia and oxidative stress in mouse testes, which induces germ cell death. Biol Reprod. 2009;80:913–9.
40. Yin Y, Hawkins KL, DeWolf WC, et al. Heat stress causes testicular germ cell apoptosis in adult mice. J Androl. 1997;18:159–65.
41. Yamamoto CM, Sinha Hikim AP, Huynh PN, et al. Redistribution of bax is an early step in an apoptotic pathway leading to germ cell death in rats, triggered by mild testicular hyperthermia. Biol Reprod. 2000;63:1683–90.
42. Cataldo L, Mastrangelo MA, Kleene KC. Differential effects of heat shock on translation of normal mRNAs in primary spermatocytes, elongated spermatids, and Sertoli cells in seminiferous tubule culture. Exp Cell Res. 1997;231:206–13.
43. Perez-Crespo M, Pintado B, Gutierrez-Adan A. Scrotal heat stress effects on sperm viability, sperm DNA integrity, and the offspring sex ratio in mice. Mol Reprod Dev. 2007;75:40–7.
44. Hadziselimovic F, Herzog B. The importance of both an early orchidopexy and germ cell maturation for fertility. Lancet. 2001;358:1156–7.
45. de Rooij DG, Okabe M, Nishimune Y. Arrest of spermatogonial differentiation in jsd/jsd, Sl17H/Sl17H, and cryptorchid mice. Biol Reprod. 1999;61:842–7.
46. Diemer T, Allen JA, Hales KH, et al. Reactive oxygen disrupts mitochondria in MA-10 tumor Leydig cells and inhibits steroidogenic acute regulatory (StAR) protein and steroidogenesis. Endocrinology. 2003;144:2882–91.
47. Ren L, Medan MS, Ozu M, et al. Effects of experimental cryptorchidism on sperm motility and testicular endocrinology in adult male rats. J Reprod Dev. 2006;52:219–28.
48. Hendricks KEM, Martins L, Hansen PJ. Consequences for the bovine embryo of being derived from a spermatozoan subjected to post-ejaculatory aging and heat shock: development to the blastocyst stage and sex ratio. J Reprod Dev. 2009;55:69–74.
49. Zhu BK, Setchell BP. Effects of paternal heat stress on the in vivo development of preimplantation embryos in the mouse. Reprod Nutr Dev. 2004;44:617–29.
50. Nakai A, Suzuki M, Tanabe M. Arrest of spermatogenesis in mice expressing an active heat shock transcription factor 1. EMBO J. 2000;19:1545–54.
51. Ohta H, Aizawa S, Nishimune Y. Functional analysis of the p53 gene in apoptosis induced by heat stress or loss of stem cell factor signaling in mouse male germ cells. Biol Reprod. 2003;68:2249–54.
52. Izu H, Inoue S, Fujimoto M, et al. Heat shock transcription factor 1 is involved in quality control mechanism in male germ cells. Biol Reprod. 2004;70:18–24.
53. Ku JH, Shim HB, Kim SW, et al. The role of apoptosis in the pathogenesis of varicocele. BJU Int. 2005;96:1092–6.
54. Shiraishi K, Takihara H, Matsuyama H. Elevated scrotal temperature, but not varicocele grade, reflects testicular oxidative stress-mediated apoptosis. World J Urol. 2010;28: 359–64.
55. Lue YH, Sinha Hikim AP, Swerdloff RS, et al. Single exposure to heat induces stage-specific germ cell apoptosis in rats: role of intratesticular testosterone on stage specificity. Endocrinology. 1999;140:1709–17.
56. Sakahira H, Enari M, Nagata S. Cleavage of CAD inhibitor in CAD activation and DNA degradation during apoptosis. Nature. 1998;391:96–9.

57. Shaha C, Tripathi R, Mishra DP. Male germ cell apoptosis: regulation and biology. Philos Trans R Soc Lond B Biol Sci. 2010;365:1501–15.
58. Lee J, Richburg JH, Shipp EB, et al. The Fas system, a regulator of testicular germ cell apoptosis, is differentially up-regulated in Sertoli cell versus germ cell injury of the testis. Endocrinology. 1999;140:852–8.
59. Crompton M. The mitochondrial permeability transition pore and its role in cell death. Biochem J. 1999;341:233–49.
60. Chance B, Sies H, Boveris A. Hydroperoxide metabolism in mammalian organs. Physiol Rev. 1979;59:527–605.
61. Nagata S, Golstein P. The Fas death factor. Science. 1995;267:1449–56.
62. Hengartner MO. The biochemistry of apoptosis. Nature. 2000;407:770–6.
63. Cory S, Adams JM. The Bcl-2 family: regulators of the cellular life-or-death switch. Nat Rev Cancer. 2002;2:647–56.
64. Maheshwari A, Misro MM, Aggarwal A, et al. Pathways involved in testicular germ cell apoptosis induced by H_2O_2 in vitro. FEBS J. 2009;276:870–81.
65. Jia Y, Castellamos J, Wang C, et al. Migotgen-activated protein kinase signaling in male germ cell apoptosis in the rat. Biol Reprod. 2009;80:771–80.
66. Johnson C, Jia Y, Wang C, et al. Role of caspase 2 in apoptotic signaling in primate and murine germ cells. Biol Reprod. 2008;79:806–14.
67. Basile A, Biziato D, Sherbet VG, et al. Hyperthermia inhibits cell proliferation and induces apoptosis: relative signaling status of P53, S100A4, and notch in heat sensitive and resistant cell lines. J Cell Biol. 2008;103:212–20.
68. Yamamoto CM, Sinha Hikim AP, Huynh PN, et al. Redistribution of Bax is an early step in an apoptotic pathway leading to germ cell death in rats, triggered by mild testicular hyperthermia. Biol Reprod. 2000;63:1683–90.
69. Shiraishi K, Naito K, Yoshida K. Vasectomy impairs spermatogenesis through germ cell apoptosis mediated by the p53-Bax pathway in rats. J Urol. 2001;166:1565–71.
70. Vydra N, Malusecka E, Jarzab M, et al. Spermatocyte-specific expression of constitutively active heat shock factor 1 induces HSP70i-resistant apoptosis in male germ cells. Cell Death Differ. 2006;13:212–22.
71. Yin Y, DeWolf WC, Morgentaler A. Experimental cyptorchidism induces testicular germ cell apoptosis by p53-dependent and -independent pathways in mice. Biol Reprod. 1998;58:492–6.
72. Yin Y, Stahl BC, DeWolf WC, et al. p53-mediated germ cell quality control in spermatogenesis. Dev Biol. 1998;204:165–71.
73. Yin Y, Stahl BC, DeWolf WC, et al. P53 and Fas are sequential mechanism of testicular germ cell apoptosis. J Androl. 2002;23:64–70.
74. Shiraishi K, Yoshida K, Fujimiya T, et al. Angiotensin II dependent testicular fibrosis and effects on spermatogenesis after vasectomy in the rat. J Urol. 2003;170:2104–8.
75. Shiraishi K, Naito K. Effects of 4-hydroxy-2-nonenal, a marker of oxidative stress, on spermatogenesis and expression of p53 protein in male infertility. J Urol. 2007;178:1012–7.
76. Schwartz D, Goldfinger N, Rotter V. Expression of p53 protein in spermatogenesis is confined to the tetraploid pachytene primary spermatocytes. Oncogene. 1993;8:1487–94.
77. Li W, Bao W, Ma J, et al. Metastasis tumor antigen 1 is involved in the resistance to heat stress-induced testicular apoptosis. FEBS Lett. 2008;582:869–73.
78. Sorimachi H, Ishiura S, Suzuki K. Structure and physiological function of calpains. Biochem J. 1997;328:721–32.
79. Shiraishi K, Naito K, Yoshida K. Inhibition of calpain but not caspase protects the testis agaisnt injury after experimental testicular torsion of rat. Biol Reprod. 2000;63:1538–48.
80. Li FJ, Kondo T, Zhao QL, et al. Enhancement of hyperthermia-induced apoptosis by a free radical initiator, 2,2′-azobis(2-amidinopropane)dihydrochloride, in human histiocytic lymphoma U937 cells. Free Radic Res. 2001;35:281–99.
81. Lizama C, Lagos CF, Lagos-Cabré R, et al. Calpain inhibitors prevent p38 MAPK activation and germ cell apoptosis after heat stress in pubertal rat testes. J Cell Physiol. 2009; 221:296–305.

82. Lu T, Xu Y, Mericle M, et al. Participation of conventional calpains in apoptosis. Biochim Biophys Acta. 2002;1590:16–26.
83. Somwaru L, Li S, Doglio L, et al. Heat-induced apoptosis of mouse meiotic cells is suppressed by ectopic expression of testis-specific calpastatin. J Androl. 2004;25:506–13.
84. Zhang XS, Zhang ZH, Jin X, et al. Dedifferentiation of adult monkey Sertoli cells through activation of extracellularly regulated kinase 1/2 induced by heat treatment. Endocrinology. 2006;147:1237–45.
85. Reuter S, Gupta SC, Chaturvedi MM, et al. Oxidative stress, inflammation, and cancer: how they linked? Free Radic Biol Med. 2010;49:1603–16.
86. Adams SH. Uncoupling protein homologs: emerging views of physiological function. J Nutr. 2000;130:711–4.
87. Zhang K, Shang Y, Liao S, et al. Uncoupling protein 2 protects testicular germ cells from hyperthermia-induced apoptosis. Biochem Biophys Res Commun. 2007;360:327–32.
88. Zhu H, Cui Y, Xie J, et al. Proteomic analysis of testis biopsies in men treated with transient scrotal hyperthermia revealed the potential targets for contraceptive development. Proteomics. 2010;10:3480–93.
89. Chowdhury DR, Small C, Wang Y, et al. Microarray-based analysis of cell-cycle gene expression during spermatogenesis in the mouse. Biol Reprod. 2010;83:663–75.
90. Tanaka H, Fujisawa M, Okada H, et al. Assessment of germ-cell kinetics in the testes of patients with varicocele using image analysis of immunostained proliferating cell nuclear antigen. Br J Urol. 1996;78:769–71.
91. Shiraishi K, Naito K. Generation of 4-hydroxy-2-nonenal modified proteins in testis predicts improvement in spermatogenesis after varicocelectomy. Fertil Steril. 2006;86:233–5.
92. Fujisawa M, Matsumoto O, Kamidono S, et al. Changes of enzymes involved in DNA synthesis in the testes of cryptorchid rats. J Reprod Fertil. 1988;84:123–30.
93. Fujisawa M, Hayashi A, Okada H, et al. Enzymes involved in DNA synthesis in the testes are regulated by temperature in vitro. Eur Urol. 1997;31:237–42.
94. Blow JJ, Tanaka TU. The chromosome cycle: coordinating replication and segregation. Second in the cycles review series. EMBO Rep. 2005;6:1028–34.
95. Wolgemuth DJ, Laurion E, Lele KM. Regulation of the mitotic and meiotic cell cycles in the male germline. Recent Prog Horm Res. 2002;57:75–101.
96. Zhang Y, Yang X, Cao H, et al. Heat stress induces Cdc2 protein decrease prior to mouse spermatogenic cell apoptosis. Acta Histochem. 2008;110:276–84.
97. King RW, Jackson PK, Kirschner MW. Mitosis in transition. Cell. 1994;79:563–71.
98. Shrivastava V, Peker M, Grosser E, et al. SUMO proteins are involved in the stress response during spermatogenesis and are localized to DNA double-strand breaks in germ cells. Reproduction. 2010;139:999–1010.
99. Zakeri ZF, Wolgemuth DJ, Hung CR. Identification and sequence analysis of a new member of the mouse HSP70 gene family and characterization of its unique cellular and developmental pattern of expression in the male germ line. Mol Cell Biol. 1988;8:2925–32.
100. Eddy EM. Role of heat shock protein HSP70-2 in spermatogenesis. Rev Reprod. 1994;4:23–30.
101. Kaushal N, Bansal MP. Dietary selenium variation-induced oxidative stress modulates CDC2/cyclin B1 expression and apoptosis of germ cells in mice testis. J Nutr Biochem. 2007;18:553–64.
102. Chandel NS, Vander Heiden MG, Thompson CB, et al. Redox regulation of p53 during hypoxia. Oncogene. 2000;19:3840–8.
103. Lindquist S. The heat-shock response. Annu Rev Biochem. 1986;55:1151–91.
104. Shimizu S, Saito M, Kinoshita Y, et al. Ischemic preconditioning and post-conditioning to decrease testicular torsion-detorsion injury. J Urol. 2009;182:1637–43.
105. Shiraishi K, Takihara H, Naito K. Testicular volume, scrotal temperature, and oxidative stress in fertile men with left varicocele. Fertil Steril. 2008;91:1388–91.
106. Bauche F, Fouchard M, Jegou B. Antioxidant system in rat testicular cells. FEBS Lett. 1994;349:392–6.

107. Gu W, Hecht NB. Developmental expression of glutathione peroxidase, catalase, and manganese superoxide dismutase mRNAs during spermatogenesis in the mouse. J Androl. 1996;17:256–62.
108. Maines MD, Ewing JF. Stress response of the rat testis: in situ hybridization and immunohistochemical analysis of heme oxygenase-1 (HSP32) induction by hyperthermia. Biol Reprod. 1996;54:1070–9.
109. Mruk DD, Silvestrini B, Mo MY, et al. Antioxidant superoxide dismutase-a review: its function, regulation in the testis, and role in male fertility. Contraception. 2002;65:305–11.
110. Fujii J, Iuchi Y, Matsuki S, et al. Cooperative function of antioxidant and redox systems against oxidative stress in male reproductive tissues. Asian J Androl. 2003;5:231–42.
111. Ishii T, Matsuki S, Iuchi Y, et al. Accelerated impairment of spermatogenic cells in SOD1-knockout mice under heat stress. Free Radic Res. 2005;39:697–705.
112. Okubo K, Nagahama K, Kamoto T, et al. GSTT1 and GSTM1 polymorphisms are associated with improvement in seminal findings after varicocelectomy. Fertil Steril. 2005;83:1579–80.
113. Zhu Y-F, Cui Y-G, Guo X-J, et al. Proteomic analysis of effect of hyperthermia on spermatogenesis in adult male mice. J Proteome Res. 2006;5:2217–25.
114. Li YC, Hu XQ, Xiao LJ, et al. An oligonucleotide microarray study on gene expression profile in mouse testis of experimental cryptorchidism. Front Biosci. 2006;11:2465–82.
115. McLean DJ, Russell LD, Griswold MD. Biological activity and enrichment of spermatogonial stem cells in vitamin A-deficient and hyperthermia-exposed testes from mice based on colonization following germ cell transplantation. Biol Reprod. 2002;66:1374–9.
116. Mines MD, Trakshel GM, Kutty RK. Characterization of two constitutive forms of rat liver microsomal heme oxygemase. Only one molecular form is inducible. J Biol Chem. 1986;261:411–9.
117. Ewing JF, Maines MD. Distribution of constitutive (HO-2) and heat inducible (HO-1) heme oxygenase isozymes in rat testis: HO-2 display stage-specific expression in spermatocytes. Endocrinology. 1995;136:2294–302.
118. Shiraishi K, Naito K. Increased expression of Leydig cell heme oxygenase-1 preserves spermatogenesis in varicocele. Hum Reprod. 2005;20:2608–13.
119. Cam K, Simsek F, Yuksel M, et al. The role of reactive oxygen species and apoptosis in the pathogenesis of varicocele in a rat model and efficiency of vitamin E treatment. Int J Androl. 2004;27:228–33.
120. Ross C, Morriss A, Khairy M, Khalaf Y, et al. A systemic review of the effect of oral antioxidants on male infertility. Reprod Biomed Online. 2010;20:711–23.
121. Ishikawa T, Fujioka H, Ishimura T, et al. Ghrelin expression in human testis and serum testosterone level. J Androl. 2006;28:320–4.
122. Dong MH, Kaunitz JD. Gastroduodenal mucosal defense. Curr Opin Gastroenterol. 2006;22:599–606.
123. Chung H, Kim E, Lee DH, et al. Ghrelin inhibits apoptosis in hypothalamin neuronal cells during oxygen-glucose deprivation. Endocrinology. 2007;148:148–59.
124. Piotrkowski B, Monzón CM, Pagotto RM, et al. Effects of heme oxygenase isozymes on Leydig cells steroidogenesis. J Endocrinol. 2009;203:155–65.
125. Allen RL, O'Brien DA, Jones CC, et al. Expression of heat shock proteins by isolated mouse spermatogenic cells. Mol Cell Biol. 1988;8:3260–6.
126. Wu C. Heat shock transcription factors: structure and regulation. Annu Rev Cell Dev Biol. 1995;11:441–69.
127. Mizzen LA, Welch WJ. Characterization of the thermotolerant cell. I. Effects on protein synthesis activity and the regulation of heat-shock protein 70 expression. J Cell Biol. 1988;106:1105–16.
128. Feder ME, Hofmann GE. Heat-shock proteins, molecular chaperones, and the stress response: evolutionary and ecological physiology. Annu Rev Physiol. 1999;61:243–82.
129. Sarge KD, Murphy SP, Morimoto RI. Activation of heat shock gene transcription by heat shock factor 1 involves oligomerization, acquisition of DNA-binding activity, and nuclear localization and can occur in the absence of stress. Mol Cell Biol. 1993;13:1392–407.

130. Beere HM. The stress of dying: the role of heat shock proteins in the regulation of apoptosis. J Cell Sci. 2004;117:2641–51.
131. Sreedhar AS, Csermely P. Heat shock proteins in the regulation of apoptosis: new strategies in tumor therapy: a comprehensive review. Pharmacol Ther. 2004;101:227–57.
132. Ogi S, Tanji N, Iseda T, et al. Expression of heat shock proteins in developing and degenerating rat testes. Arch Androl. 1999;43:163–71.
133. Rosario MO, Perkins SL, O'Brien DA, et al. Identification of the gene for the developmentally expressed 70 kDa heat-shock protein (P70) of mouse spermatogenic cells. Dev Biol. 1992;150:1–11.
134. Tsunekawa N, Matsumoto M, Tone S, Nishida T, Fujimoto H. The Hsp70 homolog gene, Hsc70t, is expressed under translational control during mouse spermiogenesis. Mol Reprod Dev. 1999;52:383–91.
135. Dix DJ. Hsp70 expression and function during gametogenesis. Cell Stress Chaperones. 1997;2:73–7.
136. Meinhardt A, Parvinen M, Bacher M, et al. Expression of mitochondrial heat shock protein 60 in distinct cell types and defined stages of rat seminiferous epithelium. Biol Reprod. 1995;52:798–807.
137. Lachance C, Fortier M, Thimon V, et al. Localization of Hsp60 and Grp78 in the human testis, epididymis and mature spermatozoa. Int J Androl. 2010;33:33–44.
138. Kumagai J, Fukuda J, Kodama H, et al. Germ cell-specific heat shock protein 105 binds to p53 in a temperature-sensitive manner in rat testis. Eur J Biochem. 2000;267:3073–8.
139. Adly MA, Assaf HA, Hussein MR. Heat shock protein 27 expression in the human testis showing normal and abnormal spermatogenesis. Cell Biol Int. 2008;32:1247–55.
140. Biggiogera M, Tanguay RM, Marin R, et al. Localization of heat shock proteins in mouse male germ cells: an immunoelectron microscopic study. Exp Cell Res. 1996;25:77–85.
141. Yamano Y, Ohyama K, Ohta M, et al. Expression of small stress protein hsp20 gene in the maturing rat testis. J Vet Med Sci. 2005;67:1181–4.
142. Morimoto RI. Regulation of the heat shock transcriptional response: cross talk between a family of heat shock factors, molecular chaperones, and negative regulators. Genes Dev. 1998;12:3788–96.
143. Nakai A, Tanabe M, Kawazoe Y, et al. HSF4, a new member of the human heat shock factor family which lacks properties of a transcriptional activator. Mol Cell Biol. 1997;17:469–81.
144. Nakai A. New aspect in the vertebrate heat shock factor system: HSF3 and HSF4. Cell Stress Chaperones. 1999;4:86–93.
145. Adachi M, Liu Y, Fujii K, et al. Oxidative stress impairs the heat stress response and delays unfolded protein recovery. PLoS One. 2009;4:e7719.
146. McMillan DR, Xiao X, Shao L, et al. Targeted disruption of heat shock transcription factor 1 abolishes thermotolerance and protection against heat-inducible apoptosis. J Biol Chem. 1998;273:7523–8.
147. Zhang Y, Huang L, Zhang J, et al. Targeted disruption of hsf1 leads to lack of thermotolerance and defines tissue-specific regulation for stress-inducible Hsp molecular chaperones. J Cell Biochem. 2002;86:376–93.
148. Wang G, Zhang J, Moskophidis D, et al. Targeted disruption of the heat shock transcription factor (hsf)-2 gene results in increased embryonic lethality, neuronal defects, and reduced spermatogenesis. Genesis. 2003;36:48–61.
149. Lecomte S, Desmots F, Le Masson F, et al. Roles of heat shock factor 1 and 2 in response to proteasome inhibition: consequence on p53 stability. Oncogene. 2010;29:4216–24.
150. Fujimoto M, Hayashida N, Katoh T, et al. A novel mouse HSF3 has the potential to activate nonclassical heat-shock genes during heat shock. Mol Biol Cell. 2010;21:106–16.
151. Ferlin A, Arredi B, Speltra E, et al. Molecular and clinical characterization of Y chromosome microdeletions in infertile men: a 10-year experience in Italy. J Clin Endocrinol Metab. 2007;92:762–70.
152. Ferlin A, Speltra E, Patassini C, et al. Heat shock protein and heat shock factor expression in sperm: relation to oligozoospermia and varicocele. J Urol. 2010;183:1248–52.

153. Koksal IT, Tefekli A, Usta M, et al. The role of reactive oxygen species in testicular dysfunction associated with varicocele. BJU Int. 2000;86:549–52.
154. Widlak W, Vydra N, Malusecka E, et al. Heat shock transcription factor 1 down-regulates spermatocyte-specicic 70 kDa heat shock protein expression prior to the induction of apoptosis in mouse testes. Genes Cells. 2007;12:487–99.
155. Rhynes WE, Ewing LL. Testicular endocrine function in Hereford bulls exposed to high ambient temperature. Endocrinology. 1973;92:509–15.
156. Wettemann RP, Desjardins C. Testicular function in boars exposed to elevated ambient temperature. Biol Reprod. 1979;20:235–41.
157. Chiba K, Ishikawa T, Yamaguchi K, et al. The efficacy of adult orchiopexy as a treatment of male infertility: our experience of 20 cases. Fertil Steril. 2009;92:1337–9.
158. Matsuki S, Iuchi Y, Ikeda Y, et al. Supression of cytochrome c release and apoptosis in tetes with heat stress by minocycline. Biochem Biophys Res Commun. 2003;312:843–9.
159. Zhang ZH, Hu ZY, Song XX, et al. Disrupted expression of intermediate filaments in the testis of rhesus monkey after experimental cryptorchidism. Int J Androl. 2004;27:234–9.
160. Guo J, Tao SX, Chen M, et al. Heat treatment induces liver receptor homolog-1 expression in monkey and rat Sertoli cells. Endocrinology. 2007;148:1255–65.
161. Chen M, Cai H, Yang JL, et al. Effect of heat stress on expression of junction-associated molecules and upstream factors androgen receptor and Wilms' tumor 1 in monkey Sertoli cells. Endocrinology. 2008;149:4871–82.
162. Meng J, Holdcraft RW, Shima JE, et al. Androgens regulate the permeability of the blood–testis barrier. Proc Natl Acad Sci USA. 2005;102:16696–700.
163. Cai H, Ren Y, Li X-X, et al. Scrotal heat stress causes a transient alteration in tight junctions and induction of TGF-β expression. Int J Androl. 2011;34:352–62.
164. Siu ER, Mruk DD, Porto CS, et al. Cadmium-induced testicular injury. Toxicol Appl Pharmacol. 2009;238:240–9.
165. Takada T, Kitamura M, Matsumiya K, et al. Infrared thermometry for rapid, noninvasive detection of reflux of spermatic vein in varicocele. J Urol. 1996;156:1652–4.
166. Jung A, Leonhardt F, Schill WB, et al. Influence of the type of undertrousers and physical activity on scrotal temperatures. Hum Reprod. 2005;20:1022–7.
167. Hjollund NH, Storgaard L, Ernst E, et al. The relation between daily activities and scrotal temperature. Reprod Toxicol. 2002;16:209–14.
168. Yamaguchi M, Sakatoku J, Takihara H. The application of intrascrotal deep body temperature measurement for the noninvasive diagnosis of varicoceles. Fertil Steril. 1989;52:295–301.
169. Takihara H, Yamaguchi M, Baba Y, et al. Deep body intrascrotal thermometer: theory and methodology. Adv Exp Med Biol. 1991;286:115–9.
170. Bengoudifa B, Mieusset R. Thermal asymmetry of the human scrotum. Hum Reprod. 2007;22:2178–82.
171. Jockenhövel F, Gräwe A, Nieschlag E. A portable digital data recorder for long-term monitoring of scrotal temperatures. Fertil Steril. 1990;54:694–700.
172. Thonneau P, Bujan L, Multigner L, et al. Occupational heat exposure and male fertility: a review. Hum Reprod. 1998;13:2122–5.
173. Ishikawa T, Fujioka H, Ishimura T, et al. Increased testicular 8-hydroxy-2′-deoxyguanosine in patients with varicocele. BJU Int. 2007;100:863–6.

Chapter 9
Cytokines and Oxidative Stress in the Germ Line

Monika Fraczek, Anna Czernikiewicz, and Maciej Kurpisz

Abstract Cytokines are important mediators of the immunologic response and involved in numerous physiological and pathological processes in the male genital tract. The same cytokines that act as elements of immunomodulation for the male gonad appear in large concentrations in semen in a number of pathological conditions, including autoimmune diseases, spinal cord injury, varicocele, or genital tract infection/inflammation. The activated macrophages and neutrophils release reactive oxygen intermediates and secrete proinflammatory cytokines, both of which can affect spermatozoa through peroxidative processes of sperm membrane components and DNA. Elucidation of these mechanisms and their interactions can be critical to develop novel diagnostic tests and treatment of male genital tract infection/inflammation. This chapter covers the current evidence about a relationship among the cytokines, antioxidants, prooxidants, and semen parameters.

Keywords Cytokines • Oxidative stress • Germ line • Spermatogenesis • Steroidogenesis • Inflammatory conditions in the gonad • Semen inflammation

9.1 Introduction

The main functions of testis are spermato- and steroidogenesis. Spermatogenesis is hormonally controlled by pituitary gonadotropins: folliculotropic hormone (FSH) and luteinizing hormone (LH) and locally produced—testosterone [1, 2]. A lot of evidence suggests that somatic cells of testis (the Sertoli, the Leydig, peritubular cells) in physiological conditions produce cytokines, such as interleukin (IL)-1,

---

M. Fraczek, PhD • A. Czernikiewicz, MSc • M. Kurpisz, MD, PhD (✉)
Institute of Human Genetics, Polish Academy of Sciences, Department of Reproductive Biology and Stem Cells, Strzeszynska 32, 60-479 Poznan, Poland
e-mail: kurpimac@man.poznan.pl

A. Agarwal et al. (eds.), *Studies on Men's Health and Fertility*, Oxidative Stress in Applied Basic Research and Clinical Practice, DOI 10.1007/978-1-61779-776-7_9,

IL-6, and nonprotein mediators of inflammation, e.g., nitric oxide (NO), which take part in the initiation of spermatogenesis and maturation of the semen [3–5].

9.2 Cytokines and Spermatogenesis

Testes are considered to be an immunoprivileged organ due to their tolerance of autoantigens secreted during sexual maturity by reproductive cells. This phenomenon supports spermatogenesis [6]. The mechanisms which protect testes against autoimmune diseases are immunological/anatomical blood–testis barrier (BTB) which protects against the antigenic leakage out of germ cells to immunological, intraparenchymal cells and against the transition of antibodies from endothelium to the lumen of seminiferous tubules; secretion of immunosuppressive factors by macrophages, the Sertoli, Leydig, and peritubular cells; and a limited presence of activated T lymphocytes (particularly that of $CD8^+$) and the presence of regulatory T_{regs} lymphocytes. Maintenance of the balance between inflammation and "immunoprivileged" gonad belongs, among other, to the function of cytokines which perform both the roles as proinflammatory mediators and immunosuppressive ones [7].

BTB consists of vascular endothelium, basal lamina of seminiferous tubules, and specialized tight junctions (TJs) between Sertoli cells. BTB creates microenvironment in which spermatogenesis takes place: meiosis, spermiogenesis, and spermiation. BTB ensures polarization of Sertoli cells and regulates intercellular transfer of water, ions, feeding substances, and biomolecules to germ cells from circulation. BTB regulation is ensured by activated TGFβ and tumor necrosis factor (TNF)-α, mitogen-activated protein kinase (MAPK), and c-Jun N-terminal kinase signaling pathways [8–11].

Ectoplasmic specialization (ES) is a junction located in two testicular areas: among postmeiotic spermatids and Sertoli cells of seminiferous tubules—apical ES is involved in migration of preleptotenic spermatocytes in stage VIII of seminiferous epithelium through BTB itself and between BTB and Sertoli cells—basal ES [12] functions with tight junctions, desmosomes, and gap junctions that are components of BTB [12–14].

Cytokines do not only play a role at directing germ cells to apoptosis, but TNF-α, TGF-β2, and TGF-β3 together with testosterone regulate spermatogenesis [9, 15].

Receptors for cytokines, i.e., TGF-β and TNF, are located on Sertoli cells which regulate migration of germ cells through the barrier to adluminal compartment [15]. One of the physiological roles of cytokines is to restructure BTB junctions and apical ES during migration of preleptotenic spermatocytes [16]. Restructuring of junctions in seminiferous epithelium depends on the changes in protein phosphorylation [17].

Cytokines may influence apical ES as well as the whole BTB through altering protein levels present in biological membranes as N-cadherin in the case of apical ES and occludin in BTB. In consequence of their action is an increase in protein endocytosis and intracellular endosome-mediated degradation [18].

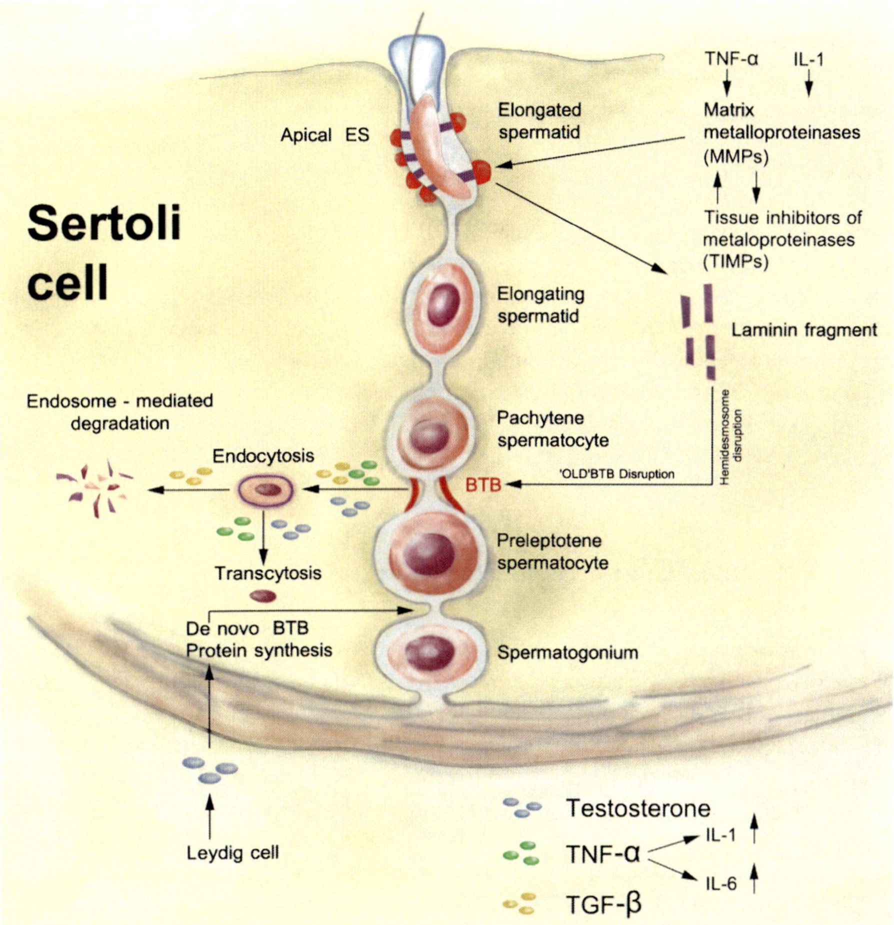

Fig. 9.1 The apical ES–blood–testis barrier (BTB)–hemidesmosome axis. Matrix proteinase-2 uncouples integrin–laminin complex and releases biologically active laminin protein fragments, which induce disruption of the "old" BTB proteins and also hemidesmosomes. The disrupted junction proteins undergo endocytosis. The endocytosed proteins are subject of endosome-mediated degradation with transforming growth factor (TGF) participation. These proteins can also be transcytosed in order to assembly new BTB in the presence of testosterone and tumor necrosis factor (TNF)

In opposition to cytokine action testosterone increases BTB integrity and enhances protein production as elements of Sertoli cells tight junctions essential for functioning and barrier stability [19, 20] (Fig. 9.1).

Testosterone and cytokines demonstrate antagonistic action on integrity of BTB within seminiferous proteins of BTB system and stabilize BTB tight junctions (Fig. 9.1). However, only testosterone directs proteins after endocytosis to renew their utilization on the membrane surface of Sertoli cells, induces "de novo" synthesis of integral proteins of BTB membrane, and promotes transcytosis through the protein relocation from the "old" to the "new" BTB side. Cytokines after endocytosis induce degradation of "old" TJ fibrils. TNF-α function is different from the remaining ones and acts on androgenic receptor, and promotes testosterone activity. In this way, transport of preleptotenic spermatocytes through BTB barrier takes place [9, 15, 21].

Coordination of restructuring of apical ES and BTB during spermatogenesis takes place at participation of apical ES–BTB–hemidesmosome functional axis. Reaction cascade begins before spermiation through action of metalloproteinase-2 which cleaves complex integrin–laminin releasing fragments of active laminin that may cause cytokine synthesis, influence endocytosis kinetics, recycling, transcytosis, or protein degeneration. It is suggested that metalloproteinase-2 influences the induction of BTB restructuring by TNF mediation. TNF also acts on production of metalloproteinase-9 in Sertoli cells which alters functioning of tight BTB junctions through cleavage of collagen [22–24].

9.3 Cytokines Influence on Steroidogenesis

Cytokines take part in gonadal steroidogenesis (Fig. 9.2). In testis, they are physiologically produced by Sertoli cells and Leydig cells and they regulate cell proliferation as well as steroidogenesis. However, increased concentrations of cytokines may adversely influence cell function. For example, IL-1α may negatively influence expression of regulatory protein steroidogenic acute regulator (StAR) as well as its phosphorylation, thus inhibiting cholesterol translocation to inner mitochondrial membrane. In physiological conditions, IL-1 stimulates Leydig cells to secretion of progesterone and testosterone [25, 26].

The Leydig cells show mRNA expression for both IL-1α and IL-1β in in vitro conditions as a response to the lipopolysaccharides (LPSs) or exogenous IL-1 activity [27, 28]. It is suggested that IL-1 inhibits the LH-dependent synthesis of testosterone by the Leydig cells, which is caused by the decrease of hCG-dependent synthesis of cAMP, lowering of the level of P450scc enzyme [29], and inhibition of P450c17 and 3β-HSD enzymes [30]. During inflammation, the presence of both isoforms of interleukin 1 may be found, also in the interstitial fluid of the testes, where they regulate the immunological response occurring in the gonad [31, 32].

The inhibition of testosterone production can be caused also by the direct influence of IL-2 on the Leydig cells as well as indirect way through activation of TNF-α and IL-1 secreted by macrophages found in the gonad. IL-2 stimulates the hypothalamic-pituitary-adrenal axis resulting in the increase in the ACTH and glycocorticosteroid levels, which in turn may have an inhibitory influence on steroidogenesis in the Leydig cells [30].

LPSs activate macrophages and monocytes, which in turn secrete TNF-α. In physiological conditions, the presence of mRNA for TNF-α is observed in spermatids and pachytene spermatocytes as well as in interstitial macrophages of testes [33]. In the Leydig cells, TNF-α causes inhibition of testosterone production

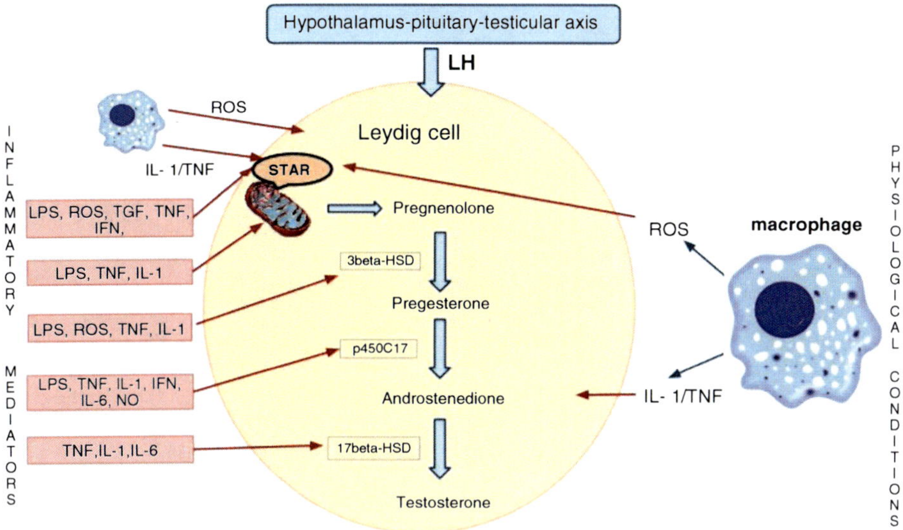

Fig. 9.2 Influence of inflammatory mediators on Leydig cell and testosterone synthesis. Inflammatory mediators, like ROS, lipopolysaccharides (LPSs), NO, or cytokines, inhibit testosterone production in Leydig cells because of their activity against steroidogenic acute regulatory (StAR) protein, 3β-hydrosteroid dehydrogenase (3β-HSD), P450c17, or 17β-hydroxysteroid dehydrogenase

stimulated by AMP through a decrease of mRNA for the cytochrome P450 enzymes: splitting the cholesterol side-chain cleavage enzyme [34] and 7α-hydroxylase/C_{17-20} lyase [33].

The Leydig cells secrete IFNs, which are in the first line of defense against viruses. They are the only cells of the gonad which produce both types of interferons (α and β). In high concentration, IFNα has an inhibitory influence on the testosterone production stimulated by LH in the Leydig cells through inhibition of expression of the StAR protein [35] and enzymes P450scc and P450c17.

The proteins of the TGFβ family are secreted in the gonad in high concentrations by the Sertoli cells, peritubular cells, and the Leydig cells. Most often, they are found in an inactive form and activation is provoked by the activity of local proteases. In physiological conditions, TGFβs function as regulators of the Leydig and reproductive cells proliferation, the Leydig cells steroidogenesis, and the testes development [2, 36, 37].

9.4 Inflammatory Conditions in the Gonad, Oxidative Stress

The oxidative stress and lipid peroxidation, which occur in the gonad, are significant factors responsible for male infertility arising from the imbalance between production of reactive forms of oxygen and the efficiency of systems which eliminate

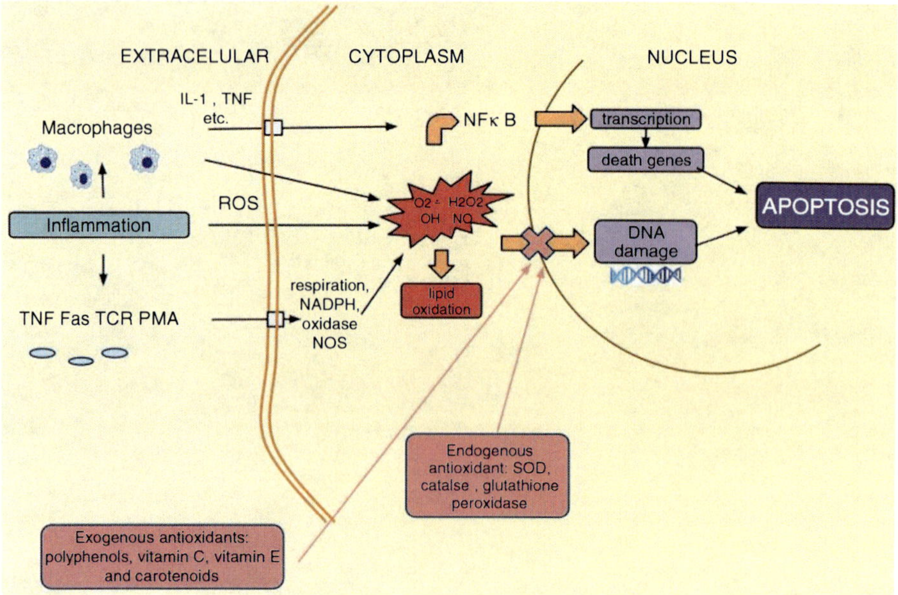

Fig. 9.3 Oxidative stress as a link between inflammation and apoptosis. The inflammatory state induces activation of macrophages and release of cytokines leading to free radicals (ROS). ROS leads to damage of plasma membranes, DNA, and other cell structures. Inflammation causes transfer of NF-κB to cell nucleus and trigger transcription of "death genes" and apoptosis or pro-proliferative genes (cytokines)

their harmful influence. To a group of oxygen radicals belong superoxide anion radical, hydroxyl radical, nitrogen oxide, and peroxide nitrite. ROS are formed as a by-product of the mitochondrial and microsomal electron transport as well as through the influence of cytochrome P450 enzymes on cholesterol and its metabolites. They are highly reactive compounds and may react with proteins, lipids, and DNA causing their damage which is prevented by the enzymatic antioxidative system, mainly superoxide dismutase (SOD), catalase, and glutathione peroxidase (GPx). However, in inflammation, the antioxidative protection is insufficient compared with the amount of radicals produced and this causes greater damage to the cell structures [38, 39] (Fig. 9.3).

Oxidative stress arises, among others, due to the exposure of the gonads to toxins, chemotherapy, ionizing radiation, bacterial inflammation, cryptorchidism, or pathology of veins in the spermatic cord (varicocele) [39]. The Leydig cells are physiologically and sometimes physically connected with macrophages, which makes possible their direct exposure to the growth factors and/or differentiation factors which are secreted by macrophages. It was shown that in physiological conditions macrophages produce and secrete 25-hydroxycholesterol which stimulates the Leydig cells to produce testosterone [40, 41]. In inflammatory conditions, however, macrophages are activated by, e.g., LPS and produce proinflammatory cytokines, such as IL-1 or TNF-α and also ROS. Therefore, beginning of the inflammatory reaction may have an influence on spermatogenesis through inhibition of steroidogenesis. LPS conditions determine an increase in the NO production by endothelial nitric oxide synthase (eNOS) in the endoepithelial cells and induce iNOS expression in macrophages. The increase in the NO concentration influences steroidogenesis in the Leydig cells in vivo; however, this mechanism has yet to be studied and has not been well recognized [41, 42].

The StAR protein takes part in the binding and transport of cholesterol to the mitochondrial matrix and, thus, steroid production. These processes depend on the mitochondrial electrochemical gradient.

Studies on MA-10 tumor Leydig cells showed that ROS may cause disturbances of the mitochondrial electrochemical gradient, and through this inhibition of cholesterol import by StAR and a decrease of the testosterone synthesis which lead to apoptosis of reproductive cells [43, 44].

The Sertoli cells protect seminiferous tubules against autoantigens and pathogens. Cytokines IL-1 and IL-6 are secreted under the influence of the inflammatory factor, such as LPS, which induces in vitro inflammatory conditions in the Sertoli cells [1, 45, 46], and ROS activity causes peroxidation of the cell membranes, which due to the high content of polyunsaturated fatty/lipid acids are susceptible to the influence of free radicals [45]. The Sertoli cells synthesize GSH, which is the source of amino acids in the process of spermatogenesis, and provides protection to spermatogenic cells against ROS [47, 48]. Reactive forms of oxygen are the reason of peroxidative damage to the sperm, which may cause infertility. Apart from the influence on peroxidation of the cell membrane lipids, ROS cause damage to DNA and may lead to apoptosis through DNA fragmentation. A large number of all divisions to which reproductive cells in the seminiferous tubules are submitted to during spermatogenesis expose their chromosomes to the harmful influence of ROS [48].

9.5 Repertoire of Cytokines in Seminal Plasma

In the male gonad, cytokines are produced physiologically and are involved in the normal function of the organ [30, 49, 50]. In this respect, they must also appear as natural components of seminal plasma [51]. A network of cytokines, chemokines, and growth factors, such as interleukins (ILs), IL-1α and -1β, IL-2, IL-4, IL-5, IL-6, IL-7, IL-8, IL-10, IL-11, IL-12, IL-13, IL-17, and IL-18, their soluble receptors and antagonists (e.g., IL-1RA, sR IL-2, sR IL-6), TNF-α, transforming growth factor (TGF) family of cytokines (TGF-α and -β), interferon gamma (IFN-γ), granulocyte colony-stimulating factor (G-SCF), granulocyte-macrophage colony-stimulating factor (GM-SCF), and macrophage inflammatory proteins alpha (MIP-1α) and beta (MIP-β), was shown to be present in human semen [51, 52]. The nature, precise origin, and regulation of the cytokines in the male genital tract are still under investigation. The main sources of cytokines occurring in the male reproductive tract are testes and, among others, testicular macrophages, although some cytokines (IL-1,

IL-6) are also produced by somatic cells in testis, including the Leydig and Sertoli cells [27]. However, some studies also indicated local production of these factors in the secondary sex glands irrespective of spermatogenesis [53, 54]. Cytokines appear to be also secreted by the epididymis and the prostate gland [55]. Prostate gland, for example, seems to be the main site of origin of IL-6 and IL-2 in the seminal plasma [54, 56–58]. Moreover, the significant correlation between IL-6 and fructose levels observed suggests that the seminal vesicles can be also included in the production of this cytokine [58]. According to some authors, in the male genital system, some cytokines (IL-1 and IL-6) can be produced and secreted not only by somatic cells, but also by germ cells [59, 60]. It is possible that cytokines may act not only as an autocrine factor, but also in a paracrine manner during spermatogenesis, sperm maturation, sperm transport, and even during the fertilization process [61]. Finally, cytokines are released by various immunocompetent cells present in the male genital tract, which are the major source of these factors produced in response to foreign antigens, pathogens, and also in semen inflammation [62]. A lot of in vivo studies on the effects of cytokines and growth factors on sperm function have provided somewhat controversial results. On the one hand, cytokines and other immune factors are intrinsically involved in normal reproductive physiology, but on the other hand their local or systemic perturbations underlie pathophysiology of sperm function. A cytokine network may play an important role in immunoregulatory effects in human semen of both fertile and infertile men. There are several topics of cytokine research in the male reproductive system: (1) presence of various cytokines in human seminal plasma, (2) demonstration of differences in cytokine concentrations between fertile and infertile men, (3) presence of negative correlations between cytokine levels and semen parameters, (4) demonstration of correlations between cytokine levels and leukocyte count, and (5) usefulness of some cytokines as clinical markers of male infertility.

9.6 Relationship Between Cytokines and Seminal Parameters

Human semen contains a variety of different cytokines and other immunological factors and their effects on semen quality and sperm function are still the subject of debate. There is increasing evidence that many of the cytokines adversely affect spermatogenesis and steroidogenesis [28]. However, there is an ongoing controversy concerning the biological role of particular cytokines in the fertilization process. A lot of studies have indicated the lack of any connection between the cytokine levels and semen quality or infertility status [53, 54, 58, 63–65]. However, a number of authors point out the observed relationship between the cytokine hypersecretion and the deterioration of semen parameters or the fertilizing ability of sperm [56, 66–78] (Table 9.1). In these studies, high levels of certain cytokines in the semen were related to the quality of semen parameters, such as sperm count, motility and morphology, or sperm–oocyte penetration rates. Also in a few in vitro studies, an inhibitory effect of some recombinant cytokines has been found in respect to sperm

Table 9.1 List of connections between various proinflammatory cytokine levels in seminal plasma and semen quality (including oxidative stress parameters) demonstrated in in vivo as well as in vitro conditions

Proinflammatory cytokine	Participation in inflammatory reaction	Consequences on semen quality	References
IL-1β	Proliferation and differentiation of B cells, chemoattraction of leukocytes to the site of inflammation, induction of neutrophils and monocytes generation, induction of apoptosis	↓ Sperm-progressive motility	[67]
		↑ ROS, ↑ MDA (in vitro conditions)	[111]
IL-2	Stimulation of growth and differentiation of T, B, NK, LAK cells, monocytes, macrophages	↓ Sperm count, motility, and morphology	[66]
IL-6	Activation and differentiation of B and T cells	↓ Sperm count	[67, 69, 74, 76, 77]
		↓ Progressive motility	[67, 69, 76, 77]
		↓ Morphology	[69]
		↓ Sperm vitality	[77]
		↑ MDA	[136]
		↑ Leukocyte count	[58, 69, 77]
		↑ ROS, ↓ TAC	[92]
IL-8	Chemoattraction of neutrophils to the site of inflammation, activation of neutrophils to phagocytosis	↓ Sperm motility	[72]
		↑ ROS	[105]
		↑ MDA (in vitro conditions)	[106, 137]
		↑ Leukocyte count	[58, 72, 104–107]
IL-12	Stimulation of proliferation, activity, and cytotoxicity of T cells and NK cells	↑ Sperm count, ↑ sperm morphology	[83]
IL-18	Stimulation of proliferation, activity, and cytotoxicity of T cells and NK cells, induction of apoptosis	↓ Sperm concentration, ↓ sperm motility	[71]

(continued)

Table 9.1 (continued)

Proinflammatory cytokine	Participation in inflammatory reaction	Consequences on semen quality	References
TNF-α	Induction of proliferation and differentiation of B cells, proliferation of T cells and NK cells, induction of apoptosis	↓ Sperm count	[76]
		↓ Sperm motility	[67, 73]
		↓ Sperm motility (in vitro conditions)	[80, 160, 161]
		↓ Sperm morphology	[73]
		↓ Integrity of the sperm membrane (in vitro conditions)	[160]
		↑ DNA fragmentation	[160]
		↑ DNA fragmentation (in vitro conditions)	[161]
		↑ ROS (in vitro conditions)	[111]
		↑ MDA (in vitro conditions)	[111, 137]
		↑ PS externalization	[161]
IFN-γ	Leukocyte attraction, growth, maturation, and differentiation of NK and B cells	↓ Sperm count, motility, morphology	[68]
		↓ Sperm motility (in vitro conditions)	[79]
		↑ MDA (in vitro conditions)	[137]

motility [79, 80] and penetration rate of human sperm to hamster oocyte [81]. However, some authors have not found a similar effect of TNF-α and IL-8 on sperm motility and the acrosome reaction [82]. In the same study, only IFN-γ showed an ability to inhibit sperm motility, but this phenomenon was observed only when high concentrations of the cytokine were applied. Thus, it appears that participation of some cytokines in the regulation of fertility is dependent upon their concentrations. For instance, the IL-12 level correlates with sperm density and morphology, which suggests a certain biological role for this interleukin in male infertility [83]. However, cytokine expression is interconnected with a variety of factors, including steroid hormones. For example, an increase in IL-1β expression in the testis during local infection or inflammation is associated with decreased testosterone production by Leydig cells and decreased intensity of spermatogenesis, probably due to apoptosis [61]. The harmful effects of cytokines are normally mediated by receptor pathways active in membrane-bound and/or soluble forms [84]. Even if the levels of cytokines in seminal plasma of fertile and infertile men are similar, possibly they may differently affect their reproductive systems because the levels of cytokine inhibitors and/or their soluble receptors may be expressed differently in fertile and infertile men [53]. Moreover, the balance of cytokines can be modulated by prostaglandins which occur in high concentrations in the semen [62].

By nature, cytokines rarely act in isolation, but rather in a network with other cytokines. For example, the overproduction of IL-18 in combination with IL-12 may be dangerous both to the cells of the immune system and other cells and tissues of the body. The toxicity of one cytokine for the sperm structures and function is also increased when other cytokines are also present. This has been suggested by different authors who observed in vivo correlations between various cytokine levels in seminal fluid, e.g., IL-1β and TNF-α [65, 85], TNF-α and IL-2 [86], IL-6 and IL-8, IL-1β and IL-8, IL-12 and IL-18 [85], and IL-18 and IFN-γ [71]. It is worth emphasizing that some of these combinations are natural interactions taking place in the inflammatory process. This synergistic effect of different cytokine combinations has been also confirmed by in vitro studies in relation to its harmful influence on sperm motility [87] or biological membranes [88].

In spite of multiple in vivo studies, the involvement of cytokines and other immunoregulatory factors in male infertility is still unclear. The same cytokines that act as testicular immunomodulatory elements and regulate the physiological function of the male gonad appear in large concentrations in the semen in a number of pathological conditions, including autoimmune diseases [89, 90], spinal cord injury [91], varicocele [92], or genital tract infection/inflammation [93, 94]. Most investigators suggest that the cytokines detected in seminal plasma are associated with the incidence of leukocytospermia but are not the cause of semen abnormalities [51, 95]. The production of ROS by leukocytes or damaged spermatozoa may be one of the mechanisms by which cellular and humoral immunity could be triggered [96]. Leukocytospermia impairs sperm function by reducing semen antioxidant activity and causing enhanced T helper 1 switch [96]. Moreover, beneficial effects of supplemented vitamin E on cell-mediated immunity and oxidative stress have also been reported [97]. Cytokines as important mediators of the immunologic response (cell activation, proliferation, growth, differentiation, mobility) are involved in numerous physiological and pathological processes in male genital tract, which include the mediation of inflammatory responses, the reproductive functions, and the regulation of gonadal steroid production and its release [65]. Although some cytokines (IL-2, IL-8) take a key role in T-cell-mediated immune responses, their participation as mediating factors between cell-mediated immunity and male infertility still seems to be controversial.

9.7 Relationship of Cytokines with Redox Status of Semen

Oxidative stress in semen has been suggested to be an important factor in the etiology of poor sperm function, especially through peroxidative damage to the sperm plasma membrane and DNA integrity in the sperm nucleus [98]. In many studies, the role of cytokines and oxidative stress in infertile patients was examined, especially together with an infection/inflammation in the genital tract [57, 85, 99, 100]. The frequent presence of leukocytes in the semen, including their major subpopulation of neutrophils, is inseparably connected with reactive oxygen intermediate (ROI) production [101, 102]. A condition in which the number of leukocytes exceeds 1×10^6/mL of semen, defined as leukocytospermia, is considered to be the threshold value above which sperm dysfunction may occur. However, rather than the leukocyte number in the semen, their activity may decide on the final effects of oxidative stress on spermatozoa [76, 103]. Many authors have observed correlations between the levels of proinflammatory cytokines and the number of leukocytes in the semen [72, 77, 104–107]. However, there have been also reports demonstrating the elevated levels of proinflammatory cytokines in the semen regardless of the presence or absence of leukocytes [108, 109]. Undoubtedly, activated macrophages and neutrophils release ROIs and secrete proinflammatory cytokines, both of which can affect spermatozoa [102, 110].

Some cytokines act as regulators of the physiological levels of ROI in the seminal plasma [57, 111]. Some authors have suggested that particular cytokines modulate the expression of genes responsible for the redox system in the semen [56, 104]. Among all the cytokines, the production of chemotactic, proinflammatory cytokine IL-8 appears to be a specific response to oxidative stress. DeForge et al. [112] have demonstrated that oxidative stress is an important regulator of IL-8 gene expression in human epithelial cells and fibroblasts. Moreover, a unifying mechanism underlying this topic may involve regulation of NF-κB by the redox status of the cell, namely, its activation by ROI and its inhibition by various antioxidants in cells stimulated with cytokines or other agents like LPS or PMA [113, 114]. An increase in ROI production by human sperm was observed after the addition of IL-1α, IL-1β, or TNF-α, the result of which was an increase in sperm membrane lipid peroxidation, as measured by the malondialdehyde (MDA) level [111]. Moreover, this process has been postulated as the mechanism by which cytokines can cause infertility. It seems likely that leukocytes mediate the induction of ROS generation by cytokines and that they have the potential to alter redox equilibrium in the semen [92]. In addition, it cannot be excluded that the oxidative stress that appears in leukocytospermia is exerted by the increased levels of cytokines themselves [105], although this has not been proved in our studies in in vitro semen inflammation conditions. However, numerous studies have demonstrated the essential interaction of proinflammatory cytokines with other mediators of inflammation to generate toxic effects at the reaction site [76, 88, 105, 115]. Some observations argue in favor of the hypothesis that interleukins do not act separately but in connection with other mediators, particularly with leukocytes [102]. In this situation, ROI (generated by leukocytes) act synergistically with proinflammatory cytokines exacerbating the destructive environment for the spermatozoa. The oxidative stress with its consequences to sperm membranes can lead to damage in sperm biological function, in spite of the fact that the semen itself consists of spermatozoa subpopulations with a different fertilizing potential.

Overproduction of reactive oxygen metabolites observed in leukocytospermia suggests a possible imbalance between ROI levels and their scavengers in the semen. Assuming that oxidative stress leads not only to an uncontrolled increase in the concentration of ROI, but also to the failure of both the enzymatic and nonenzymatic members of the antioxidant system, we cannot exclude that some cytokines act also through changes in the activities of enzymes that normally protect the sperm against harmful metabolites. Indeed, antioxidants and the glutathione precursor have been shown in the literature to downregulate cytokine transcription and biosynthesis [116, 117]. Also, interestingly, a clear relationship between increased IL-18 and decreased GPx and selenium levels in the blood serum has been noted in patients with acute pancreatitis [118]. As for TNF-α, it may be directly positively associated with zinc concentration in the regulation of endothelial function [119]. The relationship between TNF-α levels and Mn-SOD expression in human endometrial stromal cells was also demonstrated [120]. Other authors showed inducing properties of IL-6 toward mitochondrial SOD activity [121]. Direct or indirect association between antioxidant and prooxidant activity with some interleukins present in the seminal plasma has been demonstrated in patients with genital tract inflammation [76, 105, 122] and varicocele [92]. It may be hypothesized that high activities of proinflammatory cytokines in the semen influence the intensity of oxidative stress, which may then have dangerous consequences for spermatozoa, indirectly affecting their redox profile as well as their milieu. A relationship among the cytokines, antioxidants, prooxidants, and semen parameters has been diligently studied in case of the semen infection/inflammation [57, 76, 105].

9.8 The Role of Cytokines in Semen Inflammation

The activation of leukocytes at a site of the inflammatory process is strictly connected with both the oxidative burst and the initiation of an immune response directed against an infectious agent. The secretion of cytokines is one of the first signals from the innate host defense to combat inflammation. Proinflammatory cytokines are one of the mediators of the host response to inflammatory reaction. Generated by both the cells of the immune system as well as those of surrounding tissues in response to external stimuli, inflicted injury, or other cytokines, they modulate inflammatory reactions, acting as the regulatory factors in different biologic processes [84]. In literature, proinflammatory cytokines have not always correlated with the presence of bacterial agents in the semen. In some reports, seminal plasma cytokine concentrations were similar investigating samples from culture-positive and culture-negative males [72, 123]. However, in most studies regarding male genital infection/inflammation, cytokines and various soluble receptors of immunoregulatory cytokines have been expressed distinctly in the seminal plasma [71, 93, 94, 124]. Moreover, some authors reported that the concentration of some of these factors was elevated in men with pyospermia and additionally correlated with the elastase activity of polymorphonuclear white blood cells [77]. It is well known that cytokine participation in inflammation is closely connected with the accompanying leukocytospermia [57, 100, 104]. Leukocytes in the semen are the major source of ROI, and ROI production by these cells is enhanced by bacterial products or cytokines [99].

Based on the presence of bacteria and/or leukocytes in the semen, the kinetics of the inflammatory process in the male urogenital tract was studied in situ [85, 125] (Fig. 9.4). The infiltration of an infectious factor to the male reproductive system leads to migration and activation of leukocytes, mostly phagocytic cells. In the course of the inflammatory process, the excessive production of ROI and the activation of appropriate receptors and signal transduction pathways provide biologically active substances, such as proteases and inflammatory cytokines, at the same time. The released proinflammatory cytokines are the next mediators of the host response to infection/inflammation which together with ROI modulate the activities of the prooxidative and antioxidative systems to the advantage of the ROI burst. It has been suggested that the reduced total antioxidant capacity (TAC) of the seminal plasma cannot be sufficient to ensure sperm quality [126, 127]. The consequence of the locally produced mediators in the inflammatory reaction is a permanent peroxidative damage to spermatozoa and their biological function. Literature on the subject also suggests that the remnants of oxidative stress in the semen may be maintained over a long period of time, even after the infectious factor has been eliminated, and that the effects of oxidative stress accompanying genital tract inflammation (reflected by leukocytospermia) on sperm parameters might be strictly dependent on the initial antioxidant capacity. In this context, the time needed for restoring oxidative balance in semen is longer in cases of men who are infertile at the start of inflammation [85].

In spite of the fact that the semen consists of spermatozoa subpopulations with different fertilizing potential, the oxidative stress with its consequences on sperm structures can lead to a damage of sperm biological function. Most probably, the crucial role in perpetuation of the inflammatory process in the semen may belong to cytokines and for this reason these bioactive substances may constitute an important link between inflammation and male infertility [76]. The local activity of the bioactive substances released through leukocytes during the inflammatory reaction and

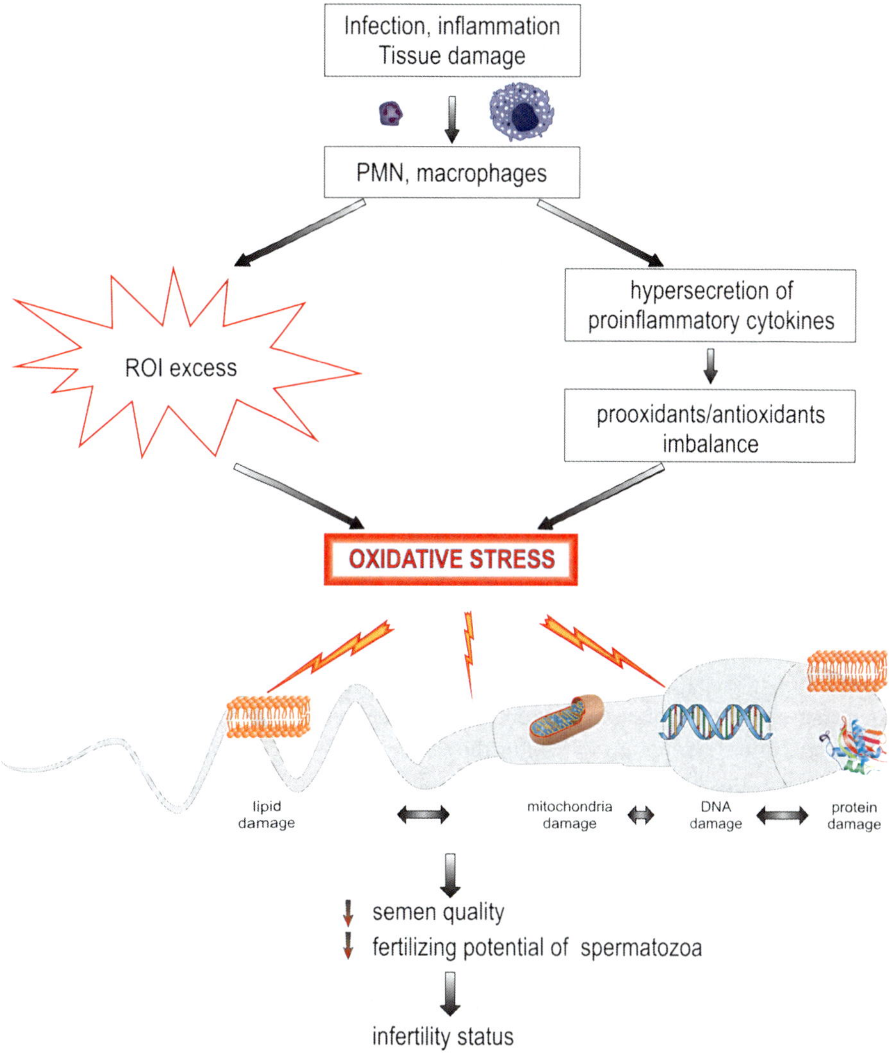

Fig. 9.4 Kinetics of inflammatory process in semen. Relationship between cytokines and oxidative stress. Bacterial invasion (phase I) or local tissue damage is accompanied by infiltrating phagocytic and mononuclear cells (phase II) connected with the production and release of large amounts of ROS and proinflammatory cytokines (phase III). The cytokines may modulate the activities of the prooxidative and antioxidative systems which also may bring enhanced secretion of ROS (phase IV). When the amount of ROS exceeds the potential of the antioxidative defense, peroxidative damage to sperm occurs, which in turn may lead to sperm dysfunction that results in fertility problems

reciprocal interactions among the particular inflammatory mediators (bacteria, leukocytes, proinflammatory cytokines) is an interesting subject, which has recently been often taken up and which has direct clinical implications. Creation of a model for in vitro inflammatory conditions enables observations of particular mediators of inflammatory reaction and their influence on the qualitative and quantitative changes in the oxygen metabolism of the semen which, in consequence, determine the intensity of toxic interactions among oxygen metabolites and the macromolecules, such as lipids, proteins, and DNA, what obviously cannot be traced in a healthy or a sick patient in in situ conditions.

9.9 Influence of Cytokines on Sperm Membrane Properties

An adverse effect of proinflammatory cytokines on sperm membrane properties (e.g., lipid peroxidation) is one of the several mechanisms by which cytokines might interfere with sperm quality. A peroxidation of the membrane structures is considered to be the first measurable destructive effect of ROI overproduction, also connected with infectious inflammation, and due to the high content of polyunsaturated fatty acids (PUFAs) spermatozoa are particularly susceptible to this process. An attack of free radicals toward the sperm membrane lipids leads to irreversible changes in membrane fluidity, enhances its nonspecific permeability, and in turn may affect the capacitation and acrosomal reactions abrogating the ability of spermatozoa to penetrate an oocyte. Our knowledge has to be broadened by systematic studies of lipid peroxidation in which the influence of lipid peroxidation on the sperm function can be evaluated [128–132]. Moreover, human semen consists of spermatozoa subpopulations, differing in the content of lipid membranes, which may determine the degree to which they are endangered to peroxidative damage [133–135]. The decrease in the PUFAs, observed in some subpopulations of ejaculated spermatozoa, may be the evidence of their previous oxidation and shedding of the products of lipid peroxidation to seminal plasma.

Many years ago, Buch et al. [111] for the first time reported the increase of the ROS production by human spermatozoa after their incubation with IL-1α, IL-1β, or TNF-α, the measurable effect of which was an increase in sperm membrane peroxidation, judged by the MDA levels. The relationship between the IL-6 or IL-8 concentrations and the intensity of sperm membrane lipid peroxidation in the semen was also suggested by other authors [106, 136]. Analysis of a wide panel of recombinant proinflammatory cytokines in pathological (observed in situ) concentrations revealed the influence of some cytokine combinations (applied together with leukocytes) on MDA levels in different spermatozoal fractions [88]. Taking into consideration that some authors [138] have documented an increase in lipid peroxidation after in vitro incubation of sperm with PMA-stimulated leukocytes which suggests a decrease in the biological value of sperm membranes in the environment of oxidative stress, it may be argued that interleukins do not act separately but in connection with other mediators of the inflammatory process which has been suggested previously [57, 95, 105, 115, 137]. It cannot be precluded that the increase of oxidative stress in leukocytospermia is modulated by the levels of some cytokines, and even more, during inflammatory conditions, produced by leukocytes ROI and proinflammatory cytokines cooperate with each other provoking a destructive effect on spermatozoa.

According to data presented in some papers, there are some differences in the rates of lipid peroxidation that are related to the sperm maturity and, presumably, sperm membrane structures [138–140]. Probably, the mediating role of leukocytes with respect to the harmful effects of secreted proinflammatory cytokines toward sperm membranes may depend on the type of sperm subpopulation [88].

Proinflammatory cytokines are characterized by their pleiotropic properties, and when they occur together they may act synergistically, additively, or antagonistically on the biological function of the target cell. This is confirmed by the synergistic effect in vitro in relation to the harmful influence of some proinflammatory cytokines on biological membranes [88]. In case of cytokines such as IL-6, IL-8, or IL-18, observations in the in vitro system appear to be complementary to evaluation of the male genital tract inflammation in vivo. The evidence of this can be high levels of IL-6, IL-8, or IL-18 observed in the seminal plasma of infertile men reported by many groups [71, 72, 74]. The extended high levels of these cytokines present during persistent infection/inflammation in the male genital tract may augment the peroxidation process and affect sperm function with a subsequent development of infertility. Many authors suggest that the measurement of IL-6, IL-8, and/or IL-18 concentration in the seminal plasma can be a sensitive marker of early diagnosis of infection/inflammation in the male genitourinary tract and a signal for fast intervention with an anti-inflammatory treatment [57, 71, 72, 76].

9.10 Influence of Cytokines on Sperm DNA and Apoptosis

The "poor quality" of sperm DNA appears to be one of the important factors affecting male reproductive ability [141]. This has been confirmed by numerous reports in which a high percentage of spermatozoa with fragmented DNA have been found in infertile men when compared to the fertile individuals [142, 143]. In case of inflammation of the genitourinary tract, the redox imbalance is probably the main etiological factor responsible for destructive effects of the inflammatory process on male gametes, associated with peroxidation of sperm macromolecules including DNA [100, 125, 144]. On the other hand, the phenomenon of oxidative stress is also associated with increased apoptosis, which leads to sperm DNA damage [145–147]. Several authors suggested an involvement of sperm apoptosis in impaired men's fertility with consequences to reduced sperm–oocyte penetration [148–150]. It is well known that some proinflammatory cytokines, such as IL-1β, TNF-α, or IL-18, take part in the regulation of the apoptotic process by the induction of the Fas/Fas ligand (FasL) system or by interaction with certain receptors, e.g., TNF-α receptor-1 [151, 152]. Participation of proinflammatory cytokines has been also suggested in the male gonad [50, 153]. It is possible that induction of sperm apoptosis can be the principal mechanism by which these proinflammatory cytokines may affect spermatozoa during the male reproductive tract infection/inflammation.

The relationship of cytokines participating in urogenital infections and human sperm apoptosis has been recently identified as a central area of interest. Of a large group of proinflammatory cytokines, IL-6 and IL-8 have been most often mentioned in the literature as the potential diagnostic markers for male genitourinary tract infections [71, 72, 76]. Additionally, combinations of proinflammatory cytokines turned out to be an important factor enhancing sperm membrane lipid peroxidation primarily caused by leukocytes in the in vitro studies [88]. However, the proapoptotic properties of IL-6 or IL-8 are still unclear. Nevertheless, there are some data on induction of the lymphocyte apoptosis by proinflammatory cytokines, including IL-6, in an acute African Swine Fever infection, thus suggesting the possibility of proapoptotic activity of IL-6 exerted on the cells of the immune system [154]. However, the premises that IL-6 and/or IL-8 contribute to apoptosis also in ejaculated spermatozoa have not been finally clarified.

The influence of IL-18 on apoptosis through the increase in the FasL expression has been previously documented in somatic cells. High levels of IL-18 were found, e.g., in liver inflammation, which suggests that this cytokine may regulate apoptosis of hepatocytes and may be a reason of liver damage. There were also some reports on the possibility of IL-18-exerted apoptosis on endothelial cells and this effect was mediated by members of TNF family or the Fas/FasL system [155]. Some reports have described the IL-18's harmful effects on semen quality in infertile men with urogenital infections [71]. In in vitro conditions, a combination of IL-18 with IL-12 used together with leukocytes was linked with a significant increase in MDA concentrations [88]. It is possible that these inflammatory cytokines may cooperate with each other in apoptosis induction of mature spermatozoa (own unpublished data).

Among the various inflammatory cytokines, TNF-α is most often presented as inducer of apoptosis in human spermatozoa. Due to participation in activation of several transduction pathways, this cytokine leads to regulation of the testicular expression of several genes involved in spermatogenesis [156]. The proapoptotic effect of TNF-α can be mediated through ROI production [157, 158]. Recently, it has also been suggested that TNF-α might alter sperm function via increase in nitric oxide production [159]. TNF-α-induced apoptosis in the ejaculated spermatozoa measured by the increase of the percentage of the spermatozoa with phosphatidylserine (PS) externalization on the cell membrane surface and/or by the increase of the TUNEL-positive spermatozoa has been confirmed in some experimental and clinical studies [160–162]. Moreover, these proapoptotic effects were reversed in the presence of a selective anti-TNF-α antibody [160]. The proapoptotic effect of this cytokine, one of the major cytokines produced during inflammation, may explain the reduced fertilizing ability of spermatozoa obtained from men with urogenital tract inflammation.

To summarize, results of the coordinated cytokines and leukocytes action are the qualitative and quantitative changes in the oxygen metabolism of spermatozoa, which determine the magnitude of toxic interactions of oxygen with cellular macromolecules (such as lipids or DNA). The course of the inflammatory process and its influence on the function of sperm depends on the type of initiating factor, time of exposition of inflammatory mediators, and the initial condition of the antioxidative system in the semen. Evaluation of the damage to cellular membranes or the integrity of sperm DNA seems to be an extremely important information for the prognosis of the male gamete fertilizing ability and may be an important step toward establishing new, additional diagnostic algorithms, which will pave the way to conservative treatment of male infertility (by antioxidants) as well as better qualification and selection of assisted reproduction techniques.

9.11 Cytokines as Potential Diagnostic Markers of Male Infertility

The clinical role of evaluation of cytokines participating in urogenital infections/inflammations has been recently found to be the important area of interest. The present routine diagnostics of inflammation in the male reproductive tract is mainly based on the determination of leukocyte concentration and microbiological evaluation of the semen. Because the harmful effect of cytokines on spermatozoa is closely connected with the accompanying leukocytospermia, the evaluation of the leukocyte count in the semen still remains an important but insufficient factor to the diagnosis and treatment monitoring of male genital tract infection/inflammation and its particular stages or kinetics. Consideration of additional biomarkers of inflammation, especially in cases of inflammatory conditions which occur without clinical symptoms, seems to be helpful in establishing the proper algorithm [163]. Determination of the levels of proinflammatory cytokines in the seminal plasma using a quantitative immunosorbent test (ELISA) may be one of the basic biochemical markers of the inflammatory process in the ejaculate, at the same time reflecting the inflammatory activity of leukocytes. In multiple studies, the measurement of cytokine concentrations in seminal plasma, such as IL-6 [63, 72, 77], IL-8 [72, 94, 164], IL-18 [71], TNF-α [75, 99], IFN-γ [99], IL-2 [68], or sIL-6R [165], can be considered as a potential marker of male infertility caused or complicated by infection/inflammation in the reproductive tract. Data from many independent groups support the need for the examination of IL-6 and IL-8 as excellent markers for early diagnosis of the inflammatory process in the male reproductive system and the rapid initiation of anti-inflammatory treatment before the reproductive potential of the sperm is inhibited [72, 76, 77, 93, 94, 164]. However, other authors reported that the concentration of IL-6 and IL-8 is closely correlated with the number of leukocytes; thus, their determination does not provide any additional information for, e.g., the diagnosis of male accessory gland infection [65, 70, 166].

Numerous reports have appeared in which the limited value of standard semen analysis has been emphasized while assessing males fertilizing potential. Moreover, the number of studies on the clinical significance of the detection of cytokines in the seminal plasma, especially in the context of genital tract infection/inflammation, is increasing. Despite these premises, the measurement of cytokine concentrations is still not a routine clinical practice. Undoubtedly, controlled prospective studies, including a large number of patients, a wide range of analyzed semen parameters, patient characteristics, and their reproductive outcome, are urgently needed to answer the question about the future of the novel biomarkers produced by activated leukocytes, such as granulocyte elastase, proinflammatory cytokines, or ROI secreted to seminal plasma.

9.12 Concluding Remarks

In spite of the quite often contradictory results from multiple clinical as well as experimental studies regarding cytokines and their effects on sperm structure and function, there is no doubt that various cytokines are involved in the regulation of gonads and sperm function. Cytokines detected in the seminal plasma seem to be mainly associated with the leukocytospermia. Moreover, the cytokines may play a crucial role in perpetuation of oxidative stress in semen, especially during the chronic inflammatory process. Due to the involvement of the mentioned bioactive factors deepening the harmful effects of oxidative stress to spermatozoa, they may be suspected to constitute an important link between inflammation and male infertility.

Acknowledgements: Financed by grants no. NN407283539, NR13006606.

References

1. Stephan JP, Syed V, Jegou B. Regulation of Sertoli cell IL-1 and IL-6 production in vitro. Mol Cell Endocrinol. 1997;134:109–18.
2. Guazzone VA, Jacobo P, Theas MS, et al. Cytokines and chemokines in testicular inflammation: a brief review. Microsc Res Tech. 2009;72:620–8.
3. O'Bryan MK, Schlatt S, Phillips DJ, et al. Bacterial lipopolysaccharide-induced inflammation compromises testicular function at multiple levels in vivo. Endocrinology. 2000;141: 238–46.
4. O'Bryan MK, Hedger MP. Inflammatory networks in the control of spermatogenesis: chronic inflammation in an immunologically privileged tissue? Adv Exp Med Biol. 2008;636: 92–114.
5. Hedger M, Klug J, Fröhlich S, et al. Regulatory cytokine expression and interstitial fluid formation in the normal and inflamed rat testis are under Leydig cell control. J Androl. 2005;26:379–86.
6. Fijak M, Bhushan S, Meinhardt A. Immunoprivileged sites: the testis. Methods Mol Biol. 2011;677:459–70.
7. Jacobo P, Guazzone VA, Theas MS, et al. Testicular autoimmunity. Autoimmun Rev. 2011;10:201–4.
8. Cheng CY, Mruk DD. Regulation of spermiogenesis, spermiation and blood-testis barrier dynamics: novel insights from studies on Eps8 and Arp3. Biochem J. 2011;435(3):553–62.
9. Cheng CY, Mruk DD. A local autocrine axis in the testes that regulates spermatogenesis. Nat Rev Endocrinol. 2010;6:380–95.
10. Wong CH, Cheng CY. The blood–testis barrier: its biology, regulation and physiological role in spermatogenesis. Curr Top Dev Biol. 2005;7:263–96.
11. Su L, Mruk DD, et al. Drug transporters, the blood–testis barrier, and spermatogenesis. J Endocrinol. 2011;208(3):207–23.

12. Yan HH, Mruk DD, Lee WM, Cheng CY. Ectoplasmic specialization: a friend or a foe of spermatogenesis? Bioessays. 2007;29(1):36–48.
13. Mruk DD, Cheng CY. Sertoli-Sertoli and Sertoli-germ cell interactions and their significance in germ cell movement in the seminiferous epithelium during spermatogenesis. Endocr Rev. 2004;25(5):747–806.
14. Xia W, Wong CH, et al. Disruption of Sertoli-germ cell adhesion function in the seminiferous epithelium of the rat testis can be limited to adherens junctions without affecting the blood-testis barrier integrity: an in vivo study using an androgen suppression model. J Cell Physiol. 2005;205(1):141–57.
15. Li MW, Mruk DD, et al. Cytokines and the junction restructuring events during spermatogenesis in the testis: an emerging new concept of regulation. Cytokine Growth Factor Rev. 2009;20(4):329–38.
16. Li MW, Mruk DD, et al. "Unlocking" the blood-testis barrier and the ectoplasmic specialization by cytokines during spermatogenesis: emerging targets for male contraception. Immunol Endocr Metab Agents Med Chem. 2008;8:20–7.
17. Siu MK, Wong CH, et al. Sertoli-germ cell anchoring junction dynamics in the testis are regulated by an interplay of lipid and protein kinases. J Biol Chem. 2005;280(26):25029–47.
18. Xia W, Mruk DD, et al. Cytokines and junction restructuring during spermatogenesis—a lesson to learn from the testis. Cytokine Growth Factor Rev. 2005;16(4–5):469–93.
19. Meng J, Holdcraft RW, et al. Androgens regulate the permeability of the blood–testis barrier. Proc Natl Acad Sci USA. 2005;102(46):16696–700.
20. Yan HH, Mruk DD, et al. Blood-testis barrier dynamics are regulated by testosterone and cytokines via their differential effects on the kinetics of protein endocytosis and recycling in Sertoli cells. FASEB J. 2008;22(6):1945–59.
21. Su L, Mruk DD, et al. Differential effects of testosterone and TGF-β3 on endocytic vesicle-mediated protein trafficking events at the blood-testis barrier. Exp Cell Res. 2010;316(17):2945–60.
22. Cheng CY, Mruk DD. An intracellular trafficking pathway in the seminiferous epithelium regulating spermatogenesis: a biochemical and molecular perspective. Crit Rev Biochem Mol Biol. 2009;44(5):245–63.
23. Yan HH, Mruk DD, et al. An autocrine axis in the testis that coordinates spermiation and blood–testis barrier restructuring during spermatogenesis. Proc Natl Acad Sci USA. 2008;105(26):8950–5.
24. Cheng CY, Wong EW, et al. Regulation of spermatogenesis in the microenvironment of the seminiferous epithelium: new insights and advances. Mol Cell Endocrinol. 2010;315(1–2):49–56.
25. Bornstein R, Rutkowski H. Cytokines and steroidogenesis. Mol Cell Endocrinol. 2004;215(1–2):135–41.
26. Guzmán C, Hernández-Bello R, et al. Regulation of steroidogenesis in reproductive, adrenal and neural tissues by cytokines. Open Neuroendocrinol J. 2010;3:161–9.
27. Cudicini C, Lejeune H, Gomez E, et al. Human Leydig cells and Sertoli cells are producers of interleukins-1 and -6. J Clin Endocrinol Metab. 1997;82:1426–33.
28. Hedger MP, Meinhardt A. Cytokines and the immune-testicular axis. J Reprod Immunol. 2003;58:1–26.
29. Rozwadowska N, Fiszer D, Kurpisz M. Function of the interleukin-1 gene system in immunomodulation, apoptosis and proliferation in the male gonad. Postepy Hig Med Dosw. 2005;59:56–67.
30. Hales DB, Diemer T, Hales KH. Role of cytokines in testicular function. Endocrine. 1999;10:201–17.
31. Khan SA, Schmidt K, Hallin P, et al. Human testis cytosol and ovarian follicular fluid contain high amounts of interleukin-1-like factor(s). Mol Cell Endocrinol. 1998;58:221–30.
32. Stéphan JP, Syed V, Jégou B. Regulation of Sertoli cell IL-1 and IL-6 production in vitro. Mol Cell Endocrinol. 1997;134:109–18.
33. Lysiak JJ. The role of tumor necrosis factor-alpha and interleukin-1 in the mammalian testis and their involvement in testicular torsion and autoimmune orchitis. Reprod Biol Endocrinol. 2004;2:9.

34. Lin T, Wang D, Nagpal ML, et al. Interleukin-1 inhibits cholesterol side-chain cleavage cytochrome P450 expression in primary cultures of Leydig cells. Endocrinology. 1991;129: 1305–11.
35. Kanzaki M, Morris PL. Growth hormone regulates steroidogenic acute regulatory protein expression and steroidogenesis in Leydig cell progenitors. Endocrinology. 1999;140:1681–6.
36. Bourguiba S, Chater S, Delalande C, et al. Regulation of aromatase gene expression in purified germ cells of adult male rats: effects of transforming growth factor β, tumor necrosis factor α, and cyclic adenosine 3′,5′-monophosphate. Biol Reprod. 2003;69:592–601.
37. O'Bryan MK, Gerdprasert O, Nikolic-Paterson DJ, et al. Cytokine profiles in the testes of rats treated with lipopolysaccharide reveal localized suppression of inflammatory responses. Am J Physiol Regul Integr Comp Physiol. 2005;288:1744–55.
38. Turner T, Lysiak J. Oxidative stress: a common factor in testicular dysfunction. J Androl. 2008;29:488–98.
39. Fujii J, Iuchi Y, Matsuki S, et al. Cooperative function of antioxidant and redox systems against oxidative stress in male reproductive tissues. Asian J Androl. 2003;5:231–42.
40. Lukyanenko YO, Chen JJ, Hutson JC. Production of 25-HC by testicular macrophages and its effects on Leydig cells. Biol Reprod. 2001;64:790–6.
41. Hales DB. Testicular macrophage modulation of Leydig cell steroidogenesis. J Reprod Immunol. 2002;57:3–18.
42. Thiemermann C, Szabo C, Peretti M, et al. Role of tumor necrosis factor in the induction of nitric oxide in a rat model of endotoxin. Br J Pharmacol. 1993;110:177–82.
43. Allen JA, Diemer T, Janus P, et al. Bacterialendotoxin lipopolysaccharide and reactive oxygen species inhibit Leydig cell steroidogenesis via perturbation of mitochondria. Endocrine. 2004;25:265–75.
44. Diemer T, Allen JA, Hales KH, et al. Reactive oxygen disrupts mitochondria in MA-10 tumor Leydig cells and inhibits steroidogenic acute regulatory (StAR) protein and steroidogenesis. Endocrinology. 2003;144:2882–91.
45. Aly HA, Lightfoot DA, El-Shemy HA. Bacterial lipopolysaccharide-induced oxidative stress in adult rat Sertoli cells in vitro. Toxicol In Vitro. 2010;24:1266–72.
46. Kern S, Robertson SA, et al. Cytokine secretion by macrophages in the rat testis. Biol Reprod. 1995;53(6):1407–16.
47. Fujii J, Iuchi Y, et al. Cooperative function of antioxidant and redox systems against oxidative stress in male reproductive tissues. Asian J Androl. 2003;5(3):231–42.
48. Maiorino M, Ursini F. Oxidative stress, spermatogenesis and fertility. Biol Chem. 2002;383:591–7.
49. Soder O, Sultana T, Jonsson C, et al. The interleukin-1 system in the testis. Andrologia. 2000;32:52–5.
50. Rozwadowska N, Fiszer D, Jedrzejczak P, et al. Interleukin-1 superfamily genes expression in normal or impaired human spermatogenesis. Genes Immun. 2007;8:100–7.
51. Maegawa M, Kamada M, Irahara M, et al. A repertoire of cytokines in human seminal plasma. J Reprod Immunol. 2002;54:33–42.
52. Politch JA, Tucker L, Bowman FP, et al. Concentrations and significance of cytokines and other immunologic factors in semen of healthy fertile men. Hum Reprod. 2007;22:2928–35.
53. Huleihel M, Lunenfeld E, Levy A, et al. Distinct expression levels of cytokines and soluble cytokine receptors in seminal plasma of fertile and infertile men. Fertil Steril. 1996;66:135–9.
54. Matalliotakis I, Kiriakou D, Fragouli I, et al. Interleukin-6 in seminal plasma of fertile and infertile men. Arch Androl. 1998;41:43–50.
55. Twillie DA, Eisenberger MA, Carducci MA, et al. Interleukin-6: a candidate mediator of human prostate cancer morbidity. Urology. 1995;45:542–9.
56. Naz RK, Kaplan P. Increased levels of interleukin-6 in seminal plasma of infertile men. J Androl. 1994;15:220–7.

57. Depuydt CE, Bosmans E, Zalata AA, et al. The relation between reactive oxygen species and cytokines in andrological patients with or without male accessory gland infection. J Androl. 1996;17:699–707.
58. Friebe K, Bohring C, Skrzypek J, et al. Levels of interleukin-6 and interleukin-8 in seminal fluid of men attending an andrological clinic. Andrologia. 2003;35:126–9.
59. Huleihel M, Lunenfeld E, Horowitz S, et al. Involvement of serum and lipopolysaccharide in the production of interleukin-1- and interleukin-6-like molecules by human sperm cells. Am J Reprod Immunol. 2000;43:41–6.
60. Huleihel M, Lunenfeld E, Horowitz S, et al. Production of interleukin-1-like molecules by human sperm cells. Fertil Steril. 2000;73:1132–7.
61. Huleihel M, Lunenfeld E. Regulation of spermatogenesis by paracrine/autocrine testicular factors. Asian J Androl. 2004;6:259–68.
62. Ochsendorf RF. Infections in the male genital tract and reactive oxygen species. Hum Reprod Update. 1999;5:399–420.
63. Comhaire F, Bosmans E, Ombelet W, et al. Cytokines in semen of normal men and of patients with andrological diseases. Am J Reprod Immunol. 1994;31:99–103.
64. Matalliotakis I, Kyriakou D, Fragouli Y, et al. Determination of interleukin-11 in seminal plasma and elevated IL-11 in seminal plasma of infertile patients with urogenital infection. Arch Androl. 1998;41:177–83.
65. Papadimas J, Goulis DG, Sotiriades A, et al. Interleukin-1 beta and tumor necrosis factor-alpha in normal/infertile men. Arch Androl. 2002;48:107–13.
66. Paradisi R, Capelli M, Mandini M, et al. Interleukin-2 in seminal plasma of fertile and infertile men. Arch Androl. 1995;35:35–41.
67. Gruschwitz MS, Brezinschek R, Brezinschek HP. Cytokine levels in the seminal plasma of infertile males. J Androl. 1996;17:158–63.
68. Paradisi R, Capelli M, Mandini M, et al. Increased levels of interferon-gamma in seminal plasma of infertile men. Andrologia. 1996;28:157–61.
69. Paradisi R, Mancini R, Bellavia E, et al. T-helper 2 type cytokine and soluble interleukin-2 receptor levels in seminal plasma of infertile men. Am J Reprod Immunol. 1997;38:94–9.
70. Dousset B, Hussenet F, Daudin M, et al. Seminal cytokine concentrations (IL-1beta, IL-2, IL-6, sR IL-2, sR IL-6), semen parameters and blood hormonal status in male infertility. Hum Reprod. 1997;12:1476–9.
71. Matalliotakis IM, Cakmak H, Fragouli Y, et al. Increased IL-18 levels in seminal plasma of infertile men with genital tract infections. Am J Reprod Immunol. 2006;55:428–33.
72. Eggert-Kruse W, Boit R, Rohr G, et al. Relationship of seminal plasma interleukin (IL) -8 and IL-6 with semen quality. Hum Reprod. 2001;16:517–28.
73. Kocak I, Yenisey C, Dundar M, et al. Relationship between seminal plasma interleukin-6 and tumor necrosis factor alpha levels with semen parameters in fertile and infertile men. Urol Res. 2002;30:263–7.
74. Furuya Y, Akashi T, Fuse H. Soluble Fas and interleukin-6 and interleukin-8 levels in seminal plasma of infertile men. Arch Androl. 2003;49:449–52.
75. Paulis G, Conti E, Voliani S, et al. Evaluation of the cytokines in genital secretions of patients with chronic prostatitis. Arch Ital Urol Androl. 2003;75:179–86.
76. Sanocka D, Jedrzejczak P, Szumala-Kakol A, et al. Male genital tract inflammation: the role of selected interleukins in regulation of pro-oxidant and antioxidant enzymatic substances in seminal plasma. J Androl. 2003;24:448–55.
77. Kopa Z, Wenzel J, Papp GK, et al. Role of granulocyte elastase and interleukin-6 in the diagnosis of male genital tract inflammation. Andrologia. 2005;37:188–94.
78. Bezold G, Politch JA, Kiviat NB, et al. Prevalence of sexually transmissible pathogens in semen from asymptomatic male infertility patients with and without leukocytospermia. Fertil Steril. 2007;87:1087–97.
79. Hill JA, Haimovici F, Politch JA, et al. Effects of soluble products of activated lymphocytes and macrophages (lymphokines and monokines) on human sperm motion parameters. Fertil Steril. 1987;47:460–5.

80. Eisermann J, Register KB, Strickler RC, et al. The effect of tumor necrosis factor on human sperm motility in vitro. J Androl. 1989;10:270–4.
81. Hill JA, Cohen J, Anderson DJ. The effects of lymphokines and monokines on human sperm fertilizing ability in the zona-free hamster egg penetration test. Am J Obstet Gynecol. 1989;160:1154–9.
82. Fedder J, Ellerman-Eriksen S. Effect of cytokines on sperm motility and ionophore-stimulated acrosome reaction. Arch Androl. 1995;35:173–85.
83. Naz RK, Evans L. Presence and modulation of interleukin-12 in seminal plasma of fertile and infertile men. J Androl. 1998;19:302–7.
84. Feldmann M, Saklatvala J. Proinflammatory cytokines. In: Oppenheim JJ, Feldman M, editors. Cytokine reference. New York: Academic; 2001. p. 291–305.
85. Sanocka D, Fraczek M, Jedrzejczak P, et al. Male genital tract infection: an influence of leukocytes and bacteria on semen. J Reprod Immunol. 2004;62:111–24.
86. Hussenet F, Dousset B, Cordonnier JL, et al. Tumour necrosis factor alpha and interleukin 2 in normal and infected human seminal fluid. Hum Reprod. 1993;8:409–11.
87. Estrada LS, Champion HC, Wang R, et al. Effect of tumour necrosis factor-alpha (TNF-alpha) and interferon-gamma (IFN-gamma) on human sperm motility, viability and motion parameters. Int J Androl. 1997;20:237–42.
88. Fraczek M, Sanocka D, Kamieniczna M, et al. Proinflammatory cytokines as an intermediate factor enhancing lipid sperm membrane peroxidation in in vitro conditions. J Androl. 2008;29:85–92.
89. Motrich RD, Maccioni M, Molina R, et al. Reduced semen quality in chronic prostatitis patients that have cellular autoimmune response to prostate antigens. Hum Reprod. 2005;20:2567–72.
90. Theas MS, Rival C, Jarazo-Dietrich S, et al. Tumour necrosis factor-alpha released by testicular macrophages induces apoptosis of germ cells in autoimmune orchitis. Hum Reprod. 2008;23:1865–72.
91. Basu S, Aballa TC, Ferrell SM, et al. Inflammatory cytokine concentrations are elevated in seminal plasma of men with spinal cord injuries. J Androl. 2004;25:250–4.
92. Nallella KP, Allamaneni SS, Pasqualotto FF, et al. Relationship of interleukin-6 with semen characteristics and oxidative stress in patients with varicocele. Urology. 2004;64:1010–3.
93. Martínez-Prado E, Camejo Bermúdez MI. Expression of IL-6, IL-8, TNF-alpha, IL-10, HSP-60, anti-HSP-60 antibodies, and anti-sperm antibodies, in semen of men with leukocytes and/or bacteria. Am J Reprod Immunol. 2010;63:233–43.
94. Kokab A, Akhondi MM, Sadeghi MR, et al. Raised inflammatory markers in semen from men with asymptomatic chlamydial infection. J Androl. 2010;31:114–20.
95. Eggert-Kruse W, Kiefer I, Beck C, et al. Role for tumor necrosis factor alpha (TNF-alpha) and interleukin 1-beta (IL-1beta) determination in seminal plasma during infertility investigation. Fertil Steril. 2007;87:810–23.
96. Omu AE, Al-Qattan F, Al-Abdul-Hadi FM, et al. Seminal immune response in infertile men with leukocytospermia: effect on antioxidant activity. Eur J Obstet Gynecol Reprod Biol. 1999;86:195–2002.
97. Lee CY, Man-Fan Wan J. Vitamin E supplementation improves cell-mediated immunity and oxidative stress of Asian men and women. J Nutr. 2000;130:2932–7.
98. Aitken RJ, Baker MA, De Iuliis GN, et al. New insights into sperm physiology and pathology. Handb Exp Pharmacol. 2010;198:99–115.
99. Wang A, Fanning L, Anderson DJ, et al. Generation of reactive oxygen species by leukocytes and sperm following exposure to urogenital tract infection. Arch Androl. 1997;39:11–7.
100. Comhaire FH, Mahmoud AM, Depuydt CE, et al. Mechanisms and effects of male genital tract infection on sperm quality and fertilizing potential: the andrologist's viewpoint. Hum Reprod Update. 1999;5:393–8.
101. Wolff H, Anderson DJ. Immunohistologic characterization and quantitation of leukocyte subpopulations in human semen. Fertil Steril. 1988;49:497–504.
102. Wolff H. The biologic significance of white blood cells in semen. Fertil Steril. 1995;63:1143–57.

103. Kovalski N, de Lamirande E, Gagnon C. Reactive oxygen species generated by human neutrophils inhibit sperm motility: protective effect of seminal plasma and scavengers. Fertil Steril. 1992;58:809–16.
104. Shimoya K, Matsuzaki N, Tsutsui T, et al. Detection of interleukin-8 (IL-8) in seminal plasma and elevated IL-8 in seminal plasma of infertile patients with leukospermia. Fertil Steril. 1993;59:885–8.
105. Rajasekaran M, Hellstrom WJ, Naz RK, et al. Oxidative stress and interleukins in seminal plasma during leukocytospermia. Fertil Steril. 1995;64:166–71.
106. Liu J, Wang YX, Wu YL, et al. Measurement of the reactive oxygen species and cytokines in the seminal plasma of leukocytospermic patients. Zhonghua Nan Ke Xue. 2003;9:103–5.
107. Jedrzejczak P, Fraczek M, Szumala-Kakol A, et al. Consequences of semen inflammation and lipid peroxidation on fertilization capacity of spermatozoa in in vitro conditions. Int J Androl. 2005;28:275–83.
108. Alexander RB, Ponniah S, Hasday J, et al. Elevated levels of proinflammatory cytokines in the semen of patients with chronic prostatitis/chronic pelvic pain syndrome. Urology. 1998;52:744–9.
109. Naz RK, Evans L, Armstrong JS, et al. Decreased levels of interleukin-12 are not correlated with leukocyte concentration and superoxide dismutase activity in semen of infertile men. Arch Androl. 1998;41:91–6.
110. Aitken RJ, Fisher H. Reactive oxygen species generation and human spermatozoa: the balance of benefit and risk. Bioessays. 1994;16:259–67.
111. Buch JP, Kolon TF, Maulik N, et al. Cytokines stimulate lipid membrane peroxidation of human sperm. Fertil Steril. 1994;62:186–8.
112. DeForge LE, Preston AM, Takeuchi E, et al. Regulation of interleukin 8 gene expression by oxidant stress. J Biol Chem. 1993;268:25568–76.
113. Kikumori T, Kambe F, Nagaya T, et al. Activation of transcriptionally active nuclear factor-kappaB by tumor necrosis factor-alpha and its inhibition by antioxidants in rat thyroid FRTL-5 cells. Endocrinology. 1998;139:1715–22.
114. Boulares AH, Giardina C, Inan MS, et al. Acetaminophen inhibits NF-kappaB activation by interfering with the oxidant signal in murine Hepa 1-6 cells. Toxicol Sci. 2000;55:370–5.
115. Das UN. Interaction(s) between essential fatty acids, eicosanoids, cytokines, growth factors and free radicals: relevance to new therapeutic strategies in rheumatoid arthritis and other collagen vascular diseases. Prostaglandins Leukot Essent Fatty Acids. 1991;44:201–10.
116. Barrett EG, Johnston C, Oberdörster G, et al. Silica-induced chemokine expression in alveolar type II cells is mediated by TNF-alpha-induced oxidant stress. Am J Physiol. 1999;276:L979–88.
117. Seo JY, Kim H, Seo JT, et al. Oxidative stress induced cytokine production in isolated rat pancreatic acinar cells: effects of small-molecule antioxidants. Pharmacology. 2002; 64:63–70.
118. Wereszczynska-Siemiatkowska U, Mroczko B, Siemi tkowski A, et al. The importance of interleukin 18, glutathione peroxidase, and selenium concentration changes in acute pancreatitis. Dig Dis Sci. 2004;49:642–50.
119. Hennig B, Meerarani P, Toborek M, et al. Antioxidant-like properties of zinc in activated endothelial cells. J Am Coll Nutr. 1999;18:152–8.
120. Karube-Harada A, Sugino N, Kashida S, et al. Induction of manganese superoxide dismutase by tumour necrosis factor-alpha in human endometrial stromal cells. Mol Hum Reprod. 2001;7:1065–72.
121. Isoherranen K, Peltola V, Laurikainen L, et al. Regulation of copper/zinc and manganese dismutase by UVB irradiation, oxidative stress and cytokines. J Photochem Photobiol B. 1997;40:288–93.
122. Omu AE, Al-Qattan F, Al-Abdul-Hadi FM, et al. Seminal immune response in infertile men with leukocytospermia: effect on antioxidant activity. Eur J Obstet Gynecol Reprod Biol. 1999;86:195–202.

123. Pannekoek Y, Trum JW, Bleker OP, et al. Cytokine concentrations in seminal plasma from subfertile men are not indicative of the presence of ureaplasma urealyticum or mycoplasma hominis in the lower genital tract. J Med Microbiol. 2000;49:697–700.
124. Matalliotakis I, Arici A, Goumenou A, et al. Distinct expression pattern of cytokines in semen of men with genital infection and oligo-terato-asthenozoospermia. Am J Reprod Immunol. 2002;48:170–5.
125. Fraczek M, Kurpisz M. Inflammatory mediators exert toxic effects of oxidative stress on human spermatozoa. J Androl. 2007;28:325–33.
126. Sharma RK, Pasqualotto FF, Nelson DR, et al. The reactive oxygen species-total antioxidant capacity score is a new measure of oxidative stress to predict male infertility. Hum Reprod. 1999;14:2801–7.
127. Agarwal A, Saleh RA, Bedaiwy MA. Role of reactive oxygen species in the pathophysiology of human reproduction. Fertil Steril. 2003;79:829–43.
128. Alvarez JG, Touchstone JC, Blasco L, et al. Spontaneous lipid peroxidation and production of hydrogen peroxide and superoxide in human spermatozoa. J Androl. 1987;8:338–48.
129. Rao B, Soufir JC, Martin M, et al. Lipid peroxidation in human spermatozoa as related to midpiece abnormalities and motility. Gamete Res. 1989;24:127–34.
130. Aitken RJ, Harkiss D, Buckingham DW. Analysis of lipid peroxidation mechanisms in human spermatozoa. Mol Reprod Dev. 1993;35:302–15.
131. Engel S, Schreiner T, Petzoldt R. Lipid peroxidation in human spermatozoa and maintenance of progressive sperm motility. Andrologia. 1999;31:17–22.
132. Fraczek M, Szkutnik D, Sanocka D, et al. Peroxidation components of sperm lipid membranes in male infertility. Ginekol Pol. 2001;72:73–9.
133. Zalata AA, Christophe AB, Depuydt CE, et al. White blood cells cause oxidative damage to the fatty acid composition of phospholipids of human spermatozoa. Int J Androl. 1998;21:154–62.
134. Fraczek M, Sanocka D, Kurpisz M. Interaction between leucocytes and human spermatozoa influencing reactive oxygen intermediates release. Int J Androl. 2004;27:69–75.
135. Fraczek M, Szumala-Kakol A, Jedrzejczak P, et al. Bacteria trigger oxygen radical release and sperm lipid peroxidation in in vitro model of semen inflammation. Fertil Steril. 2007;88:1076–85.
136. Camejo MI, Segnini A, Proverbio F. Interleukin-6 (IL-6) in seminal plasma of infertile men, and lipid peroxidation of their sperm. Arch Androl. 2001;47:97–101.
137. Martinez P, Proverbio F, Camejo MI. Sperm lipid peroxidation and pro-inflammatory cytokines. Asian J Androl. 2007;9:102–7.
138. Zalata AA, Hafez T, Comhaire F. Evaluation of the role of reactive oxygen species in male infertility. Hum Reprod. 1995;10:1444–51.
139. Aitken RJ, Krausz C, Buckingham D. Relationship between biochemical markers for residual sperm cytoplasm, reactive oxygen species generation, and the presence of leukocytes and precursor germ cells in human sperm suspensions. Mol Reprod Dev. 1994;39:268–79.
140. Huszar G, Vigue L. Correlation between the rate of lipid peroxidation and cellular maturity as measured by creatine kinase activity in human spermatozoa. J Androl. 1994;15:71–7.
141. Hughes M, Lewis SE, McKelvey-Martin VJ, et al. A comparison of baseline and induced DNA damage in human spermatozoa from fertile and infertile men, using a modified comet assay. Mol Hum Reprod. 1996;2:613–9.
142. Baccetti B, Collodel G, Piomboni P. Apoptosis in human ejaculated sperm cells (notulae seminologicae 9). J Submicrosc Cytol Pathol. 1996;28:587–96.
143. Lopes S, Sun JG, Jurisicova A, et al. Sperm deoxyribonucleic acid fragmentation is increased in poor-quality semen samples and correlates with failed fertilization in intracytoplasmic sperm injection. Fertil Steril. 1998;69:528–32.
144. Aitken RJ, Baker MA. Oxidative stress, sperm survival and fertility control. Mol Cell Endocrinol. 2006;250:66–9.
145. Aitken RJ, Krausz C. Oxidative stress, DNA damage and the Y chromosome. Reproduction. 2001;122:497–506.

146. Wang X, Sharma RK, Sikka SC, et al. Oxidative stress is associated with increased apoptosis leading to spermatozoa DNA damage in patients with male factor infertility. Fertil Steril. 2003;80:531–5.
147. Agarwal A, Said TM. Oxidative stress, DNA damage and apoptosis in male infertility: a clinical approach. BJU Int. 2005;95:503–7.
148. Oosterhuis GJ, Mulder AB, Kalsbeek-Batenburg E, et al. Measuring apoptosis in human spermatozoa: a biological assay for semen quality? Fertil Steril. 2000;74:245–50.
149. Said T, Agarwal A, Grunewald S, et al. Selection of nonapoptotic spermatozoa as a new tool for enhancing assisted reproduction outcomes: an in vitro model. Biol Reprod. 2006;74:530–7.
150. Grunewald S, Said TM, Paasch U, et al. Relationship between sperm apoptosis signalling and oocyte penetration capacity. Int J Androl. 2008;31:325–30.
151. Dinarello CA. Interleukin-18, a proinflammatory cytokine. Eur Cytokine Netw. 2000;11: 483–6.
152. Chen G, Goeddel DV. TNF-R1 signaling: a beautiful pathway. Science. 2002;296:1634–5.
153. Riccioli A, Salvati L, D'Alessio A, et al. The Fas system in the seminiferous epithelium and its possible extra-testicular role. Andrologia. 2003;35:64–70.
154. Salguero FJ, Sánchez-Cordón PJ, Núñez A, et al. Proinflammatory cytokines induce lymphocyte apoptosis in acute African swine fever infection. J Comp Pathol. 2005;132:289–302.
155. Mariño E, Cardier JE. Differential effect of IL-18 on endothelial cell apoptosis mediated by TNF-alpha and Fas (CD95). Cytokine. 2003;22:142–8.
156. Li MW, Mruk DD, Lee WM, et al. Cytokines and junction restructuring events during spermatogenesis in the testis: an emerging concept of regulation. Cytokine Growth Factor Rev. 2009;20:329–38.
157. Barroso G, Morshedi M, Oehninger S. Analysis of DNA fragmentation, plasma membrane translocation of phosphatidylserine and oxidative stress in human spermatozoa. Hum Reprod. 2000;15:1338–44.
158. Ollero M, Gil-Guzman E, Lopez MC, et al. Characterization of subsets of human spermatozoa at different stages of maturation: implications in the diagnosis and treatment of male infertility. Hum Reprod. 2001;16:1912–21.
159. Lampiao F, du Plessis SS. TNF-alpha and IL-6 affect human sperm function by elevating nitric oxide production. Reprod Biomed Online. 2008;17:628–31.
160. Said TM, Agarwal A, Falcone T, et al. Infliximab may reverse the toxic effects induced by tumor necrosis factor alpha in human spermatozoa: an in vitro model. Fertil Steril. 2005;83:1665–73.
161. Perdichizzi A, Nicoletti F, La Vignera S, et al. Effects of tumour necrosis factor-alpha on human sperm motility and apoptosis. J Clin Immunol. 2007;27:152–62.
162. Allam JP, Fronhoffs F, Fathy A, et al. High percentage of apoptotic spermatozoa in ejaculates from men with chronic genital tract inflammation. Andrologia. 2008;40:329–34.
163. Haidl G, Allam JP, Schuppe HC. Chronic epididymitis: impact on semen parameters and therapeutic options. Andrologia. 2008;40:92–6.
164. Penna G, Mondaini N, Amuchastegui S, et al. Seminal plasma cytokines and chemokines in prostate inflammation: interleukin 8 as a predictive biomarker in chronic prostatitis/chronic pelvic pain syndrome and benign prostatic hyperplasia. Eur Urol. 2007;51:524–33.
165. Matalliotakis I, Goumenou A, Fragouli Y, et al. Soluble IL-6 receptor levels in the seminal plasma of infertile patients with accessory gland infection. Arch Androl. 2000;44:237–42.
166. Krause W, Bohring C, Gueth A, et al. Cellular and biochemical markers in semen indicating male accessory gland inflammation. Andrologia. 2003;35:279–82.

Chapter 10
Metabolic Strategy in Mammalian Spermatozoa and Oxidative Stress

Juan G. Alvarez

Abstract The ability of sperm to display motion and to propel itself in vivo and in vitro is highly dependent on its capability to produce ATP. Maintenance of sperm motility is not only dependent on the normal function of the contractile proteins of the axoneme and its ability to generate ATP, but also on the availability of metabolic substrate in the female reproductive tract. Metabolic enzymatic activities in sperm have kinetic properties similar to those of muscle cells. Although ATP-utilizing enzymes in these two types of cells are very different at a molecular level, they both produce contractile force. This suggests that sperm's metabolic machinery is designed to produce energy for the contractile work of motility with only minor amounts being consumed in other reactions, including those involved in the process of fertilization. The preferential conversion of glucose to lactate through glycolysis may be an important evolutionary feature of sperm, perhaps intended to minimize the accumulation of reducing equivalents in the mitochondria and subsequent increase in oxygen radical production.

Keywords Metabolic strategy • Pyruvate kinase • Oxygen radicals • Glycolysis • ATP

J.G. Alvarez, MD, PhD (✉)
Department of Male Infertility, Centro Androgen, C/Fernando Macias 8, 1°C, 15004 La Coruña, La Coruña, Spain
e-mail: jalvarez@androgen.es

A. Agarwal et al. (eds.), *Studies on Men's Health and Fertility*, Oxidative Stress in Applied Basic Research and Clinical Practice, DOI 10.1007/978-1-61779-776-7_10,

10.1 Metabolic Strategy of Human Spermatozoa

10.1.1 Introduction

The primary function of the spermatozoon is to carry and introduce the intact DNA into the oocyte. For that purpose, after ejaculation the spermatozoon must propel itself from the vagina into the uterus in order to meet the oocyte at the site of fertilization in the ampullar region of the oviduct. Prior to oocyte fertilization, sperm must undergo in the isthmic region of the oviduct a series of biochemical changes known as capacitation that allow sperm to bind and penetrate the oocyte in vivo. Therefore, the spermatozoon could be considered as a self-propelled specialized DNA carrier. However, in order to fully accomplish its function of delivering the paternal genome to the oocyte, the fertilizing spermatozoon, through a phospholipase C_{zeta} that carries in its nuclear membrane, must first activate the oocyte and trigger the expression of homeobox genes that encode proteins responsible for binding specific DNA sequences of the zygote that ultimately determines embryo and fetal development.

The ability of sperm to display motion and to propel itself in vivo and in vitro is highly dependent on its capability of producing ATP which is used as the substrate by dynein ATPase to transduce chemical energy into mechanical work by the contractile proteins of the axoneme. The ability of sperm to maintain its characteristic motility pattern is not going to depend only on the normal function of the contractile proteins of the axoneme and its ability to generate ATP, but also on the availability of metabolic substrate in the female reproductive tract to produce ATP.

In addition to the biochemical changes that experiment during the process of capacitation, sperm are exposed to significant environmental changes as they move from the testis to the distal portion of the oviduct where fertilization normally takes place. Exposure of epididymal sperm to the seminal plasma during ejaculation reinitiates motility and sharply increases sperm energy metabolism. Further environmental changes occur in the female reproductive tract as sperm move out from the seminal plasma into the uterine and oviductal fluids. Under these conditions, sperm survival is going to depend on the efficient utilization and control of metabolic of nutrients present in the extracellular medium of the female reproductive tract.

The purpose of this chapter is to examine the metabolic strategy of human spermatozoa with special emphasis on glycolysis and oxidative metabolism and the impact of this strategy on sperm motility and fertilization.

10.1.2 Glycolysis

Most of the energy that sperm need to maintain its characteristic motility pattern is obtained through glycolysis, also known as the Embden–Meyerhof pathway, which is highly active under both aerobic and anaerobic conditions. The effects of substrate, substrate concentration and changes in medium composition on the rate of glycolysis in human sperm were first reported by Peterson and Freund [1]. In this study, the conditions under which fructose and glucose levels limit glycolysis were described. They also found that the transport of carbohydrates into sperm was not the rate-limiting step of glycolysis and that the sites of glycolytic control are downstream hexokinase. These sites are related to the regulation of the activity of glycolytic enzymes by a key cofactor: inorganic phosphate. Phosphofructokinase is involved in this critical control given its low intracellular concentration in sperm and its sensitivity to adenine nucleotides. Inorganic phosphate stimulates phosphofructokinase activity in cell-free extracts and glycolysis in intact sperm [2, 3]. However, this stimulation by inorganic phosphate only accounts for a 30% increase in the rate of glycolysis, even at phosphate concentrations up to 40 mM, thus suggesting that phosphate control of glycolysis is perhaps less important than control by other cofactors.

The enzymatic profile of human spermatozoa is not typical of cells that exhibit a high rate of aerobic glycolysis. This implies that certain metabolic changes must take place after ejaculation. This hypothesis was postulated based on the similarities in the enzymatic profile between human and bull spermatozoa. Early work by Lardy and Parks indicated that glycolysis in epididymal bull sperm sharply decreased when sperm were exposed to oxygen and increased after ejaculation, thus suggesting that a metabolic regulator may be released at the time of ejaculation that uncouples oxidative phosphorylation which, in turn, decreases the respiratory inhibition of glycolysis [3].

Human sperm contain glucose-6-phosphate dehydrogenase and 6-phosphogluconate dehydrogenase, the entry enzymes into the hexose monophosphate shunt. The shunt pathway has four major functions: (1) to provide reducing power in the form of NADPH required to drive biosynthetic reactions; (2) to provide the pentose needed for nucleic acid biosynthesis; (3) to provide an additional mechanism for the oxidative generation of ATP; and (4) to provide NADPH to maintain operative the glutathione reductase/glutathione peroxidise protective system for reduction of hydrogen peroxide and lipid hydroperoxides [4]. However, based on the fact that there is no biosynthetic activity in human sperm after their release into the seminiferous tubules, it is highly unlikely that the first two functions be of any relevance on sperm function. In addition, sperm are unable to metabolize ribose-5-phosphate, adenosine, or uridine, which are pentose shunt metabolites usually converted to lactic acid at rapid rates in cells where the shunt pathway supports biosynthesis. However, the physiological relevance of the third function indicated above is supported by the work of Sarkar et al. [5], who found that sperm incubated in the presence of NADP and glucose-6-phosphate or 6-phosphogluconate produce high levels of NADPH.

At this point is worth mentioning the significance of the low levels of α-glycerophosphate dehydrogenase in human sperm. This enzyme plays an important role in oxidative metabolism because it catalyzes the conversion of dehydroxyacetone phosphate to α-glycerophosphate which can be directly oxidized by mitochondria. Furthermore, this enzyme competes with lactate dehydrogenase for NADH, the reduced form of NAD and, therefore, can provide additional NAD for the reduction of pyruvate to lactate, a key step in sperm's metabolic strategy. Therefore, differences in the activity α-glycerophosphate dehydrogenase could explain, at least in part, some of the differences observed in oxidative metabolism between human sperm and other mammalian species. For example, higher levels of this enzyme in ram spermatozoa account for the higher rates of glucose oxidation observed in this species and also for the higher rates of sorbitol and glycerol oxidation.

The catalytic activity of glycolytic enzymes in mammalian spermatozoa exceeds the rate of glycolysis, thus indicating that glycolysis is not limited by the concentration of any particular enzyme. In sharp contrast, phosphofructokinase is found in very low concentration in sperm, suggesting a potential regulatory role for this enzyme in sperm metabolism. Furthermore, the activity of phosphofructokinase in sperm homogenates is significantly affected by cofactors known to be involved in other cell types. High concentrations of ATP inhibit phosphofructokinase activity, while AMP and inorganic phosphate stimulate its activity. These observations imply that phosphafructokinase plays an important role as a rate-limiting step in glycolysis control.

The overall enzymatic profile of epididymal sperm is characteristic of cells that exhibit a pronounced *Pasteur effect*. That is, oxygen inhibition of glycolysis. An important feature of ejaculated human sperm, however, is the absence of any detectable *Pasteur effect*. Addition of dinitrophenol, known to release the inhibitory effect of oxidative metabolism on glycolysis, to sperm suspensions barely stimulates the rate of aerobic glycolysis. The absence of a significant *Pasteur effect* in human sperm also provides an explanation for the relatively small effect of inorganic phosphate on the rate of glycolysis.

The rate of conversion of glucose to lactate by motile sperm suspended in seminal plasma-free medium is constant over a 30-fold range in glucose concentration (1–30 mM), whereas lactate conversion from fructose increases less than twofold over the same concentration range. The observation that glucose-6-phosphate enters the sperm cell without prior dephosphorylation and is metabolized to lactate at the same rate as fructose and glucose provides further evidence that glycolysis in human sperm is not substrate-limited. This also indicates that the fructose levels usually present in semen are more than enough to sustain a high rate of glycolysis, even at very low concentrations, as is the case in the female reproductive tract.

Rabbit spermatozoa convert more than 70% of the glucose consumed to lactate under aerobic conditions through the Embden–Meyerhof pathway, as shown by Murdoch and White [6]. These investigators used both the measurement of glucose uptake and lactate production of $^{14}CO_2$ produced from glucose labeled at the 1- and 6-positions. One of the two enzymes in the Embden–Meyerhof pathway that produce net ATP is pyruvate kinase [7, 8], which can be readily measured in hypotonically treated epididymal spermatozoa [9]. Based on kinetic studies on their substrates, cofactors, and effectors [10, 11], two major isozymes of pyruvate kinase have been identified. One is the type showing allosteric control by fructose-1,6-biphosphate. An isoenzyme of this type, designated L [12], is found in yeast [7] and hepatocytes [13]. Other isozymes of this type are M2 from liver nonparenchymal cells [13–15] and A from kidney [16]. The control characteristics of this type of enzyme and, in particular its allosteric activation by fructose-1,6-biphosphate, are those expected for an enzyme operating in a pathway in which the flux of substrate is closely controlled [8]. The biosynthetic pathways functioning in liver parenchymal cells require maximal utilization of ATP with minimal utilization of glucose [17]. That is, the glucose used is effectively converted to either H_2O and CO_2 or to precursors to be used in anabolic pathways. The same consideration applies to kidney cells that have the gluconeogenic pathway and the A isozyme of pyruvate kinase.

The second type of pyruvate kinase, characteristic of the muscle isozyme M [18], does not have this kind of allosteric control. It is, therefore, well suited for handling higher fluxes of glucose for conversion into lactate through the Embden–Meyerhof pathway [8]. The efficiency of ATP production, taken as moles of ATP produced per mole of glucose, is lower, but it can support a high rate of ATP production. These control characteristics of the muscle enzyme are those required for conversion of metabolic energy into mechanical work in which efficiency of glucose utilization is subordinated to maximal rate of ATP production [17]. The flux of glucose through the Embden–Meyerhof pathway is usually high in order to maintain speed and force of muscle contraction. Therefore, the kinetic properties of the pyruvate kinase in any given cell type are useful in terms of inferring the regulatory characteristics of glycolysis in any particular cell type.

10.1.3 Oxidative Metabolism

10.1.3.1 Krebs Cycle

The existence of oxidative metabolism in human sperm has long been established [2, 19]. It has also been shown that oxidative phosphorylation can support sperm motility in the absence of glycolysis [20]. However, given the idiosyncrasy of sperm, oxidative metabolism in human spermatozoa is less effective than glycolysis in maintaining intracellular levels of ATP along the length of the flagellum and in supporting sperm motility. The inability of oxidative metabolism to generate high levels of ATP may explain, at least in part, the absence of a significant *Pasteur effect* in human sperm.

The finding by Peterson and Freund [21] that phosphofructokinase activity in human sperm homogenates is markedly inhibited by ATP and that this inhibition can be prevented or reversed by the addition of inorganic phosphate is consistent with this notion. However, the addition of inorganic phosphate to intact spermatozoa results in a discrete increase in aerobic glycolysis which appears to reflect more a simple ionic requirement than glycolytic control [1]. This suggests that ATP generated aerobically in intact spermatozoa is insufficient to cause significant inhibition of enzymatic activity at the phosphofructokinase level.

With the exception of succinate, Krebs cycle intermediates, glucose and pyruvate, do not significantly increase respiration over the endogenous rate. Possible causes for this low oxidation rates include: (1) the pathway to oxygen is common at coenzyme Q for the oxidation of succinate and pyridine nucleotide-linked substrates. The high rate of succinate oxidation, therefore, eliminates this portion of the electron-transfer chain by limiting the oxidation of other substrates. This assumes that succinate-stimulated respiration is not caused by exposure and activation of succinate dehydrogenase after sperm washing. Hamner and Williams [22] reported that the level of glucose oxidation by human sperm could be increased to the level of succinate oxidation following the addition of bicarbonate. This could be explained by the fact that bicarbonate speeds the turnover rate of the Krebs cycle and of acetyl CoA oxidation by providing increasing amounts of a rate-limiting component: oxaloacetate via CoA fixation to pyruvate. Murdoch and White reported that addition of 6 mM bicarbonate slightly stimulated sperm respiration. These observations were also confirmed by Peterson and Freund using polarographic measurements. When tested in the presence of 1–10 mM sodium bicarbonate, the rate of oxidation of succinate was consistently two to fourfold higher than the rate of glucose oxidation [2, 23]. Furthermore, the respiration rates in semen, where bicarbonate levels are relatively high, are relatively low but can be stimulated by succinate. This indicates that factors other than bicarbonate limit the oxidation of pyridine nucleotide-linked substrates. On this regard, it should be pointed out that the concentration of phosphate acceptors, such as ADP and AMP, do not usually limit the oxidation of pyridine nucleotide-linked substrates. The increased rate of oxidation in the presence of succinate alone suggests that this is so because the succinate pathway bypasses two of the three potential rate-limiting steps involved in ATP biosynthesis.

Citrate and other Krebs cycle intermediates are readily formed from pyruvate in human sperm, provided that a source of oxaloacetate is present. Activation of the Krebs cycle in mammalian spermatozoa appears to be limited by oxaloacetate. The conversion of acetate to citrate, however, is not increased in the presence of substrates that increase oxaloacetate levels. This is in agreement with the report by Terner [19] who found that in human spermatozoa acetate was oxidized at much lower rates than glucose or pyruvate. Although an increased rate of citrate formation may increase the rate of oxidation of a particular substrate, the overall rate of respiration is not increased by stimulating citrate synthesis and, therefore, other substrate must regulate the control of oxidation of pyridine nucleotide-linked substrates. The high rate of succinate oxidation indicates that most Krebs cycle enzymes are present in more than adequate concentrations to handle the substrate load that sperm may usually encounter. Therefore, oxidation may be limited by either the catalytic capacity of mitochondrial pyridine nucleotide dehydrogenase or some form of stringent cofactor control. Several metabolic peculiarities of human sperm are known to contribute to the near absence of a *Pasteur effect*. First, α-glycerophosphate dehydrogenase activity is very low [24]. This would tend to increase the NADH available to lactate dehydrogenase for pyruvate reduction and, therefore, lower the acetyl CoA available for mitochondrial oxidation through the Krebs cycle. Second, high ATP/ADP ratios are not found in human spermatozoa perhaps due to the weak coupling of respiration to phosphorylation. This would tend to lower feedback inhibition of glycolysis at several points in the glycolytic pathway.

10.1.3.2 Fatty Acid Oxidation

Long-chain fatty acids must first be converted to the acyl esters of CoA in the cytosolic compartment before they can be oxidized by mitochondria. Acyl esters of CoA are then converted to the carnitine esters by the corresponding transferase located on the outside of the inner mitochondrial membrane [25–28]. In addition to oxidation of the aliphatic chain, human spermatozoa have been also shown to produce ATP through the conversion of the glycerol moiety of the lipid molecule to L-3-glycerophosphate [29, 30].

The mitochondrial oxidative activity profile of mammalian spermatozoa differs significantly between the different mammalian species. The profile of human spermatozoa falls between the full set of enzyme activity typical somatic cell mitochondria, like that of bull sperm mitochondria, and the limited set displayed by rabbit spermatozoa [31]. For example, mouse sperm lack the necessary enzymes needed to utilize the malate/aspartate shuttle [32] for transfer of these equivalents, as shown by their inability to oxidize glutamate in the presence of malate [33]. It is likely that the lactate/pyruvate shuttle, common to most mammalian spermatozoa, plays a major role in transferring NADH equivalents from the cytosol to the mitochondria in mouse sperm. On the other hand, mouse sperm can oxidize fatty acids directly from the CoA esters. These CoA esters provide mouse sperm with a supply of endogenous substrate which would enable them to remain motile for more than 4 h in a simple solution containing NaCl, tris–HCl, and $CaCl_2$ [34, 35]. Sperm from BL/6 mice maintain their high initial motility for more than 6 h in a defined saline medium buffered with phosphate and bicarbonate and lacking oxidizable substrates [36]. The differences in mitochondrial activities of human, bovine, mouse, and rabbit spermatozoa suggest that bovine gametes are more self-sufficient with regard to the ample set of substrates that they could possibly utilize, in addition to their endogenous reserves of fatty acids. If the hypothesis postulated by Storey that sperm mitochondrial activities can be used to predict the concentration of nutrients (from which energy can be derived) in the oviduct is correct, then the bovine oviductal lumen should have low concentrations of these nutrients. Data regarding substrate concentration in the lumen of bovine oviduct support this hypothesis [37–39].

10.2 Impact of Sperm Metabolic Strategy on Sperm Motility

The energy metabolism of a cell type should be related to its function. This relation has been termed as the metabolic strategy of the cell. In the case of sperm, the question that may be asked is: is sperm's metabolic strategy geared to the series of reactions related to the process of fertilization including motility, or does the maintenance of motility dominate the metabolic strategy of the spermatozoon? These questions can be clarified by a close look at the kinetic properties of pyruvate kinase in sperm, as compared with that of liver and muscle, as well as by looking at the metabolic coupling of pyruvate kinase with flagellar ATPase. Pyruvate kinase has access to its substrates in permeabilized epididymal sperm while remaining bound to the sperm cytoskeleton and suffering minimal perturbation from the pemeabilization procedure [9]. Epididymal sperm are usually immotile but motility is restored by suspending sperm in KCl medium containing Mg^{2+} upon addition of 3 mM ATP, according to the method of Gibbons and Gibbons [40], and adding 50 mM phosphoenolpyruvate to maintain ATP levels [41]. Under these conditions, phosphoenolpyruvate is converted to pyruvate. This conversion is readily measured, so that the activity of flagellar ATPase in permeabilized epididymal sperm can be quantified and putative regulators assessed.

Flagellar ATPase in mammalian spermatozoa has an activity comparable to that of somatic cell pyruvate kinase. Sperm flagellar ATPase requires Mg^{2+} for activity [41]. However, if Mg^{2+} is omitted from the medium, then sperm remain immotile even in the presence of 3 mM ATP and 50 mM phosphoenolpyruvate. Motility is restored following addition of 3 mM Mg^{2+}. The calculated K_m values for ATP and Mg^{2+} at 37°C were 0.22 and 0.25 mM, respectively, thus suggesting that the substrate for flagellar ATPase is MgATP, as previously shown by Tibbs for sperm tails from perch [42] and by Gibbons and Gibbons [40] and Hayasi for sea urchin axonemes [43].

Pyruvate kinase in mammalian spermatozoa is kinetically very similar to that of muscle and it is affected by the same inhibitors and activators as in the case of the muscle enzyme. The question of whether or not this similarity extends to the protein structure of the enzyme remains to be elucidated. The difficulty of extracting and purifying this and other enzymes from sperm without damaging their structure [44, 45] makes its characterization difficult using immunological methodology. Immunological cross-reactions can occur among some of the pyruvate kinase isozymes [46] and, therefore, protein characterization is rendered a near impossible task. For the purpose of metabolic analysis, however, the similarity of the kinetic properties and control characteristics of the sperm and muscle pyruvate kinases implies that control of the glycolytic pathway in sperm operates in a similar fashion to that of skeletal muscle. Both cells require ATP for mechanical work by contractile proteins. Both cells have access to glucose reserves: the muscle cell through glycogenolysis and the sperm cell through an inexhaustible supply of glucose in the oviduct [47]. In mature spermatozoa, the energy-producing and energy-utilizing pathways seem fairly well balanced. The activities of pyruvate kinase and flagellar ATPase and lactate dehydrogenase are all of the same order of magnitude and higher than the activity of mitochondrial lactate or pyruvate oxidation system [9]. The metabolic strategy of mature sperm depends on a high rate of glycolysis to provide ATP for motility rather than maximal efficiency through the Krebs cycle, as in the case of skeletal muscle for contractibility [8, 17].

10.3 Evolutionary Features of the Metabolic Strategy of Mammalian Spermatozoa

The enzymatic activities of mammalian spermatozoa have kinetic properties and respond to regulation in a manner that closely resembles those of muscle cells. Although ATP-utilizing enzymes in these two types of cells are very different at a molecular level, they both produce contractile force, an activity to which the cellular metabolism is geared. This suggests that sperm's metabolic machinery is geared towards the maintenance of energy for the contractile work of motility; only minor amounts of metabolic energy appear to be consumed in other reactions, including those involved in the process of fertilization.

The preferential conversion of glucose to lactate under aerobic conditions through the Embden–Meyerhof pathway may be an important evolutionary feature of sperm, perhaps intended to minimize the accumulation of reducing equivalents in the mitochondria and the subsequent increase in oxygen radical production by the mitochondria. It is well known that sperm produce oxygen radicals, that the bulk of these radicals are produced by the mitochondria, and that oxygen radical-induced damage could result in loss of sperm motility, loss of acrosomal contents, and oxidative DNA damage [48–53]. Conversion of glucose to lactate under aerobic conditions would (1) decrease production of mitochondrial NADH and FADH-reducing equivalents by the Krebs cycle; and (2) decrease electron flow in the inner mitochondrial membrane thereby downregulating oxygen radical formation. This metabolic feature of sperm is especially important outside the protective environment of the epididymis where oxygen radical-induced damage is minimized by a lower temperature and lower pO_2.

There is another evolutionary feature related to the metabolic strategy of mammalian spermatozoa worth mentioning. If we ask the question: is maximal efficiency of ATP production subordinated to maximal rate of ATP utilization in the axoneme simply because it would reduce NADH and FADH-reducing equivalents and oxygen radical production by the mitochondria? Is there another important evolutionary feature related to this strategy? The answer would be that both apply. Although ATP production by the mitochondria is far more efficient than ATP production by glycolysis in terms of moles of ATP produced per mole of glucose, should the mitochondria be the sole source of ATP for sperm, given the dimensions of the sperm flagellum and the almost lack of cytosol, diffusion of ATP to the more distal segments of the axoneme would be virtually impossible. The ATP generated in the mitochondria could only effectively diffuse to the midpiece and proximal

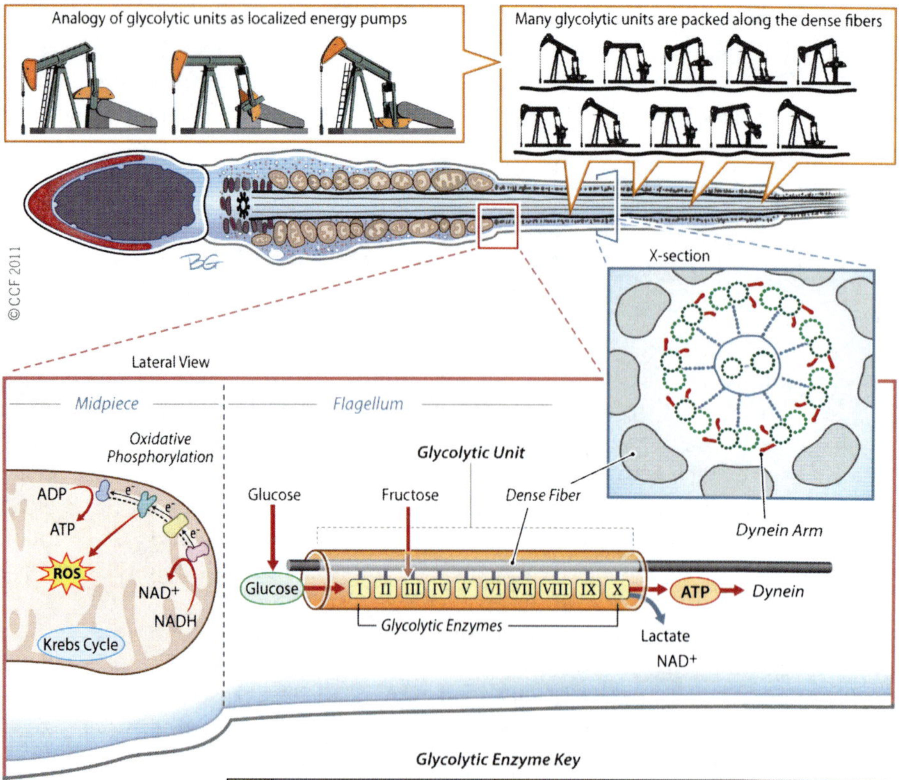

Glycolytic Enzyme Key

I	Hexokinase	VI	Glyceraldehyde Phosphate Dehydrogenase
II	Phosphoglucose Isomerase	VII	Phosphogluconate Kinase
III	Phosphofructokinase	VIII	Phosphoglycerate Mutase
IV	Aldolase	IX	Pyruvate Kinase
V	Triosephosphate Isomerase	X	Lactate Dehydrogenase

Fig. 10.1 Diagram that depicts the role of the glycolytic units in the axoneme in energy production in mammalian spermatozoa

region of the axoneme for appropriate maintenance of sperm motility. It has been also proposed that in order to optimize ATP utilization in the axoneme, glycolysis should be organized in the form of glycolytic units (Fig. 10.1). That is, rather than being in solution in the cytosol, given the fact that spermatozoa are highly compact cells with very little cytosol, glycolytic enzymes should be bound to cytoskeleton of the axoneme, i.e., fibrous sheath, and organized as structural and functional units that would span from hexokinase to lactate dehydrogenase. Rather than utilizing the "energy plant" of the mitochondria to produce ATP, sperm would utilize multiple "pumps" of ATP distributed along the length of the flagellum, pumping ATP at very high rates. The net ATP generated by these glycolytic units would be readily delivered to the flagellar ATPase. Recently, Young-Hwan et al. have revisited this concept of compartmentalization of glycolytic enzymes in the flagellum providing further support to this hypothesis. They found colocalization of a sperm flagellar energy carrier, designated as SFEC, and glycolytic enzymes attached to the fibrous sheath, thus supporting a growing literature that the principal piece of the flagellum is capable of generating and regulating ATP independently from mitochondrial oxidative phosphorylation in the midpiece. A model is proposed by these investigators by which the fibrous sheath represents a highly ordered complex, analogous to the electron transport chain of the mitochondria, in which adjacent enzymes in the glycolytic pathway are assembled to permit efficient flux of energy substrates and products with SFEC serving to mediate energy-generating and energy-consuming processes in the distal flagellum, possibly as a nucleotide shuttle between flagellar glycolysis, protein phosphorylation, and mechanisms of motility [54].

10.4 Conclusions

It appears that the metabolic strategy of sperm is similar to that of muscle cells in that (1) it depends on a high rate of glycolysis and low utilization of the Krebs cycle to provide ATP for motility; and (2) the hexose monophosphate shunt is almost exclusively devoted to produce NADPH to maintain functional the glutathione reductase/glutathione peroxidase protective system for reduction of hydrogen peroxide and lipid hydroperoxides. Sperm's metabolic strategy, therefore, is both geared to maintain high rates of ATP production/utilization for motility and to ensure that the spermatozoon destined to fertilize the egg can propel itself to reach the site of fertilization while avoiding the omnipresent threat of free radical-induced membrane and DNA damage that would interfere with sperm's primary function: *to activate the oocyte and deliver the intact DNA to the oocyte for embryo and fetal development.*

References

1. Peterson RN, Freund M. Glycolysis by washed suspensions of human spermatozoa. Effect of substrate, substrate concentration and changes in medium composition on the rate of glycolysis. Biol Reprod. 1969;1:238–46.
2. Murdoch RN, White IG. Studies on the metabolism of human spermatozoa. J Reprod Fertil. 1968;16:351–61.
3. Lardy HA, Parks RE. Phosphofructokinase. In: Gaehler O, editor. Enzymes: units of biological structure and function. New York: Academic; 1956. p. 584–92.
4. Storey BT, Alvarez JG, Thompson KA. Human sperm glutathione reductase activity in situ reveals limitation in the glutathione antioxidant defense system due to supply of NADPH. Mol Reprod Dev. 1998;49:400–7.
5. Sarkar S, Nelson AJ, Jones OW. Glucose-6-phosphate dehydrogenase activity in human sperm. J Med Genet. 1977;14:250–5.
6. Murdoch RN, White IG. The metabolism of labelled glucose by rabbit spermatozoa after incubation in vitro. J Reprod Fertil. 1967;14:213–23.
7. Hess B, Brand K. Enzyme and metabolic profiles. In: Chance B, Estabrook RW, Williamson JR, editors. Control of energy metabolism. New York: Academic; 1965. p. 111–21.

8. Newsholme EA, Start C. Regulation of metabolism. New York: Wiley; 1973. p. 88–145.
9. Storey BT, Kayne FJ. Energy metabolism of spermatozoa. V. The Embden-Meyerhof pathway of glycolysis: activities of pathway enzymes in hypotonically-treated rabbit epididymal spermatozoa. Fertil Steril. 1975;26(12):1275–65.
10. Susor WA, Ratter WJ. Some distinctive properties of pyruvate kinase purified from rate liver. Biochem Biophys Res Commun. 1968;18:527–36.
11. Carminatti HL, Jimenez de Asua L, Leiderman B, Rozengurt E. Allosteric properties of skeletal muscle pyruvate kinase. J Biol Chem. 1971;246:7284–8.
12. Seubert W, Schoner W. The regulation of pyruvate kinase. Curr Top Cell Regul. 1971;3:237–67.
13. Crisp DM, Pogson CI. Glycolytic and gluconeogenic enzyme activities in parenchymal and nonparenchymal cells from mouse liver. Biochem J. 1972;126:1009–23.
14. Jimenez de Asua L, Leiderman B, Rozengurt E, Carminatti HL. Some kinetic differences between the M isozymes of pyruvate kinase from liver and muscle. Biochim Biophys Acta. 1971;135:326–34.
15. Imamura K, Taniuchi K, Tanaka T. Multimolecular forms of pyruvate kinase. II. Purification of M2 type pyruvate kinase from Yoshida ascitis hepatoma 130 cells and comparative studies on the enzymological properties of the three types of pyruvate kinase L, M1 and M2. J Biochem. 1972;72:1001–15.
16. Berglund L, Humble E. Kinetic properties of pig pyruvate kinases type A from kidney and type M from muscle. Arch Biochem Biophys. 1979;195:347–61.
17. Scrutton MC, Utter MF. The regulation of glycolysis and gluconeogenesis in animal tissues. Annu Rev Biochem. 1968;37:249–302.
18. Kayne FJ. Pyruvate kinase. In: Boyer PD, editor. The enzymes, vol. 8. New York: Academic; 1973. p. 353–50.
19. Terner C. Oxidation of exogenous substrate by isolated human spermatozoa. Am J Physiol. 1960;198:48–50.
20. Nevo A. Relation between motility and respiration in human spermatozoa. J Reprod Fertil. 1966;11:19–26.
21. Peterson RN, Freund M. Profile of glycolytic enzyme activities in human spermatozoa. Fertil Steril. 1970;21:151–8.
22. Hamner CE, Williams WL. Identification of sperm stimulating factor of rabbit oviduct fluid. Proc Soc Exp Biol Med. 1964;117:240–3.
23. Peterson RN, Freund M. ATP synthesis and oxidative metabolism in human spermatozoa. Biol Reprod. 1970;3:47–54.
24. Peterson RN, Freund M. Glycolysis by human spermatozoa: levels of glycolytic intermediates. Biol Reprod. 1971;5:221–7.
25. Fritz IB, Yue KTN. Long-chain carnitine acyl-transferase and the role of acylcarnitine derivatives in the catalytic increase of fatty acid oxidation induced by carnitine. J Lipid Res. 1963;4:279–89.
26. Bode C, Klingenberg M. [Die veratmung von fettsauren in isolierten mitochondrien]. Biochem Z. 1965;341:271–89.
27. Haddock BA, Yates DW, Garland PB. The localization of some coenzyme A dependent enzymes in rat liver mitochondria. Biochem J. 1970;119:565–73.
28. Yates DW, Garland PB. Carnitine palmitoyl-transferase activities of rat liver mitochondria. Biochem J. 1970;119:547–52.
29. Mohri H, Mohri T, Ernster L. Isolation and enzymic properties of the midpiece of bull spermatozoa. Exp Cell Res. 1965;38:217–46.
30. Mohri H, Masaki J. Glycerokinase and its possible role in glycerol metabolism of bull spermatozoa. J Reprod Fertil. 1967;14:179–94.
31. Carey JE, Olds-Clark P, Storey BT, Storey BT. Oxidative metabolism of spermatozoa from inbred and random bred mice. J Exp Zool. 1981;216:285–92.
32. Borst P. Hydrogen transport and transport of metabolites. In: Karlson O, editor. Funktionelle und Morphologisshe Organization der Zelle. Berlin: Springer; 1963. p. 135–58.

33. LaNoue KF, Wadajtys EI, Williamson JR. Regulation of glutamate metabolism and interactions with the citric acid cycle in rat heart mitochondria. J Biol Chem. 1973;248:7171–83.
34. Saling PM, Storey BT, Wolf DP. Calcium-dependent binding of mouse epididymal spermatozoa to the zona pellucida. Dev Biol. 1978;65:515–25.
35. Heffner LJ, Saling PM, Storey BT. Separation of calcium effects on motility and zona binding ability in mouse spermatozoa. J Exp Zool. 1980;212:53–9.
36. Carey J, Olds-Clark P. Differences in sperm function in vitro but not in vivo between inbred and random-bred mice. Gamete Res. 1980;3:9–15.
37. Hamner C. Oviductal fluid-composition and physiology. In: Greep RO, editor. Handbook of physiology, section 7 (endocrinology), vol. 2, part 2. Washington, DC: American Physiological Society; 1973. p. 141–52.
38. Brackett BG, Mastroianni L. Composition of oviductal fluid. In: Johnson AD, Foley CW, editors. The oviduct and its functions. New York: Academic; 1974. p. 133–59.
39. Blandau RJ, Brackett B, Brenner RM, et al. The oviduct. In: Greep RO, Koblinsky MA, editors. Frontiers in reproduction and fertility control, part 2. Cambridge, MA: MIT; 1977. p. 132–45.
40. Gibbons BH, Gibbons IR. Flagellar movement and adenosine triphosphatase activity in sea urchin sperm extracted with Triton X-100. J Cell Biol. 1972;54:75–97.
41. Keyhani E, Storey BT. Energy conservation capacity and morphological integrity of mitochondria in hypotonically-treated rabbit epididymal spermatozoa. Biochim Biophys Acta. 1973;305:557–69.
42. Tibbs J. Adenosine triphosphate and acetylcholinesterase in relation to sperm motility. In: Bishop DW, editor. Spermatozoa motility. Washington, DC: American Association for the Advancement of Science; 1962. p. 233–50.
43. Hayasi M. Kinetic analysis of axoneme and dynein ATPase from sea urchin sperm. Arch Biochem Biophys. 1974;165:288–96.
44. Calvin HI. Isolation and subfractionation of mammalian sperm heads and tails. In: Prescott DM, editor. Methods in cell biology, vol. 13. New York: Academic; 1976. p. 85–104.
45. Hrudka F. A morphological and cytochemical study of isolated sperm mitochondria. J Ultrastruct Res. 1978;63:1–19.
46. Osterman J, Fritz PF. Pyruvate kinase isozymes from rat intestinal mucosa. Characterization and the effect of fasting and refeeding. Biochemistry. 1974;13:1731–6.
47. Holmdahl TH, Mastroianni L. Continuous collection of rabbit oviductal secretions at low temperature. Fertil Steril. 1965;16:587–95.
48. Alvarez JG, Storey BT. Spontaneous lipid peroxidation in rabbit epididymal spermatozoa. Biol Reprod. 1982;27:1102–8.
49. Holland MK, Alvarez JG, Storey BT. Production of superoxide dismutase and activity of superoxide dismutase in rabbit epididymal spermatozoa. Biol Reprod. 1982;27:1109–18.
50. Alvarez JG, Holland MK, Storey BT. Spontaneous lipid peroxidation in rabbit spermatozoa: a useful model for the reaction of O_2 metabolites with cells. In: Lubbers DW, Acker H, Leninger-Follert E, Goldstick TK, editors. Oxygen transport to tissue—V. New York: Plenum; 1984. p. 433–43.
51. Alvarez JG, Storey BT. Assessment of cell damage caused by spontaneous lipid peroxidation in rabbit spermatozoa. Biol Reprod. 1984;30:323–32.
52. Alvarez JG, Touchstone JC, Blasco L, Storey BT. Spontaneous lipid peroxidation and production of hydrogen peroxide and superoxide in human spermatozoa. Superoxide dismutase as major enzyme protectant against oxygen toxicity. J Androl. 1987;8:338–48.
53. Fraga CG, Motchnik PA, Shigenaga MK, Helbock HJ, Jacob RA, Ames BN. Ascorbic acid protects against endogenous oxidative DNA damage in human sperm. Proc Natl Acad Sci USA. 1991;88:11003–6.
54. Kim YH, Haidl G, Schaefer M, Egner U, Herr JC. Compartmentalization of a unique ADP/ATP carrier protein SFEC (sperm flagellar energy carrier, AAC4) with glycolytic enzymes in the fibrous sheath of the human sperm flagellar principal piece. Dev Biol. 2007;302:463–76.

Chapter 11
Role of Protamine Disulphide Cross-Linking in Counteracting Oxidative Damage to DNA

Juan G. Alvarez and Jaime Gosalvez

Abstract Free radical (oxygen and nitrogen-derived)-induced sperm DNA damage may take place during the process of spermatogenesis, during sperm transit through the epididymis, in the distal seminal ducts of the male genital tract and in vitro during sperm processing. Free radical-induced DNA damage during spermatogenesis may result in damage of the DNA strands and of highly sensitive telomeric DNA sequences. Post-testicular sperm DNA damage in the epididymis is considered one of the main causes of sperm DNA damage. Release of oxygen radical-producing immature spermatozoa into seminiferous tubules may result in damage of mature spermatozoa during co-migration spermatozoa through the epididymis. Post-testicular free radical-induced sperm DNA damage may also take place in the proximal portion of the epididymis through free radicals such as the superoxide anion that leak from redox recycling mechanisms involved in disulphide bond cross-linking of protamines and flagellar proteins; in the cauda epididymis through free radicals produced by the epithelial cells; and during sperm processing.

Keywords Protamine disulphide cross-linking • Counteracting oxidative damage to DNA • Oxygen radical-induced DNA damage • Candidate DNA sequences • Varicocele

---

J.G. Alvarez, MD, PhD (✉)
Department of Male Infertility, Centro Androgen, La Coruña 15004, La Coruña, Spain
e-mail: jalvarez@androgen.es

J. Gosalvez, BSc, PhD
Department of Biology, University Autónoma Madrid, Madrid 28049, Spain

A. Agarwal et al. (eds.), *Studies on Men's Health and Fertility*, Oxidative Stress in Applied Basic Research and Clinical Practice, DOI 10.1007/978-1-61779-776-7_11,

11.1 Introduction

Free radical-induced sperm DNA damage may take place during the process of spermatogenesis, during sperm transit through the epididymis [1, 2] and in vitro during sperm processing [3, 4]. When the fine balance between oxygen radical (reactive oxygen species [ROS]) production and antioxidant defences is tilted towards net ROS presence in the intracellular and extracellular medium, oxidative stress develops. This can trigger apoptotic mechanisms and cause direct damage of critical cell structures such as the plasma and acrosomal membranes and the chromatin. Exceedingly high levels of ROS may also cause necrosis [5, 6]. Oxygen radicals, including free radicals and peroxides, are particularly deleterious to the cell through direct, indirect or synergistic mechanisms. Alternatively, ROS, through oxidoreduction reactions, may be converted into more reactive radical species capable of causing more extensive cell damage [7]. Although the short-term effects of excessive ROS production are mainly expressed as loss of overall cell viability, once the DNA molecule is affected, these effects may be long-lasting, since they can be inherited [8]. If sperm DNA is damaged, given the lack of DNA repair mechanisms in the male germ line during sperm maturation, there is the risk of transmitting DNA mutations to the oocyte after fertilization [9]. Although some of these mutations could be repaired at the pronuclear stage embryo, if the transmitted DNA damage affects constitutive genes or DNA motifs related with gene regulation of epigenetic control, embryo viability could be greatly compromised.

Oxygen radicals are among the most damaging molecules capable of modifying cell components. These radicals include the superoxide anion ($O_2^{-\bullet}$), hydrogen peroxide (H_2O_2), the hydroxyl radical ($OH^{-\bullet}$), peroxynitrite, organic hydroperoxides (ROOH), alkoxy and peroxy radicals and hypochlorous acid (HOCl). Those that are liposoluble, such as H_2O_2, HOCl or ROOH, may alter membrane stability and, in the particular case of sperm motility. Free radicals such as $O_2^{-\bullet}$, may cross-react to produce Fe^{2+} using proteins as a target and generate H_2O_2 or be the precursors for the metal-catalyzed $OH^{-\bullet}$ formation. Although the biological consequences of many DNA base modifications as products of oxidative stress are known, attack of only OH free radicals, is particularly effective in modifying, for example, the C4–5 double bond of pyrimidine and the OH radical of purines. This generates a spectrum of oxidative stable conformations such as 8-OHdG, 8-OHdA, formamidopyrimidines thymine glycol, uracil glycol, urea residue, 5-OHdU, 5-OHdC, hydantoin (Fig. 11.1), which may cause unrepaired DNA lesions. Thus, for example, thymine glycol is able to block DNA replication and subsequently is potentially lethal to cells. The case of unrepaired 8-oxo-dG mismatching with dA is a well-known situation which increases G to T transition mutations with the subsequent consequences for gene expression if affecting structural genes. Although little is known about the potential negative effects of nitric oxide-derived oxidative processes, they have been shown to decrease sperm motility [10], but also to increase it [11]. Chemical studies suggest that nitric oxide may produce single-stranded DNA breaks (SSBs) and abasic sites (AP sites [12, 13]), predominately in regions with guanine residues, producing base modifications such as 8-Oxo-dG and 8-Oxo-nitro-G [14].

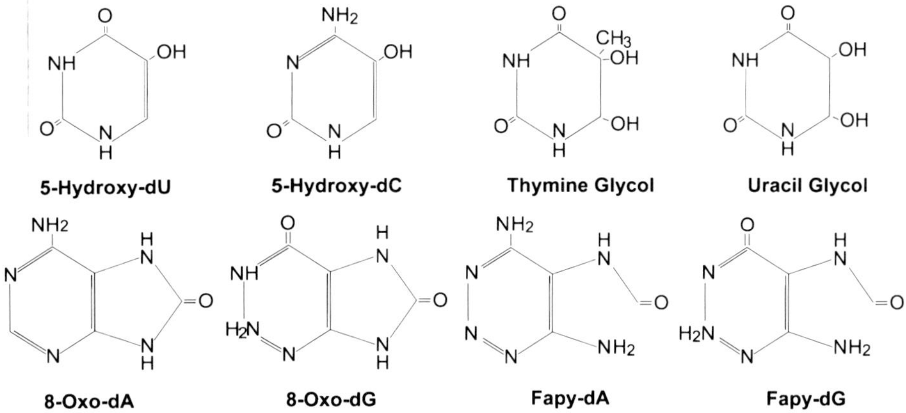

Fig. 11.1 Possible nucleotide modifications caused by attack of OH radicals. First raw, pyrimidine modifications: 5-hydroxy-dU, 5-hydroxy-dC, thymine glycol, uracil glycol. Second raw, purine modifications: 8-Oxo-dA, 8-Oxo-dG, Fapy-dA (N4-(2-deoxy-α,β-D-erytro-pentofuranosyl)-4,6-diamino-5-formamidopyrimidine) and Fapy-dG (N6-(2-deoxy-α,β-D-erytro-pentofuranosyl)-2,6-diamino-4-hydroxy-5-formamido-pyrimidine)

11.1.1 Oxygen Radical-Induced DNA Damage: Candidate DNA Sequences to Be Affected

The large variability of functional and non-functional DNA sequences in the genome is determined by a combination of A, T, C and G nucleotides. However, the mutation sensitivity of different DNA motifs made up of combinations of these four base pairs differs at different genome domains. For example, when human peripheral blood leukocytes are exposed to X-ray radiation for the analysis of the initial level of DNA breakage, using FISH and targeting for different satellite DNA sequences, such as alphoid, satellite 1, 5-bp classical satellite or telomeric DNA sequences (TEL-DNA), differences in sensitivity are observed. Irradiation of nucleoids obtained after protein removal show that alkaline unwinding solutions generate about half the amount of signal when the DNA breaks are present in the 5-bp classical DNA satellites than when the same number of breaks are present in the whole genome. Furthermore, the signal is slightly stronger when the breaks are within the alphoids or the satellite 1 sequence. Therefore, chromatin containing the 5-bp classical satellite arrangement proved to be more sensitive to breakage than the overall genome, whereas DNA in the chromatin corresponding to alphoids or satellite 1 show sensitivity similar to that of the whole genome. Interestingly, TEL-DNA sequences appear to be maximally labelled, even in unirradiated cells [15].

Similarly to the DNA damage induced by exposure to X-ray radiation, oxygen radical-induced damage may result in random DNA damage leading to whole DNA damage during the process of spermatogenesis affecting structural genes (introns or exons) or DNA sequences not directly involved in RNA production but regulating epigenetic functions. Particularly, the telomeres (TEL-DNA) seem to be highly susceptible to DNA damage [16]. Oxygen radical-induced DNA damage may not only lead to single strand DNA breaks, with the subsequent production of apurinic or apirimidinic sites (AP-sites), but also predispose to apoptosis. The particular DNA damage experimented by the telomeres of the germ line is largely unknown, but should not be overlooked, since the transmission and size maintenance of the TEL-DNA for each species is tissue specific. Large variations in the copy number of TEL-DNA sequences among male and female gametes are not expected, since this would compromise the continuity of the species. In fact, in some insects it has been shown that the largest TEL-DNA sequences are harboured in germ line tissues although a certain level of heterozygosity between homologous chromosomes has been reported. This is interesting because during meiosis, synapsis of the bivalents takes place through the telomeres and the presence of heterozygosity may disrupt bivalent synapsis. Moreover, large differences in the size of TEL-DNA between homologous chromosomes may lead to asynchrony in the replication timing assigned to these particular regions. This could be critical during the first stages of embryo development. Studies performed in fibroblasts have shown that oxygen radicals can induce telomere damage leading to defects in synapsis [17]. The highly mutagenic capacity of TEL-DNA is also found when the interstitial telomere-like DNA sequence arrays of Chinese hamster Don cells are tested for the effects of certain damaging agents. In this regard, it has been reported that around 40% of the exchanges involved a telomere-like block of DNA sequences within the rearrangement site [16]. Interestingly, this effect was independent of the DNA-damaging agent. This chromosomal behaviour suggests a general recombination capacity of interstitial telomere-like DNA sequence repeats that does not seem to be related to the initial mechanism of DNA damage, although its consequences for cell arrest, senescence or apoptotic triggering have been largely demonstrated [18]. Nitric oxide-induced DNA damage is particularly effective in inducing intragenomic heterogeneous damage since, for example, a higher density of DNA damage has been found in the telomeric DNA sequence repeats region, since they are particularly enriched in guanine residues [19]. That is precisely why it has been suggested that chronic exposure to nitric oxide in vivo may lead to premature ageing and neoplastic development. In fact, it has been shown that sperm subpopulations selected by swim-up are not homogeneous for telomere length [20]. Telomerase activity is low in mature oocytes and in cleavage stage embryos, but high in blastocysts. After fertilization, the presence of critically shortened telomeres, coming from either the sperm or the oocyte, may contribute to abnormal cleavage and development, as shown in knock-out mice. However, telomere lengthening is remarkable during the early cleavage cycles through a recombination-based mechanism [21].

It is also worth mentioning that ejaculated spermatozoa are not homogeneous in terms of telomere size [22] and that the routine sperm selection techniques used in ART, such as density gradient centrifugation and swim-up, unintentionally allow for the selection of spermatozoa with the largest TEL-DNA repeats. The question is why and where this heterogeneity for these particular chromosome domains is produced? Are they the product of DNA replicative arrested processes before meiosis or are they "allowed mutations" of telomere shortening produced during spermatogenesis and spermiogenesis?

Oxygen radical-induced telomere damage and synaptic anomalies may also occur in germ cells leading to chromosomal defects in ejaculated spermatozoa. After fertilization, these synaptic anomalies could result in defects in chromosomal segregation during embryo division, leading to embryo aneuploidy and to a reduction in pregnancy rate and an increase in miscarriage.

11.1.2 Oxidative Stress in the Seminiferous Tubules and Epididymis

As previously indicated, oxygen radical-induced DNA damage may take place in germ cells during the process of spermatogenesis leading to sperm DNA fragmentation and telomere damage. This may be the result of either an increase in oxygen radical production, a decrease in the expression of antioxidant enzymes or both. Although in most mammalian species the process of spermatogenesis results in the release of mature sperm into the seminiferous tubules, in some species, including the human, a significant proportion of the spermatozoa released have not completed their process of maturation and are therefore considered immature. This process of incomplete maturation that has been designated as *defective spermiogenesis* is characterized by the release into the seminiferous tubules of immature spermatozoa with proximal cytoplasm retention that produce very high levels of oxygen radicals [23, 24]. During spermiogenesis, round spermatids experiment a process of cell remodelling that involves the release of excess membranes from the midpiece and loss of docosahexaenoic acid (DHA) from the plasma membrane [25]. Those stages, at which elongating spermatids with retention of excess cytoplasm and high content of phospholipid-bound DHA are found in the seminiferous tubules epithelium, entail a high risk for the occurrence of intratesticular oxygen radical-induced germ cell DNA damage. During the normal process of spermiogenesis, there are several stages at which some specific elongating spermatids, given its particular content in NADPH, NADPH oxidase, NADH oxidase, superoxide dismutase (SOD), catalase, mitochondria, DHA and high rate of production of oxygen radicals, become an oxidative liability to the seminiferous tubules epithelium [26]. In fact, it has been shown in the rat model that during those stages at which these elongating spermatids with this particular enzymatic profile and high DHA content are found (stages III–VI in the rat), there is a significant increase in the expression of SOD mRNA in the Sertoli cell, most likely in an effort to protect the seminiferous tubules epithelium against the potential oxidative damage that could be produced during these stages [27]. However, in species such as the human, a significant proportion of these elongating spermatids does not complete the process of maturation and are released prematurely into the lumen of the seminiferous tubules. Should that occur, co-migration of oxygen radical-producing immature spermatozoa with mature sperm through the epididymis may lead to DNA damage in mature sperm. Since (1) sperm are highly packed in the epididymis and therefore immature sperm are in close contact with mature sperm and (2) the lifespan of oxygen radicals in the extracellular medium is of the order of nano- to microseconds, the close contact between mature and immature spermatozoa in the epididymis facilitates the occurrence of oxygen radical-induced DNA damage.

Recent studies have shown that the sperm membrane and DNA damage observed in ejaculated spermatozoa is produced, for the most part, in the epididymis [1, 2]. This is supported by the fact that in 90% of the cases, the levels of DNA damage observed in ejaculated spermatozoa are significantly higher than those observed in testicular sperm [1, 28–30]. This DNA damage may be induced directly through the impact of the free radicals on the DNA strands or indirectly through the activation of sperm caspases and/or endonucleases [2]. Direct DNA damage induced by free radicals may take place through three main mechanisms: (1) *DNA damage induced by oxygen radicals produced by immature sperm.* As previously indicated, immature sperm, especially those with proximal cytoplasm retention in the midpiece, produce very high levels of ROS and may produce cross cell damage of mature sperm [23, 24, 26]; (2) *DNA damage induced by oxygen radicals utilized in the oxidation of disulphide bridges of protamines in the sperm chromatin.* In the final segment of the caput epididymis, a recycling mechanism of oxygen radicals utilized in the oxidation of disulphide bridges has been described [31]. Alteration in the normal balance of this recycling with leakage of oxygen radicals could lead to oxidative damage of sperm DNA; and (3) *DNA damage induced by oxygen radicals produced by the epithelial cells of the epididymis.* Another mechanism that could lead to sperm DNA damage in the epididymis would be that related to the production of free radicals by epithelial cells from the epididymis coupled to the low levels of antioxidant enzymes in both the epithelium and the lumen of the epididymis [31]. Antioxidant enzymes such as the different isozymes of glutathione peroxidase (GPX) play a central role in the protection of both the epididymal epithelium and spermatozoa during their passage through the epididymis [31]. It has been shown that this damage progressively increases from the caput to the cauda epididymis and is related to a decrease in the levels of the isozyme GPX-5 [31].

In a study carried out in infertility patients, it was found that co-incubation of sperm suspensions with phytohaemagglutinin-activated polymorphonuclear leukocytes in vitro produces a significant increase in the production of oxygen radicals by immature sperm [32]. The authors of this study conclude that proinflammatory factors produced by activated leukocytes amplify oxygen radical production by immature sperm by 2–3 orders of magnitude. Therefore, in patients in whom there may be an increase in the levels of proinflammatory factors in the epididymis, e.g. inflammatory or infectious processes [33], an increase in DNA damage may take place. In addition, since it has been suggested that the epididymis is in a chronic pseudoinflammatory state [31], proinflammatory factors may play an important role in the pathophysiology of oxygen radical-induced sperm DNA damage in the epididymis. This explains, at least in part, the beneficial effects of antibiotics in patients with high levels of sperm DNA damage in ejaculated sperm [33].

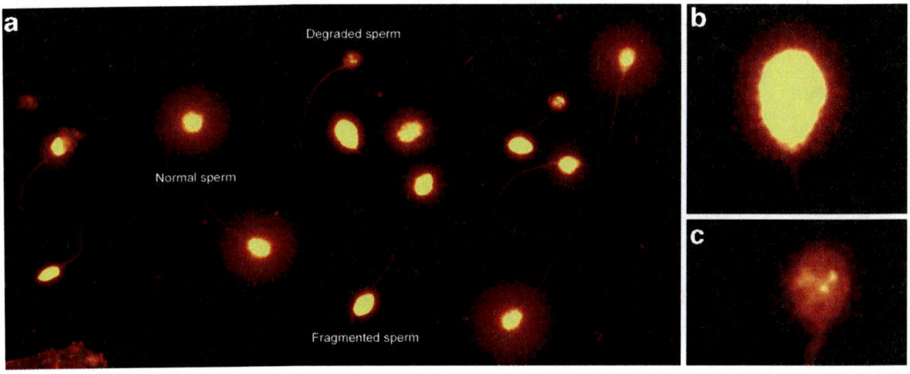

Fig. 11.2 Different patterns of sperm DNA damage in a patient with varicocele. Samples were assessed for sperm DNA damage using the Halosperm test. (**a**) General view of a sample processed using the Halosperm test and stained with Gel Red. Note the different protein/DNA removal on sperm containing fragmented DNA (small halo or no halo) when compared with a normal sperm head (large halo of chromatin dispersion). (**b**) Standard spermatozoa presenting fragmented DNA; (**c**) Sperm with "degraded" chromatin

11.1.3 Clinical Conditions of Exacerbated Oxidative Stress That Lead to Sperm DNA Damage

11.1.3.1 Varicocele

Clinical varicocele has been associated with an increase in free radical production including both oxygen and nitrogen-derived radical species [34, 35]. Nitric oxide production is proportional to the severity of varicocele [36] and patients with varicocele have increased levels of 8-oxo-guanosine (8-OHdG) in leukocyte DNA indicating oxidative damage [37]. Saleh et al., using the Sperm Chromatin Structure Assay (SCSA) test, reported a significant increase in the extent of sperm DNA damage in infertile patients with varicocele [38] and Smith et al., reported that the presence of varicocele was associated with high levels of sperm DNA damage, as assessed by the SCSA and TUNEL tests, even in the presence of normal sperm parameters. Furthermore, DNA damage was correlated with oxygen radical levels [39]. In addition, the fact that varicocelectomy is associated with a significant decrease in both oxygen radical production and sperm DNA damage, this further supports the role of oxidative stress in the pathophysiology of varicocele [40, 41]. Enciso et al. found that mean sperm DNA fragmentation levels in patients with varicocele were 32.4%. This value was 2.6 times higher than that observed in fertile patients. In addition, an even more significant difference was found when the different patterns of sperm DNA damage were evaluated, i.e. the contribution of spermatozoa with massive DNA-nuclear damage displaying the so called "degraded sperm" pattern. This particular sperm subpopulation, visualized using the sperm chromatin dispersion test (SCD) and characterized by a highly protein depletion, retaining very low amounts of DNA after massive protamine removal (Fig. 11.2a, c) when

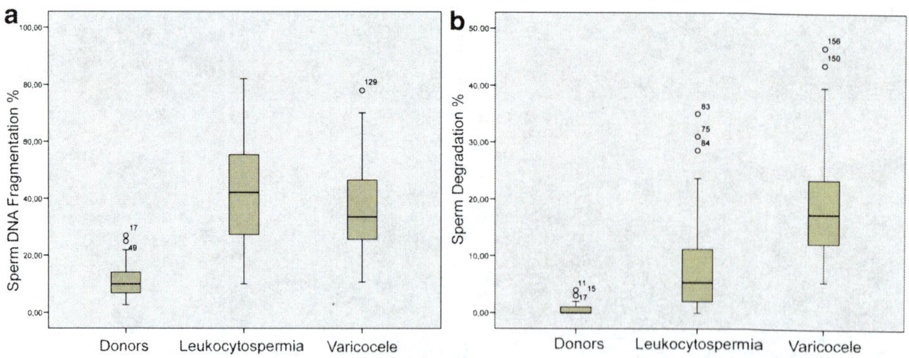

Fig. 11.3 Sperm DNA fragmentation levels in patients with leukocytospermia and varicocele. Box-and-whisker plots to compare sperm DNA fragmentation (**a**) and sperm degradation (**b**) in two different cohorts of patients (leukocytospermia and varicocele) and of sperm donors

compared with spermatozoa with fragmented DNA (Fig. 11.2b), are frequently found in patients with varicocele. Patients diagnosed with varicocele tend to present higher basal levels of SDF (Fig. 11.3a) and also higher levels of degraded sperm (Fig. 11.3b) when compared with donors. The ratio of the degraded sperm pattern to the total of sperm population containing fragmented DNA in these patients was 1–4.2 or higher. This proportion is significantly lower in normozoospermic patients (1–10 or lower). The observed increase in this degraded sperm subpopulation may be related to an increase in ROS and the resulting oxidative stress, which causes lipid peroxidation of sperm plasma membrane and nuclear DNA damage. Nitric oxide released by the endothelial cells from the dilated spermatic veins and peroxynitrite generated by its reaction with the superoxide anion are probably responsible, at least in part, for the observed oxidative damage [34]. DNA fragmentation could be either a direct expression of this damage or a consequence of the triggering of an apoptotic-like process by ROS overproduction. These sperm nuclear abnormalities, such as the degraded sperm pattern, were reproduced in an experimental model of varicocele in rats with enhanced ROS production and an increase in sperm DNA damage. Thus, the degraded sperm pattern most likely corresponds to the highest degree of nuclear damage also compromising the nuclear matrix as a consequence of an intense and prolonged exposure to DNA nuclear-damaging factors where not only the DNA but also the proteins, especially those from the nuclear matrix, are affected, leading to an advanced lytic stage. The nature of the nuclear damage in the degraded type deserves further investigation although it is anticipated that it represents a specific sperm subpopulation collapsed by the massive presence of single and double strand DNA breaks.

An enhancement in oxidative stress, in addition to be determined by an increase in ROS production, may also be determined by a decrease in antioxidant defences [42] or both [43], as it has been reported in men with varicocele. Moreover, the increase in seminal ROS levels seems to be correlated with varicocele grade [44]. This is a well-known factor that may induce DNA fragmentation either in vivo or in vitro [45]. Nitric oxide and peroxynitrite are produced in high concentrations in the dilated spermatic veins thus contributing to the high level of oxidative stress observed in men with varicocele [34, 35, 46]. In addition to the dilated veins, ROS may be released into the seminiferous tubules by immature spermatozoa with proximal cytoplasm retention [1, 24, 26], which are frequently found in semen samples from patients with varicocele.

11.1.3.2 Leukocytospermia

Leukocytospermia is defined as a concentration of leukocytes in semen above one million/mL. Leukocytospermia is often the result of an inflammatory process in the male reproductive tract including orchitis, epididymitis, prostatitis, or inflammation of the seminal vesicles. It is also associated with infertility and reduced semen quality affecting sperm motility and sperm morphology. Leukocytospermia is frequently associated to bacteriospermia. In general, leukocytospermia leads to an oxidative stress environment where ROS are largely present as a result of leukocyte activation [32]. As previously mentioned, co-incubation of sperm suspensions with phytohaemagglutinin-activated polymorphonuclear leukocytes in vitro results in a significant increase in the production of oxygen radicals by immature sperm from infertility patients [32]. One of the possible explanations for this effect is that proinflammatory factors released by activated leukocytes amplify oxygen radical production by immature sperm by 2–3 orders of magnitude.

Are there any differences in the pattern of sperm DNA damage in patients with varicocele or leukocytospermia compared to those observed in normozoospermic patients or sperm donors? In order to throw some light into this particular issue and assuming that in both clinical situations sperm are exposed to high levels of oxidative stress, we have reanalyzed some of our existing data compiling 51 donors, 90 varicocele patients and 72 patients with leukocytospermia. The results for sperm DNA fragmentation distribution and the presence of degraded sperm are shown in Fig. 11.3. Patients diagnosed with leukocytospermia tend to present higher basal levels of SDF (Fig. 11.3a) and also higher levels of degraded sperm (Fig. 11.3b) when compared with donors. There were significant differences in the level of sperm DNA damage when the three groups were compared (Kruskal–Wallis test Chi-square: 103.3; $P<0.0001$). However, for whole sperm DNA fragmentation, these differences are not so clear when leukocytospermia is compared with varicocele (U de Mann–Whitney, Z −2.049: $P=0.04$). Although there were significant differences in the prevalence of the degraded sperm pattern when the three groups were compared (Kruskal–Wallis test Chi-square: 126.04; $P<0.0001$) the prevalence of the degraded sperm pattern was significantly higher in the varicocele compared to the leukocytospermia group (U de Mann–Whitney, Z −7,632: $P=0.0001$). This implies that, even assuming that there are similar levels of DNA damage in both groups of patients, the pattern of DNA damage is different and more aggressive in the case of the clinical varicocele. Subsequently, the effects of oxidative stress on sperm DNA integrity render different results in both clinical situations suggesting that in the particular case of the varicocele either the intensity or the type of damage is different compared to the case of leukocytospermia.

11.1.4 Dynamic Sperm DNA Damage

The comparative analysis of the dynamics of sperm DNA damage in different mammalian species indicates that the resistance of the chromatin to external stressors is different for each species once the sperm is incubated ex vivo for ART. In other words, the extent of sperm DNA damage as a function of time is different for each species when the sperm in handled ex vivo, accumulating DNA impacts of different intensities at different times. While in species such as the boar or deer, the calculated rate of sperm DNA damage is of the order of 0.05 per hour; this value is 15.2 in the case of the ram, which is close to a 300-fold higher [47]. Of course, before ejaculation and when mature spermatozoa are retained in the epididymis, these rates of sperm DNA damage are not assumable. Therefore, in light of this scenario, the concept of sperm DNA damage needs to be redefined, since in all mammalian species we may discriminate between two different types of sperm DNA damage. The *constitutive sperm DNA damage* corresponding to that observed in each individual right after ejaculation, mostly determined by chromatin remodelling and protamine cross-linking during the process of spermiogenesis and passage through the epididymis; and the so-called *inducible or iatrogenic sperm DNA damage*, which appears a posteriori, as a function of incubation time after sperm liquefaction and whose intensity is directly related to (1) the time and conditions of sperm handling and (2) the species-specific resistance of the chromatin. The latter aspect of the sperm chromatin behaviour seems to be linked to the design of each specific genome.

11.1.5 Oxidative Stress, DNA Damage and Protamines

The molecular structure of the human protamine–DNA complex is still poorly understood [48–51]. Most of the models proposed to date are a mere approximation of the actual structure. There is general consensus that the sperm chromatin responds differentially to stressing factors. The reason for this differential response still belongs to the field of speculation. One of the explanations for this differential sensitivity could be related to the fact that the protamines facilitate an efficient packing of the DNA molecule. Two main types of protamine families have been described in mammalian species. The P1 family has been reported in all species of vertebrates studied to date [52–54], while the family of P2 proteins is only present in primates and most of the rodents. The sperm of stallions, lagomorphs and proboscides have also been found to contain processed protamine P2 [52, 53, 55–57]. But in the case of bull and boar sperm, two species that present the lowest rate to iatrogenic sperm DNA damage, the gene for P2 is present, but it appears to be dysfunctional or that produces an aberrant protein [56]. Among the putative functions ascribed to protamines [57, 58] are those related to imprinting of the paternal genome during spermatogenesis and the control of transcription factors to allow its reprogramming by the oocyte. However, one of the functions that should be underlined is its primary role in configuring a compact and hydrodynamic nucleus which plays a crucial role in protecting the paternal genome during migration through the male and female reproductive tract. The quasi-paracrystalline packaging of the sperm chromatin renders the DNA molecule virtually inaccessible to nucleases or mutagens that may be present in the internal or in the external media where the sperm have to endure for variable periods of time. Variations in the levels of protamination by an unbalanced P1/P2 ratio may result in male infertility. In infertile humans, where these aspects have been more deeply analyzed, these unbalance differentially impacts on the integrity of the DNA and in the reproductive outcome of these couples [59, 60]. It has been recently reported that the P1/P2 ratio in human sperm correlates with the levels of sperm DNA fragmentation and also with the rate of sperm DNA damage after ejaculation. Statistical differences were found between fertile controls and patients with different types of pathologies [61]. An altered P1/P2 ratio is associated with an increased amount of histones and intermediate proteins thus supporting the role of a balance between both fractions in order to produce a stable chromatin. Incomplete protamination makes the sperm chromatin more vulnerable to attack by endogenous and exogenous agents, such as mutagenic compounds and disulphide bond reductors [62] or oxygen radicals [63, 64]. Therefore, it appears that the protamine ratio determines the ability of the DNA molecule to reach the oocyte in an intact manner. This situation is not present in those species lacking P2 in the sperm chromatin. Thus, mammalian species such as boar or bull sperm, which only show one type of protamine for the assembling of the sperm chromatin, appear to elude the negative effects of a P1/P2 unbalance.

Additionally, we should bare in mind the role of –SS– disulphide bonds, as well as the number of arginine-lysine residues present in the protamines, that play a decisive role in the highly efficient organization of the sperm chromatin. P1 protamines of eutherian mammals are characterized by the presence of a variable number (6–9) of cysteine (Cys) residues per molecule. During spermiogenesis, SH groups are oxidized to –SS– to form a three-dimensional network of disulphide bridges between and within protamine molecules in the sperm chromatin, conferring the sperm chromatin high stability [65, 66]. In fact, once the sperm chromatin has been tightly packed with these Cys residues, an effective sperm chromatin disorganization is only possible by a reduction in the extent of –SS– cross-linking using, for example, reducing agents such as dithiothreitol or beta-mercaptoethanol [62, 67] and it has been suggested that the differential resistance of the sperm chromatin to relaxation is linked to the protective effect of the Cys residues to produce intra- or inter-protamine cross-linking [68]. The model proposed by Vilfan et al., applies to most of the mammalian species known because is mainly based in the simultaneous implication of one of the two Cys at position 5–6 and 38–39 in intermolecular cross-linking, while the other participate in a intermolecular connection (see Figure 2a in [68]). Moreover, a specific sperm nuclear glutathione peroxidase (snGPx) with properties similar to that of phospholipid hydroperoxide glutathione peroxidase (PHGPx) and identified as a 34-kDa selenoenzyme, acts as a protamine thiol peroxidase and is directly involved in the stabilization of the condensed chromatin by specific cross-linked protamine disulphide bridges [69]. With regard to the possible disulphide bonding occurring in P1, variations in the number and position varies from one species to another. However, the number of Cys residues by itself does not offer a full convincing explanation for the differential rates of sperm iatrogenic DNA damage. For example, boar containing ten Cys residues shows a low rate of sperm DNA fragmentation, while human presenting six Cys shows a relatively high rate. However, boar and bull sperm, both exhibiting a low rate of sperm DNA damage, reveal different number of Cys residues. Moreover, identical number and distribution of Cys residues among different species (stallion, ram and bull) result in large differences in the rate of sperm DNA damage.

In a recent study, the chromatin resistance to the effect of oxidative stress has been compared using sperm from *Homo sapiens*, *Mus musculus* and *Sus domesticus* and three metatherian species (*Vombatus ursinus*, *Phascolarctos cinereus*, *Macropus giganteus*). In this experiment, semen samples were exposed to increasing concentrations of hydrogen peroxide. The main result was that sperm DNA of marsupial species was significantly more sensitive to oxidative stress than spermatozoa of eutherian species [69]. It is known that protamines of eutherian species are cysteine-rich and are extensively cross-linked by disulphide bonds during epididymal transit, whereas the protamines of most marsupial species lack cysteine residues. Therefore, this observed differential susceptibility is consistent with the lack of disulphide cross-linking in marsupial sperm chromatin and suggests that the oxidation of thiols to disulphides for chromatin condensation during epididymal transit in eutherian mammals is likely to be important in order to provide stability and protect spermatozoa against the genotoxic effects of hostile environments. In fact, some studies have revealed a higher vulnerability of marsupial sperm when exposed to different treatments [70–72].

References

1. Ollero M, Gil-Guzman E, Lopez MC, Sharma RK, Agarwal A, Larson K, Evenson D, Thomas Jr AJ, Alvarez JG. Characterization of subsets of human spermatozoa at different stages of maturation: implications in the diagnosis and treatment of male infertility. Hum Reprod. 2001;16:1912–21.
2. Sakkas D, Alvarez JG. Sperm DNA fragmentation: mechanisms of origin, impact on reproductive outcome, and analysis. Fertil Steril. 2010;93:1027–36.
3. Gosálvez J, Cortés-Gutiérrez EI, Nuñez R, Fernández JL, Caballero P, López-Fernández C, Holt WV. A dynamic assessment of sperm DNA fragmentation versus sperm viability in proven fertile human donors. Fertil Steril. 2009;92:1915–9.
4. Toro E, Fernández S, Colomar A, Casanovas A, Alvarez JG, López-Teijón M, Velilla E. Processing of semen can result in increased sperm DNA fragmentation. Fertil Steril. 2009;92:2109–12.
5. Lennon SV, Martin SJ, Cotter TG. Dose-dependent induction of apoptosis in human tumour cell lines by widely diverging stimuli. Cell Prolif. 1991;24:203–14.
6. Lee YJ, Shacter E. Oxidative stress inhibits apoptosis in human lymphoma cells. J Biol Chem. 1999;274:19792–8.
7. Valko M, Morris H, Cronin MT. Metals, toxicity and oxidative stress. Curr Med Chem. 2005;12:1161–208.

8. Evans MD, Cooke MS. Factors contributing to the outcome of oxidative damage to nucleic acids. Bioessays. 2004;26:533–42.
9. Marchetti F, Wyrobek AJ. DNA repair decline during mouse spermiogenesis results in the accumulation of heritable DNA damage. DNA Repair. 2008;7:572–81.
10. Creech MM, Arnold EV, Boyle B, Muzinich MC, Montville C, Bohle DS, Atherton RW. Sperm motility enhancement by nitric oxide produced by the oocytes of fathead minnows, *Pimephelas promelas*. J Androl. 1998;19:667–74.
11. Hellstrom WJC, Bell M, Wang R, Sikka SC. Effect of sodium nitroprusside on sperm motility, viability, and lipid peroxidation. Fertil Steril. 1994;61:117–22.
12. Caulfield JL, Whishnok JS, Tannenbaum SR. Nitric oxide-induced deamination of cytosine and guanine in deoxynucleosides and oligonucleotides. J Biol Chem. 1998;273:12689–95.
13. Burney S, Caulfield JL, Niles C, Whishnok JS, Tannenbaum SR. The chemistry of DNA damage from nitric oxide and peroxynitrite. Mutat Res. 1999;424:37–49.
14. Grishko VI, Druzhyna N, LeDoux SP, Wilson GL. Nitric oxide-induced damage to mtDNA and its subsequent repair. Nucleic Acids Res. 1999;27:4510–6.
15. Vázquez-Gundín F, Riveroa M, Gosálvez J, Fernández JL. Radiation-induced DNA breaks in different human satellite DNA sequence areas, analyzed by DNA breakage detection—fluorescence in situ hybridization. Radiat Res. 2002;157:711–20.
16. Fernández JL, Gosálvez J, Goyanes V. High frequency of mutagen-induced chromatid exchanges at interstitial telomere-like DNA sequence blocks of Chinese hamster cells. Chromosome Res. 1995;3:281–4.
17. Favetta LA, Madan P, Mastromonaco GF, St. John EJ, King WA, Betts DH. The oxidative stress adaptor p66shc is required for permanent embryo arrest in vitro. BMC Dev Biol. 2007;7:132.
18. Donate LE, Blasco MA. Telomeres in cancer and ageing. Philos Trans R Soc Lond B Biol Sci. 2011;366:76–84.
19. Mosquera A, Gosálvez J, Sabatier L, Fernandez JL. Hamster cells are hypersensitive to nitric oxide damage, and DNA-PKcs has a specific local role in its repair. Genes Chromosomes Cancer. 2005;44:76–84.
20. Tamayo M, Mosquera A, Regoc I, Francisco J, Blanco FJ, Gosálvez J, Fernández JL. Decreased length of telomeric DNA sequences and increased numerical chromosome aberrations in human osteoarthriticchondrocytes. Mutat Res. 2011;708:50–8.
21. Liu L, Blasco M, Trimarchi J, Keefe D. An essential role for functional telomeres in mouse germ cells during fertilization and early development. Dev Biol. 2002;249:74–84.
22. Santiso R, Tamayo M, Gosálvez J, Meseguer M, Garrido N, Fernández JL. Swim-up procedure selects spermatozoa with longer telomere length. Mutat Res. 2010;688:88–90.
23. Aitken J, Krausz C, Buckingham D. Relationships between biochemical markers for residual sperm cytoplasm, reactive oxygen species generation, and the presence of leukocytes and precursor germ cells in human sperm suspensions. Mol Reprod Dev. 1994;39:268–79.
24. Gil-Guzman E, Ollero M, Lopez MC, Sharma RK, Alvarez JG, Thomas Jr AJ, Agarwal A. Differential production of reactive oxygen species by subsets of human spermatozoa at different stages of maturation. Hum Reprod. 2001;16:1922–30.
25. Ollero M, Powers D, Alvarez JG. Variation of docosahexaenoic acid content in subsets of human spermatozoa at different stages of maturation: implications for sperm lipoperoxidative damage. Mol Reprod Dev. 2000;55:326–34.
26. Gomez E, Buckingham DW, Brindle J, Lanzafame F, Irvine DS, Aitken RJ. Development of an image analysis system to monitor the retention of residual cytoplasm by human spermatozoa: correlation with biochemical markers of the cytoplasmic space, oxidative stress, and sperm function. J Androl. 1996;17:276–87.
27. Jow WW, Schlegel PN, Cichon Z, Phillips D, Goldstein M, Bardin CW. J Androl. 1993;14:439–47.
28. Steele EK, McClure N, Maxwell RJ, Lewis SE. A comparison of DNA damage in testicular and proximal epididymal spermatozoa in obstructive azoospermia. Mol Hum Reprod. 1999;5:831–5.

29. Greco E, Scarselli F, Iacobelli M, Rienzi L, Ubaldi F, Ferrero S, Franco G, Anniballo N, Mendoza C, Tesarik J. Efficient treatment of infertility due to sperm DNA damage by ICSI with testicular spermatozoa. Hum Reprod. 2005;20:226–30.
30. Moskovtsev SI, Jarvi K, Mullen JB, Cadesky KI, Hannam T, Lo KC. Testicular spermatozoa have statistically significant lower DNA damage compared with ejaculated spermatozoa in patients with unsuccessful oral antioxidant treatment. Fertil Steril. 2010;93:1142–6.
31. Drevet JR. The antioxidant glutathione peroxidase family and spermatozoa: a complex story. Mol Cell Endocrinol. 2006;250:70–9.
32. Alvarez JG, Sharma RK, Ollero M, Saleh RA, Lopez MC, Thomas Jr AJ, Evenson DP, Agarwal A. Increased DNA damage in sperm from leukocytospermic semen samples as determined by the sperm chromatin structure assay. Fertil Steril. 2002;78:319–29.
33. Gallegos G, Ramos B, Santiso R, Goyanes V, Gosálvez J, Fernández JL. Sperm DNA fragmentation in infertile men with genitourinary infection by *Chlamydia trachomatis* and mycoplasma. Fertil Steril. 2008;90:328–34.
34. Mitropoulos D, Deliconstantinos G, Zervas A, et al. Nitric oxide synthase and xanthine oxidase activities in the spermatic vein of patients with varicocele: a potential role for nitric oxide and peroxynitrite in sperm dysfunction. J Urol. 1996;156:1952.
35. Romeo C, Ientile R, Santoro G, et al. Nitric oxide production is increased in the spermatic veins of adolescents with left idiopathic varicocele. J Pediatr Surg. 2001;36:389.
36. Ozbek E, Turkoz Y, Gokdeniz R, et al. Increased nitric oxide production in the spermatic vein of patients with varicocele. Eur Urol. 2000;37:172.
37. Chen SS, Huang WJ, Chang LS, et al. 8-Hydroxy-2′-deoxyguanosine in leukocyte DNA of spermatic vein as a biomarker of oxidative stress in patients with varicocele. J Urol. 2004;172:1418.
38. Saleh RA, Agarwal A, Sharma RK, Said TM, Sikka SC, Thomas Jr AJ. Evaluation of nuclear DNA damage in spermatozoa from infertile men with varicocele. Fertil Steril. 2003;80:1431–6.
39. Smith R, Kaune H, Parodi D, Madariaga M, Rios R, Morales I, Castro A. Increased sperm DNA damage in patients with varicocele: relationship with seminal oxidative stress. Hum Reprod. 2006;21:986–93.
40. Dada R, Venkatesh S, Kumar K, Shamsi MB. Decreased sperm DNA fragmentation after surgical varicocelectomy is associated with increased pregnancy rate. J Urol. 2010;184:1577–82.
41. Smit M, et al. Decreased sperm DNA fragmentation after surgical varicocelectomy is associated with increased pregnancy rate. J Urol. 2010;183(1):270–4.
42. Barbieri ER, Hidalgo ME, Venegas A, Smith R, Lissi EA. Varicocele-associated decrease in antioxidant defenses. J Androl. 1999;20:713–7.
43. Hendin BN, Kolettis PN, Sharma RK, Thomas Jr AJ, Agarwal A. Varicocele is associated with elevated spermatozoal reactive oxygen species production and diminished seminal plasma antioxidant capacity. J Urol. 1999;161:1831–4.
44. Allamaneni SS, Naughton CK, Sharma RK, Thomas Jr AJ, Agarwal A. Increased seminal reactive oxygen species levels in patients with varicoceles correlate with varicocele grade but not with testis size. Fertil Steril. 2004;82:1684–6.
45. Agarwal A, Saleh RA, Bedaiwy MA. Role of reactive oxygen species in the pathophysiology of human reproduction. Fertil Steril. 2003;79:829–43.
46. Türkyilmaz Z, Gülen S, Sönmez K, Karabulut R, Dinçer S, Can Başaklar A, Kale N. Increased nitric oxide is accompanied by lipid oxidation in adolescent varicocele. Int J Androl. 2004;27:183–7.
47. Gosálvez J, López-Fernández C, Fernández JL, Gouraud A, Holt WV. The extent of iatrogenic DNA damage in spermatozoa as a species-specific characteristic. Mol Reprod Dev. Mol Reprod Dev. 2011;78:951–61.
48. Ward WS, Coffey DS. DNA packaging and organization in mammalian spermatozoa: comparison with somatic cells. Biol Reprod. 1991;44:569–74.

49. Ward WS. The structure of the sleeping genome: implications of sperm DNA organization for somatic cells. J Cell Biochem. 1994;55:77–82.
50. Wykes SM, Krawetz SA. The structural organization of sperm chromatin. J Biol Chem. 2003;278:29471–7.
51. Biegeleisen K. The probable structure of the protamine-DNA complex. J Theor Biol. 2006;241:533–40.
52. McKay DJ, Renaux BS, Dixon GH. The amino acid sequence of human sperm protamine P1. Biosci Rep. 1985;5:383–91.
53. Mengual L, Ballesca JL, Ascaso C, Oliva R. Marked differences in protamine content and P1/P2 ratios in sperm cells from Percoll fractions between patients and controls. J Androl. 2003;24:438–47.
54. Yoshii T, Kuji N, Komatsu S, Iwahashi K, Tanaka Y, Yoshida H, Wada A, Yoshimura Y. Fine resolution of human sperm nucleoproteins by two-dimensional electrophoresis. Mol Hum Reprod. 2005;11:677–81.
55. Balhorn R, Gledhill BL, Wyrobek AJ. Mouse sperm chromatin proteins: quantitative isolation and partial characterization. Biochemistry. 1977;16:4074–80.
56. Balhorn R. A model for the structure of chromatin in mammalian sperm. J Cell Biol. 1982;93:298–305.
57. Oliva R, Dixon GH. Vertebrate protamine genes and the histone-to-protamine replacement reaction. Prog Nucleic Acid Res Mol Biol. 1991;40:25–94.
58. Oliva R. Protamines and male infertility. Hum Reprod Update. 2006;12:417–35.
59. de Yebra L, Ballescà JL, Vanrell JA, Corzett M, Balhorn R, Oliva R. Detection of P2 precursors in the sperm cells of infertile patients who have reduced protamine P2 levels. Fertil Steril. 1998;69:755–9.
60. Aoki VW, Moskovtsev SI, Willis J, Liu LH, Mullen JBM, Carrell DT. DNA integrity is compromised in protamine-deficient human sperm. J Androl. 2005;26:741–8.
61. García-Peiró A, Martínez-Heredia J, Oliver-Bonet M, Abad C, Amengua JM, Navarro J, Jones C, Coward K, Gosálvez J, Benet J. Protamine P1/P2 ratio correlates with dynamic aspects of DNA fragmentation in human sperm. Fertil Steril. 2011;95(1):105–9.
62. Szczygiel MA, Ward WS. Combination of dithiothreitol and detergent treatment of spermatozoa causes paternal chromosomal damage. Biol Reprod. 2002;67:1532–7.
63. Irvine DS, Twigg JP, Gordon EL, Fulton N, Milne PA, Aitken RJ. DNA integrity in human spermatozoa: relationships with semen quality. J Androl. 2000;21:33–44.
64. Aitken RJ, De Luliis GN, McLachlan RI. Biological and clinical significance of DNA damage in the male germ line. Int J Androl. 2008;32:46–56.
65. Bedford JM, Calvin HI. Changes in –S–S– linked structures of the sperm tail during epididymal maturation, with comparative observations in sub-mammalian species. J Exp Zool. 1974;187:181–204.
66. Yanagimachi R. Stability of the mammalian sperm nucleus. Zygote. 1994;2:383–4.
67. Ahmadi A, Soon-Chye NG. Destruction of protamine in human sperm inhibits sperm binding and penetration in the zona-free hamster penetration test but increases sperm head decondensation and male pronuclear formation in the hamster–ICSI assay. J Assist Reprod Genet. 1999;16:128–32.
68. Vilfan ID, Conwell CC, Hud NV. Formation of native-like mammalian sperm cell chromatin with folded bull protamine. J Biol Chem. 2004;279:20088–95.
69. Enciso M, Johnston SD, Gosalvez J. Differential resistance of mammalian sperm chromatin to oxidative stress as assessed by a two-tailed comet assay. Reprod Fertil Dev. 2011;23(5):633–7.
70. Cummins JM. Decondensation of sperm nuclei of Australian marsupials: effects of air drying and of calcium and magnesium. Gamete Res. 1980;3:351–67.
71. Balhorn R. Molecular biology of chromosome function. In: Adolph KV, Adolph KV, editors. Mammalian protamines: structure and molecular interactions. New York: Springer; 1989. p. 366–95.
72. Bennetts LE, Aitken RJ. A comparative study of oxidative DNA damage in mammalian spermatozoa. Mol Reprod Dev. 2005;71:77–87.

Chapter 12
Role of Caspase, PARP, and Oxidative Stress in Male Infertility

Tamer M. Said and Fariba Khosravi

Abstract The etiology and pathogenesis of male infertility remain not well defined. Oxidative stress has been identified as one of the major causes of male infertility. Integrity of the sperm DNA, which could be caused by oxidative stress, also plays an important role in vivo and in vitro male fertility. Poly (ADP-ribose) polymerase (PARP) is a DNA repair enzyme that has been demonstrated in mature spermatozoa and fertile men. PARP becomes activated in an attempt to repair oxidative DNA strand breaks. Higher levels of cleaved PARP (cPARP) have been reported in infertile men. Caspases play an important role in mediating and upregulating apoptosis. The cleavage of PARP by caspase 3 inactivates PARP in two segments and inhibits the function of PARP as a DNA repair enzyme. Therefore, high levels of cPARP may be considered as a new marker of apoptosis in sperm of infertile men. In this chapter, we define the relationship between caspases, PARP, and oxidative stress and their role in male infertility.

Keywords Caspase • Poly (ADP-ribose) polymerase • Oxidative stress • Male infertility • Semen parameters • DNA fragmentation

T.M. Said, MD, PhD, HCLD (ABB) (✉)
Andrology Laboratory and Reproductive Tissue Bank, The Toronto Institute for Reproductive Medicine, 56 Aberfoyle Crescent, Toronto, ON, Canada M8X2W4
e-mail: tsaid@repromed.ca

F. Khosravi, MSc
ReproMed, Department of Andrology, 56 Aberfoyle Crescent, Suite 300, Toronto, ON, Canada M8X2W4

A. Agarwal et al. (eds.), *Studies on Men's Health and Fertility*, Oxidative Stress in Applied Basic Research and Clinical Practice, DOI 10.1007/978-1-61779-776-7_12,

12.1 Introduction

A male factor is present in about half of all cases of infertility [1]. This could be attributed to endogenous or exogenous factors [2], which may impose a genotoxic effect on male germ cells leading to alteration of spermatogenesis and subsequently male infertility. Among the endogenous agents, overproduction of reactive oxygen species (ROS) causes oxidative stress (OS), which is one of the causes of male infertility [3]. OS induces sperm alteration such as decrease in motility, viability, morphology, peroxidative damage to sperm membrane, and DNA fragmentation [4]. ROS can cause single- and double-strand DNA breaks [5], base oxidation [6], and chromatin cross-linking [7]. Exogenous elements such as physical and environmental agents can inflict damage to sperm DNA that could take the form of base alteration. DNA integrity has recently been the focus of attention due to concerns about the transmission of defective paternal DNA to the embryo in the era of assisted reproductive technologies.

Increased ROS production leads to damage of the inner and outer sperm mitochondrial membrane, resulting in the release of the cytochrome c protein and the activation of caspase and apoptotic signals [8]. In addition, ROS cause the release of apoptosis-inducing factor (AIF), which leads to high frequency of a single- and double-stranded DNA strand breaks and DNA fragmentation [8]. Therefore, it is thought that oxidative stress in sperm may mediate apoptosis in a caspase-independent pathway.

Poly (ADP-ribose) polymerases (PARP) include a large number of 18 proteins, which have a crucial role in many biological structures such as DNA repair and maintenance of genomic stability, apoptosis, and necrosis [9]. PARP and its homologues have been recently detected in both mature and immature human sperm cell [10]. It has been also demonstrated that infertile men have lower PARP levels in their sperm [2]. These findings confirm the relationship of PARP with male infertility. One of the most important roles of PARP is sperm DNA repair mechanisms. PARP is activated when OS or apoptosis occurs to repair DNA strand breaks [11]. High levels of PARP have been reported in fertile men [2]. It has been reported that caspase can cleave PARP and inactivate its DNA repair function. High level of cleaved PARP (cPARP) in the sperm of infertile men has been also reported [2].

12.2 The Caspases System

12.2.1 Caspase Activation

Apoptosis or program cell death was first introduced by Kerr, Wyllie, and Currie in 1972 as one of the homeostatic mechanism events that occur during many physiological processes such as cell proliferation, differentiation, and inflammatory response. The central component of this pathway is a family of aspartic acid-directed cysteine protease called caspases (CP, cysteinyl aspartate-specific proteinase). Until now, 14 caspases have been introduced in the human apoptotic pathway. At first, they are expressed as inactive form or proenzyme (pCP); after processing they are converted into active caspase form (aCP) [12]. The process of caspase activation starts with the removal of the inhibitory factor or binding of the cofactors either by a protease or by autocatalysis. During proteolysis, the prodomain part is cleaved and the association of the large and small subunits to form a heterodimer follows resulting into caspase activation. Caspases play a crucial role in cells that are destined for programmed death. Most of the caspases are present in cytoplasm; however, some can be found at the Golgi system (CP12) and in the mitochondria (CP2, CP3, and CP9) [13].

There are three different apoptosis pathways; each has a different caspase as an initiator: CP8 for type I or receptor-mediated apoptosis, CP9 for type II or mitochondrial-mediated apoptosis, and CP12 for type III or endoplasmic reticulum-mediated apoptosis [14]. Receptor-mediated apoptosis is triggered by attaching the members of the death receptor superfamily such as CD95 (FAS) or tumor necrosis factor (TNF) superfamily to its corresponding ligand. This complex induces a FADD (Fas-associated death domain protein) in the cytoplasm to activate procaspase-8. Activated caspase-8 induces procaspase-3, leading to production of caspase-3 and apoptotic substrate. The mitochondrial pathway occurs mostly in response to internal insult such as DNA damage and p53 activation. Pro- and anti-apoptotic members of the Bcl-2 family compete to regulate cytochrome c by a mechanism that has not been yet identified. Cytochrome c release into cytosol is followed by its association with Apaf-1 and then procaspase-9 to form the apoptosome [15]. Endoplasmic reticulum-mediated pathway is a novel caspase activation pathway which CP12 functions as the initiator caspase following endoplasmic reticulum stress [16].

12.2.2 Caspases and Spermatogenesis

During spermatogenesis, the number and quality of germ cells are balanced via apoptosis. In adults, close to 75% of spermatogonia and to lesser extent spermatocyte and spermatid are lost [17, 18]. It has been shown that germ cell apoptosis at the cellular level is controlled by endocrinal and local factors and at the intracellular level by p53 and caspases [19, 20]. Although procaspase-3, -6, and -7 and also the IAP caspase inhibitor have been found in normal histology of testis, activated forms of caspase-3, -6, and -7 were not found in normal testis using western blotting or immunohistochemistry methods. Procaspase-3 expression has been shown in testes with normal spermatogenesis and also in testis with sperm maturation arrest or sertoli cell only (SCO). Kim et al. [21] found a diffuse signal for activated caspase-3 in sertoli and germ cell and also leydig cell in testes with sperm maturation arrest; however, the expression of activated caspase-3 has never been seen in testes with normal spermatogenesis [22]. Research suggests that spermatogenic arrest could be related to increased apoptotic rate in testes and increase of activated caspase-3 [22]. In contrast, varicocele patient [23] and hypospermatogenesis and SCO are associated with low level of apoptosis [24]. In the testes, the expression of CD95 and its receptor, which have been introduced as key regulators for germ cell apoptosis [25, 26], is localized to the germ and Sertoli cells [27]. The downstream factors of CD95 activation include various caspases, among which the caspase-8 has the most significant role [28].

12.2.3 Caspase and Capacitation

Capacitation is a prerequisite for sperm fertilization; it can be determined by hyperactivation and membrane changes. It has been demonstrated that capacitation reduced significantly the proportion of sperm with initiator caspase-9 and activator caspase-3. A significant negative correlation between capacitation and caspase-9 and -3 activation [29] has been also reported. It has been shown that spermatozoa with externalized phosphatidylserine have a significantly decreased potential to undergo capacitation and acrosome reaction [30]. There are two enzymes in primate sperm that are located between the sperm membrane and the acrosome; calpain and its natural inhibitor calpastatin [31]. Further studies showed that calpain activation is related to capacitation and the acrosome reaction [32]. Caspase-1 cleaves calpastatin and enhances the activity of calpains, which indicates a potential role for apoptosis signaling and caspase-1 in capacitation [33]. Prolactin has a prosurvival effect through its receptors in postacrosomal region, neck, mid piece, and principal piece of tail in sperm [34]. It has been demonstrated that a significant correlation exists between suppression of caspase activity and capacitation following use of prolactin.

12.2.4 Caspases and Male Infertility

12.2.4.1 Semen Parameters and Caspase

A significant association has been reported between caspase-3 activity and sperm morphology and motility [35]. It has been shown that nonapoptotic sperm that are negative for caspase-3 present with higher quality in terms of normal morphology [36]. Furthermore, spermatozoa of infertile patient have shown an increase in caspase activity [37]. Similarly, a significant negative correlation between sperm motility and viability and caspase-3 activation has been reported [38–43]. In a study that evaluates caspase enzymatic activity, semen fractions of infertile patients had higher caspase enzymatic activity compared to fractions in fertile controls [44]. Although there are significant correlation between caspase activation and the sperm quality regarding motility and vitality values, other contradictory results show no significant correlation between sperm morphology and caspase-3 activation [41].

12.2.4.2 DNA Fragmentation and Caspases

There are different mechanisms for sperm DNA fragmentation including damage by caspases. It appears that sometimes caspase activation is the main factor for DNA strand break during apoptosis in spermatogenesis [14, 45]. There are two forms of DNA strand breaks, single-strand DNA breaks that may be caused by oxygen radicals and could be repaired, and double-strand DNA break, which may be mediated by caspase and usually could not be repaired [46]. These findings highlight the possibility of an association between caspase activation and in vivo and in vitro male infertility. In support, it has been demonstrated that Caspase activity in epididymal and testis sperm has negatively correlated with ICSI outcome [45, 47]. It has been proposed that selecting spermatozoa using specific preparation techniques could decrease the extent of sperm with DNA fragmentation and caspase expression [48].

12.2.4.3 Varicocele and Caspases

Apoptosis may play a role in the pathogenesis of infertility in men with varicocele. Although comparison of apoptosis-related proteins between fertile and varicocele patients by immunoblotting did not show any difference for caspase-3 and Bcl2 expression, the amount of p53, PARP, and BAK expression in varicocele patient were significantly higher than fertile group [49]. It has been shown that hypoxia results in overexpression of Bcl2 and down-regulation of caspase-9 in internal spermatic vein of varicocele patients. Therefore, intrinsic pathway for apoptosis in varicocele patient could be hypothesized [50]. There was an increased expression of active caspase-3 in varicocele patients compared to fertile men. On the other hand, caspase-9 was detected in spermatogonia, spermatocytes, and spermatids in varicocele patients, but not seen in these cells in the young fertile groups [51]. Indeed, both varicocele and aging are associated with high levels of ROS which cause DNA damage. PARP-1 is responsible for efficient repair of DNA and increased ROS can cause cessation of DNA repair and initiation of apoptosis [51].

12.2.4.4 Testicular Torsion and Caspases

Testicular torsion is a medical emergency that requires surgical intervention. Restoration of blood supply after torsion causes testicular ischemia-reperfusion injury which may lead to infertility [52]. It has been shown that testicular ischemia-reperfusion injury results in a marked increase in BAX, caspase-3, and caspase-9 following 24 h after reperfusion injury [53]. The organ damage also involves MAPK3/MAPK1 activation of the apoptotic machinery [53].

12.2.5 Methodology for Analysis of Caspase in Human Sperm

12.2.5.1 Detection of mRNA Caspases

The presence of transcripts in human spermatozoa has been initially established using reverse transcription (RT)-PCR and in situ hybridization (ISH). Sperm pellets are suspended in medium and incubated with P-labeled uridine triphosphate (UTP) that can easily permeate the plasma membrane. The purity and quantity of RNA could be determined via the measurement of the absorbance at 280 and 260 nm, respectively. RNA could be visualized with UV irradiation after ethidium bromide staining in loading buffer and the total RNA from each sample should be used for RT-PCR protocol. Primary human dermal fibroblast could be used as controls. Using this approach, Grunewald et al. have demonstrated caspase-3 mRNA in human spermatozoa of fertile donors [54, 55].

12.2.5.2 Fluorometric Assays

Caspases can be detected in living spermatozoa by fluorescence-labeled inhibitors of caspases (FLICA). This method could be used to show spermatozoa with caspase activation in a semen sample. FLICA includes a green fluorescent label (FAM) and an amino acid peptide inhibitor that attaches to the active caspase. These inhibitors are cell permeable and nontoxic [56]. High specificity of FLICA for detection of caspase activation has been shown [57]. Fluorescence microscopy or flow cytometry in spermatozoa has been used to detect FLICA. The application is easy to perform and varying fluorescence inhibitors of caspases are available; it has been proven that the evaluation of active caspase-3 is possible up to 10 days after staining of human sperm [58].

12.2.5.3 Colorimetric Assays

Activated caspases can be detected on the same principles of fluorescence-labeled inhibitors. A chromophore is attached to the caspase inhibitor, and after cleavage by the active protease, it becomes fluorescent. This color could be evaluated by spectrophotometer at 405 nm. This method is not suitable for single cell, but it has been used for the seminal ejaculate [14].

12.2.5.4 Western Blotting

This method is suitable for detailed analysis of caspase proenzyme, inactivated and activated caspase. Caspases-1, -3, -7, -8, -9, and -12 could be detected using western blotting. This approach requires an appropriate number of sperm for protein extraction. This may be the reason for the lack of detection of caspase-3 in human sperm in some reports [59].

12.2.5.5 Immunohistochemistry

Detection of caspases in testes was accomplished by immunohistochemistry methods [22, 26, 60–63]. Using this technique, caspases activation was detected during physiologic conditions such as germ cell proliferation and maturation [64] as well as environmental factors such as heat stress, irradiation, toxicants, or supraphysiological levels of prolactin [12].

12.3 Poly (ADP-Ribose) Polymerase

12.3.1 PARP Structure and Function

Sperm DNA integrity is critical for successful embryo development and the transmission of intact genetic material to the offspring [2]. During spermatogenesis, protection of the sperm DNA is achieved by tight packaging into a condensed chromatin. Incomplete chromatin condensation and persistence of DNA strand breaks during spermatogenesis are associated with DNA damage and male infertility [65]. PARP, a DNA damage repair enzyme, has been first introduced by Chambon et al. in 1963. It is associated with chromatin and specifically found in the nucleolus [66]. To date, 18 different PARP homologues have been identified, but the structure and function of some are not known yet [67]. Each PARP family member has a catalytic part which contains 50 amino acids that act as the "PARP signature" [67]. PARP family members have other domains, such as DNA-binding domains (DBDs), ankyrin repeats, WWEs domain, macro domains, and BRCA-1 domain with C terminus which is activated following DNA damage [68]. PARP proteins detect DNA strand breaks and have a particular role in both the base excision repair (BER) and nucleotide repair pathways [69]. They have been shown in testicular germ cells [69, 70].

Members of the PARP family have been categorized according to the functional domain. The first category includes PARP1 and PARP2 is activated in response to DNA strands break. Tankyrases 1 and 2 are in second group and have different functions such as telomere regulation and mitotic segregation. The third group consists of PARP12, PARP13, and TCDD-inducible PARP, which contain CCCH-type zinc fingers. Finally, the fourth group includes PARP9, PARP14, and PARP15 which have 1–3 macro domains connected to a PARP domain. They have WWE domain and a catalytic function. For other PARPs such as PARP8, 11, 16, and PARP6, no domains have been found, except a WWE domain in PARP11, therefore it is difficult to define any possible functional role for them [71].

PARP family members were recently classified on the basis of their catalytic domain sequences by Hassa and Hottiger [72]. They were divided into three groups: group 1 including PARP1, PARPb (short PARP1), PARP2, and PARP3; group 2 consists of PARP4; and group 3 consists of two PARP members, tankyrase-1, tankyrase

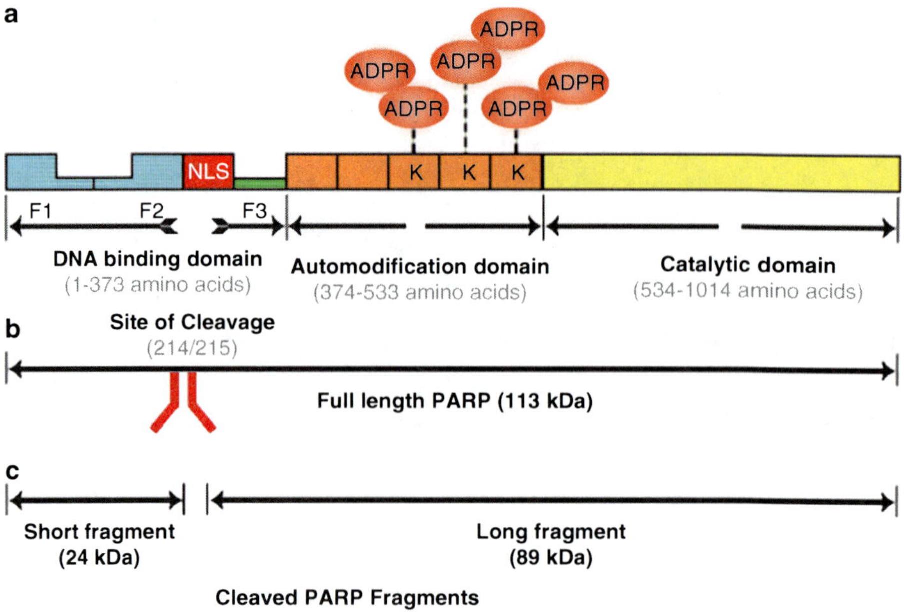

Fig. 12.1 Structural domains of PARP and its fragments. (**a**) DNA-binding domain containing Zinc fingers (F1–3) for nucleosome binding and nuclear localization (NLS) segment; Automodification domain responsible for adding ADPR (ADP-ribose) polymers through binding with Lysine (K) amino acid and catalytic domain has the PARP signature and PARP enzymatic activity. (**b**) Full-length PARP1 113 kDa molecule with a mark on the site of cleavage (214/215 amino acids). (**c**) PARP cleavage by caspase showing short (24 kDa) and long (89 kDa) cleaved PARP fragments [2]

2a, and its isoform tankyrase-2b, known as PARP5 and PARP6a/b [72]. The prototype enzyme of the PARP family is PARP1, a 113 kDa enzyme encoded by the ADP-ribosyl transferase (ADPRT) gene that sited on chromosome 1 in humans [67].

12.3.2 Structural Domains of PARP1

The protein structure of PARP1 has been identified. It contains four functional domains: (1) DBD, (2) nuclear localization signal (NLS), (3) automodification domain (AMD), and (4) Catalytic domains (CD). Figure 12.1 shows the PARP1 structure domains. The DBD also known as zinc fingers restrains to DNA breaks. The NLS has the role of ensuring the PARP1 is localized in the nucleus and is the site of caspase-3 cleavage. The AMD is the site of ADP-ribose polymers addition to PARP [73].

Several forms of molecules have been recognized as PARP activators, including histones and metal ions. PARP1 can be activated by histone H1 and H3. On the other hand, Sirtuin1 (SIRT-1), a histone deacetylase enzyme, is responsible for regulation of PARP1 activity. Cell death mediated by AIF occurs whenever SIRT-1 is absent and PARP is deregulated. In addition, some metal ions such as magnesium and calcium, which play a role in OS, can activate PARP activity thus illustrating the relationship between OS and PARP activity [74, 75]. OS causes calcium influx into the cytoplasm from the environment and also from endoplasmic reticulum. Rising of calcium in the cytoplasm leads to calcium influx into mitochondria and nucleus, which causes disruption of normal metabolism and PARP activity [74]. PARP activity may be inhibited by many endogenous and exogenous inhibitors such as hypoxanthine (endogenous) and tetracycline derivatives (exogenous) [76, 77]. Phosphorylation of PARP1 by extracellular signal-regulated kinases 1/2 (ERK1, 2) is necessary for PARP activation following DNA damage [78]. In addition, PARP phosphorylation may be induced by DNA-dependent protein kinase (DNA-PK), which is a major protein involved in repairing DNA double-strand breaks [79]. PARP1 can recruit other repair enzymes that are essential for repairing DNA damage and preserve DNA integrity. It has been shown that lack of PARP1 can inhibit BER activity; furthermore, PARP2 actively participates in BER pathways [80, 81].

12.3.3 Role of PARP in Germ Cell Death and Spermatogenesis

PARP1 is involved in programmed cell death (apoptosis) and necrosis as well [82] (Fig. 12.2). PARP1 is cleaved in two parts by caspase 3, a 25 kDa fragment consisting of DBD N-terminal and an 85 kDa fragment consisting of the AMD and CD C-terminal fragment. As shown in Fig. 12.1, when the DNA-binding domain separates from the automodification and catalytic domains, PARP1 becomes inactive and consequences of apoptosis occur. On the other hand, the short fragment, N-terminal, inhibits other uncleaved PARP, then restrains the activation of PARP1 by cPARP resulting in more consumption of NAD molecule, energy depletion, and necrosis [83]. As shown in Fig. 12.2, exogenous agents such as ROS or a genotoxin may cause DNA damage in a cell. PARP may act in one of three ways: (1) if low DNA damage occurs, PARP can gather the other repair enzymes and repair DNA damage, (2) if high DNA damage exists, PARP is overexpressed and results in ATP/NAD depletion and necrosis, and (3) after DNA damage and initiation of apoptosis event and caspase-3 activation, PARP cleavage occurs [2].

Hikim et al. proposed that exogenous agents can activate caspase-dependent cell death pathway. They showed that by increasing scrotal temperature in rats over time, the cascade signals in the mitochondria-dependent cell death pathway including relocation of Bax, translocation of cytochrome C, caspase activation, and PARP cleavage were demonstrated [84]. In support of the association between PARP and caspases, Codelia et al. reported that using a caspase-8 inhibitor and a pan-caspase inhibitor, cPARP decreases and a reduction of germ cells apoptosis takes place [85].

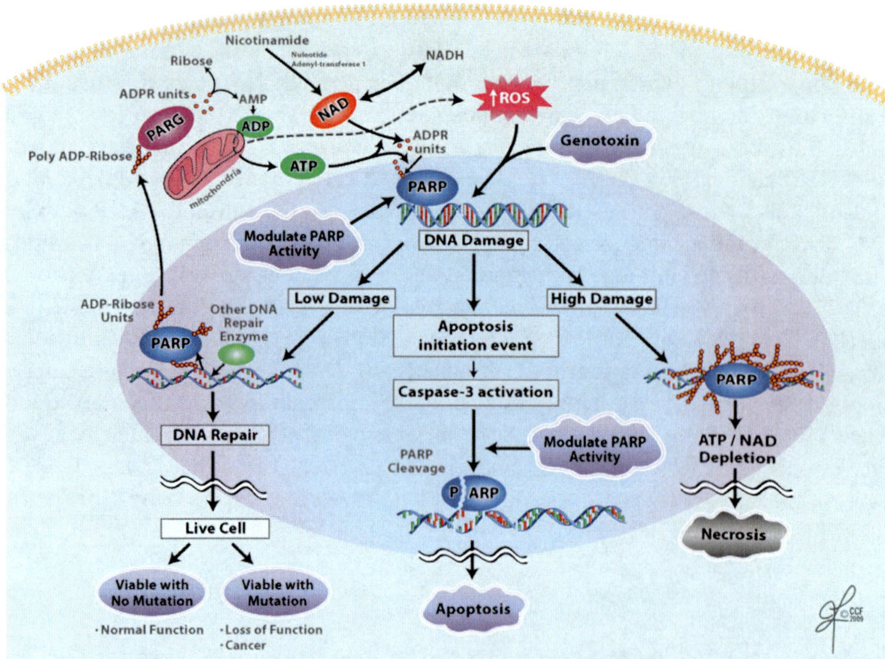

Fig. 12.2 Possible role of PARP in cell death in the event of DNA damage caused by ROS or a genotoxin; PARP targets the damaged site. If high damage occurs, PARP may become overactivated resulting in ATP/NAD depletion and necrosis. Apoptosis can also occur through caspase-3 activation and PARP cleavage. If low damage occurs, PARP can recruit other repair enzymes and DNA repair can occur [2]

12.3.4 PARP in Ejaculated Human Spermatozoa

Jha et al. recently reported several isoforms of PARP such as PARP1, PARP2, and PARP9 in ejaculated spermatozoa. PARP is located near the acrosomal regions of the sperm head and higher amounts of PARP1, PARP2, and PARP9 were detected in fertile men semen compared to infertile men [10]. A decline in level of PARP in the sperm of infertile men was proposed due to increased sperm DNA damage. In addition, sperm maturity correlated with PARP since higher levels of PARP1, PARP2, and PARP9 were found in mature sperm. It is of interest that the localization of caspase enzymes (caspase-3, -8 and -1) is also in the postacrosomal part [86]. cPARP correlates with activated caspase 3 and therefore could be considered as a new apoptotic marker in ejaculated spermatozoa [87]. Aziz et al. demonstrated that sperm deformity index scores are associated with cPARP as an early markers of apoptosis [88].

12.3.5 Potential Therapeutic Role of PARP in Male Infertility

As PARPs play an important role in cell death and apoptosis, they could be good candidates in therapeutic regimens. It was hypothesized that PARP inhibitor causes prevention of DNA repair of malignant cells following exposure to chemotherapy [89]. In cancer patients, the antitumor properties of PARP can preserve fertility even after chemotherapy or radiotherapy [2]. Jha et al. proposed that PARP inhibition can protect sperm against chemically induced injury in vitro [10]. In a recent study by Chang et al., it was shown that the levels of PARP are higher in varicocele patients compared with normal controls [49]. Manipulation of PARP in testicular cancer and other conditions such as infections and inflammatory diseases could be a future therapeutic approach.

The role of PARP in male infertility could be advocated. In a study by Wang et al., men with idiopathic infertility showed a significant positive correlation between ROS levels and caspase-9 and caspase-3, which suggests possible DNA damage and subsequently PARP in these cases [90]. In men with azoospermia, the expression of PARP-1 (enzyme) and poly(ADP-ribose) (PAR) (an indicator for PARP activity) as two markers has been localized in round and elongating spermatid. Also, PAR activity has been shown in all of the spermatocytes in maturation arrest patients at the spermatocyte level [91].

12.3.6 Methodology for Determination PARP Homologues in Human Sperm

12.3.6.1 Immunoblotting Analysis

One of the most important methods for detection of PARP homologues is immunoblotting. At first, sperm proteins are extracted, denatured, and heated at 95°C for 5 min with Laemmli buffer [92]. Thereafter, equal amount of protein should be loaded on each well and separated on a 12% SDS-PAGE for protein profile verification. All protein could be electroblotted onto nitrocellulose membranes. The blotted membranes are incubated with nonfat milk and anti-PARP1 followed by anti-goat horseradish peroxidase-conjugated dilutions. Following incubation, primary and secondary antibodies are added. PARP immunopositive bands can be verified by Totallan 100 software [10].

12.3.6.2 Peptide Mass Fingerprinting

PARP1 antibodies are used in western blotted lanes in order to detect the PARP1 immunopositive bands. The corresponding bands of PARP1 immunopositive in the SDS-PAGE could be then excised and digested by Trypsin In-Gel Digest Kit. Later, the tryptic peptides can be detected by MALDI-Tof. Bioinformatics analyses may be used to identify the corresponding protein based on peptide mass fingerprinting [10].

12.3.6.3 Flowcytometry Staining for Cleaved PARP

cPARP can be detected by an FITC-conjugated anti-PARP cleavage site-specific antibody (CSSA) kit. Sperm pellets should be first washed with phosphate-buffered saline (PBS) and sperm pellets could be fixed in IC fix buffer and finally incubated with FITC-conjugated anti-PARP CSSA. Another wash cycle should be performed in IC Perm buffer and in PBS for FACS analysis. All fluorescence signals of labeled spermatozoa could be analyzed by the flow cytometer FACScan [87].

12.4 Oxidative Stress

12.4.1 Relationship Between OS and DNA Damage in Sperm

Sperm DNA is protected from oxidative stress by two mechanisms: tight packaging of DNA through spermatogenesis and the existence of antioxidant system in semen. OS causes single and double DNA strand breaks. Furthermore, ROS could cause apoptosis by activating the cytochrome c and caspases 9 and 3 that causes more DNA single- and double-strand breaks [12]. High levels of ROS were identified as a cause of the DNA fragmentation in spermatozoa more frequently seen in infertile men [93]. It has been also shown that oral antioxidant administration can decrease DNA fragmentation in infertile men [94].

12.4.2 PARP and Oxidative Stress

ROS and nitric oxide that are produced by activated macrophages are involved in DNA damage leading to PARP activation [66]. In addition, PARP would regulate inflammation by expression-inducible nitric oxide synthase (iNOS) [66]. X-ray repair cross-complementing 1 (XRCC1), which is a DNA repair protein, is recruited by PARP1 to damaged sites following DNA damage caused by ROS [95]. Besides the effect of OS on DNA damage, it can also impair histones. Ullrich et al. explained that the 20s proteosome which is involved in degrading oxidized damaged histones can be activated by PARP. When ROS exceeds the level of the repair capacity, cleaved PARP1 is catalyzed by caspase 3 activation and initiates apoptosis [96].

Shiraishi et al. demonstrated that elevated scrotal temperature is associated with OS, impaired spermatogenesis, and apoptosis of germ cells in human testis with varicocele [97].

El-Domyati et al. showed higher level of caspase 3 and cleaved PARP1 in varicocele patients compared to fertile group expressed [51]. In another research by Chang et al., it has been shown that expression of active PARP was more in varicocele patient than in normal fertile donors [49]. Tekcan et al. showed that after varicocele induction in rat testis, PAR and PARP1 and 4-HNE expressions in germ cells increased and explained that increased level of oxidative stress and PARP overexpression caused testicular dysfunction, which is associated with varicocele [98].

12.5 Conclusion

Caspases, PARP, and OS are interlinked and play a role in the pathogenesis of male infertility. PARP expression has demonstrated a parallel relation with spermatozoa maturity and fertility potential. Decline in the level of PARP in the sperm of infertile men is associated with increased sperm DNA damage. PARP1 is involved in preventing damage to mature sperm, while PARP2 may play a role in OS which induces apoptosis. Caspase-dependent apoptosis is initiated in cytoplasmic droplet or in the mitochondria and impacts the nuclear integrity of sperm. cPARP is related to the activation of caspase 3. In addition, cPARP may be used as an apoptotic marker to discriminate between fertile from infertile men. The use of PARP inhibitor in vitro may be a useful treatment in condition-induced damage in sperm DNA.

12.6 Key Points

- Spermatozoa with damaged membranes found in infertile patients present with an increase in caspase activity.
- Nuclear and mitochondrial abnormalities induce caspase 3 activation during spermiogenesis.
- Higher levels of PARP homologues were detected in infertile men semen compared to fertile men.
- High levels of cPARP have been reported in infertile men.
- Calcium ions have an important role in the pathophysiology of oxidative stress and can play a role in the relationship between oxidative stress and PARP activity.
- Infertile male patients have higher levels of caspases and increased sperm DNA damage by ROS, which shows positive relationship between apoptosis and male factor infertility.

References

1. Schulte RT, Ohl DA, Sigman M, Smith GD (2010) Sperm DNA damage in male infertility: etiologies, assays, and outcomes. J Assist Reprod Genet 27(1):3–12
2. Agarwal A, Mahfouz RZ, Sharma RK, Sarkar O, Mangrola D, Mathur PP (2009) Potential biological role of poly (ADP-ribose) polymerase (PARP) in male gametes. Reprod Biol Endocrinol 7:143

3. Pasqualotto FF, Sharma RK, Nelson DR, Thomas AJ, Agarwal A (2000) Relationship between oxidative stress, semen characteristics, and clinical diagnosis in men undergoing infertility investigation. Fertil Steril 73(3):459–464
4. Agarwal A, Makker K, Sharma R (2008) Clinical relevance of oxidative stress in male factor infertility: an update. Am J Reprod Immunol 59(1):2–11
5. Chiu SM, Xue LY, Friedman LR, Oleinick NL (1995) Differential dependence on chromatin structure for copper and iron ion induction of DNA double-strand breaks. Biochemistry 34(8):2653–2661
6. Dizdaroglu M (1992) Oxidative damage to DNA in mammalian chromatin. Mutat Res 275(3–6):331–342
7. Altman SA, Zastawny TH, Randers-Eichhorn L et al (1995) Formation of DNA-protein cross-links in cultured mammalian cells upon treatment with iron ions. Free Radic Biol Med 19(6):897–902
8. Makker K, Agarwal A, Sharma R (2009) Oxidative stress & male infertility. Indian J Med Res 129(4):357–367
9. Burkle A (2005) Poly(ADP-ribose). The most elaborate metabolite of NAD+. FEBS J 272(18):4576–4589
10. Jha R, Agarwal A, Mahfouz R et al (2009) Determination of poly (ADP-ribose) polymerase (PARP) homologues in human ejaculated sperm and its correlation with sperm maturation. Fertil Steril 91(3):782–790
11. El-Domyati MM, Al-Din AB, Barakat MT, El-Fakahany HM, Xu J, Sakkas D (2009) Deoxyribonucleic acid repair and apoptosis in testicular germ cells of aging fertile men: the role of the poly(adenosine diphosphate-ribosyl)ation pathway. Fertil Steril 91(5 Suppl): 2221–2229
12. Said TM, Paasch U, Glander HJ, Agarwal A (2004) Role of caspases in male infertility. Hum Reprod Update 10(1):39–51
13. Chowdhury I, Tharakan B, Bhat GK (2008) Caspases—an update. Comp Biochem Physiol B Biochem Mol Biol 151(1):10–27
14. Grunewald S, Sharma R, Paasch U, Glander HJ, Agarwal A (2009) Impact of caspase activation in human spermatozoa. Microsc Res Tech 72(11):878–888
15. Hengartner MO (2000) The biochemistry of apoptosis. Nature 407(6805):770–776
16. Morishima N, Nakanishi K, Takenouchi H, Shibata T, Yasuhiko Y (2002) An endoplasmic reticulum stress-specific caspase cascade in apoptosis. Cytochrome c-independent activation of caspase-9 by caspase-12. J Biol Chem 277(37):34287–34294
17. Blanco-Rodriguez J (1998) A matter of death and life: the significance of germ cell death during spermatogenesis. Int J Androl 21(5):236–248
18. Sinha Hikim AP, Swerdloff RS (1999) Hormonal and genetic control of germ cell apoptosis in the testis. Rev Reprod 4(1):38–47
19. Russell LD, Warren J, Debeljuk L et al (2001) Spermatogenesis in Bclw-deficient mice. Biol Reprod 65(1):318–332
20. Yan W, Suominen J, Samson M, Jegou B, Toppari J (2000) Involvement of Bcl-2 family proteins in germ cell apoptosis during testicular development in the rat and pro-survival effect of stem cell factor on germ cells in vitro. Mol Cell Endocrinol 165(1–2):115–129
21. Kim SK, Yoon YD, Park YS, Seo JT, Kim JH (2007) Involvement of the Fas-Fas ligand system and active caspase-3 in abnormal apoptosis in human testes with maturation arrest and Sertoli cell-only syndrome. Fertil Steril 87(3):547–553
22. Bozec A, Amara S, Guarmit B et al (2008) Status of the executioner step of apoptosis in human with normal spermatogenesis and azoospermia. Fertil Steril 90(5):1723–1731
23. Fujisawa M, Hiramine C, Tanaka H, Okada H, Arakawa S, Kamidono S (1999) Decrease in apoptosis of germ cells in the testes of infertile men with varicocele. World J Urol 17(5): 296–300
24. Tanaka H, Fujisawa M, Okada H, Kamidono S (2002) Apoptosis-related proteins in the testes of infertile men with varicocele. BJU Int 89(9):905–909

25. Lee J, Richburg JH, Shipp EB, Meistrich ML, Boekelheide K (1999) The Fas system, a regulator of testicular germ cell apoptosis, is differentially up-regulated in Sertoli cell versus germ cell injury of the testis. Endocrinology 140(2):852–858
26. Pentikainen V, Erkkila K, Dunkel L (1999) Fas regulates germ cell apoptosis in the human testis in vitro. Am J Physiol 276(2 Pt 1):E310–E316
27. Lee J, Richburg JH, Younkin SC, Boekelheide K (1997) The Fas system is a key regulator of germ cell apoptosis in the testis. Endocrinology 138(5):2081–2088
28. Siegel RM (2006) Caspases at the crossroads of immune-cell life and death. Nat Rev Immunol 6(4):308–317
29. Grunewald S, Kriegel C, Baumann T, Glander HJ, Paasch U (2009) Interactions between apoptotic signal transduction and capacitation in human spermatozoa. Hum Reprod 24(9): 2071–2078
30. Grunewald S, Baumann T, Paasch U, Glander HJ (2006) Capacitation and acrosome reaction in nonapoptotic human spermatozoa. Ann N Y Acad Sci 1090:138–146
31. Yudin AI, Goldberg E, Robertson KR, Overstreet JW (2000) Calpain and calpastatin are located between the plasma membrane and outer acrosomal membrane of cynomolgus macaque spermatozoa. J Androl 21(5):721–729
32. Ozaki Y, Blomgren K, Ogasawara MS et al (2001) Role of calpain in human sperm activated by progesterone for fertilization. Biol Chem 382(5):831–838
33. Wang KK, Posmantur R, Nadimpalli R et al (1998) Caspase-mediated fragmentation of calpain inhibitor protein calpastatin during apoptosis. Arch Biochem Biophys 356(2):187–196
34. Pujianto DA, Curry BJ, Aitken RJ (2010) Prolactin exerts a prosurvival effect on human spermatozoa via mechanisms that involve the stimulation of Akt phosphorylation and suppression of caspase activation and capacitation. Endocrinology 151(3):1269–1279
35. Almeida C, Cardoso MF, Sousa M et al (2005) Quantitative study of caspase-3 activity in semen and after swim-up preparation in relation to sperm quality. Hum Reprod 20(5): 1307–1313
36. Aziz N, Said T, Paasch U, Agarwal A (2007) The relationship between human sperm apoptosis, morphology and the sperm deformity index. Hum Reprod 22(5):1413–1419
37. Paasch U, Grunewald S, Fitzl G, Glander HJ (2003) Deterioration of plasma membrane is associated with activated caspases in human spermatozoa. J Androl 24(2):246–252
38. Liu CH, Tsao HM, Cheng TC et al (2004) DNA fragmentation, mitochondrial dysfunction and chromosomal aneuploidy in the spermatozoa of oligoasthenoteratozoospermic males. J Assist Reprod Genet 21(4):119–126
39. Marchetti C, Obert G, Deffosez A, Formstecher P, Marchetti P (2002) Study of mitochondrial membrane potential, reactive oxygen species, DNA fragmentation and cell viability by flow cytometry in human sperm. Hum Reprod 17(5):1257–1265
40. Pena FJ, Johannisson A, Wallgren M, Rodriguez-Martinez H (2003) Assessment of fresh and frozen-thawed boar semen using an Annexin-V assay: a new method of evaluating sperm membrane integrity. Theriogenology 60(4):677–689
41. Said TM, Grunewald S, Paasch U et al (2005) Advantage of combining magnetic cell separation with sperm preparation techniques. Reprod Biomed Online 10(6):740–746
42. Shen HM, Dai J, Chia SE, Lim A, Ong CN (2002) Detection of apoptotic alterations in sperm in subfertile patients and their correlations with sperm quality. Hum Reprod 17(5): 1266–1273
43. Weng SL, Taylor SL, Morshedi M et al (2002) Caspase activity and apoptotic markers in ejaculated human sperm. Mol Hum Reprod 8(11):984–991
44. Taylor SL, Weng SL, Fox P et al (2004) Somatic cell apoptosis markers and pathways in human ejaculated sperm: potential utility as indicators of sperm quality. Mol Hum Reprod 10(11):825–834
45. Said T, Agarwal A, Grunewald S et al (2006) Selection of nonapoptotic spermatozoa as a new tool for enhancing assisted reproduction outcomes: an in vitro model. Biol Reprod 74(3):530–537

46. Sakkas D, Alvarez JG (2010) Sperm DNA fragmentation: mechanisms of origin, impact on reproductive outcome, and analysis. Fertil Steril 93(4):1027–1036
47. Brugnon F, Janny L, Artonne C, Sion B, Pouly JL, Grizard G (2009) Activated caspases in thawed epididymal and testicular spermatozoa of patients with congenital bilateral absence of the vas deferens and intracytoplasmic sperm injection outcome. Fertil Steril 92(2):557–564
48. Kotwicka M, Filipiak K, Jedrzejczak P, Warchol JB (2008) Caspase-3 activation and phosphatidylserine membrane translocation in human spermatozoa: is there a relationship? Reprod Biomed Online 16(5):657–663
49. Chang FW, Sun GH, Cheng YY, Chen IC, Chien HH, Wu GJ (2010) Effects of varicocele upon the expression of apoptosis-related proteins. Andrologia 42(4):225–230
50. Lee JD, Lu LY, Cheng WH, Jeng SY (2009) Dysregulated apoptosis through the intrinsic pathway in the internal spermatic vein of patients with varicocele. J Formos Med Assoc 108(8):612–618
51. El-Domyati MM, Al-Din AB, Barakat MT et al (2010) The expression and distribution of deoxyribonucleic acid repair and apoptosis markers in testicular germ cells of infertile varicocele patients resembles that of old fertile men. Fertil Steril 93(3):795–801
52. Ringdahl E, Teague L (2006) Testicular torsion. Am Fam Physician 74(10):1739–1743
53. Minutoli L, Antonuccio P, Polito F et al (2009) Mitogen-activated protein kinase 3/mitogen-activated protein kinase 1 activates apoptosis during testicular ischemia-reperfusion injury in a nuclear factor-kappaB-independent manner. Eur J Pharmacol 604(1–3):27–35
54. Grunewald S, Paasch U, Glander HJ, Anderegg U (2005) Mature human spermatozoa do not transcribe novel RNA. Andrologia 37(2–3):69–71
55. Ostermeier GC, Miller D, Huntriss JD, Diamond MP, Krawetz SA (2004) Reproductive biology: delivering spermatozoan RNA to the oocyte. Nature 429(6988):154
56. Ekert PG, Silke J, Vaux DL (1999) Caspase inhibitors. Cell Death Differ 6(11):1081–1086
57. Pozarowski P, Huang X, Halicka DH, Lee B, Johnson G, Darzynkiewicz Z (2003) Interactions of fluorochrome-labeled caspase inhibitors with apoptotic cells: a caution in data interpretation. Cytometry A 55(1):50–60
58. Grunewald S, Rasch M, Reinhardt M, Baumann T, Paasch U, Glander HJ (2008) Stability of fluorochrome based assays to measure subcellular sperm functions. Asian J Androl 10(3): 455–459
59. de Vries KJ, Wiedmer T, Sims PJ, Gadella BM (2003) Caspase-independent exposure of aminophospholipids and tyrosine phosphorylation in bicarbonate responsive human sperm cells. Biol Reprod 68(6):2122–2134
60. Giampietri C, Petrungaro S, Coluccia P et al (2005) Germ cell apoptosis control during spermatogenesis. Contraception 72(4):298–302
61. Hikim AP, Lue Y, Yamamoto CM et al (2003) Key apoptotic pathways for heat-induced programmed germ cell death in the testis. Endocrinology 144(7):3167–3175
62. Tesarik J, Martinez F, Rienzi L et al (2002) In-vitro effects of FSH and testosterone withdrawal on caspase activation and DNA fragmentation in different cell types of human seminiferous epithelium. Hum Reprod 17(7):1811–1819
63. Zheng S, Turner TT, Lysiak JJ (2006) Caspase 2 activity contributes to the initial wave of germ cell apoptosis during the first round of spermatogenesis. Biol Reprod 74(6):1026–1033
64. Abraham MC, Shaham S (2004) Death without caspases, caspases without death. Trends Cell Biol 14(4):184–193
65. Meyer-Ficca ML, Lonchar J, Credidio C et al (2009) Disruption of poly(ADP-ribose) homeostasis affects spermiogenesis and sperm chromatin integrity in mice. Biol Reprod 81(1):46–55
66. Decker P, Muller S (2002) Modulating poly (ADP-ribose) polymerase activity: potential for the prevention and therapy of pathogenic situations involving DNA damage and oxidative stress. Curr Pharm Biotechnol 3(3):275–283
67. Ame JC, Spenlehauer C, de Murcia G (2004) The PARP superfamily. Bioessays 26(8): 882–893

68. Wacker DA, Frizzell KM, Zhang T, Kraus WL (2007) Regulation of chromatin structure and chromatin-dependent transcription by poly(ADP-ribose) polymerase-1: possible targets for drug-based therapies. Subcell Biochem 41:45–69
69. Flohr C, Burkle A, Radicella JP, Epe B (2003) Poly(ADP-ribosyl)ation accelerates DNA repair in a pathway dependent on Cockayne syndrome B protein. Nucleic Acids Res 31(18): 5332–5337
70. Meyer-Ficca ML, Scherthan H, Burkle A, Meyer RG (2005) Poly(ADP-ribosyl)ation during chromatin remodeling steps in rat spermiogenesis. Chromosoma 114(1):67–74
71. Schreiber V, Dantzer F, Ame JC, de Murcia G (2006) Poly(ADP-ribose): novel functions for an old molecule. Nat Rev Mol Cell Biol 7(7):517–528
72. Hassa PO, Hottiger MO (2008) The diverse biological roles of mammalian PARPS, a small but powerful family of poly-ADP-ribose polymerases. Front Biosci 13:3046–3082
73. Woodhouse BC, Dianova II, Parsons JL, Dianov GL (2008) Poly(ADP-ribose) polymerase-1 modulates DNA repair capacity and prevents formation of DNA double strand breaks. DNA Repair (Amst) 7(6):932–940
74. Ermak G, Davies KJ (2002) Calcium and oxidative stress: from cell signaling to cell death. Mol Immunol 38(10):713–721
75. Kun E, Kirsten E, Mendeleyev J, Ordahl CP (2004) Regulation of the enzymatic catalysis of poly(ADP-ribose) polymerase by dsDNA, polyamines, Mg2+, Ca2+, histones H1 and H3, and ATP. Biochemistry 43(1):210–216
76. Bader M, Arama E, Steller H (2010) A novel F-box protein is required for caspase activation during cellular remodeling in *Drosophila*. Development 137(10):1679–1688
77. Baker MA, Aitken RJ (2005) Reactive oxygen species in spermatozoa: methods for monitoring and significance for the origins of genetic disease and infertility. Reprod Biol Endocrinol 3:67
78. Barbonetti A, Vassallo MR, Fortunato D, Francavilla S, Maccarrone M, Francavilla F (2010) Energetic metabolism and human sperm motility: impact of CB receptor activation. Endocrinology 151(12):5882–5892
79. Ariumi Y, Masutani M, Copeland TD et al (1999) Suppression of the poly(ADP-ribose) polymerase activity by DNA-dependent protein kinase in vitro. Oncogene 18(32):4616–4625
80. Dantzer F, de La Rubia G, Menissier-De Murcia J, Hostomsky Z, de Murcia G, Schreiber V (2000) Base excision repair is impaired in mammalian cells lacking poly(ADP-ribose) polymerase-1. Biochemistry 39(25):7559–7569
81. Schreiber V, Ame JC, Dolle P et al (2002) Poly(ADP-ribose) polymerase-2 (PARP-2) is required for efficient base excision DNA repair in association with PARP-1 and XRCC1. J Biol Chem 277(25):23028–23036
82. Brugnon F, Ouchchane L, Verheyen G et al (2009) Fluorescence microscopy and flow cytometry in measuring activated caspases in human spermatozoa. Int J Androl 32(3):265–273
83. D'Amours D, Sallmann FR, Dixit VM, Poirier GG (2001) Gain-of-function of poly(ADP-ribose) polymerase-1 upon cleavage by apoptotic proteases: implications for apoptosis. J Cell Sci 114(Pt 20):3771–3778
84. Hikim AP, Vera Y, Vernet D et al (2005) Involvement of nitric oxide-mediated intrinsic pathway signaling in age-related increase in germ cell apoptosis in male Brown-Norway rats. J Gerontol A Biol Sci Med Sci 60(6):702–708
85. Codelia VA, Cisterna M, Alvarez AR, Moreno RD (2010) p73 participates in male germ cells apoptosis induced by etoposide. Mol Hum Reprod 16(10):734–742
86. Paasch U, Grunewald S, Agarwal A, Glandera HJ (2004) Activation pattern of caspases in human spermatozoa. Fertil Steril 81(Suppl 1):802–809
87. Mahfouz RZ, Sharma RK, Poenicke K et al (2009) Evaluation of poly(ADP-ribose) polymerase cleavage (cPARP) in ejaculated human sperm fractions after induction of apoptosis. Fertil Steril 91(5 Suppl):2210–2220
88. Aziz N, Sharma RK, Mahfouz R, Jha R, Agarwal A (2011) Association of sperm morphology and the sperm deformity index (SDI) with poly (ADP-ribose) polymerase (PARP) cleavage inhibition. Fertil Steril 95(8):2481–2484

89. de la Lastra CA, Villegas I, Sanchez-Fidalgo S (2007) Poly(ADP-ribose) polymerase inhibitors: new pharmacological functions and potential clinical implications. Curr Pharm Des 13(9):933–962
90. Wang X, Sharma RK, Sikka SC, Thomas AJ Jr, Falcone T, Agarwal A (2003) Oxidative stress is associated with increased apoptosis leading to spermatozoa DNA damage in patients with male factor infertility. Fertil Steril 80(3):531–535
91. Maymon BB, Cohen-Armon M, Yavetz H et al (2006) Role of poly(ADP-ribosyl)ation during human spermatogenesis. Fertil Steril 86(5):1402–1407
92. Shukla S, Jha RK, Laloraya M, Kumar PG (2005) Identification of non-mitochondrial NADPH oxidase and the spatio-temporal organization of its components in mouse spermatozoa. Biochem Biophys Res Commun 331(2):476–483
93. Moustafa MH, Sharma RK, Thornton J et al (2004) Relationship between ROS production, apoptosis and DNA denaturation in spermatozoa from patients examined for infertility. Hum Reprod 19(1):129–138
94. Greco E, Iacobelli M, Rienzi L, Ubaldi F, Ferrero S, Tesarik J (2005) Reduction of the incidence of sperm DNA fragmentation by oral antioxidant treatment. J Androl 26(3):349–353
95. El-Khamisy SF, Masutani M, Suzuki H, Caldecott KW (2003) A requirement for PARP-1 for the assembly or stability of XRCC1 nuclear foci at sites of oxidative DNA damage. Nucleic Acids Res 31(19):5526–5533
96. Boulares AH, Yakovlev AG, Ivanova V et al (1999) Role of poly(ADP-ribose) polymerase (PARP) cleavage in apoptosis. Caspase 3-resistant PARP mutant increases rates of apoptosis in transfected cells. J Biol Chem 274(33):22932–22940
97. Shiraishi K, Takihara H, Matsuyama H (2010) Elevated scrotal temperature, but not varicocele grade, reflects testicular oxidative stress-mediated apoptosis. World J Urol 28(3):359–364
98. Tekcan M, Koksal IT, Tasatargil A, Kutlu O, Gungor E, Celik-Ozenci C (2012) Potential role of poly(ADP-ribose) polymerase activation in the pathogenesis of experimental left varicocele. J Androl 33(1):122–132

Part II
Clinical Practice

Chapter 13
Methods for the Detection of ROS in Human Sperm Samples

David Benjamin, Rakesh K. Sharma, Arozia Moazzam, and Ashok Agarwal

Abstract Male-factor significantly contributes to infertility in couples of reproductive age. Several studies have shown that oxidative stress (OS) is involved in the pathophysiology of male infertility. It is caused by an imbalance between the formation of reactive oxygen species (ROS) and the ability of the antioxidants to scavenge them. There are several methods to measure seminal ROS in the clinical setting, most notably among them is the chemiluminescence assay. This technique measures the global ROS, i.e., both the intra- and extracellular ROS. The measurement of seminal ROS employing the chemiluminescence technique in clinical andrology labs is explained in this chapter. Through a deeper understanding of ROS and its measurement, clinical andrology labs can better assist patients to achieve increased rates of fertility and pregnancy.

Keywords Use of chemiluminescence assay • Measurement of seminal reactive oxygen levels • Clinical andrology labs • Free radicals • Sources of ROS

13.1 Introduction

Infertility currently affects approximately 15% of couples of reproductive age and the incidence is growing, with male-factor contributing to as much as 50% of infertility issues [1, 2]. As infertility continues to develop into a growing concern, there is a need to better understand and identify the causes of male-factor infertility.

D. Benjamin, BA • R.K. Sharma, PhD • A. Moazzam, MD
Center for Reproductive Medicine, Cleveland Clinic, Cleveland, OH 44195, USA
e-mail: sharmar@ccf.org

A. Agarwal, PhD (✉)
Center for Reproductive Medicine, Cleveland Clinic, Lerner College of Medicine,
9500 Euclid Avenue, Cleveland, OH 44195, USA
e-mail: agarwaa@ccf.org

A. Agarwal et al. (eds.), *Studies on Men's Health and Fertility*, Oxidative Stress in Applied Basic Research and Clinical Practice, DOI 10.1007/978-1-61779-776-7_13,

Suboptimal sperm quality is a major concern for male-factor infertility, and one of the underlying causes of suboptimal sperm quality is attributed to the presence of reactive oxygen species (ROS) [3]. Although the human body naturally produces free oxygen radicals and ROS in physiological amount, essential for capacitation, hyperactivation, acrosome reaction, and oocyte fusion, an excess of ROS can be harmful to male gametes. Several studies have demonstrated that an excess of ROS, which leads to the state of oxidative stress, affects spermatozoa motility and viability, oocytes, as well as rates of fertilization and pregnancy [3–6]. There are several techniques currently used in the clinical andrology laboratory to quantify the amount of ROS in a patient's semen sample, including the chemiluminescence assay. Several recent studies have identified reference values for ROS in order to diagnose male infertility, and these reference values continue to be refined with the publication of new studies. Measurement of ROS can be used as a diagnostic tool in identifying the underlying etiology of idiopathic or unexplained infertility. In order for clinical andrology labs to better assist patients experiencing issues with infertility, it is important to understand ROS and the techniques, such as the chemiluminescence assay, which are employed to detect ROS.

13.2 What Are Free Radicals?

Free radicals are a group of atoms or molecules that are highly reactive due to having one or more unpaired electrons [7]. As a result of having an incomplete outer valance shell, these molecules attempt to react with other molecules in their vicinity in order to gain one or more electrons. However, once a molecule loses an electron to a free radical, a chain reaction is created, as now the former molecule becomes a free radical itself (Fig. 13.1). The chain reaction created from free radicals can have catastrophic effects for living cells. ROS are a collection of radicals and nonradical derivatives of oxygen. Free radicals derived from nitrogen are called reactive nitrogen species (Table 13.1). The most common ROS that is produced by spermatozoa is the superoxide anion radical; this in turn forms hydrogen peroxide (strong oxidizer) on its own or by the action of superoxide dismutase [8].

Plasma membrane of the spermatozoa is rich in polyunsaturated fatty acids (PUFA). Because their cytoplasm contains low concentrations of scavenging enzymes, they are particularly susceptible to the damage induced by excessive ROS [9, 10]. The seminal plasma, however, contains two different types of antioxidants to minimize free radical-induced damage: enzymatic and nonenzymatic antioxidants. Enzymatic antioxidants are: superoxide dismutase, catalase, glutathione reductase, and peroxidase. Nonenzymatic antioxidants are comprised of vitamins (vitamin C, E), proteins (albumin, transferrin, haptoglobin, and ceruloplasmin), and other molecules (glutathione, pyruvate, and ubiquinol). Both enzymatic and nonenzymatic antioxidants can help minimize free radical-induced damage. Excessive generation of ROS damages cells, tissues, and organs and is involved in the pathogenesis of a wide range of diseases.

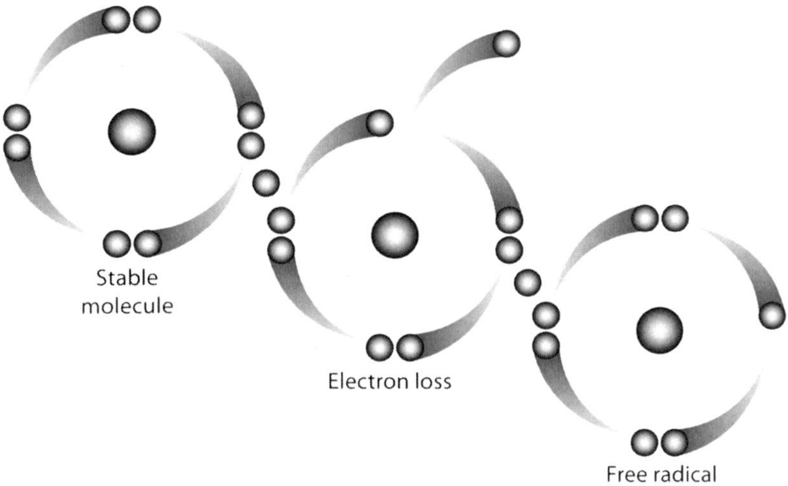

Fig. 13.1 Illustration is showing the oxygen molecule and the generation of a free radical

Table 13.1 Examples of free radicals

Reactive oxygen species	Reactive nitrogen species
Superoxide anion ($O_2^{\bullet-}$)	Nitric oxide ($NO^{\bullet}$)
Hydrogen peroxide (H_2O_2)	Nitric dioxide ($NO_2^{\bullet}$)
Hydroxyl radical ($OH^{\bullet}$)	Peroxynitrite ($ONOO^-$)

13.3 Sources of ROS

Morphologically defective spermatozoa and leukocytes found in semen are the primary causes of ROS production [11]. Several studies have shown that human spermatozoa can produce ROS, especially those with an excess of cytoplasm, also known as cytoplasmic droplets [12–15]. ROS production due to excess cytoplasm occurs due to the cytosolic enzyme glucose-6-phosphate-dehydrogenase (G6PD). Spermatozoa may generate ROS in two ways: first, through the NADPH-oxidase system found in the sperm plasma membrane, and secondly, through the NADH-dependent oxidoreductase in mitochondria [16]. Peroxidase-positive leukocytes, which are derived from the prostate and seminal vesicles, are also major contributors to ROS production [16–19]. Activation of leukocytes, stemming from inflammation and infection, can lead to ROS production as a result of increased NADPH production [3, 20]. Leukocytospermia ($>1 \times 10^6$ WBC/mL of semen) causes sperm damage by ROS production, and removal of seminal plasma during sperm preparation for assisted production can further deprive the sperm of important nutritional support (antioxidants) and further cause sperm damage through ROS production [17, 18].

13.4 Importance of Oxidative Stress and ROS in Male Infertility

Sperm dysfunction is one of the leading causes of infertility. Despite much research, there is limited knowledge about the causes of sperm dysfunction. However, oxidative stress (OS) has been identified as one of the causes of abnormal sperm function. OS refers to the state in which free radicals overwhelm the body's antioxidant capacity [21]; an excess of ROS can have harmful effects on the tissues and organs [9, 10]. ROS has been shown to negatively affect sperm quality and function [22, 23]. For normal male fertility, a fine balance between the generation of ROS and its removal is very important. Excess of ROS also has a negative impact on the pregnancy outcome as well as the assisted reproductive techniques [24, 25]. Moreover, excess ROS can damage DNA in spermatozoa, induce cell apoptosis, and cause lipid peroxidation, which leads to morphological abnormalities, decrease in fertility, and increased sperm membrane permeability [26–28]. It is important to effectively detect the amount of ROS in a semen sample in order to better treat subfertile male patients.

13.5 Methods for Detecting ROS

There are several techniques to detect ROS. These include the direct and indirect methods (Table 13.2). Direct methods include cytochrome c reduction, electron spin resonance, and nitroblue tetrazolium technique (NBT, and xyenol orange-based assay and flow cytometry [7, 22]). Indirect methods of measurement include the Endtz test, redox potential (GSH/GSSG), measurement of lipid peroxidation products levels, chemokines, measurement of oxidative DNA damage, and measurement of reactive nitrogen species by Greiss reaction and fluorescence spectroscopy [7].

13.6 Chemiluminescence Measurement of Extracellular and Intracellular ROS

Chemiluminescence depends on the measurement of light emitted after reagents are added to a sample of human spermatozoa, causing a reaction. The chemiluminescence assay is one of the most commonly used techniques to measure seminal ROS levels in clinical andrology laboratories [7, 12, 29, 30]. The two major probes used to measure ROS generation in the chemiluminescence assay are luminol (5-amino-2,3-dihydro-1,4-phthalazinedione and 3-aminophthalic hydrazide) and lucigenin (*N*,*N*′-dimethyl-9,9′-biacridinium dinitrate). Both H_2O_2 and $O_2^{\bullet-}$ are involved in luminol-dependent chemiluminescence because both catalase and SOD can disrupt the luminol signal very efficiently. The uncharged luminol molecule is membrane-permeant and can

Table 13.2 Tests for detection of reactive oxygen species (direct) or their oxidized products (indirect)

Assay	Probe	Extracellular/intracellular
Direct measurement		
Tetrazolium nitroblue	Ferricytochrome C	Extracellular
Chemiluminescence	Luminol	Both
	Lucigenin	Extracellular
Indirect measurement		
Lipid peroxidation levels	Thiobarbituric	Measures oxidized component in the body fluids
Antioxidants, micronutrients, vitamins	High-performance liquid chromatography	Serum and seminal plasma
Ascorbate	High-performance liquid chromatography	Seminal plasma
Antioxidants enzymes	Superoxide dismutase	Seminal plasma
	Catalase	Seminal plasma
	Glutathione peroxidase	Spermatozoa
	Glutathione reductase	Spermatozoa
Chemokines	ELISA	Seminal plasma
Antioxidant-pro-oxidant status	Total antioxidant	Low chain-breaking levels

react in the form of luminol or as a univalently oxidized luminol radical with a variety of ROS, including $O_2^{\bullet-}$, H_2O_2, and $OH^{\bullet}$ [12, 30]. Luminol is sensitive to H_2O_2 and this sensitivity can be greatly increased by addition of horseradish peroxidase. The luminol signal generated by human spermatozoa is initiated by a one-electron oxidative event that is mediated by H_2O_2. A luminescent signal is produced with luminol through a one-electron oxidative event mediated by H_2O_2 and either endogenous peroxidase or by addition of HRP [31]. Oxidation of luminol (one-electron) leads to the creation of a radical species which interacts with ground state oxygen to produce $O_2^{\bullet-}$, which participates in the oxygenation of luminol radical species to create an unstable endoperoxide, which breaks down and leads to light emission. In this, the $O_2^{\bullet-}$ is an essential intermediate for the luminol-dependent chemiluminescence. Also, the redox cycling activity associated with this probe allows the significant amplification of the signal and allows easy measurement of H_2O_2.

Lucigenin works through a one-electron reduction unlike luminol which requires one-electron oxidation that creates a radical from lucigenin; this radical gives up its electron to the grounds state oxygen to create $O_2^{\bullet-}$, thus returning the lucigenin to its parent state [12, 30, 31]. The luminol probe is more advantageous than the lucigenin probe for several reasons. Luminol measures both intracellular and extracellular ROS such as hydrogen peroxide, superoxide anion, and hydroxyl radical; on the other hand, lucigenin measures only the extracellular ROS, and in particular, superoxide anion [22, 30]. Lucigenin is an excellent probe for evaluating $O_2^{\bullet-}$ production as a nonspecific redox marker for the enhanced electron transfer activity associated with defective sperm function. Both the sensitivity and specificity of this probe are enhanced by its redox cycling activity [7, 29, 30].

13.7 Measurement of ROS by Chemiluminescence Assay

Herein we describe the details of measurement of ROS in unprocessed, i.e., liquefied seminal ejaculate without any further processing by chemiluminescence assay.

1. Equipment and material
   (a) Disposable polystyrene tubes with caps (15-mL)
   (b) Eppendorf pipettes (5-, 10-, 50-, and 1,000-μL)
   (c) Serological pipettes (1-, 2-mL)
   (d) Desk-top centrifuge
   (e) Disposable MicroCell slides
   (f) Dimethyl sulfoxide (DMSO; Catalog # D8779, Sigma Chemical Co., St. Louis, MO)
   (g) Luminol (5-amino-2,3 dihydro-1,4 phthalazinedione; Catalog # A8511, Sigma Chemical Co., St. Louis, MO)
   (h) Polystyrene Round bottom tubes (6-mL)
   (i) Luminometer (Model: LKB Autoplus 953)
   (j) Dulbecco's Phosphate-buffered saline solution 1× (PBS-1×; Catalog #9235, Irvine Scientific, Santa Ana, CA)
2. Preparation of reagents
   (a) *Stock luminol (100 mM)*: Weigh 177.09 mg of luminol and add it to 10 mL of DMSO solution in a polystyrene tube. The tube must be covered in an aluminum foil due to the light sensitivity of the luminol. It can be stored at room temperature in the dark until the expiration date.
   (b) *Working luminol (5 mM)*: Mix 20 μL luminol stock solution with 380 μL DMSO in a foil-covered polystyrene tube. This must be done prior to every use. Store the solution at room temperature in the dark until needed.
   (c) *DMSO solution*: Provided ready to use; store at room temperature until the expiration date.
3. Specimen preparation

   Upon arrival of the semen specimen, allow it to liquefy in the incubator at 37°C for approximately 20 min. A complete record of patient's name, allocated identity number, period of sexual abstinence, and date and time of collection is noted. Manual semen analysis is performed for assessment of sperm concentration and motility.

   First, allow the semen sample to undergo liquefaction for 20 min in a 37°C incubator. Then record the initial physical characteristics such as volume, pH, and color and manually verify sperm count and motility. After complete liquefaction is ensured, set up the luminometer for ROS measurement.
4. ROS measurement by luminometer

   This procedure is performed in a dark room. Set up the luminometer and the computer attached to it (Fig. 13.2a–c).

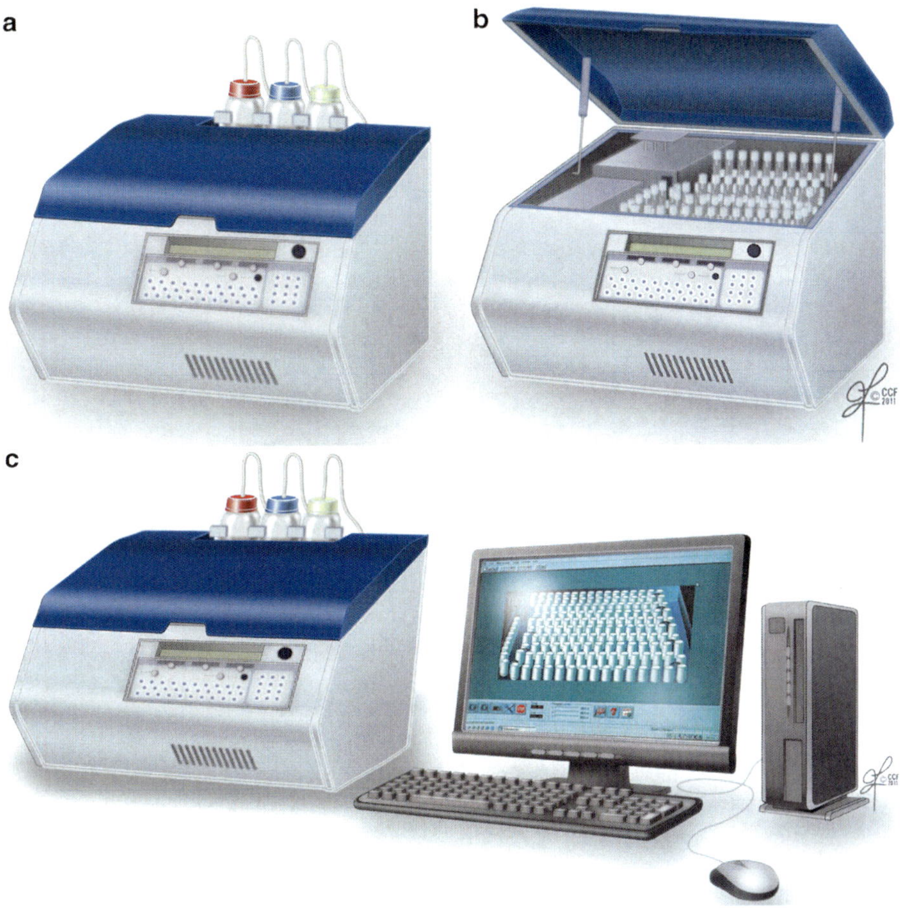

Fig. 13.2 Autolumat 953 plus luminometer used in the measurement of ROS by chemiluminescence assay. (**a**) External view and (**b**) internal view. Multiple tubes can be loaded simultaneously for measuring ROS. (**c**) The luminometer can be connected with the computer and a monitor and all the steps can be observed on the screen

Table 13.3 Setup for the measurement of ROS

Labeled tubes (no.)	PBS-1× (μL)	Specimen volume (μL)	Probe luminol (5 mM) (μL)	Hydrogen peroxide (μL)
Blank (tubes S1–3)	400	–		–
Negative control (tubes S4–6)	400	–	10	–
Patient (tubes S78)	–	400	10	–
Positive control (tubes S9–11)	400		10	50

(a) Label 11 Falcon tubes (12 × 75 mm) in duplicates and add the following reagents as indicated in Table 13.3 (Fig. 13.3).

Note: To avoid contamination, change pipette tips after each addition.

(b) Gently vortex the tubes to mix the aliquots uniformly.

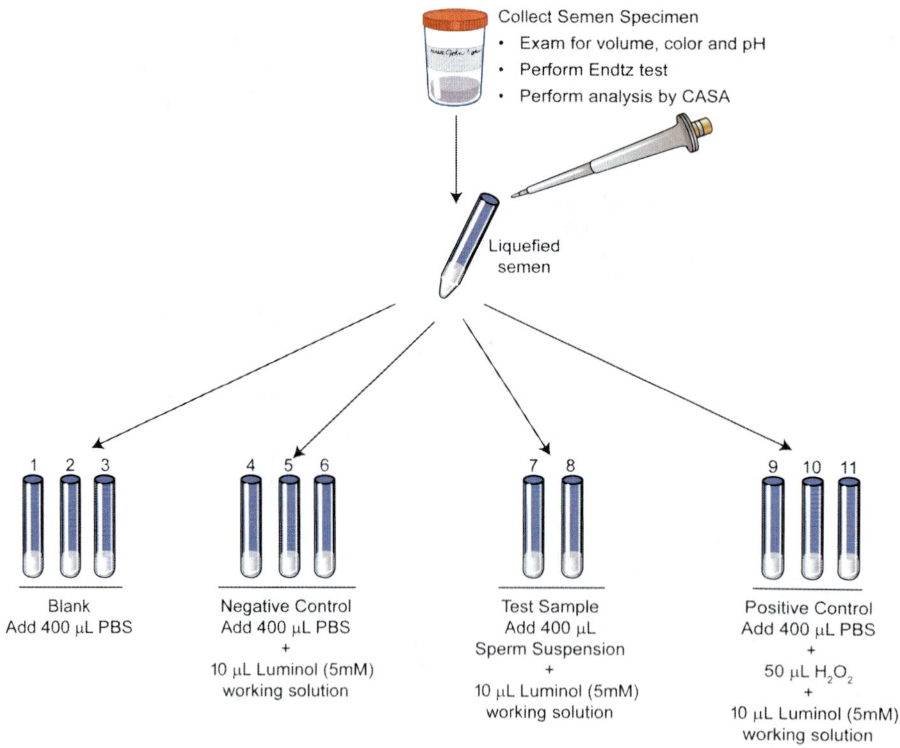

Fig. 13.3 Preparing the tubes for the ROS measurement. A total of 11 tubes are labeled from S1 to 12: Blank, negative control, patient sample, and positive control. Luminol is added only to all tubes except blank. Hydrogen peroxide is added only to the positive control

(c) Place all the labeled tubes in the luminometer in the following order: Blank (tubes labeled 1–3).

(d) Negative control (tubes labeled 4–6), patient sample (tubes labeled 7–8), and positive control (tubes labeled 9–11) (Fig. 13.3).

5. Instrument setup

(a) Turn on the instrument and the computer. From the desktop, click on "Berthold tube" master icon to start the program.

(b) From the "Setup menu," select "Measurement Definition" and then "New Measurement." You will be prompted to the following:
   - "Measurement Name" (Initials, Date, Analyte, and Measurement).
   - It will show "Measurement Definition-on the 'Tool bar.'"
   - Click "Luminometer Measurement" protocol and from the drop menu click on "Rep. assay."

– Define each "Parameter" as follows:

Read time	1 s
Background read time	0 s
Total time	900 s
Cycle time	30 s
Delay "Inj M read (s)"	0 s
Injector M (μL)	0 s
Temperature (°C)	37°C
Temperature control (0 = OFF)	1 = ON

– Press "Save"

(c) From the "Setup" menu, select "Assay Definition" and then "New Assay": It will ask for the following:

– "Assay Name" (Initials, Date, Analyte, Assay). Click "OK."
– Select "Measurement Method" and from the drop-down menu select the measurement from Step 2a above.
– Go to "Column Menu." Hide everything except the following:
Sample ID
Status
RLU mean
Read date
Read time
– Go to "Sample Type" menu and select "Normal."
– Press "OK"
– Go to file, "New" click "Workload" Press "OK."

(d) Save your "Work Load" (Date, Initial, Sample or experiment ID) in "Work Load" file.
(e) Click "File name"
(f) After saving the "Work Load," the name of the file will show in the "Title Bar."
(g) The specimens are ready to be analyzed.

6. Analyzing the samples

(a) Load the tubes into the instrument and click "Start." It will start scanning for tubes.
(b) After scanning, it will show how many tubes are detected by the instrument in each batch, press "Next."
(c) Select the "Assay Type" and type file name and then click "Finish."
(d) The "Excel spreadsheet" will open, measurement of the tubes will start.
(e) Do not touch or change the screen, wait (3–5 min) to make sure everything is working fine.
(f) After finishing measurements, it will ask for "Save Excel Spread Sheet," save it in the "My Document" under "ExcelSheet" folder.

Time	S1	S2	S3	S4	S5	S6	S7	S8	S9	S10	S11
0	6	4	9	10	13	10	19	12	1943	1719	2386
42.3	6	7	7	6	12	10	19	12	1857	1657	2248
72.3	9	7	7	10	7	12	18	9	1786	1576	2179
102.3	6	9	4	9	10	10	19	12	1742	1565	2068
132.3	9	9	7	10	7	10	16	12	1701	1493	2003
162.3	6	6	6	10	12	10	19	15	1614	1486	1943
192.3	6	7	4	9	7	15	21	13	1611	1462	1845
222.3	6	9	7	10	6	7	16	13	1582	1442	1812
252.3	7	9	10	9	10	12	16	9	1506	1408	1770
282.3	6	6	6	10	9	9	18	12	1462	1377	1729
312.3	7	9	6	12	10	12	16	12	1462	1358	1691
342.3	7	7	6	10	10	10	13	12	1391	1344	1657
372.3	4	7	7	7	9	9	18	10	1393	1330	1623
402.3	7	6	4	12	9	9	12	10	1372	1307	1583
432.3	7	10	7	9	9	15	13	10	1361	1319	1567
462.3	7	7	6	7	9	12	12	9	1319	1276	1534
492.3	6	9	7	12	9	12	9	7	1300	1275	1462
522.3	6	7	7	9	7	12	12	9	1278	1269	1437
552.3	6	12	6	10	6	13	13	10	1266	1238	1402
582.3	7	6	6	9	6	10	10	15	1244	1225	1424
612.3	10	7	6	9	9	10	10	9	1223	1216	1384
642.3	9	6	4	9	10	9	10	18	1182	1231	1344
672.3	6	9	9	7	9	13	9	27	1191	1207	1338
702.3	7	7	12	10	7	10	15	9	1169	1207	1310
732.3	9	7	7	9	7	10	12	38	1151	1157	1294
762.3	6	4	7	9	13	9	12	12	1163	1169	1287
792.3	12	7	9	7	7	9	10	7	1100	1156	1282
822.3	6	4	6	10	7	15	13	10	1114	1132	1212
852.3	3	9	10	9	9	9	15	10	1097	1142	1219
882.3	7	6	4	10	10	12	12	12	1085	1125	1234
812.3	7	9	9	9	13	18	13	10	1135	1117	1195

Fig. 13.4 A representative display of the readings showing the number of signals generated in each of the above 12 tubes (S1–11). The measurement is for a total of 900 s. As seen here, blanks have the least number of ROS produced and the positive control to which hydrogen peroxide is added has the highest number of ROS generated

   (g) Select "Excel Sheet," name the file, and save type as "Measurement Files" (*.txr). Save the "Excel Sheet" (Fig. 13.4).

7. Printing ROS results
   (a) Print Excel as well as the "chart 1" (Figs. 13.4 and 13.5).
   (b) Close the "Excel sheet"
   (c) Print the "Work Load" sheet (Fig. 13.6), save, and close it.

8. Calculating results
   (a) Calculate the "average RLU" for Negative control, Samples, and Positive control.
   (b) Calculate sample ROS by subtracting its average from negative control average.
   (c) Sample ROS = Average "RLU mean" for sample – Average "RLU mean" for negative control.
   (d) Correct the sample ROS by dividing it with "Sperm concentration/mL"

   Corrected sample ROS

   Calculated sample ROS/sperm concentration = XX (RLU/s/$\times 10^6$ sperm)

   A typical example of calculating ROS values is illustrated in Fig. 13.6.

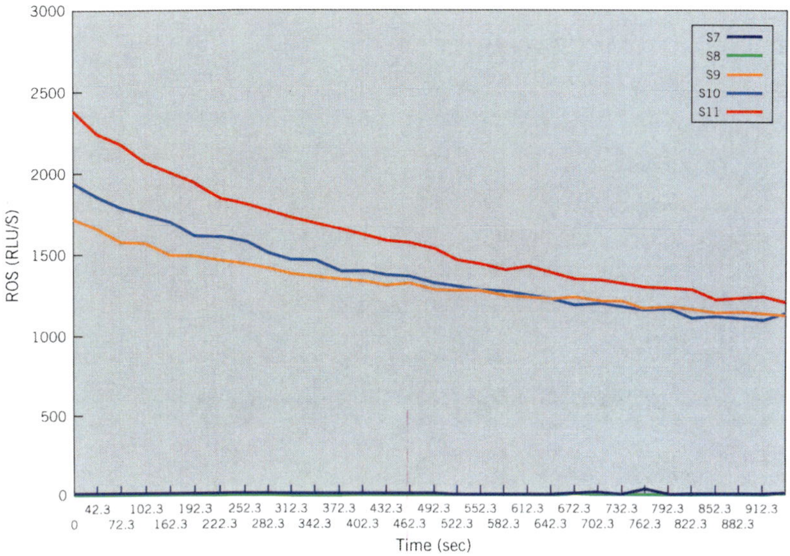

Fig. 13.5 A typical graph showing the ROS levels in the 11 tubes (S1–11). As seen here, only positive controls have significantly higher levels of ROS. Those producing low levels (tubes S1–8) of ROS are seen very close to the *X* axis

Sample	Sample ID	Status	RLU Mean	Read Date	Read Time
1	Blank	Done	6269	11/17/2010	1:35:54 PM
2		Done	6713	11/17/2010	1:35:56 PM
3		Done	6189	11/17/2010	1:35:57 PM
4	Negative Control	Done	8454	11/17/2010	1:35:59 PM
5		Done	8104	11/17/2010	1:36:00 PM
6		Done	9993	11/17/2010	1:36:02 PM
7	Test Sample	Done	12954	11/17/2010	1:36:03 PM
8		Done	11368	11/17/2010	1:36:05 PM
9	Positive Control	Done	1261225	11/17/2010	1:36:06 PM
10		Done	1207794	11/17/2010	1:36:08 PM
11		Done	1458674	11/17/2010	1:36:10 PM

Example

Patient average (P_{av}) = 12161 RLU / sec   Sperm count = 12.6 x 10^6 /mL

Negative Control average (NC_{av}) = 8850.3 RLU / sec

Corrected value = P_{av} - NC_{av}

= 3310.7 RLU / sec

Corrected ROS = $\frac{3310.7}{12.6}$ = 262.7 RLU / sec / x 10^6 sperms

Result = ROS positive

Fig. 13.6 A representative print out generated after the ROS measurement is complete. The chart shows the 11 tubes (S1–11), each representative of the type of sampler and the mean RLU value for each tube. At the *bottom* is also an example illustrating how to calculate the final amount of ROS generated in a given test (patient) sample

Reference values
Normal range: <20 RLU/s/× 10^6 sperm
Critical Values: >20 RLU/s/× 10^6 sperm

9. Quality control

   (a) *Criteria for acceptance*: control reads 20 RLU/s/× 10^6 sperm.
   (b) *Criteria for rejection*: control reads >20 RLU/s/× 10^6 sperm. The assay must be repeated and results shown to the Director.
   (c) The reagent lot numbers and expiration dates are recorded in the assay Quality Control book located in the laboratory.

13.8 Types of Samples for ROS Measurement

ROS measurement can be performed in various types of samples such as:

1. *Neat or unprocessed, whole seminal ejaculate.* In this method, the seminal plasma is not removed. All components of the seminal ejaculate are intact and ROS levels indicate the actual de novo levels in the ejaculate. This is the ideal setup for assessing the clinical relevance of ROS in the seminal ejaculate and is indicative of oxidative stress and sperm dysfunction.
2. *Processed sample*: Liquefied semen specimens are centrifuged at 300×*g* for 7 min and seminal plasma is removed. The sperm pellet is washed and resuspended to 1 mL volume in PBS. This is also called the simple "wash and resuspend" method. With this method, the seminal plasma that confers protection to the sperm and other dissolved components is removed. However, all the cellular components such as debris, round cells, white blood cells, and leukocytes are still present in the sample.
3. *Sperm preparation by the swim-up procedure*: This method is used to measure ROS in highly motile sperm prepared by swim up. After liquefaction, an aliquot of specimen is mixed with sperm wash media using a sterile Pasteur pipette. It is centrifuged at 330×*g* for 10 min. The supernatant is carefully aspirated and the pellet resuspended in 3 mL of fresh sperm wash media. The resuspended sample is carefully transferred in equal parts to two 15-mL sterile round-bottom test tubes and centrifuged at 330×*g* for 5 min. Motile sperm are allowed to swim up during the incubation of test tubes at a 45° angle in 5% CO_2 at 37°C for 1 h. Supernatant is aspirated into a clean test tube and centrifuged at 330×*g* for 7 min. The final supernatant is aspirated and the sperm pellet is resuspended in sperm wash media and is used for measurement of ROS.
4. *Sperm preparation by density gradients*: This method is used to measure ROS in immature (morphologically abnormal, poor motility) and mature (highly motile and morphologically normal) sperm. A double density gradient (40% "Upper phase" and 80% "Lower phase") is used. Both the density gradient and the sperm wash media is brought to 37°C or room temperature. Using a sterile pipette, 2.0 mL of the "Lower phase" is transferred into a 15-mL conical centrifuge tube.

Similarly, 2.0 mL of the "Upper phase" is carefully placed on the top of the lower layer. The liquefied semen sample (1–2 mL) is placed on top of the upper layer, and the tube is centrifuged for 20 min at 330 × *g* or 1,600 revolutions per minute (rpm). The uppermost layer containing clear seminal plasma and dead white cells and other debris are discarded. For separating the immature sperm, the interphase between the upper and lower layer is carefully transferred to a conical tube. For separating the mature sperm, the pellet is carefully aspirated and transferred into another centrifuge tube. Using a transfer pipette, 2–3 mL of sperm wash media is added, and the immature and mature fractions are centrifuged for 7 min at 330 × *g* or 1,600 rpm. The supernatant is discarded, and the pellet is suspended in 1.0 mL of sperm wash media. Sperm count, motility, and ROS levels are measured in the recovered fractions.

13.9 Luminometers

Several types of luminometers, varying ranging in features, design, and pricing, can be used in measuring the emitted light from the chemiluminescence assay reaction (Table 13.4) [32]. The measurements from the chemiluminescence assay are expressed as either counted photons per minute (cpm) or as relative light units (RLU). There are two different kinds of processing designs for luminometers. While direct current luminometers measure electric current, photon counting luminometers count individual photons [33]. Moreover, there are currently three types of luminometers available for commercial uses. First is single and double tube luminometers, which measure one or two samples at a time. These are inexpensive and are typically used by small research laboratories. Secondly, multiple tube luminometers that measure several tubes at a time. They are more expensive than single and double tube luminometers and are used by research centers that depend on chemiluminescence for research projects. Lastly, plate luminometers measure multiple tubes at a given time and are typically used by core research laboratories and commercial enterprises.

Table 13.4 Commercially available luminometers shown by type, sensitivity, and manufacturer

Model	Type	Sensitivity and dynamic range	Manufacturer
GloMax 20/20	Single tube	0.1 fg luciferase	Promega Corporation (Sunnyvale, CA, USA)
FB-12	Single tube	1,000 molecules of luciferase	Zylux Corporation (Oak Ridge, TN, USA)
Triathler	Single tube	<10 amol ATP/vial	Hidex (Turku, Finland)
Optocomp-2	Multiple tube	0.1 pg ATP	MGM Instruments (Hamden, CT, USA)
Autolumat Auto Plus LB 953	Multiple tube	5 amol of ATP	Berthold Technologies (Oak Ridge, TN, USA)

13.10 Factors Affecting Chemiluminescent Reactions

It is important to have a reliable and valid method of detection of ROS [29]. The luminol assay is robust; however, there are various factors that can affect ROS detection by the chemiluminescent reactions [7, 30, 34–36]. Both the probes luminol and lucigenin have great sensitivities, although the specificity is limited because of the interference of other components such as infiltrating leukocytes. Some of these confounding factors are:

1. The luminometer instrumentation, its calibration, determination of sensitivity, dynamic range, and units used.
2. The concentration and the type of probe used.
3. The concentration and volume of the semen sample, use of reagent, and temperature of the luminometer.
4. Semen age, i.e., time to analysis after sample, is collected to the time of ROS measurement.
5. Viscous samples and poor liquefaction interferes with chemiluminescent signals.
6. Repeated centrifugation: Artificial increase in chemiluminescent signal because of the shearing forces generated by centrifugation [35].
7. Serum albumin. Use of media that contain bovine serum albumin can generate spurious signals in the presence of human seminal plasma.
8. Medium pH. Luminol is sensitive to pH changes.
9. Nonspecific interference: Many compounds can artificially increase (cysteine, thiol-containing compounds) or decrease (ascorbate, uric acid) the chemiluminescent signal generated by the spermatozoa. Hence, it is necessary to run sperm-free controls as integral part of the chemiluminescent assay.

13.11 Diagnostic Application of Chemiluminescence Assays

Chemiluminescent assays provide clinically significant results that correlate with the quality of the ejaculate and the fertilizing potential of the spermatozoa both in vitro and in vivo. This is especially true in patients presenting with poor sperm quality such as in cases of oligozoospermia, or in samples where low-density immature spermatozoa recovered following sperm preparation using a density gradient technique. Defective sperm populations such as those exhibiting presence of excess residual cytoplasmic are associated with high redox activity [6, 13, 36–38]. This assay can also be used as an important diagnostic screening tool in the evaluating patients with unexplained or idiopathic infertility as these patients demonstrate high ROS levels in their seminal ejaculates.

13.12 Conclusion

Oxidative stress is implicated in the etiology of male infertility and sperm dysfunction resulting in induction of lipid peroxidation, abnormal sperm function, and increase in mitochondrial and nuclear DNA damage. Therefore, there is an intense interest in the development and evaluation of simple, convenient techniques for monitoring the generation of ROS by human spermatozoa. The chemiluminescence assay provides andrology laboratories with a reliable, heavily-tested and used method in detecting ROS levels in patient semen samples. As infertility rates continue to rise, and male-factor plays a key role in contributing to infertility in couples of reproductive age, it is critical for andrology laboratories to fully understand the chemiluminescence assay in order to provide better care for infertile patients. While there are only two probes that can be used in the assay for different measurements of ROS, there is a multitude of luminometers that are available for commercial purchase and suitable for differing laboratory settings. Andrology laboratories must review the latest literature in order to learn about the latest studies that establish more precise and accurate reference levels for ROS in semen samples. Chemiluminescent probes can be used to monitor redox activity associated with sperm suspensions and differentiate this from the redox activity associated with contaminating leukocytes as these can disrupt fertilization and DNA integrity in the spermatozoa exposed after sperm preparation and increasing their susceptibility to oxidative stress. These chemiluminescent signals are also linked to the presence of excessive residual cytoplasm in defective spermatozoa and are indicative of aberrant spermatogenic process. The redox activity measured in defective spermatozoa by chemiluminescent increases our understanding of the etiology of male infertility. The chemiluminescence assay is a key technique in the andrology laboratory that will continue to develop and be improved upon and used in testing for suboptimal sperm quality.

References

1. Stephen E, Chandra A (1998) Updated projection of infertility in the United States: 1995–2025. Fertil Steril 70:30–34
2. Bhasin S, de Kretser DM, Baker HW (1994) Clinical review 64: pathophysiology and natural history of male infertility. J Clin Endocrinol Metab 79:1525–1529
3. Pasqualotto FF, Sharma RK, Nelson DR, Thomas AJ, Agarwal A (2000) Relationship between oxidative stress, semen characteristics, and clinical diagnosis in men undergoing infertility investigation. Fertil Steril 73:459–464
4. de Lamirande E, Gagnon C (1992) Reactive oxygen species and human spermatozoa. II. Depletion of adenosine triphosphate plays an important role in the inhibition of sperm motility. J Androl 13:379–386
5. Mammoto A, Masumoto N, Tahara M et al (1996) Reactive oxygen species block sperm-egg fusion via oxidation of sperm sulfhydryl proteins in mice. Biol Reprod 55:1063–1068

6. Aitken RJ, Irvine DS, Wu FC (1991) Prospective analysis of sperm-oocyte fusion and reactive oxygen species generation as criteria for the diagnosis of infertility. Am J Obstet Gynecol 164:542–551
7. Agarwal A, Cocuzza M, Abdelrazik H, Sharma RK (2008) Oxidative stress measurement in patients with male or female factor infertility. In: Popov I, Lewin G (eds) Handbook of chemiluminescent methods in oxidative stress assessment. Tranworld Research Network, Trivandrum, Kerala, India, pp 195–218
8. Alvarez JG, Touchstone JC, Blasco L et al (1987) Spontaneous lipid peroxidation and production of hydrogen peroxide and superoxide in human spermatozoa. Superoxide dismutase as major enzyme protectant against oxygen toxicity. J Androl 8:338–348
9. Aitken RJ, Baker HW (1995) Seminal leukocytes: passengers, terrorists or good samaritans? Hum Reprod 10:1736–1739
10. Pasqualotto FF, Sharma RK, Kobayashi H, Nelson DR, Thomas AJ, Agarwal A (2001) Oxidative stress in normospermic men undergoing infertility evaluation. J Androl 22:316–322
11. Kessopoulou E, Tomlinson MJ, Barrat CLR et al (1992) Origin of reactive oxygen species in human semen spermatozoa or leukocytes. J Reprod Fertil 94:463–470
12. Aitken RJ, Buckingham DW, West KM (1992) Reactive oxygen species and human spermatozoa: analysis of the cellular mechanisms involved in luminol- and lucigenin-dependent chemiluminescence. J Cell Physiol 151:466–477
13. Gil-Guzman E, Ollero M, Lopez MC et al (2001) Differential production of reactive oxygen species by subsets of human spermatozoa at different stages of maturation. Hum Reprod 16:1922–1930
14. Hendin B, Kolettis P, Sharma RK et al (1999) Varicocele is associated with elevated spermatozoal reactive oxygen species production and diminished seminal plasma antioxidant capacity. J Urol 161:1831–1834
15. Said TM, Agarwal A, Sharma RK, Thomas AJ, Sikka SC (2005) Impact of sperm morphology on DNA damage caused by oxidative stress induced by beta-nicotinamide adenine dinucleotide phosphate. Fertil Steril 83:95–103
16. Aitken RJ, West KM (1990) Analysis of the relationship between reactive oxygen species production and leukocyte infiltration in fractions of human semen separated on Percoll gradients. Int J Androl 3:433–451
17. Ochsendorf FR (1999) Infections in the male genital tract and reactive oxygen species. Hum Reprod 5:399–420
18. Shekarriz M, Sharma RK, Thomas AJ et al (1995) Positive myeloperoxidase staining (Endtz Test) as an indicator of excessive reactive oxygen species formation in semen. J Assist Reprod Genet 12:70–74
19. Wolff H (1995) The biologic significance of white blood cells in semen. Fertil Steril 63:1143–1157
20. Blake DR, Allen RE, Lunec J (1987) Free radical in biological systems—a review oriented to inflammatory processes. Br Med Bull 43:371–385
21. Sharma RK, Pasqualotto FF, Nelson DR, Thomas AJ, Agarwal A (1999) The reactive oxygen species-total antioxidant capacity score is a new measure of oxidative stress to predict male infertility. Hum Reprod 14:2801–2807
22. Sharma RK, Agarwal A (1996) Role of reactive oxygen species in male infertility. Urology 48:835–850
23. Sikka SC, Rajasekaran M, Hellstrom WJG (1995) Role of oxidative stress and antioxidants in male infertility. J Androl 16:464–468
24. Kefer JC, Agarwal A, Sabanegh E (2009) Role of antioxidants in the treatment of male infertility. Int J Urol 16:449–457
25. Lanzafame FM, La Vignera S, Vicari E, Calogero AE (2009) Oxidative stress and medical antioxidant treatment in male infertility. Reprod Biomed Online 19:638–659

26. Tremellen K (2008) Oxidative stress and male infertility—a clinical perspective. Hum Reprod Update 14:243–258
27. Saleh RA, Agarwal A (2002) Oxidative stress and male infertility: from research bench to clinical practice. J Androl 23:737–752
28. Agarwal A, Said TM (2003) Role of sperm chromatin abnormalities and DNA damage in male infertility. Hum Reprod Update 9:331–345
29. Kobayashi H, Gil-Guzman E, Mahran AM et al (2001) Quality control of reactive oxygen species measurement by luminol-dependent chemiluminescence assay. J Androl 22:568–574
30. Aitken RJ, Baker MA, O'Bryan M (2004) Shedding light on chemiluminescence: the application of chemiluminescence in diagnostic andrology. J Androl 25:455–465
31. Faulkner K, Fridovich I (1993) Luminol and lucigenin as detectors for $O_2^{-\bullet}$. Free Radic Biol Med 15:447–451
32. Stanley PE (1999) Commercially available fluorometers, luminometers and imaging devices for low-light level measurements and allied kits and reagents: survey update 6. Luminescence 14:201–213
33. Agarwal A, Allamaneni S, Said T (2004) Chemiluminescence technique for measuring reactive oxygen species. Reprod Biomed Online 9:466–468
34. Berthold R, Herick K, Siewe RM (2000) Luminometer design and low light detection. Methods Enzymol 305:62–87
35. Shekarriz M, DeWire DM, Thomas AJ Jr, Agarwal A (1995) A method of human semen centrifugation to minimize the iatrogenic sperm injuries caused by reactive oxygen species. Eur Urol 28:31–35
36. Aitken RJ, Clarkson JS (1987) Cellular basis of defective sperm function and its association with the genesis of reactive oxygen species by human spermatozoa. J Reprod Fertil 81:459–469
37. Aitken RJ, Clarkson JS, Hargreave TB, Irvine DS, Wu FC (1989) Analysis of the relationship between defective sperm function and the generation of reactive oxygen species in cases of oligozoospermia. J Androl 10:214–220
38. Ollero M, Gil-Guzman E, Lopez MC, Sharma RK, Agarwal A, Larson K, Evenson D, Thomas AJ Jr, Alvarez JG (2001) Characterization of subsets of human spermatozoa at different stages of maturation: implications in the diagnosis and treatment of male infertility. Hum Reprod 16:1912–1921

Chapter 14
Direct Methods for the Detection of Reactive Oxygen Species in Human Semen Samples

R. John Aitken, Geoffry N. De Iuliis, and Mark A. Baker

Abstract The accurate, selective measurement of reactive oxygen species (ROS) production by human spermatozoa is still a work-in-progress. Traditionally, chemiluminescence approaches have been employed using luminol or lucigenin as probes to provide a sensitive readout of the redox status of human sperm populations. A particular concern with these chemiluminescent probes is that they can also generate spurious signals as a consequence of redox cycling activity, particularly in the case of lucigenin which only needs to experience a one-electron reduction at the hands of a cellular oxidoreductase to generate intense but artificial redox signals. Most importantly, the measurement of ROS in sperm suspensions is confounded by the presence of leukocytes. Alternatives are to either physically remove the leukocytes using magnetic beads coated with a monoclonal antibody against the common leukocyte antigen (CD45) or to use fluorescent probes in combination with fluorescence microscopy or flow cytometry to focus the analysis on individual cells rather than cell suspensions.

Keywords Chemiluminescence techniques • Direct methods • Detection of reactive oxygen species • Semen • Chemiluminescent assays

R.J. Aitken, PhD, ScD (✉) • G.N. De Iuliis, PhD, BSc (Hons)
Department of Biological Sciences, University of Newcastle, Callaghan, NSW 2308, Australia
e-mail: john.aitken@nerwcastle.edu.au

M.A. Baker, PhD
Department of Environmental and Life Science, University of Newcastle, Callaghan, NSW, Australia

A. Agarwal et al. (eds.), *Studies on Men's Health and Fertility*, Oxidative Stress in Applied Basic Research and Clinical Practice, DOI 10.1007/978-1-61779-776-7_14,

14.1 Introduction

In 1943, MacLeod [1] first noted that spermatozoa incubated under high oxygen tensions rapidly lost their motility. He suggested that such loss was due to the cellular production of hydrogen peroxide (H_2O_2) because addition of catalase (a hydrogen peroxide scavenger) under the same conditions restored sperm movement. The notion that actively respiring spermatozoa can produce factors that inhibit cell metabolism can be traced back to studies on the phenomenon of allelostasis in populations of sea urchin spermatozoa [2]. Allelostasis refers to the increase in oxygen uptake that accompanies the dilution of sperm populations which, it was thought, might have been due to the "autointoxication" of spermatozoa as a result of oxygen metabolites generated in densely packed populations of these cells. The first clear demonstration that spermatozoa can in fact generate reactive oxygen species (ROS) was presented in a paper published in *Nature* by Tosic and Walton [3]. These authors identified H_2O_2 as an inhibitor of bovine sperm respiration and used a very simple benzidine-peroxidase reaction to quantify the generation of this powerful oxidant. In 1950, Tosic and Walton [4] produced a second more extensive paper describing the autointoxication of bovine sperm populations, which seemed to generate a potent inhibitor of respiration and motility when incubated under aerobic conditions. The inhibition was reversed by catalase and peroxidase and was chemically identified as H_2O_2. The particular toxicity of this oxidant towards spermatozoa was not only demonstrated in this paper, but has been subsequently rediscovered many times in several different species including man, bull, dog, fowl, ram, mouse, rabbit, and rat [5–9]. Thus, the importance of cellular ROS generation is well established in the sperm biology literature and predates the discovery of this activity in phagocytic leukocytes by several decades.

The vulnerability of spermatozoa to ROS exposure stems from several important, and even unique, attributes of these cells. For example, spermatozoa are very well endowed with substrates for free radical attack, particularly in the form of polyunsaturated fatty acids, notably docosahexaenoic acid (22:6), which readily undergoes lipid peroxidation [10, 11]. Spermatozoa also possess additional targets for oxidative stress in the form of proteins, which can be modified by ROS to generate a wide variety of products. For example, ROS may oxidize amino acid side chains into ketone or aldehyde derivatives or induce the formation of nitrotyrosine residues, advanced glycation end products, or protein adducts via their interaction with the electrophilic products of lipid peroxidation, such as 4 hydroxynonenal [12–14]. Of course, another major target for oxidative stress in the male germ line is the DNA in the sperm nucleus and mitochondria [15]. Where and how such oxidative attacks arise is the subject of some speculation; however, the consequences of this stress, in terms of the functionality of the spermatozoa and their capacity to support normal embryonic development, are beyond reasonable doubt [16, 17]. In light of these considerations, it is clearly important that we develop reliable, robust methods for detecting the generation of ROS by mammalian spermatozoa. The issues that have to be addressed in developing such diagnostic tests vary somewhat between species, so at the outset it is important to define the context in which such diagnostic procedures will be applied—before discussing their technical idiosyncrasies. For the purpose of this discussion, we shall focus exclusively on the diagnosis of ROS generation by human spermatozoa.

14.2 Chemiluminescence Techniques

14.2.1 The Significance of Leukocyte Contamination

When it comes to the measurement of cellular ROS generation, human semen presents some significant problems that might not be encountered in other biological materials. Thus, one of the major drawbacks encountered in analyzing human semen from a ROS perspective is the presence of relatively high levels of leukocyte contamination. Every human semen sample is contaminated with leukocytes, the median concentration being around 24,900/mL or 91,000/ejaculate [18]. Although this may not seem like a large number of leukocytes, particularly given the concomitant presence of hundreds of millions of spermatozoa, it should be recognized that most of these contaminating cells (>80%) are granulocytes.

The dominance of granulocytes in the leukocytic profile is profoundly important because these cells are professional generators of ROS. Furthermore, at least some of the leukocytes contaminating human semen are in an activated state, generating a very high correlation coefficient ($r=0.8$) between chemiluminescent measures of ROS generation and leukocyte concentration in semen (Fig. 14.1a, b) [18, 19]. Importantly, this correlation between leukocyte numbers and ROS generation extends across the entire range of leukocyte concentrations found in human semen (typically 10^4–10^5), far below the leukocyte concentration deemed as pathological by the World Health Organization [20] of 1×10^6/mL. Thus, even in semen samples that would not be classified as leukocytospermic by conventional criteria, the presence of leukocytes is still a significant source of ROS. Although the presence of a few thousand leukocytes may seem trivial, in terms of the measurement of ROS generation by spermatozoa it is of major importance because each leukocyte is log orders of magnitude more active than a spermatozoon in the generation of ROS. Even though some authors may assert that the semen samples subjected for analysis are not significantly contaminated with leukocytes from a clinical perspective (i.e., are not leukocytospermic), from a biochemical standpoint the presence of such contaminating cells is a major confounding factor in the measurement of ROS generation by spermatozoa that can render such determinations worthless.

The key to generating meaningful data when assessing ROS generation by entire sperm suspensions, as when chemiluminescence or amplex red are used, is to prepare the cells adequately. If unselected sperm populations are employed which have been prepared by repeated cycles of centrifugation and resuspension in a simple defined culture medium, the outcome is going to be a ROS signal that is dominated by the presence of contaminating leukocytes, dwarfing the relatively weak signal

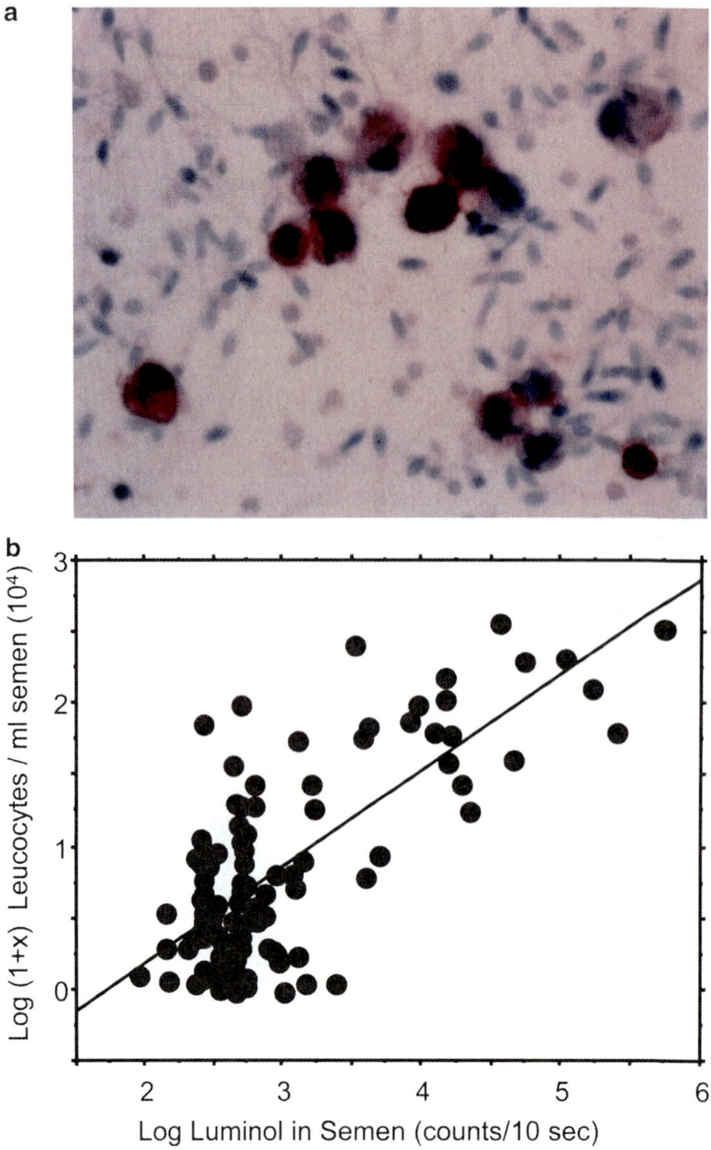

Fig. 14.1 Leukocytes are a conspicuous feature of the free radical generating profile of human spermatozoa. (**a**) Immunocytochemical detection of leukocytes in human semen samples using a monoclonal antibody against the common leukocyte antigen (CD45). (**b**) In unfractionated semen samples, the ROS signal measured by luminol-dependent chemiluminescence is strongly correlated with leukocyte concentration measured using CD45 immunocytochemistry. Redrawn from Aitken et al. [18], by permission of Oxford University Press

produced by the spermatozoa [21]. In order to affect the removal of contaminating leukocytes, some form of sperm fractionation procedure is required. This may take the form of a swim-up or swim-down procedure or, most commonly, discontinuous gradient centrifugation [21]. When sperm suspensions are subjected to such cell purification procedures, the ROS measurements obtained on these cell populations are significantly, and invariably, diminished (Fig. 14.2) [21, 22].

Very few authors have engaged the process of carefully dissecting the contribution made by leukocytes to the ROS signals generated by purified human sperm suspensions. However, where such studies have been conducted, the results are unequivocal in demonstrating that while leukocytes are the major source of ROS in such samples, defective spermatozoa are also capable of this activity [22, 23].

One of the earlier studies was conducted on unselected semen donors and examined ROS generation following fractionation of human semen samples on discontinuous Percoll gradients [22]. This study employed three variations of the luminol assay comprising: (1) luminol alone, (2) luminol plus horse radish peroxide to focus attention on H_2O_2 generation, and (3) luminol/peroxidase-dependent chemiluminescence before and after the addition of catalase in order to secure a precise measure of extracellular H_2O_2 levels, as indicated in Fig. 14.2. The results of this analysis again demonstrated that the low-density Percoll fractions exhibited significantly higher ROS signals than the purified high-density fractions, but that the latter still exhibited an impressive range of chemiluminescent responses [22, 23] (Fig. 14.2). Whether these signals were generated by spermatozoa, precursor germ cells, or residual leukocytes was then investigated using a monoclonal antibody against the common leukocyte antigen (CD45) to quantify the levels of white cell contamination. The results of one such analysis, illustrated in Fig. 14.3, indicated that the high-density Percoll fractions contained very few leukocytes, but whenever leukocytes were present a chemiluminescent signal was evident (Fig. 14.3). These samples also contained a low number of germ cells (0.1×10^4 per 10^7 spermatozoa), but the presence of these cells showed no relationship with the generation of ROS. As a result, when leukocyte concentrations are found to be low, much of the variation in ROS generation within these high-density Percoll fractions can be ascribed to the spermatozoa. Furthermore, these chemiluminescent signals were significantly higher in the spermatozoa of oligozoospermic patients than in normozoospermic controls (Fig. 14.3). An analysis focusing on high-density Percoll fractions from oligozoospermic samples, from which all detectable leukocyte contamination had been excluded, clearly demonstrated that the spermatozoa of these patients were significantly more active than controls in generating ROS [23].

Intriguingly, not only did the purified spermatozoa of oligozoospermic specimens generate higher levels of ROS, but the low-density fractions invariably contained significantly more leukocytes than normozoospermic control specimens (Fig. 14.3) [23]. This link between defective sperm function and leukocytic infiltration into the ejaculate has been observed subsequently [24, 25] and may represent a physiological leukocytic response to the presence of defective, moribund spermatozoa in the male reproductive tract, possibly triggered by apoptotic markers appearing on the surface of these cells [25].

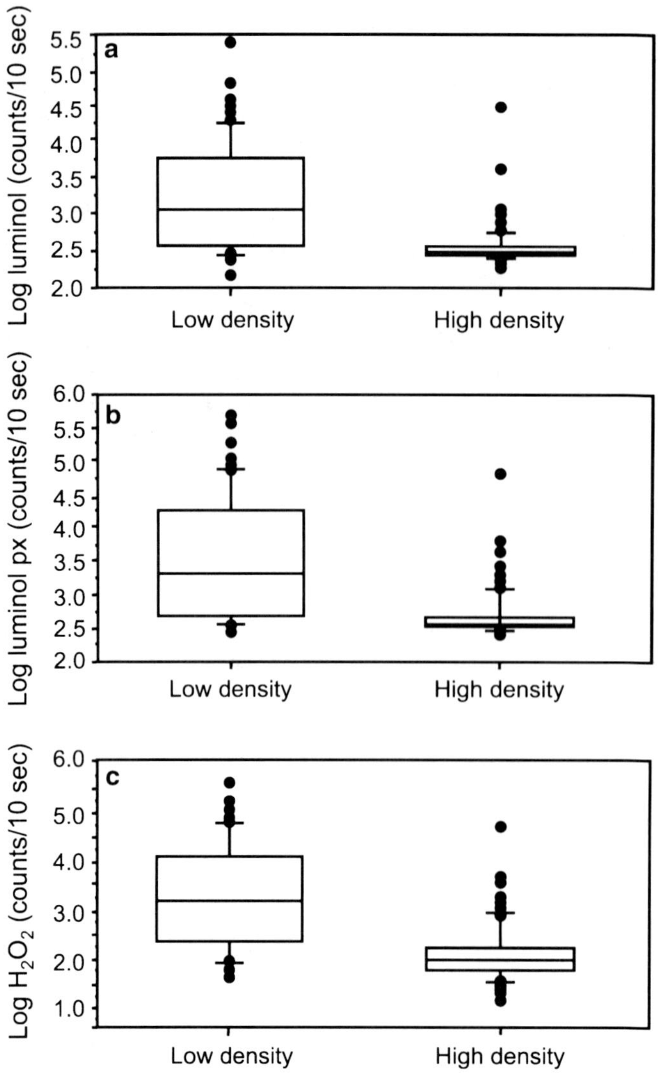

Fig. 14.2 Comparison of the ROS-generating potential of spermatozoa recovered from the high and low-density regions of Percoll gradients. Purification of these cell populations in order to diminish the level of leukocyte contamination leads to a significant reduction in the ROS signal as measured by: (**a**) luminol; (**b**) luminol and peroxidase, and (**c**) the change in luminol and peroxidase chemiluminescence observed following the addition of catalase, in order to focus attention of extracellular H_2O_2. *Boxes* indicate 25th and 75th percentiles while the *horizontal lines* through the box represent the 50th percentile (median). *Vertical lines* indicate the 10th and 90th percentile limits of the data, while extreme results outside of these limits are represented as single data points. Redrawn from Aitken and West [22] by permission of John Wiley and Sons

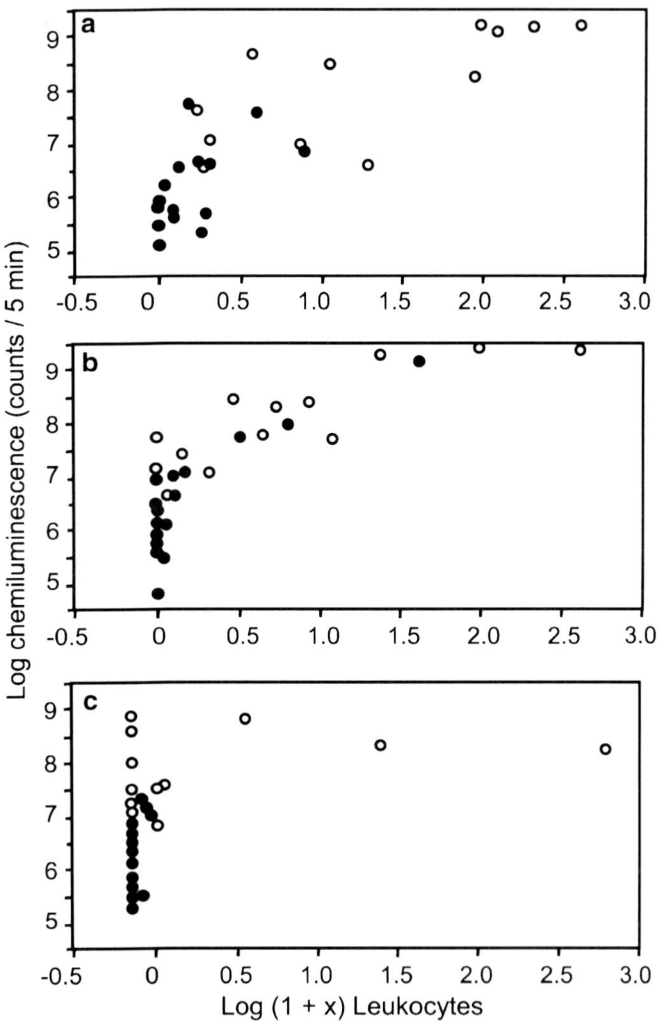

Fig. 14.3 Relationship between leukocyte concentration and ROS signals measured following treatment of the spermatozoa with the 12-myristate, 13-acetate phorbol ester. (**a**) Cells from the Percoll/semen interface; (**b**) cells from the low-density/high-density Percoll interface and (**c**) samples recovered in the high-density Percoll pellet. *Closed circles* represent fertile donors while *open circles* represent oligozoospermic patients. Whenever leukocytes are present, significant ROS signals are recorded. More leukocytes are present in the oligozoospermic specimens, particularly in the low-density Percoll fractions. However, when samples were generated lacking detectable leukocyte contamination (as in **c**), it is clear that the spermatozoa exhibited a range of ROS-generating activity with the oligozoospermic specimens being particularly active. Redrawn from Aitken et al. [23], by permission from BioScientifica

Such data clearly suggest that the spermatozoa are a source of ROS in cases of defective sperm function; however, this contribution cannot be detected if leukocytes are present in the same cell suspension. Any assessments of ROS generation by human sperm suspensions that do not rigorously exclude a contribution from infiltrating leukocytes are therefore seriously compromised. There are two major ways in which this problem can be solved. First, if whole sperm populations are to be analyzed, then the only solution is to physically remove the leukocytes from the suspension. This can be done using Percoll gradient centrifugation or electrophoretic sperm isolation [26] to achieve the initial purification of the sperm suspension followed by a cell extraction procedure employing magnetic beads or ferrofluids coated in a monoclonal antibody targeting the common leukocyte antigen [27]. Using this technique, human sperm suspensions can be rapidly cleared of contaminating leukocytes allowing assessments of ROS generation by sperm suspensions that are entirely focused on the gametes.

Furthermore, the purity of the sperm suspensions generated using such cell purification techniques can be validated using simple leukocyte provocation assays employing either FMLP (formyl methionyl leucyl phenylalanine) or opsonized zymosan as the stimulant. The basic principle behind these assays is that following the addition of chemiluminescence reagents such as luminol/peroxidase, an agonist is added to the sperm suspension that can only trigger ROS generation by phagocytic leukocytes. FMLP was the first stimulant to be used for the detection of leukocytes in human sperm suspensions [28, 29], but more recently opsonized zymosan has been employed for the same purpose [26]. The signals elicited by these two probes are highly correlated, both with each other and with the size of the contaminating leukocyte population [29].

If such techniques are not applied, there is a strong possibility that the responses recorded are more reflective of the level of leukocyte contamination than abnormal redox activity on the part of the spermatozoa. For example, the claim that NGF (nerve growth factor) will stimulate ROS generation in suspensions of human spermatozoa [30] is almost certainly a measure of the level of leukocyte contamination in these preparations [29].

14.2.2 The Chemical Principles Underpinning Chemiluminescence

The two major probes that have been used to assess ROS generation by human spermatozoa are luminol (5 amino-2,3-dihydro-1,4-phthalazinedione) and lucigenin (10,10'-dimethyl-9,9'-biacridinium dinitrate). Because lucigenin carries a positive ionic charge, it is generally thought to be membrane-impermeant and to respond to ROS, particularly superoxide ($O_2^{-\bullet}$) anion in the extracellular space. In contrast, the uncharged luminol is thought to be relatively membrane-permeant and to react with a variety of ROS including $O_2^{-\bullet}$, H_2O_2, and $OH^\bullet$, both in the intra- and extracellular spaces. However, the sensitivity of this probe towards extracellular H_2O_2 can be greatly accentuated by the addition of horseradish peroxidase [31, 32]. While these probes are extremely sensitive and convenient for diagnostic purposes, the data they generate have to be interpreted with care and should be underpinned by knowledge of the chemical principles responsible for their reactivity. As emphasized above, a variety of confounding factors can influence the chemiluminescence responses observed with these probes in addition to the inevitable presence of contaminating leukocytes including incubation time, medium composition, pH, presence of seminal plasma, and albumin source, all of which can have a profound impact on the signals obtained [29].

14.2.2.1 Luminol

It has been presumed that both H_2O_2 and $O_2^{-\bullet}$ are involved in luminol-dependent chemiluminescence because both catalase and superoxide dismutase (SOD) can disrupt the signal with great efficiency (Fig. 14.4a–c). The luminol signal generated by human spermatozoa is initiated by a one-electron oxidative event mediated by H_2O_2 and either endogenous peroxidase [32–34] or, in order to sensitize the assay for extracellular H_2O_2, by the addition of horse radish peroxidase to the medium [32, 35]. The one-electron oxidation of luminol leads to the creation of a radical species ($L^\bullet$). The latter interacts with ground state oxygen to produce $O_2^{-\bullet}$ that then participates in the oxygenation of $L^\bullet$ to create an unstable endoperoxide, which breaks down with the release of light (Fig. 14.4a). According to this scheme, $O_2^{-\bullet}$ is an essential intermediate in the manifestation of luminol-dependent chemiluminescence and it is for this reason that SOD is such an effective inhibitor of this probe. However, the biochemical activity of SOD should never be taken to indicate the primary production of $O_2^{-\bullet}$ by human spermatozoa, as sometimes suggested; $O_2^{-\bullet}$ is simply an artificially created intermediate that is essential for luminol-dependent chemiluminescence [32]. Indeed, any univalent oxidant has the potential to generate $O_2^{-\bullet}$, and hence chemiluminescence, in the presence of luminol, including ferricyanide, persulfate, hypochlorite, $ONOO^-$, and xanthine oxidase (Fig. 14.4a).

H_2O_2 lies upstream of $O_2^{-\bullet}$ in the reaction scheme depicted in Fig. 14.4a and its involvement in the initial oxidation of luminol partly accounts for the inhibitory effects of catalase (Fig. 14.4b). In addition, H_2O_2 will also react directly with the azaquinone (L^+) and thereby contribute to the formation of excited aminophthalic acid, the chemiluminescent species [36]. In some species (rat and mouse but not human), secondary radical species are created in the presence spermatozoa and luminol/peroxidase that generate very intense chemiluminescent signals over prolonged periods of time. These responses may reflect the nonenzymatic generation of NO and ONOO as a consequence of H_2O_2-mediated attacks on arginine [37].

One of the most important points to emphasize about Fig. 14.4 is the opportunity this chemiluminescence detection system presents for redox cycling. All that is needed is a source of H_2O_2 (or alternative oxidizing species) to initiate the one-electron oxidation of luminol, and an azaquinone reductase, such as diaphorase, to reduce L^+ back to the parent luminol (L). The remaining elements of the chemiluminescent cascade can be generated by the detection system itself ($O_2^{-\bullet}$ by the interaction of $L^\bullet$ with ground state oxygen, H_2O_2 by the SOD-induced dismutation of $O_2^{-\bullet}$, L^+ by the dismutation of $L^\bullet$).

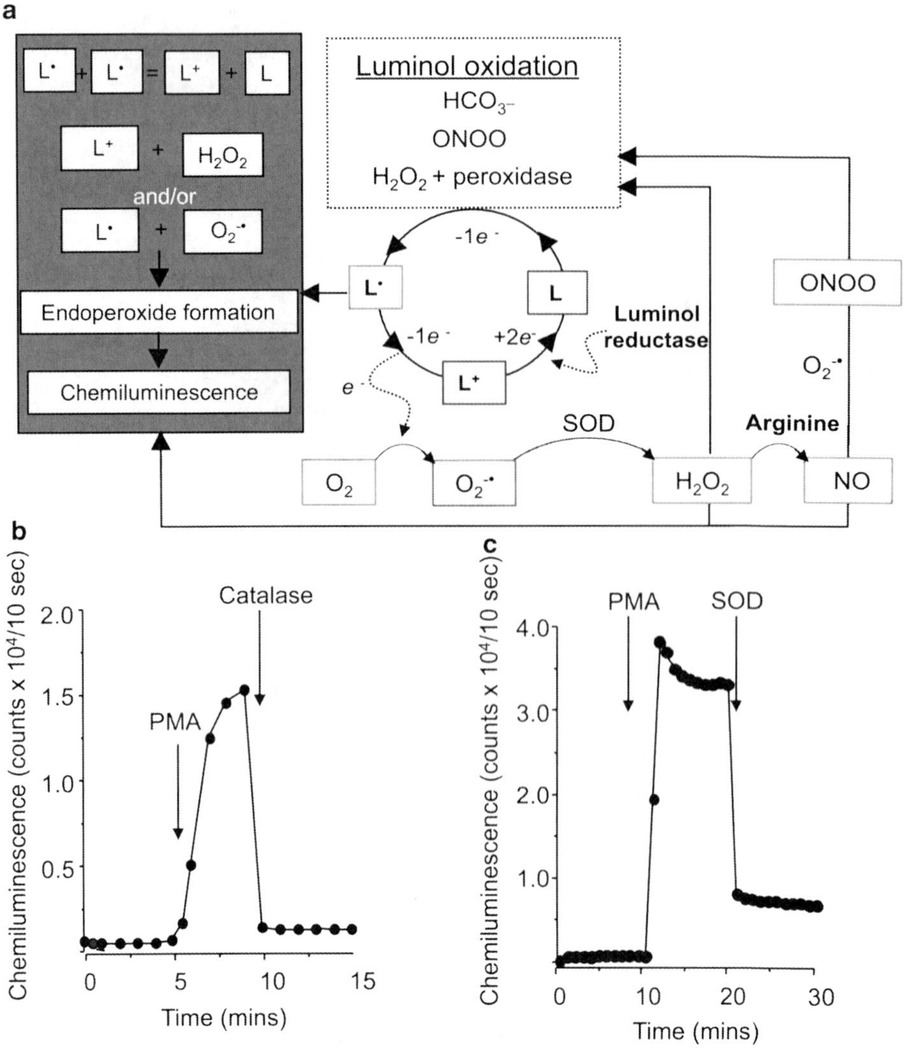

Fig. 14.4 Luminol-dependent chemiluminescence. (**a**) Schematic representation of the underlying chemistry; L=luminol; L•=a luminol radical created by the one-electron oxidation of L. L^+=azaquinone formed by the further one-electron oxidation of L• by oxygen, generating $O_2^{-\bullet}$ as a by-product. The reaction of L• with $O_2^{-\bullet}$ or L^+ with H_2O_2 generates an unstable endoperoxide, the decomposition of which leads to production of the chemiluminescence species, an electronically excited aminophthalate. Redox cycling of the probe could result if human spermatozoa possessed an appropriate reductase to convert L^+ back to the parent L. Any reactant that can achieve the univalent oxidation of luminol will generate chemiluminescence in this assay including H_2O_2/peroxidase and $ONOO^-$. (**b**) PMA-induced chemiluminescence quenched by catalase. (**c**) PMA-induced chemiluminescence quenched by SOD. Reproduced from Aitken et al. [29], with kind permission from American Society of Andrology

While the scheme presented in Fig. 14.4a is a major simplification of the chemistry involved in luminol-dependent chemiluminescence, from a diagnostic andrology point of view there are several points to emerge from this analysis that should be emphasized: (1) the redox cycling activity characteristic of this probe will lead to a significant amplification of the signal and may explain why alternative methods for measuring H_2O_2 have failed to detect this oxidant in purified suspensions of human spermatozoa, even though such activity has been clearly established in other species; (2) the complexity of this redox chemistry is such that we cannot state with certainty what is being measured with the luminol assay, it is certainly not just H_2O_2 [29]; (3) in particular, luminol assays may be measuring redox activity characterized by the cellular generation of oxidizing species capable of creating $L^{\bullet}$; (4) notwithstanding the reservations that might be expressed concerning the specificity of this probe, the luminol assay is robust and generates results that are highly correlated with sperm function (see below).

14.2.2.2 Lucigenin

This probe is thought to be sensitive to the cellular generation of $O_2^{-\bullet}$ largely because of the ability of SOD to suppress lucigenin-dependent cellular signals [32]. However, the reservations that apply to the use of luminol to detect specific ROS also apply to lucigenin. In the case of lucigenin, activation of the probe requires a one-electron reduction, rather than the one-electron oxidation associated with luminol-dependent chemiluminescence [33]. This one-electron reduction creates a radical ($LucH^{+\bullet}$) from lucigenin (Luc^{2+}) that rapidly gives up its electron to ground state oxygen to create $O_2^{-\bullet}$ and return the lucigenin to its parent state (Fig. 14.5). The $LucH^{+\bullet}$ generated from the one-electron reduction of lucigenin then combines with $O_2^{-\bullet}$ to produce the dioxetane (Fig. 14.5), which in turn decomposes with the generation of light (chemiluminescence). The $O_2^{-\bullet}$ involved in the last reaction could come from an independent cellular source, such as an NADPH oxidase, in which case the chemiluminescence recorded would reflect the generation of $O_2^{-\bullet}$, as originally proposed for both leukocytes [38] and spermatozoa [32, 39]. However, an unknown proportion of the $O_2^{-\bullet}$ involved in this reaction is an artifact created by the reaction between $LucH^{+\bullet}$ and ground state oxygen. Chemiluminescence created by the cellular generation of $O_2^{-\bullet}$ or the redox cycling of lucigenin cannot be readily distinguished, since both sources of ROS are suppressible by SOD. When an agonist such as PMA (12-myristate, 13-acetate phorbol ester) is used to stimulate chemiluminescence through the activation of protein kinase C, then it is probable that the cellular production of $O_2^{-\bullet}$ is being measured (Fig. 14.5b). However, when chemiluminescence is generated by the addition of NAD(P)H, then the system may also be detecting the presence of any oxidoreductase capable of effecting the one-electron reduction of lucigenin [40]. In the case of spermatozoa, we have clearly demonstrated that lucigenin chemiluminescence in the presence of NADH or NADPH does not represent $O_2^{-\bullet}$ production, but rather the respective abilities of cytochrome *b*5 reductase and cytochrome P450 reductase to reduce Luc^{2+} to $LucH^{+\bullet}$ and artificially trigger a redox cycle that generates $O_2^{-\bullet}$ as a by-product [41, 42].

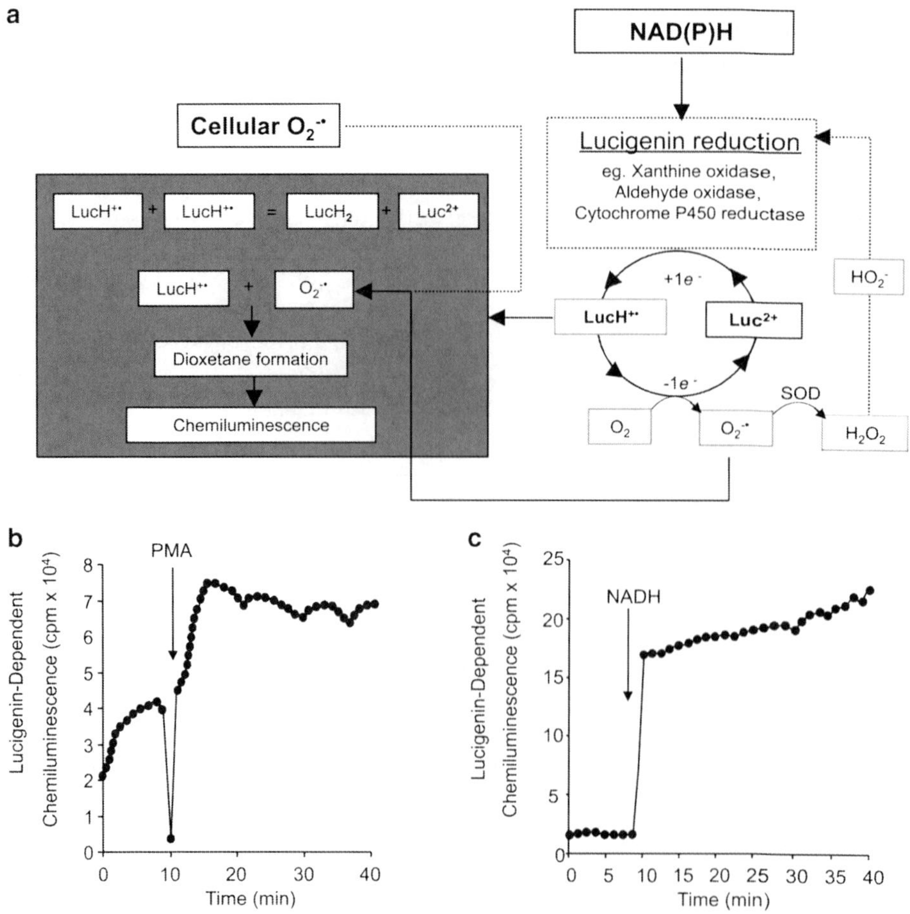

Fig. 14.5 Lucigenin-dependent chemiluminescence. (**a**) Schematic representation of the underlying chemistry; Luc^{2+} = lucigenin; $LH^{+\bullet}$ = a lucigenin radical created by the one-electron reduction of Luc^{2+}. The reaction of $LH^{+\bullet}$ with oxygen generates $O_2^{-\bullet}$. The latter then participates in an oxygenation reaction with $LH^{+\bullet}$ generating a dioxetane that decomposes with the generation of chemiluminescence. Any entity that can affect the one-electron reduction of lucigenin will, in the presence of oxygen, create a redox cycle that produces high levels of $O_2^{-\bullet}$ and chemiluminescence. It is impossible to distinguish the relative contribution of such probe-dependent and cell-dependent chemiluminescence. (**b**) PMA-induced signal is likely to involve the primary cellular production of $O_2^{-\bullet}$. (**c**) In contrast, the chemiluminescence generated by the addition of exogenous NADH (or NADPH) is likely to involve secondary $O_2^{-\bullet}$ production following the univalent reduction of the probe by NAD(P)H-dependent oxidoreductases. Reproduced from Aitken et al. [29], with kind permission from American Society of Andrology

Notwithstanding the deficiencies in lucigenin as a probe for evaluating $O_2^{-\bullet}$ production, it does have value as a nonspecific redox marker for the enhanced electron transfer activity associated with defective sperm function [43]. In many ways, the sensitivity and diagnostic value of the probe is enhanced by its redox cycling activity, rather than diminished. If specific quantification of $O_2^{-•}$ production is required, then there are alternative chemiluminescent probes that do not redox cycle including the Cypridina luciferin analog, MCLA [44], and coelenterazine [45].

14.2.3 Laboratory Application of Chemiluminescent Assays

From the above descriptions, it will be evident that while chemiluminescent probes such as luminol and lucigenin cannot yield specific data on ROS generation by human spermatozoa, they are nevertheless sensitive, semiquantitative indicators of redox activity in these cells and many publications have shown that the chemiluminescent activity detected in their presence is inversely related to sperm function and significantly elevated in the cases of male infertility [23, 46–54]. A unique prospective study, conducted in the days before assisted conception therapy was readily available, went so far as to demonstrate an excellent relationship between spontaneous pregnancy rates over a 4-year follow-up period and measurements made at the beginning of the study of luminol-dependent chemiluminescent signals generated by the spermatozoa in couples characterized by a normal female factor [55]. The predictive power of the luminol-dependent assay used in this study was all the more impressive because within this data set the conventional criteria of semen quality (count, morphology, and motility) were of no diagnostic significance whatsoever.

As discussed above, the major caveat with the clinical interpretation of such studies is that if the level of leukocyte contamination has not been meticulously assessed, then the source and significance of the enhanced chemiluminescent activity are open to question. Much will depend on when, where, and how the leukocytes were activated [56]. Another clinically significant issue with chemiluminescence as a diagnostic tool is that the readout of individual luminometers depends heavily upon the response characteristics of the photomultipliers used to monitor the chemiluminescent response. Since every individual luminometer is different in this respect, any attempt to describe a threshold level of chemiluminescence as a universal reference point for normal sperm function, though desirable, is simply not achievable [57].

14.2.3.1 Chemical Interference with Chemiluminescence Assays

While the sensitivity of chemiluminescent probes is extremely valuable, it renders such systems very susceptible to interference in a manner that may distort their diagnostic information content. Some of these confounding factors are described below:

1. *Time to analysis*. Chemiluminescent activity tends to decline with time following the isolation of spermatozoa from seminal plasma [46]. As a result, such assays are best conducted within 1 h of sperm isolation [58].

2. *Bovine serum albumin*. Supplementation of culture media with bovine serum albumin (BSA) has the potential to generate spurious chemiluminescent signals in the presence of human seminal plasma. These signals are generated because most commercial BSA preparations are heavily contaminated with polyamine oxidase(s), which will generate luminol-dependent chemiluminescence on contact with the polyamines (spermine and spermidine) in human seminal plasma and/or coated onto the surface of human spermatozoa [29, 59].
3. *Medium pH*. Chemiluminescence systems are sensitive to changes in pH; for example, a change of 1 pH unit from pH 7.4 to 8.4 has a dramatic effect on the quantum yield of luminol [29].
4. In the luminol-peroxidase system, the presence of electron donors such as NADPH or cysteine will generate massive chemiluminescent signals without any need for spermatozoa; it is entirely a chemical artifact. Alternatively, many small molecular mass-free radical scavengers, such as ascorbate or uric acid, will quench cellular chemiluminescence as will the common pH indicator, phenol red [29]. In light of this susceptibility to nonspecific interference, it is always advisable to run cell-free controls as an integral part of chemiluminescence assays.

14.2.4 Chemiluminescence and the Antioxidant Properties of Seminal Plasma

Another variant on the chemiluminescence theme is to use this technology to monitor both ROS generation by the ejaculate and the antioxidant potential of the seminal plasma, with a view to generating a readout reflecting the balance between these two opposing forces. There can be no doubt that seminal plasma is a very powerful antioxidant medium that serves to protect spermatozoa during their perilous journey from the male to the female reproductive tract. This fluid is rich in antioxidant enzymes such as SOD, glutathione peroxidase, and some catalase as well as small molecular mass scavengers, including ascorbate, pyruvate, urate, alpha tocopherol, glutathione, albumin, beta carotene, hypotaurine, and ubiquinol [10, 60, 61]. The ability of this fluid to protect spermatozoa from the emission of free radicals by contaminating leukocytes has been clearly demonstrated, as have the adverse consequences of removing this antioxidant protection when leukocytes are present in the sperm suspension [18, 28]. Significantly, seminal plasma has been shown to protect spermatozoa from DNA damage inflicted as a consequence of leukocytic infiltration [62].

Although the antioxidant properties of semen can be assessed by measuring the individual constituents of the system such as SOD [63] alpha-tocopherol or ascorbate [64], a more convenient approach is to assess the sum total of all antioxidant activities in the semen using a TRAP assay. The latter measures the ability of seminal plasma to extinguish a free radical signal typically generated by a compound such as 2,2-azobis-(2-amidinopropane) (ABAP), which generates alkylperoxyl radicals at 37°C. In the presence of luminol, a steady state chemiluminescent signal is generated with ABAP and the ability of seminal plasma to inhibit this output gives an indication of total antioxidant capacity. Using a related compound [2,2' Azino-*bis*(3-ethylbenzothiazoline-6-sulfonic acid)] as the radical source in a postaddition assay, Rhemrev et al. [65] demonstrated that the total TRAP activity of seminal plasma is about ten times higher than that of blood plasma. These authors also differentiated (1) a fast TRAP activity detectable in 10 s, 37% of which could be attributed to vitamin C, uric acid, and tyrosine, while proteins and polyphenolic compounds contributed a further 57% and (2) a slow TRAP activity detectable in 300 s attributable to vitamin C (1%), uric acid (2%), tyrosine (15%), and to proteins and polyphenolic compounds (33%). It was not possible to account for the remaining 49%.

An alternative ROS source (H_2O_2 in the presence of horse radish peroxidase linked to immunoglobulin) has also been used in combination with luminol as the basis for an alternative TRAP assay [66]. This approach would focus such antioxidant assessments on H_2O_2-scavenging activity. Whether this is an advantage is not known at the present time because the relationship between the results of this assay and conventional ABAP-based TRAP assays has never been assessed. Indeed, there are several variants on the TRAP assay that use a variety of free radical/oxidant sources and detection systems. Systematic evaluation of these assays for their relative ability to detect differences in the antioxidant activity of human seminal plasma would be a worthwhile exercise.

Many authors have found an inverse relationship between ROS generation and seminal antioxidant activity in semen [62, 66]. This inverse relationship probably reflects the fact that quantitatively, a majority of the free radicals detected in human semen samples are generated by contaminating leukocytes. Since the leukocytes release ROS into the extracellular space, they will be rapidly intercepted by the free radical scavengers in seminal plasma leading to the observed negative correlation between ROS generation and TRAP values. Viewed in this light, the loss of antioxidant activity might be considered a *consequence* of excess ROS generation by leukocytes entering the seminal fluid at the moment of ejaculation. Alternatively, antioxidant activity in seminal plasma may be a reflection of the redox status of the individual and be a *cause* of oxidative stress in the germ line. Thus, the oxidative DNA damage seen in the spermatozoa of heavy smokers is thought to reflect a systemic loss of antioxidant protection precipitated by the free radicals present in cigarette smoke [67]. Under such circumstances, it is not clear how a loss of antioxidant protection could induce an increase in ROS generation by spermatozoa, unless oxidative stress itself triggers ROS production by these cells [68].

Simultaneous chemiluminescence assessments of ROS generation and antioxidant protection are potentially valuable and have the possibility of being expressed as a ratio of activities, thereby circumventing the difficulties inherent in calibrating chemiluminescence assays. However, interpretation of the results requires a knowledge of the leukocytic contribution to the ROS signal, which could be readily acquired using leukocyte-specific agonists (opsonized zymosan) as part of the chemiluminescence protocol possibly in combination with sperm isolation protocols that effectively remove contaminating leukocytes such as swim-up or electrophoresis.

14.3 Use of Flow Cytometry to Analyze ROS Generation by Spermatozoa

A possible solution to the weaknesses inherent in the chemiluminescence approach to measuring ROS in human sperm suspensions can be found in a variety of redox-sensitive fluorescence probes which can be loaded into spermatozoa and monitored by flow cytometry. With this technology platform, the readout for such probes can be focused specifically on the sperm population via the appropriate setting of gates. Of course, such assays are again only semiquantitative because they just indicate the percentage of cells exhibiting a high level of activity, without indicating the concentration or cellular content of the metabolites being evaluated. Nevertheless, they are sensitive, easy to use, and clinically valuable.

14.3.1 Dihydroethidium

The first probe to be assessed in detail in this context was dihydroethidium or hydroethidine (DHE) [69]. This probe is nonfluorescent in its own right, but on oxidation will produce DNA-sensitive fluorochromes that will stain the mitochondrial and nuclear DNA (Fig. 14.6). The application of this probe is not straightforward, however, and its interpretation has to be considered with care. One of the major issues with DHE is that it is generated by the chemical reduction of ethidium bromide (Et^+), a general DNA-sensitive fluorochrome. As a result, every commercial preparation of DHE will be contaminated with Et^+, which will, in turn, stain every nonviable cell in the sperm suspension with a bright red fluorescence. Consequently, this probe cannot be used in isolation, but has to be coupled with a cell vitality marker. If a cell vitality marker is not used, then it will be impossible to distinguish free radical generating cells from those that are simply dead. SYTOX green is an appropriate vitality stain in this context because its emission spectrum (500–530 nm) is completely different from Et^+ (>560 nm). This combination of probes therefore allows flow cytometry to be used to rapidly screen human spermatozoa in order to identify those cells that are alive and generating ROS ([69]; Fig. 14.6).

A second problem with DHE as a probe is that it can be nonspecifically oxidized to generate Et^+ by a variety of metabolites and xenobiotic chemical reagents including hydrogen peroxide, hypochlorous acid, peroxynitrite as well as a certain quinone species including *o*-chloranil or *o*-naphthoquinone ([69, 70]; Fig. 14.6). In order to be certain that the probe is capable of measuring superoxide anion ($O_2^{-\bullet}$), it is necessary to demonstrate that the cellular oxidation of DHE generates

a

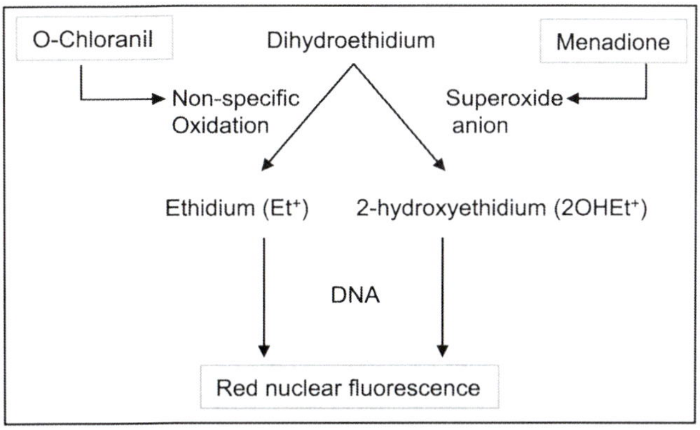

b

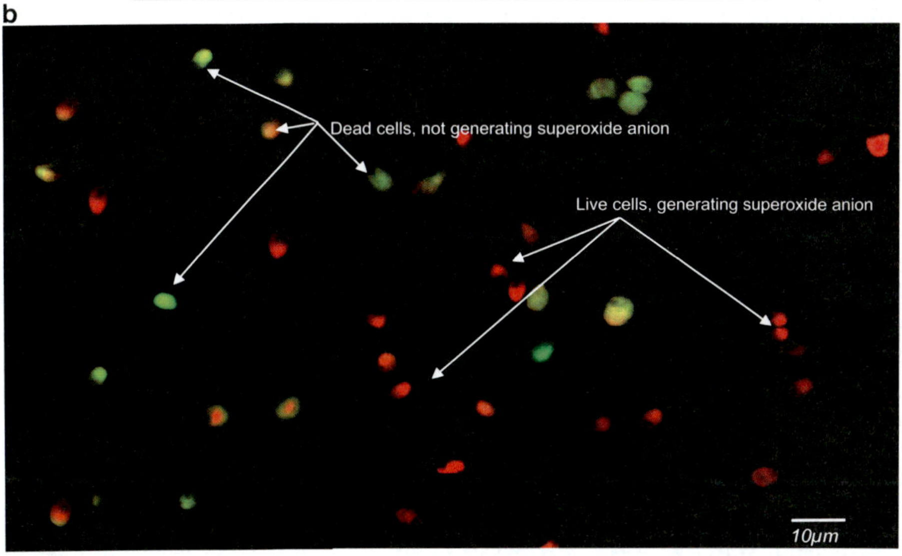

Fig. 14.6 DHE as a probe for ROS generation by human spermatozoa. (**a**) Diagrammatic representation of the chemistry underlying the use of DHE as a probe for ROS generation. DHE may either undergo a nonspecific 2-electron oxidation to generate Et^+ or react with $O_2^{-\bullet}$ to generate the specific reaction product, $2OHET^+$. Both of these fluorochromes will interact with DNA in the sperm nucleus to generate a red fluorescence. In order to be certain that spermatozoa were producing $O_2^{-\bullet}$, it was essential to isolate the putative $2OHET^+$ by HPLC and verify the chemical identity of the product using a variety of analytical tools including NMR spectroscopy, mass spectrometry, and spectrofluorimetry. In the course of these studies, *o*-chloranil was used to drive the nonspecific oxidation of DHE to Et^+, while the closely related compound, menadione, was used to stimulate the $O_2^{-\bullet}$-induced formation of $2OHET^+$. (**b**) Confocal image of human spermatozoa stained with DHE and SYTOX® Green; overlay image of red and green fluorescence. Nonviable cells are *green* or *orange* while viable cells that are generating ROS are red. Scale bar = 10 μm. Reproduced from De Iuliis et al. [69], by kind permission from The Endocrine Society

2-hydroxyethidium ($2OHET^+$), a unique reaction product that can only be created when DHE is oxidized by $O_2^{-\bullet}$ [71]. Using HPLC, two reaction products were found to be formed when free radical generating spermatozoa were treated with DHE; mass spectrometry revealed these peaks to be Eth^+ and $2OHET^+$ [69]. Appropriately, the $2OHET^+$ signal was found to be greatly augmented when $O_2^{-\bullet}$ production by human spermatozoa, was artificially stimulated with menadione (vitamin K) in concert with a dramatic decline in sperm motility and a corresponding increase in the incidence of DNA damage [69]. Furthermore, using the DHE:Sytox Green assay to monitor $O_2^{-\bullet}$ generation in viable cells, a clear negative correlation was observed between ROS generation and sperm motility. The fact that the mitochondrial inhibitor CCCP (carbonyl cyanide *m*-chlorophenylhydrazone, a proton ionophore which collapses the mitochondrial membrane potential] had no impact on the DHE signal generated by human spermatozoa, prompted us to propose that the $O_2^{-\bullet}$ detected in these cells was not of mitochondrial origin. This, it turned out, was not true. Sperm mitochondria are very active generators of $O_2^{-\bullet}$; however, they are unusual in that the production of ROS, at least in the short term, does not depend on the mitochondria membrane potential [72].

The most commonly employed probe for measuring mitochondrial ROS generation is MitoSOX Red (MSR). This probe is based on DHE but carries an additional charge, which results in its selective accumulation within the mitochondria. Following reaction with $O_2^{-\bullet}$, MSR produces DNA-sensitive fluorochromes that generate a red fluorescence when excited at 510 nm that can be detected by flow cytometry. As with DHE, it is important to ensure that only live cells are imaged when MSR fluorescence is being monitored. Failure to incorporate this step into the protocol may produce spurious results as a consequence of dead cells staining with the residual Eth^+ contaminating commercial MSR preparations. In order to simultaneously monitor mitochondrial ROS and cell vitality, we again advocate the use of a Sytox green/MSR combination in a flow cytometry-based protocol [72].

Using MSR, we have found that the mitochondria of normal human spermatozoa can be induced to generate $O_2^{-\bullet}$ with inhibitors of complex III in the mitochondrial electron transport chain (ETC). Thus, while normal human spermatozoa exhibit extremely low rates of spontaneous ROS generation, addition of antimycin A and myxothiazol, both of which act at complex III of the ETC, resulted in a dramatic increase in redox activity (Fig. 14.7). Myxothiazol binds close to the bL heme of this complex, allowing ubiquinol (QH_2) to access the Rieske iron sulfur center in order to undergo a one-electron oxidation to create the semiquinone radical, $Q^{-\bullet}$, which is unstable and rapidly reverts to the parent quinone (Q) with the release of an electron which is avidly taken up by oxygen to create $O_2^{-\bullet}$. Antimycin A treatment also leads to the generation of $Q^{-\bullet}$ by inhibiting the reoxidation of heme bL through its capacity to disrupt electron transfer from heme bH to $Q^{-\bullet}$ [72]. As a consequence of these interactions, both myxothiazol and antimycin A stimulate the generation of $Q^{-\bullet}$, which then reduces O_2 to $O_2^{-\bullet}$ in the intramembranous space. This $O_2^{-\bullet}$ then dismutates to H_2O_2 under the influence of SOD and escapes to the outside of the cell, where it can be detected by the luminol-peroxidase monitoring system. Rotenone (an inhibitor of electron transfer from FeSN-2 cluster to ubiquinone) was found to have a minor stimulatory effect on H_2O_2 release at 10 μM, possibly because the $O_2^{-\bullet}$ generated with this

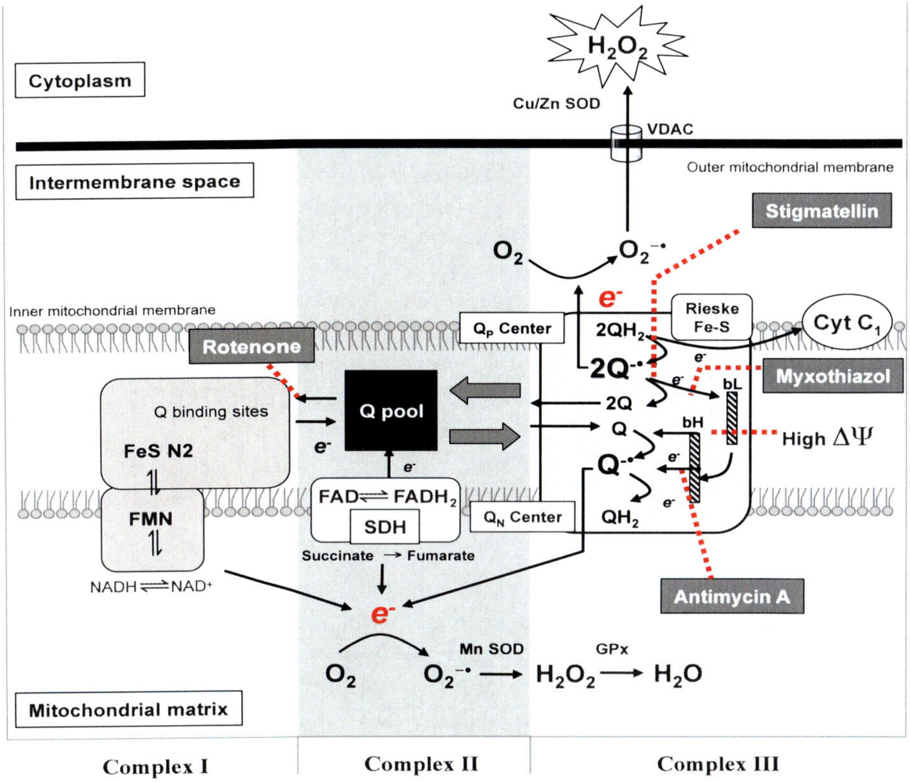

Fig. 14.7 Schematic diagram showing the main pathways of electron flux through the mitochondrial electron transport chain. Sites of action for the mitochondrial inhibitors used in this study are also indicated. *VDAC* voltage-dependent anion channel; *SDH* succinate dehydrogenase; *Q* ubiquinone; *QH2* ubiquinol; *Q*$^{-\bullet}$ ubisemiquinone; *SOD* superoxide dismutase; *GPx* glutathione peroxidase; *FMN* flavin mononucleotide; *NADH* nicotinamide adenine dinucleotide; *FAD* flavin adenine dinucleotide. With antimycin and myxothiazol, electron leakage leads to the generation of $O_2^{-\bullet}$ in the intermembranous space, while rotenone leads to $O_2^{-\bullet}$ formation in the mitochondrial matrix. It is primarily the generation of ROS in the matrix that appears to damage spermatozoa. Reproduced from Koppers et al. [72], with kind permission from The Endocrine Society

compound is formed in the mitochondrial matrix and may be neutralized by the antioxidant enzymes (SOD and glutathione peroxidase) that abound at this site (Fig. 14.7). Stigmatellin (10 μM), which inhibits the transfer of electrons from ubiquinol (QH_2) to the Rieske iron sulfur cluster and prevents semiquinone ($Q^{-\bullet}$) formation (Fig. 14.7), failed to induce mitochondrial ROS generation. The generation of ROS with these mitochondrial electron transport inhibitors is not suppressed by collapsing the mitochondrial membrane potential with CCCP; indeed, the addition of these reagents actually enhances ROS generation [72].

Biologically, the spontaneous generation of mitochondrial ROS is associated with a significant decline in sperm motility as a result of lipid peroxidation occurring in the midpiece of the cell [72]. On the basis of the damaging effects observed with rotenone, it has been suggested that mitochondrial ROS are particularly damaging to sperm function if they are generated in the mitochondrial matrix, rather than in the intermembranous space [72]. In the latter position, the $O_2^{-\bullet}$ appears to escape rapidly to the cytosol where it dismutates to H_2O_2 and leaves the cell. However, when the ROS are generated in the mitochondrial matrix, they do not escape so readily and precipitate the peroxidative damage that leads to motility loss [72]. Since the MSR probe is targeted to the mitochondrial matrix, it is in the perfect position to detect ROS formation in this strategically important location.

The triggers for mitochondrial ROS generation are varied, including electromagnetic radiation, the local availability of lipid peroxides, and polyunsaturated fatty acids [73, 74]. The fact that the spontaneous generation of mitochondrial ROS is highly correlated with the polyunsaturated fatty acid content of human spermatozoa strongly suggests that the etiology of oxidative stress in these cells involves a lipid imbalance generated by errors in metabolism or, possibly, diet [75].

14.4 Other ROS Detection Systems

The foregoing discussion highlights the major strength and weaknesses of protocols for assessing the generation of ROS by human spermatozoa based around the chemiluminescent analysis of redox activity and detection of $O_2^{-\bullet}$ using flow cytometry. There are other fluorescent probes in use for measuring ROS generation by cells particularly dichlorofluorescin diacetate (DCFH-DA) [68]. This probe becomes fluorescent on oxidation and is purported to measure cellular H_2O_2 production. However, in reality this oxidant has no effect on DCFH-DA fluorescence unless it is accompanied by peroxidase activity which in a highly compartmentalized, cytoplasm-deficient cell such as the spermatozoon may not always be the case. It is also possible that the DCFH/H_2O_2/peroxidase detection system may actually generate H_2O_2 in a scheme whereby H_2O_2 reacts with peroxidase to form compound 1, which then oxidizes DCFH to the $DCF^{\bullet-}$ semiquinone-free radical. The latter then decays to fluorescent DCF with the release of an electron to oxygen, generating $O_2^{-\bullet}$ which will rapidly dismutate to H_2O_2 and continue the cycle. Other ROS such as the peroxynitrite, hypochlorous acid, and the hydroxyl radical can also oxidize this probe and might make significant contributions to the positive signals observed in defective human spermatozoa [68, 76]. Finally, the dye, nitroblue tetrazolium (NBT), has been used to detect ROS generation by human spermatozoa. NBT is held to be reduced by the donation of electrons from $O_2^{-\bullet}$ to generate a detectable chemical signal; in this case, blue-black deposits of formazan [77]. The problem with this probe is that it is not specific for spermatozoa, and unless the sperm suspensions are carefully purified, will generate signals dominated by the presence of contaminating leukocytes [78]. Furthermore, this probe is not specific for $O_2^{-\bullet}$ since it can be activated by cytochrome P450-reductase oxidoreductases which can effect a one-electron reduction of the probe [42] to produce unstable tetrazoinyl radical intermediates, which then give up their electrons to reduce O_2 to $O_2^{-\bullet}$ in a reversible process. SOD, by removing $O_2^{-\bullet}$ displaces this oxidation to the right and thus prevents production of the formazan. For this reason, many aerobic tetrazolium reductions are inhibitable by SOD even though $O_2^{-\bullet}$ was not actually being generated in the system in the absence of NBT [79].

The above discussion covers most of the methods that are currently being used to assess ROS production by human spermatozoa. Clearly, this is still an unresolved issue. We are still not certain which particular ROS are damaging to spermatozoa, how they originate, and which of the above methods is the most appropriate for their routine diagnostic assessment. Clearly, the field needs to undertake detailed comparative studies where the readout of selected assays is correlated with defective sperm function, in the form of lost motility or oxidative DNA damage, in order to determine which is the most sensitive procedure to incorporate into routine diagnostic analyses of semen quality.

14.5 Key Points

- We still do not have the perfect method to measure ROS generation in human sperm suspensions.
- Methods that target the entire sperm suspension such as chemiluminescence or NBT reduction are sensitive but are only semiquantitative, difficult to calibrate, easily distorted by the activity of cellular oxidoreductases, and too easily dominated by the presence of leukocytes.
- If measures are taken to either quantify the degree of leukocyte contamination or to purify the spermatozoa to the point that leukocytes are no longer present, then these forms of measurement are valuable and of diagnostic significance.
- Flow cytometry may be used to measure the fluorescence of probes that are responsive to the presence of specific ROS. Both DHE and MSR have been validated for the generation of $O_2^{-\bullet}$ in different compartments of the spermatozoon, but the ROS responsible for DCFH fluorescence still await resolution.
- The diagnostic application of the DHE and MSR assays requires the simultaneous determination of cell viability.
- The advantage of flow cytometry is that the analysis can be focused directly on the spermatozoa and that acceptably large numbers of cells can be analyzed with ease.
- The major problem with these probes is that they require relatively sophisticated, expensive hardware in the form of flow cytometers and the results do not quantify the target ROS but simply indicate that the percentage of cells are particularly active in their creation. Nevertheless, at present they represent the methods of choice for assessing ROS production by human spermatozoa.

References

1. MacLeod J (1943) The role of oxygen in the metabolism and motility of human spermatozoa. Am J Physiol 138:512–518
2. Lord R (1951) Sea urchin spermatozoa. Biol Rev 26:1–27
3. Tosic J, Walton A (1946) Formation of hydrogen peroxide by spermatozoa and its inhibitory effect on respiration. Nature 158:485
4. Tosic J, Walton A (1950) Metabolism of spermatozoa. Formation of hydrogen peroxide by spermatozoa and its effects on motility and survival. Biochem J 47:199–212
5. Wales RG, White IG, Lamond DR (1959) The spermicidal activity of hydrogen peroxide in vitro and in vivo. J Endocrinol 18:236–244
6. Oehninger S, Blackmore P, Mahony M, Hodgen G (1995) Effects of hydrogen peroxide on human spermatozoa. J Assist Reprod Genet 12:41–47
7. Aitken RJ, Buckingham D, Harkiss D (1993) Use of a xanthine oxidase oxidant generating system to investigate the cytotoxic effects of reactive oxygen species on human spermatozoa. J Reprod Fertil 97:441–450
8. Hsu PC, Hsu CC, Guo YL (1999) Hydrogen peroxide induces premature acrosome reaction in rat sperm and reduces their penetration of the zona pellucida. Toxicology 139:93–101
9. Lozano GM, Bejarano I, Espino J et al (2009) Relationship between caspase activity and apoptotic markers in human sperm in response to hydrogen peroxide and progesterone. J Reprod Dev 55:615–621
10. Jones R, Mann T, Sherins R (1979) Peroxidative breakdown of phospholipids in human spermatozoa, spermicidal properties of fatty acid peroxides, and protective action of seminal plasma. Fertil Steril 31:531–537
11. Ollero M, Gil-Guzman E, Lopez MC et al (2001) Characterization of subsets of human spermatozoa at different stages of maturation: implications in the diagnosis and treatment of male infertility. Hum Reprod 16:1912–1921
12. Vignini A, Nanetti L, Buldreghini E et al (2006) The production of peroxynitrite by human spermatozoa may affect sperm motility through the formation of protein nitrotyrosine. Fertil Steril 85:947–953
13. Chigurupati S, Son TG, Hyun DH et al (2008) Lifelong running reduces oxidative stress and degenerative changes in the testes of mice. J Endocrinol 199:333–341
14. Mallidis C, Agbaje IM, Rogers DA et al (2009) Advanced glycation end products accumulate in the reproductive tract of men with diabetes. Int J Androl 32:295–305
15. Aitken RJ, De Iuliis GN (2010) On the possible origins of DNA damage in human spermatozoa. Mol Hum Reprod 16:3–13
16. Aitken RJ, Curry BJ (2011) Redox regulation of human sperm function: from the physiological control of sperm capacitation to the etiology of infertility and DNA damage in the germ line. Antioxid Redox Signal 14:367–381
17. Aitken RJ, Koopman P, Lewis SE (2004) Seeds of concern. Nature 432:48–52
18. Aitken RJ, Buckingham D, Brindle J, Gomez E, Baker G, Irvine DS (1995) Analysis of sperm movement in relation to the oxidative stress created by leucocytes in washed sperm preparations and seminal plasma. Hum Reprod 10:2061–2071
19. Aitken RJ, West K, Buckingham D (1994) Leukocytic infiltration into the human ejaculate and its association with semen quality, oxidative stress, and sperm function. J Androl 15:343–352
20. World Health Organization (1999) WHO laboratory manual for the examination of human semen and sperm-cervical mucus interaction, 4th edn. Cambridge University Press, Cambridge, UK
21. Aitken RJ, Clarkson JS (1988) Significance of reactive oxygen species and antioxidants in defining the efficacy of sperm preparation techniques. J Androl 9:367–376
22. Aitken RJ, West KM (1990) Analysis of the relationship between reactive oxygen species production and leucocyte infiltration in fractions of human semen separated on Percoll gradients. Int J Androl 13:433–451

23. Aitken RJ, Buckingham D, West K, Wu FC, Zikopoulo K, Richardson DW (1992) Differential contribution of leucocytes and spermatozoa to the high levels of reactive oxygen species recorded in the ejaculates of oligozoospermic patients. J Reprod Fertil 94:451–462
24. Tremellen K, Tunc O (2010) Macrophage activity in semen is significantly correlated with sperm quality in infertile men. Int J Androl 33(6):823–831
25. Aitken RJ, Findlay JK, Hutt KJ, Kerr JB (2011) Apoptosis in the germ line. Reproduction 141:139–150
26. Ainsworth C, Nixon B, Aitken RJ (2005) Development of a novel electrophoretic system for the isolation of human spermatozoa. Hum Reprod 20:2261–2270
27. Aitken RJ, Buckingham W, West K, Brindle J (1996) On the use of paramagnetic beads and ferrofluids to assess and eliminate the leukocytic contribution to oxygen radical generation by human sperm suspensions. Am J Reprod Immunol 35:541–551
28. Krausz C, Mills C, Rogers S, Tan SL, Aitken RJ (1994) Stimulation of oxidant generation by human sperm suspensions using phorbol esters and formyl peptides: relationships with motility and fertilization in vitro. Fertil Steril 62:599–605
29. Aitken RJ, Baker MA, O'Bryan M (2004) Shedding light on chemiluminescence: the application of chemiluminescence in diagnostic andrology. J Androl 25:455–465
30. Weese DL, Peaster ML, Hernandez RD, Leach GE, Lad PM, Zimmern PE (1993) Chemoattractant agents and nerve growth factor stimulate human spermatozoal reactive oxygen species generation. Fertil Steril 59:869–875
31. Cormier MJ, Pritchard PM (1968) An investigation of the mechanism of the luminescent peroxidation of luminol by stopped flow techniques. J Biol Chem 243:4706–4714
32. Aitken RJ, Buckingham DW, West KM (1992) Reactive oxygen species and human spermatozoa: analysis of the cellular mechanisms involved in luminol- and lucigenin-dependent chemiluminescence. J Cell Physiol 151:466–477
33. Faulkner K, Fridovich I (1993) Luminol and lucigenin as detectors for $O_2^{-\bullet}$. Free Radic Biol Med 15:447–451
34. Gil-Guzman E, Ollero M, Lopez MC, Sharma RK, Alvarez JG, Thomas AJ Jr, Agarwal A (2001) Differential production of reactive oxygen species by subsets of human spermatozoa at different stages of maturation. Hum Reprod 16:1922–1930
35. Gomez E, Irvine DS, Aitken RJ (1998) Evaluation of a spectrophotometric assay for the measurement of malondialdehyde and 4-hydroxyalkenals in human spermatozoa: relationships with semen quality and sperm function. Int J Androl 21:81–94
36. Nakamura M, Nakamura S (1998) One- and two-electron oxidations of luminol by peroxidase systems. Free Radic Biol Med 24:537–544
37. Aitken RJ, Ryan AL, Baker MA, McLaughlin EA (2004) Redox activity associated with the maturation and capacitation of mammalian spermatozoa. Free Radic Biol Med 36:994–1010
38. Gyllenhammar H (1987) Lucigenin chemiluminescence in the assessment of neutrophil superoxide production. J Immunol Methods 97:209–213
39. McKinney KA, Lewis SE, Thompson W (1996) Reactive oxygen species generation in human sperm: luminol and lucigenin chemiluminescence probes. Arch Androl 36:119–125
40. Aitken RJ, Fisher H, Fulton N, Knox W, Lewis B (1997) Reactive oxygen species generation by human spermatozoa is induced by exogenous NADPH and inhibited by the flavoprotein inhibitors diphenylene iodonium and quinacrine. Mol Reprod Dev 47:468–482
41. Baker MA, Krutskikh A, Curry BJ, Hetherington L, Aitken RJ (2005) Identification of cytochrome-b5 reductase as the enzyme responsible for NADH-dependent lucigenin chemiluminescence in human spermatozoa. Biol Reprod 73:334–342
42. Baker MA, Krutskikh A, Curry BJ, McLaughlin EA, Aitken RJ (2004) Identification of cytochrome P450-reductase as the enzyme responsible for NADPH-dependent lucigenin and tetrazolium salt reduction in rat epididymal sperm preparations. Biol Reprod 71:307–318
43. Aitken RJ, Ryan AL, Baker MA, McLaughlin EA (2004) Redox activity associated with the maturation and capacitation of mammalian spermatozoa. Free Radic Biol Med 36:994–1010
44. de Lamirande E, Gagnon C (1995) Capacitation-associated production of superoxide anion by human spermatozoa. Free Radic Biol Med 18:487–495

45. Tarpey MM, White CR, Suarez E, Richardson G, Radi R, Freeman BA (1999) Chemiluminescent detection of oxidants in vascular tissue. Lucigenin but not coelenterazine enhances superoxide formation. Circ Res 84:1203–1211
46. Aitken RJ, Clarkson JS (1987) Cellular basis of defective sperm function and its association with the genesis of reactive oxygen species by human spermatozoa. J Reprod Fertil 81:459–469
47. Aitken RJ, Clarkson JS, Hargreave TB, Irvine DS, Wu FC (1989) Analysis of the relationship between defective sperm function and the generation of reactive oxygen species in cases of oligozoospermia. J Androl 10:214–220
48. D'Agata R, Vicari E, Moncada ML et al (1990) Generation of reactive oxygen species in subgroups of infertile men. Int J Androl 13:344–351
49. Ochsendorf FR, Thiele J, Fuchs J et al (1994) Chemiluminescence in semen of infertile men. Andrologia 26:289–293
50. Zalata A, Hafez T, Comhaire F (1995) Evaluation of the role of reactive oxygen species in male infertility. Hum Reprod 10:1444–1451
51. Alkan I, Sim ek F, Haklar G et al (1997) Reactive oxygen species production by the spermatozoa of patients with idiopathic infertility: relationship to seminal plasma antioxidants. J Urol 157:140–143
52. Henkel R, Ichikawa T, Sánchez R, Miska W, Ohmori H, Schill WB (1997) Differentiation of ejaculates showing reactive oxygen species production by spermatozoa or leukocytes. Andrologia 29:295–301
53. Hendin BN, Kolettis PN, Sharma RK, Thomas AJ Jr, Agarwal A (1999) Varicocele is associated with elevated spermatozoal reactive oxygen species production and diminished seminal plasma antioxidant capacity. J Urol 161:1831–1834
54. Said TM, Agarwal A, Sharma RK, Mascha E, Sikka SC, Thomas AJ Jr (2004) Human sperm superoxide anion generation and correlation with semen quality in patients with male infertility. Fertil Steril 82:871–877
55. Aitken RJ, Irvine DS, Wu FC (1991) Prospective analysis of sperm-oocyte fusion and reactive oxygen species generation as criteria for the diagnosis of infertility. Am J Obstet Gynecol 164:542–551
56. Yumura Y, Iwasaki A, Saito K, Ogawa T, Hirokawa M (2009) Effect of reactive oxygen species in semen on the pregnancy of infertile couples. Int J Urol 16:202–207
57. Desai N, Sharma R, Makker K, Sabanegh E, Agarwal A (2009) Physiologic and pathologic levels of reactive oxygen species in neat semen of infertile men. Fertil Steril 92:1626–1631
58. Kobayashi H, Gil-Guzman E, Mahran AM, Rakesh, Nelson DR, Thomas AJ Jr, Agarwal A (2001) Quality control of reactive oxygen species measurement by luminol-dependent chemiluminescence assay. J Androl 22:568–574
59. Quinn P, Whittingham DG, Stanger JD (1982) Interaction of semen with ova in vitro. Arch Androl 8:189–198
60. Holmes RP, Goodman HO, Shihabi ZK, Jarow JP (1992) The taurine and hypotaurine content of human semen. J Androl 13:289–292
61. Smith R, Vantman D, Ponce J, Escobar J, Lissi E (1996) Total antioxidant capacity of human seminal plasma. Hum Reprod 11:1655–1660
62. Twigg J, Irvine DS, Houston P, Fulton N, Michael L, Aitken RJ (1998) Iatrogenic DNA damage induced in human spermatozoa during sperm preparation: protective significance of seminal plasma. Mol Hum Reprod 4:439–445
63. Aitken RJ, Buckingham DW, Carreras A, Irvine DS (1996) Superoxide dismutase in human sperm suspensions: relationship with cellular composition, oxidative stress, and sperm function. Free Radic Biol Med 21:495–504
64. Kao SH, Chao HT, Chen HW, Hwang TI, Liao TL, Wei YH (2008) Increase of oxidative stress in human sperm with lower motility. Fertil Steril 89:1183–1190
65. Rhemrev JP, van Overveld FW, Haenen GR, Teerlink T, Bast A, Vermeiden JP (2000) Quantification of the nonenzymatic fast and slow TRAP in a postaddition assay in human

seminal plasma and the antioxidant contributions of various seminal compounds. J Androl 21:913–920
66. Sharma RK, Pasqualotto FF, Nelson DR, Thomas AJ Jr, Agarwal A (1999) The reactive oxygen species-total antioxidant capacity score is a new measure of oxidative stress to predict male infertility. Hum Reprod 14:2801–2807
67. Fraga CG, Motchnik PA, Wyrobek AJ, Rempel DM, Ames BN (1996) Smoking and low antioxidant levels increase oxidative damage to sperm DNA. Mutat Res 351:199–200
68. Mahfouz RZ, du Plessis SS, Aziz N, Sharma R, Sabanegh E, Agarwal A (2010) Sperm viability, apoptosis, and intracellular reactive oxygen species levels in human spermatozoa before and after induction of oxidative stress. Fertil Steril 93:814–821
69. De Iuliis GN, Wingate JK, Koppers AJ, McLaughlin EA, Aitken RJ (2006) Definitive evidence for the nonmitochondrial production of superoxide anion by human spermatozoa. J Clin Endocrinol Metab 91:1968–1975
70. Hughes LM, Griffith R, Carey A et al (2009) The spermostatic and microbicidal actions of quinones and maleimides: toward a dual-purpose contraceptive agent. Mol Pharmacol 76:113–124
71. Zhao H, Kalivendi S, Zhang H et al (2003) Superoxide reacts with hydroethidine but forms a fluorescent product that is distinctly different from ethidium: potential implications in intracellular fluorescence detection of superoxide. Free Radic Biol Med 34:1359–1368
72. Koppers AJ, De Iuliis GN, Finnie JM, McLaughlin EA, Aitken RJ (2008) Significance of mitochondrial reactive oxygen species in the generation of oxidative stress in spermatozoa. J Clin Endocrinol Metab 93:3199–3207
73. Aitken RJ, Wingate JK, De Iuliis GN, Koppers AJ, McLaughlin EA (2006) Cis-unsaturated fatty acids stimulate reactive oxygen species generation and lipid peroxidation in human spermatozoa. J Clin Endocrinol Metab 91:4154–4163
74. De Iuliis GN, Newey RJ, King BV, Aitken RJ (2009) Mobile phone radiation induces reactive oxygen species production and DNA damage in human spermatozoa in vitro. PLoS One 4:e6446
75. Koppers AJ, Garg ML, Aitken RJ (2010) Stimulation of mitochondrial reactive oxygen species production by unesterified, unsaturated fatty acids in defective human spermatozoa. Free Radic Biol Med 48:112–119
76. Myhre O, Andersen JM, Aarnes H, Fonnum F (2003) Evaluation of the probes 2',7'-dichlorofluorescin diacetate, luminol, and lucigenin as indicators of reactive species formation. Biochem Pharmacol 65:1575–1582
77. Tunc O, Thompson J, Tremellen K (2010) Development of the NBT assay as a marker of sperm oxidative stress. Int J Androl 33:13–21
78. Esfandiari N, Sharma RK, Saleh RA, Thomas AJ Jr, Agarwal A (2003) Utility of the nitroblue tetrazolium reduction test for assessment of reactive oxygen species production by seminal leukocytes and spermatozoa. J Androl 24:862–870
79. Fridovitch I (1997) Superoxide anion radical ($O_2^{-\bullet}$), superoxide dismutases, and related matters. J Biol Chem 272:18515–18517

Chapter 15
ROS and Semen Quality

Ralf Henkel

Abstract Infertility affects an estimate of 50–80 × 10^6 (=7–15% of men at reproductive age) men globally and annually. Among these men, a large number of patients (25–40%) suffer from male genital tract infection/inflammation, conditions where reactive oxygen species (ROS) and oxidative stress in the semen play prominent roles in the pathogenesis of infertility. Considering the extraordinary high amount of polyunsaturated fatty acids present in the plasma membrane, spermatozoa are exceptionally prone to oxidative damage. This damage may include the plasma membrane including the flagellar structure as well as the DNA. Thereby, sperm functions are damaged. As natural protective mechanisms, several antioxidants, enzymatic, non-enzymatic and preventive, are available in the seminal plasma. However, considering that certain essential physiological sperm functions are dependent on a limited amount of ROS, the balance between oxidation and reduction is of utmost importance. Clinically, this balance should be determined by estimating the seminal oxidative stress and antioxidant capacity.

Keywords ROS • Semen quality • Leukocytes • Male germ cells • Leukocytospermia • Spermatozoa • Detrimental effects of ROS • Impact of varicoceles

15.1 Introduction

According to the World Health Organization (WHO), infertility is defined as a condition where a couple is unable to conceive spontaneously despite having unprotected, regular sexual intercourse for more than 1 year [1]. Infertility, particularly male infertility, is a major matter of concern, as it is a clinical challenge of increasing significance.

R. Henkel, PhD (✉)
Department of Medical Bioscience, University of the Western Cape,
Modderdam Road, Bellville 7535, Western Cape, South Africa
e-mail: rhenkel@uwc.ac.za

A. Agarwal et al. (eds.), *Studies on Men's Health and Fertility*, Oxidative Stress in Applied Basic Research and Clinical Practice, DOI 10.1007/978-1-61779-776-7_15,

Worldwide, an estimate of 50–80 × 10^6 couples are infertile, which equals 7–15% of all couples at reproductive age (15–45 years of age). These numbers include both primary (couples have never conceived) and secondary infertility (couples have difficulties to conceive after having conceived before). However, due to high rates of infection-induced infertility in certain regions in the so-called "infertility belt" of sub-Saharan Africa, the prevalence of secondary infertility is amounting to up to 30%. Among all involuntary childless couples, male infertility accounts for approximately 30–50% of the cases [2]. This means that more than 7% of men are affected by infertility during their reproductive lifetime. Thus, the prevalence of male infertility is even higher than that for diabetes mellitus type I and II, which, with an overall estimate of 2.8% in the year 2000 and 4.4% in 2030 [3], is considered as a common disease [4].

Standard semen analysis is limited and only examines parameters such as sperm concentration, viability, normal sperm morphology and motility. In only very few laboratories, advanced sperm tests are performed to determine the functionality of the acrosome, chromatin condensation or DNA fragmentation. It has also been shown that standard semen parameters do not detect sperm abnormalities resulting in infertility in about 20% of infertile men [5]. Consequently, high numbers of idiopathic infertility are observed [6]. For a number of years, few groups have started investigating the causes of sperm DNA damage and identified oxidative stress is defined as an major cause [7] with a reported prevalence between 25 and 40% in unselected infertile men [8, 9] and up to 96% in patients suffering from spinal cord injuries [10].

15.2 Oxidative Stress and Reactive Oxygen Species

Oxidative stress is defined as an imbalance between oxidation and reduction towards the oxidative status [11], which can potentially result in cellular or genetic damage. This condition is triggered by so-called reactive oxygen species (ROS), which are chemical intermediates deriving from oxygen, most of them having one or more unpaired electrons (radicals) causing electronic instability and therefore very short half-life times in the nanosecond (10^{-9} s) to millisecond range (10^{-3} s) with high reactivity of the molecules [12]. Practically, these radicals react at the site of generation. The most relevant examples of ROS in Andrology are the hydroxyl radical ($^{\bullet}OH$), superoxide anion ($^{\bullet}O_2^-$) and hydrogen peroxide (H_2O_2). On the other hand, one has to distinguish between ROS, which can be radicals, but need not to be always (exception: H_2O_2), and radicals. Radicals are any form of molecule exhibiting one or more unpaired electrons.

15.3 Origin of ROS in Semen

In an ejaculate, there are two major sources of ROS, namely, leukocytes and the male germ cells themselves. Leukocytes physiologically generate high amounts of ROS, since this process plays a major role in infections, inflammations and cellular defence mechanisms to kill pathogens. If activated, leukocyte ROS generation can be up to 1,000-times higher than that of spermatozoa [9, 13]. In addition, ROS and oxidative stress in semen can also be caused by numerous other factors (for review see [14, 15]). Among them are exogenous sources such as environmental pollution by heavy metals [16] or other chemical compounds [17], and even lifestyle factors such as obesity [18], smoking [19], alcohol consumption [20] and certain medical conditions such as spinal cord injuries [21] or varicoceles [22].

Moreover, in a recent report, Cocuzza et al. [23] described a significant increase in seminal ROS production in men older than 40 years of age. Despite these higher seminal ROS levels in older men, no increase in the seminal leukocyte concentration was found. The latter observation was also described by Henkel et al. [24]. Yet, these authors demonstrated a significantly higher prevalence of male genital tract infections in older men as determined by means of the concentration of polymorphonuclear elastase. This interpretation would be consistent with the fact that age-related changes in leukocyte activity include increased secretion of pro-inflammatory cytokines [25] which, in turn, stimulate increased ROS generation in spermatozoa with subsequent damage of membrane function by lipid peroxidation (LPO) [26].

15.3.1 Leukocytes as ROS Producers

Leukocytes frequently appear in ejaculates, even in those from fertile men [27]. The WHO, in the latest edition of the laboratory manual [28] recommended a cut-off value for leukocytospermia, an excessive amount of leukocytes in the ejaculate, of 10^6/mL. Yet, this cut-off value is not unquestioned, and it has been shown that leukocyte concentrations as low as 4×10^4/mL can inflict damage to sperm functions and the DNA [29–31]. Therefore, several groups [32–36] suggested that the cut-off value for leukocytospermia should be lowered. In contrast, other groups found that the detection of leukocytospermia is of no diagnostic value for the identification of men with actual microbial infections [37]. Since no humoral immunological response was detected, bacteria found in semen cultures were only regarded as contaminants from the genital tract [38].

A recent study (Mupfiga and Henkel, unpublished data) showed that, as expected, the seminal leukocyte concentration is correlated not only significantly with the seminal ROS production but also with the production of superoxide, the activation of the effector caspases-3/7 in spermatozoa and the percentage of sperm with disrupted mitochondrial membrane potential. These observations give evidence that a high seminal concentration of activated leukocytes, as it is the case in male genital tract infections/inflammation, can trigger programmed cell death, apoptosis, by means of activating the caspase system and consequently leading to male infertility.

15.3.2 Spermatozoa as ROS Producers

In spermatozoa, like in other living cell, energy is produced to a large extent by means of oxidative phosphorylation in the mitochondria as well as through oxidation of hydrogen in the form of nicotinamide adenine dinucleotide (NADH). During the process of energy production in the electron transfer chain (ETC), elementary oxygen (O_2) takes up four electrons and is thus reduced to highly reactive free radicals as intermediate products with water (H_2O) as end product. Nevertheless, this process is not efficient enough to convert all the consumed oxygen into energy. As a result, 1–5% of the consumed oxygen is converted into free radicals [39]. ROS that are produced via this mechanism are regarded as cytotoxic by-products involved in the aetiology of disease and ageing [40], and spermatozoa are no exception. Therefore, male germ cells are competent producers of ROS such as superoxide and H_2O_2 (reviewed in [41]).

Particularly, morphologically abnormal spermatozoa exhibiting excess residual cytoplasm with its content of glucose-6-phosphate dehydrogenase [42], which fuels ROS production [43, 44], are deemed immature. As a result, such spermatozoa have impaired fertilizing capacity [45]. Although lower in the total amount as compared to the ROS production by activated leukocytes, the male germ cells' own ROS production appears to be of clinical importance. This is due to the so-called "intrinsic" ROS production in spermatozoa, which correlates considerably stronger with different sperm parameters such as DNA fragmentation than "extrinsic" ROS production by leukocytes [34].

15.4 Sensitivity of Spermatozoa to Oxidative Damage

Spermatozoa are very special cells. They are not only the smallest but also the most polarized cells in the body that fulfil their functions even outside the body in a different individual, the female genital tract. To maintain this extreme polarization, spermatozoa exhibit a specially composed plasma membrane containing an extraordinary amount of polyunsaturated fatty acids (PUFA) [46, 47]. This high PUFA content is essential for normal sperm function, as it is the foundation for the sperm plasma membrane's high fluidity [48]. In turn, membrane fluidity is directly related to normal sperm functions [47, 49] which can actually be regarded as membrane functions [50, 51].

The total content of PUFA in human spermatozoa accounts for about 50% of the content of fatty acids. Docosahexaenoic acid (DHA), a PUFA containing six double bonds per molecule [46] is of particular importance as its content represents about 21.5% of the total fatty acid and 43.0% of the unsaturated fatty acid content of human spermatozoa (reviewed in [41]). The importance of the PUFA content of the sperm plasma membrane in infertile patients with oligozoospermia and/or asthenozoospermia was debated, as contradictory data were presented by different groups. While Zalata et al. [46], Lenzi et al. [52] and Tavilani et al. [53] showed low PUFA levels in patients with poor sperm motility, Ollero et al. [54] as well as Khosrowbeygi and Zarghami [55] observed higher PUFA concentrations, especially DHA, in patients with poor sperm motility than in normozoospermic samples. Apparently, this obvious discrepancy is due to aberrant transformation of defective germ cells during their morphogenesis leaving these cells with too high PUFA levels triggering mitochondrial ROS production and causing oxidative stress [56]. The aforementioned discrepancies are interpreted by these authors as result of different end points used in the different studies.

Furthermore, spermatozoa are characterized by their very special morphological features, which comprise not only the extreme polarization but also the dramatic loss of most of the cytoplasm during spermatogenesis. As a result, the male germ cells exhibit an inevitable lack of intrinsic antioxidative protection by ROS scavengers such as catalase, glutathione peroxidase (GPx) or superoxide dismutase (SOD) as well as non-enzymatic molecules such as vitamin C, vitamin E or glutathione. Both these factors, the intrinsic lack of antioxidant protection together with the extraordinary high content of PUFA in the plasma membrane, make the male germ cell extremely vulnerable to oxidative stress. Since developing spermatozoa have a very limited ability to DNA repair, replenishment and regeneration of glutathione, this oxidative stress does lead not only to disturbed sperm functions but also to damaged DNA [57].

15.5 Detrimental Effects of ROS on Sperm Function

In view of the extraordinary high amount of PUFAs in the sperm plasma membranes and the extremely low concentration of antioxidants available in the male germ cell itself, ROS can trigger serious damage to various sperm functions, including motility or DNA integrity. However, ROS do not deteriorate sperm functions in an isolated manner, because ROS have direct and indirect influence on the whole sperm cell. Therefore, oxidative stress has been established as a major ethiological cause of male infertility. The causes for seminal oxidative stress are manifold and among these varicocele (18.1%) and male genital tract infections (35%) are the most frequent [24, 58]. Both causes, particularly infections, are potentially correctable using appropriate antibiotic and anti-inflammatory treatment to relieve the consequences of the infection, obstruction of the excurrent genital ducts [59]. However, for the treatment of varicoceles contradictory data have been published.

15.5.1 Effects of ROS on Membranes

As mentioned before, the sperm plasma membrane is particularly susceptible to oxidative stress because of its extraordinary content of double-bonds bearing membrane lipids, which can easily be oxidized by excessive ROS levels produced by both, the sperm cells as well as leukocytes. The process by which these plasma membrane lipids are oxidatively damaged is called "lipid peroxidation" (LPO).

Eventually, this process results in decreased membrane fluidity. As a result, receptor-mediated signal transduction as well as ion gradients derail [60] leading to loss of sperm membrane function and therefore functional capacity to fertilize oocytes.

LPO can be subdivided into three phases, initiation, propagation and termination. During the initiation phase, ROS attack the PUFA at carbon atoms adjacent to the double bonds leading to hydrogen abstraction from neighbouring methylene groups, which are especially reactive. Eventually, this process creates a lipid radical and water. The free electron is transferred to the lipid creating a new radical, which is subsequently stabilized by delocalization of the free electron over the whole molecule resulting in an energetically more stable structure than the initiating free radical. Considering that the resulting lipid radical is also not a very stable molecule, it will spontaneously react with molecular oxygen to form lipid peroxide. The propagation of the reaction is characterized by a reaction of this lipid peroxide radical molecule with a neighbouring fatty acid creating another fatty acid radical in a process called "radical chain reaction". Ultimately, this process causes damage to numerous molecules and up to 60% of the unsaturated fatty acid contend present in the plasma membrane can be oxidized [15]. The propagation of LPO terminates when one radical reacts with another radical, thus producing a non-radical, stable product whereby the two free electrons from the two radicals form a covalent bond.

With regard to the impact of oxidative stress on membrane lipids Khosrowbeygi and Zarghami [55] showed that spermatozoa from patients with asthenozoospermia, asthenoteratozoospermia or oligoasthenoteratozoospermia had significantly higher PUFA levels in their plasma membranes than normozoospermic men and were therefore more susceptible to oxidative stress and LPO. Additionally, significantly higher malondialdehyde (MDA) concentrations, which correlated negatively with the sperm count, were detected in samples from infertile patients [61]. Another study revealed that the amount of MDA present in spermatozoa correlated negatively with fertilization rates in an IVF programme [62].

Essentially, sperm exposure to oxidative stress cause by leukocytes induces a decrease in membrane fluidity, which is directly related to the membrane function and its fusogenic capacity [46]. This loss of membrane fluidity and functionality might lead to a rapid loss of intracellular ATP causing decreases in motility with axonemal damages and viability [63]. Consequently, not only sperm motility but also capacitation [64], acrosomal function [65] and acrosin activity are impaired [66].

15.5.2 Effect of ROS on DNA

During the process of LPO, numerous stable carbonyl-containing by-products such as MDA and 4-hydroxy-2-alkenals such as 4-hydroxy-nonenal (HNE), resulting from ω6-fatty acids such as DHA are formed. These by-products themselves are either highly mutagenic (MDA) or genotoxic (HNE) [67]. Accordingly, these by-products pose the danger for further DNA damage by adduct formation on spermatozoa [68]. In addition to the higher release of MDA in semen of men with seminal oxidative stress, protein carbonyl and sialic acid are elevated [61] as an indication of oxidative damage to proteins and possible protective response of the body to such oxidative stress, respectively [69].

Among other factors, the different physico-chemical behaviour of different ROS (H_2O_2 vs. $^{\bullet}OH$ or $^{\bullet}O_2^-$) is the reason for a differentiated action of membrane permeable and non-membrane permeable ROS on different sperm functions such as motility or DNA integrity. Also, the location of the production, extrinsic by leukocytes or intrinsic by the male germ cells themselves, appears to play a role as extrinsic ROS produced by leukocytes rather impairs sperm motility, while intrinsic ROS production seems to preferentially affect sperm nuclear DNA (nDNA) fragmentation [34]. Such nDNA damage has repeatedly and unequivocally been shown to be associated with fertilization and/or pregnancy failure after intrauterine insemination [70], in vitro fertilization [71–74] and ICSI [75–78] and is therefore predictive of the success of assisted reproduction.

Only until recently, investigations on sperm DNA damage merely focused on the nDNA. nDNA is surrounded and protected by nuclear proteins, namely, histones in somatic cells and protamines in spermatozoa. However, any eukaryotic cell contains a second type of DNA, mitochondrial DNA (mtDNA), which is, in contrast to nDNA, not protected. Furthermore, mtDNA replicates very fast without proper proof-reading and has only a very basic repair mechanism [79], thus making certain regions of the mitochondrial genome up to 100-times more susceptible to damage and mutations [80]. As a result, mtDNA is exceptionally susceptible to mutations and numerous diseases including male infertility such as asthenozoospermia or oligoasthenozoospermia [81, 82].

mtDNA encodes for 13 polypeptides that are essential for the ETC on the inner mitochondrial membrane and 22 tRNAs and 2 rRNAs that are necessary for the translation of these polypeptides. These polypeptides are intimately involved in oxidative phosphorylation and ATP production in the mitochondria. During this process, mitochondria continuously oxidize different substrates while, at the same time, reducing oxygen to water [83]. However, as a by-product of energy production, the mitochondria also generate most of the endogenous ROS of the cell, and these damage the mitochondria, mtDNA and the cell. Consequently, mtDNA damages will result in decreased mitochondrial membrane potential ($\Delta\psi_m$) and defective mitochondrial function. Yet, the latter is essential for sperm motility [84] and is negatively correlated with seminal oxidative stress [85]. In turn, mitochondrial dysfunction has repeatedly been shown to cause an increased release of mitochondrial ROS early events of apoptosis [86, 87]. Therefore, this parameter has been suggested as being a highly sensitive parameter [88].

Ruiz-Pesini et al. [89] and Hoshi et al. [90] found a correlation between the quality of the semen and the functionality of the respiratory chain in sperm mitochondria. Moreover, it has been demonstrated that mtDNA point mutations, mtDNA single nucleotide polymorphisms and mtDNA haplogroups can significantly influence semen quality [91–94]. Thus, due to the high sensitivity of mtDNA to damage and the essential role mitochondria play in the cells' energy production, it is plausible that seminal oxidative stress may lead to mtDNA damage. This may result in dysfunction of the mitochondrial respiratory chain including further stimulation of mitochondrial ROS production and oxidative damage due to derailment of the ETC causing a vicious cycle [95].

15.5.3 Impact of ROS on Flagellar Structure

Recent data by El-Taieb et al. [96] show that the oxidative stress-related seminal protein carbonyl concentration is positively correlated with axonemal abnormalities that are associated with poor sperm motility. This ROS-related damage of structural components of the flagellum might arise in the epididymis during sperm epididymal maturation, which has been shown to reflect on sperm ultrastructure and morphology not only for the sperm head but also for the flagellum [97].

15.5.4 Infections/Inflammations

Both, male genital tract infections and inflammations are medical conditions that are seriously affecting spermatogenesis and sperm transit during ejaculation in terms of an obstruction of the relevant ducts. Their effects can be seen in clinical findings in cases of oligozoospermia (decreased number of sperm), asthenozoospermia (decreased sperm motility) or azoospermia (absence of sperm in the ejaculate) [98, 99]. Moreover, they are also the cause of dysfunctional male accessory glands [59] and significantly impaired sperm functions [100, 101]. Reportedly, the prevalence of male genital tract infection related infertility amounts to up to 35% of patients consulting for infertility [24].

According to Monga and Roberts [102], the damages to the male fertility potential can be triggered by direct action of the pathogens on spermatozoa and sperm functions or indirectly by inducing inflammatory processes in the seminal tract by activating leukocytes [103]. As a result of the latter, seminal ROS and cytokine (IL-6, IL-8 or TNF-α) levels increase and exert their detrimental effects [34, 104].

Both, ROS as well as cytokines, which in turn can trigger the production of excessive ROS [26] in spermatozoa, have been shown to be associated with the impairment of various sperm functions such as motility, acrosome reaction or sperm DNA integrity through oxidative stress [34, 65, 71, 105, 106]. The mechanism by which this damage takes place involves oxidation of sperm membranes by means of LPO as well as direct oxidation of the DNA [107, 108].

15.5.5 Detrimental Effects of ROS in Patients with Prostatitis

In patients suffering from prostatitis, a negative association with sperm motility and morphology has repeatedly been shown because this condition is also associated with increased leukocyte infiltration in semen. [109, 110]. Accordingly, significantly increased seminal oxidative stress has been observed in patients with prostatitis categories NIH I, NIH II, NIH IIIA and NIH IV [111, 112]. However, even in prostatitis categories NIH III (chronic abacterial prostatitis) and NIH IIIB (non-inflammatory chronic pelvic pain syndrome), elevated ROS concentrations accompanied by decreased antioxidant levels are present in the semen causing oxidative stress irrespective of the presence of leukocytes [113]. This oxidative stress could be triggered by cytokines [26] and is thought to prompt acrosomal dysfunction in terms of significantly reduced inducibility of the acrosome reaction [59], possibly via destabilization of the sperm plasma membrane by LPO.

15.5.6 Effects of Leukocytes and Infections on Chromatin Packaging

The events of sperm chromatin condensation can be separated into two phases, a testicular phase where the exchange of histones by protamines takes place [114], and an epididymal where the final protamine structure is fixed by the formation of disulphide-bridges [115]. Considering that the latter process involves redox reactions it can be affected by an infection or inflammatory process. ROS produced by spermatozoa and seminal leukocytes correlate positively with poor chromatin condensation as determined by means of the chromomycin A_3 stain, suggesting that the histone replacement by protamines and the occurrence of sperm DNA damage are separate processes [116].

15.5.7 Impact of Varicoceles

A varicocele is a tortuous distension of intra-scrotal veins of the pampiniform plexus, and is regarded as one of the leading causes of male infertility with a prevalence of about 15–20% among the general adult men population and up to 40% in infertile males [117]. The aetiology of a varicocele is likely to be multifactorial [118]. Pathophysiologically, clinical varicoceles are characterized by hyperthermia of the scrotum and the affected testicle leading to spermatogenic damage [119] possibly induced by oxidative stress mediated apoptosis [120] with an increased expression of apoptosis-related proteins in spermatozoa [121]. Moreover, endocrine and paracrine imbalances compromising Sertoli cell function have been described [118].

Due to the release of nitric oxide synthase and xanthin oxidase significantly higher levels of oxidative stress were detected in the spermatic vein [122–124]. Additionally, Mostafa et al. [125] found that, apart from increased levels of oxidative stress as determined by means of the seminal concentrations of MDA, hydrogen peroxide and nitric oxide, levels of antioxidants (SOD, catalase, GPx, vitamin C) were significantly reduced in spermatic venous blood as compared to peripheral blood. This finding is consistent with the observation by Shamsi and Dada (2010; personal communication) that seminal concentrations of MDA and antioxidants significantly and positively correlate with those in the blood. In turn, if there is a positive correlation between genital oxidative stress with that found in the blood, this raises the question if this could possibly also be the case if there is a systemic infection or generally too high systemic oxidative stress with high levels of oxidants. The latter might then cause increased oxidative stress in the genital tract system, leading to infertility.

It is also not surprising that the seminal concentrations of ROS and total antioxidant capacity (TAC) were increased and decreased, respectively [126]. Consequently, varicocele not only does affect spermatogenesis leading to poor sperm count, morphology and DNA integrity [127] but also leads to deteriorated sperm functionality by seminal oxidative stress. The latter could possibly be induced by the inflammatory effect caused by varicocele and which is resulting in significantly elevated levels of IL-6 or IL-8, comparable to those found in patients with genital tract infections [128].

On the contrary, another research group [129] presented data questioning an impact of clinical varicocele on testicular size and seminal ROS levels on male fertility, at least in fertile men. For patients with unknown fertility status, these authors conclude that increased ROS levels may be indicative of an early sign for a decline in fertility if the condition is untreated. Thus, these authors differentiate regarding varicocele-associated fertility problems between fertile and infertile men.

Varicocelectomy, in turn, has been shown to significantly attenuate seminal oxidative stress including the resultant sperm DNA damage 3 months after the operation [130, 131]. Recently, data supporting the clinical value of varicocelectomy have been reported in a randomized controlled study [132].

15.6 Antioxidants in the Semen

In view of the male germ cell's high vulnerability to extrinsic and intrinsic ROS as well as lack of its own protection, spermatozoa have to receive protection against oxidative stress through relevant scavengers by the male and/or female reproductive tracts. Therefore, in the male, the testis and the accessory sex glands that are producing the seminal plasma have to provide substances and mechanisms for such protection. Indeed, seminal plasma is the biological fluid containing highest concentrations of antioxidant substances, even more than in any other physiological fluid. Basically, this antioxidative protection can be separated into two systems, interceptive and preventive. Among the scavenging antioxidants another distinction between enzymatic and non-enzymatic systems can be made [133].

15.6.1 Enzymatic Antioxidant Protection

The enzymatic system consists of SOD [134], catalase [135] and GPx [136]. SOD is the most important system in seminal fluid and scavenges intra- and extracellular superoxide, thus preventing LPO of membrane lipids. Considering that this reaction results in the production of hydrogen peroxide, another, but less reactive oxidant, SOD should act together with catalase or GPx [137]. Catalase and GPx detoxify intra- and extracellular hydrogen peroxide to produce water and oxygen [138]. Data by the Zini group suggest that the seminal catalase and SOD activities do not derive from the testis or epididymis, but rather from seminal vesicle or prostate [139].

Recent studies revealed that seminal MDA, nitric oxide and SOD levels were significantly altered in infertile patients as compared to fertile controls [140, 141], confirming earlier observations by Kobayashi et al. [134] and Calamera et al. [142]. Since no difference in the seminal concentration of catalase between fertile and infertile patients could be found, it appears that seminal SOD activity plays a bigger role for the maintenance of sperm motility and the sperm cells' protection against ROS than catalase [143]. However, the usefulness of the determination of SOD levels to diagnose male infertility is also questioned by Hsieh et al. [144] who could not find any significant correlation with motility of sperm concentration, neither for the intracellular, nor for the seminal SOD levels. The relationship of seminal SOD levels on various sperm functions such as acrosome reaction, however, did not appear so clear [140, 141, 145].

For the seminal activity of GPx, Hsieh et al. [146] established a positive, but not significant association with sperm motility. However, the intracellular GPx levels appear to be more important as intracellular components of the glutathione system, including GPx and glutathione, seem to be altered in infertile men where it is particularly linked to poor sperm morphology [147]. On the other hand, there is also evidence that intra-spermatozoal expression of phospholipid hydroperoxide glutathione (PHGPx) peroxidase is decreased in about 25% of oligo-asthenozoospermic men. Apparently, this lower expression of the enzyme is not caused by mutations of the PHGPx gene [148].

15.6.2 Non-enzymatic Protection

Non-enzymatic antioxidative systems consists of co-enzyme Q10 (radical scavenger), dietary vitamins such as vitamins A (β-carotene; singlet oxygen quencher), C (ascorbate; diverse antioxidant functions) and E (α-tocopherol; chain-breaking compound) or other substances such as urate (radical scavenger) or aliphatic polyamines such as spermidine and spermine. Vitamin E is the most efficient compound in the lipid phase [149]. All these substances have in common that they detoxify free radicals [150]. Hence, it is not surprising to find the most important natural antioxidants such as vitamin C and E [151, 152], uric acid [153], glutathione [154] in highest concentrations in the seminal fluid. Reportedly, seminal vitamin C and E concentrations are lower in infertile men than in normal patients [155, 156]. In addition, patients with lower seminal ascorbate concentrations showed a higher, though not significant, percentage of sperm DNA fragmentation [155].

Moreover, other substances such as the aliphatic polyamines spermidine and spermine are present in seminal plasma and act directly as a free radical scavenger to inhibit LPO [157]. Spermine, for instance, is present at high concentrations of about 0.6 mg/mL suggesting an important influence in the protection of spermatozoa against free radicals. On the other hand, there is also evidence that the ROS scavenging capacity of polyamines is only of marginal relevance because the rate constants for their reaction are relatively low [158]. The antioxidative functions of polyamines are thought to be mainly due to their metal-chelating properties. In contrast, Allen and Roberts [159] showed significant cytotoxic effects of seminal plasma and spermine on lymphocyte cultures. Thus, the physiological function of spermine is still unclear.

The ROS scavenging activity of the seminal fluid not only does protect the male germ cell against oxidative damage but also regulates sperm functions, namely, capacitation. In this regard, it has been shown by de Lamirande et al. [160] that semenogelin, the main protein in the human seminal coagulum that is deriving from the seminal vesicle, or one of its degradation products, is scavenging superoxide at concentrations much lower than those of semen. This is an important observation, since superoxide triggers sperm hyperactivation and capacitation [161]. Apparently, semenogelin and the element zinc, which is derived in large concentrations from the prostate, are important modulators of sperm capacitation by fine tuning the sperm cells' own generation of superoxide and nitric oxide [162].

15.6.3 Preventive Antioxidant Systems

Among the preventive antioxidants metal chelators such as albumin, transferrin, metallothionein, lactoferrin or coeruloplasmin are to mention. The antioxidative activity is achieved by eliminating transition metal ions from the active centre of pro-oxidative enzymes and thereby detoxifying free radicals [150, 163].

15.7 Determination of Oxidative Stress in Seminal Plasma

In order to determine seminal oxidative stress, two different approaches can be chosen, the direct measurement of ROS by a chemiluminescent method using luminol [8, 164] or other chemiluminescent substances, or the indirect assessment of oxidative stress by means of the measurement of MDA [165], which is a stable end product of LPO [107]. Luminescent methods are fast and highly sensitive as light can be detected at very low emission rates. However, specialized equipment (luminometer) is necessary and not every Andrology laboratory may be in possession thereof. In the indirect method, MDA, as thiobarbituric acid reactive substance (TBARS) is measured by means of standard photometry, which is available in almost every laboratory. Thus, determination of TBARS is factually a measure of the degree of LPO, of which semen samples derived from infertile men were shown to have elevated levels [166]. Furthermore, in abnormal semen samples TBARS levels correlate negatively with the sperm count [61], while in other studies the TBARS concentration was negatively correlated with the male germ cells' acrosin activity [66] as well as with fertilization rates in an IVF programme [62]. These data indicate that the determination of seminal TBARS levels can be indicative of male infertility. However, proper cut-off values have not been calculated yet.

For the direct measurement of ROS, the Agarwal group [167] has performed a receiver operating characteristics (ROC) analysis to establish a cut-off value of 0.145×10^6 cpm/20×10^6 spermatozoa to distinguish between fertile and infertile men with a sensitivity and specificity of the test of 87 and 57%, respectively. Although outputs from different luminometers cannot be directly compared because of different response characteristics of the photomultipliers, Henkel et al. (unpublished data) came to a similar result of 0.229×10^3 cpm/20×10^6 spermatozoa after ROC analysis for fertilization.

15.8 Determination of the Seminal Antioxidant Capacity

The antioxidant activity in biological fluids can be measured either by determination of the concentrations of the individual antioxidants or by the TAC [168, 169]. Among the different methodologies, which include the determination of the oxygen radical absorbance capacity [170], ferric reducing ability [171] or the phycoerythrin fluorescence-based assay [172], an enhanced chemiluminescent assay described by Whitehead et al. [173] is most commonly used. The principle of this assay is that the chemiluminescence emitted by a chemiluminescent substance such as luminol is inhibited by the antioxidant activity and then compared and standardized with that of Trolox® (6-hydroxy-2,5,7,8-tetramethylchroman-2-carboxylic acid), a water-soluble vitamin E analogue, and measured as molar Trolox® equivalent. More recently, a reliable, accurate, more rapid and simple colorimetric method has been developed [174].

The determination of TAC in seminal plasma revealed significant differences between fertile and infertile subjects with TAC being higher in fertile men [168, 175, 176]. Said et al. [174] developed a colorimetric assay, which is simple, relatively cheap and easy to perform, and was as reliable as the luminometric test system. More recently, Mahfouz et al. [177] have established a clinical cut-off value for the TAC of 1,420 μM and thus demonstrated its diagnostic value. Yet, proper clinical evaluation including the calculation of cut-off values by means of ROC still has to be carried out.

15.9 Balance Between Oxidation and Reduction

On the one hand, ROS and oxidative stress are detrimental to spermatozoa and sperm functions because of the very special membrane lipid composition of the male germ cells. On the other hand, however, ground-breaking studies by de Lamirande and Gagnon [178, 179] show that superoxide not only is detrimental to sperm but also plays an essential role in triggering cellular events such as capacitation and hyperactivation, which are crucial requirements for sperm zona pellucida penetration [180]. Aitken et al. [181, 182] as well as Griveau and Le Lannou [183] confirmed this function of ROS. Moreover, it could be shown that besides capacitation and hyperactivation, acrosome reaction, sperm zona binding and oocyte fusion are also stimulated by various low levels of ROS even in other species [43, 50, 184].

Conversely, high levels of activated, ROS-producing leukocytes are without question harmful to sperm functions and male fertility. Nevertheless, this detrimental effect of oxidative stress is also depending on the TAC of the ejaculate. Thus, it is not enough to determine only one parameter, either seminal ROS or TAC. Therefore, a combination of both parameters as a ROS-TAC score as proposed by Sharma et al. [185] appears to be superior to ROS or TAC in discriminating between fertile and infertile men. This suggestion seems reasonable because the balance between oxidants (ROS) and antioxidants (TAC) is essential for normal sperm function.

Since the ROS-TAC score decreases as oxidative stress increases caused by leukocytes, it is difficult, if not impossible, to determine a proper cut-off for leukocyte contamination [32, 186]. For instance, a patient's fertility might not be compromised even if high numbers of non-activated leukocytes producing only small amounts of ROS are present in his ejaculate if this patient shows high seminal levels of TAC at the same time. In contrast, another patient might have compromised antioxidative seminal protection with low numbers of activated leukocytes, which produce high ROS levels. Because of this, the latter patient might be infertile. This example shows that the system of seminal oxidants and antioxidants as a whole has to be finely balanced in order to function properly [41]. Essentially, spermatozoa have to "walk a tightrope" and they will not be functionally competent, if the system of oxidants and antioxidants as a whole deviates to either side; it has to be finely balanced in order to function properly. Therefore, to be on the safe side in a clinical set-up, the presence of seminal leukocytes should rather be lower than higher and the cut-off for leukocytospermia of 1×10^6 leukocytes/mL as recommended by the WHO [28], should be lowered as previously suggested [32–34].

15.10 Conclusions

A very fine balance between a physiological oxidative stress and the antioxidative capacity of the ejaculate is of critical importance for the physiology as well as the pathophysiology of the male germ cells [15, 187]. Thus, a derailment of this balance has an impact on both male fertility and andrological diagnostics. Owing to these complex interactions between oxidation and antioxidative protection, it is indisputably not sufficient to determine a male's fertility status by simply relying on a standard semen analysis with some additional testing of sperm functions. The determination of the seminal redox status is the way forward to improve andrological diagnostic. However, this would then have to include the determination of ROS levels as well as the antioxidant capacity present in the ejaculate in terms of a ROS-TAC score [168, 177]. This concept of oxidative stress also explains the discrepancies in the literature about the impact and importance of ROS on sperm functions as well as that of leukocytes.

With regard to the determination of this balance, one would have to test for both, ROS and TAC, in order to determine the true oxidative status in the semen of patients. Unfortunately, a ROS-TAC score as suggested by Sharma et al. [185] has not been clinically evaluated yet. This is regrettable, not only since it is a necessary completion of andrological diagnostic but also because many patients look for treatment with antioxidants. However, it is simply not sufficient to treat patient with antioxidants without knowing the redox status of the semen. Since too much of antioxidant treatment might also cause harm [188], the redox status should be detected before the prescription of an antioxidative treatment.

References

1. Rowe PJ, Comhaire FH, Hargreave TB, Mahmoud AMA. WHO manual for the standardized investigation, diagnosis and management of the infertile male. Cambridge: Cambridge University Press; 2000.
2. Hull MG, Glazener CM, Kelly NJ, Conway DI, Foster PA, Hinton RA, Coulson C, Lambert PA, Watt EM, Desai KM. Population study of causes, treatment, and outcome of infertility. Br Med J (Clin Res Ed). 1985;291:1693–7.
3. Wild S, Roglic G, Green A, Sicree R, King H. Global prevalence of diabetes: estimates for the year 2000 and projections for 2030. Diabetes Care. 2004;27:1047–53.
4. Nieschlag E, Behre HM. Andrology. Male reproductive health and dysfunction. 2nd ed. Berlin: Springer; 2000.
5. Romeo C, Ientile R, Santoro G, Impellizzeri P, Turiaco N, Impala P, Cifala S, Cutroneo G, Trimarchi F, Gentile C. Nitric oxide production is increased in the spermatic veins of adolescents with left idiophatic varicocele. J Pediatr Surg. 2001;36:389–93.
6. Hendin BN, Kolettis PN, Sharma RK, Thomas Jr AJ, Agarwal A. Varicocele is associated with elevated spermatozoal reactive oxygen species production and diminished seminal plasma antioxidant capacity. J Urol. 1999;161:1831–4.
7. Agarwal A, Said TM. Oxidative stress, DNA damage and apoptosis in male infertility: a clinical approach. BJU Int. 2005;95:503–7.
8. Iwasaki A, Gagnon C. Formation of reactive oxygen species in spermatozoa of infertile patients. Fertil Steril. 1992;57:409–16.
9. de Lamirande E, Gagnon C. Capacitation-associated production of superoxide anion by human spermatozoa. Free Radic Biol Med. 1995;18:487–95.
10. Padron OF, Brackett NL, Sharma RK, Lynne CM, Thomas Jr AJ, Agarwal A. Seminal reactive oxygen species and sperm motility and morphology in men with spinal cord injury. Fertil Steril. 1997;67:1115–20.
11. Sies H. Oxidative stress: introductory remarks. In: Sies H, editor. Oxidative stress. London: Academic; 1985. p. 1–8.
12. Halliwell B, Gutteridge JMC. Free radicals in biology and medicine. 2nd ed. Oxford: Clarendon Press; 1989.
13. Plante M, de Lamirande E, Gagnon C. Reactive oxygen species released by activated neutrophils, but not by deficient spermatozoa, are sufficient to affect normal sperm motility. Fertil Steril. 1994;62:387–93.
14. Varghese AC, du Plessis SS, Agarwal A. Male gamete survival at stake: causes and solutions. Reprod Biomed Online. 2008;17:866–80.

15. Kothari S, Thompson A, Agarwal A, du Plessis S. Free radicals: their beneficial and detrimental effects on sperm function. Indian J Exp Biol. 2010;48:425–35.
16. Acharya UR, Acharya S, Mishra M. Lead acetate induced cytotoxicity in male germinal cells of Swiss mice. Ind Health. 2003;41:291–4.
17. Kasahara E, Sato EF, Miyoshi M, Konaka R, Hiramoto K, Sasaki J, Tokuda M, Nakano Y, Inoue M. Role of oxidative stress in germ cell apoptosis induced by di(2-ethylhexyl)phthalate. Biochem J. 2002;365:849–56.
18. Tunc O, Bakos HW, Tremellen K. Impact of body mass index on seminal oxidative stress. Andrologia. 2011;43:121–8.
19. Saleh RA, Agarwal A, Sharma RK, Nelson DR, Thomas Jr AJ. Effect of cigarette smoking on levels of seminal oxidative stress in infertile men: a prospective study. Fertil Steril. 2003;78:491–9.
20. Koch OR, Pani G, Borrello S, Colavitti R, Cravero A, Farrè S, Galeotti T. Oxidative stress and antioxidant defense in ethanol-induced cell injury. Mol Aspects Med. 2004;25:191–8.
21. Linsenmeyer TA, Perkash I. Infertility in men with spinal cord injury. Arch Phys Med Rehabil. 1991;72:747–54.
22. Dada R, Shamsi MB, Venkatesh S, Gupta NP, Kumar R. Attenuation of oxidative stress and DNA damage in varicocelectomy: implications in infertility management. Indian J Med Res. 2010;132:728–30.
23. Cocuzza M, Athayde KS, Agarwal A, Sharma R, Pagani R, Lucon AM, Srougi M, Hallak J. Age-related increase of reactive oxygen species in neat semen in healthy fertile men. Urology. 2008;71:490–4.
24. Henkel R, Maaß G, Jung A, Haidl G, Schill WB, Schuppe HC. Age-related changes in seminal polymorphonuclear elastase in men with asymptomatic inflammation of the genital tract. Asian J Androl. 2007;9:299–304.
25. Ortega E, Garcia JJ, De La Fuente M. Ageing modulates some aspects of the non-specific immune response of murine macrophages and lymphocytes. Exp Physiol. 2000;85: 519–25.
26. Buch JP, Kolon TF, Maulik N, Kreutzer DL, Das DK. Cytokines stimulate lipid membrane peroxidation of human sperm. Fertil Steril. 1994;62:186–8.
27. Wolff H. The biologic significance of white blood cells in semen. Fertil Steril. 1995;63:1143–57.
28. World Health Organization. WHO laboratory manual for the examination and processing of human semen. 5th ed. Geneva: World Health Organization; 2010.
29. Aitken RJ, Buckingham DW, Brindle J, Gomez E, Baker HW, Irvine DS. Analysis of sperm movement in relation to the oxidative stress created by leukocytes in washed sperm preparations and seminal plasma. Hum Reprod. 1995;10:2061–71.
30. Alvarez JG, Sharma RK, Ollero M, Saleh RA, Lopez MC, Thomas AJ, Evenson DP, Agarwal A. Increased DNA damage in sperm from leukocytospermic semen samples as determined by the sperm chromatin structure assay. Fertil Steril. 2002;78:319–29.
31. Erenpreiss J, Hlevicka S, Zalkalns J, Erenpreisa J. Effect of leukocytospermia on sperm DNA integrity: a negative effect in abnormal semen samples. J Androl. 2002;23:717–23.
32. Sharma RK, Pasqualotto AE, Nelson DR, Thomas Jr AJ, Agarwal A. Relationship between seminal white blood cell counts and oxidative stress in men treated at an infertility clinic. J Androl. 2001;22:575–83.
33. Punab M, Loivukene K, Kermes K, Mändar R. The limit of leucocytospermia from the microbiological viewpoint. Andrologia. 2003;35:271–8.
34. Henkel R, Kierspel E, Stalf T, Mehnert C, Menkveld R, Tinneberg HR, Schill WB, Kruger TF. Effect of reactive oxygen species produced by spermatozoa and leukocytes on sperm functions in non-leukocytospermic patients. Fertil Steril. 2005;83:635–42.
35. Gdoura R, Kchaou W, Chaari C, Znazen A, Keskes L, Rebai T, Hammami A. *Ureaplasma urealyticum*, *Ureaplasma parvum*, *Mycoplasma hominis* and *Mycoplasma genitalium* infections and semen quality of infertile men. BMC Infect Dis. 2007;7:129.

36. Oborna I, Fingerova H, Novotny J, Brezinova J, Svobodova M, Aziz N. Reactive oxygen species in human semen in relation to leukocyte contamination. Biomed Pap Med Fac Univ Palacky Olomouc Czech Repub. 2009;153:53–7.
37. Trum JW, Mol BWJ, Pannekoek Y, Spanjaard L, Wertheim P, Bleker OP, van der Veen F. Value of detecting leukocytospermia in the diagnosis of genital tract infection in subfertile men. Fertil Steril. 1998;70:315–9.
38. Lackner JE, Herwig R, Schmidbauer J, Schatzl G, Kratzik C, Marberger M. Correlation of leukocytospermia with clinical infection and the positive effect of antiinflammatory treatment on semen quality. Fertil Steril. 2006;86:601–5.
39. Chance B, Sies H, Boveris H. Hydroperoxide metabolism in mammalian organs. Physiol Rev. 1979;59:527–605.
40. Raha S, Robinson BH. Mitochondria, oxygen free radicals, disease and ageing. Trends Biochem Sci. 2000;25:502–8.
41. Henkel R. Leukocytes and oxidative stress: dilemma for sperm function and male fertility. Asian J Androl. 2011;13:43–52.
42. Gomez E, Buckingham DW, Brindle J, Lanzafame F, Irvine DS, Aitken RJ. Development of an image analysis system to monitor the retention of residual cytoplasm by human spermatozoa: correlation with biochemical markers of the cytoplasmic space, oxidative stress, and sperm function. J Androl. 1996;17:276–87.
43. Aitken RJ, Fisher HM, Fulton N, Gomez E, Knox W, Lewis B, Irvine S. Reactive oxygen species generation by human spermatozoa is induced by exogenous NADPH and inhibited by the flavoprotein inhibitors diphenylene iodonium and quinacrine. Mol Reprod Dev. 1997;47:468–82.
44. Aziz N, Saleh RA, Sharma RK, Lewis-Jones I, Esfandiari N, Thomas Jr AJ, Agarwal A. Novel association between sperm reactive oxygen species production, sperm morphological defects, and the sperm deformity index. Fertil Steril. 2004;81:349–54.
45. Keating J, Grundy CE, Fivey PS, Elliott M, Robinson J. Investigation of the association between the presence of cytoplasmic residues on the human sperm midpiece and defective sperm function. J Reprod Fertil. 1997;110:71–7.
46. Zalata AA, Christophe AB, Depuydt CE, Schoonjans F, Comhaire FH. The fatty acid composition of phospholipids of spermatozoa from infertile patients. Mol Hum Reprod. 1998;4:111–8.
47. Force A, Grizard G, Giraud MN, Motta C, Sion B, Boucher D. Membrane fluidity and lipid content of human spermatozoa selected by swim-up method. Int J Androl. 2001;24:327–34.
48. Flesch FM, Gadella BM. Dynamics of the mammalian sperm plasma membrane in the process of fertilization. Biochim Biophys Acta. 2000;1469:197–235.
49. Gulaya NM, Margitich VM, Govseeva NM, Klimashevsky VM, Gorpynchenko II, Boyko MI. Phospholipid composition of human sperm and seminal plasma in relation to sperm fertility. Arch Androl. 2001;46:169–75.
50. Aitken RJ, Clarkson JS, Hargreave TB, Irvine DS, Wu FCW. Analysis of the relationship between defective sperm function and the generation of reactive oxygen species in cases of oligozoospermia. J Androl. 1989;10:214–20.
51. Aitken RJ, Irvine DS, Wu FC. Prospective analysis of sperm-oocyte fusion and reactive oxygen species generation as criteria for the diagnosis of infertility. Am J Obstet Gynecol. 1991;164:542–51.
52. Lenzi A, Gandini L, Maresca V, Rago R, Sgro P, Dondero F, Picardo M. Fatty acid composition of spermatozoa and immature germ cells. Mol Hum Reprod. 2000;6:226–31.
53. Tavilani H, Doosti M, Abdi K, Vaisiraygani A, Joshaghani HR. Decreased polyunsaturated and increased saturated fatty acid concentration in spermatozoa from asthenozoospermic males as compared with normozoospermic males. Andrologia. 2006;38:173–8.
54. Ollero M, Gil-Guzman E, Lopez MC, Sharma RK, Agarwal A, Larson K, Evenson D, Thomas Jr AJ, Alvarez JG. Characterization of subsets of human spermatozoa at different stages of maturation: implications in the diagnosis and treatment of male infertility. Hum Reprod. 2001;16:1912–21.

55. Khosrowbeygi A, Zarghami N. Fatty acid composition of human spermatozoa and seminal plasma levels of oxidative stress biomarkers in subfertile males. Prostaglandins Leukot Essent Fatty Acids. 2007;77:117–21.
56. Koppers AJ, Garg ML, Aitken RJ. Stimulation of mitochondrial reactive oxygen species production by unesterified, unsaturated fatty acids in defective human spermatozoa. Free Radic Biol Med. 2010;48:112–9.
57. Bauche F, Fouchard MH, Jegou B. Antioxidant system in rat testicular cells. FEBS Lett. 1994;349:392–6.
58. Nieschlag E, Behre H. Andrology. Male reproductive health and dysfunction. Berlin: Springer; 1997.
59. Weidner W, Krause W, Ludwig M. Relevance of male accessory gland infection for subsequent fertility with special focus on prostatitis. Hum Reprod Update. 1999;5:421–32.
60. Sikka SC, Rajasekaran M, Hellstrom WJ. Role of oxidative stress and antioxidants in male infertility. J Androl. 1995;16:464–8.
61. Saraniya A, Koner BC, Doureradjou P, Selvaraj N, Shivagourou V. Altered malondialdehyde, protein carbonyl and sialic acid levels in seminal plasma of microscopically abnormal semen. Andrologia. 2008;40:56–7.
62. Zabludovsky N, Eltes F, Geva E, Berkovitz E, Amit A, Barak Y, Har-Even D, Bartoov B. Relationship between human sperm lipid peroxidation, comprehensive quality parameters and IVF outcome. Andrologia. 1999;31:91–8.
63. Bansal AK, Bilaspuri GS. Impacts of oxidative stress and antioxidants on semen functions. Vet Med Int. 2010;pii:686137.
64. Sikka SC. Oxidative stress and role of antioxidants in normal and abnormal sperm function. Front Biosci. 1996;1:e78–86.
65. Henkel R, Ludwig M, Schuppe HC, Diemer T, Schill WB, Weidner W. Chronic pelvic pain syndrome/chronic prostatitis affect the acrosome reaction in human spermatozoa. World J Urol. 2006;24:39–44.
66. Zalata AA, Ahmed AH, Allamaneni SSR, Comhaire FH, Agarwal A. Relationship between acrosin activity of human spermatozoa and oxidative stress. Asian J Androl. 2004;6:313–8.
67. Luczaj W, Skrzydlewska E. DNA damage caused by lipid peroxidation products. Cell Mol Biol Lett. 2003;8:391–413.
68. Esterbauer H. Cytotoxicity and genotoxicity of lipid-oxidation products. Am J Clin Nutr. 1993;57(5 Suppl):779S–85.
69. Eguchi H, Ikeda Y, Ookawara T, Koyota S, Fujiwara N, Honke K, Wang PG, Taniguchi N, Suzuki K. Modification of oligosaccharides by reactive oxygen species decreases sialyl lewis x-mediated cell adhesion. Glycobiology. 2005;15:1094–101.
70. Duran EH, Morshedi M, Taylor S, Oehninger S. Sperm DNA quality predicts intrauterine insemination outcome: a prospective cohort study. Hum Reprod. 2002;17:3122–8.
71. Henkel R, Hajimohammad M, Stalf T, Hoogendijk C, Mehnert C, Menkveld R, Gips H, Schill W-B, Kruger TF. Influence of deoxyribonucleic acid damage on fertilization and pregnancy. Fertil Steril. 2004;81:965–72.
72. Sun JG, Jurisicova A, Casper RF. Detection of deoxyribonucleic acid fragmentation in human sperm: correlation with fertilization in vitro. Biol Reprod. 1997;56:602–7.
73. Host E, Lindenberg S, Smidt-Jensen S. DNA strand breaks in human spermatozoa: correlation with fertilization in vitro in oligozoospermic men and in men with unexplained infertility. Acta Obstet Gynecol Scand. 2000;79:189–93.
74. Morris ID, Ilott S, Dixon L, Brison DR. The spectrum of DNA damage in human sperm assessed by single cell gel electrophoresis (Comet assay) and its relationship to fertilization and embryo development. Hum Reprod. 2002;17:990–8.
75. Lopes S, Sun JG, Jurisicova A, Meriano J, Casper RF. Sperm deoxyribonucleic acid fragmentation is increased in poor-quality semen samples and correlates with failed fertilization in intracytoplasmic sperm injection. Fertil Steril. 1998;69:528–32.
76. Henkel R, Kierspel E, Hajimohammad M, Stalf T, Hoogendijk C, Mehnert C, Menkveld R, Schill WB, Kruger TF. DNA fragmentation of spermatozoa and assisted reproduction technology. Reprod Biomed Online. 2003;7:477–84.

77. Lewis SEM, O'Connell M, Stevenson M, Thompson-Cree L, McClure N. An algorithm to predict pregnancy in assisted reproduction. Hum Reprod. 2004;19:1385–94.
78. Tesarik J, Greco E, Mendoza C. Late, but not early, paternal effect on human embryo development is related to sperm DNA fragmentation. Hum Reprod. 2004;19:611–5.
79. Croteau DL, Stierum RH, Bohr VA. Mitochondrial DNA repair pathways. Mutat Res. 1999;434:137–48.
80. Pesole G, Gissi C, De Chirico A, Saccone C. Nucleotide substitution rate of mammalian mitochondrial genomes. J Mol Evol. 1999;48:427–34.
81. Folgero T, Bertheussen K, Lindal S, Torbergsen T, Oian P. Mitochondrial disease and reduced sperm motility. Hum Reprod. 1993;8:1863–8.
82. Lestienne P, Reynier P, Chretien MF, Penisson-Besnier I, Malthiery Y, Rohmer V. Oligoasthenospermia associated with multiple mitochondrial DNA arrangements. Mol Hum Reprod. 1997;3:811–4.
83. Chance B, Williams GR. The respiratory chain and oxidative phosphorylation. Adv Enzymol Biochem. 1956;17:65–134.
84. Kasai T, Ogawa K, Mizuno K, Nagai S, Uchida Y, Ohta S, Fujie M, Suzuki K, Hirata S, Hoshi K. Relationship between sperm mitochondrial membrane potential, sperm motility, and fertility potential. Asian J Androl. 2002;4:97–103.
85. Wang X, Sharma RK, Gupta A, George V, Thomas AJ, Falcone T, Agarwal A. Alterations in mitochondria membrane potential and oxidative stress in infertile men: a prospective observational study. Fertil Steril. 2003;80 Suppl 2:844–50.
86. Quillet-Mary A, Jaffrezou JP, Mansat V, Bordier C, Naval J, Laurent G. Implication of mitochondrial hydrogen peroxide generation in ceramide-induced apoptosis. J Biol Chem. 1997;272:21388–95.
87. Heerdt BG, Houston MA, Anthony GM, Augenlicht LH. Mitochondrial membrane potential (delta psi(mt)) in the coordination of p53-independent proliferation and apoptosis pathways in human colonic carcinoma cells. Cancer Res. 1998;58:2869–75.
88. Marchetti C, Obert G, Deffosez A, Formstecher P, Marchetti P. Study of mitochondrial membrane potential, reactive oxygen species, DNA fragmentation and cell viability by flow cytometry in human sperm. Hum Reprod. 2002;17:1257–65.
89. Ruiz-Pesini E, Lapena AC, Diez C, Alvarez E, Enriquez JA, Lopez-Perez MJ. Seminal quality correlates with mitochondrial functionality. Clin Chim Acta. 2001;300:97–105.
90. Hoshi K, Kasai T, Ogawa K, Mizuno K, Nagai S, Uchida Y, Ohta S, Fujie M, Suzuki K, Hirata S. Relationship between sperm mitochondrial membrane potential, sperm motility and fertility potential. Asian J Androl. 2002;4:97–103.
91. Holyoake AJ, Sin IL, Benny PS, Sin FYT. Association of a novel human mtDNA ATPase6 mutation with immature sperm cells. Andrologia. 1999;31:339–45.
92. Holyoake AJ, McHugh P, Wu M, O'Carroll S, Benny PS, Sin IL, Sin FY. High incidence of single nucleotide substitutions in the mitochondrial genome is associated with poor semen parameters in men. Int J Androl. 2001;24:175–82.
93. Ruiz-Pesini E, Lapena AC, Diez-Sanchez C, Perez-Martos A, Montoya J, Alvarez E, Diaz M, Urries A, Montoro L, Lopez-Perez MJ. Human mtDNA haplogroups associated with high or reduced spermatozoa motility. Am J Hum Genet. 2000;67:682–96.
94. Surtano JM, Cummins JM, Greef J, Lymbery AJ. Mitochondrial DNA polymorhisms and fertility in beef cattle. Theriogenology. 2002;57:1603–10.
95. Henkel R. ROS and sperm DNA integrity—implications of male accessory gland infections. In: Giwercman A, Tournaye H, Björndahl L, Weidner W, editors. Clinical andrology. London: Informa Healthcare (incorporating Parthenon Publishing, Marcel Dekker, and Taylor & Francis Medical); 2010. p. 324–8.
96. El-Taieb MA, Herwig R, Nada EA, Greilberger J, Marberger M. Oxidative stress and epididymal sperm transport, motility and morphological defects. Eur J Obstet Gynecol Reprod Biol. 2009;144 Suppl 1:S199–203.
97. Gil-Guzman E, Ollero M, Lopez MC, Sharma RK, Alvarez JG, Thomas Jr AJ, Agarwal A. Differential production of reactive oxygen species by subsets of human spermatozoa at different stages of maturation. Hum Reprod. 2001;16:1922–30.

98. Weidner W, Colpi GM, Hargreave TB, Papp GK, Pomerol JM, Ghosh C; EAU Working Group on Male Infertility. EAU guidelines on male infertility. Eur Urol. 2002;42:313–22.
99. Schuppe HC, Meinhardt A, Allam JP, Bergmann M, Weidner W, Haidl G. Chronic orchitis: a neglected cause of male infertility? Andrologia. 2008;40:84–91.
100. Henkel R, Schill WB. Sperm separation in patients with urogenital infections. Andrologia. 1998;30 Suppl 1:91–7.
101. Sanocka-Maciejewska D, Ciupinska M, Kurpisz M. Bacterial infection and semen quality. J Reprod Immunol. 2005;67:51–6.
102. Monga M, Roberts JA. Sperm agglutination by bacteria: receptor-specific interactions. J Androl. 1994;15:151–6.
103. Eggert-Kruse W, Kiefer I, Beck C, Demirakca T, Strowitzki T. Role for tumor necrosis factor alpha (TNF-alpha) and interleukin 1-beta (IL-1beta) determination in seminal plasma during infertility investigation. Fertil Steril. 2007;87:810–23.
104. Comhaire FH, Mahmoud AM, Depuydt CE, Zalata AA, Christophe AB. Mechanisms and effects of male genital tract infection on sperm quality and fertilizing potential: the andrologist's viewpoint. Hum Reprod Update. 1999;5:393–8.
105. Alvarez JG, Sharma RK, Ollero M, Saleh RA, Lopez MC, Thomas AJ, Evenson DP, Agarwal A. Increased DNA damage in sperm from leukocytospermic semen samples as determined by the sperm chromatin structure assay. Fertil Steril. 2002;78:319–29.
106. Kocak I, Yenisey C, Dündar M, Okyay P, Serter M. Relationship between seminal plasma interleukin-6 and tumor necrosis factor alpha levels with semen parameters in fertile and infertile men. Urol Res. 2002;30:263–7.
107. Aitken RJ, Clarkson JS, Fishel S. Generation of reactive oxygen species, lipid peroxidation, and human sperm function. Biol Reprod. 1989;41:183–97.
108. Martinez R, Proverbio F, Camejo MI. Sperm lipid peroxidation and pro-inflammatory cytokines. Asian J Androl. 2007;9:102–7.
109. Krieger JN, Berger RE, Ross SO, Rothman I, Muller CH. Seminal fluid findings in men with nonbacterial prostatitis and prostatodynia. J Androl. 1996;17:310–8.
110. Menkveld R, Huwe P, Ludwig M, Weidner W. Morphological sperm alterations in different types of prostatitis. Andrologia. 2003;35:288–93.
111. Zhou JF, Xiao WQ, Zheng YC, Dong J, Zhang SM. Increased oxidative stress and oxidative damage associated with chronic bacterial prostatitis. Asian J Androl. 2006;8:317–23.
112. Kullisaar T, Türk S, Punab M, Korrovits P, Kisand K, Rehema A, Zilmer K, Zilmer M, Mändar R. Oxidative stress in leucocytospermic prostatitis patients: preliminary results. Andrologia. 2008;40:161–72.
113. Pasqualotto FF, Sharma RK, Potts JM, Nelson DR, Thomas AJ, Agarwal A. Seminal oxidative stress in patients with chronic prostatitis. Urology. 2000;55:881–5.
114. Poccia D. Remodeling of nucleoproteins during gametogenesis, fertilization and early development. Int Rev Cytol. 1986;105:1–65.
115. Haidl G, Badura B, Schill WB. Function of human epididymal spermatozoa. J Androl. 1994;15(Suppl):23S–7.
116. Henkel R, Bastiaan HS, Schuller S, Hoppe I, Starker W, Menkveld R. Leukocytes and intrinsic ROS production may be factors compromising sperm chromatin condensation status. Andrologia. 2010;42:69–75.
117. Schoor RA, Elhanbly SM, Niederberger C. The pathophysiology of varicocele-associated male infertility. Curr Urol Rep. 2001;2:432–6.
118. Skoog SJ, Roberts KP, Goldstein M, Pryor JL. The adolescent varicocele: what's new with an old problem in young patients? Pediatrics. 1997;100:112–21.
119. Goldstein M, Eid JF. Elevation of intratesticular and scrotal skin surface temperature in men with varicocele. J Urol. 1989;142:743–5.
120. Shiraishi K, Takihara H, Matsuyama H. Elevated scrotal temperature, but not varicocele grade, reflects testicular oxidative stress-mediated apoptosis. World J Urol. 2010;28:359–64.
121. Chang FW, Sun GH, Cheng YY, Chen IC, Chien HH, Wu GJ. Effects of varicocele upon the expression of apoptosis-related proteins. Andrologia. 2010;42:225–30.

122. Romeo C, Ientile R, Santoro G, Impellizzeri P, Turiaco N, Impala P, Cifala S, Cutroneo G, Trimarchi F, Gentile C. Nitric oxide production is increased in the spermatic veins of adolescents with left idiophatic varicocele. J Pediatr Surg. 2001;36:389–93.
123. Mostafa T, Anis T, Imam H, El-Nashar AR, Osman IA. Seminal reactive oxygen species-antioxidant relationship in fertile males with and without varicocele. Andrologia. 2009; 41:125–9.
124. Yoon CJ, Park HJ, Park NC. Reactive oxygen species in the internal spermatic and brachial veins of patients with varicocele-induced infertility. Korean J Urol. 2010;51:348–53.
125. Mostafa T, Anis TH, Ghazi S, El-Nashar AR, Imam H, Osman IA. Reactive oxygen species and antioxidants relationship in the internal spermatic vein blood of infertile men with varicocele. Asian J Androl. 2006;8:451–4.
126. Pasqualotto FF, Sundaram A, Sharma RK, Borges Jr E, Pasqualotto EB, Agarwal A. Semen quality and oxidative stress scores in fertile and infertile patients with varicocele. Fertil Steril. 2007;89:602–7.
127. Smith R, Kaune H, Parodi D, Madariaga M, Rios R, Morales I, Castro A. Increased sperm DNA damage in patients with varicocele: relationship with seminal oxidative stress. Hum Reprod. 2006;21:986–93.
128. Moretti E, Cosci I, Spreafico A, Serchi T, Cuppone AM, Collodel G. Semen characteristics and inflammatory mediators in infertile men with different clinical diagnoses. Int J Androl. 2009;32:637–46.
129. Cocuzza M, Athayde KS, Agarwal A, Pagani R, Sikka SC, Lucon AM, Srougi M, Hallak J. Impact of clinical varicocele and testis size on seminal reactive oxygen species levels in a fertile population: a prospective controlled study. Fertil Steril. 2008;90:1103–8.
130. Mostafa T, Anis TH, El-Nashar AR, Imam H, Othman IA. Varicocelectomy reduces reactive oxygen species levels and increases antioxidant activity of seminal plasma from infertile men with varicocele. Int J Androl. 2001;24:261–5.
131. Dada R, Shamsi MB, Venkatesh S, Gupta NP, Kumar R. Attenuation of oxidative stress & DNA damage in varicocelectomy: implications in infertility management. Indian J Med Res. 2010;132:728–30.
132. Abdel-Meguid TA, Al-Sayyad A, Tayib A, Farsi HM. Does varicocele repair improve male infertility? An evidence-based perspective from a randomized, controlled trial. Eur Urol. 2011;59:455–61.
133. Sies H. Oxidative stress: oxidants and antioxidants. Exp Physiol. 1997;82:291–5.
134. Kobayashi T, Miyazaki T, Natori M, Nozawa S. Protective role of superoxide dismutase in human sperm motility: superoxide dismutase activity and lipid peroxide in human seminal plasma and spermatozoa. Hum Reprod. 1991;6:987–91.
135. Jeulin C, Soufir JC, Weber P, Laval-Martin D, Calvayrac R. Catalase activity in human spermatozoa and seminal plasma. Gamete Res. 1989;24:185–96.
136. Drevet JR. The antioxidant glutathione peroxidase family and spermatozoa: a complex story. Mol Cell Endocrinol. 2006;250:70–9.
137. Agarwal A, Prabakaran S. Mechanism, measurement and prevention of oxidative stress in male reproductive physiology. Indian J Exp Biol. 2005;43:963–74.
138. Ochsendorf FR. Infections in the male genital tract and reactive oxygen species. Hum Reprod Update. 1999;5:399–420.
139. Zini A, Fischer MA, Mak V, Phang D, Jarvi K. Catalase-like and superoxide dismutase-like activities in human seminal plasma. Urol Res. 2002;30:321–3.
140. Ben Abdallah F, Dammak I, Attia H, Hentati B, Ammar-Keskes L. Lipid peroxidation and antioxidant enzyme activities in infertile men: correlation with semen parameter. J Clin Lab Anal. 2009;23:99–104.
141. Ramya T, Misro MM, Sinha D, Nandan D, Mithal S. Altered levels of seminal nitric oxide, nitric oxide synthase, and enzymatic antioxidants and their association with sperm function in infertile subjects. Fertil Steril. 2011;95:135–40.
142. Calamera J, Buffone M, Ollero M, Alvarez J, Doncel GF. Superoxide dismutase content and fatty acid composition in subsets of human spermatozoa from normozoospermic, asthenozoospermic, and polyzoospermic semen samples. Mol Reprod Dev. 2003;66:422–30.

143. Zini A, Garrels K, Phang D. Antioxidant activity in the semen of fertile and infertile men. Urology. 2000;55:922–6.
144. Hsieh YY, Sun YL, Chang CC, Lee YS, Tsai HD, Lin CS. Superoxide dismutase activities of spermatozoa and seminal plasma are not correlated with male infertility. J Clin Lab Anal. 2002;16:127–31.
145. Komori K, Tsujimura A, Okamoto Y, Matsuoka Y, Takao T, Miyagawa Y, Takada S, Nonomura N, Okuyama A. Relationship between substances in seminal plasma and Acrobeads Test results. Fertil Steril. 2009;91:179–84.
146. Hsieh YY, Chang CC, Lin CS. Seminal malondialdehyde concentration but not glutathione peroxidase activity is negatively correlated with seminal concentration and motility. Int J Biol Sci. 2006;2:23–9.
147. Garrido N, Meseguer M, Alvarez J, Simon C, Pellicer A, Remohi J. Relationship among standard semen parameters, glutathione peroxidase/glutathione reductase activity, and mRNA expression and reduced glutathione content in ejaculated spermatozoa from fertile and infertile men. Fertil Steril. 2004;82 Suppl 3:1059–66.
148. Diaconu M, Tangat Y, Böhm D, Kühn H, Michelmann HW, Schreiber G, Haidl G, Glander HJ, Engel W, Nayernia K. Failure of phospholipid hydroperoxide glutathione peroxidase expression in oligoasthenozoospermia and mutations in the PHGPx gene. Andrologia. 2006;38: 152–7.
149. Burton GW, Joyce A, Ingold KU. Is vitamin E the only lipid-soluble, chain-breaking antioxidant in human blood plasma and erythrocyte membranes? Arch Biochem Biophys. 1983;221: 281–90.
150. Sies H. Strategies of antioxidant defense. Eur J Biochem. 1993;215:213–9.
151. Chow CK. Vitamin E and oxidative stress. Free Radic Biol Med. 1991;11:215–32.
152. Niki E. Action of ascorbic acid as a scavenger of active and stable oxygen radicals. Am J Clin Nutr. 1991;54:1119S–24.
153. Grootveldt M, Halliwell B. Measurement of allantoin and uric acid in human body fluids. Biochem J. 1987;242:803–8.
154. Li TK. The glutathione and thiol content of mammalian spermatozoa and seminal plasma. Biol Reprod. 1975;12:641–6.
155. Song GJ, Norkus EP, Lewis V. Relationship between seminal ascorbic acid and sperm DNA integrity in infertile men. Int J Androl. 2006;29:569–75.
156. Patel S, Panda S, Nanda R, Mangaraj M, Mohapatra PC. Influence of oxidants and antioxidants on semen parameters in infertile males. J Indian Med Assoc. 2009;107:78–80, 82.
157. Lovaas E, Carlin G. Spermine: an anti-oxidant and anti-inflammatory agent. Free Radic Biol Med. 1991;11:455–61.
158. Bors W, Langebartels C, Michel C, Sandermann JH. Polyamines as radical scavengers against ozone damage. Phytochemistry. 1989;27:1589–95.
159. Allen RD, Roberts TK. Role of spermine in the cytotoxic effects of seminal plasma. Am J Reprod Immunol Microbiol. 1987;13:4–8.
160. de Lamirande E, Yoshida K, Yoshiike TM, Iwamoto T, Gagnon C. Semenogelin, the main protein of semen coagulum, inhibits human sperm capacitation by interfering with the superoxide anion generated during this process. J Androl. 2001;22:672–9.
161. de Lamirande E, Gagnon C. A positive role for the superoxide anion in triggering hyperactivation and capacitation of human spermatozoa. Int J Androl. 1993;16:21–5.
162. de Lamirande E, Lamothe G. Levels of semenogelin in human spermatozoa decrease during capacitation: involvement of reactive oxygen species and zinc. Hum Reprod. 2010;25: 1619–30.
163. Sanocka D, Kurpisz M. Reactive oxygen species and sperm cells. Reprod Biol Endocrinol. 2004;2:12.
164. Henkel R, Ichikawa T, Sanchez R, Miska W, Ohmori H, Schill W-B. Differentiation of ejaculates showing reactive oxygen species production by spermatozoa or leukocytes. Andrologia. 1997;29:295–301.

165. Laudat A, Lecourbe K, Guechot J, Palluel AM. Values of sperm thiobarbituric acid-reactive substance in fertile men. Clin Chim Acta. 2002;325:113–5.
166. Nakamura H, Kimura T, Nakajima A, Shimoya K, Takemura M, Hashimoto K, Isaka S, Azuma C, Koyama M, Murata Y. Detection of oxidative stress in seminal plasma and fractionated sperm from subfertile male patients. Eur J Obstet Gynecol Reprod Biol. 2002;105:155–60.
167. Allamaneni SS, Agarwal A, Nallella KP, Sharma RK, Thomas Jr AJ, Sikka SC. Characterization of oxidative stress status by evaluation of reactive oxygen species levels in whole semen and isolated spermatozoa. Fertil Steril. 2005;83:800–3.
168. Lewis SEM, Boyle PM, McKinney KA, Young IS, Thompson W. Total antioxidant capacity of seminal plasma is different in fertile and infertile men. Fertil Steril. 1985;64:868–70.
169. Smith R, Vantman D, Ponce J, Escobar J, Lissi E. Total antioxidant capacity of human seminal plasma. Hum Reprod. 1996;11:1655–60.
170. Cao G, Prior RL. Comparison of different analytical methods for assessing total antioxidant capacity of human serum. Clin Chem. 1998;44(6 Pt 1):1309–15.
171. Benzie IF, Strain JJ. The ferric reducing ability of plasma (FRAP) as a measure of "antioxidant power": the FRAP assay. Anal Biochem. 1996;239:70–6.
172. Glazer AN. Phycoerythrin fluorescence-based assay for reactive oxygen species. Methods Enzymol. 1990;100:161–8.
173. Whitehead TP, Thorpe GHG, Maxwell SRJ. Enhanced chemiluminescent assay for antioxidant capacity in biological fluids. Anal Chim Acta. 1992;266:265–77.
174. Said TM, Kattal N, Sharma RK, Sikka SC, Thomas Jr AJ, Mascha E, Agarwal A. Enhanced chemiluminescence assay vs colorimetric assay for measurement of the total antioxidant capacity of human seminal plasma. J Androl. 2003;24:676–80.
175. Khosrowbeygi A, Zarghami N. Levels of oxidative stress biomarkers in seminal plasma and their relationship with seminal parameters. BMC Clin Pathol. 2007;7:6.
176. Lewis SEM, Sterling ESL, Young IS, Thompson W. Comparison of individual antioxidants of sperm and seminal plasma in fertile and infertile men. Fertil Steril. 1997;67:142–7.
177. Mahfouz R, Sharma R, Sharma D, Sabanegh E, Agarwal A. Diagnostic value of the total antioxidant capacity (TAC) in human seminal plasma. Fertil Steril. 2009;91:805–11.
178. de Lamirande E, Gagnon C. Human sperm hyperactivation and capacitation as parts of an oxidative process. Free Radic Biol Med. 1993;14:157–66.
179. de Lamirande E, Gagnon C. A positive role for the superoxide anion in triggering hyperactivation and capacitation of human spermatozoa. Int J Androl. 1993;16:21–5.
180. Stauss CR, Votta TJ, Suarez SS. Sperm motility hyperactivation facilitates penetration of the hamster zona pellucida. Biol Reprod. 1995;53:1280–5.
181. Aitken RJ, Paterson M, Fisher H, Buckingham DW, van Duin M. Redox regulation of tyrosine phosphorylation in human spermatozoa and its role in the control of human sperm function. J Cell Sci. 1995;108:2017–25.
182. Aitken RJ, Harkiss D, Knox W, Paterson M, Irvine DS. A novel signal transduction cascade in capacitating human spermatozoa characterised by a redox-regulated, cAMP-mediated induction of tyrosine phosphorylation. J Cell Sci. 1998;111:645–56.
183. Griveau JF, Le Lannou D. Reactive oxygen species and human spermatozoa: physiology and pathology. Int J Androl. 1997;20:61–9.
184. Bize I, Santander G, Cabello P, Driscoll D, Sharpe C. Hydrogen peroxide is involved in hamster sperm capacitation in vitro. Biol Reprod. 1991;44:398–403.
185. Sharma RK, Pasqualotto FF, Nelson DR, Thomas Jr AJ, Agarwal A. The reactive oxygen species—total antioxidant capacity score is a new measure of oxidative stress to predict male infertility. Hum Reprod. 1999;14:2801–7.
186. Esfandiari N, Sharma RK, Saleh RA, Thomas Jr AJ, Agarwal A. Utility of the nitroblue tetrazolium reduction test for assessment of reactive oxygen species production by seminal leukocytes and spermatozoa. J Androl. 2003;24:862–70.
187. Aitken J, Fisher H. Reactive oxygen species generation and human spermatozoa: the balance of benefit and risk. Bioessays. 1994;16:259–67.
188. Halliwell B. The antioxidant paradox. Lancet. 2000;355(9210):1179–80.

Chapter 16
Oxidative Stress and Male Infertility: A Clinical Perspective

Kelton Tremellen

Abstract Studies have now conclusively linked impaired sperm function with reactive oxygen species attack on sperm—a process more commonly referred to as "oxidative stress". While research studies suggest that up to 80% of infertile men have evidence of oxidative stress, the majority of fertility specialists do not test their patients for oxidative stress, nor offer empirical antioxidant treatment. This is primarily due to a lack of availability of oxidative stress testing in the majority of clinical andrology laboratories, and a general ignorance of the importance of oxidative stress as a cause of male infertility amongst treating physicians. The first aim of this chapter is to outline the causes of oxidative stress so that they may be identified by simple history and examination. Sentinel signs of oxidative stress seen on routine semen analysis are then reviewed, together with a discussion of the benefits and pitfalls of direct testing for sperm oxidative stress. Finally, the management of sperm oxidative stress is outlined, including improvements in lifestyle, better management of underlying chronic illness, antioxidant medical therapy and surgical approaches to oxidative stress.

Keywords Oxidative stress • Clinical perspective • Identification of sperm oxidative stress • Routine semen analysis • Oral antioxidant therapy • Clinical examinations • Direct laboratory assessment

K. Tremellen, MBBS (Hons), PhD, FRANZCOG, CREI (✉)
School of Pharmacy and Medical Science, University of South Australia, Adelaide, SA, Australia

Repromed, 180 Fullarton Road, Dulwich, SA 5065, Australia
e-mail: ktremellen@repromed.com.au

A. Agarwal et al. (eds.), *Studies on Men's Health and Fertility*, Oxidative Stress in Applied Basic Research and Clinical Practice, DOI 10.1007/978-1-61779-776-7_16,

16.1 Introduction

Infertility is a condition that affects one in six couples, with impaired semen quality playing a significant role in at least half of all cases of infertility. Even though sperm dysfunction is a common underlying cause of infertility, many fertility clinics around the world have lost interest in investigating the cause of their patient's impaired sperm function, instead relying on mechanical treatments such as IVF-ICSI. While it is acknowledged that the advent of IVF-ICSI has been a huge advance in the treatment of the infertile male, an over-reliance on this type of "generic" treatment is potentially flawed. Evidence now suggests that reactive oxygen species (ROS) mediated damage to sperm ("oxidative stress") is a significant contributing pathology in 30–80% of cases of male infertility [1–6]. ROS produce infertility by two principal mechanisms [6]. First, ROS damage the sperm membrane, which in turn reduces the sperm's motility and ability to fuse with the oocyte (impaired fertilisation). Second, ROS directly damage the sperm DNA, compromising the paternal genomic contribution to the embryo. While IVF-ICSI undoubtedly overcomes any oxidative impairment of fertilisation, it has no therapeutic effect on the quality of the paternal genome. Therefore, the injection of sperm with oxidatively damaged DNA may result in impaired blastocyst development, an increased risk of miscarriage and the birth of a child with a sub-optimal paternal genetic compliment, potentially leading to disease in later life [7]. For all of these reasons, it is imperative that clinicians treat the underlying causes of sperm dysfunction, not just become practitioners of IVF.

The preceding chapters have conclusively established the science behind oxidative stress being a significant cause of male sub-fertility. Therefore, the outstanding issues facing the clinician managing patients with male factor infertility are three-fold. First, the treating physician must be able to identify the presence of oxidative stress in his male patients. Second, the clinician must identify the underlying cause(s) of sperm oxidative stress and where possible, attempt to reverse or ameliorate these pathological processes responsible for damaging sperm. Finally, the clinician may initiate appropriate oral antioxidant therapy to "neutralise" any residual excess ROS left over after treating the underlying oxidative pathology. This chapter discusses each of these basic clinical issues in turn, critically analysing the available data along the way.

16.2 Identification of Sperm Oxidative Stress

Evidence for oxidative damage to sperm is present in 30–80% of infertile men's sperm, depending upon the techniques used to measure oxidative stress and the patient population studied [6]. Since oxidative stress is such a common pathology, it would appear logical that all infertile men should be screened for its presence. Surprisingly, however, this is not current standard clinical practice. The reasons for this are multiple but include a general ignorance amongst treating physicians of the importance of oxidative stress as a cause of male infertility, a perceived lack of cost-effective assays for sperm oxidative stress and the absence of a universally accepted treatment regime. Chapters 13 and 14 have already outlined the various assays for sperm oxidative stress. The Luminol assay is probably the most commonly used direct measure of oxidative stress, but more than 30 other assays currently exist [6]. The Luminol assay is the only assay of sperm oxidative stress described in the current WHO semen analysis manual, but unfortunately has the disadvantage of requiring expensive specialised equipment (Luminometer) not commonly present in clinical andrology laboratories and can be quite technically demanding from a quality control perspective [8]. In my own fertility clinic, we measure sperm oxidative stress with the Nitro Blue Tetrazolium (NBT) assay, since this is easy to perform with limited specialised equipment and is relatively inexpensive, making it an ideal assay for a clinical andrology laboratory [9]. It will be left up to the discretion of individual clinician to decide on what oxidative stress assay they should adopt based on the relative merits of each assay described in Chaps. 13 and 14. However, for the purposes of this discussion it will be assumed that none of these direct or indirect assays of sperm oxidative stress are available, as is the most common current clinical scenario, with the diagnosis of presumptive oxidative stress being made solely based on clinical history, physical examination and interpretation of the routine WHO semen analysis.

16.3 Identification of Sperm Oxidative Stress from Clinical History

The various causes of male infertility are outlined in Fig. 16.1, with the majority of these sources of sperm dysfunction being relatively easy to identify through history alone. The mnemonic "Testicular" is a helpful aide-memoire for recalling the underlying causes of male infertility, with sperm oxidative stress playing a pivotal role in the majority of these male infertility pathologies.

Exposure to *toxins* such as heavy metals through the manufacture of lead acid batteries or soldering fumes is known to create sperm oxidative stress [10, 11] and lead to sperm DNA damage, infertility and miscarriage [12, 13]. Several environmental pollutants have been linked with testicular oxidative stress. Pesticides such as lindane [14], methoxychlor [15] and the herbicide dioxin-TCDD [16] have all been linked with testicular oxidative stress in rodent models. The commonly used preservative sulphur dioxide has also been shown to produce testicular oxidative stress in laboratory animals [17]. Air pollutants such as diesel particulate matter act as potent stimuli for leukocyte ROS generation [18, 19]. While no study has directly linked airborne pollutants with testicular oxidative stress, it is possible that this oxidative insult is responsible for the increase in sperm DNA damage seen following periods of airborne pollution [20]. Phthalates are chemicals used as a plastics softener and are contained in a wide range of food packaging and personal care products. Exposure to phthalates can occur via dietary consumption, dermal absorption or inhalation and

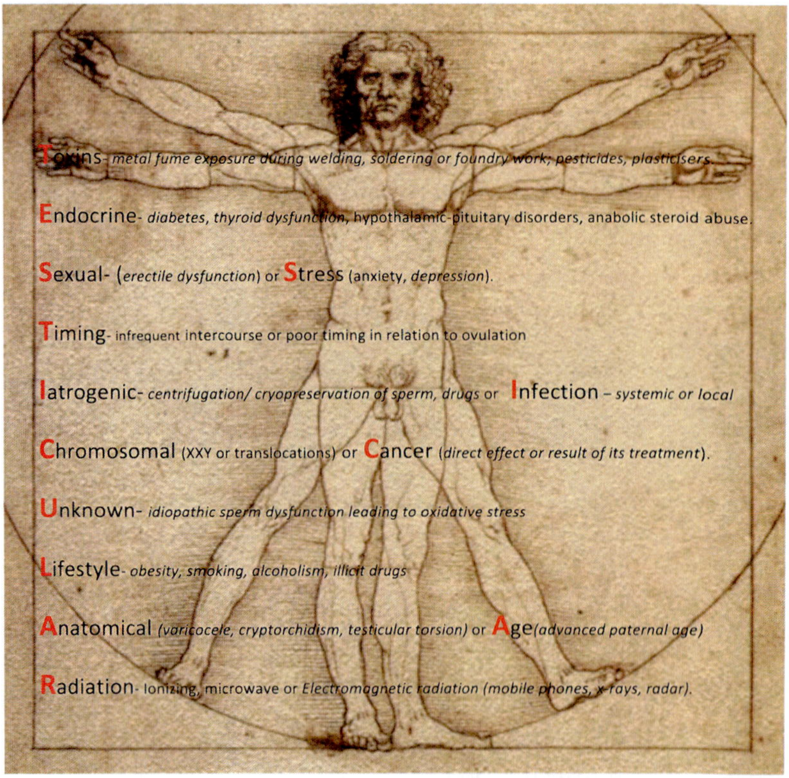

Fig. 16.1 The "Testicular" mnemonic summarising the common causes for male factor infertility. Those aetiologies in which oxidative stress plays a significant role in causing sperm dysfunction are highlighted in *italics*

has been linked with impaired spermatogenesis and increased sperm DNA damage [21–24]. Oral administration of phthalate esters to rats is reported to increase the generation of ROS within the testis and a concomitant decrease in antioxidant levels, culminating in impaired spermatogenesis [25].

Diabetes is an ever increasing *endocrine* cause of sperm oxidative stress as obesity levels increase in the Western population. Poor glycaemic control has been linked with excessive systemic production of ROS and an increase in sperm DNA fragmentation levels [26]. Further studies by this group confirmed that type I diabetes is associated with an increase in the oxidative DNA adduct 8-OHdG in sperm, confirming that diabetes induced oxidative stress is at least partially responsible for the decline in sperm DNA quality seen in diabetic men [27]. Furthermore, animal studies using the streptozotocin-induced diabetic rat model have found a significant increase in testicular oxidative stress within 6 weeks of initiation of the diabetic state [28], while the antioxidant quercetin is capable of significant amelioration of the negative effects of diabetes on sperm quality [29]. This all suggests that oxidative stress is a major mediator of the decline in sperm function seen in diabetic men.

Thyroid dysfunction may also be an *endocrine* cause of sperm oxidative stress as both hyperthyroidism and hypothyroidism have been linked with systemic oxidative stress [30–32]. This systemic state of oxidative stress appears to extend to the testicle, since animal studies using experimentally induced hyperthyroidism and hypothyroidism have shown increased levels of lipid peroxidation products and reduced antioxidant status within the testicle tissue [33, 34]. Furthermore, a recent study examining the link between thyroid hormones and antioxidant capacity in seminal plasma of infertile men reported a significant positive correlation between seminal plasma antioxidant capacity and serum free thyroxin levels [35]. Therefore, screening of infertile men for thyroid dysfunction may be useful, especially since returning thyroid hormone status to normal has been shown to reduce systemic levels of oxidative stress within 1–2 months [30].

Oxidative stress is believed to be a major cause of erectile dysfunction, a common *sexual* cause of male infertility in the older man. Of some concern is the observation that as many as one in ten men aged in their 40s will also experience significant erectile dysfunction [36]. Penile erection is dependent upon vascular smooth muscle relaxation in erectile tissue and penile arteries, the principal mediator of relaxation being nitric oxide (NO). Evidence from basic scientific studies indicates that oxidative stress may be central to impaired cavernosal function in erectile dysfunction (ED) [37, 38]. Increased inactivation of NO by the superoxide ROS results in impaired penile NO transmission and smooth muscle relaxation. Furthermore, propagation of endothelial dysfunction by ROS may result in chronic impairment of penile vascular function, a process analogous to early atherogenesis. Furthermore, both animal models and human studies have shown that supplementation of the diet with antioxidants may significantly improve intracavernosal blood flow and erectile activity, supporting a pivotal role for oxidative stress in erectile dysfunction [39–41].

Psychological *stress* is known to produce a decline in sperm function and a state of systemic oxidative stress. Two prospective studies have linked a period of psychological stress with a reduction in sperm quality mediated by an increase in seminal plasma ROS generation and a reduction in antioxidant protection [42, 43]. Interestingly, after the removal of the psychological stressor (medical student exams in these observational studies), sperm function improved, suggesting that effective management of emotional stress may improve sperm function.

The processing of sperm by centrifugation for later use in assisted reproductive treatment (IVF, IUI) is an example of *iatrogenic* generation of sperm oxidative stress. Centrifugation is known to enhance the production of free radicals by sperm [1, 4], while also bringing sperm in close proximity to ROS producing leukocytes in the cellular pellet. This, together with the fact that centrifugation removes sperm from their protective bath of antioxidants contained in seminal plasma, is a potent inducer of sperm oxidative damage [44]. In addition, cryopreservation of sperm, another commonly used technique in ART, is associated with an increase in sperm oxidative stress [45, 46]. Furthermore, medications prescribed to patients are another potential cause of iatrogenic oxidative stress. Drugs such as aspirin and paracetamol (acetaminophen) can produce oxidative stress by increasing cytochrome P450 activity, thereby boosting ROS generation [47]. The SSRI class of anti-depressants has been linked with reduced libido and impaired ejaculatory frequency [48]. This results in sperm being stored for prolonged periods in the epididymis where they are potentially exposed to oxidative attack, resulting in a decline in sperm DNA integrity [49].

Infective causes for sperm oxidative stress include local infections such as Male Accessory Gland Infection (MAGI) or systemic infections such as Hepatitis, HIV, TB and Malaria. Leukocytes are professional producers of free radicals, releasing ROS at relatively high concentrations to destroy infective pathogens. Therefore, it is not surprising that activation of the immune system within the male reproductive tract is likely to result in sperm oxidative damage. Up to 50% of men will experience prostatitis at some point in their lives, with prostatitis becoming chronic in 10% of men [50]. Bacteria responsible for prostate infection may originate from the urinary tract or can be sexually transmitted [51]. Typical non-STD pathogens include streptococci (*Streptococcus viridans* and *S. pyogens*), coagulase-negative staphylococci (*Staphylococcus epidermidis, S. haemolyticus*), gram-negative bacteria (*Escherichia coli, Proteus mirabilis*) and atypical mycoplasma strains (*Ureaplasma urealyticum, Mycoplasma hominis*). All of these pathogens will create an acute inflammatory response with an influx of leukocytes into the genital tract and a resulting increase in ROS production [52–55]. Men prone to recurrent genitourinary tract infections, such as paraplegics, have been confirmed to have high degrees of sperm oxidative pathology [56, 57]. Current or past *Chlamydia* infection has also been linked with an increase in oxidative damage to sperm [58]. Viral infections such as Herpes may initiate oxidative damage to sperm. Herpes simplex DNA is found in 4–50% of infertile men's semen [59, 60], with acute HSV infection being associated with a tenfold increase in the rate of leukospermia [61, 62]. Given the well-recognised link between leukospermia and seminal ROS levels, together with the observation of a reduction in sperm motility in men positive for seminal HSV DNA [59], it is likely that HSV is a viral pathogen involved in oxidative stress.

Several chronic systemic infections have been linked with increased oxidative stress throughout the body. Human immunodeficiency virus (HIV) infection is associated with an increase in leukocyte number and activation within semen [63]. Hepatitis B and C infection has been also correlated with significant hepatic oxidative stress [64, 65]. At present, it is unknown if this oxidative stress extends to the semen, but impaired sperm motility seen in Hepatitis B and C patients [66, 67] makes this likely. Finally, chronic infections such as Tuberculosis [68], Leprosy [69], Malaria [70] and Chagas Disease [71] have been all linked with elevated degrees of systemic oxidative stress. While no study has directly linked these chronic infectious diseases with sperm oxidative stress, it is unlikely that the male reproductive tract would be spared from this systemic oxidative insult.

Similarly, non-infective *inflammatory* stimuli such as occurs following a vasectomy reversal will produce sperm oxidative stress. Following a vasectomy, sperm are able to cross the blood–testis immune privilege barrier and activate an inflammatory response [72]. Even after successful reanastamosis of the vas, there is ongoing immune recognition of sperm resulting in a sterile inflammatory response which can harm sperm function [73, 74]. Chronic non-bacterial prostatitis (NIH Category III), a chronic inflammation of the prostate in the absence infection, has been reported by several groups to be associated with elevated levels of seminal oxidative stress [75–77]. Chronic non-bacterial prostatitis affects 10% of men during their lifetime [50] and is believed to be caused by an adverse autoimmune response to seminal or prostate antigens, leading to an increase in pro-inflammatory cytokines and activated ROS producing leukocytes within the semen [78, 79].

Chronic inflammation and oxidative stress are highly prevalent in patients with chronic kidney disease and end-stage renal disease [80]. Surprisingly, even when uraemia is reversed by haemodialysis, a persisting state of chronic inflammation and oxidative stress persists [81, 82]. Furthermore, renal transplant patients with stable renal function and no obvious signs of immune rejection of their graft also have elevated levels of oxidative stress [83]. This systemic state of oxidative stress may at least account for some of the observed impaired sperm function seen in renal patients. Furthermore, patients with haemoglobinopathies such as beta-thalassaemia major have high degrees of systemic oxidative stress [84], with this oxidative damage confirmed to involve sperm [85]. The likely cause of oxidative stress is iron overload from multiple blood transfusions. Iron is a potent pro-oxidant capable of redox cycling when not safely bound to transferrin in the blood or stored as ferritin in tissue. Likewise, men with haemochromatosis are likely to have oxidative damage to their sperm caused by excess iron stores.

Cancer and its treatment by chemotherapy and radiation are known to be a potent systemic oxidative insult, potentially impairing sperm function. Drugs such as the chemotherapy agent cyclophosphamide have been linked with sperm oxidative stress. Administration of cyclophosphamide to animals is reported to increase testicular malondialdehyde (MDA) levels and produce a fall in testicular catalase, implying the presence of oxidative stress [86, 87]. Similarly, radiation exposure has been shown to cause a systemic inflammatory reaction and increase in oxidative stress in both the irradiated tissue and non-irradiated bystander normal tissue [88, 89]. DNA fragments of apoptotic irradiated cancer cells are released into the intercellular space and interact with the DNA-binding receptors of the bystander non-irradiated cells, initiating activation of lymphocyte signalling pathways associated with synthesis of reactive oxygen and nitrogen species, thereby inducing secondary oxidative stress [90]. Finally, the patient with untreated cancer often exhibits oxidative stress as a part of the normal malignant metabolic response [91–94]. This may account for the usual impairment in semen quality often seen in these patients when they attend fertility clinics for semen cryopreservation prior to initiation of potentially sterilising cancer therapy [95].

In the vast majority of infertile men, no clear cause for sperm dysfunction is found on routine history and examination—making idiopathic infertility a common finding. However, recent research suggests that infertility of *unknown* origin often has an oxidative stress basis. The ability of sperm to produce ROS inversely correlates with their maturational state. During spermatogenesis, there is a loss of cytoplasm to allow the sperm to form its condensed, elongated form. Immature teratozoospermic sperm are often characterised by the presence of excess cytoplasmic residues in the mid-piece. These residues are rich in the enzyme glucose-6-phosphate dehydrogenase (G6PD), an enzyme which controls the rate of glucose flux and intracellular production of beta-nicotinamide adenine dinucleotide phosphate (NADPH) through the hexose monophosphate shunt. NADPH is used to fuel the generation of ROS via NADPH oxidase located within the sperm membrane [96]. As a result, teratozoospermic sperm produce increased amounts of ROS compared to morphologically normal sperm. Interestingly, however, even normozoospermic men in an infertile relationship produce higher amounts of ROS than fertile men [97]. Why this occurs is not fully understood but may include subtle metabolic defects in the sperm leading to excessive production of ROS from the mitochondrial respiratory chain [98].

Lifestyle causes of oxidative stress such as obesity, smoking and alcohol abuse are likely to be a common cause of oxidative male infertility. Obesity has recently been linked with excessive production of ROS in semen, possibly mediated by the systemic inflammatory response observed in the obese state [99–101]. Furthermore, accumulation of adipose tissue within the groin region results in heating of the testicle which has been linked with oxidative stress and reduced sperm quality [102–104]. Animal studies suggest that high fat diets may also cause sperm themselves to produce excess ROS through an as yet unidentified process [105]. Finally, a "fast food" diet low in vegetables and fresh fruits, all too common in today's overweight Western society, is likely to be deficient in dietary antioxidants and therefore place these men at increased risk of sperm oxidative stress. Dietary deficiencies have been linked with sperm oxidative damage by several research groups. The AGES study examined the self-reported dietary intake of various antioxidants and nutrients (vitamins C and E, beta-carotene, folate and zinc) in a group of healthy non-smokers and observed a significant correlation between vitamin C intake and sperm concentration and between vitamin E intake and total progressively motile sperm [106]. These observations are also consistent with earlier reports of a significant link between seminal plasma vitamin E levels and an increase in percentage of motile sperm [107] and low seminal plasma vitamin C levels with increased levels of sperm DNA damage [108].

Exposure to cigarette smoke is a well-established cause of both testicular and systemic oxidative stress and a clear cause of male sub-fertility. Smoking results in a 48% increase in seminal leukocyte concentrations and a 107% increase in seminal ROS levels [109]. Smokers have decreased levels of seminal plasma antioxidants such as vitamin E [110] and vitamin C [111], placing their sperm at additional risk of oxidative damage. This has been confirmed by the finding of a significant increase in levels of 8-OHdG within smoker's seminal plasma [110].

Excessive alcohol consumption causes an increased in systemic oxidative stress as ethanol stimulates the body's production of ROS, while many alcohol abusers have diets deficient in protective antioxidants [112]. A study of 46 alcoholic men of reproductive age has confirmed the presence of ethanol induced oxidative stress within the testicle by reporting a significant increase in serum lipid peroxidation by-products plus a drop in antioxidants [113].

The use of illicit drugs such as cocaine and MDMA (ecstasy) may produce sperm oxidative stress, as both have been linked with a systemic state of oxidative stress

[114, 115], although only animal studies to date have confirmed the ability of these illicit drugs to impair sperm function [116, 117].

The presence of a varicocele, present in up to 40% of infertile men, is a common *anatomical* cause for sperm oxidative stress. Oxidative stress is now widely believed to be the principal underlying pathology linking varicocele with male infertility [118–125]. While it is not currently understood exactly how the presence of a varicocele produces sperm oxidative stress, suggested mechanism includes an elevation in scrotal temperature [126], genital tract inflammation [127] and acidification of seminal plasma inhibiting antioxidant enzyme activity [128]. The increase in varicocele-related ROS production is strongly correlated with a reduction in sperm DNA integrity when assessed by either TUNEL [122] or 8-hydroxy-2′-deoxyguanosine DNA oxidative metabolite levels [129].

Cryptorchidism is another common anatomical cause for male factor infertility in which the primary pathology is hypo-spermatogenesis due to deficient maturation of gonocytes to type A spermatogonia [130]. However, it has been reported that even in men with cryptorchidism surgically treated with orchidopexy early in life, there is a markedly elevated production of ROS by sperm and an associated increase in sperm DNA fragmentation compared to fertile controls [124]. Finally, testicular oxidative stress initiated from torsion of the spermatic cord is believed to be a cause of male infertility. It is reported that oxidative stress is precipitated by an inflammatory response to the ischemia–reperfusion injury in both the torted and contralateral testis. A prolonged period of ischemia, followed by surgical or spontaneous restoration of blood flow, leads to an influx of activated leukocytes into both testes [131] and a consequent increase in generation of free radicals [132]. Oxidative stress then leads to necrosis of the germinal cells with resulting sub-fertility or infertility.

Advanced paternal *age* is another potential cause of sperm oxidative stress. It has been known for a long time that sperm quality slowly declines with advancing age [133]. Several large studies have shown that systemic oxidative stress increases with advancing age [134], but only recently has it been reported that this age-related oxidative insult definitely extends to the sperm themselves [135]. Animal studies using the Brown Norway rat, an established model of male reproductive ageing, confirm that older animal's sperm produce more free radicals than young animals and have a reduced enzymatic antioxidant activity, resulting in an increase in ROS mediated sperm DNA damage [136, 137]. Several studies have reported that sperm DNA damage increases with advancing age in both fertile [138] and infertile men [139, 140]. It is possible that an increase in oxidative sperm DNA damage is one of the underlying pathologies behind this age-related decline in the paternal genome.

Finally, electromagnetic *radiation* in the form of mobile phone exposure has recently been shown to produce sperm oxidative stress [141, 142]. As the vast majority of men in industrialised societies now use a mobile phone, this cause of sperm oxidative stress may become an endemic potential cause of male sub-fertility. Pulsed microwave radiation exposure, as occurs in workers who are in close proximity to radar equipment, has also been linked with a systemic state of oxidative stress [143] and a significant decline in sperm quality [144]. Low dose exposure to ionising radiation is commonly seen in medical workers in the radiology sector, nuclear facility employees, and some miners and aircraft personnel. Such low dose ionising radiation exposure has also been linked with a systemic state of oxidative stress [145] and a potential decline in semen quality [146].

16.4 Identification of Sperm Oxidative Stress from Clinical Examination

Physical examination of the infertile male should focus on the salient features of disease processes that are linked with sperm oxidative stress. Measurement of height, weight and abdominal circumference is a useful measure of the degree of obesity and oxidative stress associated with the so called "metabolic syndrome". The diagnosis of metabolic syndrome is made in the presence of at least three of the following clinical features—waist circumference > 102 cm (Caucasian men), BP > 130/85, abnormal glucose profile and adverse lipid profile (low HDL or high triglycerides) [147]. Excess adipose tissue releases pro-inflammatory chemicals which activate the immune system and result in a systemic increase in production of ROS, with a resulting damage to sperm function. The presence of a BMI in excess of 30 kg/m^2 has been linked with sperm oxidative stress [100] and a significant decline in sperm function and increase in sperm DNA damage by several investigators [148, 149].

In addition, "dipstick" testing of the urine for protein and glucose is an integral part of the routine physical examination. Since diabetes has been linked with sperm oxidative stress, the finding of a BMI > 30, an abdominal circumference > 102 cm or the presence of glucose in the urine should prompt screening for diabetes with at least the measurement of a fasting Blood Sugar level (BSL) (ideal fasting BSL < 6 mmol/L) or ideally a glucose tolerance test.

Physical inspection of the patient may identify likely causes of sperm oxidative stress. Patients are often less than truthful about their smoking and alcohol intake. Yellow stained fingertips or the odour of tobacco may indicate a significant smoking habit. Palmar erythemia, spider naevi, gynaecomastica and tender hepatomegaly may suggest alcohol-related liver damage. The presence of tattoos or needle tract scars in the cubital fossa places the patient at increased risk of Hepatitis B/C and HIV, all linked with sperm oxidative stress.

Palpation of the scrotal contents can provide useful clues to the cause of sperm oxidative damage. Varicoceles, most commonly present on the left side, are a well-established cause of oxidative stress. A tender or hardened epididymis on palpation may suggest the presence of acute or past epididymal infection respectively. Rectal examination with palpation of an enlarged boggy or tender prostate may signify MAGI. This type of intrusive examination is best reserved for men with likely MAGI on history and examination of the semen. Trans-rectal ultrasound of the accessory sex glands, in combination with microbiological examination of the semen will help confirm the presence of MAGI and tailor effective treatment. Testicular cancer is present in 0.3% of infertile men [150], and therefore, the presence of any irregularity in testicular surface texture on palpation mandates the ordering of a scrotal ultrasound to exclude the presence of testicular cancer.

16.5 Identification of Sperm Oxidative Stress from Routine Semen Analysis

Semen analysis is a cornerstone of assessment of the infertile male. While direct measurement of sperm oxidative stress is not presently part of the routine WHO semen analysis [151], the clinician can often make a presumptive diagnosis of oxidative stress from a routine semen analysis (Table 16.1). As described above, MAGI may cause sperm oxidative stress. A diagnosis of past or present MAGI can be made when there is a alteration in normal semen viscosity (increase or decrease), discolouration of the semen or an increase in semen pH above the normal range (pH > 8). Hyperviscosity of seminal plasma is associated with increased levels of seminal plasma MDA [152] and reduced seminal plasma antioxidant status [153], making impaired viscosity a reasonable surrogate marker of oxidative stress. Infection of the semen with *U. urealyticum* is associated with increased seminal plasma viscosity [154] and an increase in ROS production [55]. It is possible that these infections may damage the prostate and seminal vesicle, altering the substrates required for creation of normal semen viscosity.

Furthermore, the presence of large numbers of round cells may suggest oxidative stress caused by leukocytospermia [73]. However, round cells may also be immature sperm rather than leukocytes, so formal identification of leukocytes requires ancillary tests such as the peroxidase test, CD45 staining or measurement of seminal elastase [151, 155, 156]. Poor sperm motility, especially in the absence of any features of antisperm antibody generation (sperm "clumping"), is highly suggestive of the presence of sperm oxidative damage. Sperm with abnormal morphology, especially those with excess residual cytoplasm and cytoplasmic "droplets" are cardinal features of pathological sperm producing high levels of ROS. In addition, the more severe the semen defect is, the higher the chance that sperm oxidative stress will be present. Finally, poor sperm membrane integrity assessed by the hypo-osmolar swelling test (HOST) has been linked with the presence of sperm oxidative stress [157].

Advanced ancillary semen assays may also indirectly suggest the presence of oxidative stress. High levels of the neutrophil activation marker elastase in seminal plasma have been linked with sperm oxidative stress [155, 156]. More recently, the

Table 16.1 Cardinal signs of sperm oxidative stress seen in a routine semen analysis

- Altered semen viscosity
- Discolouration of semen
- Raised semen pH
- High number of round cells in semen
- Leukocytospermia on the Endtz or immuno-cytochemical analysis
- Poor sperm motility, especially in the absence of anti-sperm antibodies
- High number of morphologically abnormal sperm (especially cytoplasmic droplets)
- Severe OAT "triple semen defect"
- Low hypo-osmolar swelling test (HOST) results

presence of the macrophage activity marker neopterin in seminal plasma has been linked with sperm oxidative stress [158]. Sperm DNA integrity tests such as TUNEL or SCSA may suggest the possibility of sperm oxidative stress if high levels of DNA fragmentation are recorded [6, 47]. However, as sperm DNA fragmentation can also be caused by non-oxidative processes (abortive apoptosis, DNA nicks made during DNA remodelling), neither is conclusive evidence for the presence of sperm oxidative damage.

16.6 Direct Laboratory Assessment of Sperm Oxidative Stress

A full description of the assays used to directly measure sperm oxidative stress is outlined in Chaps. 13 and 14. At present, there are over 30 such assays, but only two are in common usage [6]. One of the oldest and most widely used methods of assessing sperm membrane peroxidation is the measurement of MDA levels in sperm or seminal plasma with the thiobarbituric acid assay (TBARS). MDA levels in sperm are quite low and therefore require the use of sensitive HPLC equipment or the use of iron-based promoters and spectrofluorometry measurement. Seminal plasma levels of MDA are five to tenfold higher than sperm, making measurement on standard spectrophotometers possible [6]. Measurement of MDA appears to be of some clinical relevance, since its concentration within both seminal plasma and sperm is elevated in infertile men with excess ROS production, compared to fertile controls or normozoospermic individuals [6, 157]. Furthermore, in vitro impairment of motility, sperm DNA integrity and sperm–oocyte fusion capacity by ROS is accompanied by an increase in MDA concentration [6].

Chemiluminescence assays using luminol are probably the most commonly described technique to detect ROS production within semen. This probe is very sensitive and has the advantage of relatively well-established reported ranges for both the fertile and infertile population [3–5, 8, 118, 120]. However, its general uptake by clinical andrology laboratories has been hampered by expensive equipment (luminometer) and difficulties with quality control created by assay confounders such as incubation time, leukocyte contamination and presence of seminal plasma contamination [8]. Measurement of total antioxidant capacity (TAC) within seminal plasma can also be measured using luminol, with the ability of seminal plasma to inhibit chemiluminescence elicited by a constant source of ROS (horseradish peroxidase) being the most commonly used technique. The TAC is usually quantified against a vitamin E analogue (Trolox) and expressed as a ROS-TAC score [109]. At the present moment the ROS-TAC score appears to be the best-established method for analysing the balance between free radical production (ROS) and antioxidant protection (TAC) of sperm. However, since all assays of sperm oxidative stress are relatively expensive compared to routine semen analysis, many treating physicians will continue to avoid directly testing their patients for oxidative stress and instead offer empirical therapy antioxidant treatment which is relatively inexpensive and without side effects.

16.7 Treatment of the Underlying Cause of Sperm Oxidative Stress

> No man is an island, entire of itself; every man is a piece of the continent, a part of the main.
>
> John Donne (1572–1631)

The seventeenth century English poet and clergyman John Donne wrote this quote when reflecting upon the basic human need for interaction with others, not to act alone in isolation. Likewise, this quote is a useful poetic reminder to the clinician to not see sperm as an isolated testicular cell, but rather as part of a rich interlocking network of body systems, all capable of modifying sperm function. Therefore, when identifying and treating the underlying cause of sperm oxidative stress the physician must look beyond the sperm or testicle, and treat the patient in a holistic fashion.

Lifestyle and environmental factors should first be addressed when treating oxidative stress-related male infertility. Patients in the overweight category should be encouraged to lose weight, even though this has not yet been shown to improve sperm function by any properly conducted trial. A recent randomised control trial has shown that weight reduction created by a combination of exercise and calorie restriction does result in a significant reduction in levels of systemic oxidative stress [159], and there is no reason to believe that this improvement would not also be extended to the testicular micro-environment. Setting realistic goals for weight loss (1 kg/week for the first few weeks and then 0.5 kg/week thereafter) through a combination of dieting and exercise is important. Vigorous exercise regimes have been linked with the generation of oxidative stress [160], so it is probably best to limit exercise to a moderate amount three to four times per week to minimise any exercise pro-oxidant effect. Furthermore, exercise which significantly elevates scrotal temperature (e.g. prolonged bicycle training) is best avoided as this may also induce oxidative stress [161]. In the massively obese patient, pharmacological interventions (orlistat, sibutramine) and lap-band surgery probably have the highest chance of success.

The cessation or a reduction in smoking can best be achieved by a combination of pharmacological support (nicotine replacement therapy, bupropion) and supportive counselling. Similarly, a reduction in alcohol intake may be achieved by supportive psychotherapy.

The removal or reduction of environmental toxin exposure may require significant changes to the patient's work environment, including changing jobs where occupational exposure is unavoidable. However, simple measures such as adequate ventilation and the use of protective clothing may effectively reduce toxin exposure and improve sperm quality. Exposure to electromagnetic radiation from mobile phones can be minimised by preferably using landline phones when possible, avoiding long calls on the mobile phone and storing the mobile phone well away from the testicular area when not in use. Finally, a diet rich in antioxidant minerals (zinc, selenium) and antioxidant compounds such as vitamin C, vitamin E, carotenoids and flavonoids, all commonly present in fruit and vegetables, are likely to help augment sperm function.

While as yet untested by clinical trials, the effective treatment of systemic diseases linked with oxidative stress (diabetes, Hepatitis B/C, HIV, malaria, haemochromatosis, haemoglobinopathies, inflammatory bowel disease, psoriasis, rheumatoid arthritis, depression) is likely to reduce overall oxidative stress in the body and benefit sperm function. It is therefore ideal that patients delay conception until after these systemic diseases are under effective control, unless the medications used to achieve control have a detrimental effect on sperm function (e.g. methotrexate treatment of inflammatory conditions).

Several investigators have reported that surgical treatment of a varicocele can reduce seminal ROS levels and improve sperm DNA integrity [162–167]. While some of these studies suggest that this type of surgical treatment may result in increased rates of natural conception, this has not yet been conclusively proven by randomised controlled trials. However, as surgical ligation of a varicocele is a relatively simple low-risk procedure, it makes reasonable clinical sense to offer this therapy if lifestyle modification and simple oral antioxidants have not been successful in achieving conception.

Antibiotic therapy for men with MAGI may reduce the inflammatory stimulus causing neutrophils and macrophages to release ROS in close proximity of sperm. Two studies have now confirmed the ability of antibiotic treatment to reduce sperm oxidative stress and subsequently improve sperm quality [168, 169]. One relatively large and well-conducted study randomised men with *Chlamydia* or *Ureaplasma* infection to either 3 months of antibiotics or no treatment [169]. Compared to the controls, the antibiotic treated group exhibited a significant fall in seminal leukocytes and ROS production at 3 months, an improvement in sperm motility and a significant improvement in natural conception. A smaller study using only 10 days of antibiotic treatment did not produce any significant decline in seminal leukocyte count or improvement in motility [62]. While this study did not measure ROS production in semen, it is likely that prolonged courses of antibiotics (3 months) are required to completely irradiate difficult to treat MAGIs and reverse oxidative pathology. In addition to antibiotic treatment, non-steroidal anti-inflammatory (NSAID) drugs may also reduce seminal leukocytes production of free radicals. In one study men with antibiotic treated *Chlamydia* or *Ureaplasma* infection were randomised to either a NSAID or carnitine antioxidant and monitored for improvements in sperm quality over the next 4 months [170]. Those men treated with 2 months of NSAID followed by 2 months of carnitine had the most significant reduction in seminal ROS production and improvement in sperm motility/viability. A 1 month course of a COX-2 anti-inflammatory has been shown to significantly reduce sperm leukocyte count, while improving sperm motility, morphology and viability [171]. It would therefore appear that a combination of antibiotics followed by a course of anti-inflammatory medication is the preferred treatment path in infection-related oxidative stress.

If initial semen analysis suggests the presence of MAGI, formal semen culture and PCR analysis for *Chlamydia* and gonorrhoea should be performed to allow for tailored anti-microbial therapy. Of course, the presence of a confirmed sexually transmitted disease should trigger the screening and appropriate treatment of all sexual partners to prevent reinfection of the male patient.

Testicular sperm extraction is a new and controversial treatment for sperm oxidative stress damage. It is generally recognised that the primary site at which sperm oxidative attack occurs is while sperm are being stored in the epididymis. Here, the sperm sit for many days before ejaculation, unprotected from oxidative attack by supportive Sertoli cells or seminal plasma derived antioxidants. It has been proposed that by surgically collecting sperm direct from the testicle, it is possible to still obtain sperm that have yet to be damaged by ROS attack as they are "fresh off the sperm production line". Three studies have now shown that such a surgical approach can result in improved sperm DNA integrity and pregnancy outcomes [172–174]. However, surgical sperm aspiration can only be performed in conjunction with IVF-ICSI and does of course have some potential adverse effects (haemorrhage, infection, pain). Therefore, it is our practice to only offer this type of treatment when the use of ejaculate sperm has resulted in uniformly poor quality embryos despite our best conservative efforts to ameliorate sperm oxidative stress.

As previously outlined, sperm preparation using density gradient centrifugation (DGC) can exacerbate underlying sperm oxidative stress. Because of this we have moved away from routine use of DGC as a means of preparing sperm for IVF. Instead, we use a low g force centrifugation wash to remove seminal plasma, followed by a simple sperm "swim up" to isolate motile good quality sperm, as this creates less of an oxidative insult to sperm. Furthermore, sperm should be processed and stored in media containing antioxidant compounds [175] before being added to the oocytes for in vitro fertilisation. Finally, sperm should not be stored for long periods at atmospheric concentrations of oxygen as this has been shown to exacerbate oxidative stress-related DNA damage [176, 177]. A 5% oxygen environment is more than adequate to meet the sperm's metabolic needs, while minimising levels of oxidative stress.

16.8 Oral Antioxidant Therapy

As outlined in Chap. 23, many studies have been conducted examining the effect of a wide variety of different antioxidant combinations on sperm function and pregnancy outcome. Table 16.2 summarises the outcomes of the various antioxidant treatment regimes tested in a placebo controlled setting. In January 2011, the Cochrane collaboration, widely perceived as the definitive authority on evidence-based medicine, published a review on the effect of oral antioxidant therapy on sperm quality and pregnancy outcomes [178]. This review identified 34 randomised placebo controlled studies examining the effect of antioxidants on sperm quality and pregnancy outcomes in couples seeking fertility assistance. While 15 trials examined the effect of antioxidants on pregnancy rates, unfortunately only 3 studies examined live birth rate, the most meaningful clinical outcome. Further compounding the effective performance of a meta-analysis was the fact that dozens of different antioxidant combinations for various periods of time were used in these studies. Despite these drawbacks, the Cochrane review concluded that there is good evidence suggesting

Table 16.2 Placebo controlled studies examining the effect of antioxidant therapy on male reproductive heath

Study reference	Therapy used per day	Duration of therapy (months)	Oxidative stress as an inclusion criteria	Positive changes in semen quality	Positive changes in sperm OS end points	Positive changes in reproductive outcomes
[183]	Vitamin E 300 mg	6	No	↑ Motility	↓ MDA	Pregnancy 17% active group vs. 0% placebo
[184]	Vitamin E 300 mg	3	No	Nil	Nil	Improved sperm zona binding
[185]	Vitamin C 200 or 1,000 mg	1	No	↑ Motility, morph and viability	Not tested	Not reported
[186]	Vitamin E 800 mg, vitamin C 1,000 mg	2	No	Nil	Not tested	None
[187]	Vitamin E and C 1,000 mg each	2	No	Nil	↓ Sperm DNA damage	Not reported
[188]	Vitamin E 10 mg, vitamin C 5 mg, zinc 200 mg	3	No	Nil	Trend for ↓ MDA	Not reported
[189]	Coenzyme Q10 300 mg	6	No	↑ Conc and motility	Not tested	No difference in pregnancy rates
[190]	Sn 100 mg, vitamin A 1 mg, vitamin C 10 mg, vitamin E 15 mg	3	No	↑ Motility	Not tested	No difference in pregnancy rates (11% vs. 0% placebo)
[191]	Sn 200 mg, NAC 600 mg	6	No	↑ Conc, motility and morph	Not tested	Not reported
[192]	Glutathione 600 mg	2	No	↑ Motility and morph	Not tested	Not reported
[193]	L-Carnitine 2 g	2	No	Nil (raw data analysis)	No change in MDA	No difference
[194]	NAC 600 mg	3	No	↑ Sperm concentration	Not tested	Not reported
[195]	NAC 600 mg	3	No	↑ Motility	Not tested	Not reported
[196]	Astaxanthin 16 mg	3	No	↑ Motility	↓ Semen ROS	↑ Natural + IUI conceptions
[197]	Vitamin E 400 mg, Sn 225 mg	3	No	↑ Motility	↓ MDA	Not reported
[198]	Vitamin C 30 mg, vitamin E 5 mg, beta-glucan 20 mg, papaya 50 mg, lactoferrin 97 mg	3	No	↑ Motility and morph	No change in DNA quality	Not reported
[199]	Menevit (vitamin C, vitamin E, Sn, lycopne, folate, zinc and garlic oil)	3	No	Not reported	Not tested	↑ IVF-ICSI conceptions on active antioxidant (38.5% vs. 16% placebo)

MDA malondialdehyde; *ROS* reactive oxygen species; *Sn* selenium; *NAC* *N*-acetyl cysteine; *IUI* intra-uterine insemination

that oral antioxidant therapy can boost both natural and IVF assisted pregnancy and live birth rates. As antioxidant therapy is relatively inexpensive compared to other fertility treatments, and there is no definitive evidence that antioxidants will cause harm to patients, this positive Cochrane review should result in a more widespread uptake of oral antioxidant therapy for male infertility in the future.

To date there have been no trials clearly showing one antioxidant formulation to be superior to another. Therefore, when faced with the dilemma of imperfect information the treating clinician must make his/her own decision on what is the ideal antioxidant combination.

Vitamin E is an essential fat soluble vitamin, with alpha-tocopherol being the most common form of vitamin E available in food. Vitamin E is major chain breaking antioxidant that directly neutralises superoxide anions, hydrogen peroxide and the hydroxyl radical. As sperm membranes contain abundant phospholipids which are prone to oxidative damage, it is believed that vitamin E plays a critical role in protecting cellular structures from damage caused by free radicals and reactive products of lipid peroxidation. Second, vitamin E exhibits some anti-inflammatory activity and therefore may reduce leukocyte initiated sperm oxidative stress. The Recommended Dietary Allowance (RDA) for vitamin E is suggested to be 15 mg (equivalent to 22.4 IU) of alpha-tocopherol per day for adult men, with the tolerable upper intake being suggested as 1,000 mg (1,500 IU) by the US National Institute of Health [179]. However, a meta-analysis of 19 clinical trials using long-term vitamin E supplementation in patients with chronic disease has reported that at dosages of 400 IU or greater per day, vitamin E may actually increase overall mortality compared to placebo [180]. Studies using dosages of 200–300 mg vitamin E (300–450 IU) have been shown to produce a significant fall in sperm lipid peroxidation (Table 16.2) so it would appear that 400 IU of vitamin E per day is a safe and effective therapy for sperm oxidative stress. However, as vitamin E is known to inhibit platelet aggregation and has been linked with an increased risk of haemorrhagic stroke, its use in infertile men on anticoagulants or at risk of serious haemorrhagic illness is probably contraindicated [180].

Vitamin C (ascorbic acid) is an important water soluble antioxidant that competitively protects lipoproteins from peroxyl radical attack while also enhancing the antioxidant activity of vitamin E by assisting in its recycling. Seminal plasma vitamin C levels are tenfold higher than serum [181], suggesting a very important protective role for vitamin C in the male reproductive tract. The RDA for vitamin C in the adult male is 75 mg, with the tolerable upper intake limit being suggested as 2,000 mg/day [179]. While some trials have used vitamin C supplementation at doses as high as 1,000 mg/day, lower-dose supplementation such as 100 mg/day is probably more preferable, since vitamin C can act as a pro-oxidant at high concentrations in the presence of iron [182]. Furthermore, high-dose vitamin C may also lead to the development of kidney stones and cause side effects such as nausea, abdominal cramps and diarrhoea.

Phytochemicals with known powerful antioxidant and anti-inflammatory action include lycopene, garlic and astaxanthin, with all being shown to various degrees to reduce oxidative attack on sperm and improve male reproductive performance

Table 16.3 Recommended Dietary Allowance (RDA) and Tolerable Upper Intake Levels (UL) of common dietary antioxidants*

Antioxidant nutrient	RDA (mg)	UL (mg)
Vitamin C	90	2,000
Vitamin E	15	1,000
Selenium	55	400
Zinc	11	40

RDA = The average daily dietary nutrient intake sufficient to meet the nutritional requirements of the majority of healthy individuals
UL = The highest daily intake level that is likely to pose no risk of adverse heath effects to the majority of individuals in the general population
*US National Academy of Sciences, Washington, USA, 2005

(Table 16.2). Finally, minerals such as zinc and selenium both have antioxidant properties and play an important role in protamination packaging of sperm DNA, helping protect the sperm DNA from ROS mediated damage (Table 16.2). It would therefore appear logical that an ideal fertility promoting antioxidant formulation would contain these two minerals.

The current recommendations for the Recommended Dietary Allowances (RDA) for various antioxidant compounds issued by the National Academy of Sciences (USA) [179] are presented in Table 16.3 as they will hopefully guide the clinicians in deciding the safe and effective dose of an antioxidant to prescribe to their own infertile patients.

The various studies examining the effect of oral antioxidants on sperm function have used antioxidant therapy for between 4 and 26 weeks duration, with the majority of studies being of 2–3 months duration (Table 16.2). Since sperm production takes on average 70 days, it would appear logical that at least 3 months of antioxidant therapy is most likely to have a beneficial effect on sperm function. It is our own clinical practice to start men on antioxidant therapy several months before the female partner commences her fertility treatment.

16.9 Conclusion

Following more than 70 years of research on the topic, it has now been irrefutably proven that sperm oxidative stress is a common underlying cause of sub-fertility in a large proportion of infertile men. It is hoped that this chapter will make it easier for clinicians to identify and successfully treat sperm oxidative stress, thereby maximising the reproductive potential of their patients before resorting to more complex and expensive treatments such as IVF-ICSI. Lifestyle modifications, removal of toxins from the environment, treatment of male genital tract infections and the effective treatment of illness linked with systemic oxidative stress is the first line of management in all infertile individuals. Once seminal ROS production has been reduced to the lowest possible baseline levels, oral antioxidant therapy can then be initiated to further enhance sperm function. It is believed that this type of coordinated "holistic" approach to the management of sperm oxidative stress is most likely to result in a positive pregnancy outcome for the couple.

16.10 Key Points

- Oxidative stress is a common pathology, effecting up to 80% of infertile men.
- Common lifestyle causes of sperm oxidative stress include obesity, smoking and alcohol abuse.
- Occupational exposure to metal fumes such as occurs with welding, soldering, manufacturing of lead acid batteries and foundry work all can cause oxidative stress.
- Systemic diseases characterised by a general state of oxidative stress such as diabetes, thyroid dysfunction, chronic viral infections (Hepatitis and HIV), haemoglobinopathies (thalassaemia, sickle cell disease), haemochromatosis and depression are all likely to negatively impact on sperm function.
- Features in the routine semen analysis suggestive of the presence of oxidative stress include abnormal sperm morphology with "cytoplasmic droplets", poor sperm motility, low HOST results, large numbers of round cells or leukocytes and altered semen viscosity or raised pH.
- First line therapy for the management of sperm oxidative stress is to remove any lifestyle or environmental triggers (decrease weight, stop smoking, removal of environmental toxins), followed by the amelioration of any systemic disease processes linked with oxidative stress.
- The Cochrane Collaboration review on oral antioxidant therapy for male infertility concludes that antioxidants are effective in improving both natural and IVF assisted pregnancy rates.
- No universally accepted antioxidant formulation has been agreed upon for the management of male infertility. However, based on the available data a combinational formulation containing vitamins C, vitamin E, zinc, selenium, and antioxidant phytochemicals such as lycopene, garlic, astaxanthin or resveratrol would make a reasonable choice. As the sperm production cycle is 70 days, maximal effect of antioxidant therapy is likely to take 2–3 months.

References

1. Iwasaki A, Gagnon C. Formation of reactive oxygen species in spermatozoa of infertile patients. Fertil Steril. 1992;57:409–16.
2. Zini A, de Lamirande E, Gagnon C. Reactive oxygen species in semen of infertile patients: levels of superoxide dismutase- and catalase-like activities in seminal plasma and spermatozoa. Int J Androl. 1993;16:183–8.

3. Ochsendorf FR, Thiele J, Fuchs J, Schuttau H, Freisleben HJ, Buslau M, Milbradt R. Chemiluminescence in semen of infertile men. Andrologia. 1994;26:289–93.
4. Shekarriz M, Thomas Jr AJ, Agarwal A. Incidence and level of seminal reactive oxygen species in normal men. Urology. 1995;45:103–7.
5. Agarwal A, Sharma RK, Nallella KP, Thomas Jr AJ, Alvarez JG, Sikka SC. Reactive oxygen species as an independent marker of male factor infertility. Fertil Steril. 2006;86:878–85.
6. Tremellen K. Oxidative stress and male infertility—a clinical perspective. Hum Reprod Update. 2008;14(3):243–58.
7. Aitken RJ, Baker MA, Sawyer D. Oxidative stress in the male germ line and its role in the aetiology of male infertility and genetic disease. Reprod Biomed Online. 2003;7:65–70.
8. Aitken RJ, Baker MA, O'Bryan M. Shedding light on chemiluminescence: the application of chemiluminescence in diagnostic andrology. J Androl. 2004;25:455–65.
9. Tunc O, Thompson J, Tremellen K. Development of the NBT assay as a marker of sperm oxidative stress. Int J Androl. 2010;33(1):13–21.
10. Xu DX, Shen HM, Zhu QX, Chua L, Wang QN, Chia SE, Ong CN. The associations among semen quality, oxidative DNA damage in human spermatozoa and concentrations of cadmium, lead and selenium in seminal plasma. Mutat Res. 2003;534:155–63.
11. Naha N, Chowdhury AR. Inorganic lead exposure in battery and paint factory: effect on human sperm structure and functional activity. J UOEH. 2006;28:157–71.
12. Gennart JP, Buchet JP, Roels H, Ghyselen P, Ceulemans E, Lauwerys R. Fertility of male workers exposed to cadmium, lead, or manganese. Am J Epidemiol. 1992;135:1208–19.
13. Bonde JP. The risk of male subfecundity attributable to welding of metals. Studies of semen quality, infertility, fertility, adverse pregnancy outcome and childhood malignancy. Int J Androl. 1993;16 Suppl 1:1–29.
14. Chitra KC, Sujatha R, Latchoumycandane C, Mathur PP. Effect of lindane on antioxidant enzymes in epididymis and epididymal sperm of adult rats. Asian J Androl. 2001;3:205–8.
15. Latchoumycandane C, Mathur PP. Induction of oxidative stress in the rat testis after short-term exposure to the organochlorine pesticide methoxychlor. Arch Toxicol. 2002;76:692–8.
16. Latchoumycandane C, Chitra KC, Mathur PP. 2,3,7,8-Tetrachlorodibenzo-p-dioxin (TCDD) induces oxidative stress in the epididymis and epididymal sperm of adult rats. Arch Toxicol. 2003;77:280–4.
17. Meng Z, Bai W. Oxidation damage of sulfur dioxide on testicles of mice. Environ Res. 2004;96:298–304.
18. Gonzalez-Flecha B. Oxidant mechanisms in response to ambient air particles. Mol Aspects Med. 2004;25:169–82.
19. Alaghmand M, Blough NV. Source-dependent variation in hydroxyl radical production by airborne particulate matter. Environ Sci Technol. 2007;41:2364–70.
20. Rubes J, Selevan SG, Evenson DP, Zudova D, Vozdova M, Zudova Z, Robbins WA, Perreault SD. Episodic air pollution is associated with increased DNA fragmentation in human sperm without other changes in semen quality. Hum Reprod. 2005;20:2776–83.
21. Agarwal DK, Maronpot RR, Lamb JCT, Kluwe WM. Adverse effects of butyl benzyl phthalate on the reproductive and hematopoietic systems of male rats. Toxicology. 1985;35:189–206.
22. Srivastava SP, Srivastava S, Saxena DK, Chandra SV, Seth PK. Testicular effects of di-n-butyl phthalate (DBP): biochemical and histopathological alterations. Arch Toxicol. 1990;64:148–52.
23. Kasahara E, Sato EF, Miyoshi M, Konaka R, Hiramoto K, Sasaki J, Tokuda M, Nakano Y, Inoue M. Role of oxidative stress in germ cell apoptosis induced by di(2-ethylhexyl)phthalate. Biochem J. 2002;365:849–56.
24. Hauser R, Meeker JD, Singh NP, Silva MJ, Ryan L, Duty S, Calafat AM. DNA damage in human sperm is related to urinary levels of phthalate monoester and oxidative metabolites. Hum Reprod. 2007;22:688–95.
25. Lee E, Ahn MY, Kim HJ, Kim IY, Han SY, Kang TS, Hong JH, Park KL, Lee BM, Kim HS. Effect of di(n-butyl) phthalate on testicular oxidative damage and antioxidant enzymes in hyperthyroid rats. Environ Toxicol. 2007;22:245–55.

26. Agbaje IM, Rogers DA, McVicar CM, McClure N, Atkinson AB, Mallidis C, Lewis SE. Insulin dependant diabetes mellitus: implications for male reproductive function. Hum Reprod. 2007;22:1871–7.
27. Agbaje IM, McVicar CM, Schock BC, McClure N, Atkinson AB, Rogers D, Lewis SE. Increased concentrations of the oxidative DNA adduct 7,8-dihydro-8-oxo-2-deoxyguanosine in the germ-line of men with type 1 diabetes. Reprod Biomed Online. 2008;16(3):401–9.
28. Shrilatha B, Muralidhara DR. Occurrence of oxidative impairments, response of antioxidant defences and associated biochemical perturbations in male reproductive milieu in the streptozotocin-diabetic rat. Int J Androl. 2007;30(6):508–18.
29. Khaki A, Fathiazad F, Nouri M, Khaki A, Maleki NA, Khamnei HJ, Ahmadi P. Beneficial effects of quercetin on sperm parameters in streptozotocin-induced diabetic male rats. Phytother Res. 2010;24(9):1285–91.
30. Erdamar H, Demirci H, Yaman H, Erbil MK, Yakar T, Sancak B, Elbeg S, Bibero lu G, Yetkin I. The effect of hypothyroidism, hyperthyroidism, and their treatment on parameters of oxidative stress and antioxidant status. Clin Chem Lab Med. 2008;46(7):1004–10.
31. Torun AN, Kulaksizoglu S, Kulaksizoglu M, Pamuk BO, Isbilen E, Tutuncu NB. Serum total antioxidant status and lipid peroxidation marker malondialdehyde levels in overt and subclinical hypothyroidism. Clin Endocrinol (Oxf). 2009;70(3):469–74.
32. Santi A, Duarte MM, Moresco RN, Menezes C, Bagatini MD, Schetinger MR, Loro VL. Association between thyroid hormones, lipids and oxidative stress biomarkers in overt hypothyroidism. Clin Chem Lab Med. 2010;48(11):1635–9.
33. Chattopadhyay S, Choudhury S, Roy A, Chainy GB, Samanta L. T3 fails to restore mitochondrial thiol redox status altered by experimental hypothyroidism in rat testis. Gen Comp Endocrinol. 2010;169(1):39–47.
34. Sakr SA, Mahran HA, Nofal AE. Effect of selenium on carbimazole-induced testicular damage and oxidative stress in albino rats. J Trace Elem Med Biol. 2011;25(1):59–66.
35. Mancini A, Festa R, Silvestrini A, Nicolotti N, Di Donna V, La Torre G, Pontecorvi A, Meucci E. Hormonal regulation of total antioxidant capacity in seminal plasma. J Androl. 2009;30(5):534–40.
36. Moreira Jr ED, Lbo CF, Diament A, Nicolosi A, Glasser DB. Incidence of erectile dysfunction in men 40 to 69 years old: results from a population-based cohort study in Brazil. Urology. 2003;61(2):431–6.
37. Jones RW, Rees RW, Minhas S, Ralph D, Persad RA, Jeremy JY. Oxygen free radicals and the penis. Expert Opin Pharmacother. 2002;3(7):889–97.
38. Meldrum DR, Gambone JC, Morris MA, Ignarro LJ. A multifaceted approach to maximize erectile function and vascular health. Fertil Steril. 2010;94(7):2514–20.
39. Azadzoi KM, Schulman RN, Aviram M, Siroky MB. Oxidative stress in arteriogenic erectile dysfunction: prophylactic role of antioxidants. J Urol. 2005;174(1):386–93.
40. Stanislavov R, Nikolova V, Rohdewald P. Improvement of erectile function with Prelox: a randomized, double-blind, placebo-controlled, crossover trial. Int J Impot Res. 2008;20(2):173–80.
41. Zhang Q, Radisavljevic ZM, Siroky MB, Azadzoi KM. Dietary antioxidants improve arteriogenic erectile dysfunction. Int J Androl. 2011;34(3):225–35.
42. Eskiocak S, Gozen AS, Taskiran A, Kilic AS, Eskiocak M, Gulen S. Effect of psychological stress on the L-arginine-nitric oxide pathway and semen quality. Braz J Med Biol Res. 2006;39:581–8.
43. Eskiocak S, Gozen AS, Yapar SB, Tavas F, Kilic AS, Eskiocak M. Glutathione and free sulphydryl content of seminal plasma in healthy medical students during and after exam stress. Hum Reprod. 2005;20:2595–600.
44. Potts RJ, Notarianni LJ, Jefferies TM. Seminal plasma reduces exogenous oxidative damage to human sperm, determined by the measurement of DNA strand breaks and lipid peroxidation. Mutat Res. 2000;447:249–56.
45. Watson PF. The causes of reduced fertility with cryopreserved semen. Anim Reprod Sci. 2000;60–61:481–92.

46. Thomson LK, Fleming SD, Aitken RJ, De Iuliis GN, Zieschang JA, Clark AM. Cryopreservation-induced human sperm DNA damage is predominantly mediated by oxidative stress rather than apoptosis. Hum Reprod. 2009;24(9):2061–70.
47. Agarwal A, Said TM. Oxidative stress, DNA damage and apoptosis in male infertility: a clinical approach. BJU Int. 2005;95:503–7.
48. Corona G, Ricca V, Bandini E, Mannucci E, Lotti F, Boddi V, Rastrelli G, Sforza A, Faravelli C, Forti G, Maggi M. Selective serotonin reuptake inhibitor-induced sexual dysfunction. J Sex Med. 2009;6(5):1259–69.
49. Tanrikut C, Feldman AS, Altemus M, Paduch DA, Schlegel PN. Adverse effect of paroxetine on sperm. Fertil Steril. 2010;94(3):1021–6.
50. Schaeffer AJ. Epidemiology and demographics of prostatitis. Andrologia. 2003;35:252–7.
51. Fraczek M, Kurpisz M. Inflammatory mediators exert toxic effects of oxidative stress on human spermatozoa. J Androl. 2007;28:325–33.
52. Mazzilli F, Rossi T, Marchesini M, Ronconi C, Dondero F. Superoxide anion in human semen related to seminal parameters and clinical aspects. Fertil Steril. 1994;62:862–8.
53. Depuydt CE, Bosmans E, Zalata A, Schoonjans F, Comhaire FH. The relation between reactive oxygen species and cytokines in andrological patients with or without male accessory gland infection. J Androl. 1996;17:699–707.
54. Ochsendorf FR. Infections in the male genital tract and reactive oxygen species. Hum Reprod Update. 1999;5:399–420.
55. Potts JM, Sharma R, Pasqualotto F, Nelson D, Hall G, Agarwal A. Association of *Ureaplasma urealyticum* with abnormal reactive oxygen species levels and absence of leukocytospermia. J Urol. 2000;163:1775–8.
56. Padron OF, Brackett NL, Sharma RK, Lynne CM, Thomas Jr AJ, Agarwal A. Seminal reactive oxygen species and sperm motility and morphology in men with spinal cord injury. Fertil Steril. 1997;67:1115–20.
57. Brackett NL, Cohen DR, Ibrahim E, Aballa TC, Lynne CM. Neutralization of cytokine activity at the receptor level improves sperm motility in men with spinal cord injuries. J Androl. 2007;28(5):717–21.
58. Segnini A, Camejo MI, Proverbio F. *Chlamydia trachomatis* and sperm lipid peroxidation in infertile men. Asian J Androl. 2003;5:47–9.
59. Kapranos N, Petrakou E, Anastasiadou C, Kotronias D. Detection of herpes simplex virus, cytomegalovirus, and Epstein-Barr virus in the semen of men attending an infertility clinic. Fertil Steril. 2003;79 Suppl 3:1566–70.
60. Bezold G, Politch JA, Kiviat NB, Kuypers JM, Wolff H, Anderson DJ. Prevalence of sexually transmissible pathogens in semen from asymptomatic male infertility patients with and without leukocytospermia. Fertil Steril. 2007;87:1087–97.
61. Krause W, Herbstreit F, Slenzka W. Are viral infections the cause of leukocytospermia? Andrologia. 2002;34:87–90.
62. Krause W, Bohring C, Gueth A, Horster S, Krisp A, Skrzypek J. Cellular and biochemical markers in semen indicating male accessory gland inflammation. Andrologia. 2003;35:279–82.
63. Umapathy E, Simbini T, Chipata T, Mbizvo M. Sperm characteristics and accessory sex gland functions in HIV-infected men. Arch Androl. 2001;46:153–8.
64. Seronello S, Sheikh MY, Choi J. Redox regulation of hepatitis C in nonalcoholic and alcoholic liver. Free Radic Biol Med. 2007;43(6):869–82.
65. Chen J, Siddiqui A. Hepatitis B virus X protein stimulates the mitochondrial translocation of Raf-1 via oxidative stress. J Virol. 2007;81:6757–60.
66. Vicari E, Arcoria D, Di Mauro C, Noto R, Noto Z, La Vignera S. Sperm output in patients with primary infertility and hepatitis B or C virus; negative influence of HBV infection during concomitant varicocele. Minerva Med. 2006;97:65–77.
67. Durazzo M, Premoli A, Di Bisceglie C, Bertagna A, Faga E, Biroli G, Manieri C, Bo S, Pagano G. Alterations of seminal and hormonal parameters: an extrahepatic manifestation of HCV infection? World J Gastroenterol. 2006;12:3073–6.

68. Srinivasan S, Pragasam V, Jenita X, Kalaiselvi P, Muthu V, Varalakshmi P. Oxidative stress in urogenital tuberculosis patients: a predisposing factor for renal stone formation—amelioration by vitamin E supplementation. Clin Chim Acta. 2004;350:57–63.
69. Vijayaraghavan R, Suribabu CS, Sekar B, Oommen PK, Kavithalakshmi SN, Madhusudhanan N, Panneerselvam C. Protective role of vitamin E on the oxidative stress in Hansen's disease (leprosy) patients. Eur J Clin Nutr. 2005;59:1121–8.
70. Guha M, Kumar S, Choubey V, Maity P, Bandyopadhyay U. Apoptosis in liver during malaria: role of oxidative stress and implication of mitochondrial pathway. FASEB J. 2006;20:1224–6.
71. Macao LB, Filho DW, Pedrosa RC, Pereira A, Backes P, Torres MA, Frode TS. Antioxidant therapy attenuates oxidative stress in chronic cardiopathy associated with Chagas' disease. Int J Cardiol. 2007;123(1):43–9.
72. Filippini A, Riccioli A, Padula F, Lauretti P, D'Alessio A, De Cesaris P, Gandini L, Lenzi A, Ziparo E. Control and impairment of immune privilege in the testis and in semen. Hum Reprod Update. 2001;7:444–9.
73. Sharma RK, Pasqualotto AE, Nelson DR, Thomas Jr AJ, Agarwal A. Relationship between seminal white blood cell counts and oxidative stress in men treated at an infertility clinic. J Androl. 2001;22:575–83.
74. Nandipati KC, Pasqualotto FF, Thomas Jr AJ, Agarwal A. Relationship of interleukin-6 with semen characteristics and oxidative stress in vasectomy reversal patients. Andrologia. 2005;37:131–4.
75. Shahed AR, Shoskes DA. Oxidative stress in prostatic fluid of patients with chronic pelvic pain syndrome: correlation with gram positive bacterial growth and treatment response. J Androl. 2000;21:669–75.
76. Pasqualotto FF, Sharma RK, Potts JM, Nelson DR, Thomas AJ, Agarwal A. Seminal oxidative stress in patients with chronic prostatitis. Urology. 2000;55:881–5.
77. Potts JM, Pasqualotto FF. Seminal oxidative stress in patients with chronic prostatitis. Andrologia. 2003;35:304–8.
78. Motrich RD, Maccioni M, Molina R, Tissera A, Olmedo J, Riera CM, Rivero VE. Reduced semen quality in chronic prostatitis patients that have cellular autoimmune response to prostate antigens. Hum Reprod. 2005;20:2567–72.
79. Motrich RD, Mackern-Oberti JP, Maccioni M, Rivero VE. Effects of autoimmunity to the prostate on the fertility of the male rat. Fertil Steril. 2009;91(5 Suppl):2273–80.
80. Oberg BP, McMenamin E, Lucas FL, McMonagle E, Morrow J, Ikizler TA, Himmelfarb J. Increased prevalence of oxidant stress and inflammation in patients with moderate to severe chronic kidney disease. Kidney Int. 2004;65:1009–16.
81. Danielski M, Ikizler TA, McMonagle E, Kane JC, Pupim L, Morrow J, Himmelfarb J. Linkage of hypoalbuminemia, inflammation, and oxidative stress in patients receiving maintenance hemodialysis therapy. Am J Kidney Dis. 2003;42:286–94.
82. Pupim LB, Himmelfarb J, McMonagle E, Shyr Y, Ikizler TA. Influence of initiation of maintenance hemodialysis on biomarkers of inflammation and oxidative stress. Kidney Int. 2004;65:2371–9.
83. Moreno JM, Ruiz MC, Ruiz N, Gomez I, Vargas F, Asensio C, Osuna A. Modulation factors of oxidative status in stable renal transplantation. Transplant Proc. 2005;37:1428–30.
84. Livrea MA, Tesoriere L, Pintaudi AM, Calabrese A, Maggio A, Freisleben HJ, D'Arpa D, D'Anna R, Bongiorno A. Oxidative stress and antioxidant status in beta-thalassemia major: iron overload and depletion of lipid-soluble antioxidants. Blood. 1996;88:3608–14.
85. Carpino A, Tarantino P, Rago V, De Sanctis V, Siciliano L. Antioxidant capacity in seminal plasma of transfusion-dependent beta-thalassemic patients. Exp Clin Endocrinol Diabetes. 2004;112:131–4.
86. Das UB, Mallick M, Debnath JM, Ghosh D. Protective effect of ascorbic acid on cyclophosphamide-induced testicular gametogenic and androgenic disorders in male rats. Asian J Androl. 2002;4:201–7.

87. Ghosh D, Das UB, Misro M. Protective role of alpha-tocopherol-succinate (provitamin-E) in cyclophosphamide induced testicular gametogenic and steroidogenic disorders: a correlative approach to oxidative stress. Free Radic Res. 2002;36:1209–18.
88. Wolfram RM, Budinsky AC, Palumbo B, Palumbo R, Sinzinger H. Radioiodine therapy induces dose-dependent in vivo oxidation injury: evidence by increased isoprostane 8-epi-PGF(2 alpha). J Nucl Med. 2002;43(9):1254–8.
89. Zhao W, Robbins ME. Inflammation and chronic oxidative stress in radiation-induced late normal tissue injury: therapeutic implications. Curr Med Chem. 2009;16(2):130–43.
90. Little JB. Cellular radiation effects and the bystander response. Mutat Res. 2006; 597(1–2):113–8.
91. Yalçin O, Karataş F, Erulaş FA, Ozdemir E. The levels of glutathione peroxidase, vitamin A, E, C and lipid peroxidation in patients with transitional cell carcinoma of the bladder. BJU Int. 2004;93(6):863–6.
92. Sharma M, Rajappa M, Kumar G, Sharma A. Oxidant-antioxidant status in Indian patients with carcinoma of posterior one-third of tongue. Cancer Biomark. 2009;5(6):253–60.
93. Badjatia N, Satyam A, Singh P, Seth A, Sharma A. Altered antioxidant status and lipid peroxidation in Indian patients with urothelial bladder carcinoma. Urol Oncol. 2010;28(4): 360–7.
94. Engin AB, Karahalil B, Engin A, Karakaya AE. Oxidative stress, *Helicobacter pylori*, and OGG1 Ser326Cys, XPC Lys939Gln, and XPD Lys751Gln polymorphisms in a Turkish population with colorectal carcinoma. Genet Test Mol Biomarkers. 2010;14(4):559–64.
95. Said TM, Tellez S, Evenson DP, Del Valle AP. Assessment of sperm quality, DNA integrity and cryopreservation protocols in men diagnosed with testicular and systemic malignancies. Andrologia. 2009;41(6):377–82.
96. Gomez E, Buckingham DW, Brindle J, Lanzafame F, Irvine DS, Aitken RJ. Development of an image analysis system to monitor the retention of residual cytoplasm by human spermatozoa: correlation with biochemical markers of the cytoplasmic space, oxidative stress, and sperm function. J Androl. 1996;17:276–87.
97. Said TM, Agarwal A, Sharma RK, Thomas Jr AJ, Sikka SC. Impact of sperm morphology on DNA damage caused by oxidative stress induced by beta-nicotinamide adenine dinucleotide phosphate. Fertil Steril. 2005;83:95–103.
98. Koppers AJ, De Iuliis GN, Finnie JM, McLaughlin EA, Aitken RJ. Significance of mitochondrial reactive oxygen species in the generation of oxidative stress in spermatozoa. J Clin Endocrinol Metab. 2008;93(8):3199–207.
99. Singer G, Granger DN. Inflammatory responses underlying the microvascular dysfunction associated with obesity and insulin resistance. Microcirculation. 2007;14:375–87.
100. Tunc O, Bakos HW, Tremellen K. Impact of body mass index on seminal oxidative stress. Andrologia. 2011;43(2):121–8.
101. Iyer A, Fairlie DP, Prins JB, Hammock BD, Brown L. Inflammatory lipid mediators in adipocyte function and obesity. Nat Rev Endocrinol. 2010;6(2):71–82.
102. Ishii T, Matsuki S, Iuchi Y, Okada F, Toyosaki S, Tomita Y, Ikeda Y, Fujii J. Accelerated impairment of spermatogenic cells in SOD1-knockout mice under heat stress. Free Radic Res. 2005;39:697–705.
103. Banks S, King SA, Irvine DS, Saunders PT. Impact of a mild scrotal heat stress on DNA integrity in murine spermatozoa. Reproduction. 2005;129:505–14.
104. Perez-Crespo M, Pintado B, Gutierrez-Adan A. Scrotal heat stress effects on sperm viability, sperm DNA integrity, and the offspring sex ratio in mice. Mol Reprod Dev. 2007;75:40–7.
105. Bakos HW, Mitchell M, Setchell BP, Lane M. The effect of paternal diet-induced obesity on sperm function and fertilization in a mouse model. Int J Androl. 2011;34(5 Pt 1):402–10.
106. Eskenazi B, Kidd SA, Marks AR, Sloter E, Block G, Wyrobek AJ. Antioxidant intake is associated with semen quality in healthy men. Hum Reprod. 2005;20:1006–12.
107. Therond P, Auger J, Legrand A, Jouannet P. alpha-Tocopherol in human spermatozoa and seminal plasma: relationships with motility, antioxidant enzymes and leukocytes. Mol Hum Reprod. 1996;2:739–44.

108. Fraga CG, Motchnik PA, Shigenaga MK, Helbock HJ, Jacob RA, Ames BN. Ascorbic acid protects against endogenous oxidative DNA damage in human sperm. Proc Natl Acad Sci USA. 1991;88:11003–6.
109. Saleh RA, Agarwal A, Sharma RK, Nelson DR, Thomas Jr AJ. Effect of cigarette smoking on levels of seminal oxidative stress in infertile men: a prospective study. Fertil Steril. 2002;78:491–9.
110. Fraga CG, Motchnik PA, Wyrobek AJ, Rempel DM, Ames BN. Smoking and low antioxidant levels increase oxidative damage to sperm DNA. Mutat Res. 1996;351:199–203.
111. Mostafa T, Tawadrous G, Roaia MM, Amer MK, Kader RA, Aziz A. Effect of smoking on seminal plasma ascorbic acid in infertile and fertile males. Andrologia. 2006;38:221–4.
112. Koch OR, Pani G, Borrello S, Colavitti R, Cravero A, Farre S, Galeotti T. Oxidative stress and antioxidant defenses in ethanol-induced cell injury. Mol Aspects Med. 2004;25:191–8.
113. Maneesh M, Dutta S, Chakrabarti A, Vasudevan DM. Alcohol abuse-duration dependent decrease in plasma testosterone and antioxidants in males. Indian J Physiol Pharmacol. 2006;50:291–6.
114. Moritz F, Monteil C, Isabelle M, Bauer F, Renet S, Mulder P, Richard V, Thuillez C. Role of reactive oxygen species in cocaine-induced cardiac dysfunction. Cardiovasc Res. 2003;59(4):834–43.
115. Song BJ, Moon KH, Upreti VV, Eddington ND, Lee IJ. Mechanisms of MDMA (ecstasy)-induced oxidative stress, mitochondrial dysfunction, and organ damage. Curr Pharm Biotechnol. 2010;11(5):434–43.
116. Barroso-Moguel R, Méndez-Armenta M, Villeda-Hernández J. Testicular lesions by chronic administration of cocaine in rats. J Appl Toxicol. 1994;14(1):37–41.
117. Barenys M, Macia N, Camps L, de Lapuente J, Gomez-Catalan J, Gonzalez-Linares J, Borras M, Rodamilans M, Llobet JM. Chronic exposure to MDMA (ecstasy) increases DNA damage in sperm and alters testes histopathology in male rats. Toxicol Lett. 2009;191(1):40–6.
118. Hendin BN, Kolettis PN, Sharma RK, Thomas Jr AJ, Agarwal A. Varicocele is associated with elevated spermatozoal reactive oxygen species production and diminished seminal plasma antioxidant capacity. J Urol. 1999;161:1831–4.
119. Barbieri ER, Hidalgo ME, Venegas A, Smith R, Lissi EA. Varicocele-associated decrease in antioxidant defenses. J Androl. 1999;20:713–7.
120. Saleh RA, Agarwal A, Nada EA, El-Tonsy MH, Sharma RK, Meyer A, Nelson DR, Thomas AJ. Negative effects of increased sperm DNA damage in relation to seminal oxidative stress in men with idiopathic and male factor infertility. Fertil Steril. 2003;79 Suppl 3:1597–605.
121. Nallella KP, Allamaneni SS, Pasqualotto FF, Sharma RK, Thomas Jr AJ, Agarwal A. Relationship of interleukin-6 with semen characteristics and oxidative stress in patients with varicocele. Urology. 2004;64:1010–3.
122. Smith R, Kaune H, Parodi D, Madariaga M, Rios R, Morales I, Castro A. Increased sperm DNA damage in patients with varicocele: relationship with seminal oxidative stress. Hum Reprod. 2006;21:986–93.
123. Agarwal A, Prabakaran S, Allamaneni SS. Relationship between oxidative stress, varicocele and infertility: a meta-analysis. Reprod Biomed Online. 2006;12:630–3.
124. Smith GR, Kaune GH, Parodi D, Madariaga AM, Morales DI, Rios SR, Castro GA. Extent of sperm DNA damage in spermatozoa from men examined for infertility. Relationship with oxidative stress. Rev Med Chil. 2007;135:279–86.
125. Ishikawa T, Fujioka H, Ishimura T, Takenaka A, Fujisawa M. Increased testicular 8-hydroxy-2'-deoxyguanosine in patients with varicocele. BJU Int. 2007;100(4):863–6.
126. Shiraishi K, Takihara H, Matsuyama H. Elevated scrotal temperature, but not varicocele grade, reflects testicular oxidative stress-mediated apoptosis. World J Urol. 2010;28(3):359–64.
127. Moretti E, Cosci I, Spreafico A, Serchi T, Cuppone AM, Collodel G. Semen characteristics and inflammatory mediators in infertile men with different clinical diagnoses. Int J Androl. 2009;32(6):637–46.

128. Ghabili K, Shoja MM, Agutter PS, Agarwal A. Hypothesis: intracellular acidification contributes to infertility in varicocele. Fertil Steril. 2009;92(1):399–401.
129. Chen SS, Huang WJ, Chang LS, Wei YH. 8-Hydroxy-2'-deoxyguanosine in leukocyte DNA of spermatic vein as a biomarker of oxidative stress in patients with varicocele. J Urol. 2004;172:1418–21.
130. Huff DS, Hadziselimovic F, Snyder III HM, Blyth B, Duckett JW. Early postnatal testicular maldevelopment in cryptorchidism. J Urol. 1991;146:624–6.
131. Turner TT, Bang HJ, Lysiak JL. The molecular pathology of experimental testicular torsion suggests adjunct therapy to surgical repair. J Urol. 2004;172:2574–8.
132. Filho DW, Torres MA, Bordin AL, Crezcynski-Pasa TB, Boveris A. Spermatic cord torsion, reactive oxygen and nitrogen species and ischemia-reperfusion injury. Mol Aspects Med. 2004;25:199–210.
133. Kidd SA, Eskenazi B, Wyrobek AJ. Effects of male age on semen quality and fertility: a review of the literature. Fertil Steril. 2001;75(2):237–48.
134. Junqueira VB, Barros SB, Chan SS, Rodrigues L, Giavarotti L, Abud RL, Deucher P. Aging and oxidative stress. Mol Aspects Med. 2004;25:5–16.
135. Cocuzza M, Athayde KS, Agarwal A, Sharma R, Pagani R, Lucon AM, Srougi M, Hallak J. Age-related increase of reactive oxygen species in neat semen in healthy fertile men. Urology. 2008;71(3):490–4.
136. Zubkova EV, Wade M, Robaire B. Changes in spermatozoal chromatin packaging and susceptibility to oxidative challenge during aging. Fertil Steril. 2005;84 Suppl 2:1191–8.
137. Weir CP, Robaire B. Spermatozoa have decreased antioxidant enzymatic capacity and increased reactive oxygen species production during aging in the Brown Norway rat. J Androl. 2007;28L:229–40.
138. Wyrobek AJ, Eskenazi B, Young S, Arnheim N, Tiemann-Boege I, Jabs EW, Glaser RL, Pearson FS, Evenson D. Advancing age has differential effects on DNA damage, chromatin integrity, gene mutations, and aneuploidies in sperm. Proc Natl Acad Sci USA. 2006; 103:9601–6.
139. Singh NP, Muller CH, Berger RE. Effects of age on DNA double-strand breaks and apoptosis in human sperm. Fertil Steril. 2003;80:1420–30.
140. Moskovtsev SI, Willis J, Mullen JB. Age-related decline in sperm deoxyribonucleic acid integrity in patients evaluated for male infertility. Fertil Steril. 2006;85:496–9.
141. Makker K, Varghese A, Desai NR, Mouradi R, Agarwal A. Cell phones: modern man's nemesis? Reprod Biomed Online. 2009;18(1):148–57.
142. De Iuliis GN, Newey RJ, King BV, Aitken RJ. Mobile phone radiation induces reactive oxygen species production and DNA damage in human spermatozoa in vitro. PLoS One. 2009;4(7):e6446.
143. Garaj-Vrhovac V, Gajski G, Pažanin S, Saroli A, Domijan AM, Flajs D, Peraica M. Assessment of cytogenetic damage and oxidative stress in personnel occupationally exposed to the pulsed microwave radiation of marine radar equipment. Int J Hyg Environ Health. 2011;214(1):59–65.
144. Weyandt TB, Schrader SM, Turner TW, Simon SD. Semen analysis of military personnel associated with military duty assignments. Reprod Toxicol. 1996;10(6):521–8.
145. Engin AB, Ergun MA, Yurtcu E, Kan D, Sahin G. Effect of ionizing radiation on the pteridine metabolic pathway and evaluation of its cytotoxicity in exposed hospital staff. Mutat Res. 2005;585(1–2):184–92.
146. Bartoov B, Zabludovsky N, Eltes F, Smirnov VV, Grischenko VI VI, Fischbein A. Semen quality of workers exposed to ionizing radiation in decontamination work after the chernobyl nuclear reactor accident. Int J Occup Environ Health. 1997;3(3):198–203.
147. Alberti KG, Eckel RH, Grundy SM, Zimmet PZ, Cleeman JI, Donato KA, Fruchart JC, James WP, Loria CM, Smith Jr SC. Harmonizing the metabolic syndrome: a joint interim statement of the International Diabetes Federation Task Force on Epidemiology and Prevention; National Heart, Lung, and Blood Institute; American Heart Association; World Heart

Federation; International Atherosclerosis Society; and International Association for the Study of Obesity. Circulation. 2009;120(16):1640–5.
148. Kort HI, Massey JB, Elsner CW, Mitchell-Leef D, Shapiro DB, Witt MA, Roudebush WE. Impact of body mass index values on sperm quantity and quality. J Androl. 2006;27(3): 450–2.
149. Chavarro JE, Toth TL, Wright DL, Meeker JD, Hauser R. Body mass index in relation to semen quality, sperm DNA integrity, and serum reproductive hormone levels among men attending an infertility clinic. Fertil Steril. 2010;93(7):2222–31.
150. Raman JD, Nobert CF, Goldstein M. Increased incidence of testicular cancer in men presenting with infertility and abnormal semen analysis. J Urol. 2005;174(5):1819–22.
151. World Health Organization. Laboratory manual for the examination of human semen and sperm-cervical mucous interaction. 5th ed. New York: Cambridge University Press; 2010.
152. Aydemir B, Onaran I, Kiziler AR, Alici B, Akyolcu MC. The influence of oxidative damage on viscosity of seminal fluid in infertile men. J Androl. 2008;29(1):41–6.
153. Siciliano L, Tarantino P, Longobardi F, Rago V, De Stefano C, Carpino A. Impaired seminal antioxidant capacity in human semen with hyperviscosity or oligoasthenozoospermia. J Androl. 2001;22:798–803.
154. Wang Y, Liang CL, Wu JQ, Xu C, Qin SX, Gao ES. Do *Ureaplasma urealyticum* infections in the genital tract affect semen quality? Asian J Androl. 2006;8:562–8.
155. Zorn B, Sesek-Briski A, Osredkar J, Meden-Vrtovec H. Semen polymorphonuclear neutrophil leukocyte elastase as a diagnostic and prognostic marker of genital tract inflammation—a review. Clin Chem Lab Med. 2003;41:2–12.
156. Kopa Z, Wenzel J, Papp GK, Haidl G. Role of granulocyte elastase and interleukin-6 in the diagnosis of male genital tract inflammation. Andrologia. 2005;37:188–94.
157. Dandekar SP, Nadkarni GD, Kulkarni VS, Punekar S. Lipid peroxidation and antioxidant enzymes in male infertility. J Postgrad Med. 2002;48:186–9.
158. Tremellen K, Tunc O. Macrophage activity in semen is significantly correlated with sperm quality in infertile men. Int J Androl. 2010;33(6):823–31.
159. Esposito K, Di Palo C, Maiorino MI, Petrizzo M, Bellastella G, Siniscalchi I, Giugliano D. Long-term effect of Mediterranean-style diet and calorie restriction on biomarkers of longevity and oxidative stress in overweight men. Cardiol Res Pract. 2010;2011:293916.
160. Powers SK, Nelson WB, Hudson MB. Exercise-induced oxidative stress in humans: cause and consequences. Free Radic Biol Med. 2011;51(5):942–50.
161. Paul C, Teng S, Saunders PT. A single, mild, transient scrotal heat stress causes hypoxia and oxidative stress in mouse testes, which induces germ cell death. Biol Reprod. 2009;80(5):913–9.
162. Mostafa T, Anis TH, El-Nashar A, Imam H, Othman IA. Varicocelectomy reduces reactive oxygen species levels and increases antioxidant activity of seminal plasma from infertile men with varicocele. Int J Androl. 2001;24:261–5.
163. Zini A, Blumenfeld A, Libman J, Willis J. Beneficial effect of microsurgical varicocelectomy on human sperm DNA integrity. Hum Reprod. 2005;20:1018–21.
164. Hurtado de Catalfo GE, Ranieri-Casilla A, Marra FA, de Alaniz MJ, Marra CA. Oxidative stress biomarkers and hormonal profile in human patients undergoing varicocelectomy. Int J Androl. 2007;30(6):519–30.
165. Werthman P, Wixon R, Kasperson K, Evenson DP. Significant decrease in sperm deoxyribonucleic acid fragmentation after varicocelectomy. Fertil Steril. 2008;90(5):1800–4.
166. Chen SS, Huang WJ, Chang LS, Wei YH. Attenuation of oxidative stress after varicocelectomy in subfertile patients with varicocele. J Urol. 2008;179(2):639–42.
167. Sakamoto Y, Ishikawa T, Kondo Y, Yamaguchi K, Fujisawa M. The assessment of oxidative stress in infertile patients with varicocele. BJU Int. 2008;101(12):1547–52.
168. Omu AE, al-Othman S, Mohamad AS, al-Kaluwby NM, Fernandes S. Antibiotic therapy for seminal infection. Effect on antioxidant activity and T-helper cytokines. J Reprod Med. 1998;43:857–64.

169. Vicari E. Effectiveness and limits of antimicrobial treatment on seminal leukocyte concentration and related reactive oxygen species production in patients with male accessory gland infection. Hum Reprod. 2000;15:2536–44.
170. Vicari E, La Vignera S, Calogero AE. Antioxidant treatment with carnitines is effective in infertile patients with prostatovesiculoepididymitis and elevated seminal leukocyte concentrations after treatment with nonsteroidal anti-inflammatory compounds. Fertil Steril. 2002;78:1203–8.
171. Gambera L, Serafini F, Morgante G, Focarelli R, De Leo V, Piomboni P. Sperm quality and pregnancy rate after COX-2 inhibitor therapy of infertile males with abacterial leukocytospermia. Hum Reprod. 2007;22(4):1047–51.
172. O'Connell M, McClure N, Lewis SE. Mitochondrial DNA deletions and nuclear DNA fragmentation in testicular and epididymal human sperm. Hum Reprod. 2002;17:1565–70.
173. Greco E, Scarselli F, Iacobelli M, Rienzi L, Ubaldi F, Ferrero S, Franco G, Anniballo N, Mendoza C, Tesarik J. Efficient treatment of infertility due to sperm DNA damage by ICSI with testicular spermatozoa. Hum Reprod. 2005;20:226–30.
174. Moskovtsev SI, Jarvi K, Mullen JB, Cadesky KI, Hannam T, Lo KC. Testicular spermatozoa have statistically significantly lower DNA damage compared with ejaculated spermatozoa in patients with unsuccessful oral antioxidant treatment. Fertil Steril. 2010;93(4):1142–6.
175. Zini A, San Gabriel M, Libman J. Lycopene supplementation in vitro can protect human sperm deoxyribonucleic acid from oxidative damage. Fertil Steril. 2010;94(3):1033–6.
176. Griveau JF, Le Lannou D. Influence of oxygen tension on reactive oxygen species production and human sperm function. Int J Androl. 1997;20:195–200.
177. Dalzell LH, McVicar CM, McClure N, Lutton D, Lewis SE. Effects of short and long incubations on DNA fragmentation of testicular sperm. Fertil Steril. 2004;82(5):1443–5.
178. Showell MG, Brown J, Yazdani A, Stankiewicz MT, Hart RJ. Antioxidants for male subfertility. Cochrane Database Syst Rev. 2011;(1):CD007411.
179. National Academy of Sciences (USA). Dietary reference intakes for energy, carbohydrate, fiber, fat, fatty acids, cholesterol, protein and amino acids. Washington, DC: National Academy of Sciences (USA); 2005.
180. Miller III ER, Pastor-Barriuso R, Dalal D, Riemersma RA, Appel LJ, Guallar E. Meta-analysis: high-dosage vitamin E supplementation may increase all-cause mortality. Ann Intern Med. 2005;142(1):37–46.
181. Jacob RA, Pianalto FS, Agee RE. Cellular ascorbate depletion in healthy men. J Nutr. 1992;122(5):1111–8.
182. Wayner DD, Burton GW, Ingold KU. The antioxidant efficiency of vitamin C is concentration-dependent. Biochim Biophys Acta. 1986;884(1):119–23.
183. Suleiman SA, Ali ME, Zaki ZM, El-Malik EM, Nasr MA. Lipid peroxidation and human sperm motility: protective role of vitamin E. J Androl. 1996;17(5):530–7.
184. Kessopoulou E, Powers HJ, Sharma KK, Pearson MJ, Russell JM, Cooke ID, Barratt CL. A double-blind randomized placebo cross-over controlled trial using the antioxidant vitamin E to treat reactive oxygen species associated male infertility. Fertil Steril. 1995;64(4):825–31.
185. Dawson EB, Harris WA, Rankin WE, Charpentier LA, McGanity WJ. Effect of ascorbic acid on male fertility. Ann N Y Acad Sci. 1987;498:312–23.
186. Rolf C, Cooper TG, Yeung CH, Nieschlag E. Antioxidant treatment of patients with asthenozoospermia or moderate oligoasthenozoospermia with high-dose vitamin C and vitamin E: a randomized, placebo-controlled, double-blind study. Hum Reprod. 1999;14(4):1028–33.
187. Greco E, Iacobelli M, Rienzi L, Ubaldi F, Ferrero S, Tesarik J. Reduction of the incidence of sperm DNA fragmentation by oral antioxidant treatment. J Androl. 2005;26(3):349–53.
188. Omu AE, Al-Azemi MK, Kehinde EO, Anim JT, Oriowo MA, Mathew TC. Indications of the mechanisms involved in improved sperm parameters by zinc therapy. Med Princ Pract. 2008;17(2):108–16.
189. Safarinejad MR. Efficacy of coenzyme Q10 on semen parameters, sperm function and reproductive hormones in infertile men. J Urol. 2009;182(1):237–48.

190. Scott R, MacPherson A, Yates RW, Hussain B, Dixon J. The effect of oral selenium supplementation on human sperm motility. Br J Urol. 1998;82(1):76–80.
191. Safarinejad MR, Safarinejad S. Efficacy of selenium and/or *N*-acetyl-cysteine for improving semen parameters in infertile men: a double-blind, placebo controlled, randomized study. J Urol. 2009;181(2):741–51.
192. Lenzi A, Culasso F, Gandini L, Lombardo F, Dondero F. Placebo-controlled, double-blind, cross-over trial of glutathione therapy in male infertility. Hum Reprod. 1993;8(10):1657–62.
193. Lenzi A, Lombardo F, Sgrò P, Salacone P, Caponecchia L, Dondero F, Gandini L. Use of carnitine therapy in selected cases of male factor infertility: a double-blind crossover trial. Fertil Steril. 2003;79(2):292–300.
194. Paradiso Galatioto G, Gravina GL, Angelozzi G, Sacchetti A, Innominato PF, Pace G, Ranieri G, Vicentini C. May antioxidant therapy improve sperm parameters of men with persistent oligospermia after retrograde embolization for varicocele? World J Urol. 2008;26(1):97–102.
195. Ciftci H, Verit A, Savas M, Yeni E, Erel O. Effects of *N*-acetylcysteine on semen parameters and oxidative/antioxidant status. Urology. 2009;74(1):73–6.
196. Comhaire FH, El Garem Y, Mahmoud A, Eertmans F, Schoonjans F. Combined conventional/antioxidant "Astaxanthin" treatment for male infertility: a double blind, randomized trial. Asian J Androl. 2005;7(3):257–62.
197. Keskes-Ammar L, Feki-Chakroun N, Rebai T, Sahnoun Z, Ghozzi H, Hammami S, Zghal K, Fki H, Damak J, Bahloul A. Sperm oxidative stress and the effect of an oral vitamin E and selenium supplement on semen quality in infertile men. Arch Androl. 2003;49(2):83–94.
198. Piomboni P, Gambera L, Serafini F, Campanella G, Morgante G, De Leo V. Sperm quality improvement after natural anti-oxidant treatment of asthenoteratospermic men with leukocytospermia. Asian J Androl. 2008;10(2):201–6.
199. Tremellen K, Miari G, Froiland D, Thompson J. A randomised control trial examining the effect of an antioxidant (Menevit) on pregnancy outcome during IVF-ICSI treatment. Aust N Z J Obstet Gynaecol. 2007;47(3):216–21.

Chapter 17
Oxidative Stress and Testicular Torsion

Dikmen Dokmeci

Abstract Testicular torsion is a medical urologic syndrome that constitutes a surgical emergency affecting newborns, children, and adolescent boys. The main pathophysiological event in testicular torsion is ischemia (torsion) followed by reperfusion (detorsion) of the testis, and multifactorial mechanisms seem to mediate this condition. Testicular damage after ischemia/reperfusion (I/R) is related to the duration of ischemia and to the severity of the torsion. I/R injury is associated with activation of neutrophils, inflammatory cytokines, and adhesion molecules with increased thrombogenicity, release of massive intracellular calcium, and overgeneration of reactive oxygen species and reactive nitrogen species. Although the results are not conclusive and the molecular mechanism by which antioxidants control male fertility have not yet been clearly identified, several antioxidant enzymes and antioxidant drugs have been studied to prevent such I/R injury in testis. Antioxidant therapy may represent a new option within a broader therapeutic strategy in testicular torsion with oxidative stress-mediated infertility.

Keywords Antioxidant • İschemia/reperfusion • Lipid peroxidation • Male infertility • Oxidative stress • Reactive oxygen/nitrogen species • Testicular torsion

Abbreviations

8-OHdG	8-Hydroxydeoxyguanosine
AP	Activating protein
ATP	Adenosine triphosphate

D. Dokmeci, MD, PhD (✉)
Department of Pharmacology, Faculty of Medicine,
Trakya University, 22030 Edirne, Turkey
e-mail: ddokmeci@hotmail.com

A. Agarwal et al. (eds.), *Studies on Men's Health and Fertility*, Oxidative Stress in Applied Basic Research and Clinical Practice, DOI 10.1007/978-1-61779-776-7_17,

CD	Conjugated diene
CT	Computed tomography
DNA	Deoxyribonucleic acid
EFS	Electrical field stimulation
EGF	Epidermal growth factor
eNOS	Endothelial nitric oxide synthase
ERK	Extracellular signal-regulated kinase
FDA	Food and Drug Administration
FSH	Follicule-stimulating hormone
GSH	Glutathione
GSH-Px	Glutathione peroxidase
HIF	Hypoxia-inducible factor
HSP	Heat shock protein
I/R	Ischemia/reperfusion
ICAM	Intercellular adhesion molecule
IGF	Insulin-like growth factor
IL	Interleukin
iNOS	Inducible nitric oxide synthase
JNK	c-Jun-NH_2-terminal kinase
L-NAME	L-nitro arginine methyl ester
L-NMMA	L-nitro monomethyl arginine
MAPK	Mitogen-activated protein kinase
MDA	Malondialdehyde
MPO	Myeloperoxidase
MRI	Magnetic resonance imaging
NADPH	Nicotinamid adenine dinucleotide phosphate
NF-κB	Nuclear factor kappa B
nNOS	Neuronal nitric oxide synthase
NO	Nitric oxide
PAF	Platelet-activating factor
PARP	Poly (adenosine triphosphate-ribose) polymerase
PPAR	Peroxisome proliferator-activated receptor
RNA	Ribonucleic acid
RNS	Reactive nitrogen species
ROS	Reactive oxygen species
SOD	Superoxide dismutase
TBARS	Thiobarbituric acid reactive substances
TNF	Tumor necrosis factor
VCAM	Vascular cellular adhesion molecule
VEGF	Vascular endothelial growth factor
VIP	Vasoactive intestinal polypeptide

17.1 Introduction

Testicular torsion or spermatic cord torsion is the most common genitourinary tract emergency of childhood and probably the second most common surgical emergency in the adolescent age group after acute appendicitis [1–3]. This clinical entity is commonly referred to as "the acute scrotum." Of the diseases that cause an acute scrotum, spermatic cord torsion is undoubtedly the most important. Misdiagnosis and inappropriate treatment lead to male factor infertility. Thus, a child presenting with an acute scrotum is clinically one of the most difficult situation for pediatricians, pediatric surgeons, and pediatric radiologist [3, 4].

Tissue ischemia is a major cause of morbidity and mortality. Cell death following ischemic injury is a clinically important process involved in a number of human diseases, including stroke, heart disease, renal failure, and cancer. The primary pathophysiologic event in testicular torsion is ischemia followed by reperfusion; thus, testicular torsion/detorsion is an ischemia/reperfusion (I/R) injury to the testis. The torsion must be treated promptly to avoid loss of function of ipsilateral and contralateral testis. This syndrome often leads to subfertility or infertility of the ipsilateral (torted) and contralateral (not torted) testis, but the mechanisms of cellular injury remain incompletely understood [5].

Mammalian testes are highly sensitive to oxidative stress and particularly to lipid peroxidation due to their high concentration of polyunsaturated fatty acids in the plasma membrane. The fatty acids are an essential requirement for the male germ cell to maintain sperm functions. Some detrimental factors such as I/R injury in testicular torsion may cause deoxyribonucleic acid (DNA) damage, inhibition of protein synthesis, and corruption of the sperm formation cycle, inducing in abnormal spermatogenesis [6]. On the other hand, unilateral testicular torsion reduces contralateral testicular blood flow, which gradually increases after the detorsion procedure. Ischemic damage leads to breakdown of the blood–testis barrier [7, 8]. I/R injury is also associated with activation of neutrophils, inflammatory cytokines, and adhesion molecules with increased thrombogenicity, release of massive intracellular calcium, and overgeneration of reactive oxygen species (ROS) and reactive nitrogen species (RNS). They damage several cellular components by peroxidation of cell membrane lipids [9, 10]. In addition, experimental studies have suggested that unilateral testicular torsion results in increase in lipid peroxidation products in the contralateral untwisted testis. Several studies have shown that applying substances to prevent the formation of oxygen free radicals or scavenging them before testicular detorsion significantly reduces damage in the ipsilateral and/or contralateral testis [11–14].

Testicular damage after testicular torsion is related to the duration of ischemia and to the severity of the torsion [1, 3, 15, 16]. Early diagnosis and definitive management are important to avoid testicular loss [17]. Anamnesis, physical examination, scrotal Doppler ultrasonography, and testicular scintigraphy are the main diagnostic methods. The diagnosis is particularly clinical and the management is emergency surgical untwisting and bilateral fixation.

To date, a number of chemicals, drugs, and physical methods such as intratesticular testosterone, melatonin, carnitine, ibuprofen, rosiglitazon, immunosuppression with cyclosporine and prednisone, pentoxifylline, vasoactive intestinal polypeptide (VIP), zinc aspartate, biocompatible surfactant (tetronic 1107), hyperbaric oxygen therapy, and chemical sympathectomy have been used to prevent I/R injury in testicular torsion and detorsion. They were found to be effective in preventing testicular damage, but none have been implemented in clinical practice. In addition, none have been tested in clinical trials and they cannot be applied in patients because of severe adverse effects [18, 19], apart from cooling the scrotum [20, 21]. It is very crucial that long-term effects of testicular torsion and its treatment should be observed in adulthood in terms of reduced fertility and impairment of spermatogenesis.

17.2 Testicular Torsion

17.2.1 Definition

Torsion of the testis was first described by Delasiauve [22]. The condition was first reported in the newborn in 1897 by Taylor [23]. In 1893, Nash first described manipulative detorsion of the testis in 1893 [24]. Testicular torsion is a medical urologic syndrome mainly caused by torsion of the spermatic cord that constitutes a surgical emergence.

17.2.2 Incidence

The annual incidence of testicular torsion is between one in 4,000 males and one in 158 males younger than 25 years [25, 26].

17.2.3 Laterality

Torsion is usually unilateral. Bilateral torsion occurs with only 2% of the patients and is an extremely tragic condition especially in neonates [27]. For instance, Bagci et al. recently reported the case of a patient with bilateral perinatal testicular torsion [28]. The left side is more commonly affected [1].

17.2.4 Age

Testicular torsion can occur at any age, but there are two peaks in incidence, the largest around puberty (accounting for 65% of all torsions) with another much smaller peak in the first year of life [1, 29]. Testicular torsion can also be present in adult man, the oldest age at which torsion has been reported is 69 years [30], but testicular torsion is rare in the adults [31, 32].

17.2.5 Etiology

The risk factors for acute testicular torsion are poorly understood. Torsion usually occurs in the absence of any precipitating event [16]. Environmental factors have been implicated by some authors and discredited by others. Srinivasan et al. [33] speculated that an increased incidence of testicular torsion is seen with decreasing atmospheric temperature humidity. Some reports demonstrated that no statistically significant differences were seen with regard to the seasonal or monthly occurrence of testicular torsion [34–36]. However, there is a cold weather seasonal trend for the torsion [37], and higher incidence rates of torsion have been reported during the colder months, especially December [36]. It has been reported that 40 of 46 cases of testicular torsion occurred when the ambient temperature was less than 2°C (35.6°F) and the torsion occurred from a cold-induced contraction of the cremasteric muscles [38, 39].

The child's age is the first clue to the etiology of the testicular torsion, since spermatic cord torsion is more common in adolescents and newborns, whereas torsion of the appendix testes/epididymis more commonly presents in prepubertal boys [40]. Marulaiah et al. [41] analyzed retrospectively the pathology results of all testicular and paratesticular 502 specimens (474 patients) between August 1995 to September 2007, and testicular torsion was found in 11.2%, with bimodal peak ages of <1 year and 13–14 years; the mean age was 9.4 years. Interestingly, it occurred more on the left side testis. Similar results are reported in other studies [42, 43]. Other predisposing factors are an increase in testicular volume (often associated with puberty), testicular trauma, tumor, testicles with horizontal lie, a history of cryptorhidism, and a spermatic cord with a long intrascrotal portion [16].

Familiar testicular torsion has been infrequently reported in the literature. Only a few cases have been reported so far. Shteynshlyuger and Freyle [44] have identified 11 families with familial testicular torsion, with 30 probands and 37 testicular torsions among them. They reported the largest number of first-degree relatives affected by testicular torsion in three consecutive generations of one family. Okeke and Ikuerowo [45] reported a familial series of a man and his three children all suffering from left testicular torsion. Family history in first-degree relatives may be helpful in the evaluation of a patient with testicular torsion. There is a strong predisposition in familial testicular torsion history.

17.2.6 Types [46–53]

Based on anatomy and age, testicular torsion is divided into two types: extravaginal and intravaginal torsion.

17.2.6.1 Intravaginal Torsion

This is the most common type of testicular torsion. Normally, tunica vaginalis invests the epididymis and posterior surface of the testicle, which fixes it to the scrotum and prevents it from twisting. If the tunica vaginalis attaches in a more proximal position on the spermatic cord, the testicle and epididymis hang free in the scrotum and can twist around the longitudinal axis of the spermatic cord within the tunica vaginalis. The testis usually undergoes torsion on the last few centimeters of the spermatic cord. The predisposing anatomical factors are the spiral arrangement and low insertion of the fibers of the cremasteric muscle, the tunica vaginalis, which extends proximally around the spermatic cord (the bell-clapper deformity), and the abnormality of the junction of the epididymis with the testis, forming a mesorchium. Bell-clapper deformity is the most common type, found 12% in cadaveric studies and often bilateral. The intravaginal spermatic cord may rotate and cause infarction of the testis and epididymis in the bell-clapper deformity.

Intravaginal testicular torsion is commonly seen in adolescents and older males; the peak incidence is 13 years and the left testis is more frequently involved. This type commonly occurs during sleep. It is possibly the result of a strong cremasteric reflex and spasm associated with nocturnal erections [1, 54]. The other initiating forces may be trauma, cold weather, or vigorous exercise such as cycling, ice-skating, swimming, parachuting, football, and rugby. Hormonal causes, rapid growth, and the increase of the size and vascularity of the testis during puberty may also be precursors to intravaginal torsion. Furthermore, testicular torsion is ten times more likely in undescended testes [1, 16, 29, 36].

17.2.6.2 Extravaginal Torsion

This type of testicular torsion includes approximately 5% of all torsions. Up to 20% of cases are synchronous, and 3% are asynchronous bilateral. Extravaginal torsion is usually seen in the prenatal or postnatal (neonatal) period. Prenatal torsion usually occurs in utero around 32 weeks and is present at birth. The pathogenesis is unknown and there isn't generally any anatomic defect. The other is postnatal testicular torsion occurring within the first 30 days of life. High birth weight and trauma during difficult delivery or breech presentation are important for torsion. The difference is important since the postnatal torsion requires emergent exploration and treatment with detorsion and fixation.

17.2.7 Pathogenesis

Torsion results from twisting of the spermatic cord, which causes ischemic changes depending on the degree and duration of twisting. Torsional twisting usually occurs away from the midline, probably secondary to the direction of the cremasteric muscle fibers. The degree of the torsion varies from 180° to more than 1,440°. This variation in torsion contributes to the variant presentations of acute torsion from severe to subacute and chronic torsion. Torsion blocks both arterial supply and venous drainage. This contributes to edema, congestion, swelling, hemorrhage, ecchymosis, and cellulitis. The edema further results in altered blood flow dynamics to the testis and accentuates arterial blockage, hypoxia, degeneration, infarction, necrosis, apoptosis, and gangrene of the testis [1, 15, 55, 56]. Ischemia may cause ultrastructural changes in the contralateral testis [13] and induce antitesticular or antisperm antibody production [57–60].

17.2.8 Clinical Features

A sudden onset and short duration of severe pain, abdominal discomfort, nausea and vomiting, high position of the testis, and abnormal cremasteric reflex are associated with increased risk of testicular torsion [53, 61–63].

17.2.9 Diagnosis

Testicular torsion is an emergent condition and timely recognition by almost all physicians, such as pediatricians, pediatric surgeons, urologist, neonatologist, emergency medicine physicians, and primary care physicians, is of paramount importance for early diagnosis and management to prevent testicular loss [31, 46].

There is currently no diagnostic test to unequivocally establish the diagnosis of torsion. The most important aspect in discerning the correct diagnosis is the history and physical examination. Information such as the duration of symptoms, character and quality of pain, time of onset of symptoms, activity at the time of initial symptoms, and the patient response to the symptoms are all important. Genital, inguinal, and abdominal examination should be done carefully. Testicular examination should include for lie (low or high) and axis (vertical or horizontal), comparing the torsione and untorsione side. Tender, elevated, transversely located testis with loss of cremasteric reflex are important in the diagnosis [64–66]. Laboratory examination should include a complete blood count and urinalysis which are usually normal. Rarely pyuria may be seen. Leukocytosis and pyuria are more common in infectious diseases, such as epididymitis or epididymoorchitis [1, 31].

Good imaging studies, however, can certainly be helpful in diagnosis. Gray-scale imaging, color Doppler, power Doppler and pulsed Doppler ultrasound, scintigraphy, nuclear magnetic resonance imaging (MRI), and computed tomography (CT) evaluate anatomy and perfusion. The ultrasound results differ depending on the duration and degree of the torsion. Ultrasonographic diagnosis is established by color or power Doppler, when there is not a detectable flow within the testicular parenchyma and cannot evaluate the duration of torsion and cannot predict testicular viability. Color Doppler ultrasound has a sensitivity of 88.9% and a specificity of 98.8 and 1% false positive results [67, 68], but is technically difficult in infants. Ideally, both pulsed and color Doppler ultrasound should be used. In complete torsion, the blood flow is completely absent on Doppler imaging. In incomplete torsion, the blood flow may still be seen. Color Doppler is less sensitive from power Doppler ultrasound. In addition, ultrasound can also be used to diagnose prenatal torsion [69–74].

In the diagnosis of testicular torsion, nuclear MRI can also be used [75, 76]. MRI has a sensitivity rate of 93% and a specificity rate of 100% in diagnosis of testicular torsion [77]. Nuclear MRI has also been used to diagnose torsion and to differentiate torsion from other acute scrotum conditions. Its use, however, is limited, because of the need for sedation in young children and the fact that other, less expensive imaging modalities are available [29]. CT scans provide detailed anatomical information; on the other hand, contains the risk of exposure to radiation and gives no information on blood flow to the testes. Testicular radionuclide scintigraphy, using Tc-99m, Xe-133, and RP-30A, has been reported to have sensitivity of 100%, while specificity varies from 80 to 100%. Spermatic cord torsion is associated with poor radionuclide uptake and appears as cold spots on the sintigram. Radionuclide scintigraphy requires an intravenous process and this procedure usually takes a longer time compared with ultrasound imaging [31, 64, 78, 79].

Interestingly, the effectiveness of using infrared thermography [80, 81], testicular oximetry [82], or the testis rigidity tester [83] for the noninvasive diagnosis of testicular torsion has been reported.

The procalcitonin assay is an easy, fast, noninvasive, inexpensive, and useful diagnostic maker for infectious disease; its level in healthy humans is 0.1 ng mL^{-1}. During infections, it rises to over 0.5 ng mL^{-1}. Yamis et al. [84] used this assay in differential diagnosis of torsion and epididymoorchitis. The increase in the epididymoorchitis group could help investigator to differentiate the epididymoorchitis from testicular torsion, and patients with high levels of procalcitonin should be accepted as epididymoorchitis and treated primarily by suitable antibiotics.

Ischemia-modified albumin is a new and sensitive biomarker of ischemia and oxidative stress. The U.S. Food and Drug Administration (FDA) recently licensed it for diagnostic use in suspected myocardial ischemia. Kutlu et al. [85] compared with sham and I/R testis tissue in terms of the malondialdehyde (MDA), glutathione (GSH), and myeloperoxidase (MPO) levels. They determined a high level of ischemia-modified albumin in testis torsion, indicating a potential value for testicular torsion diagnosis.

Inhibin B, a gonadal peptide regulating follicule-stimulating hormone (FSH) secretion, is an established marker of Sertoli cell function and spermatogenesis in adults. In contrast to the other hormones of the hypothalamo-pituitarygonadal axis, inhibin B is also secreted in detectable amounts during childhood. Inhibin B secretion depends on the interaction between Sertoli cells and germ cells [86]. Reduced levels of inhibin B have been described in adolescent boys after testicular torsion, particularly after orchidectomy [87]. Ozkan et al. [88] compared serum inhibin B levels to histopathologic parameters (Johnsen's score) on contralateral testicular damage after unilateral testicular torsion. They suggest that measurement of inhibin B levels to evaluate contralateral testicular damage after unilateral testicular torsion is more effective than histopathologic examination and this parameter evaluates complete testicular function.

17.2.10 Differential Diagnosis

Differential diagnoses include testicular trauma, tumor, polyorchidism, idiopathic scrotal edema, orchitis, epididymitis, inguinal hernias, haematocele, adrenal hemorrhages, meconium peritonitis, scrotal-periscrotal abscess, hydrocele, varicocele, appendicitis, acute hydronephrosis, funiculitis, fat necrosis, Henoch-Schonlein purpura, polyarteritis nodosa, and torsion of the testicular appendages [1, 31, 89, 90].

Patients usually present with hemiscrotal pain which is acute and severe. The differential diagnosis of scrotal pain includes torsion of appendix testis, epididymitis, and epididymoorchitis. Torsion of the appendix may present with a tender nodule and a bluish black spot (*blue dot*) on the upper pole of the testicle, epididymitis with localized tenderness of the epididymis, and edematous scrotal skin like orange peel and a normal testicle. Epididymoorchitis is characterized by scrotal and testicular pain, tenderness, edema, and erythema [50, 64].

Radiologic studies may differentiate torsion from other acute scrotum conditions. Scintigraphy also evaluate perfusion and anatomy. Typically, normal flow is present with torsion of appendices, normal or increased flow in epididymitis. Increased radiotracer is seen in inflamed testicle, whereas decreased perfusion is found in ischemic testes [64].

17.2.11 Treatment

Manual detorsion of testicular torsion was first described by Nash [24]. Van der Poel later reported a 25-year-old physician suffering recurrent testicular torsion who managed himself manual testicular detorsion successfully [91]. Adequate intravenous sedation or spermatic cord block may be necessary. The torsed testis usually twists from the medial to lateral side. This procedure is probably possible in early salvageable cases only [1, 16].

The aim of the surgical intervention is to restore the blood flow to the testis as soon as possible after the onset of symptoms. Once the diagnosis has been established, during exploration, the clinician faces the dilemma of whether to remove a testicle whose viability seems questionable and, in the absence of objective criteria, can only rely upon their own empirical experience. This decision is usually taken during surgical exploration, and the problem resides in the fact that no objective criteria exist to assess testicular viability [55]. Thus, surgery for torsion should occur immediately to minimize I/R injury and oxidative stress. Testicular salvage is estimated at 90% if detorsion occurs less than 6 h from the onset of symptoms, but it decreases to 50% after 12 h and to less than 10% after 24 h. Orchidectomy may be required if the testicle is necrotic, nonviable and perfusion not observed. Orchidopexy is performed on the affected testes and the contralateral testis (as a preventive measure) if the testes are found to be viable and perfusion is reestablished. Orchidopexy involves suturing the tunica albuginea to the dartos muscle using nonabsorbable sutures. Fixation may be one-, two-, three-, or four-points [16, 31, 54, 87, 92–96].

Based on experimental animal models, the mode of surgery plays an important role in the respect of the fate of the contralateral testis. However, clinical studies' data comparing testicular function after orchidectomy and detorsion and orchidopexy are still uncertain [87]. It has been suggested that, after a prolonged period of torsion, preserving surgery of the affected testis is even more harmful to the contralateral testis when compared to orchidectomy of affected testis [7]. On the other hand, in another study, Cimador et al. [55] speculated that as for the diagnosis and therapy of testicular torsion, there is no one history, physical, laboratory, or radiological finding that might predict testicular salvageability which can only be determined at surgical exploration.

Anderson et al. [97] evaluated 16 patients with regard to semen quality, endocrine parameters, and the presence or absence of semen antisperm antibodies. Nine testicular torsion patients were treated with detorsion and bilateral orchidopexy and seven were treated with ipsilateral orchidectomy and contralateral orchidopexy. Their study demonstrated that testicular injury (changes in semen quality and/or serum FSH, luteinizing hormone and testosterone) was occurred in the bilateral testis, regardless of treatment modality. However, with early intervention by detorsion and testicular salvage, subsequent semen quality remained within normal limits. Late surgical intervention, even with removal of the necrotic testis, resulted in significant impairment of semen quality.

Ichikawa et al. [98] evaluated 24 patients following spermatic cord torsion for semen quality and endocrine parameters after spermatic cord torsion. Of patients, 12 were treated with bilateral orchiodopexy, and 12 were treated with ipsilateral orchidectomy and contralateral orchidopexy. The average sperm density and the average total sperm count in the orchidectomy group were significantly lower than those in the orchidopexy group. The mean serum FSH level in the orchidectomy group was significantly higher than that in the orchidopexy group. These results suggested a significant decrease in testicular function in the orchidectomy group. All the patients in the orchidopexy group demonstrated a normal semen quality and endocrine parameters during follow-up. Four of the eight patients in the orchidectomy group whose duration of follow-up was more than 2 years still demonstrated oligozoospermia. The average age at operation of these four patients with abnormal semen quality was significantly higher than that of the other four patients with normal semen quality, whereas no significant difference in duration of torsion preceding surgical therapy was observed between these two groups. These findings suggested that subsequent semen quality remained within normal limits with early surgical treatment by bilateral orchidopexy. Ipsilateral orchidectomy was occurred less damage of the contralateral testis in the younger generation than in the older generation.

Arap et al. [57] analyzed the records of 24 patients who underwent surgical treatment (group I: orchidectomy + contralateral orchidopexy and group II: detorsion + bilateral orchidopexy) for testicular torsion. All patients were assessed by semen analysis, endocrine profile (levels of follicle-stimulating hormone FSH, luteinizing hormone, and testosterone), and seminal antisperm antibody levels. Median ischemia time in group 1 (48 h) was significantly higher than in group II (7 h). The investigators reported that patients treated for testicular torsion show mainly morphologic semen abnormalities and endocrine profiles within the normal range. Testicular fate did not have any correlation with the formation of antisperm antibodies. In addition, no significant correlation was found between antisperm antibody levels and age at torsion, ischemia time, seminal parameters, or surgical treatment type. Patients treated with orchidectomy presented better motility and morphology compared with the detorsion group. These results demonstrated that the maintenance of a severely ischemic testicle may impair seminal parameters; thus, the best procedure is to remove the affected testis.

In the other retrospective study, Taskinen et al. [87] investigated 17 patients operated on for testicular torsion. Testicular volume, serum FSH, testosterone, and inhibin B levels were measured early (median 36 days) and/or late (median 1.1 years) after operation. One third of patients with testicular torsion were found to have a reduced fertility potential according to serum FSH and inhibin B levels. It seems that the fertility prognosis is better after testis preservation than after orchidectomy if the testis is not obviously necrotic. Surprisingly, the onset of symptoms does not necessarily cause complete testicular atrophy and sometimes the testis can survive despite a long history of symptoms. In addition, preserving surgery can also sometimes be attempted after delayed diagnosis.

Shafik et al. [99] applied to transcutaneous electro-orchidogram in eight patients with testicular torsion. Slow waves were absent during torsion and after detorsion in seven patients. Semen analysis was done 1, 3, and 6 months after detorsion. Slow waves and semen normalized in one patient 6 months after detorsion, but semen analysis was not performed in the acute stage of the disease as the patients could not tolerate this test. Authors suggested that the electro-orchidogram might be an investigative tool in the diagnosis and follow-up testicular torsion, thus further investigations are needed to determine long-term effects of this tools.

Baker et al. [100] reported a case of missed bilateral testicular torsion with resulting primary testicular failure and hypogonadism with erectile dysfunction. The patient was seen for infertility and erectile dysfunction 10 years after the testicular torsion, semen analysis revealed azoospermia, serum total testosterone was at the castration level, and his infertility was irreversible. His erectile dysfunction was managed successfully with testosterone replacement and sildenafil therapy.

In another study, 25 patients were reexamined 1–12 years after the surgery. Age of torsion, duration of symptoms, and operative findings were reevaluated. It has been speculated that results of testicular atrophy correlated with duration of symptoms and operative findings. In all patients of surgical detorsion in which torsion lasted more than 24 h and testicular viability was questionable, subsequent atrophy was rule [101].

Moreover, it has been reported that hypothermia induced by the application of ice packs probably delays the production of oxygen-free radicals and conserves cellular energy sources and decrease oxygen consumption following testicular torsion/detorsion in rats, which in turn can inhibit lipid peroxidation and increase the survivability of the torsional testis [15, 102–107]. Another similar surgical approach to decrease reperfusion injury is ischemic pre- and postconditioning, which is a variation of controlled reperfusion. It is a simple method that provides a new tool to protect testes against I/R injury [108–113]. Nevertheless, Taneli et al. [114] didn't find effective mesh bioprosthesis. Woodruff et al. [115] suggested that the use of cryopreservation must always be offered to these patients regardless of the situation, even after surgery and possibly if patient becomes anorchic.

17.3 Testicular Torsion, I/R Injury in Testis, and Oxidative Stress

The mechanism of tissue damage due to I/R is common to other organs such as the brain, heart, kidneys, and testis. The primary pathophysiologic event in testicular torsion is ischemia followed by reperfusion; thus, testicular torsion/detorsion is I/R injury to the testis [7]. I/R also induce morphological and biochemical changes by both ischemia and reperfusion of the tissues and induce germ cell-specific apoptosis through cell membrane lipid peroxidation, protein denaturation, and DNA damage. These changes lead to testicular apoptosis, atrophy, necrosis, the loss of spermatogenesis, altered hormone production, and male subfertility or infertility. Testicular infarction begins after 2 h of ischemia, becomes irreversible after 6 h, and complete infarction is established after 24 h [116–118].

Even though the presence of some earlier reports about no injury related to testicular torsion/detorsion, the general acceptance is that of bilateral injuries during testicular I/R process. However, the exact mechanism of the bilateral, especially contralateral injury, is obscure. Torsion of the spermatic cord causes a reduction in the testicular blood flow and leads to testicular ischemia on the affected side. The injury in ipsilateral testis is related to tissue hypoxia due to ischemia and overgeneration of ROS and RNS and necrosis of germinal cells [11]. In terms of the contralateral injury, several theories have been postulated, including blood flow alterations, increased adhesion molecule expression, leukocyte migration and damage to the spermatogenic epithelium, autoimmune response, triggered by blood–testis barrier breakdown, subclinical attacks of contralateral testicular torsion, secondary to ischemic damage leading to exposition of antigenic material and formation of antibodies against testicular elements, reflex vasoconstriction in contralateral testicular vessels mediated by sympathetic nerves, release of acrosomal enzymes, paired neuroendocrine or vasomotor response during torsion, presence of an underlying defect in spermatogenesis, underlying testicular defect, and presence of an inherent gonadal abnormality [8, 119–121].

Decrease in blood flow to a tissue causes hypoxia, which results in elevated levels of lipid peroxidation products such as lactic acid, hypoxanthine, and thiobarbituric acid reactive products in ischemic tissue. Increase in blood flow after lipid peroxidation leads to the formation of large amounts of oxygen and/or nitrogen-derived free radicals. Indeed, this causes further damage in the ischemic tissue and is known as reperfusion injury. Reperfusion after ischemia causes oxidative stress, which is characterized by imbalance between ROS and antioxidative defense system, such as superoxide dismutase (SOD), catalase, and glutathione peroxidase (GSH-Px). Although reinstitute of blood flow and oxygenation are very important to salvage affected testis, reperfusion may cause further damage paradoxically in the ischemic tissue. Thus, reperfusion injury may play an important role in testicular damage related to testicular torsion [108]. It has been postulated that gradual detorsion of torsioned testis has a tendency to decrease the degree of testicular reperfusion injury [108, 113].

Torsion (ischemia) and detorsion (reperfusion) of testis is an important phenomenon that has not been completely defined from the perspective of the signaling pathways. I/R is associated with activation of neutrophils, inflammatory cytokines, and adhesion molecules with increased thrombogenicity, release of massive intracellular calcium, and generation of oxygen-derived free radicals [122]. I/R of testis also results in stimulation of proinflammatory cytokines, secretion of tumor necrosis factor (TNF) alpha (α) and interleukin (IL) 1 beta (β), and activation of nitric oxide synthase (NOS) [123, 124]. Hence, this I/R injury is associated with overgeneration of ROS and RNS [125].

It has been also demonstrated that lipid peroxidation activated mitogen-activated protein kinases (MAPKs), which were of vital importance for signal transduction pathways of germ cell apoptosis [126, 127] and MAPKs family plays an important role in the pathogenesis of testis I/R damage. Active MAPKs are responsible for the phosphorylation of several transcription factors and the production of proinflammatory cytokines. These cytokines has multifunctional effect such as proinflammatory response, immunoregulatory response, apoptosis, and certain testicular pathologies, especially testicular torsion. These findings suggest that a block of MAPKs activation might present a potential therapeutic approach in the treatment of testicular torsion [128].

During the detorsion process, with the resumption of blood flow, a huge amount of molecular oxygen is supplied to the tissues and abundant amounts of free oxygen radicals are produced. Ischemia followed by reperfusion leads to a burst of ROS such as superoxide anions, hydrogen peroxide, hydroxyl radical, nitric oxide (NO), and peroxy nitrite, which cause lipid peroxidation and oxidative stress [18, 108, 111, 117]. High levels of ROS endanger sperm function and viability in reproductive tract. Oxidative stress can arise as a consequence of excessive production of ROS or impaired antioxidant defense systems in semen. Thus, oxidative stress causes a range of pathologies that are thought to affect the male reproductive system. ROS-mediated damage to sperm may account for the defective sperm function observed in a high proportion of infertility patients. Production of high levels of ROS in semen have been correlated with reduced sperm motility and damage to sperm nuclear DNA [129]. Some of the experimental studies to determine oxidative stress in animal models of testicular torsion have measured lipid peroxidation.

There are several strategies for decreasing the effect of ROS after I/R. The first strategy involves prevention of ROS formation. Another strategy for reducing ROS effects involves eliminating ROS once they have formed. Free radical scavengers such as SOD and catalase convert toxic ROS to water and oxygen [117]. The ROS are difficult to quantify directly in tissue because of their high reactivity and short half-life. MDA, a stable end-product of lipid peroxidation generated by ROS, is usually used as a good indicator of the degree of lipid peroxidation. MDA levels in testicular tissue increase after testicular injury [130]. GSH is a key component in cell growth, differentiation, and protection. MPO is stored in the primary granules of neutrophils and the enzyme activity is a common measure of neutrophil accumulation.

Testicular torsion has been used in the laboratory animals to reproduce the clinical situation and to study the biological effects of ischemia on both testes and fertility. The end-points include testicular size, weight, histopathology, testicular biopsy score index (Johnsen score), biochemical enzyme parameters, semen characteristics, apoptosis, and general endocrine changes [131–134].

Testicular size correlates primarily with germinal epithelial mass. It has been known that the ratio of ischemia determines testicular salvage. Endocrine function (testosterone production from the Leydig cells) appears to be unaffected if the testis remains viable. Nevertheless, exocrine function (spermatogenesis) is unusually abnormal because ipsilateral ischemia has affected the contralateral testis. Spermatogenesis is significantly impaired in most patients after testicular torsion. Antisperm antibodies have been implicated as a result of the exposure of testicular tissue to the bloodstream after ischemic event. Detection of antisperm antibodies after torsion is variable. Many reports have confirmed abnormal semen analysis after acute testicular torsion [29].

To avoid testicular loss and eventual impaired fertility, prompt diagnosis and immediate surgery are the most important factors in patients for the treatment of testicular torsion. It has been postulated that unilateral testicular torsion causes damage to the ipsi- and contralateral testis and reduces future fertility. There are lots of experimental studies in animals about the short- and long-term (for example 60 days) [5] impacts of testicular torsion on fertility potentials in terms of semen analysis and fertility rate [135]. Moreover, animal models usually show histologic lesions and biochemical changes and, abnormal levels of antisperm antibodies after experimental testicular torsion. However, in human studies such an effect has not been fully proven in terms of long-term impacts by histopathologic examination or other conventional assays of spermatogenesis. Abnormal semen results have been noted in 40 to >60% of patients after unilateral testicular torsion [136]. Clinical problems, medicolegal issues for surgeon, ethical approach, and biopsy are limited to long-term and prospective clinical studies. Testicular ipsi- and/or contralateral biopsies are traumatic management and can reduce the future fertility of humans. Thus, there are case reports and retrospective analysis studies on testicular torsion in humans. Surprisingly few follow-up studies have been reported on semen analysis and fertility issues in humans after testicular torsion. In the light of these findings, it can be said that testicular torsion is an I/R injury and lead to oxidative stress. Thus, it affects semen quality and lead to infertility.

It has been demonstrated that testicular venous serum testosterone concentrations have not been affected after experimental torsion repair in rats. The resting testicular vein testosterone concentrations or those in response to in vivo ED_{50} or ED_{100} luteinizing hormone administration were little changed after repair of 1 h, 720° torsion in the rat, whether at acute (1 day) or chronic (30 day) periods [137]. However, another reports showed that reperfusion-induced oxidative stress in testicular torsion might play a role in Leydig cell dysfunction, as well as by acting directly in germ cell apoptosis [10].

Recently, Yang et al. [138] reported a 20-year retrospective study (between 1990 and 2010) in a single institution. They advocated immediate surgical exploration with suspected testicular torsion. Long-term hormonal levels are within the normal range regardless of the fate of the testis. They suggest that further follow-up studies are needed to confirm fertility after testicular torsion.

Numerous studies have used testicular torsion in rats as a model to study the effects of I/R on testicular tissue. Administration of many antioxidant agents and ROS scavengers provides significant rescue in experimental model of testicular torsion; however, none of these agents have been used as adjunctive therapy to torsion repair in humans [139].

17.4 Experimental Testicular Torsion and Antioxidant Therapy

Under normal condition, free radicals are produced and their effects are counterbalanced by the endogenous antioxidant system. When ROS generation exceeds the defense mechanisms' capacity to control, oxidative stress is occurred and contributes to reversible or irreversible cell damage [7, 140]. All living aerobic cells are normally exposed to some ROS, but if ROS levels rise, oxidative stress may occur, which results in oxygen and oxygen-derived oxidants, and in turn increases the rates of cellular damage. Oxidative stress has been shown to be a major cause of male infertility; a large proportion of infertile men have elevated levels of seminal ROS [141]. ROS can directly damage spermatozoa by inducing peroxidation of the lipid-containing sperm plasma membrane, which decreases its integrity, and may also affect sperm motility by damaging the axonemal structure [142–144].

Like all cells living under aerobic conditions, spermatozoa produce ROS, mostly originating from normal metabolic activity. High concentrations of ROS cause sperm pathology such as adenosine triphosphate (ATP) depletion leading to insufficient axonemal phosphorylation, lipid peroxidation, and loss of motility and viability. However, spermatozoa and seminal plasma contain a battery of ROS scavengers, including enzymes, such as SOD, catalase, and GSH peroxidase/reductase system, and also a variety of substances with SOD- or catalase-like activities, such as ascorbic acid, GSH, pyruvate, taurine, hypotaurine, albumin, and carnitine. The fine balance between ROS production and scavenging as well as the right timing for ROS production is of paramount importance for the acquisition of fertilizing ability by spermatozoa. Excessive ROS generation that overcomes the ROS scavenging ability of human spermatozoa appears to be related to male infertility, such as testicular torsion [141, 144–146].

There are two available antioxidant strategies. In the first strategy, the superoxide radical and hydrogen peroxide are removed using specific enzymes such as SOD, catalase, and GSH peroxidase (Px), either by administration of these enzymes or by increasing their in vivo activities. In the second strategy, radical generation is prevented. However, these systems are not always fully operative [118]. Normally, a balance exists between concentrations of ROS and antioxidant scavenging systems. One of the rational strategies to counteract the oxidative stress is to increase the scavenging capacity of seminal plasma with antioxidants.

Experimental spermatic cord torsion has been widely studied under several different aspects, including the effect of ischemia on testicular structure, I/R, contralateral testis injury, therapy with pharmacologic agents, and measures to avoid or reduce these injuries. There are conflicting results because of several factors such as animal type and species, age, model of ischemic injury, ischemia and/or reperfusion time, and technique for the evaluation of testicular injury [26]. Several studies have claimed that the contralateral testis is not affected by unilateral torsion [147–150], while others favored the opposite view [12, 151–153]. However, animal studies have demonstrated that significant testicular damage occurs in both testes, and contralateral testicular damage may be avoided by early removal of the torsed testis and pretreatment with antioxidant therapy. Experimental models of testicular torsion are valuable tools to evaluate the relationship between the degree and duration of torsion with the blood flow and resultant damage to the torsed and/or untorsed testis. The results, however, have been conflicting. With 720° of torsion, several studies have reported reduction in testicular blood flow varying from 61.7 to 100%, with ischemic injury after periods of time as diverse as 1 and 8 h [26]. Numerous studies have evaluated the efficacy of antioxidants in male infertility [146, 154, 155]. Furthermore, pretreatment with antioxidants and ROS scavengers has been shown to prevent reperfusion injury in testes.

Korean red ginseng is a potent antioxidant and free radical scavenger. Kim et al. [156] investigated the effects of it on testicular damage in a rat testicular I/R injury model. Researchers have evaluated superoxide generation (by using a lucigenin-enhanced chemiluminescence assay) and the blood levels of ROS (by using the free oxygen radical test-FORT). The FORT provides an indirect measure of hydroperoxides, which are a useful measure of oxidative stress because they indicate the presence of intermediate oxidative products of lipids, amino acids, and peptides. In this study, Korean red ginseng attenuated the increase in the testis FORT level and recovered the testis dysfunction caused by ischemia and subsequent reperfusion in the rat testis by suppressing superoxide production.

Coenzyme Q10 is an essential component for electron transport in oxidative phosphorylation of mitochondria. It is a potent antioxidant, a membrane stabilizer, and cofactor in the production of ATP by oxidative phosphorylation. It can reverse endothelial dysfunction by preventing oxidative and nitrative stress, scavenging free radicals, and inhibiting lipid peroxidation and inflammation. Erol et al. [157] demonstrated that coenzyme Q10 administration before reperfusion period of testicular torsion provided a significant reduction in the level of testicular MDA and expression of inducible nitric oxide synthase (iNOS), endothelial NOS (eNOS), and germ cell-specific apoptosis.

Aktoz et al. [158] showed a significant reduction in the activity of TUNEL, eNOS, and a rise in the expression of testosterone in testes tissue of I/R treated with quercetin therapy. The TUNEL method detects fragmentation of DNA in the nucleus during apoptotic cell death in situ. Quercetin prevented oxidant injury and cell death by scavenging oxygen radicals, protecting against lipid peroxidation, and chelating metal ions.

Montelukast is a new anti-inflammatory drug with antioxidant properties that interferes directly with leukotriene production (5-lipoxygenase inhibitors) and/or reception (leukotriene receptor antagonists). Ozturk et al. [159] investigated whether montelukast (MK-0476) could protect the testis from injury associated with testicular torsion and detorsion. To accomplish this, they measured the MDA, GSH, and MPO levels and performed for histological examination. Montelukast significantly reduced the I/R-induced elevation in testes tissue. MDA and MPO levels and testicular GSH levels were significantly higher when compared with the I/R/untreated rats.

In an experimental study by Shirazi et al. [160], the comparison of the protective effects of papaverine, lidocain, and verapamil on sperm quality of the testis after I/R-induced damage was reported. In conclusion, verapamil and lidocaine, as antioxidants, but not papaverine, as vasodilator, had beneficial effects on semen analysis parameters after testicular torsion. Gao et al. [103] investigated whether verapamil and hypothermia protect spermatogenesis of torsioned testes; they showed that either verapamil or local hypothermia can enhance testicular resistance against I/R injury and the combination of two can more efficiently prevent the germ cells from apoptosis.

Yang et al. [161] investigated antioxidant and lipid-lowering agent simvastatin on testicular I/R damage in rats. Simvastatin protected testes from torsion/detorsion injury in a dose-dependent manner. They speculated that this effect may involve attenuating nuclear factor kappa B (NF-κB) activation and decreasing oxidative stress induced by torsion/detorsion.

Apoptosis (programmed cell death) is characterized by a variety of changes resulting in the recognition and phagocytosis of apoptotic cells. Caspases (cysteinyl aspartate-specific proteinases) play a central role in the regulation of apoptosis in the human seminiferous epithelium, sperm differentiation, and testicular maturity. However, caspases have been implicated in the pathogenesis of multiple andrological pathologies such as impaired spermatogenesis, decreased sperm motility, increased levels of sperm DNA fragmentation, and apoptosis in testicular torsion [162]. In recent years, poly (adenosine triphosphate-ribose) polymerase (PARP) has been implicated in the process of apoptotic cell death and acts as a "death substrate" for caspases. Several PARP inhibitors such as nicotinamide, 3-aminobenzamide, 1,5-dihydroxyisoquinoline, or 4-amino-1,8-naphthalimide have been examined in testicular torsion/detorsion model [163–165]. Kar et al. [166] wanted to show biochemical, histopathological, and apoptotic changes caused by unilateral spermatic cord torsion in ipsilateral and contralateral testis and the effect of the PARP inhibitor, nicotinamide, on these changes in early and late periods. I/R injury-related changes were assessed by levels of MDA and total and free GSH in the serum. Rats were divided into two major groups as early and late periods. Bilateral orchidectomy was performed by the end of the fourth hour in early and 2 months after I/R in late groups. Nicotinamide decreased tubular damage and apoptosis in early and late periods in both testicles.

Lysiak et al. [19] reported that I/R of the testis results in germ cell-specific apoptosis, followed by a reduction in testis weight and daily sperm production. They said that this has been associated with an increase in the adhesion of neutrophils to testicular subtunical venules and an increase in ROS. In the study, they demonstrated a partial rescue with the catalase, Cu-Zn SOD, catalase plus SOD, and M40403 (a nonpeptide mimic of SOD) by infusing right femoral vein on 30 days after I/R of the testis, and the administration of ROS scavengers significantly reduced the I/R injury. Moon et al. [167] investigated the expression of tyrosine kinase receptors A and B, and p75 growth factor receptor in I/R rat testis. They reported that tyrosine kinase receptors A and B, but not p75 growth factor receptor, are involved differently in the survival of testicular cells during acute I/R injury. Tyrosine kinase receptor A increases especially related to germ cell survival.

Recently published studies demonstrated that MAPK family is of vital importance for signal transduction pathways and belong to the extracellular signal-regulated kinase family and are serine/threonine protein kinases activated by a variety of cell surface receptors. Indeed, the MAPK, MAPK3/MAPK1 (also named extracellular signal-regulated kinase [ERK] 1/2), MAPK8 (also named c-Jun-NH_2-terminal kinase [JNK]), and MAPK14 (also named p-38), has a role in the pathogenesis of testicular I/R injury [126, 128, 168–170]. Active MAPKs are responsible for the phosphorylation of a variety of effector proteins, including several transcription factors, such as NF-κB, activating protein (AP) 1, and the production of proinflammatory cytokines, including TNF-α. These cytokines have multifunctional effect such as proinflammatory response, immunoregulatory response, apoptosis, and certain testicular pathologies, especially testicular torsion [169]. NF-κB, a nuclear transcription factor, controls a number of cellular processes including the immune response, inflammation, proliferation, apoptosis, and calcium homeostasis and also plays a role in the testicular I/R damage. Minutoli et al. [168] studied the involvement of MAPKs (ERK 1/2, JNK and p38) activation in NF-κB knockout mice in testicular I/R model. In addition, it has been demonstrated that PD98059 is a flavonoid and inhibits MAPK3/MAPK1, blunted MAPK3/MAPK1 and MAPK8, and decreased TNF expression and improved the testicular damage caused by I/R injury in both testes [169]. Recently, in the other study, same researchers reported that testicular I/R may cause two different pathways of organ damage: (1) an early pathway involving NF-κB, MAPK3/MAPK1, MAPK8, MAPK14 and TNF-α which produces an inflammatory derangement of the testis and (2) a more delayed response involving MAPK3/MAPK1 only, TNF-α, BAX, and caspase-3 and -9 that likely activates apoptosis [128]. Moon et al. [170] demonstrated that both the Akt/protein kinase B and ERK1/2 activation increased in damaged seminiferous tubules, suggesting that these signaling pathways contribute to the survival of testicular cells after testicular I/R injury.

Hypoxia-inducible factor (HIF)-1 is a transcription factor that regulates response to hypoxia and oxygen homeostasis in many tissues involving testis. Powell et al. [171] showed that HIF-1 played a role in the cellular response to hypoxia after torsion.

Al-Maghrebi et al. [172] recently reported that increased survivin expression in I/R injury was correlated with oxidative stress, apoptosis, and spermatogenic damage. Down-regulation of testicular survivin expression is a potential new target for the prevention of germ cell death during testicular torsion.

Sildenafil, vardenafil, and tadalafil are the phosphodiesterase type 5 inhibitors. They are new class of vasoactive drugs that have been developed for the treatment of erectile dysfunction. The beneficial effect of intraperitoneal vardenafil [173] and sildenafil [174, 175] on I/R injury in a rat model of testicular torsion has been investigated. Verdanafil treatment exerted antiischemic effect, increased total antioxidant enzymes levels, and decreased MDA levels, germ cell apoptosis, and eNOS and iNOS levels. Similarly, catalase and SOD activities are increased, and MDA concentrations are decreased with sildenafil treatment. However, the effect of oral vardenafil [176] and oral sildenafil and vardenafil [177] has been investigated and they are found not to have protective effect on testicular I/R injury.

Regulation of testicular function was involved in several growth factors, such as epidermal growth factor (EGF), insulin-like growth factor (IGF) I, transforming growth factor-β, fibroblast growth factor, insulin, and erythropoietin. These factors are able to influence Leydig cell steroidogenesis by binding to specific plasma membrane receptors [178]. In addition, pretreatment with some growth factors, such as EGF, vascular endothelial growth factor (VEGF), and IGF, has been shown to prevent I/R damage in testis. Uguralp et al. [179] investigated the effects of sustained and local administration of EGF on improving bilateral testicular tissue after torsion. MDA levels were significantly lower in the EGF groups than in the control groups. GSH-Px levels in the control group were significantly higher than in the EGF group. Their study showed that local and sustained EGF release after testicular torsion improves bilateral testicular injury. VEGF, an angiogenic peptide, mediates angiogenesis and vasculogenesis and promotes endothelial cell survival. VEGF expresses in testis, prostate, and seminal vesicles and presents in high concentration in semen. Thus, it could play an important role in male reproductive system. Recently, it has been suggested that VEGF has the protective effect on different I/R injury models. Hashimoto et al. [180] reported that the spontaneous increase in VEGF in I/R testis may lead to a reduction of ischemic testicular damage. Tunçkiran et al. [181] evaluated the effect of VEGF injection into the testis, especially on spermatogenesis and apoptosis. They measured mean seminiferous tubular diameter, germinal epithelial cell thickness, mean testicular biopsy score, and apoptosis (caspase-3-positive cells). Researchers found that VEGF administration caused a significant decrease in testicular caspase-3-positive cells compared with I/R group. However, administering VEGF before reperfusion might have the potential to reduce the long-term histopathologic damage after testicular torsion. Other growth factors are IGF-1 and IGF-2. They are locally produced in the testis and may regulate the balance between pro- and antiapoptotic proteins at a cellular level. It was interesting that IGF-1 levels decreased in the contralateral testis following testicular torsion [182] and IGF-1 administration seems to lower the levels of germ cell apoptosis, which may be important for protecting the testes from torsion/detorsion injury [183].

Previous studies have demonstrated that NF-κB played a critical role in I/R injury. NF-κB transcriptionally activates many genes involved in the inflammatory process, including intercellular adhesion molecule 1 (ICAM-1), vascular cellular adhesion molecule 1 (VCAM-1), and E-selectin (ELAM-1), which are important mediators of the inflammatory process in reperfusion injury. Apoptosis always occurs in the ischemic-reperfused tissue, and the functions of the tissue never remain unchanged. Many factors, including ROS, IL-1, and TNF-α, cause IκB phosphorylation. Sulfasalazine acts as a potent inhibitor of NF-κB by inhibiting I kappa B phosphorylation, thereby preventing its translocation into the nucleus and decreasing adhesion molecule expression. Zhao et al. [184, 185] concluded that sulfosalazine prevented apoptosis in spermatogenic cells after the testicular torsion inhibiting NF-κB activation.

Tumor necrosis-α is produced by round spermatids, pachytene spermatocytes, and testicular macrophages in the testis. The type 1 TNF receptor has been found on Sertoli and Leydig cells and numerous studies suggest a paracrine mode of action for TNF-α in the normal testis. IL-1α has been reported to be produced by Sertoli cells, testicular macrophages, and possibly postmeiotic germ cells. IL-1 receptors have been reported on Sertoli cells, Leydig cells, testicular macrophages, and germ cells suggesting both autocrine and paracrine functions. While these proinflammatory cytokines have important roles in normal testicular homeostasis, an elevation of their expression can lead to testicular dysfunctions. A pivotal role for IL-1β in the pathology of testicular torsion has been recently described whereby an increase in IL-1β production after reperfusion of the testis is correlated with the activation of the stress-related kinase, JNK, and ultimately resulting in neutrophil recruitment to the testis and germ cell apoptosis [186].

Several studies have suggested that the bcl-2 family of proteins, including pro- (bax, bak, bid, bad, and bim) and antiapoptotic (bcl-2, bcl-xs) molecules, plays a crucial role in the regulation of germ cell apoptosis. In addition, bax/bcl-2 system is important for the evolution of normal spermatogenesis. Sukhotnik et al. [187] reported that in ipsilateral testis, bax/bcl-2 ratio did not change significantly, and the elevation of germ cell apoptosis was not marked; in the contralateral testis, germ cell apoptosis increased after 6 h of detorsion, achieved statistical significance after 24 h, and decreased after 72 h of detorsion and was initiated by decreased bcl-2 messenger ribonucleic acid (RNA) levels and elevated bax/bcl-2 ratio within the first 6 h of detorsion.

I/R of the testis results in germ cell-specific apoptosis and can lead to decrease spermatogenesis. Germ cell-specific apoptosis after I/R of the testis has been shown to be correlated with and dependent on neutrophil recruitment to the testis after I/R. In experimental animals models, cell adhesion molecules such as E- and P-selectins, ICAMs, CD44 play an important role in neutrophil recruitment to the testis after torsion-/detorsion and resulting germ cell-specific apoptosis. After torsion/detorsion of the testis, cytokines (both TNF-α and IL-1β) increase and they regulate selectins expression. Therefore, selectins expression in the testis may be a diagnostic indicator of I/R of the testes and blockade of selectins function may have therapeutic values [188–190]. Studies that used E-selectin-deficient mice have demonstrated that E-selectin expression is critical for neutrophil recruitment to subtunical venules in the testis after I/R and for the resultant germ cell-specific apoptosis. Celebi and Paul [188, 189] administered a function-blocking monoclonal anti-mouse E-selectin or P-selectin antibody (FBMABs), and these FBMABs inhibited significantly inflammation and neutrophil migration to the I/R-induced testis. They suggested that anti E-selectin antibody alone or combination of anti E- and P-selectin antibody have been a new point of view by antiadhesion therapy to the clinical management. In addition, Moon et al. [191] showed increased expression of CD44 in rat testis I/R model.

Lysiak et al. [192] demonstrated that the proinflammatory cytokines, TNF-α and IL-1β, were stimulated after I/R as was the phosphorylation of JNK. The downstream transcription factors of JNK, ATF-2, and c-Jun were also phosphorylated at specific times after I/R of the testis. Activation of the JNK stress-related kinase pathway was correlated with an increase in E-selectin expression and neutrophil recruitment to the testis after I/R. Intratesticular injection of IL-1β also caused JNK phosphorylation and neutrophil recruitment to the testis. These results speculated that testicular I/R injury stimulated IL-1β expression, which lead to activation of the JNK signaling pathway and ultimately E-selectin expression and neutrophil recruitment to the testis. This provided the first evidence of a cytokine/stress-related kinase signaling pathway to E-selectin expression in vivo.

Recently, Tsounapi et al. [21] investigated the protective effects of sivelestat sodium hydrate, a neutrophil elastase inhibitor, on ipsilateral and contralateral testes after unilateral testicular I/R injury in rats. They demonstrated that unilateral testicular I/R resulted in marked increases in bilateral testicular MDA levels, MPO activity, and enhanced expression levels of heat shock protein (HSP) 70. Sivelestat ameliorated the I/R-induced ipsilateral testicular damage bilaterally through its ability to inhibit neutrophil elastase. By inhibiting neutrophil elastase, it has a direct action to the accumulated and activated leucocytes, offering efficient protection against the production of oxygen radicals and cytokines.

Trapidil is an antianginal drug which is a phosphodiesterase and platelet-derived growth factor inhibitor. It has effects of nitroglycerine-like vasodilatation, inhibition of platelet aggregation via blockage of thromboxane A_2, reduction of lipid peroxidation, IL-6 and IL-12, and procoagulant activity by inhibiting the CD40 pathway in monocytes. Bozlu et al. [20] investigated the protective effects of trapidil on long-term (60 days after I/R) histologic damage; and Somuncu et al. [193] also investigated on MDA, NO, and total sulfhydryl levels. These studies demonstrated that trapidil decreased free oxygen radical formation and attenuated histopathological damage testes.

There is a consistent basal expression of erythropoietin mRNA, and this expression is increased by hypoxia; this expression has been also detected in Sertoli and peritubular myoid cells in testis. It has also been observed that testosterone production has been stimulated by erythropoietin in rat Leydig cell and that it influences steroidogenesis. Testosterone production has been increased by intravenous erythropoietin administration in patients with renal failure. It has been suggested that this effect of erythropoietin might act directly on human Leydig cell function, which requires activation of a protein kinase-C-dependent pathway [178]. Moreover, erythropoietin has free radical scavenger, antiapoptotic and anti-inflammatory activity. In addition, recently published studies have shown a protective role of erythropoietin on experimental I/R injury in testis [178, 194, 195]. Moreover, one of the recently published studies has reported that administration of darbepoetin-α, a novel erythropoietin protein, caused reduction of MDA and NO levels and an increase of GSH levels in testis torsion [196].

Ozturk et al. [197] demonstrated that platelet-activating factor (PAF) antagonist BN-52021 decreased MDA values and the testicular injury score and increased SOD, catalase, and GSH-Px values in the BN-52021-treated group compared to in the I/R group in rats with reperfusion damage due to unilateral testicular torsion.

Another experimental study was designed to evaluate the effects of VIP on lipid peroxidation and histopathology in both testes after unilateral testicular torsion and detorsion. It has antioxidant, anti-inflammatory, antiapoptotic, neurotrophic, and immunomodulatory actions. Immunohistochemical studies have demonstrated the presence of VIP and its receptors in the testis, as well as other parts of the male urogenital tract of various animal species and humans. Can et al. [198] demonstrated that VIP could protect testicular tissue from detorsion injury.

Recent studies showed that ethyl pyruvate (anti-inflammatory, antioxidant, and antiapoptotic) [127], angiotensin-converting enzyme inhibitor (lisinopril), angiotensin II type 1 receptor blocker (losartan) [199], and PAF antagonist BN-52021 (reduced ICAM-1 expression) [197] reduced germ cell-specific apoptosis and oxidative stress on testicular torsion/detorsion model in rats. Sozubir et al. [200] reported that Insl3 mutant mice predisposes to testicular pathologies such as torsion, atrophy. Thus, Insl3 is a candidate signaling molecule in human testicular torsion.

NO, a water- and lipid-soluble, freely diffusible gaseous molecule with a short half-life, is formed from L-arginine and molecular oxygen by a family of NOS. It has been demonstrated that the three isoforms of NOS (iNOS, eNOS, and neuronal nitric oxide synthase [nNOS]) involved in the pathogenesis of testicular I/R injury in association with germ cell death, through either necrosis or apoptosis. Expression of eNOS in testes has been demonstrated in Leydig cell, Sertoli cell, spermatocyte, spermatids, and endothelial cells [122, 201, 202]. Although several investigators have reported that NO which has dual effects on cell survival and death is an important signal transduction molecule in I/R injury. Nevertheless, there are conflicting results in testicular torsion. Dokucu et al. [203] reported the protective effect of NO on contralateral testis in unilateral testicular torsion. In Barlas and Hatibo lu's study [204], L-arginine, precursor of NO, was used to increase NO synthesis; this caused a decrease in reperfusion injury. L-nitro arginine methyl ester (L-NAME), a competitive inhibitor of NO synthase, was used to reduce NO formation; this concluded a negative effects on testicular I/R injury. However, Ozokutan et al. [205] also used L-arginine and (L-nitro monomethyl arginine) L-NMMA, a competitive inhibitor of NO synthase. Inhibition of NO synthesis with L-NMMA significantly improved I/R injury, increasing NO production with L-arginine-increased testicular damage. In Dokucu's study [206], they suggested that molsidomine has a protective effect against I/R injury in rat testes. Molsidomine, a precursor of NO, reduced MDA and histopathologic score and increased expression of Sonic hedgehog and HIF α-1 in testes I/R injury. L-NAME, the NOS inhibitor, reversed the protective effect of molsidomine against I/R injury. The effect of molsidomine may be related to reducing the effects of oxidative stress in testes. Also, Ozturk et al. [207] demonstrated that NO regulated adhesion molecules expression such as tenascin, lectin, and ICAM-1, which was the proof of inflammation in torted testicle; furthermore, molsidomine played an important role in the prevention of I/R injury.

Gabexate mesilate, a synthetic serine protease inhibitor, has anticoagulant activity. It inhibits trypsin, plasmin, kallikrein, and thrombin; it has been used for the treatment of acute pancreatitis and disseminated intravascular coagulation and for anticoagulation during hemodialysis. Gezici et al. [208] found that the treatment with gabexate decreased MDA levels significantly when compared to the I/R/untreated group. In addition, gabexate protected the levels of antioxidant enzymes such as SOD, catalase, and GSH-Px. In addition, the Tc-99m pertechnetate uptake ratio and the perfusion index were significantly increased after gabexate treatment.

So far, it has been found that capsinoids possess the biological properties of antitumor, antioxidant, and antiobesity. Chili peppers are the major source of nature capsaicinoids, which consist of capsaicin, dihydrocapsaicin, etc. Sarioglu et al. [209] showed that capsaicin effectively prevented apoptosis in the contralateral testis after ipsilateral torsion.

Rosuvastatin is one of the synthetic statins, which is used for hyperlipidemia treatment. It has been experimentally shown that it has antioxidant and anti-inflammatory effect, decreasing leukocyte adhesion and thrombocyte aggregation. Recently, Karakaya et al. [210] measured testis basal blood flow with laser Doppler flowmeter and they reported that rosuvastatin could protect tissue perfusion in the experimental testicular torsion model.

Romeo et al. [211] aimed to evaluate the effects of raxofelast, a vitamin E-like antioxidant, on lipid peroxidation and histopathology in both testes after unilateral testicular torsion and detorsion. Conjugated dienes (CD) levels, an index of lipid peroxidation, and testis histopathology were evaluated. Testicular I/R in untreated rats produced high testicular levels of CD. Furthermore, histological examination revealed marked damage to the testis interstitium with severe hemorrhage and edema. The administration of raxofelast lowered CD levels and significantly reduced histological damage. Antonuccio et al. [126] supported this study evaluating testicular JNK, ERK, and TNF-α activation by Western blot analysis, and mRNA expression and CD using a spectrophotometer technique. These data suggested that the hydrophilic vitamin E-like antioxidants are good candidates for testicular torsion.

However, Turan et al. [212] investigated the role of lipid peroxidation in ipsilateral and contralateral testicular reperfusion injury following unilateral testicular torsion and the effect of vitamin E in the management of this injury. There were no significant differences between right and left testes within groups or between right or left testicular MDA values in different groups. The results suggested that vitamin E given before or after detorsion of testes is not useful in preventing testicular reperfusion injury.

Recently, surprising results were observed in the coadministration of vitamin C and dopamine on I/R injury after experimental testicular torsion in rats. Investigators could not find more beneficial effects from the combined administration of vitamin C and dopamine than vitamin C alone, although it was better than the dopamine administration alone [213].

Lycopene is a potent antioxidant. It was reported that lycopene treatment in men with idiopathic infertility provided an improvement in male infertility, especially in sperm motility, concentration, and morphology. In addition, regular lycopene intake for 9 months in oligospermia patients improved the sperm characteristics. Moreover, testicular I/R model in rats, lycopene increased the motility in bilateral testes and decreased the rate of abnormal sperm in ipsilateral testis, but did not rise sperm concentration in bilateral testes. Lycopene also restored antioxidant enzyme activities, but not reduced glutathione level [6].

Ginkgo biloba (EGB 761), a free radical scavenger, is investigated by Akgül et al. [214]. Ginkgo biloba decreased MDA, nitrate–nitrite concentrations on unilateral testicular torsion model. Moreover in another study, treatment with Ginkgo biloba showed significant decrease in apoptotic cells, eNOS, and iNOS activities in testes [215].

L-carnitine, a naturally occurring enzymatic antioxidant, is a necessary factor in the utilization of long chain fatty acids to produce energy. L-carnitine exhibits a wide range of biological activities including anti-inflammatory, antiapoptotic, neuroprotective, cardioprotective, and gastroprotective properties. Furthermore, these effects are attributed to its antioxidative and free radical scavenging activity [18, 216–219]. In addition, it plays a pivotal role in the maturation of spermatozoa within the male reproductive tract. Epididymal plasma contains the highest levels of L-carnitine found in the human body, and initiation of sperm motility occurs in parallel to L-carnitine increase in the epididymal lumen. It is known that L-carnitine prevents the formation of ROS and scavenges free radicals and it protects cells from peroxidative stress. Moreover, it plays a key role in sperm metabolism by providing readily available energy for use by spermatozoa, which positively affects sperm motility, maturation, and the spermatogenic process [146]. The use of L-carnitine and its derivatives in therapy has been proposed in recent years for treatment of male infertility, and a number of controlled and uncontrolled human and animal studies published to indicate their possible application. In several clinical studies with infertile patients, seminal plasma total carnitine level was found to be low and carnitine supplementation improved reproductive function [220–222]. An increase in spermatozoa motility has been observed in treated patients affected with idiopathic forms of oligospermia, asthenozoospermia, and oligoasthenoteratozoospermia [223–225] or affected with bacterial inflammation of the accessory sexual gland [226, 227]. However, the mechanisms by which carnitine controls male fertility has not yet been clearly identified. In our study, this is the first time we showed that, L-carnitine has a protective effect in experimental testicular torsion/detorsion model in rats; L-carnitine decreased MDA levels and improved histological damage in testicular tissue [228]. In addition, it has been reported that L-carnitine has the protective effect on apoptosis I/R injury in testis [229–231]. Recently, the combined use of L-carnitine (an antioxidant and free radical scavenger) and meloxicam (COX-2 inhibitor) was investigated in the treatment of cellular damage caused by testicular torsion. Meloxicam caused the most dominant inhibitory effect on the expression of specific genes of inflammation as well as combination therapy. Because the effect of these inflammatory genes was still evident 4 days after detorsion, combination therapy using these agents could be administered until the late postoperative period to achieve the most reproductive treatment of testicular torsion at the cellular level. This process would prevent the autoimmune activity against sperm cells and protect the innocent contralateral testis from the insult of the antisperm antibodies [232].

N-acetylcysteine is a small molecule containing a thiol group. It was first introduced as a mucolytic drug in the 1960s, after that *N*-acetylcysteine has been found to have an antioxidant effect, acting as a free radical scavenging agent and a precursor of GSH. Recently, Akta et al. [233] suggested that early administration of *N*-acetylcysteine may have a protective effect in the rat experimental testicular torsion/detorsion models. *N*-acetylcysteine significantly reduced thiobarbituric acid reactive substances (TBARS) levels and histological parameters of spermatogenesis and improved in bilateral testes when compared with torsion and torsion/detorsion groups. In the other two studies [234, 235] *N*-acetylcysteine prevented the oxidative effects of reperfusion injury after torsion, regulating antioxidant enzyme activities and germ cell apoptosis. Another study showed protective effect of erdosteine in testis I/R injury [236]. Also, erdosteine is a mucolytic agent, and owing to the presence of two sulfhydryl group in its metabolites, can act as a free radical scavenger and this component of the mechanism of action is likely to be involved in the testicular injury induced by I/R.

Curcumin is a major active component of turmeric, which is extracted from the powdered dry rhizomes of *Curcuma longa*. Curcumin exhibits a wide range of pharmacological activities, including antioxidant, anti-inflammatory, and antitumoral effects. It reduces production of proinflammatory cytokines and activation of NOS, cyclooxygenase-2, lipooxygenase, AP-1, and NF-κB. There are a few studies investigating curcumin on testis I/R injury. Basaran et al. [237] demonstrated the protective effect of curcumin on testicular I/R injury for the first time. They showed that the curcumin treatment diminished iNOS and eNOS immunoreactivity in ipsilateral and contralateral testes. However, it had no effect on testicular MDA levels. Later, Wei et al. [238] reported that the rats treated with curcumin had significant decreases in MDA level and xanthine oxidase activity and had significant increases in heme oxygenase-1 protein expression level and testicular spermatogenesis in ipsilateral testes, compared with I/R group.

Nigella sativa L. and thymoquinone are effective free radical scavengers. They have antioxidant activity and protect against the damage caused by free radicals. Gökçe et al. [239] showed that thymoquinone administration reduced histological damage, MDA, total oxidative stress, and oxidative stress index values, but did not affect the total antioxidant capacity and MPO activity in the I/R testes.

Edaravone, a free radical scavenger and anti-inflammatory agent, is used for the clinical treatment of cerebral ischemia, protecting neurons. Its effect is especially on hydroxyl radicals. To induce testicular I/R, Tamamura et al. [240] clamped testicular artery in the rat testis for 30 min. Blood flow and NO release were monitored with a laser Doppler flowmeter and an NO-selective electrode, respectively. Tissue MDA, nitrate-nitrite, 8-hydroxydeoxyguanosine (8-OHdG), MPO, and HSP 70 and its mRNA were measured. In this research, investigators showed that testicular I/R resulted in marked increases in nitrate-nitrite, MDA, 8-OHdG and MPO levels, and in the expression levels of HSP 70 and its mRNA in rat testes. Edaravone ameliorated the I/R-induced testicular damage through its ability to scavenge free radicals and anti-inflammation.

Uguralp et al. [241] evaluated the effects of resveratrol on testicular I/R injury. Resveratrol, a nonflavinoid polyphenol found in great amounts in grapes, mulberries, peanuts, and in red wine, exhibits a wide range of biological activities such as antioxidant, anti-inflammatory, and antitumor properties. In the study, the testicular tissue levels of MDA and GSH and histological changes were determined. In rats treated with resveratrol, MDA levels were significantly decreased, GSH levels were significantly increased, and the mean testicular tissue injury score in the resveratrol group was significantly lower than in torsion and detorsion groups. This study demonstrated that intraperitoneal administration of resveratrol in rats may protect testis against injury associated with reperfusion. Moreover, in the other study resveratrol decreased germ cell apoptosis in the ipsilateral testes [242]. In addition, tea polyphenols had a protective effect against testicular torsion [243].

Another experimental study was designed to evaluate the effects of garlic extract on testicular tissue after testicular torsion/detorsion. Garlic extract, an antioxidant, attenuated the generation of toxic-free radicals and inhibited the xanthine oxidase-mediated I/R injury cascade [244].

α-Lipoic acid acts as antioxidant not only through free radical quenching, but also indirectly through recycling of other cellular antioxidants. Free lipoic acid has not been detected in human beings because it is bound to proteins. However, after therapeutic applications, free lipoic acid can be found in the circulation. Guimarães et al. [140] studied the protective effects of α-lipoic acid in experimental testicular torsion. α-Lipoic acid pretreatment promoted decreased TBARS concentrations and increased GSH levels in testes tissue and total antioxidant power in plasma.

Etensel et al. [245, 246] investigated the protective effects of dexpanthenol after torsion at the first, fifth minute, and first hour and testicular atrophy at 60th day on testicular I/R injury model. Dexpanthenol (provitamin B5) is the biologically active alcohol of pantothenic acid (vitamin B5). Dexpanthenol has approved FDA and it is safe, cost-effective, and readily available agent. They reported that dexpanthenol attenuated lipid peroxidation and tissue damage and significantly prevented testicular atrophy at 60th day of testicular torsion.

Avlan et al. [247] investigated the effects of selenium on ipsilateral and contralateral testicular damage after unilateral testicular torsion/detorsion. The highest MDA and the lowest SOD values were determined in both testes in torsion/detorsion group. There were statistically significant differences in MDA levels and SOD enzyme activities in torsion/detorsion group compared with control group. These results suggested that I/R injury occurred in both testes after unilateral testicular torsion/detorsion and that selenium administration before detorsion prevented reperfusion injury in testicular torsion. These findings were also confirmed histopathologically.

Caffeic acid phenethyl ester (CAPE) is an active component of honeybee propolis extracts and has been used as a folk medicine for many years. CAPE was known to have antioxidant, anti-inflammatory, antimicrobial, immunomodulatory, and antitumoral effects. It has been investigated that the effects of CAPE in rats subjected to testicular torsion/detorsion [248–250]. Pretreatment with CAPE attenuated the testicular injury, as well as the increase in the tissue levels of MPO, TBARS, NO levels, and iNOS activity caused by torsion/detorsion in the testis. Testis tissues showed a significant increase in GSH-Px activity compared to the torsion group when CAPE was applied. In conclusion, investigators concluded that CAPE treatment exerts a protective effect on testicular I/R.

Adivarekar et al. [251] showed that administration of lomodex (low molecular weight dextran) and $MgSO_4$ prior to detorsion resulted in prolonged testicular salvage with a potential of subsequent improvement in semen quality, fertility, and reduction in long-term morbidity and showed preservation of tubular morphology. In another study, Abes et al. [252] reported that ATP-$MgCl_2$ administration before or after detorsion may prevent reperfusion injury in testicular torsion.

Metal aspartates including zinc aspartate are inhibitors of ROS. Their inhibitory activities are a consequence of both the scavenging of the free radicals and the inhibition of xanthine oxidase and nicotinamide adenine dinucleotide phosphate (NADPH) oxidase activities. Ozkan et al. [253] demonstrated that zinc aspartate reduces I/R injury by its antioxidant effects after unilateral testicular torsion/detorsion and affects the antioxidant enzyme systems such as SOD, catalase.

Another experimental I/R study was designed by Wei et al. [254]. They presented the beneficial effects of taurine on I/R testis model. Taurine is a potent antioxidant and antiapoptotic and also exerts cytoprotective effect on I/R injury of other organ. Taurine has been found in Leydig cells, vascular endothelial cells, and other interstitial cells in the testis. In addition, sperm and seminal fluid are rich in taurine. It has also been shown to be involved in sperm motility, capacitation, acrosome reaction, and osmoregulation. Taurine acts as an antioxidant and prevents sperm lipid peroxidation. Wei and collaborates demonstrated that taurine decreased ROS generation by diminishing the neutrophil recruitment to the testis.

Ibuprofen is one of the most useful nonsteroidal anti-inflammatory agents available to humans. It is known to inhibit cyclooxygenase, modulate channel activities, and activate peroxisome proliferator-activated receptor (PPAR), an orphan nuclear receptor that acts as a ligand-activated transcription factor, thereby inhibiting proinflammatory cytokine production. In addition, ibuprofen has antiradical and antioxidant effects and scavenges ROS. It protects the lipids of biological membranes from oxidation and consequently inhibits the accumulation of lipid peroxidation products such as MDA. Several studies have showed that ibuprofen could scavenge hydroxyl and superoxide radicals and that it possessed radioprotective [255] and neuroprotective effects such as Alzheimer disease [256, 257]. Recently, ibuprofen was successfully used to decrease I/R injury in multiple organ systems, including the retina, liver, heart, and brain. Nonetheless, to our knowledge, the effect of ibuprofen on testicular I/R injury had not been previously reported. Our results showed that because of its anti-inflammatory and antioxidant effects, ibuprofen pretreatment may have protective effects in the experimental testicular torsion/detorsion model in rats [258].

The peroxisome proliferator-activated receptor-gamma (PPAR-γ) is a member of the nuclear hormone receptor superfamily that is involved in several physiological processes, such as glucose homeostasis, cellular differentiation, regulation of lipid, and lipoprotein metabolism, as well as in pathological states, including atherosclerosis, inflammation, cancer, infertility, and demyelination. Thiazolidinediones are synthetic PPAR-γ ligands used in the control of type 2 diabetes, whereby rosiglitazone is the most potent and selective agent in this class, which binds the receptor with a higher affinity than other thiazolidinediones, such as pioglitazone or ciglitazone. In recent studies, it was demonstrated that rosiglitazone ameliorated the lesions associated with I/R of kidney, heart, and gut, possibly through its anti-inflammatory and antioxidant properties. Inan et al. [259] aimed to evaluate the effects of rosiglitazone on lipid peroxidation, eNOS immunoreactivity, and the histopathology in both testes after unilateral I/R in rats. Investigators reported rosiglitazone prevented I/R injury in both testes. In addition, Mogilner et al. [260] found that diclofenac did not change histologic parameters of spermatogenesis, but decreased germ cell apoptosis in both testes following testicular I/R. It has been shown that PPAR beta/delta (β/δ) is expressed in testis. Minutoli et al. [261] tested PPAR-β/δ agonist (L-165,041) and antagonist (GW9662) and evaluated testicular ERK, TNF-α, and IL-6, and also investigated PPAR-β/δ activation, mRNA expression, and organ damage. They found that L-165,041 administration increased the PPAR-β/δ message and protein, inhibited ERK, TNF-α, and IL-6 expression, and decreased histological damage. Concomitant administration of GW9662 reversed the protection exerted by PPAR-β/δ agonist.

Salmasi et al. [262] reported that morphine increased the ipsilateral intratesticular antioxidant markers (SOD, catalase, GSH-Px) during the reperfusion phase after unilateral testicular torsion. The ipsilateral MDA levels in the morphine group were significantly lower than in the I/R group. The observed effects of morphine may have been a result of its impact on neutrophils and its direct scavenging action against peroxynitrite.

Pentoxifylline, a methyl xanthine derivative, has antioxidant effect. It has been used for its regulatory effects on microvascular blood flow. It was reported that unilateral testicular torsion and detorsion caused an increase in the MDA levels of both testes. Pentoxifylline decreased MDA levels on both side and attenuated reperfusion damage on both side, possibly with its effects on blood flow and neutrophils [263–265].

The effect of methylene blue on the experimental testis torsion/detorsion models has been investigated. Both Greenstein [266] and Inan [267] suggested that methylene blue didn't protect testes from histological damage in case of ipsilateral testis after I/R injury. Nevertheless, surprisingly Inan et al. reported that methylene blue had harmful effect and increased MDA levels on the contralateral testis.

Propofol, which is widely used as an intravenous anesthetic, has been shown to have an antioxidant activity on several tissues. Unsal et al. [268] investigated the prevention of reperfusion injury with propofol after testicular torsion. They assessed tissue MDA level, xanthine oxidase, catalase, and GSH-Px activities. Pretreatment with propofol prevented a further increase in MDA levels and significantly decreased catalase activity following detorsion. GSH-Px activities were not affected either by torsion/detorsion or propofol pretreatment. Histologically, torsion caused some separation between germinal cells in the seminiferous tubules, which became much more prominent in the detorsion group and attenuated with propofol pretreatment. In brief, propofol decreased free radical formation and attenuated histopathological damage in the testis after reperfusion. Yagmurdur et al. [269] suggested that propofol prevented testicular damage by scavenging reactive oxygen and nitrogen species and inhibiting lipid peroxidation and attenuated germ cell-specific apoptosis [270]. Recently, Harvey et al. [271] used propofol sedation performing manual testicular detorsion to a 14-year-old adolescent boy. Patient's testicular pain is completely reduced.

Nezami et al. [272] stated that the use of immunophilin ligands (cyclosporine and FK 506-tacrolimus) had significantly decreased testicular torsion damage. The results of biochemical studies suggest that reduction of oxidative stress along with attenuated neutrophil accumulation by immunophilin ligands may have a major role in their cytoprotective effects and their potent biologic properties in inhibiting programmed cell death and necrosis.

Mexiletine is a congener of lidocaine and safe antiarrhythmic agent, preventing I/R injury in myocardium. Also, mexiletine is a potent antioxidant agent, most of the reports about its protective effects on I/R injury on heart. Ozen et al. [273] investigated its effects on the electrical field stimulation (EFS)-induced contractions in rabbit vasa deferentia after torsion/detorsion. However, mexiletine had no preventive effect on this inhibition.

Allopurinol has been shown to have preventive effects on testicular I/R injury [274–276]. Kehinde et al. [276] tried five antioxidant (acetyl salicylic acid, ascorbic acid, allopurino, quercetin, and SOD), but they found that only allopurinol had a beneficial effect at 3 months after experimental torsion in a rabbit model. Moreover, in Abasiyanik and Dağdönderen's study [274], allopurinol treatment prevented only MDA increase, but no histopathologic results. However, Silva et al. [277] reported that allopurinol didn't protect the testes on 60 days after I/R injury.

Melatonin is one of the most powerful endogenous antioxidants known. Antioxidant effects of melatonin can occur by either a direct or an indirect mechanism. Melatonin itself exerts direct antioxidant effects via scavenging the hydroxyl radical, peroxyl radical, singlet oxygen, peroxynitrite anion, and superoxide anion. Additionally, it acts as an indirect antioxidant by stimulating several antioxidative enzymes, including GSH-Px, GSH reductase, glucose-6-phosphate dehydrogenase, and SOD. Conversely, it inhibits a prooxidative enzyme, NOS. Barun et al. [278] investigated whether EFS-evoked biphasic contractions are altered in ipsilateral and contralateral rat vasa deferentia obtained from animals exposed to the unilateral testicular torsion/detorsion procedure. They also evaluated the effects of melatonin on these contractile responses. The investigators suggested that melatonin produces an inhibition on EFS-evoked biphasic twitch responses in the ipsilateral and contralateral rat vasa deferentia following unilateral testicular torsion/detorsion in the rat. In another study, Abasiyanik and Da dönderen [274] compared beneficial effects of melatonin and allopurinol in experimental testicular torsion. Their results showed that melatonin treatment prevented testes from injury both biochemically and histopathologically. In addition, Ozturk [279], Duru [280], and Yurtçu et al. [281, 282] showed preventive effects of melatonin administration on reperfusion damage in experimental testis I/R. In their studies, it was demonstrated that MDA levels were lower in both torsioned and contralateral testes. Furthermore, Yurtçu et al. [281] reported that melatonin increased serum inhibin B levels which reflect Sertoli cell function and the state of spermatogenesis. Jeong et al. [283] compared the preventive effects of cyclosporine A combined with prednisolone and melatonin on damage to contralateral testis after ipsilateral torsion/detorsion between pubertal and adult rats. Their study stated that the preventive effects of cyclosporine A combined with prednisolone on contralateral testicular damage were characteristic only in pubertal rats, while the preventive effects of melatonin were characteristic in pubertal and adult rats. These results suggested that damage to the contralateral testis induced by an immunological mechanism may be more significant during puberty than during adulthood. Kanter [284] confirmed these experimental studies with immunohistochemistry. They reported that melatonin treatment increased the immunoexpression of proliferating cell nuclear antigen and testosterone and decreased germ cell apoptosis in I/R testis. Recently, Kucer et al. [285] found that melatonin treatment improved sperm morphology and epididymal sperm quality in I/R injury.

Yazawa et al. [286] demonstrated the suppressing effect of dexamethasone, a potent synthetic glucocorticoid hormone, on apoptosis of testicular germ cells and vascular neutrophil adhesion after I/R in testis. This inhibition was suppressed by intravenous administration of mifepriston, a glucocorticoid receptor antagonist. In recent years, Cho et al. [287] suggested that medical treatment, such as drug treatment and surgery treatment should be performed as early as 2 h after testicular torsion. Because apoptosis occurs at least partially via the phosphoinositide-dependent protein kinase-1/serum- and glucocorticoid-inducible kinase 1/forkhead transcriptional factor FOXO3a signaling cascade, these cascades impair the testosterone secretion capacity of the testes after only 2 h of torsion.

Yapanoglu [288] and Aksoy [289] studied about the effects of dehydroepiandrosterone on testicular torsion damage in rats. They found that dehydroepiandrosterone may be a protective agent for preventing biochemical, apoptotic, and histopathologic changes related to oxidative stress in testicular injury. Unal et al. [290] investigated the preventive effects of trimetazidine on I/R injury in rats. They also reported that trimetazidine may be a protective agent for preventing biochemical and histopathologic changes related to oxidative stress in testicular injury. Savaş et al. [291] reported that human chorionic gonadotropin treatment improved contralateral histopathologic injury and increased the serum testosterone levels. In addition, the positive effect of hyperbaric oxygen therapy [292] and surfactant tetronic 1107 [293] on testicular torsion model in rats has been shown.

17.5 Conclusion

In recent years, experimental and clinical studies have suggested that the injured testis may result in damage to the contralateral testis after unilateral testicular torsion. The result of a I/R injury to the testis may be related to significant increase in germ cell apoptosis due to high testicular oxidative stress following the reperfusion. However, the specific mechanisms still need to be clarified. Impaired spermatogenesis has been reported in animals and humans after testicular torsion. Biochemical, hormonal, immunological, histopathological, and vasculogenic studies have been performed, but the results are still controversial. To avoid testicular loss and eventual impaired fertility, prompt diagnosis and immediate surgery are the most important issues for the treatment of these patients.

Many antioxidants and free radical scavengers have been proposed in recent years for treatment of testicular torsion-induced subfertility or infertility, and a number of animal studies published to indicate their possible application. However, the molecular mechanism by which antioxidants control testicular torsion-induced male fertility has not yet been clearly identified. As a result, antioxidant therapy such as carnitine, melatonin, selenium, ibuprofen, and resveratrol may represent a new nonhormonal option within a broader therapeutic strategy. Antioxidant therapy reestablishes adequate oxidative balance in terms of the ratio of prooxidants to scavenger factors in men with ROS-mediated testicular torsion-induced infertility. Experimentally tested drugs or methods of preventing testicular damage caused by torsion still await clinical application. Identification of pharmacologic agents, administered as adjunctive therapy to surgical repair for rescuing the testis from I/R injury, is a clinically important goal. In brief, additional works are being performed in this area to further assess the late effects of testicular torsion and antioxidant therapy.

References

1. Heyns CF, Visser AJ. Testicular torsion. In: Schill WB, Comhaire FR, Hargreave TB, editors. Andrology for the clinician. Berlin: Springer; 2006. p. 134–61.
2. Rampaul MS, Hosking SW. Testicular torsion: most delay occurs outside hospital. Ann R Coll Surg Engl. 1998;80:169–72.
3. Saxena AK, Höllwarth ME. Acute Scrotum. In: Puri P, Höllwarth M, editors. Pediatric surgery: diagnosis and management. Berlin: Springer; 2009. p. 927–34.
4. Baldisserotto M. Scrotal emergencies. Pediatr Radiol. 2009;39:516–21.
5. Vigueras RM, Reyes G, Rojas-Castañeda J, et al. Testicular torsion and its effects on the spermatogenic cycle in the contralateral testis of the rat. Lab Anim. 2004;38:313–20.
6. Hekimoglu A, Kurcer Z, Aral F, et al. Lycopene, an antioxidant carotenoid, attenuates testicular injury caused by ischemia/reperfusion in rats. Tohoku J Exp Med. 2009;218:141–7.
7. Filho DW, Torres MA, Bordin AL, et al. Spermatic cord torsion, reactive oxygen and nitrogen species and ischemia-reperfusion injury. Mol Aspects Med. 2004;25:199–210.
8. Visser AJ, Heyns CF. Testicular function after torsion of the spermatic cord. BJU Int. 2003;92:200–3.

9. Sun J, Liu GH, Zhao HT, et al. Long-term influence of prepubertal testicular torsion on spermatogenesis. Urol Int. 2006;77:275–8.
10. Turner TT, Bang HJ, Lysiak JJ. Experimental testicular torsion: reperfusion blood flow and subsequent testicular venous plasma testosterone concentrations. Urology. 2005;65:390–4.
11. Pampal A, Ozen IO, Ekingen G, et al. The morphological evaluation of ipsilateral and contralateral vasa deferentia in a rat model of unilateral spermatic cord torsion. Pediatr Surg Int. 2010;26:287–92.
12. Saba M, Morales CR, De Lamirande E, et al. Morphological and biochemical changes following acute unilateral testicular torsion in prepubertal rats. J Urol. 1997;157:1149–54.
13. Savas C, Ozogul C, Karaoz E, et al. Ischemia, whether from ligation or torsion, causes ultrastructural changes on the contralateral testis. Scand J Urol Nephrol. 2002;36:302–6.
14. Yao QS, Ye ZQ, Wang XK, et al. Experimental study of contralateral testicular changes after unilateral testicular torsion in rats. Zhonghua Nan Ke Xue. 2003;9:586–8.
15. Pentyala S, Lee J, Yalamanchili P, et al. Testicular torsion: a review. J Low Genit Tract Dis. 2001;5:38–47.
16. Ringdahl E, Teague L. Testicular torsion. Am Fam Physician. 2006;74:1739–43.
17. Cosentino MJ, Nishida M, Rabinowitz R, et al. Histopathology of prepubertal rat testes subjected to various durations of spermatic cord torsion. J Androl. 1986;7:23–31.
18. Dokmeci D. Testicular torsion, oxidative stress and the role of antioxidant therapy. Folia Med (Plovdiv). 2006;48:16–21.
19. Lysiak JJ, Nguyen QA, Turner TT. Peptide and nonpeptide reactive oxygen scavengers provide partial rescue of the testis after torsion. J Androl. 2002;23:400–9.
20. Bozlu M, Acar D, Cayan S, et al. Protective effect of trapidil on long-term histologic damage in a rat model of testicular ischemia-reperfusion injury. World J Urol. 2009;27:117–22.
21. Tsounapi P, Saito M, Dimitriadis F, et al. Protective effect of sivelestat, a neutrophil elastase inhibitor, on ipsilateral and contralateral testes after unilateral testicular ischaemia-reperfusion injury in rats. BJU Int. 2010;107:329–36.
22. Delasiauve LJF. Descente tardive du testicule gauche, prisé pour une hernie etrangleé. Rev Med Franc Etrang. 1840;1:363.
23. Taylor MR. A case of testicular strangulation at birth. Castration, recovery. Br Med J. 1897;1:458.
24. Nash W. Acute torsion of spermatic cord: reduction: immediate relief. Br Med J. 1893;1:742.
25. Cuckow PM, Frank JD. Torsion of the testis. BJU Int. 2000;86:349–53.
26. Romero FR, Gomes RP, Lorenzini F, et al. Ipsilateral testicular necrosis and atrophy after 1,080-degree torsion of the spermatic cord in rats. Acta Cir Bras. 2009;24:118–23.
27. Zampieri N, Cecchetto M, Ottolenghi A, et al. Bilateral testicular torsion: an extremely tragic rare condition. Minerva Pediatr. 2009;61:455–7.
28. Bagci S, Bachour H, Woelfle JF, Mueller A, et al. Bilateral perinatal testicular torsion in an infant: a rare neonatal emergency. Pediatr Int. 2010;52:e227–8.
29. Cilento BG, Atala A. Cryptorchidism, testicular torsion, and torsion of the testicular appendages. In: Kandeel FR, editor. Male reproductive dysfunction: pathophysiology and treatment. New York: Informa Healthcare; 2007. p. 165–71.
30. Davol P, Simmons J. Testicular torsion in a 68-year-old man. Urology. 2005;66:195.
31. Kapoor S. Testicular torsion: a race against time. Int J Clin Pract. 2008;62:821–7.
32. Lazarov R, Stapper G, Lock MT, et al. Testicular torsion can also be present in adult men. Ned Tijdschr Geneeskd. 2003;147:1389–93.
33. Srinivasan AK, Freyle J, Gitlin JS, et al. Climatic conditions and the risk of testicular torsion in adolescent males. J Urol. 2007;178:2585–8.
34. Mabogunje OA. Testicular torsion and low relative humidity in a tropical country. Br Med J (Clin Res Ed). 1986;292:363–4.
35. Preshaw RM. Seasonal frequency of testicular torsion. Can J Surg. 1984;27:404–5.
36. Williams CR, Heaven KJ, Joseph DB. Testicular torsion: is there a seasonal predilection for occurrence? Urology. 2003;61:638–41.

37. Molokwu CN, Somani BK, Goodman CM. Outcomes of scrotal exploration for acute scrotal pain suspicious of testicular torsion: a consecutive case series of 173 patients. BJU Int. 2011;107:990–3.
38. Lyronis ID, Ploumis N, Vlahakis I, et al. Acute scrotum—etiology, clinical presentation and seasonal variation. Indian J Pediatr. 2009;76:407–10.
39. Shukla RB, Kelly DG, Daly L, et al. Association of cold weather with testicular torsion. Br Med J (Clin Res Ed). 1982;285:1459–60.
40. Gatti JM, Patrick Murphy J. Current management of the acute scrotum. Semin Pediatr Surg. 2007;16:58–63.
41. Marulaiah M, Gilhotra A, Moore L, et al. Testicular and paratesticular pathology in children: a 12-year histopathological review. World J Surg. 2010;34:969–74.
42. Murphy FL, Law H, Mushtaq I, et al. Testicular and paratesticular pathology in infants and children: the histopathological experience of a tertiary paediatric unit over a 17 year period. Pediatr Surg Int. 2007;23:867–72.
43. Mushtaq I, Fung M, Glasson MJ. Retrospective review of paediatric patients with acute scrotum. ANZ J Surg. 2003;73:55–8.
44. Shteynshlyuger A, Freyle J. Familial testicular torsion in three consecutive generations of first-degree relatives. J Pediatr Urol. 2011;7:86–91.
45. Okeke LI, Ikuerowo OS. Familial torsion of the testis. Int Urol Nephrol. 2006;38:641–2.
46. Al-Salem AH. Intrauterine testicular torsion: a surgical emergency. J Pediatr Surg. 2007;42: 1887–91.
47. Cuervo JL, Grillo A, Vecchiarelli C, et al. Perinatal testicular torsion: a unique strategy. J Pediatr Surg. 2007;42:699–703.
48. Djahangirian O, Ouimet A, Saint-Vil D. Timing and surgical management of neonatal testicular torsions. J Pediatr Surg. 2010;45:1012–5.
49. Eifinger F, Ahrens U, Wille S, et al. Neonatal testicular infarction—possibly due to compression of the umbilical cord? Urology. 2010;75:1482–4.
50. Hutson JM. Testicular torsion. In: Grosfeld JL, O'Neill JA, Coran AG, Fonkalsrud EW, editors. Pediatric surgery. 6th ed. Philadelphia: Mosby Elsevier; 2006. p. 1205–7.
51. Jung YJ, Chung JM. Testicular torsion in the inguinal region in an extremely low birth weight infant. Korean J Pediatr. 2010;53:852–4.
52. Kyriazis ID, Dimopoulos J, Sakellaris G, et al. Extravaginal testicular torsion: a clinical entity with unspecified surgical anatomy. Int Braz J Urol. 2008;34:617–23.
53. Wan J, Bloom DA. Male genital tract. In: Oldham KT, Colombani PM, Foglia RP, Skinner MA, editors. Principles and practice of pediatric surgery. Philadelphia: Lippincott Williams & Wilkins; 2005. p. 1602–3.
54. Kessler CS, Bauml J. Non-traumatic urologic emergencies in men: a clinical review. West J Emerg Med. 2009;10:281–7.
55. Cimador M, DiPace MR, Castagnetti M, et al. Predictors of testicular viability in testicular torsion. J Pediatr Urol. 2007;3:387–90.
56. Moon C, Shin T. Involvement of macrophages in germ cell death in the rat testis with acute experimental testicular torsion. Korean J Vet Res. 2004;44:329–34.
57. Arap MA, Vicentini FC, Cocuzza M, et al. Late hormonal levels, semen parameters, and presence of antisperm antibodies in patients treated for testicular torsion. J Androl. 2007; 28:528–32.
58. Fu GB, Qian LX, Cui YG, et al. Antisperm-antibodies induced by testicular torsion and its influence on testicular function. Zhonghua Nan Ke Xue. 2006;12:988–91.
59. Koşar A, Küpeli B, Alçigir G, et al. Immunologic aspect of testicular torsion: detection of antisperm antibodies in contralateral testicle. Eur Urol. 1999;36:640–4.
60. Rybkiewicz M. Long-term and late results of treatment in patients with a history of testicular torsion. Ann Acad Med Stetin. 2001;47:61–75.
61. Beni-Israel T, Goldman M, Bar Chaim S, et al. Clinical predictors for testicular torsion as seen in the pediatric ED. Am J Emerg Med. 2010;28:786–9.

62. Shehzad KN, Riaz AA. Unusual cause of a painful right testicle in a 16-year-old man: a case report. J Med Case Reports. 2011;5:27.
63. Wiener SL. Testicular pain. In: Walker HK, Hall WD, Hurst JW, editors. Clinical methods: the history, physical, and laboratory examinations. 3rd ed. Boston: Butterworths; 1990. p. 853–6.
64. Chan JL, Knoll JM, Depowski PL, et al. Mesorchial testicular torsion: case report and a review of the literature. Urology. 2009;73:83–6.
65. Schmitz D, Safranek S. Clinical inquiries. How useful is a physical exam in diagnosing testicular torsion? J Fam Pract. 2009;58:433–4.
66. Somani BK, Watson G, Townell N. Testicular torsion. BMJ. 2010;341:c3213. doi:10.1136/bmj.c3213.
67. Baker LA, Sigman D, Mathews RI, et al. An analysis of clinical outcomes using color Doppler testicular ultrasound for testicular torsion. Pediatrics. 2000;105:604–7.
68. Dudea SM, Ciurea A, Chiorean A, et al. Doppler applications in testicular and scrotal disease. Med Ultrason. 2010;12:43–51.
69. Ahmed SJ, Kaplan GW, DeCambre ME. Perinatal testicular torsion: preoperative radiological findings and the argument for urgent surgical exploration. J Pediatr Surg. 2008;43:1563–5.
70. Chmelnik M, Schenk JP, Hinz U, et al. Testicular torsion: sonomorphological appearance as a predictor for testicular viability and outcome in neonates and children. Pediatr Surg Int. 2010;26:281–6.
71. Kwon JW, Kim WS, Cheon JE, et al. Evaluation of testicular viability by power Doppler ultrasonography in experimentally induced acute testicular torsion. Invest Radiol. 2005;40:682–7.
72. Mernagh JR, Caco C, De Maria J. Testicular torsion revisited. Curr Probl Diagn Radiol. 2004;33:60–73.
73. Prando D. Torsion of the spermatic cord: the main gray-scale and Doppler sonographic signs. Abdom Imaging. 2009;34:648–61.
74. Waldert M, Klatte T, Schmidbauer J, et al. Color Doppler sonography reliably identifies testicular torsion in boys. Urology. 2010;75:1170–4.
75. Gotto GT, Chang SD, Nigro MK. MRI in the diagnosis of incomplete testicular torsion. Br J Radiol. 2010;83:e105–7.
76. Terai A, Yoshimura K, Ichioka K, et al. Dynamic contrast-enhanced subtraction magnetic resonance imaging in diagnostics of testicular torsion. Urology. 2006;67:1278–82.
77. Watanabe Y, Nagayama M, Okumura A, et al. MR imaging of testicular torsion: features of testicular hemorrhagic necrosis and clinical outcomes. J Magn Reson Imaging. 2007;26:100–8.
78. Amini B, Patel CB, Lewin MR, et al. Diagnostic nuclear medicine in the ED. Am J Emerg Med. 2011;29:91–101.
79. Ovali GY, Yilmaz O, Tarhan S, et al. Perfusion CT evaluation in experimentally induced testicular torsion. Can Urol Assoc J. 2009;3:383–6.
80. Capraro GA, Coughlin BF, Mader TJ, et al. Testicular cooling associated with testicular torsion and its detection by infrared thermography: an experimental study in sheep. J Urol. 2008;180:2688–93.
81. Capraro GA, Mader TJ, Coughlin BF, et al. Feasibility of using near-infrared spectroscopy to diagnose testicular torsion: an experimental study in sheep. Ann Emerg Med. 2007;49:520–5.
82. Kram HB, Miyamoto E, Rajfer J, et al. Testicular oximetry: a new method for the assessment of tissue perfusion and viability following torsion and detorsion. J Pediatr Surg. 1989;24:1297–302.
83. Mirilas P, Kontis E, Mentessidou A, et al. Testis rigidity tester: a prototype device for intraoperative detection of changes in testicular pressure and/or elasticity.First experimental experiences. J Pediatr Urol. 2010;6:277–81.

84. Yamis S, Gedik A, Sahin H, et al. Is procalcitonin valuable in the differential diagnosis of testicular torsion and epididymo-orchitis. Saudi Med J. 2010;31:170–4.
85. Kutlu O, Mentese A, Turkmen S, et al. Investigation of the possibility of using ischemia-modified albumin in testicular torsion: an experimental study. Fertil Steril. 2011;95(4):1333–7.
86. Kubini K, Zachmann M, Albers N, et al. Basal inhibin B and the testosterone response to human chorionic gonadotropin correlate in prepubertal boys. J Clin Endocrinol Metab. 2000;85:134–8.
87. Taskinen S, Taskinen M, Rintala R. Testicular torsion: orchidectomy or orchiopexy? J Pediatr Urol. 2008;4:210–3.
88. Ozkan KU, Küçükaydin M, Muhtaroğlu S, et al. Evaluation of contralateral testicular damage after unilateral testicular torsion by serum inhibin B levels. J Pediatr Surg. 2001;36:1050–3.
89. Moore CP, Marr JK, Huang CJ. Cryptorchid testicular torsion. Pediatr Emerg Care. 2011;27:121–3.
90. Phuwapraisirisan S, Lim M, Suwanthanma W. Surgical reduction in a delayed case of traumatic testicular dislocation. J Med Assoc Thai. 2010;93:1317–20.
91. Van der Poel J. Strangulation of the testis and epididymis from torsion of the spermatic cord. Med Rec. 1895;47:737.
92. Bah OR, Roupret M, Guirassy S, et al. Clinical aspects and management of the torsion of the spermatic cord: study of 27 cases. Prog Urol. 2010;20:527–31.
93. Bhagra A, Suravaram S, Schears RM. Testicular torsion—a common surgical emergency. Int J Emerg Med. 2008;1:147.
94. Cavusoglu YH, Karaman A, Karaman I, et al. Acute scrotum—etiology and management. Indian J Pediatr. 2005;72:201–3.
95. Guerra LA, Wiesenthal J, Pike J, et al. Management of neonatal testicular torsion: which way to turn? Can Urol Assoc J. 2008;2:376–9.
96. Soccorso G, Ninan GK, Rajimwale A, et al. Acute scrotum: is scrotal exploration the best management? Eur J Pediatr Surg. 2010;20:312–5.
97. Anderson MJ, Dunn JK, Lipshultz LI, et al. Semen quality and endocrine parameters after acute testicular torsion. J Urol. 1992;147:1545–50.
98. Ichikawa T, Kitagawa N, Shiseki Y, et al. Testicular function after spermatic cord torsion. Hinyokika Kiyo. 1993;39:243–8.
99. Shafik A, El-Sibal O, Shafik I. Electro-orchidogram: a non-invasive diagnostic tool in testicular pathologies. Med Sci Monit. 2006;12:MT51–5.
100. Baker BW, Surratt JK, Run W. Missed testicular torsion presenting with male infertility and erectile dysfunction. Zhonghua Nan Ke Xue. 2005;11:615–6.
101. Tryfonas G, Violaki A, Tsikopoulos G, et al. Late postoperative results in males treated for testicular torsion during childhood. J Pediatr Surg. 1994;29:553–6.
102. Chen HT, Zheng XM, Yang ZW, et al. Effects of hypothermia on antioxidant capacity of rat testes after testicular torsion. Zhonghua Nan Ke Xue. 2005;11:426–7.
103. Gao DJ, Xuan XJ, Wang YJ, et al. Verapamil and hypothermia protect spermatogenesis of torsioned testes in rats. Zhonghua Nan Ke Xue. 2009;15:796–800.
104. Haj M, Shasha SM, Loberant N, et al. Effect of external scrotal cooling on the viability of the testis with torsion in rats. Eur Surg Res. 2007;39:160–9.
105. Power RE, Scanlon R, Kay EW, et al. Long-term protective effects of hypothermia on reperfusion injury post-testicular torsion. Scand J Urol Nephrol. 2003;37:456–60.
106. Qian HJ, Du XJ, Zhang C, et al. Cold ischemia time influences spermatogenesis in a testicular ischemia/reperfusion injury model. Transplant Proc. 2010;42:1610–3.
107. Sarica K, Bakir K. Semiquantitative evaluation of testicular histology after testicular torsion: protective effect of external cooling. Urol Int. 1999;63:110–4.
108. Akçora B, Altuğ ME, Balci A, et al. Gradual detorsion of torsioned rat testis attenuates ischemia reperfusion injury. J Pediatr Surg. 2008;43:1879–84.
109. Duru FI, Olabiyi O, Noronha CC, et al. Brief ischaemia reduces testicular lipid peroxidation following subsequent ischaemia: an evidence for ischaemic preconditioning. Nig Q J Hosp Med. 2008;18:149–52.

110. Ozkisacik S, Yazici M, Gursoy H, et al. The effects of short-interval postconditioning in preventing testicular ischemia-reperfusion injury in rats. J Pediatr Surg. 2011;46:546–50.
111. Sahinkanat T, Ozkan KU, Tolun FI, et al. The protective effect of ischemic preconditioning on rat testis. Reprod Biol Endocrinol. 2007;5:47.
112. Shimizu S, Saito M, Dimitriadis F, et al. Protective effect of ischaemic post-conditioning on ipsilateral and contralateral testes after unilateral testicular ischaemia-reperfusion injury. Int J Androl. 2011;34(3):268–75. doi:10.1111/j.1365-2605.2010.01077.x.
113. Shimizu S, Saito M, Kinoshita Y, et al. Ischemic preconditioning and post-conditioning to decrease testicular torsion-detorsion injury. J Urol. 2009;182:1637–43.
114. Taneli F, Aydede H, Vatansever S, et al. The long-term effect of mesh bioprosthesis in inguinal hernia repair on testicular nitric oxide metabolism and apoptosis in rat testis. Cell Biochem Funct. 2005;23:213–20.
115. Woodruff DY, Horwitz G, Weigel J, et al. Fertility preservation following torsion and severe ischemic injury of a solitary testis. Fertil Steril. 2010;94:352.
116. Henkel RR. Leukocytes and oxidative stress: dilemma for sperm function and male fertility. Asian J Androl. 2011;13:43–52.
117. Prillaman HM, Turner TT. Rescue of testicular function after acute experimental torsion. J Urol. 1997;157:340–5.
118. Sharma R, Agarwal A. Role of reactive oxygen species in gynecologic diseases. Reprod Med Biol. 2004;3:177–99.
119. Karagüzel G, Güngör F, Karagüzel G, et al. Unilateral spermatic cord torsion without ipsilateral spermatogenetic material: effects on testicular blood flow and fertility potential. Urol Res. 2004;32:51–4.
120. Otçu S, Durakoğugil M, Orer HS, et al. Contralateral genitofemoral sympathetic nerve discharge increases following ipsilateral testicular torsion. Urol Res. 2002;30:324–8.
121. Romeo C, Impellizzeri P, Arrigo T, et al. Late hormonal function after testicular torsion. J Pediatr Surg. 2010;45:411–3.
122. Shiraishi K, Naito K, Yoshida K. Nitric oxide promotes germ cell necrosis in the delayed phase after experimental testicular torsion of rat. Biol Reprod. 2001;65:514–21.
123. Ahmed A, Bello A, Mbibu NH, et al. Epidemiological and aetiological factors of male infertility in northern Nigeria. Niger J Clin Pract. 2010;13:205–9.
124. Sarkar O, Bahrainwala J, Chandrasekaran S, et al. Impact of inflammation on male fertility. Front Biosci (Elite Ed). 2011;3:89–95.
125. Cuzzocrea S, Riley DP, Caputi AP, et al. Antioxidant therapy: a new pharmacological approach in shock, inflammation, and ischemia/reperfusion injury. Pharmacol Rev. 2001; 53:135–59.
126. Antonuccio P, Minutoli L, Romeo C, et al. Lipid peroxidation activates mitogen-activated protein kinases in testicular ischemia-reperfusion injury. J Urol. 2006;176:1666–72.
127. Payabvash S, Kiumehr S, Tavangar SM, et al. Ethyl pyruvate reduces germ cell-specific apoptosis and oxidative stress in rat model of testicular torsion/detorsion. J Pediatr Surg. 2008;43:705–12.
128. Minutoli L, Antonuccio P, Polito F, et al. Mitogen-activated protein kinase 3/mitogen-activated protein kinase 1 activates apoptosis during testicular ischemia-reperfusion injury in a nuclear factor-kappaB-independent manner. Eur J Pharmacol. 2009;604:27–35.
129. Agarwal A, Saleh RA. Role of oxidants in male infertility: rationale, significance, and treatment. Urol Clin North Am. 2002;29:817–27.
130. Kehinde EO, Mojiminiyi OA, Mahmoud AH, et al. The significance of measuring the time course of serum malondialdehyde concentration in patients with torsion of the testis. J Urol. 2003;169:2177–80.
131. Ahmed FA, Jequier AM, Cummins JM, et al. Differentially expressed DNA sequences following recovery from unilateral testicular torsion in rat. Biochim Biophys Acta. 2001; 1535:192–9.
132. Ahmed FA, Whelan J, Jequier AM, et al. Torsion-induced injury in rat testes does not affect mitochondrial respiration or the accumulation of mitochondrial mutations. Int J Androl. 2000;23:347–56.

133. Andiran F, Okur DH, Kilinç A, et al. Do experimentally induced ipsilateral testicular torsion, vas deferens obstruction, intra-abdominal testis or venous obstruction damage the contralateral testis through a common mechanism? BJU Int. 2000;85:330–5.
134. Heindel RM, Pakyz RE, Reinking LN, et al. The effect of various degrees of unilateral spermatic cord torsion on fertility in the rat. J Urol. 1990;144:366–9.
135. Kamada K, Takihara H, Shirataki S, et al. Flow cytometric DNA analysis demonstrates contralateral testicular deterioration in experimental unilateral testicular torsion of prepubertal rats. Andrologia. 1993;25:239–44.
136. Nguyen L, Lievano G, Ghosh L, et al. Effect of unilateral testicular torsion on blood flow and histology of contralateral testes. J Pediatr Surg. 1999;34:680–3.
137. Baker LA, Turner TT. Leydig cell function after experimental testicular torsion despite loss of spermatogenesis. J Androl. 1995;16:12–7.
138. Yang C, Song B, Tan J, et al. Testicular torsion in children: a 20-year retrospective study in a single institution. Sci World J. 2011;11:362–8.
139. Turner TT, Bang HJ, Lysiak JL. The molecular pathology of experimental testicular torsion suggests adjunct therapy to surgical repair. J Urol. 2004;172:2574–8.
140. Guimarães SB, Santos JM, Aragão AA, et al. Protective effect of alpha-lipoic acid in experimental spermatic cord torsion. Nutrition. 2007;23:76–80.
141. Said TM, Agarwal A, Sharma RK, et al. Human sperm superoxide anion generation and correlation with semen quality in patients with male infertility. Fertil Steril. 2004;82:871–7.
142. Agarwal A, Said TM. Oxidative stress, DNA damage and apoptosis in male infertility: a clinical approach. BJU Int. 2005;95:503–7.
143. Kothari S, Thompson A, Agarwal A, et al. Free radicals: their beneficial and detrimental effects on sperm function. Indian J Exp Biol. 2010;48:425–35.
144. Saleh A, Agarwal A. Oxidative stress and male infertility: from research bench to clinical practice. J Androl. 2002;23:737–52.
145. Aitken RJ, Baker MA, De Iuliis GN, et al. New insights into sperm physiology and pathology. Handb Exp Pharmacol. 2010;198:99–115.
146. Dokmeci D. Oxidative stress, male infertility and the role of carnitines. Folia Med (Plovdiv). 2005;47:26–30.
147. Akgür FM, Kilinç K, Aktuğ T. Is ipsilateral testis mandatory for the occurrence of contralateral intratesticular biochemical changes indicative of hypoxia after unilateral spermatic cord torsion? Eur Urol. 1995;28:143–6.
148. Anderson JB, Williamson RC. Fertility after torsion of the spermatic cord. Br J Urol. 1990;65:225–30.
149. Becker Jr EJ, Turner TT. Endocrine and exocrine effects of testicular torsion in the prepubertal and adult rat. J Androl. 1995;16:342–51.
150. Turner TT. Acute experimental testicular torsion. No effect on the contralateral testis. J Androl. 1985;6:65–72.
151. Cosentino MJ, Rabinowitz R, Valvo JR, et al. The effect of prepubertal spermatic cord torsion on subsequent fertility in rats. J Androl. 1984;5:93–8.
152. Tanyel FC, Büyükpamukçu N, Hiçsönmez A. Contralateral testicular blood flow during unilateral testicular torsion. Br J Urol. 1989;63:522–4.
153. Thomas WE, Cooper MJ, Crane GA, et al. Testicular exocrine malfunction after torsion. Lancet. 1984;2:1357–60.
154. Agarwal A, Nallella K, Allamaneni S, et al. Role of antioxidants in treatment of male infertility: an overview of the literature. Reprod Biomed Online. 2004;8:616–27.
155. Agarwal A, Sekhon LH. The role of antioxidant therapy in the treatment of male infertility. Hum Fertil (Camb). 2010;13:217–25.
156. Kim YH, Kim GH, Shin JH, et al. Effect of korean red ginseng on testicular tissue injury after torsion and detorsion. Korean J Urol. 2010;51:794–9.
157. Erol B, Bozlu M, Hanci V, et al. Coenzyme Q10 treatment reduces lipid peroxidation, inducible and endothelial nitric oxide synthases, and germ cell-specific apoptosis in a rat model of testicular ischemia/reperfusion injury. Fertil Steril. 2010;93:280–2.

158. Aktoz T, Kanter M, Aktas C. Protective effects of quercetin on testicular torsion/detorsion-induced ischaemia-reperfusion injury in rats. Andrologia. 2010;42:376–83.
159. Ozturk H, Ozturk H, Gideroglu K, et al. Montelukast protects against testes ischemia/reperfusion injury in rats. Can Urol Assoc J. 2010;4:174–9.
160. Shirazi M, Noorafshan A, Karbalay-Doust S, et al. Comparison of the protective effects of papaverine, lidocaine and verapamil on the sperm quality of the testis after induced torsion-detorsion in rats. Scand J Urol Nephrol. 2010;44:133–7.
161. Yang S, Shih HJ, Chow YC, et al. Simvastatin attenuates testicular injury induced by torsion-detorsion. J Urol. 2010;184:750–6.
162. Said TM, Paasch U, Glander HJ, et al. Role of caspases in male infertility. Hum Reprod Update. 2004;10:39–51.
163. Bozlu M, Coşkun B, Cayan S, et al. Inhibition of poly(adenosine diphosphate-ribose) polymerase decreases long-term histologic damage in testicular ischemia-reperfusion injury. Urology. 2004;63:791–5.
164. Bozlu M, Eskandari G, Cayan S, et al. The effect of poly (adenosine diphosphate-ribose) polymerase inhibitors on biochemical changes in testicular ischemia-reperfusion injury. J Urol. 2003;169:1870–3.
165. Hekimoglu A, Kurcer Z, Aral F, et al. Effects of 3-aminobenzamide on unilateral testicular ischemia-reperfusion injury: what is the role of PARP inhibition? Can J Physiol Pharmacol. 2010;88:1123–9.
166. Kar A, Ozden E, Yakupoglu YK, et al. Experimental unilateral spermatic cord torsion: the effect of polypolymerase enzyme inhibitor on histopathological and biochemical changes in the early and late periods in the ipsilateral and contralateral testicles. Urology. 2010;76:507.e1–5.
167. Moon C, Yasuzumi F, Okura N, et al. Enhanced expression of tyrosine kinase receptor a in germ cells of rat testis with acute experimental testicular torsion. Urol Int. 2005;74:79–85.
168. Minutoli L, Antonuccio P, Polito F, et al. Involvement of mitogen-activated protein kinases (MAPKs) during testicular ischemia-reperfusion injury in nuclear factor-kappaB knock-out mice. Life Sci. 2007;81:413–22.
169. Minutoli L, Antonuccio P, Romeo C, et al. Evidence for a role of mitogen-activated protein kinase 3/mitogen-activated protein kinase in the development of testicular ischemia-reperfusion injury. Biol Reprod. 2005;73:730–6.
170. Moon C, Kim JS, Jang H, et al. Activation of Akt/protein kinase B and extracellular signal-regulated kinase in rats with acute experimental testicular torsion. J Vet Med Sci. 2008;70:337–41.
171. Powell JD, Elshtein R, Forest DJ, et al. Stimulation of hypoxia-inducible factor-1 alpha (HIF-1alpha) protein in the adult rat testis following ischemic injury occurs without an increase in HIF-1alpha messenger RNA expression. Biol Reprod. 2002;67:995–1002.
172. Al-Maghrebi M, Kehinde EO, Anim JT. Long term testicular ischemia-reperfusion injury-induced apoptosis: involvement of survivin down-regulation. Biochem Biophys Res Commun. 2010;395:342–7.
173. Erol B, Tokgoz H, Hanci V, et al. Vardenafil reduces testicular damage following ischemia/reperfusion injury in rats. Kaohsiung J Med Sci. 2009;25:374–80.
174. Beheshtian A, Salmasi AH, Payabvash S, et al. Protective effects of sildenafil administration on testicular torsion/detorsion damage in rats. World J Urol. 2008;26:197–202.
175. Yildiz H, Durmus AS, Simşek H, et al. Protective effect of sildenafil citrate on contralateral testis injury after unilateral testicular torsion/detorsion. Clinics (Sao Paulo). 2011;66:137–42.
176. Istanbulluoglu MO, Zor M, Celik A, et al. Effects of vardenafil on testicular torsion/detorsion damage: an experimental study in pigs. Urol Int. 2011;86:228–32.
177. Ustün H, Akgül KT, Ayyildiz A, et al. Effect of phosphodiesterase 5 inhibitors on apoptosis and nitric oxide synthases in testis torsion: an experimental study. Pediatr Surg Int. 2008;24:205–11.

178. Köseoğlu B, Yilmaz E, Ceylan K, et al. The protective effect of erythropoietin infusion on testicular torsion/detorsion: an experimental study. Int Urol Nephrol. 2009;41:85–91.
179. Uguralp S, Bay Karabulut A, Mizrak B, et al. The effect of sustained and local administration of epidermal growth factor on improving bilateral testicular tissue after torsion. Urol Res. 2004;32:323–31.
180. Hashimoto H, Ishikawa T, Yamaguchi K, et al. Experimental ischaemia-reperfusion injury induces vascular endothelial growth factor expression in the rat testis. Andrologia. 2009;41:216–21.
181. Tunçkiran A, Cayan S, Bozlu M, et al. Protective effect of vascular endothelial growth factor on histologic changes in testicular ischemia-reperfusion injury. Fertil Steril. 2005;84: 468–73.
182. Savas C, Ozguner IF, Ozguner M, et al. The effects of unilateral testicular ischemia and hemicastration on contralateral testicular IGF-1 level, histology and lipid peroxidation. Int Urol Nephrol. 2003;35:231–5.
183. Ozkurkcugil C, Yardimoglu M, Dalcik H, et al. Effect of insulin-like growth factor-1 on apoptosis of rat testicular germ cells induced by testicular torsion. BJU Int. 2004;93:1094–7.
184. Zhao YG, Zheng XM, Zhou J, et al. Changes of nuclear factor-kappa gene binding expression in and apoptosis of spermatogenic epithelial cells in the restored testis after torsion: experiment with rats. Zhonghua Yi Xue Za Zhi. 2006;86:1381–5.
185. Zhao YG, Zheng XM, Zhou J, et al. Sulfasalazine prevents apoptosis in spermatogenic cells after experimental testicular torsion/detorsion. Acta Pharmacol Sin. 2006;27:603–8.
186. Lysiak JJ. The role of tumor necrosis factor-alpha and interleukin-1 in the mammalian testis and their involvement in testicular torsion and autoimmune orchitis. Reprod Biol Endocrinol. 2004;2:1–10.
187. Sukhotnik I, Voskoboinik K, Lurie M, et al. Involvement of the bax and bcl-2 system in the induction of germ cell apoptosis is correlated with the time of reperfusion after testicular ischemia in a rat model. Fertil Steril. 2009;92:1466–9.
188. Celebi M, Paul AG. Blockade of p-selectin reduces neutrophil infiltration into the murine testis after ischemia-reperfusion-injury. Dtsch Tierarztl Wochenschr. 2008;115:457–60.
189. Celebi M, Paul AG. Blocking E-selectin inhibits ischaemia-reperfusion-induced neutrophil recruitment to the murine testis. Andrologia. 2008;40:235–9.
190. Ozturk H, Ozturk H, Dokucu AI. The role of cell adhesion molecules in ischemic epididymal injury. Int Urol Nephrol. 2008;40:137–42.
191. Moon C, Jeong CW, Kim H, et al. Expression of CD44 adhesion molecule in rat testis with ischemia/reperfusion injury. J Vet Med Sci. 2006;68:761–4.
192. Lysiak JJ, Nguyen QA, Kirby JL, et al. Ischemia-reperfusion of the murine testis stimulates the expression of proinflammatory cytokines and activation of c-Jun NH_2-terminal kinase in a pathway to E-selectin expression. Biol Reprod. 2003;69:202–10.
193. Somuncu S, Cakmak M, Erdogan S, et al. Protective effects of trapidil in ischemia-reperfusion injury due to testicular torsion and detorsion: an experimental study. Int J Urol. 2006;13:601–5.
194. Ergur BU, Kiray M, Pekcetin C, et al. Protective effect of erythropoietin pretreatment in testicular ischemia-reperfusion injury in rats. J Pediatr Surg. 2008;43:722–8.
195. Yazihan N, Ataoglu H, Koku N, et al. Protective role of erythropoietin during testicular torsion of the rats. World J Urol. 2007;25:531–6.
196. Akcora B, Altug ME, Kontas T, et al. The protective effect of darbepoetin alfa on experimental testicular torsion and detorsion injury. Int J Urol. 2007;14:846–50.
197. Ozturk H, Ozturk H, Yagmur Y. PAF antagonist BN-52021 reduces intercellular adhesion molecule-1 expression and oxidative stress in rats with reperfusion damage due to unilateral testicular torsion. Pediatr Surg Int. 2006;22:191–6.
198. Can C, Töre F, Tunçel N, et al. Protective effect of vasoactive intestinal peptide on testicular torsion-detorsion injury: association with heparin-containing mast cells. Urology. 2004; 63:195–200.

199. Gokce G, Karboga H, Yildiz E, et al. Effect of angiotensin-converting enzyme inhibition and angiotensin II type 1 receptor blockade on apoptotic changes in contralateral testis following unilateral testicular torsion. Int Urol Nephrol. 2008;40:989–95.
200. Sozubir S, Barber T, Wang Y, et al. Loss of Insl3: a potential predisposing factor for testicular torsion. J Urol. 2010;183:2373–9.
201. Moon C, Ahn M, Kim S, et al. Increased expression of both constitutive and inducible forms of nitric oxide synthase in the delayed phase of acute experimental testicular torsion. J Vet Med Sci. 2005;67:453–6.
202. Shiraishi K, Yoshida K, Naito K. Activation of endothelial nitric oxide synthase in contralateral testis during unilateral testicular torsion in rats. Arch Androl. 2003;49:179–90.
203. Dokucu AI, Oztürk H, Ozdemir E, et al. The protective effects of nitric oxide on the contralateral testis in prepubertal rats with unilateral testicular torsion. BJU Int. 2000;85:767–71.
204. Barlas M, Hatiboğlu C. The effect of nitric oxide in testicular ischemia-reperfusion injury. Int Urol Nephrol. 2002;34:81–6.
205. Ozokutan BH, Küçükaydin M, Muhtaroğlu S, et al. The role of nitric oxide in testicular ischemia-reperfusion injury. J Pediatr Surg. 2000;35:101–3.
206. Dokucu AI, Ozturk H, Ozturk H, et al. The effects of molsidomine on hypoxia inducible factor alpha and Sonic hedgehog in testicular ischemia/reperfusion injury in rats. Int Urol Nephrol. 2009;41:101–8.
207. Ozturk H, Buyukbayram H, Ozdemir E, et al. The effects of nitric oxide on the expression of cell adhesion molecules (ICAM-1, UEA-1, and tenascin) in rats with unilateral testicular torsion. J Pediatr Surg. 2003;38:1621–7.
208. Gezici A, Ozturk H, Buyukbayram H, et al. Effects of gabexate mesilate on ischemia-reperfusion-induced testicular injury in rats. Pediatr Surg Int. 2006;22:435–41.
209. Sarioglu A, Gedikoglu G, Bingol-Kologlu M, et al. Capsaicin in albino rats prevents contralateral testis from the damaging effects posed by ipsilateral testis that underwent torsion. Eur Urol. 2001;40:469–72.
210. Karakaya E, Ateş O, Akgür FM, et al. Rosuvastatin protects tissue perfusion in the experimental testicular torsion model. Int Urol Nephrol. 2010;42:357–60.
211. Romeo C, Antonuccio P, Esposito M, et al. Raxofelast, a hydrophilic vitamin E-like antioxidant, reduces testicular ischemia-reperfusion injury. Urol Res. 2004;32:367–71.
212. Turan C, Küçükaydin N, Bekerecioğlu A, et al. The effect of vitamin E on ipsilateral and contralateral testis following unilateral testicular torsion in rats. Res Exp Med (Berl). 1996;196:243–6.
213. Azizollahi S, Babaei H, Derakhshanfar A, et al. Effects of co-administration of dopamine and vitamin C on ischaemia-reperfusion injury after experimental testicular torsion-detorsion in rats. Andrologia. 2011;43:100–5.
214. Akgül T, Ayyildiz A, Nuhoğlu B, et al. Ginkgo biloba (EGb 761) usage attenuates testicular injury induced by testicular ischemia/reperfusion in rats. Int Urol Nephrol. 2008;40:685–90.
215. Akgül T, Karagüzel E, Sürer H, et al. Ginkgo biloba (EGB 761) affects apoptosis and nitric-oxide synthases in testicular torsion: an experimental study. Int Urol Nephrol. 2009;41:531–6.
216. Dokmeci D, Akpolat M, Aydogdu N, et al. L-carnitine inhibits ethanol-induced gastric mucosal injury in rats. Pharmacol Rep. 2005;57:481–8.
217. Dokmeci D, Akpolat M, Aydogdu N, et al. The modifying effect of ibuprofen on total body irradiation-induced elevation of oxidative reactions in male hamsters. Acta Med Biol. 2004;52:67–72.
218. Erkin B, Dokmeci D, Altaner S, et al. Gastroprotective effect of L-carnitine on indomethacin-induced gastric mucosal injury in rats: a preliminary study. Folia Med (Plovdiv). 2006;48:86–9.
219. Tastekin N, Aydogdu N, Dokmeci D, et al. Protective effects of L-carnitine and alpha-lipoic acid in rats with adjuvant arthritis. Pharmacol Res. 2007;56:303–10.
220. Gurbuz B, Yalti S, Ficicioglu C, et al. Relationship between semen quality and seminal plasma total carnitine in infertile men. J Obstet Gynaecol. 2003;23:653–6.

221. Lenzi A, Lombardo F, Sgro P, et al. Use of carnitine therapy in selected cases of male factor infertility: a double-blind crossover trial. Fertil Steril. 2003;79:292–300.
222. Matalliotakis I, Koumantaki Y, Evageliou A, et al. L-carnitine levels in the seminal plasma of fertile and infertile men: correlation with sperm quality. Int J Fertil Womens Med. 2000;45:236–40.
223. Costa M, Canale D, Filicori M, et al. L-carnitine in idiopathic asthenozoospermia: a multicenter study. Italian Study Group on carnitine and male infertility. Andrologia. 1994;26:155–9.
224. Lenzi A, Sgro P, Salacone P, et al. A placebo-controlled double-blind randomized trial of the use of combined l-carnitine and l-acetyl-carnitine treatment in men with asthenozoospermia. Fertil Steril. 2004;81:1578–84.
225. Vitali G, Parente R, Melotti C. Carnitine supplementation in human idiopathic asthenospermia: clinical results. Drugs Exp Clin Res. 1995;21:157–9.
226. Vicari E, Calogero AE. Effects of treatment with carnitines in infertile patients with prostato-vesiculo-epididymitis. Hum Reprod. 2001;16:2338–42.
227. Vicari E, La Vignera S, Calogero AE. Antioxidant treatment with carnitines is effective in infertile patients with prostatovesiculoepididymitis and elevated seminal leukocyte concentrations after treatment with nonsteroidal anti-inflammatory compounds. Fertil Steril. 2002;78:1203–8.
228. Dokmeci D, Inan M, Basaran UN, et al. Protective effect of L-carnitine on testicular ischaemia-reperfusion injury in rats. Cell Biochem Funct. 2007;25:611–8.
229. Cankorkmaz L, Köylüoğlu G, Ozer H, et al. The role of apoptosis and protective effect of carnitine in contralateral testicular injury in experimental unilateral testicular torsion. Ulus Travma Acil Cerrahi Derg. 2009;15:529–34.
230. Guan Y, Zheng XM, Yang ZW, et al. Apoptosis of epididymal epithelium and the content of epididymal carnitine following testicular torsion/detorsion in rats. Zhonghua Nan Ke Xue. 2006;12:636–8.
231. Guan Y, Zheng XM, Yang ZW, et al. Protective effects of L-carnitine upon testicular ischemia-reperfusion damage in rats. Zhonghua Yi Xue Za Zhi. 2009;89:1858–61.
232. Gürocak S, Yilmaz A, Alp E, et al. Inflammation and oxidative stress in testicular torsion: do they deserve intensive treatment to save both guilty and innocent testes? Urology. 2011;78(1):164–9.
233. Aktaş BK, Bulut S, Bulut S, et al. The effects of N-acetylcysteine on testicular damage in experimental testicular ischemia/reperfusion injury. Pediatr Surg Int. 2010;26:293–8.
234. Cay A, Alver A, Küçük M, et al. The effects of N-acetylcysteine on antioxidant enzyme activities in experimental testicular torsion. J Surg Res. 2006;131:199–203.
235. Payabvash S, Salmasi AH, Kiumehr S, et al. Salutary effects of N-acetylcysteine on apoptotic damage in a rat model of testicular torsion. Urol Int. 2007;79:248–54.
236. Koc A, Narci A, Duru M, et al. The protective role of erdosteine on testicular tissue after testicular torsion and detorsion. Mol Cell Biochem. 2005;280:193–9.
237. Basaran UN, Dokmeci D, Yalcin O, et al. Effect of curcumin on ipsilateral and contralateral testes after unilateral testicular torsion in a rat model. Urol Int. 2008;80:201–7.
238. Wei SM, Yan ZZ, Zhou J. Curcumin attenuates ischemia-reperfusion injury in rat testis. Fertil Steril. 2009;91:271–7.
239. Gökçe A, Oktar S, Koc A, et al. Protective effect of thymoquinone in experimental testicular torsion. Urol Int. 2010;85:461–5.
240. Tamamura M, Saito M, Kinoshita Y, et al. Protective effect of edaravone, a free-radical scavenger, on ischaemia-reperfusion injury in the rat testis. BJU Int. 2010;105:870–6.
241. Uguralp S, Mizrak B, Bay Karabulut A. Resveratrol reduces ischemia reperfusion injury after experimental testicular torsion. Eur J Pediatr Surg. 2005;15:114–9.
242. Uguralp S, Usta U, Mizrak B. Resveratrol may reduce apoptosis of rat testicular germ cells after experimental testicular torsion. Eur J Pediatr Surg. 2005;15:333–6.
243. He QX, Zhang YG, Hu X, et al. Tea polyphenols protects the testis following testicular torsion/detorsion in rats. Zhonghua Nan Ke Xue. 2010;16:615–8.

244. Unsal A, Eroglu M, Avci A, et al. Protective role of natural antioxidant supplementation on testicular tissue after testicular torsion and detorsion. Scand J Urol Nephrol. 2006;40:17–22.
245. Etensel B, Ozkisacik S, Ozkara E, et al. Dexpanthenol attenuates lipid peroxidation and testicular damage at experimental ischemia and reperfusion injury. Pediatr Surg Int. 2007; 23:177–81.
246. Etensel B, Ozkisacik S, Ozkara E, et al. The protective effect of dexpanthenol on testicular atrophy at 60th day following experimental testicular torsion. Pediatr Surg Int. 2007; 23:271–5.
247. Avlan D, Erdouğan K, Cimen B, et al. The protective effect of selenium on ipsilateral and contralateral testes in testicular reperfusion injury. Pediatr Surg Int. 2005;21:274–8.
248. Atik E, Görür S, Kiper AN. The effect of caffeic acid phenethyl ester (CAPE) on histopathological changes in testicular ischemia-reperfusion injury. Pharmacol Res. 2006;54:293–7.
249. Koltuksuz U, Irmak MK, Karaman A, et al. Testicular nitric oxide levels after unilateral testicular torsion/detorsion in rats pretreated with caffeic acid phenethyl ester. Urol Res. 2000;28:360–3.
250. Uz E, Söğüt S, Sahin S, et al. The protective role of caffeic acid phenethyl ester (CAPE) on testicular tissue after testicular torsion and detorsion. World J Urol. 2002;20:264–70.
251. Adivarekar PK, Bhagwat SS, Raghavan V, et al. Effect of Lomodex-$MgSO_4$ in the prevention of reperfusion injury following unilateral testicular torsion: an experimental study in rats. Pediatr Surg Int. 2005;21:184–90.
252. Abes M, Sarihan H, Deger O, et al. The effect of ATP-$MgCl_2$ on prevention of reperfusion injury after unilateral testicular torsion. Eur J Pediatr Surg. 2001;11:255–8.
253. Ozkan KU, Boran C, Kilinç M, et al. The effect of zinc aspartate pretreatment on ischemia-reperfusion injury and early changes of blood and tissue antioxidant enzyme activities after unilateral testicular torsion-detorsion. J Pediatr Surg. 2004;39:91–5.
254. Wei SM, Yan ZZ, Zhou J. Beneficial effect of taurine on testicular ischemia-reperfusion injury in rats. Urology. 2007;70:1237–42.
255. Dokmeci D, Akpolat M, Aydogdu N, et al. The protective effect of L-carnitine on ionizing radiation-induced free oxygen radicals. Scand J Lab Anim Sci. 2006;33:75–83.
256. Dokmeci D. Ibuprofen and Alzheimer's disease. Folia Med (Plovdiv). 2004;46:5–10.
257. Dokmeci D. The role of peroxisome proliferator-activated receptors in Alzheimer's disease. In: Welsh EM, editor. New research on Alzheimer's disease. New York: Nova Science Publishers; 2006. p. 125–63.
258. Dokmeci D, Kanter M, Inan M, et al. Protective effects of ibuprofen on testicular torsion/detorsion-induced ischemia/reperfusion injury in rats. Arch Toxicol. 2007;81:655–63.
259. Inan M, Basaran U, Dokmeci D, et al. Rosiglitazone, an agonist of peroxisome proliferator-activated receptor-gamma, prevents contralateral testicular ischaemia-reperfusion injury in prepubertal rats. Clin Exp Pharmacol Physiol. 2007;34:457–61.
260. Mogilner JG, Lurie M, Coran AG, et al. Effect of diclofenac on germ cell apoptosis following testicular ischemia-reperfusion injury in a rat. Pediatr Surg Int. 2006;22:99–105.
261. Minutoli L, Antonuccio P, Polito F, et al. Peroxisome proliferator activated receptor beta/delta activation prevents extracellular regulated kinase 1/2 phosphorylation and protects the testis from ischemia and reperfusion injury. J Urol. 2009;181:1913–21.
262. Salmasi AH, Beheshtian A, Payabvash S, et al. Effect of morphine on ischemia-reperfusion injury: experimental study in testicular torsion rat model. Urology. 2005;66:1338–42.
263. Liu ZM, Zheng XM, Yang ZW, et al. Protective effect of pentoxifylline on spermatogenesis following testicular torsion/detorsion in rats. Zhonghua Nan Ke Xue. 2006;12:323–5.
264. Sava C, Dindar H, Aras T, et al. Pentoxifylline improves blood flow to both testes in testicular torsion. Int Urol Nephrol. 2002;33:81–5.
265. Savas C, Dindar H, Bilgehan A, et al. Pentoxifylline attenuates reperfusion injury in testicular torsion. Scand J Urol Nephrol. 2002;36:65–70.
266. Greenstein A, Schreiber L, Matzkin H. The effect of methylene blue on histological damage after spermatic cord torsion in a rat model. BJU Int. 2001;88:90–2.

267. Inan M, Basaran UN, Dokmeci D, et al. Methylene blue increases contralateral testicular ischaemia-reperfusion injury after unilateral testicular torsion. Clin Exp Pharmacol Physiol. 2008;35:50–4.
268. Unsal A, Devrim E, Guven C, et al. Propofol attenuates reperfusion injury after testicular torsion and detorsion. World J Urol. 2004;22:461–5.
269. Yagmurdur H, Ayyildiz A, Karaguzel E, et al. The preventive effects of thiopental and propofol on testicular ischemia-reperfusion injury. Acta Anaesthesiol Scand. 2006;50:1238–43.
270. Yagmurdur H, Ayyildiz A, Karaguzel E, et al. Propofol reduces nitric oxide-induced apoptosis in testicular ischemia-reperfusion injury by downregulating the expression of inducible nitric oxide synthase. Acta Anaesthesiol Scand. 2008;52:350–7.
271. Harvey M, Chanwai G, Cave G. Manual testicular detorsion under propofol sedation. Case Report Med. 2009;2009:529346.
272. Nezami BG, Rahimpour S, Gholipour T, et al. Protective effects of immunophilin ligands on testicular torsion/detorsion damage in rats. Int Urol Nephrol. 2009;41:93–9.
273. Ozen IO, Vural IM, Moralioğlu S, et al. Effects of mexiletine on electrical field stimulation-induced contractile responses in the ipsilateral and contralateral vasa deferentia after unilateral testicular torsion/detorsion. Eur Surg Res. 2006;38:423–9.
274. Abasiyanik A, Dağdönderen L. Beneficial effects of melatonin compared with allopurinol in experimental testicular torsion. J Pediatr Surg. 2004;39:1238–41.
275. Akgür FM, Kilinç K, Aktuğ T, et al. The effect of allopurinol pretreatment before detorting testicular torsion. J Urol. 1994;151:1715–7.
276. Kehinde EO, Anim JT, Mojiminiyi OA, et al. Allopurinol provides long-term protection for experimentally induced testicular torsion in a rabbit model. BJU Int. 2005;96:175–80.
277. Silva AC, Ortiz V, Silva RA, et al. Effect of allopurinol on rat testicles morphology, submitted to ischaemia for spermatic cord torsion followed by reperfusion. Acta Cir Bras. 2005;20:468–72.
278. Barun S, Ekingen G, Mert Vural I, et al. The effects of melatonin on electrical field stimulation-evoked biphasic twitch responses in the ipsilateral and contralateral rat vasa deferentia after unilateral testicular torsion/detorsion. Naunyn Schmiedebergs Arch Pharmacol. 2005;371:351–8.
279. Ozturk A, Baltaci AK, Mogulkoc R, et al. The effect of prophylactic melatonin administration on reperfusion damage in experimental testis ischemia-reperfusion. Neuro Endocrinol Lett. 2003;24:170–2.
280. Duru FI, Noronha CC, Akinwande AI, et al. Effects of torsion, detorsion and melatonin on testicular malondialdehyde level. West Afr J Med. 2007;26:312–5.
281. Yurtçu M, Abasiyanik A, Avunduk MC, et al. Effects of melatonin on spermatogenesis and testicular ischemia-reperfusion injury after unilateral testicular torsion-detorsion. J Pediatr Surg. 2008;43:1873–8.
282. Yurtçu M, Abasiyanik A, Biçer S, et al. Efficacy of antioxidant treatment in the prevention of testicular atrophy in experimental testicular torsion. J Pediatr Surg. 2009;44:1754–8.
283. Jeong SJ, Choi WS, Chung JS, et al. Preventive effects of cyclosporine a combined with prednisolone and melatonin on contralateral testicular damage after ipsilateral torsion-detorsion in pubertal and adult rats. J Urol. 2010;184:790–6.
284. Kanter M. Protective effects of melatonin on testicular torsion/detorsion-induced ischemia-reperfusion injury in rats. Exp Mol Pathol. 2010;89:314–20.
285. Kucer Z, Hekimoglu A, Aral F, et al. Effect of melatonin on epididymal sperm quality after testicular ischemia/reperfusion in rats. Fertil Steril. 2010;93:1545–9.
286. Yazawa H, Sasagawa I, Suzuki Y, et al. Glucocorticoid hormone can suppress apoptosis of rat testicular germ cells induced by testicular ischemia. Fertil Steril. 2001;75:980–5.
287. Cho YM, Pu HF, Huang WJ, et al. Role of serum- and glucocorticoid-inducible kinase-1 in regulating torsion-induced apoptosis in rats. Int J Androl. 2011;34(4):379–89. doi:10.1111/j.1365-2605.2010.01091.x.

288. Yapanoglu T, Aksoy Y, Gursan N, et al. Antiapoptotic effects of dehydroepiandrosterone on testicular torsion/detorsion in rats. Andrologia. 2008;40:38–43.
289. Aksoy H, Yapanoglu T, Aksoy Y, et al. Dehydroepiandrosterone treatment attenuates reperfusion injury after testicular torsion and detorsion in rats. J Pediatr Surg. 2007;42:1740–4.
290. Unal D, Karatas OF, Savas M, et al. Protective effects of trimetazidine on testicular ischemia-reperfusion injury in rats. Urol Int. 2007;78:356–62.
291. Savaş C, Ozgüner M, Ozgüner F, et al. The effects of human chorionic gonadotropin treatment on the contralateral side in unilateral testicular torsion. Int Urol Nephrol. 2003;35: 237–45.
292. Kolski JM, Mazolewski PJ, Stephenson LL, et al. Effect of hyperbaric oxygen therapy on testicular ischemia-reperfusion injury. J Urol. 1998;160:601–4.
293. Palmer JS, Cromie WJ, Lee RC. Surfactant administration reduces testicular ischemia-reperfusion injury. J Urol. 1998;159:2136–9.

Chapter 18
Varicocele and Oxidative Stress

Armand Zini and Naif Al-hathal

Abstract A varicocele is an abnormal dilatation of the pampiniform plexus within the spermatic cord. It is the most common male infertility factor, with multiple potential etiologies involved in its development. However, despite ongoing extensive research on varicoceles, the exact mechanism(s) by which varicocele influences male fertility is not known. Recent studies have shown that infertile men with varicocele have higher levels of seminal oxidative stress (OS) markers, and/or lower seminal antioxidant levels, than do fertile men and infertile men without varicocele. The abnormally high levels of seminal OS biomarkers (e.g., reactive oxygen species, malonaldehyde) in infertile men with varicocele is clinically relevant as these markers have been associated with poor sperm function and reduced fertility potential. In addition, infertile patients with varicocele possess high levels of sperm DNA damage and the mechanism of varicocele-induced sperm DNA damage is believed to be at least in part due to OS. The observed improvement in seminal OS and sperm DNA damage after varicocele repair supports the premise that varicocele can induce seminal OS.

Keywords Varicocele • Oxidative stress • Mechanisms of disease • Pathophysiology of varicocele • Varicocelectomy • DNA damage

18.1 Introduction

A varicocele is an abnormal dilatation of the pampiniform plexus within the spermatic cord. It is generally believed to be the most common treatable cause of male infertility. Varicocele is found in approximately 15–20% of the general population,

A. Zini, MD (✉) • N. Al-hathal, MD
Department of Surgery, St. Mary's Hospital, Montréal, QC, Canada
e-mail: ziniarmand@yahoo.com

A. Agarwal et al. (eds.), *Studies on Men's Health and Fertility*, Oxidative Stress in Applied Basic Research and Clinical Practice, DOI 10.1007/978-1-61779-776-7_18,

but the prevalence increases to 40% in men with infertility and up to 80% in men with secondary infertility [1]. This chapter reviews the current concepts regarding the association between varicocele, oxidative stress, and DNA damage.

18.2 Etiology of Varicocele

The development of varicocele is believed to be multifactorial. Anatomically, the gonadal vein is longer on left than the right side. The left gonadal vein enters the left renal vein, whereas the right renal vein enters the inferior vena cava (IVC). The narrower diameter renal vein (compared to the IVC) and the perpendicular entry of the left gonadal vein into the renal vein both increase the hydrostatic pressure along the gonadal vein, especially, when one is in the upright position [2]. Furthermore, the absent or incompetent valves within the internal spermatic veins (left more so than right), as seen in autopsy series of cadavers with varicocele, seem to play a role in the etiology of varicocele [3]. Increased hydrostatic pressure within the left internal spermatic vein is also attributed to compression of left renal vein between the superior mesenteric artery and aorta, the so-called "nutcracker" phenomenon.

18.3 Mechanisms of Disease

Despite ongoing extensive research on varicoceles, the exact mechanism(s) by which varicocele induces male infertility is not known. The most plausible mechanism is that varicocele induces scrotal hyperthermia and this results in defective spermatogenesis [4–6]. Studies have shown that elevations in scrotal temperature can impair spermatogenesis in patients with or without varicocele [7]. In support of this theory, external cooling of the scrotum has been associated with improved sperm parameters [8]. An additional mechanism by which varicocele may induce male infertility is the reflux of adrenal and renal metabolites into the spermatic veins. These metabolites have been shown to have a detrimental toxic effect on the testes and high levels of catecholamines [9], prostaglandins E and F [10], and adrenomedullin [11] have been reported in the internal spermatic vein of infertile patients with varicocele. Finally, venous stasis and hypoxia has also been implicated as a potential mechanism. Venous stasis and hypoxia may induce apoptosis and reactive oxygen species (ROS) generation at the testicular level with deleterious effects on spermatogenesis [12, 13].

18.4 Pathophysiology of Varicocele

The true effect of varicocele on testicular function and male fertility remains unknown. Although numerous studies have clearly demonstrated an association between varicocele and reduced male reproductive function (e.g., poor semen parameters, infertility), most varicocele studies involve highly selected populations (e.g., infertile men), making it difficult to ascertain the role of varicoceles in male infertility. Moreover, the lack of reliable end-points for measuring fertility represents another challenge in relating varicoceles with male infertility. Conventional sperm parameters (sperm concentration, motility, and morphology) are generally monitored in varicocele studies, but these parameters exhibit a high degree of biological variability and are of modest value in predicting male fertility potential [14]. Pregnancy is also of limited value in assessing the influence of varicocele on male fertility potential because this outcome is heavily influenced by female factors [15]. Studies of noninfertility populations provide conflicting results on the relationship between varicocele and fertility. As such, a cause and effect relationship between varicocele and male infertility has not been established.

An adverse effect of varicocele on male fertility is suggested by the testicular atrophy that is frequently associated with this condition [16–22]. In men with a left varicocele, mean left testicular volume is less than right testicular volume [18, 22]. However, the relationship between varicocele grade and the degree of testicular atrophy is less clear. Zini et al. [23] found that in men with unilateral left varicocele, the loss of left testicular volume relative to the right (i.e., right minus left) increased with increasing varicocele grade, whereas Alukal et al. [24] found no such correlation between varicocele grade and volume differential.

The impact of testicular atrophy on male fertility remains to be established, even though most studies indicate that testicular atrophy is associated with poor sperm parameters. Sigman and Jarow [20] have reported that in men with left varicocele, those with testicular atrophy have poorer semen parameters than do men without atrophy. Similarly, Diamond et al. [25] have shown that in the adolescent, a volume differential greater than 10% between the normal and varicocele testis correlates with a significantly decreased sperm concentration and total motile sperm count. Nonetheless, loss of testicular volume is not clearly associated with loss of fertility [19].

Varicocele is associated with bilateral spermatogenic abnormalities and Leydig cell dysfunction [26–29]. The testicular histology in infertile men with varicocele is variable, but most studies report reduced spermatogenesis (hypospermatogenesis) [30, 31]. Recently, Santoro and Romeo [32] described abnormalities in the ultrastructure of testicular tissue of men with varicocele. They noted that histologic changes were less pronounced in adolescents than in adults, supporting the concept that an uncorrected varicocele is associated with a progressive decline in testicular function. The observed increase in germ cell apoptosis associated with varicocele is thought to occur as a result of hyperthermia and low testosterone levels in the testis [33]. Serum testosterone levels are lower in older (>30 years) compared to younger men with varicocele, a trend not seen in men without varicocele, suggesting a progressive, adverse effect of varicocele on Leydig cell function [21].

The influence of varicocele on sperm parameters has not been established conclusively. In studies of infertile men, varicoceles have been associated with abnormal sperm parameters. MacLeod [34] and other investigators [21] observed that the majority of semen samples from infertile men with varicocele have poorer sperm parameters (lower sperm counts, increased numbers of spermatozoa with abnormal forms, and decreased sperm motility) than fertile men. However, the "stress pattern" described by MacLeod (i.e., increased proportions of sperm with tapered heads and immature forms) is not a specific marker for varicocele and, therefore, is not diagnostic of this condition [35]. In studies of unselected men (i.e., noninfertile), the relationship between varicocele and sperm parameters is less clear. Johnson et al. [36] showed that in a cohort of asymptomatic military recruits, nearly 70% of men with a palpable varicocele had an abnormality on semen analysis. In contrast, Zargooshi [37] observed that most young military recruits with moderate-to-large (grade 2 and 3) varicoceles have normal semen parameters.

Although studies on the prevalence of varicocele in men with primary and secondary infertility suggest that the presence of a varicocele may cause a progressive decline in fertility, this has not been confirmed by prospective studies. Chehval and Purcell [38] conducted a prospective, uncontrolled study of untreated varicocele and observed a significant deterioration in both sperm density and motility at 9–96 months follow-up. In contrast, Lund and Larsen [39] conducted a prospective, controlled trial of untreated men with and without varicocele and found no decline in semen parameters in either group after 8 years of follow-up.

18.5 Varicocele and Oxidative Stress

Seminal oxidative stress (OS) results from an imbalance between ROS production (superoxide anions, hydrogen peroxide, hydroxyl radical, hydroperoxyl radical, and nitric oxide) and ROS scavenging by seminal antioxidants (superoxide dismutase, glutathione peroxidase, catalase, uric acid, vitamins C and E, and albumin). Seminal OS is believed to be one of the main factors in the pathogenesis of sperm dysfunction and sperm DNA damage in male infertility [40–43]. Spermatozoa are particularly susceptible to oxidative injury due to the abundance of plasma membrane polyunsaturated fatty acids [44–47]. These unsaturated fatty acids provide fluidity that is necessary for membrane fusion events (e.g., the acrosome reaction and sperm–egg interaction) and for sperm motility. However, the unsaturated nature of these molecules predisposes them to free radical attack and ongoing lipid peroxidation throughout the sperm plasma membrane. Once this process has been initiated, lipid peroxides accumulate on the sperm surface and this results in loss of sperm motility and viability. ROS can cause damage to the DNA, directly or indirectly via production and subsequent translocation of lipid peroxides [44, 46, 48–51].

Several studies have examined the levels of OS markers (e.g., ROS, lipid peroxidation, oxidative DNA damage) in the semen of infertile and fertile men with and without varicocele. These studies have shown that infertile men with varicocele have higher levels of seminal OS markers than fertile men and infertile men without varicocele [52–62] (Table 18.1). Indeed, seven studies have reported that infertile men with varicocele demonstrate higher semen ROS levels than do fertile men [53–56, 59, 60, 62]. One study reported higher semen ROS levels in men with high-grade compared to low-grade varicocele [57]. In line with these observations, other

Table 18.1 Seminal oxidative stress in men with varicocele

Study	Seminal OS marker	Subjects (*n*)	Controls	Findings
Weese et al. [52]	Stimulated ROS	Fertile men with palpable varicocele (7)	Fertile men without varicocele (7)	Higher ROS levels in men with varicocele
Hendin et al. [53]	ROS (by CL)	Infertile (21) and fertile (15) men with varicocele	Sperm donors without varicocele (17)	Higher ROS levels in infertile and fertile men with varicocele
Sharma et al. [54]	ROS (by CL)	Infertile men with varicocele (55)	Sperm donors	Higher ROS levels in men with varicocele
Pasqualotto et al. [55]	ROS (by CL)	Infertile men with varicocele (77)	Fertile men (19)	Higher ROS levels in varicocele
Pasqualotto et al. [56]	ROS (by CL)	Infertile normospermic men with varicocele (16)	Healthy donors (19)	High ROS levels in varicocele group
Allamaneni et al. [57]	ROS (by CL)	Infertile men with grades 2 and 3 left varicocele	Infertile men with grade 1 left varicocele	Higher ROS levels in grade 2–3 varicocele compared to grade 1
Hurtado de Catalfo et al. [58]	TBARS (seminal and peripheral blood)	Infertile men with varicocele (36)	Fertile men (33)	Higher ROS in infertile men with varicocele
Pasqualotto et al. [59]	ROS (by CL)	Infertile men with varicocele (21)	Fertile men (17)	Higher ROS in varicocele group
Sakamoto et al. [60]	NO	Oligospermic (15) and normospermic men (15) with varicocele	Normospermic and oligospermic infertile men (15) without varicocele	Higher NO levels in varicocele group
Cocuzza et al. [61]	ROS (by CL)	Fertile men with varicocele (33)	Fertile men without varicocele (81)	No significant difference in ROS levels
Mostafa et al. [62]	TBARS, H_2O_2	Infertile (42) and fertile (45) men with varicocele	Fertile men without varicocele (45)	High ROS levels in varicocele groups vs. fertile men

CL Chemiluminescence; *MDA* malondialdehyde; *NO* nitric oxide; *OS* oxidative stress; *ROS* reactive oxygen species; *TBARS* thiobarbituric acid reactive substances

investigators have detected significantly higher levels of lipid peroxidation (measured as thiobarbituric acid reactive substances, TBARS) and oxidative DNA damage (8-OHdG, 8-hydroxy-2′-deoxyguanosine content) in the semen of infertile men with varicocele compared to that of fertile men [58, 60, 62]. However, it is less clear whether fertile men with varicocele have higher levels of seminal OS markers than fertile men without varicocele, as one of four published studies reported no significant difference in semen OS markers between these groups [52, 53, 61, 62]. Taken together, these studies demonstrate that infertile men with varicocele have elevated semen ROS levels when compared to men without varicocele.

Serum and testicular OS markers have also been examined in men with varicocele. Studies have shown that testicular tissue from infertile men with varicocele have higher levels of OS markers (e.g., 8-OHdG) than do testicular tissue from men without varicocele and support the premise that infertile men with varicocele have testicular OS [63, 64]. A number of studies have compared the levels of testicular vein and peripheral vein serum OS markers (e.g., NO, H_2O_2) in infertile men with varicocele and have generally shown higher OS markers in the testicular venous blood compared to the peripheral blood samples [65–68]. In three of these studies, individual patients served as their own controls, and due to the absence of appropriate controls (e.g., infertile men without varicocele, fertile men), the accurate interpretation of the data is not possible. In the one study with appropriate controls (healthy men without varicocele), higher levels of OS markers in the testicular compared to the peripheral blood samples were seen in men with clinical varicocele but not men without varicocele, supporting the premise that varicocele may cause a local (testicular vein), as well as, a milder systemic OS [65].

A number of studies have examined the antioxidant capacity (e.g., catalase, superoxide dismutase, SOD, total antioxidant capacity, TAC) in the semen of infertile and fertile men with and without varicocele. These studies have generally shown that infertile men with varicocele have lower seminal antioxidant levels than do fertile men [53–56, 58–60, 62] (Table 18.2). However, one compared infertile men with and without varicocele and reported that infertile men with varicocele have higher seminal antioxidant levels than do infertile men without varicocele [60]. These studies generally support the premise that infertile men with varicocele have lower seminal antioxidant capacity than fertile men, but there is no evidence to show that the seminal antioxidant capacity is lower than in infertile men without varicocele.

To date, there are a number of studies indicating that antioxidants may be beneficial in men with varicocele [69–71]. These studies support the premise that varicocele may, in part, be caused by oxidative stress.

18.6 Oxidative Molecular and Cellular Pathways

The exact mechanism by which varicocele induces testicular failure and sperm dysfunction is still under debate; however, oxidative stress seems to play a principal role in reproductive dysfunction. In addition, the oxidative cellular and

Table 18.2 Seminal antioxidant capacity in men with varicocele

Study	Semen antioxidant	Subjects (*n*)	Controls (*n*)	Findings
Hendin et al. [53]	Ascorbate, urate, tocopherol, and glutathione	Infertile (21) and fertile men (15) with varicocele	Sperm donors without varicocele (17)	Lower TAC in men with varicocele vs. controls
Sharma et al. [54]	TAC	Infertile men with varicocele (56)	Sperm donors (24)	Lower TAC in men with varicocele vs. controls
Pasqualotto et al. [55]	TAC	Infertile men with varicocele (77)	Healthy men (19)	Lower TAC in men with varicocele vs. controls
Pasqualotto et al. [56]	TAC	Infertile normospermic men with varicocele (16)	Healthy donors (19)	Lower TAC in varicocele group vs. controls
Hurtado de Catalfo et al. [58]	Nonenzymatic antioxidants, SOD, CAT	Infertile men with varicocele (36)	Fertile men (33)	Lower antioxidant levels in infertile men with varicocele
Pasqualotto et al. [55]	TAC	Infertile (21) and fertile (15) men with varicocele	Fertile men (17)	Lower TAC in varicocele groups vs. controls
Sakamoto et al. [60]	SOD	Oligospermic (15) and normospermic men with varicocele (15)	Oligospermic and normospermic infertile men (15) without varicocele	Higher SOD activity in varicocele group vs. men without varicocele
Mostafa et al. [62]	SOD, CAT, GPx, vitamins C and E	Infertile (42) and fertile (45) men with varicocele	Fertile men without varicocele (45)	Lower antioxidant levels in varicocele groups vs. fertile men

CAT Catalase; *GPx* glutathione peroxidase; *ROS* reactive oxygen species; *SOD* superoxide dismutase; *TAC* total antioxidant capacity

molecular pathways, which may be abnormally activated in men with varicocele, are not completely understood. One possible mechanism is the compensatory release of nitric oxide (NO), in response to testicular venous congestion, with associated elevation in malondialdehyde (MDA) that together contribute to excessive lipid peroxidation and defective spermatogenesis [72]. Another hypothesis is cytokine-mediated oxidative stress. The expression of IL-1 and IL-6, both potent proinflammatory activators, has been shown to be upregulated in experimental and clinical varicocele, respectively [72, 73]. Leptin expression has been associated with oxidative stress (lipid peroxidation) in obesity [74]. Studies have shown an increased expression of leptin in Leydig cells and germ cells in men with varicocele, suggesting a possible role of leptin in inducing OS in infertile men with varicocele [72, 75]. Glial cell line-derived neurotropic factor receptor-alpha-1, a growth factor that protects neuron from OS, has been shown to play a role in spermatogenesis and DNA synthesis [76]. This neurotropic factor is expressed at lower cellular levels in spermatids and Leydig cells in patients with untreated varicocele [77], which may account for OS in the testis. Other mechanisms that could explain the association between varicocele and OS include the antagonistic effect of cadmium on zinc (an antioxidant) [78], whereby increased levels of cadmium in infertile men with varicocele [79] could reduce seminal antioxidant capacity [72]. Finally, a reduced expression of HO-isoenzyme 1 may result in an increased vulnerability of testicular germ cell to oxidative injury in the presence of varicocele [80].

18.7 Effect of Varicocelectomy on Oxidative Stress

Varicocelectomy is the most common urological procedure for male factor infertility. In most of the published studies, varicocele repair has been associated with an improvement in semen parameters and spontaneous pregnancy rates, although the data remain controversial [81, 82]. Varicocele repair may improve pregnancy outcomes after assisted reproduction technologies (ARTs) [83, 84]. There is now evidence to show that varicocelectomy may reduce seminal oxidative stress and this provides another mechanism for the improved male fertility potential following varicocelectomy, as seminal OS has been associated with poor sperm function and reduced fertility potential [85].

As discussed earlier, seminal ROS levels are higher in semen samples from infertile men with varicocele [53, 86, 87] (including men with normozoospermia [86, 88]) than in controls and there are now several reports showing that varicocelectomy is generally associated with a reduction in seminal OS. However, the effect of varicocelectomy on seminal antioxidant capacity and seminal antioxidant enzyme activity (e.g., SOD) is mixed—with some studies reporting lower and others higher antioxidant activity after varicocelectomy [58, 60, 89–93] (Table 18.3). Nevertheless, the true effect of varicocelectomy on seminal OS is not proven as most of the studies are retrospective and none are randomized, controlled trials.

Table 18.3 Summary of studies that measured seminal OS and antioxidant (AOX) capacity before and after varicocelectomy

Study	OS marker and/or AOX capacity	*n*	Before varicocelectomy	After varicocelectomy	*P*-value	Conclusion
Zini [89], 1999	% Sperm with residual cytoplasm	43	28.5%	18.1%	0.0003	Reduced OS
Mostafa [90], 2001	MDA (nm/mL)	68	MDA = 23.2 ± 2.7	At 3–6 months MDA = 15.3 ± 4.6	$P<0.0001$ except albumin and vitamin E	Reduced OS and increased AOX capacity
	H_2O_2 (µm/mL)		H_2O_2 = 45.2 ± 10.2	H_2O_2 = 39.4 ± 7.9		
	NO (nm/L)		NO = 5.8 ± 1.0	NO = 5.5 ± 0.9		
	SOD (U/mL)		SOD = 6.6 ± 1.7	SOD = 10.7 ± 2.6		
	CAT (µm/mL)		CAT = 6.1 ± 1.1	CAT = 9.4 ± 2.5		
	GPx (U/mL)		GPx = 1.3 ± 0.6	GPx = 2.7 ± 1.2		
	Vitamin C and E (µg/mL)		Vitamin C = 2.7 ± 0.6 Vitamin E = 0.36 ± 0.06	Vitamin C = 3.1 ± 0.7 Vitamin E = 0.25 ± 0.08		
	Albumin (mg/mL)		Albumin = 0.28 ± 0.04	Albumin = 0.29 ± 0.03		
Mancini [91], 2004	TAC	25	Lag time = 107 ± 8.8 s	At 10–24 months Lag time = 106 ± 8.6 s	NS	No change in TAC
Hurtado de Catalfo et al. [58], 2007	GSH and GSSG (µmol/L)	36	GSH = 1.5 ± 0.21	At 8 months GST = 2.2 ± 0.22	<0.05 only for GSH, Zn, Se	Reduced OS and reduced AOX capacity
			GSSG = 0.19 ± 0.01	GSSG = 0.1 ± 0.06		
	Zn (mg/dL)		Zn = 45 ± 16	Zn = 233 ± 11		
	Se (µg/L)		Se = 41.5 ± 4.4	Se = 58.3 ± 6.5		
	SOD		SOD = 1,703 ± 134	SOD = 1,380 ± 135		
	CAT (U/g Hb)		CAT = 157 ± 21	CAT = 124 ± 17		
Chen [92], 2008	8-OHdG (/10^5 dG)	30	8-OHdG = 10.27 ± 2.24	At 6 months 8-OHdG = 5.95 ± 1.46	$P<0.001$	Reduced OS and increased AOX capacity
	Protein thiols (nmol/mL)		Thiols = 0.77 ± 0.75	Thiols = 3 ± 1.17		
	Ascorbic acid (AA) (mg/dL)		AA = 1.87 ± 0.40	AA = 3.12 ± 0.94		

(continued)

Table 18.3 (continued)

Study	OS marker and/or AOX capacity	*n*	Before varicocelectomy	After varicocelectomy	*P*-value	Conclusion
Sakamoto et al. [60], 2008	NO (μmol/L)	30	NO = 17.1(9.1)	At 6 months NO = 7.5(4.5)	$P < 0.001$	Reduced OS sperm DNA damage and increased AOX capacity
	8-OHdG (μmol/L)		8-OHdG = 10.3(4.7)	8-OHdG = 6.2(2.5)		
	SOD (%)		SOD = 85.8(5.8)	SOD = 78.1(8.1)		
	HEL (μmol/L)		HEL = 137.3(67.9)	HEL = 90.9(28.5)		
Dada [93], 2010	ROS levels in washed (W) and neat (N) semen	11	W = 3,121,725.65	At 3–6 months W = 98,971.081	$P < 0.001$	Reduced OS
			N = 142,897.704 (RLU/20 million sperm/min)	N = 6,456.249		

8-OHdG 8-Hydroxy-2′-deoxyguanosine; *CAT* catalase; *GPx* glutathione peroxidase; *GSH* reduced glutathione; *GSSG* oxidized glutathione; *HEL* hexanoyl-lysine; *MDA* malondialdehyde; *NO* nitric oxide; *ROS* reactive oxygen species; *Se* selenium; *SOD* superoxide dismutase; *TAC* total antioxidant capacity; *TBARS* thiobarbituric acid reactive substances; *Zn* zinc

Shiraishi and Naito examined the effect of varicocelectomy on the levels of testicular 4-hydroxy-2-nonenal (4-HNE)-modified proteins, an OS marker. They found higher levels of testicular 4-HNE-modified proteins in those men who responded to varicocelectomy suggesting that varicocelectomy reduces OS in those infertile men who exhibit a high level of baseline testicular OS and that varicocelctomy is not as effective in men without OS [64]. Mostafa et al. [90] reported a significant reduction in seminal plasma oxidants (NO, H_2O_2, MDA) and a significant increase in seminal plasma antioxidant levels (SOD, catalase, glutathione peroxidase, and vitamin C) at 3 and 6 months after varicocele repair. Hurtado de Catalfo et al. [58] demonstrated a shift in the seminal glutathione status (with increased reduced glutathione, GSH and decreased oxidized glutathione, GSSG), an increase in the levels of nonenzymatic antioxidants (Zn, Se), and a paradoxical decrease in the antioxidant enzyme levels (to levels comparable to that of fertile controls) as early as 1 month after varicocele repair. Chen et al. evaluated the effect of varicocelectomy on sperm mitochondrial DNA deletions (by polymerase chain reaction), DNA oxidation (8-OHdG), and seminal plasma protein thiols and ascorbic acid levels. They reported lower sperm mitochondrial DNA deletions and DNA oxidation levels and higher seminal plasma protein thiols and ascorbic acid levels at 6 months after varicocelectomy compared to preoperatively [92]. Dada et al. [93] evaluated the effect of varicocelectomy on seminal ROS levels and sperm DNA damage and reported lower semen ROS levels as early as 1 month after surgery with a delayed improvement in sperm DNA integrity (at 3 and 6 months after surgery). Sakamoto et al. [60] evaluated the effect of varicocelectomy on markers of seminal OS (nitric oxide, 8-OHdG, hexanoyl-lysine) and seminal SOD activity and reported lower levels of semen OS markers levels and lower seminal SOD activity after surgery. In contrast to otherwise largely positive studies, Mancini et al. [91] evaluated the effect of varicocelectomy on seminal plasma TAC and reported no significant reduction in seminal TAC after varicocele repair.

One report evaluated the effect of varicocelectomy on peripheral blood plasma oxidative stress markers (C-reactive protein, TBARS, and plasma lipid peroxidation susceptibility) in 11 adolescents with left-side varicocele and ipsilateral testicular hypotrophy [94]. They reported higher levels of TBARS and a higher mean plasma peroxidation susceptibility in adolescents with varicocele compared to controls. They also observed a significant reduction in plasma TBARS levels and in the plasma peroxidation susceptibility 1 year after varicocelectomy [94].

18.8 Varicocele and DNA Damage

Spermatozoa of infertile men possess substantially more chromatin defects and DNA damage than spermatozoa of fertile men, suggesting that sperm DNA damage is predictive of male fertility potential [95–97]. The etiology of sperm DNA damage is multifactorial and most investigators have proposed that ultimately, oxidative stress, aberrant chromatin remodeling (compaction), and abortive apoptosis can result in sperm DNA damage [98–101].

Table 18.4 Prospective studies on the effect of clinical varicocele repair on mean (±SD) sperm% DNA fragmentation index (DFI)

Study	*n*	Pre-op	Post-op	*P*-value
Smit et al. [84], 2010	49	35 ± 13	30 ± 15	0.019
Zini et al. [115], 2010	25	18 ± 11	11 ± 6	0.0009

Several studies have shown that infertile patients with varicocele have high levels of sperm DNA damage [87, 88, 102–104]. The mechanism of varicocele-induced sperm DNA damage is believed to be multifactorial, although most studies have suggested that the sperm DNA damage in these men is, at least in part, due to OS [53]. Smith et al. reported high levels of sperm DNA damage, using two assays of DNA damage (sperm chromatin structure assay, SCSA and Terminal deoxynucleotidyl transferase-mediated dUTP Nick End-Labeling, TUNEL), in patients with varicocele regardless of semen profile when compared to normozoospermic donors. They also observed that semen ROS levels (as measured by chemiluminescence) were significantly associated with sperm DNA damage [87]. Similarly, Saleh et al. [88] observed that infertile men with varicocele had significantly higher levels of sperm DNA damage (measured by SCSA) and OS (higher semen ROS levels and lower antioxidant capacity) when compared to fertile controls.

Another mechanism by which varicocele may induce sperm DNA damage is by increased apoptosis in the testis. Various endogenous and exogenous factors have been linked to testicular apoptosis in men with varicocele. Factors linked to testicular apoptosis in men with varicocele include alterations in intracellular calcium [104], hypoxia [105, 106], androgen deprivation [107], heat stress [108, 109], exposure to toxins (e.g., cadmium) [79, 110], and increased levels of apoptotic factors such as IL-6 and gonadal peptides [111, 112].

Several studies have now evaluated the effect of varicocelectomy on sperm DNA damage [84, 113, 114]. An evaluation of two published prospective studies of varicocelectomy on sperm DNA damage demonstrates that varicocele repair is associated with a significant decrease in sperm DNA damage (Table 18.4) [84, 115]. These data further support the premise that varicocele reduces testicular OS if we consider that the sperm DNA damage in these men is likely due to OS.

18.9 Summary

Varicocele is a common clinical finding in infertile men seeking fertility treatment. Varicocelectomy remains the most cost-effective and reversible treatment of infertile men with varicocele. Oxidative stress biomarkers (e.g., ROS, malonaldehyde) are significantly higher in the seminal fluid of infertile men with varicocele and this is clinically relevant as oxidative stress has been associated with poor sperm function and sperm DNA damage. Most studies have shown that infertile men with varicocele have suppressed levels of seminal antioxidant capacity, although some studies have found high levels of seminal antioxidant enzyme activity in these men. Varicocele repair is associated with improvement in semen parameters, pregnancy rates, and reduced seminal oxidative stress. However, randomized controlled trials are needed to confirm the effects of varicocelectomy on seminal oxidative stress as the available data are derived from uncontrolled studies.

References

1. Naughton CK, Nangia AK, Agarwal A. Pathophysiology of varicoceles in male infertility. Hum Reprod Update. 2001;7(5):473–81.
2. Shafik A, Bedeir GA. Venous tension patterns in cord veins. I: in normal and varicocele individuals. J Urol. 1980;123(3):383–5.
3. Ahlberg NE, Bartley O, Chidekel N. Right and left gonadal veins. An anatomical and statistical study. Acta Radiol Diagn (Stockh). 1966;4(6):593–601.
4. Comhaire F, Monteyne R, Kunnen M. The value of scrotal thermography as compared with selective retrograde venography of the internal spermatic vein for the diagnosis of "subclinical" varicocele. Fertil Steril. 1976;27(6):694–8.
5. Zorgniotti AW, Macleod J. Studies in temperature, human semen quality, and varicocele. Fertil Steril. 1973;24(11):854–63.
6. Goldstein M, Eid JF. Elevation of intratesticular and scrotal skin surface temperature in men with varicocele. J Urol. 1989;142(3):743–5.
7. Mieusset R, et al. Association of scrotal hyperthermia with impaired spermatogenesis in infertile men. Fertil Steril. 1987;48(6):1006–11.
8. Jung A, Eberl M, Schill WB. Improvement of semen quality by nocturnal scrotal cooling and moderate behavioural change to reduce genital heat stress in men with oligoasthenoteratozoospermia. Reproduction. 2001;121(4):595–603.
9. Comhaire F, Vermeulen A. Varicocele sterility: cortisol and catecholamines. Fertil Steril. 1974;25(1):88–95.
10. Ito H, et al. Internal spermatic vein prostaglandins in varicocele patients. Fertil Steril. 1982;37(2):218–22.
11. Ozbek E, et al. The role of adrenomedullin in varicocele and impotence. BJU Int. 2000;86(6):694–8.
12. Kilinc F, et al. Experimental varicocele induces hypoxia inducible factor-1alpha, vascular endothelial growth factor expression and angiogenesis in the rat testis. J Urol. 2004;172(3): 1188–91.
13. Lee JD, Jeng SY, Lee TH. Increased expression of hypoxia-inducible factor-1alpha in the internal spermatic vein of patients with varicocele. J Urol. 2006;175(3 Pt 1):1045–8; discussion 1048.
14. Guzick DS, et al. Sperm morphology, motility, and concentration in fertile and infertile men. N Engl J Med. 2001;345(19):1388–93.
15. ESHRE Capri Workshop Group. Social determinants of human reproduction. Hum Reprod. 2001;16(7):1518–26.
16. Akbay E, et al. The prevalence of varicocele and varicocele-related testicular atrophy in Turkish children and adolescents. BJU Int. 2000;86(4):490–3.
17. Laven JS, et al. Effects of varicocele treatment in adolescents: a randomized study. Fertil Steril. 1992;58(4):756–62.
18. Lipshultz LI, Corriere Jr JN. Progressive testicular atrophy in the varicocele patient. J Urol. 1977;117(2):175–6.
19. Pinto KJ, Kroovand RL, Jarow JP. Varicocele related testicular atrophy and its predictive effect upon fertility. J Urol. 1994;152(2 Pt 2):788–90.

20. Sigman M, Jarow JP. Ipsilateral testicular hypotrophy is associated with decreased sperm counts in infertile men with varicoceles. J Urol. 1997;158(2):605–7.
21. World Health Organization. The influence of varicocele on parameters of fertility in a large group of men presenting to infertility clinics. Fertil Steril. 1992;57(6):1289–93.
22. Zini A, et al. The influence of clinical and subclinical varicocele on testicular volume. Fertil Steril. 1997;68(4):671–4.
23. Zini A, et al. Loss of left testicular volume in men with clinical left varicocele: correlation with grade of varicocele. Arch Androl. 1998;41(1):37–41.
24. Alukal JP, et al. Testicular hypotrophy does not correlate with grade of adolescent varicocele. J Urol. 2005;174(6):2367–70; discussion 2370.
25. Diamond DA, et al. Relationship of varicocele grade and testicular hypotrophy to semen parameters in adolescents. J Urol. 2007;178(4 Pt 2):1584–8.
26. Dubin L, Hotchkiss RS. Testis biopsy in subfertile men with varicocele. Fertil Steril. 1969;20(1):51–7.
27. Comhaire F, Vermeulen A. Plasma testosterone in patients with varicocele and sexual inadequacy. J Clin Endocrinol Metab. 1975;40(5):824–9.
28. Johnsen SG, Agger P. Quantitative evaluation of testicular biopsies before and after operation for varicocele. Fertil Steril. 1978;29(1):58–63.
29. Hudson RW. The endocrinology of varicoceles. Fertil Steril. 1988;49(2):199–208.
30. Agger P, Johnsen SG. Quantitative evaluation of testicular biopsies in varicocele. Fertil Steril. 1978;29(1):52–7.
31. Ibrahim AA, et al. Bilateral testicular biopsy in men with varicocele. Fertil Steril. 1977;28(6):663–7.
32. Santoro G, Romeo C. Normal and varicocele testis in adolescents. Asian J Androl. 2001; 3(4):259–62.
33. Lue YH, et al. Mild testicular hyperthermia induces profound transitional spermatogenic suppression through increased germ cell apoptosis in adult cynomolgus monkeys (*Macaca fascicularis*). J Androl. 2002;23(6):799–805.
34. MacLeod J. Seminal cytology in the presence of varicocele. Fertil Steril. 1965;16(6):735–57.
35. Ayodeji O, Baker HW. Is there a specific abnormality of sperm morphology in men with varicoceles? Fertil Steril. 1986;45(6):839–42.
36. Johnson DE, Pohl DR, Rivera-Correa H. Varicocele: an innocuous condition? South Med J. 1970;63(1):34–6.
37. Zargooshi J. Sperm count and sperm motility in incidental high-grade varicocele. Fertil Steril. 2007;88(5):1470–3.
38. Chehval MJ, Purcell MH. Deterioration of semen parameters over time in men with untreated varicocele: evidence of progressive testicular damage. Fertil Steril. 1992;57(1):174–7.
39. Lund L, Larsen SB. A follow-up study of semen quality and fertility in men with varicocele testis and in control subjects. Br J Urol. 1998;82(5):682–6.
40. Zini A, San Gabriel M, Baazeem A. Antioxidants and sperm DNA damage: a clinical perspective. J Assist Reprod Genet. 2009;26(8):427–32.
41. Aitken RJ, et al. Analysis of the relationships between oxidative stress, DNA damage and sperm vitality in a patient population: development of diagnostic criteria. Hum Reprod. 2010;25(10):2415–26.
42. Fraga CG, et al. Ascorbic acid protects against endogenous oxidative DNA damage in human sperm. Proc Natl Acad Sci USA. 1991;88(24):11003–6.
43. Iwasaki A, Gagnon C. Formation of reactive oxygen species in spermatozoa of infertile patients. Fertil Steril. 1992;57(2):409–16.
44. Aitken RJ, Clarkson JS. Cellular basis of defective sperm function and its association with the genesis of reactive oxygen species by human spermatozoa. J Reprod Fertil. 1987; 81(2):459–69.
45. de Lamirande E, Gagnon C. Reactive oxygen species and human spermatozoa. II. Depletion of adenosine triphosphate plays an important role in the inhibition of sperm motility. J Androl. 1992;13(5):379–86.

46. de Lamirande E, Gagnon C. Reactive oxygen species and human spermatozoa. I: effects on the motility of intact spermatozoa and on sperm axonemes. J Androl. 1992;13(5):368–78.
47. Zini A, Garrels K, Phang D. Antioxidant activity in the semen of fertile and infertile men. Urology. 2000;55(6):922–6.
48. Yang MH, Schaich KM. Factors affecting DNA damage caused by lipid hydroperoxides and aldehydes. Free Radic Biol Med. 1996;20(2):225–36.
49. Lewis SE, Aitken RJ. DNA damage to spermatozoa has impacts on fertilization and pregnancy. Cell Tissue Res. 2005;322(1):33–41.
50. Alvarez JG, et al. Spontaneous lipid peroxidation and production of hydrogen peroxide and superoxide in human spermatozoa. Superoxide dismutase as major enzyme protectant against oxygen toxicity. J Androl. 1987;8(5):338–48.
51. Twigg J, et al. Analysis of the impact of intracellular reactive oxygen species generation on the structural and functional integrity of human spermatozoa: lipid peroxidation, DNA fragmentation and effectiveness of antioxidants. Hum Reprod. 1998;13(6):1429–36.
52. Weese DL, et al. Stimulated reactive oxygen species generation in the spermatozoa of infertile men. J Urol. 1993;149(1):64–7.
53. Hendin BN, et al. Varicocele is associated with elevated spermatozoal reactive oxygen species production and diminished seminal plasma antioxidant capacity. J Urol. 1999;161(6):1831–4.
54. Sharma RK, et al. The reactive oxygen species-total antioxidant capacity score is a new measure of oxidative stress to predict male infertility. Hum Reprod. 1999;14(11):2801–7.
55. Pasqualotto FF, et al. Relationship between oxidative stress, semen characteristics, and clinical diagnosis in men undergoing infertility investigation. Fertil Steril. 2000;73(3):459–64.
56. Pasqualotto FF, et al. Oxidative stress in normospermic men undergoing infertility evaluation. J Androl. 2001;22(2):316–22.
57. Allamaneni SS, et al. Increased seminal reactive oxygen species levels in patients with varicoceles correlate with varicocele grade but not with testis size. Fertil Steril. 2004;82(6):1684–6.
58. Hurtado de Catalfo GE, et al. Oxidative stress biomarkers and hormonal profile in human patients undergoing varicocelectomy. Int J Androl. 2007;30(6):519–30.
59. Pasqualotto FF, et al. Semen quality and oxidative stress scores in fertile and infertile patients with varicocele. Fertil Steril. 2008;89(3):602–7.
60. Sakamoto Y, et al. The assessment of oxidative stress in infertile patients with varicocele. BJU Int. 2008;101(12):1547–52.
61. Cocuzza M, et al. Impact of clinical varicocele and testis size on seminal reactive oxygen species levels in a fertile population: a prospective controlled study. Fertil Steril. 2008; 90(4):1103–8.
62. Mostafa T, et al. Seminal reactive oxygen species-antioxidant relationship in fertile males with and without varicocele. Andrologia. 2009;41(2):125–9.
63. Ishikawa T, et al. Increased testicular 8-hydroxy-2'-deoxyguanosine in patients with varicocele. BJU Int. 2007;100(4):863–6.
64. Shiraishi K, Naito K. Generation of 4-hydroxy-2-nonenal modified proteins in testes predicts improvement in spermatogenesis after varicocelectomy. Fertil Steril. 2006;86(1):233–5.
65. Chen SS, Chang LS, Wei YH. Oxidative damage to proteins and decrease of antioxidant capacity in patients with varicocele. Free Radic Biol Med. 2001;30(11):1328–34.
66. Mostafa T, et al. Reactive oxygen species and antioxidants relationship in the internal spermatic vein blood of infertile men with varicocele. Asian J Androl. 2006;8(4):451–4.
67. Romeo C, et al. Nitric oxide production is increased in the spermatic veins of adolescents with left idiophatic varicocele. J Pediatr Surg. 2001;36(2):389–93.
68. Turkyilmaz Z, et al. Increased nitric oxide is accompanied by lipid oxidation in adolescent varicocele. Int J Androl. 2004;27(3):183–7.
69. Moskovtsev SI, et al. Cause-specific treatment in patients with high sperm DNA damage resulted in significant DNA improvement. Syst Biol Reprod Med. 2009;55(2):109–15.
70. Oliva A, Dotta A, Multigner L. Pentoxifylline and antioxidants improve sperm quality in male patients with varicocele. Fertil Steril. 2009;91(4 Suppl):1536–9.

71. Paradiso Galatioto G, et al. May antioxidant therapy improve sperm parameters of men with persistent oligospermia after retrograde embolization for varicocele. World J Urol. 2008; 26(1):97–102.
72. Agarwal A, et al. Role of oxidative stress in pathogenesis of varicocele and infertility. Urology. 2009;73(3):461–9.
73. Nallella KP, et al. Relationship of interleukin-6 with semen characteristics and oxidative stress in patients with varicocele. Urology. 2004;64(5):1010–3.
74. Konukoglu D, Serin O, Turhan MS. Plasma leptin and its relationship with lipid peroxidation and nitric oxide in obese female patients with or without hypertension. Arch Med Res. 2006;37(5):602–6.
75. Ishikawa T, et al. Expression of leptin and leptin receptor in the testis of fertile and infertile patients. Andrologia. 2007;39(1):22–7.
76. Chen Z, et al. Glial cell line-derived neurotrophic factor promotes survival and induces differentiation through the phosphatidylinositol 3-kinase and mitogen-activated protein kinase pathway respectively in PC12 cells. Neuroscience. 2001;104(2):593–8.
77. Akkoyunlu G, et al. Immunolocalization of glial cell-derived neurotrophic factor (GDNF) and its receptor GFR-alpha1 in varicocele-induced rat testis. Acta Histochem. 2007;109(2): 130–7.
78. Benoff S, et al. Deletions in L-type calcium channel alpha1 subunit testicular transcripts correlate with testicular cadmium and apoptosis in infertile men with varicoceles. Fertil Steril. 2005;83(3):622–34.
79. Benoff SH, et al. Bilateral increased apoptosis and bilateral accumulation of cadmium in infertile men with left varicocele. Hum Reprod. 2004;19(3):616–27.
80. Shiraishi K, Naito K. Increased expression of Leydig cell haem oxygenase-1 preserves spermatogenesis in varicocele. Hum Reprod. 2005;20(9):2608–13.
81. Schlesinger MH, Wilets IF, Nagler HM. Treatment outcome after varicocelectomy. A critical analysis. Urol Clin North Am. 1994;21(3):517–29.
82. Kim ED, et al. Varicocele repair improves semen parameters in azoospermic men with spermatogenic failure. J Urol. 1999;162(3 Pt 1):737–40.
83. Cayan S, et al. Can varicocelectomy significantly change the way couples use assisted reproductive technologies? J Urol. 2002;167(4):1749–52.
84. Smit M, et al. Decreased sperm DNA fragmentation after surgical varicocelectomy is associated with increased pregnancy rate. J Urol. 2010;183(1):270–4.
85. Aitken RJ, Irvine DS, Wu FC. Prospective analysis of sperm-oocyte fusion and reactive oxygen species generation as criteria for the diagnosis of infertility. Am J Obstet Gynecol. 1991;164(2):542–51.
86. Barbieri ER, et al. Varicocele-associated decrease in antioxidant defenses. J Androl. 1999;20(6):713–7.
87. Smith R, et al. Increased sperm DNA damage in patients with varicocele: relationship with seminal oxidative stress. Hum Reprod. 2006;21(4):986–93.
88. Saleh RA, et al. Evaluation of nuclear DNA damage in spermatozoa from infertile men with varicocele. Fertil Steril. 2003;80(6):1431–6.
89. Zini A, et al. Effect of varicocelectomy on the abnormal retention of residual cytoplasm by human spermatozoa. Hum Reprod. 1999;14(7):1791–3.
90. Mostafa T, et al. Varicocelectomy reduces reactive oxygen species levels and increases antioxidant activity of seminal plasma from infertile men with varicocele. Int J Androl. 2001;24(5):261–5.
91. Mancini A, et al. Seminal antioxidant capacity in pre- and postoperative varicocele. J Androl. 2004;25(1):44–9.
92. Chen SS, et al. Attenuation of oxidative stress after varicocelectomy in subfertile patients with varicocele. J Urol. 2008;179(2):639–42.
93. Dada R, et al. Attenuation of oxidative stress & DNA damage in varicocelectomy: implications in infertility management. Indian J Med Res. 2010;132(6):728–30.

94. Cervellione RM, et al. Effect of varicocelectomy on the plasma oxidative stress parameters. J Pediatr Surg. 2006;41(2):403–6.
95. Evenson DP, et al. Utility of the sperm chromatin structure assay as a diagnostic and prognostic tool in the human fertility clinic. Hum Reprod. 1999;14(4):1039–49.
96. Zini A, et al. Correlations between two markers of sperm DNA integrity, DNA denaturation and DNA fragmentation, in fertile and infertile men. Fertil Steril. 2001;75(4):674–7.
97. Spano M, et al. Sperm chromatin damage impairs human fertility. The Danish First Pregnancy Planner Study Team. Fertil Steril. 2000;73(1):43–50.
98. Carrell DT, Liu L. Altered protamine 2 expression is uncommon in donors of known fertility, but common among men with poor fertilizing capacity, and may reflect other abnormalities of spermiogenesis. J Androl. 2001;22(4):604–10.
99. de Yebra L, et al. Detection of P2 precursors in the sperm cells of infertile patients who have reduced protamine P2 levels. Fertil Steril. 1998;69(4):755–9.
100. Sakkas D, et al. Abnormal spermatozoa in the ejaculate: abortive apoptosis and faulty nuclear remodelling during spermatogenesis. Reprod Biomed Online. 2003;7(4):428–32.
101. Aitken RJ, De Iuliis GN. Origins and consequences of DNA damage in male germ cells. Reprod Biomed Online. 2007;14(6):727–33.
102. Enciso M, et al. Infertile men with varicocele show a high relative proportion of sperm cells with intense nuclear damage level, evidenced by the sperm chromatin dispersion test. J Androl. 2006;27(1):106–11.
103. Simsek F, et al. Role of apoptosis in testicular tissue damage caused by varicocele. Arch Esp Urol. 1998;51(9):947–50.
104. El-Domyati MM, et al. The expression and distribution of deoxyribonucleic acid repair and apoptosis markers in testicular germ cells of infertile varicocele patients resembles that of old fertile men. Fertil Steril. 2010;93(3):795–801.
105. Hsu HS, et al. Decreased blood flow and defective energy metabolism in the varicocele-bearing testicles of rats. Eur Urol. 1994;25(1):71–5.
106. Li H, et al. Effect of surgically induced varicocele on testicular blood flow and Sertoli cell function. Urology. 1999;53(6):1258–62.
107. Fujisawa M, et al. The significance of gonadotropin-releasing hormone test for predicting fertility after varicocelectomy. Fertil Steril. 1994;61(4):779–82.
108. Mieusset R, Bujan L. Testicular heating and its possible contributions to male infertility: a review. Int J Androl. 1995;18(4):169–84.
109. Zini A, et al. Varicocele is associated with abnormal retention of cytoplasmic droplets by human spermatozoa. Fertil Steril. 2000;74(3):461–4.
110. Benoff S, et al. A potential role for cadmium in the etiology of varicocele-associated infertility. Fertil Steril. 1997;67(2):336–47.
111. Marmar JL. The pathophysiology of varicoceles in the light of current molecular and genetic information. Hum Reprod Update. 2001;7(5):461–72.
112. French DB, Desai NR, Agarwal A. Varicocele repair: does it still have a role in infertility treatment? Curr Opin Obstet Gynecol. 2008;20(3):269–74.
113. Werthman P, et al. Significant decrease in sperm deoxyribonucleic acid fragmentation after varicocelectomy. Fertil Steril. 2008;90(5):1800–4.
114. Zini A, et al. Beneficial effect of microsurgical varicocelectomy on human sperm DNA integrity. Hum Reprod. 2005;20(4):1018–21.
115. Zini A, et al. Effect of microsurgical varicocelectomy on human sperm chromatin and DNA integrity: a prospective trial. Int J Androl. 2010;34(1):14–9.

Chapter 19
Recreational Drugs and ROS Production in Mammalian Spermatozoa

Fábio Firmbach Pasqualotto and Eleonora Bedin Pasqualotto

Abstract Oxidative stress is caused by an imbalance between the production of reactive oxygen species (ROS) and antioxidant capacity of the cell. Oxidative stress appears to be the major cause of DNA damage in the male germ line. Furthermore, many studies have indicated a significant correlation between DNA damage and high levels of ROS in infertile patients. Male reproductive health is already under threat from a range of environmental factors including endocrine disruptors, toxic pollutants, ionizing and lifestyle factors such as sexually transmitted infections, alcoholism, smoking, and anabolic steroid use. Further hazards such as fast food, recreational drugs, and stress levels may also impair male fertility. Recreational drugs such as alcohol, barbiturates, amphetamines, THC, PCP, cocaine, and heroin, but also includes caffeine in coffee and cola beverages, and environmental factors may increase ROS production, leading to DNA damage and further male infertility.

Keywords Recreational drugs • ROS production • Mammalian spermatozoa • Psychotropic medications • Cigarette smoking

19.1 Introduction

Oxidative stress is caused by an imbalance between the production of reactive oxygen species (ROS) and antioxidant capacity of the cell [1]. Oxidative stress appears to be the major cause of DNA damage in the male germ line [2–5]. Furthermore,

---

F.F. Pasqualotto, MD, PhD (✉)
Department of Urology, University of Caxias do Sul,
Pinheiro Machado 2569, sl 23/24, Caxias do Sul, RS 95020172, Brazil
e-mail: Fabio@conception-rs.com.br

E.B. Pasqualotto, MD, PhD
Department of Gynecology, University of Caxias do Sul, Caxias do Sul, RS, Brazil

A. Agarwal et al. (eds.), *Studies on Men's Health and Fertility*, Oxidative Stress in Applied Basic Research and Clinical Practice, DOI 10.1007/978-1-61779-776-7_19,

many studies have indicated a significant correlation between DNA damage and high levels of ROS in infertile patients [6–10]. DNA damage in the male germ line has been associated with poor sperm quality, low fertilization rates, impaired preimplantation development, increased abortion rates, and an elevated incidence of disease in the offspring, including childhood cancer. The exact causes of this DNA damage are still unclear, but the major candidates are oxidative stress and aberrant apoptosis [10]. Several investigators have now found a link between oxidative stress and sperm DNA damage [11].

At least 14% of the reproductive population worldwide is affected by infertility [12]. Male reproductive health is already under threat from a range of environmental factors including endocrine disruptors, toxic pollutants, ionizing and lifestyle factors such as sexually transmitted infections, alcoholism, smoking, and anabolic steroid use. Further hazards such as fast food, recreational drugs, and stress levels may also impair male fertility.

Recreational drug may be defined as any substance with pharmacologic effects that is taken voluntarily for personal pleasure or satisfaction rather than for medicinal purposes. The term is generally applied to alcohol, barbiturates, amphetamines, tetrahydrocannabinol (THC), PCP, cocaine, and heroin, but also includes caffeine in coffee and cola beverages.

The drugs most popular for recreational use worldwide are the following:

- *Caffeine* (from *coffee*, *tea*, and other plant sources)—Legal in all parts of the world, but not consumed by members of some religions.
- *Cannabis* (commonly known as *marijuana* (MJ); contains *cannabinoids*, primarily *THC*)—Illegal in most parts of the world.
- *Ethanol* (commonly referred to as (ethyl) *alcohol*, produced through fermentation by yeast in *alcoholic beverages* such as wine and beer)—Legal but regulated in most parts of the world, and illegal in *several Muslim countries* such as Libya, Sudan, and Saudi Arabia; not consumed by members of some religions. In chemistry *alcohol* can refer to substances other than ethyl alcohol. Methyl (wood) alcohol is poisonous.
- *Tobacco* (contains *nicotine* and *β-carboline alkaloids*)—Legal but regulated in most parts of the world and not consumed by members of some religions.
- *Opiates* and *opioids*—Generally legal by prescription only, for relief of pain. These drugs include *hydrocodone*, *oxycodone*, *morphine*, and others; some opiates are illegal in some countries but used for medical purposes in others, such as diacetylmorphine (heroin).
- *Cocaine*—A stimulant derived from the *coca plant* in South America. Use of the stimulating coca leaf, but not cocaine, is legal in Peru and Bolivia. Cocaine is illegal in most parts of the world, but its derivatives such as *lidocaine* and *novocaine* are used in medicine and dentistry for local anesthesia.
- Psychotropic drugs.

19.2 Marijuana

MJ is the most commonly used recreational drug worldwide [13–15]. It is one of the most commonly abused substances in the USA, where 3.3% of young adults 19–28 years of age use MJ on a daily basis and 54% of people between 26 and 34 have used MJ at least once [16]. Medicinal uses of cannabis date back thousands of years, and both crude smoke and the psychoactive component, delta-9-tetrahydocannabinol (D9-THC), have been used for treating migraine headache, glaucoma, nausea, and anorexia [17]. MJ contains at least 20 active cannabinoids [18], the primary psychoactive cannabinoid being D9-THC [13, 14, 18]. Some reports suggest that, over the past 10–20 years, the cannabinoid content in MJ cigarettes may have increased severalfold [19, 20].

Despite widespread use, little information is available regarding toxic effects of MJ smoke. Persistent efforts to legalize MJ and political movements advocating medicinal uses tend to promote the notion that little or no hazardous risk is associated with MJ smoking. In vitro and whole-animal studies suggest that D9-THC has a direct immunosuppressive effect on a variety of immune cells, including macrophages, natural killer cells, and T lymphocytes [21]. Habitual MJ smoking has also been shown to alter alveolar macrophage morphology, phagocytic function, fungicidal and bactericidal activity, and oxidative burst superoxide production [22–24].

The expression/activity of cannabinoid receptor 1 controls the physiological alterations of DNA packaging during spermiogenesis and epididymal transit [13–15, 22–24]. Therefore, given the deleterious effects of sperm DNA damage on male fertility, the reproductive function of MJ users may also be impaired by deregulation of the endogenous endocannabinoid system [22–24].

The endocannabinoid system regulates many functions in the human body, including reproduction. Two cannabinoid receptors have thus so far been identified—CB1 and CB2—as well as a family of lipid signaling ligands, of which anandamide, or arachidonyl ethanoamide (AEA), is the most abundant in reproductive fluids [24]. The active psychotrophic ingredient of MJ, THC, acting as an antagonist and binding to these cannabinoid receptors, may upset the endogenous cannabinoid signaling pathways and thus disrupt the normal functions of reproduction [23]. CB1 receptors have been found in rat testes and mouse vas deferens [23, 24]. Therefore, they may play a possible role in the control of spermatogenesis and male fertility [25–27].

Exposure to MJ smoke causes a dramatic increase in ROS over control levels, an increase that was as much as threefold higher than the increment resulting from exposure to a similar amount of tobacco smoke [28]. Oxidizing agents prooxidants such as H_2O_2 are responsible for some of the MJ-induced effects. The MJ-induced ROS appeared to be cannabinoid-dependent because smoke from cigarettes lacking D9-THC produces no increase in ROS compared with control [28].

Impaired sperm motility may be due to deficient oxidative phosphorylation as a result of damaged or absent proteins in the electron transport chain subunits caused by mitochondrial deletions or base substitutions [29]. Also of note, the levels of mtDNA damage in sperm are much higher than those observed in somatic cells [30–32], suggesting that sperm may be more vulnerable to damage. Owing to the lack of histone proteins to protect the DNA against attack by ROS, mitochondria have little protection against oxidative assault, so their DNA accumulates large numbers of mutations [33]. The most immediate effect of ROS damage is in the mtDNA, as it is continually exposed to a high steady-state level of ROS and free radicals within the mitochondrial matrix [34]. It is also vulnerable because of the mitochondria's lack of repair mechanisms.

Sarafian et al. [35] reported that MJ smoke containing THC is a potent source of cellular oxidative stress that could contribute to cellular damage. Because the respiratory chain has the capacity for ROS production, an upregulation by THC may cause significant oxidative damage to mtDNA [35].

THC alters another essential sperm function, that of the acrosome reaction, at higher recreational concentrations in the 90% fraction [33]. The inhibition of the acrosome reaction was greater in the 45% fraction at both recreational and therapeutic concentrations, suggesting again that THC has more pronounced effects on sperm that are of poorer quality [33]. THC may impair fertilization by modulating acrosome reactions at recreational concentrations.

At low doses of MJ, there is a significant increase in lipid peroxidation and decrease in testicular lipid content, but the effects are significantly less at higher doses and at the withdrawal of cannabis treatment (recovery dose) [33]. There is a marked decrease in antioxidant enzyme profiles (superoxide dismutase, catalase, and glutathione peroxidase) and glutathione (GSH) content at low doses, but these effects are higher at higher dose and at withdrawal of the treatment (recovery effect) [33, 35]. Testicular histology reveals significant shrinkage of tubular diameter and detrimental changes in seminiferous epithelium of testis with resulting lowered serum testosterone and pituitary gonadotropins (follicular stimulating [FSH] and luteinizing hormones [LH]) levels at low doses [33, 35]. But at higher doses and particularly after withdrawing the treatment, the regression of various germ cell layers of testes through may be observed, indicating that recovery effects on testes may due through the endogenous testicular antioxidant enzymes profiles and pituitary gonadotropins hormones feedback mechanisms [33].

The generation of ROS has several undesirable consequences, including the impairment of cellular energetic and defense systems and the promotion of malignant transformation [36–38]. Cell death induced by MJ smoke is largely necrotic. These deleterious effects of MJ smoke could have serious implications for tissues in direct contact with cannabinoid-containing smoke, including lung macrophages and surface epithelial cells in the upper aerodigestive tract and the tracheobronchial mucosa. Such effects need to be taken into consideration when evaluating risk–benefit factors associated with MJ consumption.

19.3 Cocaine

Cocaine promotes oxidation via involvement of the enzymes inducible nitric oxide synthase (iNOS) and xanthine oxidase [39].

Daily intraperitoneal injection of 20 mg/kg cocaine in rats shows a significant oxidative stress in basal conditions that may be related to their impaired learning ability [39]. It may probably mean that changes in intensity of LPO processes observed during prenatal cocaine exposure are not directly linked to NO pathway activation.

Use of cocaine within 2 years of their first semen analysis has been found to be twice as common among men with sperm counts less than 20×10^6 mL [40]. Duration of cocaine use for 5 or more years was more common in men with low sperm motility and in those with low concentrations and a large proportion of abnormal forms. This association, together with the high prevalence of cocaine use in the general male population, suggests that cocaine may now be related to male subfertility and that history of use should be ascertained during diagnostic interviews.

Human spermatozoa acutely exposed to high concentrations of cocaine initially demonstrate a decrease in two motion kinematics: straight line velocity and linearity [41]. However, overall, cocaine exposure has no significant effects on sperm motility and fertilizing capability.

19.4 Alcohol

In adults, a possible effect of alcohol exposure during spermatozoa production on semen parameters has been assessed in a limited number of studies with conflicting results: some indicate a detrimental effect, but other point toward no or even a protective effect of moderate amounts of alcohol [42]. Whether maternal alcohol consumption during pregnancy is associated with poor semen quality in the male offspring has, to the best of our knowledge, never been examined and published. If such an association exists it may well be a candidate cause of the large differences in semen quality between different populations and time periods [43]. In fact, moderate prenatal exposure to alcohol was associated with lower sperm concentration, and sons exposed to ≥4.5 drinks/week in uterus had approximately one third lower sperm concentration than sons exposed to 1 drink/week.

An ethanol-induced oxidative stress is not restricted to the liver, where ethanol is actively oxidized, but can affect various extrahepatic tissues as shown by experimental data obtained in the rat during acute or chronic ethanol intoxication. Most of these data concern the central nervous system, the heart, and the testes [44].

That alcohol abuse may lead to testicular lipid peroxidation is suggested by the fact that ethanol is a known testicular toxin and its chronic use leads to both endocrine and reproductive failure. Because testicular membranes are rich in polyenoic fatty acids that are prone to undergo peroxidative decomposition, it is reasonable to consider that lipid peroxidation may contribute to the membrane injury and gonadal dysfunction that occurs as a result of alcohol abuse and/or chronic use. Consistent with such a mechanism for putative alcohol-associated testicular toxicity are the observed reductions in the testicular content of polyenoic fatty acids and GSH content of the testes of alcohol-fed animals as compared to isocalorically fed controls [45].

The increased conversion of xanthine dehydrogenase into xanthine oxidase as well as the activation of peroxisomal acyl CoA-oxidase linked to ethanol administration could contribute to the oxidative stress. Chronic ethanol administration elicits in the testes an enhancement in mitochondrial lipid peroxidation and a decrease in the GSH level, which appear to be correlated to the gross testicular atrophy observed. It is well known that peroxidation injury can be attenuated when it occurs in association with dietary vitamin A supplementation. Thus, it is of interest to note that vitamin A, acting as an antioxidant, stabilizes testicular membranes by reducing lipid peroxidation and prevents the alcohol-induced atrophy that occurs in animals not receiving vitamin A-enriched diets. Vitamin A supplementation attenuates the changes in lipid peroxidation, glutathione, and testicular morphology [46].

Taken together, these observations suggest that the enhanced peroxidation of testicular lipids that occurs following ethanol exposure may be an important factor in the pathogenesis of alcohol-associated gonadal injury.

19.5 Cigarette Smoking

Tobacco plants, like many other plants, will invariably contain certain amounts of lead absorbed from the soil; however, lead may also be deposited on the surface of the leaves. The widespread use of lead arsenate as a pesticide on tobacco crops has, in the past, resulted in lead contamination of tobacco products. The effect of tobacco constituents on classical sperm parameters and fertility has been under investigation for decades [47].

Künzle et al. [48], when evaluating many men attending a fertility clinic, found that cigarette smoking was associated with reductions in sperm concentration, motility, and normal morphology. Gaur et al. [49] showed that motility is one of the first sperm parameters affected and asthenozoospermia may be an early indicator of reduced semen quality in light smokers. They reported significantly high teratozoospermia in heavy smokers compared with nonsmokers.

In a study conducted on fertile men, it was observed that fertile men who were smokers showed reductions in semen volume in comparison with nonsmokers, and this reduction in semen volume was in proportion to the number of cigarettes smoked [5].

The largest cross-sectional study, including 2,542 healthy males, on this issue was carried out and published by Ramlau-Hansen et al. [50] and it showed that with increasing smoking, a 20–30% reduction in sperm count, volume and motile spermatozoa were observed. A study on voluntary males of reproductive age showed that after ejaculation, sperm motility deteriorated much more rapidly in heavy smokers in comparison to controls [51].

However, the exact effect of smoking on male fertility remains controversial. Toxic metabolites in smoke, such as nicotine, carbon monoxide, recognized carcinogens and mutagens such as radioactive polonium, cadmium, benzo(a) pyrene, and others have been demonstrated to be negatively associated with the normal development of male and female gametes and embryos [52, 53].

Also, direct exposure of spermatozoa to the toxins in cigarettes smoke probably tilts the delicate balance of ROS that are produced by spermatozoa for their special functions such as capacitation and acrosome reaction. Increased quantities of ROS have been shown to be detrimental to the DNA of spermatozoa, thus producing a negative effect on the viability and morphology of spermatozoa [54].

In fact, cigarette smoking has been associated with significantly increased levels of seminal ROS, which cause oxidative stress [6]. Cigarette smoking may inversely affect sperm quality and decrease the antioxidant activity in seminal plasma [7].

The exact molecular mechanisms for these effects are not well understood [55]. Animal models using rats linked the reduction of sperm developmental processes and fertility with mutagenic components of smoke [56]. These metabolites may cause deficiencies in spermatogenesis and chromatin condensation and reductions in sperm motility and the number of germ cells [57], as well as induction of apoptosis in the genital cells of the rat testes [58], a secretory deficit of Leydig and Sertoli cells [59], or impairment of testicular histology [60].

Different groups have studied sperm DNA fragmentation in smoking and nonsmoking men from infertile couples. An increased rate of sperm DNA fragmentation in smokers was found in pre and post-swim-up samples [61]. The protamine deposition can also be incomplete in smokers, resulting in ratios of histone to protamine and of protamine 1 to protamine 2 that differ from the normal. Both types of defects in spermiogenesis are associated with subfertility or infertility [62]. Oxidative stress occurs in seminal plasma of male smokers with increased levels of seminal ROS and with increases in the concentrations of cadmium and lead and with decreases in the concentration of ascorbic acid and the activity of other components of the antioxidant defense mechanism [6, 7]. Infertile men who smoke have higher levels of seminal oxidative stress indicators than infertile nonsmokers [6]. Oxidative stress has been shown to be a major cause of male infertility; a large proportion of infertile men were shown in one study to have elevated levels of seminal ROS [4]. ROS causes oxidative damage to normal sperm DNA, proteins and lipids, which may be related to sperm abnormalities [1]. Moreover, plasma membranes of spermatozoa, whose physically properties and functional integrity largely determine motility and fertilizing ability, hold a large amount of polyunsaturated fatty acids, prone to oxidative injury, whereas the cytoplasm contains low concentrations of antioxidant enzymes [63].

Cotinine concentrations in seminal plasma are considered as a biomarker for smoking [64]. A significant negative correlation was detected with P2 levels in spermatozoa, and a significant positive correlation was detected with P1/P2 ratios [65]. This is the first study conducted to evaluate the effect of smoking on the protamination process. It demonstrates the negative effect of smoking on protamination of sperm chromatin and indicates that the reason for higher P1/P2 ratios in smokers is the underrepresentation of P2, suggesting that smoking may affect P2 expression greater than P1. Smoking affects either the structure of protamines or their binding to DNA or affects sperm DNA directly through increasing oxidative damage.

Some investigators have described seminal leukocytes as a possible source of free radicals [66, 67]. The status of ROS-induced spermatic damage depends on the location of the inflammatory reaction, on the duration of the exposure of the sperm to these products, and the ability of the spermatic cell to activate its intrinsic defense system [68].

Recently, a study evaluated the levels of plasma ascorbic acid levels in smokers. The study showed a significant decrease in the mean seminal plasma ascorbic acid levels in smokers vs. nonsmokers in all groups. Also, in fertile subjects, smokers or not, demonstrated significant higher seminal ascorbic acid levels than any infertile group. Seminal plasma ascorbic acid in smokers and nonsmokers was correlated significantly with sperm concentration, sperm motility, but negatively with sperm abnormal forms. The study concluded that seminal plasma ascorbic acid decreases significantly in smokers and infertile men vs. nonsmokers and fertile men, and correlates with the main sperm parameters: count, motility, and normal morphology. Also, cigarette smoking is associated with reduced semen main parameters that could worsen the male fertilizing potential, especially in borderline cases [69].

DNA fragmentation index, high DNA stainability and round-head sperms are increased in idiopathic infertile men; this increase is associated with cigarette smoking. These defects may be attributed to increased oxidative stress and insufficient scavenging antioxidant enzymes in the seminal fluid of infertile patients [70].

The extent of these alterations was lesser in the group administered with nicotine along with selenium [71]. Analysis of plasma revealed reduced quantity of cotinine in the group coadministered with nicotine along with selenium in comparison with the nicotine group. Nondetectable levels of nicotine were present in the coadministered group. This indicates altered metabolism of nicotine when administered along with selenium.

19.6 Psychotropic Medications

The mechanism(s) of drug action are conventionally rationalized in terms of their binding as agonists or antagonists to specific proteins. However, for compounds acting on the membrane level other modes of action are also possible, owing to the principles of biomembrane operations as many body systems composed of lipids and associated proteins [72].

Radicals such as superoxide radical anion, hydroxyl radical (•OH), and hydrogen peroxide (H_2O_2) are produced in oxidative processes of cellular metabolism and accumulate in tissues upon aging as well as under pathophysiological conditions such as inflammation, leading to a state known as oxidative stress. These compounds are highly reactive toward cellular macromolecules, unsaturated lipids in particular, and their overproduction has been linked with several major diseases such as cancer, diabetes, schizophrenia, and neurodegenerative disorders, most notably both Parkinson's and Alzheimer's disease [72].

The changes in the chemical composition of the lipid bilayer caused by phospholipid peroxidation include shortening of the acyl chains and a decrease in the degree of their unsaturation, together with the modification of the acyl chains by polar carbonyl, carboxyl, hydroxide, and peroxide moieties. The above can be readily expected to result in alterations in several key physical properties of the bilayer, including lipid–lipid interactions, the polarity profile, chain order, lateral pressure profile, as well as thermal phase behavior, lateral organization, and membrane asymmetry. These modifications of the membrane lipid matrix can be anticipated to modulate the functions and lipid interactions of the various membrane-associated proteins in addition to possible ROS-induced direct chemical modification of the proteins themselves [72].

Increased oxidative stress has been associated with the pathogenesis of tardive dyskinesia and other extrapyramidal side effects of the antipsychotic treatment of schizophrenia. Chronic exposure to chlorpromazine been shown to decrease the expression and activity of key antioxidant enzymes and to elevate the level of lipid peroxides in rat brain [72].

Selective serotonin reuptake inhibitors can impair semen quality and damage sperm DNA integrity. In men with normal semen parameters, paroxetine induced abnormal sperm DNA fragmentation in a significant proportion of subjects, without a measurable effect on semen parameters. The fertility potential of a substantial number of men on paroxetine may be adversely affected by these changes in sperm DNA integrity. In fact, it has been suggested that treatment with selective serotonin reuptake inhibitors may be associated with an increase in sperm having DNA damage, as assessed by the SCSA assay. In vitro exposure to estradiol, by-products of estrogen metabolism, or genistein increased damage to the human sperm DNA, possibly by causing oxidative damage [72].

19.7 Morphine

Morphine sulfate and other opiate/narcotic analgesics are the most effective treatments for acute and chronic severe pain. However, their clinical utility is often hampered by the development of analgesic tolerance as well as by de novo painful hypersensitivity to innocuous and noxious stimuli with such phenomena observed in both animal and human studies [73–75]. Human spermatozoa express δ-, κ-, and μ-opioid receptors. These receptors are located in different parts of the head, in the middle region, and in the tail of the sperm. The δ-receptor antagonist naltrindole significantly reduces progressive motility immediately after its addition. However, the δ-receptor agonist DPDPE has no significant effect. Finally, neither the κ-receptor agonist U50488 nor its antagonist nor-binaltorphimine significantly affected the progressive motility of human spermatozoa.

Considerable evidence implicates nitroxidative stress in the development of pain of several etiologies and importantly in opiate antinociceptive tolerance, caused by the presence of superoxide, $O_2^{-\bullet}$, nitric oxide, $^{\bullet}NO$, and more recently peroxynitrite or its protonated counterpart ONOOH that is the product of their interaction.

The mechanisms by which prolonged opiate exposure induces tolerance and hypersensitivity remain unclear, although a role for peroxynitrite-mediated nitroxidative stress has been identified [76]. Peroxynitrite is a potent proinflammatory and proapoptotic reactive species and a potent inducer of hyperalgesia (defined as augmented pain intensity in response to painful stimuli) [77–80].

Considerable evidence over the years has supported the roles of $^{\bullet}NO$ and $O_2^{-\bullet}$ as precursors of peroxynitrite, in the development of morphine antinociceptive tolerance. Since the rate of interaction between $^{\bullet}NO$ and $O_2^{-\bullet}$ to form peroxynitrite is faster than the dismutation of $O_2^{-\bullet}$ by superoxide dismutase, peroxynitrite formation from $O_2^{-\bullet}$ and $^{\bullet}NO$ is the likely signaling molecule involved in antinociceptive tolerance as in pain of several etiologies [77, 80–83]. Because studies have only recently begun to unravel the role of peroxynitrite in antinociceptive tolerance and pain, few data are available to help understand the molecular and biochemical pathways engaged by this nitrooxidative species. To date, we know that peroxynitrite contributes to peripheral and central sensitization by increasing production of proinflammatory cytokines, by activating PARP, and modulating the cyclooxygenase pathway to increase the production of proinflammatory and pronociceptive PGE_2 (activation of COX-1 and COX-2 and induction of COX-2) [80]. Peroxynitrite is also involved in neuroimmune activation, apoptosis and post-translational nitration and modification of key proteins known to be implicated in central and peripheral sensitization [82]. Additionally, nitroxidative species may be involved more subtly in central sensitization at least in part by sensitizing wide dynamic range neurons in the dorsal horn [84]. Importantly for eventual clinical management, the peroxynitrite decomposition catalysts evaluated to date apparently synergize with nonselective COX-1/COX-2 inhibitors and selective COX-2 inhibitors, and do so at greatly reduced doses. This synergism should minimize the obvious side effects of either drug class when coadministered [82]. Considering the many molecular, biochemical, and pharmacological similarities between opiate-mediated antinociceptive tolerance and the hypersensitivity associated with chronic neuropathic pain, the broader implication of our proposed studies is that peroxynitrite is a viable therapeutic target in both disease states.

Morphine dependence can induce testis cell apoptosis, an increase in testis nitric oxide synthase positive cells, a decrease in calmodulin content and the activity of superoxide dismutase and glutathione superoxidase in the testis. NG-nitro-L-arginine has the curative effect on the morphine abstinent syndrome, protects testis cells against apoptosis and improves their involved biochemical indexes.

19.8 Caffeine

Caffeine (1,3,7-trimethylxanthine) is probably the most frequently ingested pharmacologically active substance in the entire world [85]. It is found in common beverages (coffee, tea, soft drinks), in products containing cocoa and medications, including headache or pain medicines and over-the-counter stimulants [85, 86]. Because of the wide consumption at different levels by most segments of the population, the public and the scientific community have expressed interest in the potential for caffeine to produce adverse effects on human health [85–87].

Coffee is considered a nutritional and pharmaceutical drink, it is richer in minerals than isotonic drinks, and also contains niacin (vitamin B). In addition, coffee contains natural antioxidants called chlorogenic acids [88]. An average intake of three cups a day provides calcium, vitamins, and other basic nutrients for human health without causing obesity. Recent cumulative scientific evidence have helped to demystify the idea that coffee is a psychotropic plant (due to the presence of caffeine) and to characterize it as a product with nutritional and medicinal (or pharmaceutical) properties, which can prove to be highly beneficial to human health [85, 89–91].

In fact, the alkaloid caffeine and its catabolic products theobromine and xanthine exhibit antioxidant properties [86–93]. Further, all three compounds proved to be capable of binding and reducing Cu(II) to Cu(I), leading to the generation of oxygen radicals. However, such an activity requires a relatively high concentration of the compounds and copper ions. Although ingested caffeine is capable of crossing the blood–testis barrier, caffeine consumption as a factor could alter male reproduction function has not been investigated extensively. Data from in vitro studies suggested that caffeine has variable, dose-related effects on human sperm motility, number, and structure [93]. Also, Jensen et al. [94] found no association between caffeine intake and semen quality in men exposed to caffeine for an extended period at dose levels as high as >700 mg/day. However, in a different center, Pasqualotto et al. [92] demonstrated that no differences are detected in sperm concentration, motility, and sperm motion characteristics in mild, moderate, heavy, and nonconsumers of coffee.

Therefore, several recreational drugs and environmental factors may increase ROS production, leading to DNA damage and further male infertility.

References

1. Aitken RJ, Baker MA. Oxidative stress, sperm survival and fertility control. Mol Cell Endocrinol. 2006;250:66–9.
2. Aitken RJ, De Luliis GN. On the possible origins of DNA damage in human spermatozoa. Mol Hum Reprod. 2010;16:3–13.
3. Aitken RJ, Baker MA, Sawyer D. Oxidative stress in the male germ line and its role in the aetiology of male infertility and genetic disease. Reprod Biomed Online. 2003;7:65–70.
4. Sharma RK, Agarwal A. Role of reactive oxygen species in male infertility. Urology. 1996;48:835–50.
5. Pasqualotto FF, Sobreiro BP, Hallak J, Pasqualotto EB, Lucon AM. Cigarette smoking is related to a decrease in semen volume in a population of fertile men. BJU Int. 2006;97:324–6.
6. Pasqualotto FF, Umezu FM, Salvador M, Borges Jr E, Sobreiro BP, Pasqualotto EB. Effect of cigarette smoking on antioxidant levels and presence of leukocytospermia in infertile men: a prospective study. Fertil Steril. 2008;90:278–83.
7. Saleh RA, Agarwal A, Sharma RK, Nelson DR, Thomas AJ. Effect of cigarette smoking on levels of seminal oxidative stress in infertile men: a prospective study. Fertil Steril. 2002;78:491–9.
8. Aitken RJ, Baker MA. Oxidative stress and male reproductive biology. Reprod Fertil Dev. 2004;16:581–8.
9. Barroso G, Morshedi M, Oehninger S. Analysis of DNA fragmentation, plasma membrane translocation of phosphatidylserine and oxidative stress in human spermatozoa. Hum Reprod. 2000;15:1338–44.

10. Lewis SEM, Agbaje I, Alvarez J. Sperm DNA tests as useful adjuncts to semen analysis. Syst Biol Reprod Med. 2008;54:111–25.
11. Tremellen K. Oxidative stress and male infertility: a clinical perspective. Hum Reprod Update. 2008;14:243–58.
12. Fishel S, Dowell K, Thornton S. Reproductive possibilities for infertile couples: present and future. In: Bentley GR, Mascie-Taylor CGN, editors. Infertility in the modern world. Cambridge, UK: Cambridge University Press; 2000. p. 17–45.
13. Park B, McPartland JM, Glass M. Cannabis, cannabinoids and reproduction. Prostaglandins Leukot Essent Fatty Acids. 2004;70:189–97.
14. Sharpe P, Smith G. Cannabis: time for scientific evaluation of this ancient remedy. Anesth Analg. 2000;90:237–40.
15. Hall W, Solowij N. Adverse effects of cannabis. Lancet. 1998;352:1611–6.
16. Johnston LD, O'Malley PM, Bachman JG. National survey results on drug use from the monitoring the future study, 1975–1995, vol II. College students and young adults (NIH Publication No. 98-4140). Washington, DC: USDHHS; 1997. p. 35–74.
17. Grinspoon L, Bakalar JB. Marijuana: the forbidden medicine, revised and expanded edition. New Haven, CT: Yale University Press; 1997.
18. Ashton H. Pharmacology and effects of cannabis: a brief review. Br J Psychiatry. 2001;178:101–6.
19. ElSohly MA, Ross SA. Quarterly report, NIDA potency monitoring project, report no. 50. Rockville, MD: National Institute on Drug Abuse; 1994.
20. Zimmer L, Morgan JP. Marijuana myths, marijuana facts. New York, NY: Lindesmith Center; 1997. p. 135–41.
21. Burnette-Curley D, Cabral GA. Differential inhibition of RAW264.7 macrophage tumoricidal activity by D9-tetrahydrocannabinol. Proc Soc Exp Biol Med. 1995;210:64–76.
22. Davies P, Sornberger GC, Huber GL. Effects of experimental marijuana and tobacco smoke inhalation on alveolar macrophages. Lab Invest. 1979;41:220–3.
23. Beals TF, Fligiel SEG, Stuth S, Tashkin DP. Morphological alterations of alveolar macrophages from marijuana smokers. Am Rev Respir Dis. 1989;139:A336 (Abstract).
24. Sherman MP, Campbell LA, Gong Jr H, Roth MD, Tashkin DP. Respiratory burst and microbicidal characteristics of pulmonary alveolar macrophages recovered from smokers of marijuana alone, smokers of tobacco alone, smokers of marijuana and tobacco and nonsmokers. Am Rev Respir Dis. 1991;144:1351–6.
25. Gerard C, Mollereau C, Vassart G, Parmentier M. Nucleotide sequence of a human cannabinoid receptor cDNA. Nucleic Acids Res. 1990;18:7142.
26. Sugiura T, Kondo S, Sukagawa A, Tonegawa T, Nakane S, Yamashita A, et al. Enzymatic synthesis of anandamide, an endogenous cannabinoid receptor ligand, through *N*-acylphosphatidylethanolamine pathway in testis: involvement of Ca(2+)-dependent transacylase and phosphodiesterase activities. Biochem Biophys Res Commun. 1996;218:113–7.
27. Pertwee RG, Joe-Adigwe G, Hawksworth GM. Further evidence for the presence of cannabinoid CB1 receptors in mouse vas deferens. Eur J Pharmacol. 1996;296:169–72.
28. Chioccarelli T, Cacciola G, Altucci L, Lewis SE, Ricci G, Ledent C, Meccariello R, Fasano S, Pierantoni R, Cobellis G. Cannabinoid receptor 1 influences chromatin remodeling in mouse spermatids by affecting content of transition protein 2 mRNA and histone displacement. Endocrinology. 2010;151(10):5017–29.
29. Cummins JM, Jequier AM, Martin R, Mehmet D, Goldblatt J. Semen levels of mitochondrial DNA deletions in men attending an infertility clinic do not correlate with phenotype. Int J Androl. 1998;21:47–52.
30. Hayakawa M, Hattori K, Sugiyama S, Ozawa T. Age-associated oxygen damage and mutations in mitochondrial DNA in human hearts. Biochem Biophys Res Commun. 1992;189: 979–85.
31. Corral-Debrinski M, Horton T, Lott MT, Shoffner JM, McKee AC, Beal MF, et al. Marked changes in mitochondrial DNA deletion levels in Alzheimer brains. Genomics. 1994; 23:471–6.

32. Fromenty B, Carrozzo R, Shanske S, Schon EA. High proportions of mtDNA duplications in patients with Kearns-Sayre syndrome occur in the heart. Am J Med Genet. 1997;71:443–52.
33. Wiseman H, Halliwell B. Damage to DNA by reactive oxygen and nitrogen species: role in inflammatory disease and progression to cancer. Biochem J. 1996;313:17–29.
34. Wei YH, Lu CY, Lee HC, Pang CY, Ma YS. Oxidative damage and mutation to mitochondrial DNA and age-dependent decline of mitochondrial respiratory function. Ann N Y Acad Sci. 1998;854:155–70.
35. Sarafian TA, Magallanes JAM, Shau H, Tashkin DP, Roth MD. Oxidative stress produced by marijuana smoke. An adverse effect enhanced by cannabinoids. Am J Respir Cell Mol Biol. 1999;20:1286–93.
36. Wallace KB, Eelss JT, Madeira VMC, Cortopassi G, Jones DP. Mitochondria-mediated cell injury. Fundam Appl Toxicol. 1997;38:23–37.
37. Repine JE, Bast A, Lankhorst I; The Oxidative Stress Study Group. Oxidative stress in chronic obstructive pulmonary disease. Am J Respir Crit Care Med. 1997;156:341–57.
38. Bankson DD, Kestin M, Rifai N. Role of free radicals in cancer and atherosclerosis. Clin Lab Med. 1993;13:463–80.
39. Bashkatova V, Meunier J, Vanin A, Maurice T. Nitric oxide and oxidative stress in the brain of rats exposed in utero to cocaine. Ann N Y Acad Sci. 2006;1074:632–42.
40. Yelian FD, Sacco AG, Ginsburg KA, Doerr PA, Armant DR. The effects of in vitro cocaine exposure on human sperm motility, intracellular calcium, and oocyte penetration. Fertil Steril. 1994;61(5):915–21.
41. Bracken MB, Eskenazy B, Sachse K, McSharry JE, Hellenbrand K, Leo Summers L. Association of cocaine use with sperm concentration, motility and morphology. Fertil Steril. 1990;53:315–22.
42. Marinelli D, Gaspari L, Pedotti P, Taioli E. Mini-review of studies on the effect of smoking and drinking habits on semen parameters. Int J Hyg Environ Health. 2004;207:185–92.
43. Swan SH, Elkin EP, Fenster L. The question of declining sperm density revisited: an analysis of 101 studies published 1934–1996. Environ Health Perspect. 2000;108:961–6.
44. Rosenblum ER, Gavaler JS, Van Thiel DH. Lipid peroxidation: a mechanism for alcohol-induced testicular injury. Free Radic Biol Med. 1989;7:569–77.
45. Nordmann R, Ribiere C, Rouach H. Ethanol-induced induced lipid peroxidation and oxidative stress in extrahepatic tissues. Alcohol. 1990;25:231–7.
46. Nordmann R. Alcohol and antioxidant systems. Alcohol. 1994;29:513–22.
47. Soares SR, Melo M. Cigarette smoking and reproductive function. Curr Opin Obstet Gynecol. 2008;20:281–91.
48. Künzle R, Mueller MD, Hänggi W, Birkhäuser MH, Drescher H, Bersinger NA. Semen quality of male smokers and nonsmokers in infertile couples. Fertil Steril. 2003;79:287–91.
49. Gaur DS, Talekar M, Pathak VP. Effect of cigarette smoking on semen quality of infertile men. Singapore Med J. 2007;48:119–23.
50. Ramlau-Hansen CH, Thulstrup AM, Aggerholm AS, Jensen MS, Toft G, Bonde JP. Is smoking a risk factor for decreased semen quality? A cross-sectional analysis. Hum Reprod. 2007;22:188–96.
51. Suleiman SA, Elamin Ali M, Zaki EMS, EL-Malik EMA, Nasr MA. Lipid peroxidation and human sperm motility: protective role of Vitamin E. J Androl. 1996;17:530–7.
52. Kumosani TA, Elshal MF, Al-Jonaid AA, Abduljabar HS. The influence of smoking on semen quality, seminal microelements and Ca^{2+}-ATPase activity among infertile and fertile men. Clin Biochem. 2008;41:1199–203.
53. Thompson J, Bannigan J. Cadmium: toxic effects on the reproductive system and the embryo. Reprod Toxicol. 2008;25:304–15.
54. Koksal IT, Usta M, Orhan I, Abbasoglu S, Kadioglu A. Potential role of reactive oxygen species on testicular pathology associated with infertility. Asian J Androl. 2003;5:95–9.
55. Arabi M, Mosthaghi H. Influence of cigarette smoking on spermatozoa via seminal plasma. Andrologia. 2005;37:119–24.

56. Sorahan T, Prior P, Lancashire RJ, Faux SP, Hulten MA, Peck IM, Stewart AM. Childhood cancer and parental use of tobacco: deaths from 1971 to 1976. Br J Cancer. 1997;76:1525–31.
57. Yamamoto Y, Isoyama E, Sofikitis N, Miyagawa I. Effects of smoking on testicular function and fertilizing potential in rats. Urol Res. 1998;26:45–8.
58. Rajpurkar A, Jiang Y, Dhabuwala CB, Dunbar JC, Li H. Cigarette smoking induces apoptosis in rat testis. J Environ Pathol Toxicol Oncol. 2002;21:243–8.
59. Kapawa A, Giannakis D, Tsoukanelis K, Kanakas N, Baltogiannis D, Agapitos E, Loutradis D, Miyagawa I, Sofikitis N. Effects of paternal cigarette smoking on testicular function, sperm fertilizing capacity, embryonic development, and blastocyst capacity for implantation in rats. Andrologia. 2004;36:57–68.
60. Ahmadnia H, Ghanbari M, Moradi MR, Khaje-Dalouee M. Effect of cigarette smoke on spermatogenesis in rats. Urol J. 2007;4:159–63.
61. Sepaniak S, Forges T, Gerard H, Foliguet B, Bene MC, Monnier-Barbarino P. The influence of cigarette smoking on human sperm quality and DNA fragmentation. Toxicology. 2006;223:54–60.
62. Aoki VW, Emery BR, Liu L, Griffin J, Carrell DT. Protamine levels vary between individual sperm cells of infertile human males and correlate with viability and DNA integrity. J Androl. 2006;27:890–8.
63. Velando A, Torres R, Alonso-Alvarez C. Avoiding bad genes: oxidatively damaged DNA in germ line and mate choice. Bioessays. 2008;30:1212–9.
64. Viloria T, Meseguer M, Conejero JAM, O'Çonnor JE, Remohi J, Pellicer A, Garrido N. Cigarette smoking affects specific sperm oxidative defenses but does not cause oxidative DNA damage in infertile men. Fertil Steril. 2010;94:631–7.
65. Hammadeh ME, Hamad MF, Montenarh M, Fischer-Hammadeh C. Protamine contents and P1/P2 ratio in human spermatozoa from smokers and non-smokers. Hum Reprod. 2010;25:2708–20.
66. Potts JM, Pasqualotto FF. Seminal oxidative stress in patients with chronic prostatitis. Andrologia. 2003;35:304–8.
67. Tortolero I, Duarte Ojeda JM, Pamplona Casamayor M, Alvarez Gonzalez E, Arata-Bellabarba G, Regadera J, et al. The effect of seminal leukocytes on semen quality in subfertile males with and without varicocele. Arch Esp Urol. 2004;57:921–8.
68. Everaert K, Mahmoud A, Depuydt C, Maeyaert M, Comhaire F. Chronic prostatitis and male accessory gland infection—is there an impact on male infertility (diagnosis and therapy)? Andrologia. 2003;35:325–30.
69. Mostafa T, Tawadrous G, Roaia MM, Amer MK, Kader RA, Aziz A. Effect of smoling on seminal plasma ascorbic acid in infertile and fertile males. Andrologia. 2006;38(6):221–4.
70. Eshai MF, El-Sayed IH, Elsaied MA, El-Masry SA, Kumosani TA. Sperm head defects and disturbances in spermatozoal chromatin and DNA integrities in idiopathic infertile subjects: association with cigarette smoking. Clin Biochem. 2009;42:589–94.
71. Xu DX, Shen HM, Zhu QX, Chua L, Wang QN, Chia SE, Ong CN. The associations among semen quality, oxidative DNA damage in human spermatozoa and concentrations of cadmium, lead and selenium in seminal plasma. Mutat Res. 2003;534:155–63.
72. Mattila JP, Sabatini K, Kinnunen PK. Oxidized phospholipids as potential novel drug targets. Biophys J. 2007;93(9):3105–12.
73. Mao J, Price DD, Mayer DJ. Mechanisms of hyperalgesia and morphine tolerance: a current view of their possible interactions. Pain. 1995;62:259–74.
74. Arner S, Rawal N, Gustafsson LL. Clinical experience of long-term treatment with epidural and intrathecal opioids—a nationwide survey. Acta Anaesthesiol Scand. 1988;32:253–9.
75. Ossipov MH, Lai J, King T, Vanderah TW, Malan Jr TP, Hruby VJ, Porreca F. Antinociceptive and nociceptive actions of opioids. J Neurobiol. 2004;61:126–48.
76. Muscoli C, Cuzzocrea S, Ndengele MM, Mollace V, Porreca F, Fabrizi F, Esposito E, Masini E, Matuschak GM, Salvemini D. Therapeutic manipulation of peroxynitrite attenuates the development of opiate-induced antinociceptive tolerance in mice. J Clin Invest. 2007; 117:3530–9.

77. Jagtap P, Szabo C. Poly(ADP-ribose) polymerase and the therapeutic effects of its inhibitors. Nat Rev Drug Discov. 2005;4:421–40.
78. Beckman JS, Beckman TW, Chen J, Marshall PA, Freeman BA. Apparent hydroxyl radical production by peroxynitrite: implications for endothelial injury from nitric oxide and superoxide. Proc Natl Acad Sci USA. 1990;87:1620–4.
79. Salvemini D, Jensen MP, Riley DP, Misko TP. Therapeutic manipulations of peroxynitrite. Drug News Perspect. 1998;11:204–14.
80. Ndengele MM, Cuzzocrea S, Esposito E, Mazzon E, Di Paola R, Matuschak GM, Salvemini D. Cyclooxygenases 1 and 2 contribute to peroxynitrite-mediated inflammatory pain hypersensitivity. FASEB J. 2008;22(9):3154–64.
81. Muscoli C, Sacco I, Alecce W, Palma E, Nistico R, Costa N, Clementi F, Rotiroti D, Romeo F, Salvemini D, Mehta JL, Mollace V. The protective effect of superoxide dismutase mimetic M40401 on balloon injury-related neointima formation: role of the lectin-like oxidized low-density lipoprotein receptor-1. J Pharmacol Exp Ther. 2004;311:44–50.
82. Wang ZQ, Porreca F, Cuzzocrea S, Galen K, Lightfoot R, Masini E, Muscoli C, Mollace V, Ndengele M, Ischiropoulos H, Salvemini D. A newly identified role for superoxide in inflammatory pain. J Pharmacol Exp Ther. 2004;309:869–78.
83. Bezerra MM, Brain SD, Girao VC, Greenacre S, Keeble J, Rocha FA. Neutrophils-derived peroxynitrite contributes to acute hyperalgesia and cell influx in zymosan arthritis. Naunyn Schmiedebergs Arch Pharmacol. 2007;374:265–73.
84. Lee I, Kim HK, Kim JH, Chung K, Chung JM. The role of reactive oxygen species in capsaicin-induced mechanical hyperalgesia and in the activities of dorsal horn neurons. Pain. 2007;133:9–17.
85. Nawrot P, Jordan S, Eastwood J, Rotstein J, Hugenholz A, Feeley M. Effects of caffeine on human health. Food Addit Contam. 2003;20:1–30.
86. Azam S, Hadi N, Khan NU, Hadi SM. Antioxidant and prooxidant properties of caffeine, theobromine, and xantine. Med Sci Monit. 2003;9:BR325–30.
87. Christian MS, Brent RL. Teratogen update: evaluation of the reproductive and developmental risks of caffeine. Teratology. 2001;64:51–78.
88. Esknazi B. Caffeine during pregnancy: grounds for concern? JAMA. 1993;270:2973–4.
89. Ashihara H, Crozier A. Caffeine: a well-known but little mentioned compound in plant science. Trends Plant Sci. 2001;6:407–18.
90. Joesoef MR, Beral V, Rolfs RT, Aral SO, Cramer DW. Are caffeinated beverages risk factors for delayed conception? Lancet. 1990;335:136–7.
91. Olsen J, Overvad K, Frische G. Coffee consumption, birthweight, and reproductive failures. Epidemiology. 1991;2:370–4.
92. Pasqualotto FF, Lucon AM, Sobreiro BP, Hallak J, Pasqualotto EB, Arap S. Does caffeine intake impair semen quality and hormonal levels in a fertile population? Fertil Steril. 2003;80:S231.
93. Dlugosz L, Bracken MB. Reproductive effects of caffeine: a review and theoretical analysis. Epidemiol Rev. 1992;14:83–100.
94. Jensen TK, Henriksen TB, Hjollund NH, Sceike T, Kolstad H, Giwercman A, et al. Caffeine intake and fecundability: a follow-up study among 430 Danish couples planning their first pregnancy. Reprod Toxicol. 1998;12:289–95.

Chapter 20
Oxidative Stress, DNA Damage, and Apoptosis in Male Infertility

Tamer M. Said, Constanze Fischer-Hammadeh, Mohammed Hamad, Khaled Refaat, and Mohamad Eid Hammadeh

Abstract Oxidative stress, sperm DNA damage, and apoptosis are associated with male infertility as a cause and as a manifestation. Several clinical conditions and laboratory findings may be indicative of the presence of oxidative stress, sperm DNA damage, and apoptosis. The use of standardized assays for the diagnosis of oxidative stress, sperm DNA damage, and apoptosis should be limited to cases where these conditions are suspected. The identifications of the causative factor for infertility will lead to a specific and subsequently more successful treatment option.

In this chapter, we provide a description of the clinical conditions associated with oxidative stress, sperm DNA damage, and apoptosis that lead to male infertility. The different signs and results of laboratory investigations will be provided to highlight manifestations of these conditions. We also describe the impact of oxidative stress, sperm DNA damage, and apoptosis on male fertility and the different approaches that could be used for diagnosis and prevention.

Keywords Oxidative stress • Apoptosis • Male infertility • Reactive oxygen species • Spermatozoa DNA damage • Apoptosis

---

T.M. Said, MD, PhD, HCLD (ABB) (✉)
Andrology Laboratory and Reproductive Tissue Bank, The Toronto Institute for Reproductive Medicine, 56 Aberfoyle Crescent, Toronto, ON, Canada M8X2W4
e-mail: tsaid@repromed.ca

C. Fischer-Hammadeh, MD • M. Eid Hammadeh, DVM, BSC, PhD
Department of Obstetrics and Gynecology, University of Saarland, Homburg, Saarland, Germany

M. Hamad, PhD
Department of Pharmacology and Bioscience, Petra University, Amman, Jordan

K. Refaat, MD
Department of Obstetrics and Gynecology, Al Azhar University, Assuit, Egypt

A. Agarwal et al. (eds.), *Studies on Men's Health and Fertility*, Oxidative Stress in Applied Basic Research and Clinical Practice, DOI 10.1007/978-1-61779-776-7_20,

20.1 Oxidative Stress, Apoptosis, and DNA Damage

20.1.1 Reactive Oxygen Species and Oxidative Stress

A free radical is defined as any species capable of independent existence that contains one or more unpaired electrons [1, 2]. Reactive oxygen species (ROS) include free radicals, as well as other oxygen-related reactive compounds. ROS are also formed during reduction of molecular oxygen to water in cellular respiration in the mitochondrial electron transport chain, via the cyclooxygenase pathway, as well as cellular enzymes such as cytochrome P450 oxidase and xanthine oxidase [3]. The most common ROS that have potential implication in reproductive biology include superoxide (O_2^-) anion, hydrogen peroxide (H_2O_2), peroxyl (ROO^-), and hydroxyl (OH^-) radicals [4, 5]. ROS are produced by some physiological processes such as phagocytosis, arachidonic acid pathway, mitochondrial oxidative phosphorylation, and other processes. Moreover, ROS are produced from the spermatozoa themselves or from the leukocytes [6–8].

In the human ejaculate, ROS are mainly produced by leukocytes, while marginal amounts are produced by spermatozoa [9]. It is to be noted that leukocytes are more powerful generators of ROS than spermatozoa [10]. ROS were detected in seminal plasma of infertile men [11–13]. ROS are also regarded as essential participants in cell signaling and gene regulation [14–16]. However, low levels of ROS are necessary for spermatozoa to develop normally and be capable of fertilization [17, 18]. In addition, H_2O_2 and O_2^- were found to promote sperm capacitation, acrosome reaction, hyperactivation, and oocyte fusion [19]. High rates of ROS production that exceed the antioxidant capacity of the seminal plasma result in oxidative stress (OS) which is harmful to spermatozoa [20–22]. Oxidative stress also attacks the fluidity of the sperm plasma membrane and the integrity of DNA in the sperm nucleus [23]. ROS may accelerate the process of germ cell apoptosis, leading to decline in sperm counts, DNA damage, and poor ART fertility outcomes [24]. All cellular components including lipids, proteins, nucleic acids, and sugars are potential targets of OS [25].

20.1.2 Seminal Antioxidants

Depending on the solubility in water or in lipids, antioxidants can be classified into two main divisions: hydrophilic which are water-soluble or hydrophobic which are lipid-soluble. In general, water-soluble antioxidants react with oxidants in the cell cytosol and the blood plasma, while lipid-soluble antioxidants protect cell membranes from lipid peroxidation. These compounds can be formed in the body or obtained from the diet [26].

The different types of antioxidants are present at various concentrations in body fluids and tissues, with some such as glutathione or ubiquinone mostly present within cells, while others such as uric acid being more evenly distributed. Some compounds contribute to antioxidant defense by chelating transition metals and preventing them from catalyzing the production of free radicals in the cell. Particularly important is the ability to sequester iron, which is the function of iron-binding proteins such as transferrin and ferritin. Selenium and zinc are commonly referred to as antioxidant nutrients; however, these chemical elements have no antioxidant action themselves and are instead required for the activity of some antioxidant enzymes [27].

20.1.3 Apoptosis and Oxidative Stress

Apoptosis is a mode of programmed cellular death based on a genetic mechanism that induces a series of cellular, morphological, and biochemical alterations, leading the cell to suicide without eliciting an inflammatory response. Mature sperm cells have been reported to express distinct markers of apoptosis-related cell damage [28–30].

Infertile males were reported to have relatively high rates of apoptosis in their testicular biopsies [31]. In addition, the percentage of apoptotic sperm is higher in ejaculated semen samples from infertile men compared with healthy men [32]. Moreover, sperm caspases become more activated in patients with infertility than in healthy donors during cryopreservation [33]. Nevertheless, it has not been confirmed if the apoptotic markers detected in spermatozoa are due to an abortive apoptotic process started before ejaculation or whether they result from apoptosis started in the postejaculation [34, 35].

20.1.3.1 Abortive Apoptosis

During spermatogenesis, Sertoli cells are responsible for the induction of apoptosis in about 50–60% of all germ cells that enter meiosis I. These cells are determined by apoptotic markers of the Fas type [36]. However, this mechanism may not always occur efficiently and a proportion of these defective germ cells enter the process of sperm remodeling during spermiogenesis, appearing later in the ejaculate. It has been proposed that apoptosis is responsible for the process of stripping the cytoplasm in the final stages of sperm maturation. In men with oligozoospermia, the possibility that a spermatozoon with normal morphology to be aneuploid is much higher than in normozoospermic men [37].

High levels of ROS can disrupt the mitochondrial membranes, inducing the release of the cytochrome c and activating caspases and subsequently apoptosis. Apoptosis in spermatozoa may also include the activation of caspases 1, 3, 8, and 9, annexin-V binding, and the mitochondrial generation of ROS [38]. Moreover, production of the mitochondrial ROS is increased in defective spermatozoa leading to the induction of DNA damage, possibly as a component of apoptosis [39]. The exposure of mitochondria to ROS ends by releasing apoptosis-inducing factor (AIF), which directly interacts with the DNA and leads to DNA fragmentation [40, 41]. Since ROS are important in mediating apoptosis by inducing caspases 3 and 9 and cytochrome c resulting in single and double DNA strand breaks [42], a significant portion of sperm DNA damage is oxidative in nature [43, 44].

Other proteins are required to initiate apoptotic response such as caspase-activated DNase that is responsible for DNA fragmentation [45]. These protein families include Bcl-2 family (Bcl-xs, Bcl-w, Bax, Bak, Bid, Bad), the tumor suppressor p53, the nuclear factor κB (NF-κB), and the heat shock proteins (HSPs) [46]. Moreover in spermatozoa, apoptosis may be initiated by ROS-independent pathways involving the cell surface protein Fas [47]. In contrast is the inhibitor gene of apoptosis, bcl-2, which protects the cell, most likely by decreasing ROS production [48].

Wang et al. [49] reported a positive correlation between the presence of apoptosis markers, due to OS-induced apoptosis, and sperm DNA damage. It has been shown that infertile patients with higher ROS levels in their seminal plasma had a higher percentage of sperm apoptosis than normal healthy donors [50]. Oxidative stress-induced apoptosis in ejaculated spermatozoa has been recently reviewed [51, 52]. Therefore, in the context of male infertility, seminal plasma OS level, sperm DNA damage, and apoptosis are interlinked and constitute a unified pathogenic molecular mechanism.

20.1.3.2 The BCL-2 Family

The BCL-2 family members of proteins reside upstream of irreversible cellular damage and focus much of their efforts at the level of mitochondria; they play a main role in deciding whether a cell will live or die. The BCL-2 proto-oncogene was discovered at the chromosomal breakpoint oft (14;18) bearing human B-cell lymphomas. The BCL-2 family of proteins has expanded significantly and includes both pro- as well as antiapoptotic molecules. Actually, the ratio between these two component helps to determine, in part, the susceptibility of cells to a death signal [53].

This family is characterized by their ability to form homo- as well as heterodimers, suggesting neutralizing competition between these proteins. A further characteristic of probable functional significance is their ability to become integral membrane proteins. BCL-2 family members possess up to four conserved BCL-2 homology (BH) domains designated BH1, BH2, BH3, and BH4, which correspond to a-helical segments [54, 55].

20.1.4 DNA Damage and Oxidative Stress

Many theories were proposed to explain the molecular mechanism of sperm DNA damage of spermatozoa; these are: (a) apoptosis occurring mainly during spermatogenesis [56, 57] with DNA double-strand breaks (DSBs) in spermatozoa arising through an abortive apoptotic pathway [58]; (b) ROS [59–61];

(c) defective chromatin packaging; and (d) DNA fragmentation induced by endogenous endonucleases [62]. Sperm DNA damage may result from various physiological and pathological conditions including advanced age [63], varicocele [64], cancer [65], leukocytospermia [66], or high prolonged fever [67]. In addition, environmental factors can also be involved such as smoking and air pollution [59, 68], radiation [43, 69], pesticides, chemicals, heat, and hydrogen peroxide [70]. In addition, spermatozoa DNA damage may result due to lack of the antioxidant protection or excessive exposure to ROS or both. Antioxidants also might be achieved by extracellular antioxidants provided by the secretions of the male reproductive tract, e.g., vitamin C, uric acid, tryptophan, spermine, and taurine [71, 72]. A reduction in the incidence of sperm DNA fragmentation by oral antioxidant treatment with such scavengers (vitamin C and E) has been also reported [73].

It has also been shown that in vitro separation of the spermatozoa from the seminal plasma in ART setting leaves them unprotected and vulnerable to ROS damage from leukocytes [74]. Removal of leukocytes and adding antioxidants to the ART medium such as reduced glutathione may reverse the action of leukocytes and decrease the negative effect of ROS on spermatozoa DNA [75–77]. However, the presence of sperm DNA damage has been closely associated with impaired sperm function as well as male infertility [56].

20.2 Clinical Presentations

20.2.1 Varicocele

Infertile patients with varicocele present with higher levels of ROS and lower levels of seminal antioxidants leading to increased OS [78]. Published data have shown that the grade of varicocele has an impact on the levels of seminal ROS, which were proven to decline following the surgical correction of varicocele [78]. The relationship between varicocele and OS is not fully understood; however, some mechanisms have been proposed. Significant amount of nitric oxide (NO) production are released from the dilated pampiniform plexus of veins leading to OS [79]. Additionally, NO may also interact with superoxide anion resulting in the production peroxynitrate, which result in further sperm damage [80].

Patients diagnosed with infertility and varicocele had significantly higher levels of OS and sperm DNA damage compared to healthy controls [64]. This could indicate a potential pathway for sperm DNA damage resulting from OS in these patients with varicocele [64]. OS and its resulting sperm dysfunction were estimated to cause infertility in 15% of males with varicocele [81]. Another potential cause of sperm DNA damage in varicocele patients is apoptosis. Levels of apoptosis are higher in ejaculated spermatozoa from varicocele patients than in spermatozoa from healthy men [52].

20.2.2 Unexplained Infertility

Unexplained infertility is diagnosed whenever routine testing methods including semen analysis fail to identify a reason for delay in achieving pregnancy. Other testing methods such as assessment of ROS and DNA fragmentation may reveal the actual cause of infertility in these cases [82]. Higher seminal ROS levels and lower antioxidant levels were shown in men diagnosed with unexplained infertility compared to healthy controls [83]. Moreover, abnormal levels of ROS were seen in 25–40% of men diagnosed with unexplained infertility [84].

Assessment of the sperm DNA fragmentation analysis can also provide the exact etiology in cases with unexplained infertility especially in the presence of normal semen parameters [85]. Since OS is one of the causes of increased sperm DNA damage, both conditions may play a conjoint potentiating role resulting in male infertility. Also, apoptosis markers such as cytochrome c, mitochondrial membrane potential, and caspases 3, 9 have been shown in association with OS-induced sperm damage which may point to the implication of apoptosis in unexplained infertility as seen in cases with normal semen parameters [49].

Apoptosis has been approved to play a major role in male infertility [32]; however, the exact mechanisms of its involvement need to be explained. Relatively high levels of apoptosis have been reported in testicular biopsies from infertile men with different degrees of testicular insufficiency [31]. The percentage of apoptotic sperm is reported to be higher in ejaculated semen samples from infertile men compared to healthy men [32]. Although apoptosis is considered a mechanism to ensure selection of sperm cells with undamaged DNA, sperm with DNA damage that are not eliminated by apoptosis may fertilize an ovum [86]. Poor chromatin packaging and/or damaged DNA have been implicated in the failure of sperm decondensation after intracytoplasmic sperm injection (ICSI), resulting in fertilization failure [87].

20.2.3 Genitourinary Tract Infections

Infections and inflammation in the male genital tract are associated with an increase in levels of ROS, which in turn results in an imbalance between ROS and antioxidant defenses resulting in OS [88]. Increased seminal leukocyte concentration above the accepted value ($>1 \times 10^6$/mL) is termed as leukocytospermia. It has been associated with high level of DNA fragmentation in samples from men diagnosed with genitourinary tract infections [89]. Other mediators of inflammation, cytokines, such as interleukin-6, and interleukin-8 also play a role in increasing levels of ROS [89]. It is important to note that the increased levels of ROS production by leukocytes occur as a defense mechanism against pathogenic organisms [90]. However, these increased levels are also detrimental for the sperm function. The occurrence of OS in cases with genitourinary tract infections is also caused by decrease in antioxidants [91]. In these cases, semen samples showed low levels of antioxidants such as citric acid and zinc [91]. These antioxidants are also important in maintaining the sperm DNA integrity [92].

20.2.4 *Malignant Conditions*

Testicular and systemic malignant conditions have been proven to result in male infertility even before the initiation of cancer treatments [93]. Patients with testicular and systemic malignancies present with higher levels of DNA fragmentation compared to fertile controls [86]. The levels of DNA fragmentation appear to be more significant in patients diagnosed with Hodgkin's and non-Hodgkin's diseases [94]. In addition, management of malignant conditions including chemo and radiotherapy has a negative impact on male fertility. The different therapeutic agents used for cancer treatment result in sperm DNA damage, which may present with mutations, repeated miscarriages, and carcinogenesis in offspring [95].

The structural organization of sperm DNA is vital for the proper functioning of the spermatozoa [96]. It is not clear whether the cancer itself is capable of inducing changes in the genomic integrity of the spermatozoa. If the DNA is structurally intact, such patients can enroll in sperm banking and have increased success when using assisted reproductive procedures such as ICSI.

20.2.5 *Outcomes of Assisted Reproductive Techniques*

Seminal ROS levels correlate with success rates of intrauterine insemination (IUI), in vitro fertilization (IVF), and ICSI [97]. Since spermatozoa used in assisted reproductive techniques originate from an environment conducive to OS, they could be the source of introduction of high levels of ROS [98]. In support, embryo culture media showing higher levels of ROS have been correlated with low fertilization rate, decreased blastocyst and cleavage rate, and decreased embryo quality and pregnancy rates following assisted reproductive techniques [99].

Ample data document the negative impact of sperm DNA fragmentation on the success rates and outcomes of assisted reproductive techniques, specifically the embryo development and pregnancy rates [100]. These negative effects are seen whenever the extent of sperm DNA damage exceeds the repairing capabilities of the oocyte [21].

20.3 Laboratory Presentations

20.3.1 *Leukocytospermia*

Leukocytes are a normal constituent of the human seminal fluid. However, if the leukocyte concentration exceeds the normal values (1×10^6/mL), sperm dysfunction may be seen [101]. Higher leukocyte concentrations can be seen in ejaculates of men with varicocele, genitourinary tract infection, and those who heavily consume alcohol and smoke tobacco [102]. Male smokers diagnosed with infertility have an increased incidence of OS, which may be due to an increase in leukocyte concentrations by as much as 48% [103]. Sperm dysfunction due to the presence of leukocytes may be also seen at levels below 1×10^6/mL since activated leukocytes produce excessive amounts of ROS and therefore are the main cause of OS in the male genital tract [8].

Leukocytospermia has been correlated with abnormal sperm parameters which appear to be positively correlated with the leukocyte concentration [104]. Sperm DNA fragmentation has also been proven to correlate with leukocytopsermia [105]. Sperm DNA fragmentation occurs due to the presence of inflammatory cytokines leading to direct DNA structural abnormalities [106]. Alternatively, sperm DNA fragmentation may occur due to OS as documented by the increase in levels of 8-hydroxy-2′-deoxyguanosine (8-OH-dG), a marker of oxidative DNA damage, which is seen in patients with leukocytospermia [106].

20.3.2 Asthenozoospermia

One of the manifestations of ROS-induced damage is a decrease in sperm motility. This is due to ROS causing lipid peroxidation of the sperm plasma membrane. In turn, lipid peroxidation damages the sperm axonemal structure leading to decline in sperm motility. In support, leukocytes which generate ROS had a negative correlation with sperm motility [107]. Evidence shows that deregulated apoptosis may also play a role in the decline of sperm motility. Apoptosis-like manifestations in the form of increase levels of caspases were shown in ejaculated spermatozoa in athenozoospermic semen sample [108].

20.3.3 Azoospermia

Apoptosis is an ongoing physiological phenomenon that maintains the number of germ cells in the testicular vicinity within the supportive capacity of Sertoli cells. However, deregulated apoptosis may be both a sign and cause of abnormal spermatogenesis. Immature germ cells characterized by maturation arrest have signs of active apoptosis and DNA fragmentation [109]. Therefore, azoospermia may be one of the presentations of deregulated apoptosis [110]. The sperm concentration in the human ejaculate appears to be correlated with ROS levels. In men diagnosed with male infertility, sperm concentration was negatively correlated with ROS irrespective of the exact clinical diagnosis and etiology for infertility [110].

20.3.4 Teratozoospermia

One of the features seen in spermatozoa with abnormal sperm morphology is the retention of cytoplasmic droplets, which are extruded during spermiogenesis.

High levels of morphologically abnormal sperm, teratozoospermia, showing cytoplasmic droplets lead to excessive ROS generation [111]. Therefore, these patients are more prone to develop seminal OS. Sperm ROS production negatively correlates with the percentage of normal morphology. This relationship allows for the identification of cases with high seminal ROS based on the assessment of sperm morphology [112].

20.4 Management

20.4.1 Diagnosis

The diagnosis of OS, DNA damage, and apoptosis in the context of male infertility can be made by focusing on the clinical and laboratory presentations that were proven to be associated with these conditions. Additionally, there are specific tests that can be used to identify whether OS, DNA damage, and apoptosis do actually play a role in a specific case of male infertility [113]. There has been no recommendation to implement these tests in routine clinical practice to date. However, there are some specific cases that will definitely benefit from such testing [114].

Testing for OS should include the assessment of ROS and the total antioxidant capacity (TAC) of seminal plasma since it is the end result of an imbalance between the two. The chemiluminescence assay with luminol and lucigenin has been consistently used to detect ROS levels [90]. This method is the most standardized to date for the assessment of ROS and provides a clear cut-off point as threshold for normality [115]. Flow cytometry has been recently successfully used for the detection of intracellular ROS in samples with low sperm concentration [116]. The seminal TAC can be evaluated using the enhanced chemiluminescence assay. This method is accurate, but colorimetry has been identified as less cumbersome [117].

Apoptosis and apoptosis-like manifestations can be evaluated in ejaculated spermatozoa using fluorescent staining methods. Using this approach, several markers such as externalization of phosphatidylserine, mitochondrial membrane potential, and caspase activation have been identified in human spermatozoa [40]. Assessment of sperm DNA fragmentation has been an issue of controversy. Several methods are currently available; however, the lack of standardization and cut-off point between normal levels in the average fertile population and the minimal levels of sperm DNA integrity required for achieving pregnancy deprives most of these methods of clinical significance [118]. Currently, there are two methods recognized to be of a clinical benefit. The sperm chromatin structure assay (SCSA) and the terminal deoxynucleotidyl transferase-mediated deoxyuridine triphosphate-nick end labeling (TUNEL) [21, 119]. Both assays correlate well with fertility as well as with IVF outcomes [100]. Although results from both assays correlate very well, each has a different range for normal values. In addition, these different techniques determine different aspects of sperm DNA damage [120].

The clinical application of DNA fragmentation testing includes the use of testicular sperm in patients with high levels of sperm DNA fragmentation in semen. In these cases, the use of surgically retrieved testicular sperm in ART may be recommended. In addition to the use of testicular sperm, another strategy that could be used in patients with high levels of sperm DNA fragmentation is the selection of spermatozoa with low levels of DNA damage [121].

20.4.2 Prevention

Etiological factors resulting in OS, sperm DNA fragmentation, and apoptosis such as smoking, environmental toxins, and heat should be avoided. Varicocele, genitourinary tract infections, and inflammations should be treated to avoid excessive ROS and inflammatory cytokine production. There are also considerations for minimizing the levels of iatrogenic ROS during the sperm processing in vitro [98]. The various sources of ROS generation in the seminal fluid including immature spermatozoa and leukocytes could be extracted and discarded using density gradient centrifugation. This should offer some degree of protection for mature spermatozoa against ROS-induced damage. Supplementation of culture media during sperm preparation and cryopreservation with antioxidants may also help protect spermatozoa against OS-induced damage and subsequent DNA damage and apoptosis. Albumin, vitamin C supplementation, and vitamin E are the most common supplements used in vitro [89].

20.5 Key Points

- In the context of male infertility, OS, sperm DNA damage, and apoptosis are linked together leading to decrease in the sperm fertilization potential.
- Several clinical conditions have been strongly associated with seminal OS, sperm DNA damage, and apoptosis.
- Clinical symptoms and signs and laboratory presentations could be used to identify infertile males who have seminal OS, sperm DNA damage, and apoptosis as a causative factor. In these cases, specific testing is warranted for the presence of seminal OS, sperm DNA damage, and apoptosis.
- Proper management of some clinical conditions will improve sperm parameters which are initially deteriorated due to OS, DNA damage, and apoptosis.
- Several methods could be applied to decrease the occurrence of OS, DNA damage, and apoptosis in vitro.

References

1. Halliwell B, Aruoma OI. DNA damage by oxygen-derived species. Its mechanisms and measurement in mammalian systems. FEBS Lett. 1991;281:9–19.
2. Halliwell B. Free radicals, antioxidants, and human disease: curiosity, cause, or consequence? Lancet. 1994;344:721–4.

3. Janssen YMW, van Houten B, Borm PJA, Mossman BT. Cell and tissue responses to oxidative damage. Lab Invest. 1993;69:261–74.
4. Agarwal A, Gupta S, Sikka S. The role of free radicals and antioxidants in reproduction. Curr Opin Obstet Gynecol. 2006;18(3):325–32.
5. Bagchi K, Puri S. Free radicals and antioxidants in health and disease. East Mediterr Health J. 1998;4:350–60.
6. Gil-Guzman E, Ollero M, Lopez MC, Sharma RK, Alvarez JG, Thomas Jr AJ, Agarwal A. Differential production of reactive oxygen species by subsets of human spermatozoa at different stages of maturation. Hum Reprod. 2001;16:1922–30.
7. Hendin BN, Kolettis PN, Sharma RK, Thomas Jr AJ, Agarwal A. Varicocele is associated with elevated spermatozoal reactive oxygen species production and diminished seminal plasma antioxidant capacity. J Urol. 1999;161:1831–4.
8. Plante M, de Lamirande E, Gagnon C. Reactive oxygen species released by activated neutrophils, but not by deficient spermatozoa, are sufficient to affect normal sperm motility. Fertil Steril. 1994;62:87–393.
9. Aitken RJ, Baker MA. Reactive oxygen species generation by human spermatozoa: a continuing enigma. Int J Androl. 2002;25:191–4.
10. Aitken RJ, West KM. Analysis of the relationship between reactive oxygen species production and leukocyte infiltration in fractions of human semen separated on Percoll gradients. Int J Androl. 1990;13:433–51.
11. Padron OF, Brackett NL, Sharma RK, Lynne CM, Thomas Jr AJ, Agarwal A. Seminal reactive oxygen species and sperm motility and morphology in men with spinal cord injury. Fertil Steril. 1997;67:1115–20.
12. Hammadeh ME, Al Hasani S, Rosenbaum P, Schmidt W, Fischer-Hammadeh C. Reactive oxygen species (ROS), total antioxidants (TAS) concentration of seminal plasma and their effect on sperm parameters and outcome of IVF/ICSI patients. Arch Gynecol Obstet. 2008;277:515–26.
13. Hammadeh ME, Radwan M, Al-Hasani S, Micu R, Rosenbaum P, Lorenz M, Schmidt W. Comparison between reactive oxygen species (ROS) concentration in seminal plasma and semen parameters in partners of patients who became pregnant after IVF/ICSI and those who did not. Reprod Biomed Online. 2006;13(5):696–706.
14. Mossman BT. Signal transduction by oxidants: look who's talking. Free Radic Biol Med. 2000;28:1315–6.
15. Thannickal VJ, Fanburg BL. Reactive oxygen species in cell signalling. Am J Physiol Lung Cell Mol Physiol. 2000;279:L1005–28.
16. Allen RG, Tresini M. Oxidative stress and gene regulation. Free Radic Biol Med. 2000; 28:463–99.
17. Gagnon C, Iwasaki A, De Lamirande E, Kovalski N. Reactive oxygen species and human spermatozoa. Ann N Y Acad Sci. 1991;637:436–44.
18. Aitken RJ, Fisher HM, Fulton N, Gomez E, Knox W, Lewis B, Irvine S. Reactive oxygen species generation by human spermatozoa is induced by exogenous NADPH and inhibited by the flavoprotein inhibitors diphenylene iodonium and quinacrine. Mol Reprod Dev. 1997; 47:468–82.
19. de Lamirande E, Gagnon C. Human sperm hyperactivation and capacitation as parts of an oxidative process. Free Radic Biol Med. 1993;14:255–65.
20. Said TM, Aziz N, Sharma RK, Lewis-Jones I, Thomas Jr AJ, Agarwal A. A novel association between sperm deformity index and oxidative stress-induced DNA damage in infertile male patients. Asian J Androl. 2005;7:121–6.
21. Agarwal A, Said T. Role of sperm chromatin abnormalities and DNA damage in male infertility. Hum Reprod Update. 2003;9:331–45.
22. Berry EM, Kohen R. Is the biological antioxidant system integrated and regulated? Med Hypotheses. 1999;53:397–401.
23. Agarwal A, Saleh RA, Bedaiwy MA. Role of reactive oxygen species in the pathophysiology of human reproduction. Fertil Steril. 2003;79(4):829–43.

24. Agarwal A, Gupta S, Sharma RK. Role of oxidative stress in female reproduction. Reprod Biol Endocrinol. 2005;3:28–37.
25. Shiota K, Yamada S. Intrauterine environment-genome interaction and children's development (3): assisted reproductive technologies and developmental disorders. J Toxicol Sci. 2009;34 Suppl 2:SP287–91.
26. Rhee SG. Cell signaling: H_2O_2, a necessary evil for cell signaling. Science. 2006;312(5782): 1882–3.
27. Krieger-Liszkay A. Singlet oxygen production in photosynthesis. J Exp Bot. 2004;56(411): 337–46.
28. Sakkas D, Mariethoz E, St John JC. Abnormal sperm parameters in humans are indicative of an abortive apoptotic mechanism linked to the Fas-mediated pathway. Exp Cell Res. 1999;251:350–5.
29. Barroso G, Morshedi M, Oehninger S. Analysis of DNA fragmentation, plasma membrane translocation of phosphatidylserine and oxidative stress in human spermatozoa. Hum Reprod. 2000;15:1338–44.
30. Shen HM, Dai J, Chia SE, Lim A, Ong CN. Detection of apoptotic alterations in sperm in subfertile patients and their correlations with sperm quality. Hum Reprod. 2002;17: 1266–73.
31. Jurisicova A, Lopes S, Meriano J, Oppedisano L, Casper RF, Varmuza S. DNA damage in round spermatids of mice with a targeted disruption of the Pp1cgamma gene and in testicular biopsies of patients with non-obstructive azoospermia. Mol Hum Reprod. 1999;5:323–30.
32. Taylor SL, Weng SL, Fox P, Duran EH, Morshedi MS, Oehninger S, Beebe SJ. Somatic cell apoptosis markers and pathways in human ejaculated sperm: potential utility as indicators of sperm quality. Mol Hum Reprod. 2004;10:825–34.
33. Grunewald S, Paasch U, Wuendrich K, Glander HJ. Sperm caspases become more activated in infertility patients than in healthy donors during cryopreservation. Arch Androl. 2005; 51:449–60.
34. Tesarik J, Martinez F, Rienzi L, Iacobelli M, Ubaldi F, Mendoza C, Greco E. In-vitro effects of FSH and testosterone withdrawal on caspase activation and DNA fragmentation in different cell types of human seminiferous epithelium. Hum Reprod. 2002;17:1811–9.
35. Lachaud C, Tesarik J, Canadas ML, Mendoza C. Apoptosis and necrosis in human ejaculated spermatozoa. Hum Reprod. 2004;19:607–10.
36. Pentikainen V, Erkkila K, Dunkel L. Fas regulates germ cell apoptosis in the human testis in vitro. Am J Physiol. 1999;276:310–6.
37. Burrello N, Arcidiacono G, Vicari E, Asero P, Di BD, De PA, et al. Morphologically normal spermatozoa of patients with secretory oligo-asthenoteratozoospermia have an increased aneuploidy rate. Hum Reprod. 2004;19:2298–302.
38. Paasch U, Grunewald S, Agarwal A, Glandera HJ. Activation pattern of caspases in human spermatozoa. Fertil Steril. 2004;81 Suppl 1:802–9.
39. Koppers AJ, de Iuliis GN, Finnie JM, McLaughlin EA, Aitken RJ. Significance of mitochondrial reactive oxygen species in the generation of oxidative stress in spermatozoa. J Clin Endocrinol Metab. 2008;93:3199–207.
40. Paasch U, Sharma RK, Gupta AK, et al. Cryopreservation and thawing is associated with varying extent of activation of apoptotic machinery in subsets of ejaculated human spermatozoa. Biol Reprod. 2004;71:1828–37.
41. Cande C, Cecconi F, Dessen P, Kroemer G. Apoptosis-inducing factor (AIF): key to the conserved caspase-independent pathways of cell death? J Cell Sci. 2002;115:4727–34.
42. Said T, Paasch U, Glander H, Agarwal A. Role of caspases in male infertility. Hum Reprod Update. 2004;10:39–51.
43. de Iuliis GN, Thomson LK, Mitchell LA, Finnie JM, Koppers AJ, et al. DNA damage in human spermatozoa is highly correlated with the efficiency of chromatin remodelling and the formation of 8-hydroxy-29-deoxyguanosine, a marker of oxidative stress. Biol Reprod. 2009;81:517–24.

44. Sotolongo B, Huang TT, Isenberger E, Ward WS. An endogenous nuclease in hamster, mouse, and human spermatozoa cleaves DNA into loop-sized fragments. J Androl. 2005;26:272–80.
45. Nagata S. Apoptotic DNA fragmentation. Exp Cell Res. 2000;256:12–8.
46. Barkett M, Gilmore TD. Control of apoptosis by Rel/NFkB transcription factors. Oncogene. 1999;18:6910–24.
47. Lee J, Richburg JH, Younkin SC, Boekelheide K. The Fas system is a key regulator of germ cell apoptosis in the testis. Endocrinology. 1997;138:2081–8.
48. Kane DJ, Sarafian TA, Anton R, Hahn H, Gralla EB, Valentine JS, et al. Bcl-2 inhibition of neural death: decreased generation of reactive oxygen species. Science. 1993;262:1274–7.
49. Wang X, Sharma R, Sikka S, Thomas Jr A, Falcone T, Agarwal A. Oxidative stress is associated with increased apoptosis leading to spermatozoa DNA damage in patients with male factor infertility. Fertil Steril. 2003;80:531–5.
50. Moustafa MH, Sharma RK, Thornton J, Mascha E, Abdel-Hafez M, Thomas AJ, Agarwal A. Relationship between ROS production, apoptosis and DNA denaturation in spermatozoa from patients examined for infertility. Hum Reprod. 2004;19:129–38.
51. Aitken RJ, Koppers AJ. Apoptosis and DNA damage in human spermatozoa. Asian J Androl. 2011;13:36–42.
52. Agarwal A, Said TM. Oxidative stress, DNA damage and apoptosis in male infertility: a clinical approach. BJU Int. 2005;95:503–7.
53. Oltvai ZN, Milliman CL, Korsmeyer SJ. Bcl-2 heterodimerizes in vivo with a conserved homolog, Bax, that accelerates programmed cell death. Cell. 1993;74:609–19.
54. Adams JM, Cory S. The Bcl-2 protein family: arbiters of cell survival. Science. 1998;281:1322–6.
55. Kelekar A, Thompson CB. Bcl-2-family proteins: the role of the BH3 domain in apoptosis. Trends Cell Biol. 1998;8:324–30.
56. Sakkas D, Mariethoz E, Manicardi G, Bizzaro D, Bianchi PG, Bianchi U. Origin of DNA damage in ejaculated human spermatozoa. Rev Reprod. 1999;4:31–7.
57. Shen H, Ong C. Detection of oxidative DNA damage in human sperm and its association with sperm function and male infertility. Free Radic Biol Med. 2000;28:529–36.
58. Cisternas P, Moreno RD. Comparative analysis of apoptotic pathways in rat, mouse, and hamster spermatozoa. Mol Reprod Dev. 2006;73:1318–25.
59. Said TM, Agarwal A, Sharma RK, Thomas Jr AJ, Sikka SC. Impact of sperm morphology on DNA damage caused by oxidative stress induced by beta-nicotinamide adenine dinucleotide phosphate. Fertil Steril. 2005;83:95–103.
60. Twigg J, Fulton N, Gomez E, Irvine DS, Aitken RJ. Analysis of the impact of intracellular reactive oxygen species generation on the structural and functional integrity of human spermatozoa: lipid peroxidation, DNA fragmentation and effectiveness of antioxidants. Hum Reprod. 1998;13:1429–36.
61. Agarwal A, Saleh RA, Bedaiwy MA. Role of reactive oxygen species in the pathophysiology of human reproduction. Fertil Steril. 2003;79:829–43.
62. Alvarez JG. The predictive value of sperm chromatin structure assay. Hum Reprod. 2005;20:2365–7.
63. Singh NP, Muller CH, Berger RE. Effects of age on DNA double-strand breaks and apoptosis in human sperm. Fertil Steril. 2003;80:1420–30.
64. Saleh R, Agarwal A, Sharma R, Said T, Sikka S, Thomas Jr A. Evaluation of nuclear DNA damage in spermatozoa from infertile men with varicocele. Fertil Steril. 2003;80:1431–6.
65. O'Donovan M. An evaluation of chromatin condensation and DNA integrity in the spermatozoa of men with cancer before and after therapy. Andrologia. 2005;37:83–90.
66. Erenpreiss J, Hlevicka S, Zalkalns J, Erenpreisa J. Effect of leukocytospermia on sperm DNA integrity: a negative effect in abnormal semen samples. J Androl. 2002;23:717–23.
67. Evenson DP, Jost LK, Corzett M, Balhorn R. Characteristics of human sperm chromatin structure following an episode of influenza and high fever: a case study. J Androl. 2000;21:739–46.

68. Hammadeh ME, Hamad MF, Montenarh M, Fischer-Hammadeh C. Protamine contents and P1/P2 ratio in human spermatozoa from smokers and non-smokers patients. Hum Reprod. 2010;00:1–13.
69. Aitken RJ, Bennetts LE, Sawyer D, Wiklendt AM, King BV. Impact of radio frequency electromagnetic radiation on DNA integrity in the male germline. Int J Androl. 2005;28:171–9.
70. Lozano GM, Bejarano I, Espino J, González D, Ortiz A, et al. Relationship between caspase activity and apoptotic markers in human sperm in response to hydrogen peroxide and progesterone. J Reprod Dev. 2009;55:615–21.
71. Henkel RR, Schill WB. Sperm preparation for ART. Reprod Biol Endocrinol. 2003;1:108.
72. Rhemrev JP, van Overveld FW, Haenen GR, Teerlink T, Bast A, et al. Quantification of the nonenzymatic fast and slow TRAP in a postaddition assay in human seminal plasma and the antioxidant contributions of various seminal compounds. J Androl. 2000;21:913–20.
73. Greco E, Iacobelli M, Rienzi L, Ubaldi F, Ferrero S, et al. Reduction of the incidence of sperm DNA fragmentation by oral antioxidant treatment. J Androl. 2005;26:349–53.
74. Aitken RJ, Clarkson JS. Significance of reactive oxygen species and antioxidants in defining the efficacy of sperm preparation techniques. J Androl. 1988;9:367–76.
75. Baker HW, Brindle J, Irvine DS, Aitken RJ. Protective effect of antioxidants on the impairment of sperm motility by activated polymorphonuclear leukocytes. Fertil Steril. 1996;65:411–9.
76. Moskovtsev SI, Willis J, White J, Mullen JB. Leukocytospermia: relationship to sperm deoxyribonucleic acid integrity in patients evaluated for male factor infertility. Fertil Steril. 2007;88:737–40.
77. Henkel R, Kierspel E, Stalf T, Mehnert C, Menkveld R, et al. Effect of reactive oxygen species produced by spermatozoa and leukocytes on sperm functions in nonleukocytospermic patients. Fertil Steril. 2005;83:635–42.
78. Barbieri E, Hidalgo M, Venegas A, Smith R, Lissi E. Varicocele-associated decrease in antioxidant defenses. J Androl. 1999;20:713–7.
79. Ozbek E, Turkoz Y, Gokdeniz R, Davarci M, Ozugurlu F. Increased nitric oxide production in the spermatic vein of patients with varicocele. Eur Urol. 2000;37:172–5.
80. Weinberg JB, Doty E, Bonaventura J, Haney AF. Nitric oxide inhibition of human sperm motility. Fertil Steril. 1995;64:408–13.
81. Saleh R, Agarwal A, Nada E, et al. Negative effects of increased sperm DNA damage in relation to seminal oxidative stress in men with idiopathic and male factor infertility. Fertil Steril. 2003;79:79–87.
82. Vicari E, La Vignera S, Calogero AE. Antioxidant treatment with carnitines is effective in infertile patients with prostatovesiculoepididymitis and elevated seminal leukocyte concentrations after treatment with non steroidal anti-inflammatory compounds. Fertil Steril. 2002;78:1203–8.
83. Pasqualotto F, Sharma R, Kobayashi H, Nelson D, Thomas Jr A, Agarwal A. Oxidative stress in normospermic men undergoing infertility evaluation. J Androl. 2001;73:459–64.
84. Alkan I, Simsek F, Haklar G, et al. Reactive oxygen species production by the spermatozoa of patients with idiopathic infertility: relationship to seminal plasma antioxidants. J Urol. 1997;157:140–3.
85. Host E, Lindenberg S, Smidt-Jensen S. DNA strand breaks in human spermatozoa: correlation with fertilization in vitro in oligozoospermic men and in men with unexplained infertility. Acta Obstet Gynecol Scand. 2000;79:189–93.
86. Sun JG, Jurisicova A, Casper RF. Detection of deoxyribonucleic acid fragmentation in human sperm: correlation with fertilization in vitro. Biol Reprod. 1997;56:602–7.
87. Shoukir Y, Chardonnens D, Campana A, Sakkas D. Blastocyst development from supernumerary embryos after intracytoplasmic sperm injection: a paternal influence? Hum Reprod. 1998;13:1632–7.
88. Saleh R, Agarwal A, Kandirali E, et al. Leukocytospermia is associated with increased reactive oxygen species production by human sperm. Fertil Steril. 2002;78:1215–24.

89. Agarwal A, Nallella K, Allamaneni S, Said T. Role of antioxidants in treatment of male infertility: an overview of the literature. Reprod Biomed Online. 2004;8:616–27.
90. Agarwal A, Allamaneni SS, Said TM. Chemiluminescence technique for measuring reactive oxygen species. Reprod Biomed Online. 2004;9:466–8.
91. Comhaire FH, Vermeulen L, Pieters O. Study of the accuracy of physical and biochemical markers in semen to detect infectious dysfunction of the accessory sex glands. J Androl. 1989;10:50–3.
92. Kjellberg S, Bjorndahl L, Kvist U. Sperm chromatin stability and zinc binding properties in semen from men in barren unions. Int J Androl. 1992;15:103–13.
93. Said TM, Tellez S, Evenson DP, Del Valle AP. Assessment of sperm quality, DNA integrity and cryopreservation protocols in men diagnosed with testicular and systemic malignancies. Andrologia. 2009;41:377–82.
94. Kobayashi H, Larson K, Sharma R, et al. DNA damage in cancer patients before treatment as measured by the sperm chromatin structure assay. Fertil Steril. 2001;75:469–75.
95. Aitken RJ, De Iuliis GN. Origins and consequences of DNA damage in male germ cells. Reprod Biomed Online. 2007;14:727–33.
96. Ward WS, Coffey DS. DNA packaging and organization in mammalian spermatozoa: comparison with somatic cells. Biol Reprod. 1991;44:569–74.
97. Agarwal A, Makker K, Sharma R. Clinical relevance of oxidative stress in male factor infertility: an update. Am J Reprod Immunol. 2008;59(1):2–11.
98. Agarwal A, Said TM, Bedaiwy MA, Banerjee J, Alvarez JG. Oxidative stress in an assisted reproductive techniques setting. Fertil Steril. 2006;86:503–12.
99. Bedaiwy MA, Falcone T, Mohamed MS, et al. Differential growth of human embryos in vitro: role of reactive oxygen species. Fertil Steril. 2004;82:593–600.
100. Sharma R, Said T, Agarwal A. Sperm DNA damage and its clinical relevance in assessing reproductive outcome. Asian J Androl. 2004;6:139–48.
101. World Health Organization. WHO laboratory manual for the examination of human semen and sperm-cervical mucus interaction. 4th ed. Cambridge: Cambridge University Press; 2010.
102. Close CE, Roberts PL, Berger RE. Cigarettes, alcohol and marijuana are related to pyospermia in infertile men. J Urol. 1990;144:900–3.
103. Saleh R, Agarwal A, Sharma R, Nelson D, Thomas AJ. Effect of cigarette smoking on levels of seminal oxidative stress in infertile men: a prospective study. Fertil Steril. 2002;78: 491–9.
104. Sharma R, Pasqualotto A, Nelson D, Thomas Jr A, Agarwal A. Relationship between seminal white blood cell counts and oxidative stress in men treated at an infertility clinic. J Androl. 2001;22:575–83.
105. Maruyama Jr DK, Hale RW, Rogers BJ. Effects of white blood cells on the in vitro penetration of zona-free hamster eggs by human spermatozoa. J Androl. 1985;6:127–35.
106. Kodama H, Yamaguchi R, Fukuda J, Kasai H, Tanaka T. Increased oxidative deoxyribonucleic acid damage in the spermatozoa of infertile male patients. Fertil Steril. 1997;68: 519–24.
107. Lackner JE, Agarwal A, Mahfouz R, du Plessis SS, Schatzl G. The association between leukocytes and sperm quality is concentration dependent. Reprod Biol Endocrinol. 2010;8:12.
108. Weng S, Taylor S, Morshedi M, et al. Caspase activity and apoptotic markers in ejaculated human sperm. Mol Hum Reprod. 2002;8:984–91.
109. Tesarik J, Greco E, Cohen-Bacrie P, Mendoza C. Germ cell apoptosis in men with complete and incomplete spermiogenesis failure. Mol Hum Reprod. 1998;4:757–62.
110. Pasqualotto F, Sharma R, Nelson D, Thomas Jr A, Agarwal A. Relationship between oxidative stress, semen characteristics, and clinical diagnosis in men undergoing fertility investigation. Fertil Steril. 2000;73:459–64.
111. Aitken R. The Amoroso Lecture. The human spermatozoon—a cell in crisis? J Reprod Fertil. 1999;115:1–7.

112. Aziz N, Saleh R, Sharma R, et al. Novel association between sperm reactive oxygen species production, sperm morphological defects, and the sperm deformity index. Fertil Steril. 2004;81:349–54.
113. Collins JA, Barnhart KT, Schlegel PN. Do sperm DNA integrity tests predict pregnancy with in vitro fertilization? Fertil Steril. 2008;89:823–31.
114. Desai NR, Mahfouz R, Sharma R, Gupta S, Agarwal A. Reactive oxygen species levels are independent of sperm concentration, motility, and abstinence in a normal, healthy, proven fertile man: a longitudinal study. Fertil Steril. 2010;94(4):1541–3.
115. Athayde KS, Cocuzza M, Agarwal A, et al. Development of normal reference values for seminal reactive oxygen species and their correlation with leukocytes and semen parameters in a fertile population. J Androl. 2007;28:613–20.
116. Mahfouz R, Sharma R, Lackner J, Aziz N, Agarwal A. Evaluation of chemiluminescence and flow cytometry as tools in assessing production of hydrogen peroxide and superoxide anion in human spermatozoa. Fertil Steril. 2009;92:819–27.
117. Said T, Kattal N, Sharma R, et al. Enhanced chemiluminescence assay vs colorimetric assay for measurement of the total antioxidant capacity of human seminal plasma. J Androl. 2003;24:676–80.
118. Schlegel PN, Paduch DA. Yet another test of sperm chromatin structure. Fertil Steril. 2005;84:854–9.
119. Sharma RK, Sabanegh E, Mahfouz R, Gupta S, Thiyagarajan A, Agarwal A. TUNEL as a test for sperm DNA damage in the evaluation of male infertility. Urology. 2010;76:1380–6.
120. Henkel R, Hoogendijk CF, Bouic PJ, Kruger TF. TUNEL assay and SCSA determine different aspects of sperm DNA damage. Andrologia. 2010;42:305–13.
121. Sakkas D, Alvarez G. Sperm DNA fragmentation: mechanisms of origin, impact on reproductive outcome, and analysis. Fertil Steril. 2010;93:1027–37.

Chapter 21
Effect of Oxidative Stress on ART Outcome

Mohamad Eid Hammadeh, Mohammed Hamad, Khaled Refaat, Tamer M. Said, and Constanze Fischer-Hammadeh

Abstract Oxidative stress is a condition that causes cellular damage including destruction of all cellular components including lipids, proteins, nucleic acids, and sugars. It affects negatively the quality of oocytes, sperm oocyte interaction, implantation, and early embryo development which influence the success of pregnancy. Many events related to infertility may occur due to OS in the female reproductive tract, such as endometriosis, hydrosalpinx, polycystic ovarian disease, unexplained infertility, and recurrent pregnancy loss.

Oxidative stress may cause retardation in embryo development and growth due to cell-membrane damage, DNA damage, and apoptosis. Poor quality of spermatozoa due to oxidative stress has been correlated with poor fertilization rates, impaired embryo development, and increased rates of pregnancy loss.

There are many strategies that overcome OS in assisted reproductive techniques like ensuring in vitro culture under low oxygen tension, antioxidant culture media

M.E. Hammadeh, DVM, BSc, PhD (✉) • C. Fischer-Hammadeh, MD
Department of Obstetrics and Gynecology, University of Saarland, Homburg, Saarland, Germany
e-mail: mehammadeh@yahoo.de; c.fischer-hammadeh@web.de

M. Hamad, PhD
Department of Pharmacology and Bioscience, Petra University, Amman, Jordan
e-mail: moh_1968@yahoo.com

K. Refaat, MD
Department of Obstetrics and Gynecology, Al Azhar University, Assiut, Egypt
e-mail: khalidra1@yahoo.com

T. Said, MD, PhD, HCLD (ABB)
Andrology Laboratory and Reproductive Tissue Bank,
The Toronto Institute for Reproductive Medicine, 56 Aberfoyle Crescent, Toronto, ON, Canada M8X2W4
e-mail: tsaid@repromed.ca

A. Agarwal et al. (eds.), *Studies on Men's Health and Fertility*, Oxidative Stress in Applied Basic Research and Clinical Practice, DOI 10.1007/978-1-61779-776-7_21,

supplementation, control of sperm reactive oxygen species production, and reducing sperm-oocyte coincubation time.

Keywords Assisted reproduction technology • Reactive oxygen species • Oxidative stress • Spermatozoa DNA damage • Female infertility • Male infertility

21.1 Introduction

The applied assisted reproductive techniques (ART) are aimed to achieve fertilization in patients who cannot conceive naturally.

Defective sperm functions are the most prevalent causes of male infertility and a difficult condition to treat [1]. The defective sperm functions and infertility have been correlated with many environmental, physiological, and genetic factors [2, 3].

Oxidative stress (OS), among the various causes, affect the fertility status and physiology of spermatozoa [4]. Oxidative stress reflects the state of the imbalance between the production of reactive oxygen species (ROS) and the biological systems' ability to readily detoxify the reactive intermediates [5]. ROS including free radicals and peroxides are the destructive aspects of OS [6].

Low levels of ROS affect the gametes and reproductive processes such as sperm fertilization capacity, implantation of the zygote, and embryo development [7, 8]. On the contrary, antioxidants are the main defense factors against oxidative stress induced by free radicals [1].

On the other side, oxidative stress negatively affects the quality of oocytes, sperm oocyte interaction, implantation, and early embryo development which influence the success of pregnancy. This article reviews the impact of oxidative stress and ROS and its effect on the ART outcomes.

21.2 Oxidative Stress Markers

Oxidative stress is a condition resulting from overproduction of ROS that exceeds the antioxidant capacity of the seminal plasma [9]. Oxidative stress through ROS causes cellular damage through destruction of all cellular components including lipids, proteins, nucleic acids, and sugars which are potential targets of oxidative stress [4], which leads to poor sperm quality ended with infertility [10]. Also, oxidative stress is related to DNA fragmentation of spermatozoa.

21.2.1 Free Radicals

Free radicals are reactive chemical elements that carry one or more unpaired electrons [11, 12]. These electrons induce oxidation to other cellular structures as they bind with them, of these structures; nucleic acids, proteins and lipids of the cell membranes [13]. They are required by the cells as intracellular signaling necessary for many cell functions such as cell proliferation, differentiation, and migration [14, 15]. Also in reproductive system, they are important for various reproductive functions [16]. However, high levels of free radicals may alter spermiogenesis resulting in production of abnormal spermatozoa [12].

21.2.2 Reactive Oxygen Species

The most relevant free radicals in biological systems are derivatives of oxygen. The superoxide free radical anion ("superoxide") is produced as a result of oxygen reduction by transfer of a single electron to it:

$$O_2^{\bullet} + e^- \rightarrow O_2^{-\bullet}$$

While, hydrogen peroxide is generated either by two-electron reduction of oxygen:

$$O_2^{\bullet} + 2e^- - + 2H^+ \rightarrow H_2O_2$$

or via the generation of superoxide where two superoxide molecules can react together to form hydrogen peroxide and oxygen:

$$2O_2^{-\bullet} + 2H^+ \rightarrow H_2O_2 + O_2$$

The breaking down of hydrogen peroxide in the presence of transition metal ions such as iron will generate the most reactive and damaging of the oxygen free radicals, the hydroxyl radical (OH):

$$H_2O_2 + Fe^{2+} \rightarrow {}^{\bullet}OH + OH^- - + Fe^{3+}$$

In addition, hydroxyl radical produced by reaction of superoxide directly with hydrogen peroxide:

$$O_2^{-\bullet} + H_2O_2 \rightarrow {}^{\bullet}OH + OH^- - + O_2^{\bullet}$$

In biological system, ROS can be formed in three ways [17]:

1. By the homolytic cleavage of a covalent bond of a normal molecule, with each fragment retaining one of the paired electrons ($X : Y \rightarrow X^{\bullet} + Y^{\bullet}$).
2. By the loss of a single electron from a normal molecule ($X : Y \rightarrow X :^- + Y^+$).
3. By the addition of a single electron to a normal molecule ($A + e^- \rightarrow A^{-\bullet}$).

ROS are generated during the normal enzymatic reactions of inter- and intracellular signaling [15]. Spermatozoa have the ability to produce many types of ROS aerobically, such as, hydrogen peroxide (H_2O_2), the superoxide anion ($^{\bullet}O_2^-$), the hydroxyl radical ($OH^{\bullet}$), hypochlorite radical ($OHCl^{\bullet}$), nitric oxide, peroxyl, ozone, lipid peroxides, nitrous oxide, peroxynitrite, and nitroxyl ion [18]. Binding of these compounds with other cellular structures can cause oxidation and ending with cellular damage [18, 19] The presence of high ROS levels has been reported in the semen of 25–40% of infertile men [20].

21.2.3 Reactive Nitrogen Species (Nitric Oxide)

L-arginine (L-Arg) along with the molecular oxygen forms nitric oxide (NO) and citrulline and this reaction is catalyzed by nitric oxide synthase. NO also has a role in cell and tissue destruction, sterile inflammation, and formation of adhesions [21]. The two common examples of reactive nitrogen species are nitric oxide (NO) and nitrogen dioxide [22, 23]. NO is involved in the regulation of various physiological processes and is proven to be toxic if present in excess levels [21, 22, 24]. The actions of NO in a cell depend on its concentration, the cellular redox state, and the abundance of metals, protein thiols and low-molecular weight thiols (glutathione) as well as other nucleophil targets [25]. High levels of nitric oxide (NO) alter both motility and sperm competence for zona binding [4].

21.2.4 Illumination

The detrimental effects of light on fertilized oocytes and embryos that were exposed to light when cultured in vitro have been discussed in many studies. Embryos developed in vitro are exposed to environmental stress factor such as fluctuation of temperature and pH and the presence of visible light, which all have a negative effect on embryo development [26]. The use of microscopic light during in vitro production (IVP) is a common stress factor which may compromise embryo development and viability [27]. Visible light induces photodynamic stress and can cause oxidative damage to these lipids [28] which generate ROS and consequently damage the embryos.

The nonionizing irradiation of visible light (350–800 nm) passes through non-pigmented embryonic cells causing production of ROS, such as the hydroxyl radical, which are known to be toxic and/or mutagenic to mammalian cells [29]. In particular, exposure to microscopic light compromised development of two-cell rabbit and hamster embryos [30]. Reduced exposure to light or manipulating embryos in actinic light has been shown to be beneficial [19].

Visible light acts as an exogenous source of OS in that it promotes ROS production, causing cellular damage through the oxidation of DNA bases and DNA strand breaks.

The findings of a study by Beehler et al. implicated UV radiation in ROS-induced DNA base modifications [31].

Furthermore, Takenaka et al. [32] showed that smaller amounts of short-wavelength visible light are advantageous in ART as they expose oocytes and embryos to lower levels of OS than cool white light. These results suggest that minimizing the exposure of oocytes, zygotes, and embryos to visible light and near-UV light will serve to better mimic in vivo conditions, thereby yielding more successful ART outcomes [32].

Therefore, transient exposure of embryos to light of X5 min in vitro may cause ROS production [33]. Nevertheless, smaller amount of short-wavelength visible light are advantageous in ART as they expose the oocytes and embryos to lower level of OS than the cool white light and will serve to better mimic in vivo conditions, thereby yielding more successful ART outcomes [32].

21.2.5 Metallic Ions

Oxidative stress is a common feature of the mechanism of injury induced by a broad range of environmental agents [34]. Metal ions also induce production of ROS directly through the Haber–Weiss reaction. Toxicologically, a wide variety of environmental contaminants ranging from aromatic hydrocarbons [35] to heavy metal ions [6] has been shown to impair mitochondrial respiration, with ensuing production of ROS. Redox-active metals, such as iron, copper, and chromium, undergo redox cycling, whereas redox-inactive metals, such as lead, cadmium, mercury, and others, deplete cells' major antioxidants, particularly thiol-containing antioxidants and enzymes. Either redox-active or redox-inactive metals may cause an increase in production of ROS such as hydroxyl radical ($HO^{\bullet}$), superoxide radical ($O_2^{\bullet -}$), or hydrogen peroxide (H_2O_2). Consequently, it is suggested that metal-induced oxidative stress in cells can be partially responsible for the toxic effects of heavy metals [36].

Redox-active metal ions such as Fe (II) and Cu (I) are released from storage proteins by ROS and have been detected in a number of diseased tissues for example in the rheumatoid Arthritis (RA) joint [37]. These metal ions have been shown to participate in several reactions that enhance oxidative stress [38]. By switching oxidation states, these redox-active metal ions further activate species like hydrogen peroxide (H_2O_2) and superoxide ($O_2^{\bullet -}$) to the highly reactive hydroxyl radical ($^{\bullet}OH$). Nitric oxide ($NO^{\bullet}$) reacts with ($O_2^{\bullet -}$) to form the potent species peroxynitrite ($ONOO^{-\bullet}$). The reduction of peroxides by any metal species is usually called Fenton-like reaction with rather complex mechanism. Fenton-like reactions may be commonly associated with most membranous fractions including mitochondria, microsomes, and peroxisome [38].

Damage to biological components by ($ONOO^{-}$) is catalyzed by Fe (III) ions. Supplementation of transition metal ions such as Fe^{2+} to the sperm suspension results in a sudden acceleration of LPO and loss of sperm functions such as motility and viability [39].

Environmental oxidative stress can also involve the depletion of cellular antioxidant defense mechanisms or dysregulation of oxidative metabolism processes in the cell [36]. ROS are implicated in the induction of apoptosis [40] and cellular senescence [41].

Mitochondria are a known source of partially reduced oxygen species generated as a by-product of oxidative metabolism in the cell [42]. Dysregulation of mitochondrial function with metabolic inhibitors has been shown to induce the release of ROS and associated oxidative stress [43]. Addition of EDTA (ethylenediamine tetraacetic acid) or transferrin overcomes embryonic developmental arrests [44]. Besides, culture media supplemented with metal ion chelating agents may decrease the production of oxidants and help in successful embryo development and pregnancy [45].

21.2.5.1 Oxygen Concentration

Among the various culture conditions and exogenous factors that lead to elevated production of ROS in embryos, including traces of metallic cations, visible light, and amino oxidases, the in vitro oxygen tension is the most studied and the easiest to control [19].

The fertilization and embryo development in vivo takes place in an environment of low oxygen tension [46]. Oxygen tension in the oviduct is one quarter to one third of the atmospheric tension [47]. The influence of the gaseous environment is more in in vitro embryos than the in vivo ones [48]. A study conducted by Bavister et al. [49] showed that embryos cultured in an environment with atmospheric oxygen concentration and exposed to increased level of hydrogen peroxide radicals resulted in DNA fragmentation of approximately 20% of embryos, compared with DNA damage seen in 5% of embryos grown in an environment with low oxygen tension.

In vitro fertilization studies (IVF) in mice, cattle, sheep, rabbits, hamsters, rats, cows, and pigs have demonstrated that when embryos were cultured in oxygen concentrations of 5%, they present a higher viability and a better development to the blastocyst stage [50]. However, human embryos showed that cultures in vitro in atmospheric of (20%) or reduced (5%) oxygen concentrations resulted in similar fecundation and preimplantational embryo development processes.

Therefore, in vitro culture media modification such as utilizing lower oxygen tension (5%) has shown benefit in human embryos resultant improved implantation rates and increase in live birth rate [45, 51]. Recently, the impact of oxygen concentration on implantation, pregnancy, and delivery rates in IVF patients above age of 40 year with transfer of blastocysts has been investigated [50]. The implantation and pregnancy rates are significantly higher in women older than 40 years from the 5% of O_2 group, in comparison to the 20% group (25.00% vs. 2.70% and 41.38% vs. 5.56%; $P<0.05$). The deliveries rates were 13.79 and 5.56% in the 5 and 20% oxygen groups, respectively (P: NS).

In a recent publication, Kovačič et al. [52] did not find improvements in the implantation rates in older women over 40 years of age whose embryos were cultured with oxygen at 5% as compared to cultures at 20% of O_2 and embryo transfer in the third day. It is possible that effects of low oxygen concentration may be observed if embryos are transferred in the fifth or sixth days. In addition, it has been noted that embryos cultured in 20% oxygen conditions sustain a tenfold increase in intracellular H_2O_2 levels and a twofold increase in the frequency of permanent embryo arrest at the two- to four-cell stage, compared to embryos cultured in 5% oxygen tensions [53]. Interestingly, oxygen tension has recently been shown to alter gene expression in blastocysts [54] possibly through epigenetic mechanisms [55].

21.2.5.2 Gamete and Embryo Handling

With the advent of ART technologies, sperm preparation became an important procedure to obtain optimal results. It's selected the best spermatozoa available from a semen samples.

However, exposures to environmental or industrial toxins, genetics, oxidative stress, smoking, etc. are known to cause sperm DNA fragmentation (SDF) and infertility [56–58].

Centrifugation used for sperm selection can damage sperm with generation of ROS and peroxidation of sperm membranes, which can bring negative effects on sperm–oocyte fusion [59, 60]. Even washing sperm by centrifugation can also generate ROS [61, 62], which has prompted some authors to add antioxidants to sperm preparations to avoid the oxidative stress induced by centrifugation. However, the addition of vitamin E to semen samples and the addition of glutathione or/and hypotaurine to a sperm preparation medium did not show beneficial effects on the sperm progressive motility or the baseline DNA integrity [63]. Prolonged sperm-oocyte incubation (16–20 h) increases ROS generation. Coincubation of 1–2 h results in better quality embryos and improved implantation and pregnancy rates. Prospective randomized trials have recommended shorter sperm-oocyte coincubation time to improve ART outcomes [64]. In fact, a reduced exposure of oocyte to spermatozoa favors embryo viability in humans, possibly due to a decrease in potential damage from sperm metabolic waste products [65] and would create suboptimal conditions due to excessive generation of ROS [66].

21.3 Oxidative Stress and Male Infertility

21.3.1 Physiologic Role of ROS

ROS generation is mainly due to electron leakage from the mitochondrial membrane [14]. Many research groups showed that male germ cells have the ability to produce ROS at different stages. ROS may have good or harmful effects on sperm quality depending on the level of ROS, their nature, location, and length of exposure [67]. Under physiological conditions, low levels of ROS generated by spermatozoa are required for their acrosomal reaction and capacitation. Moreover, superoxide anion is important for this function [5].

21.3.2 Origin of ROS in Male Reproductive System

21.3.2.1 ROS Production by Spermatozoa

Many studies had proved that human spermatozoa are vulnerable to oxidative stress created by ROS [68]. Human spermatozoa have the ability to generate ROS, especially H_2O_2 that is important for sperm capacitation through the tyrosine phosphorylation reactions. H_2O_2 has the ability to suppress tyrosine phosphatase activity and indirectly stimulate cAMP production by adenylyl cyclase [68, 69]. cAMP is required for tyrosine phosphorylation stimulation. The effect of tyrosine phosphorylation on sperm capacitation was found in human and other species [68–70].

21.3.2.2 Role of NADPH and NADPH Oxidase Activity

Two possible mechanisms had been proposed regarding the production of ROS by human spermatozoa; these are the nicotinamide adenine dinucleotide phosphate (NADPH) oxidase system at the level of the sperm plasma membrane and nicotinamide adenine dinucleotide (NADH)-dependent oxido-reductase (diphorase) at the level of mitochondria [71]. Many research groups have reported the presence of NADPH oxidase-like activity, at the human sperm plasma membrane. This activity has subsequent consequences of induced OS on spermatozoa [72]. The presence of large number of mitochondria in spermatozoa is necessary to provide a constant source of energy for their motility. However, the increased number of damaged mitochondria in spermatozoa may elevate the production of ROS which in turn disrupt the function of the spermatozoa [73].

21.3.2.3 Retained Cytoplasma

High levels of ROS are generated by immature germ cells, damaged spermatozoa, and those with retention of residual cytoplasm [74]. In addition, ROS production by spermatozoa has been associated with midpiece abnormalities, cytoplasmic droplets, and spermatozoa immaturity [75–77]. It has been shown that the retention of residual cytoplasm in the sperm midpiece after spermeation is responsible for excess ROS generation by spermatozo [77] Immature spermatozoa are also a well-characterized source of ROS which may alter the semen quality. The major ROS type produced by immature spermatozoa are superoxide anion ($O_2^{\bullet-}$) and hydrogen peroxide (H_2O_2) [78, 79].

Stimulation of ROS generation by sperm residual cytoplasm is elevated due to high levels of NADPH produced by the enzyme glucose-6-phosphate dehydrogenase (G6PD), which is exhibited in sperm residual cytoplasm. Therefore, the retention of residual cytoplasm creates a situation in which sufficient substrate would be available to support excessive NADPH-dependent ROS generation [77, 80, 81].

Spermatozoa may be damaged during transferring from the seminiferous tubules to the epididymis due to high ROS generation by immature, morphologically abnormal spermatozoa with cytoplasmic residues such as those confronted in teratozoospermic semen [82]. However, immature spermatozoa, sperm with cytoplasmic droplets at the midpiece and leukocytes, are the major ROS producers in semen [83].

Single study [84] demonstrated that sperm morphology was not found to be associated with oxidative stress. However, retained cytoplasmic residues in the sperm may be an important source of ROS in both primary and secondary infertile men [85].

21.3.2.4 ROS Production by Leukocytes

Leukocytes are found in the male reproductive tract and human ejaculates [85]; they play an immunological role against pathogens as well as phagocytic clearance of abnormal sperm [86]. The leucocytes produce ROS as a result of their activity [59]. Male subfertility and infertility have been correlated with reproductive tract inflammation and the presence of leukocytes in ejaculate [13, 87]. Patients exhibiting leukocytospermia secondary to infection or as consequence of a paraplegia showed high level of oxidative stress in their ejaculates [39]

Plante et al. [88] reported that activated leukocytes can produce 100-fold higher amounts of ROS than nonactivated leukocytes. Inflammation and infections activate leukocytes [89]. High levels of ROS produced by leukocytes [90] or during separating spermatozoa from seminal plasma for assisted reproduction will increase the incidence of spermatozoa damage [13]. This spermatozoal damages may include chromatin alterations [91], poor motility [92], and inhibition of the mitochondrial function [93].

21.4 Oxidative Stress and Spermatozoa Functions

21.4.1 Association Between Oxidative Stress and Subfertility

ROS had been identified by MacLeod [94] who showed that sperm motility lost as they are exposed to oxygen, while the correlation between spermatozoa functions and oxidative stress had been identified at the Cambridge University [95].

Increased levels of ROS were correlated with female infertility as well as male subfertility [96]. Low levels of natural antioxidants production or excessive ROS production result in oxidative stress which may cause oxidative damage to tissues [97, 98].

The presence of high amounts of polyunsaturated fatty acids (PUFA) in the plasma membrane of spermatozoa make it susceptible to damage by oxidative stress [99]. ROS negatively affect sperm function by contributing to the occurrence of lipid peroxidation (LPO) [99, 100]. Furthermore, the severity of oxidative stress increases as the levels of antioxidants decreased [101, 102].

The effect of ROS on male fertility is dependent on their concentration in seminal plasma regardless of their sources of production. Significant opposite correlation has been established between defective sperm chromatin structure and fertility in IVF cycles [103]. However, this condition can be overcome by applying intracytoplasmic sperm injection (ICSI), in which normal fertilization and pregnancy rates can be achieved with cells that have not completed spermatogenesis, such as epididymal and testicular spermatozoa [104].

Hammadeh et al. [105] found in patients undergoing IVF or ICSI therapy that the concentration of ROS and total antioxidants (TAS) in seminal plasma did not significantly differ; however, they showed a negative correlation between ROS concentration in seminal plasma and sperm vitality, membrane integrity, sperm density, chromatin condensation, and DNA single-stand breaks in both IVF and ICSI groups [105].

Oxidative stress affects the mean semen parameters (count, motility, morphology) and therefore, asthenozoospermia is probably the best indicator for oxidative stress in a routine semen analysis.

21.4.2 Oxidative Stress and Sperm Motility

High levels of ROS negatively affect sperm motility. Kao et al. [106] reported a highly significant correlation between oxidation of sperm DNA and reduced motility. In addition, ROS production has been related to the outcome of in vitro sperm mucus penetration tests. A significant negative correlation was found between ROS and number of progressively motile spermatozoa present in the mucus [92].

21.4.3 Oxidative Stress and Sperm Capacitation and Acrosomal Reaction

Spermatozoal capacitation and acrosomal reaction are generally promoted by certain ROS such as H_2O_2, nitric oxide, and superoxide anion ($O_2^{\bullet-}$) [100, 107]. In human spermatozoa, NADPH-dependent ROS generation seems to regulate AR by tyrosine phosphorylation [108]. Aitken et al. [109] showed that capacitation of spermatozoa may thus occur by different ROS, specifically by H_2O_2 following an increase in cAMP, activation of protein kinase A, and downstream tyrosine kinase activation.

Moreover, many research groups reported in vitro induction of spermatozoa capacitation by ROS, such as, superoxide anion ($O_2^{\bullet-}$), hydrogen peroxidase (H_2O_2), and nitric oxide NO [110, 111].

21.4.4 *Oxidative Stress and Sperm Egg Binding and Fusion*

The ability of human spermatozoa to bind to zona pellucida is enhanced by low levels of ROS [112]. Moreover, addition of NO^--releasing compounds to capacitating medium increased the number of spermatozoa bound to the hemizona [113]. Also, production of low levels of hydrogen peroxides by spermatozoa plays a role in the signaling events controlling capacitation and sperm–oocytes fusion [109, 114].

On the contrary, studies revealed that elevated ROS concentrations are associated with poor sperm–oocytes fusion, IVF experiment [75], and standard IVF [115, 116]. Also, increased generation of lipoperoxidase has been proposed by Aitken et al. [117] as a reason of high failure rate of sperm-oocyte fusion bioassay [117].

21.4.5 *Oxidative Stress and Sperm Fertilization Ability*

Oxidative stress of spermatozoa may affect the generation of phosphorylation and adenosine triphosphate (ATP) which will affect the fertilizing potential of the spermatozoa. Spermatozoa chromatin structure affects its fertilization and subsequent development [8]. Increased DNA damage and apoptosis in infertile patients due to high level of ROS alter sperm-fertilizing capacity [67]. In addition, spermatozoa with damaged DNA could not participate in the fertilization process because of collateral peroxidative damage to the sperm plasma membrane [118]. Moreover, a negative correlation between ROS and embryo development to the blastocyst stage has been observed in ICSI program [119].

21.4.6 *Oxidative Stress, Sperm DNA Damage, and Reproductive Outcomes*

Spermatozoa DNA integrity may be used as a predictive factor for male infertility and pregnancy outcomes. Male with high levels of sperm DNA damage have very low potential for natural fertility [120]. During natural fertilization, the chance of spermatozoa with better DNA integrity to bind the oocytes is higher than those with damaged DNA. The ART bypass this by direct injection of the spermatozoa in the oocytes despite their impaired DNA integrity [121]. Many studies examined the effect of Spermatozoa DNA integrity on IVF/ICSI outcomes. Some of these studies reported that no consistent correlation between sperm DNA damage and fertilization rates during IVF or IVF/ICSI. In addition, no consistent correlation between sperm DNA damage and embryo quality after ICSI was reported. However, high levels of sperm DNA damage are inversely related to pregnancy rates in most, but not all, studies [122, 123].

Besides that, IVF/ICSI outcomes are correlated with high risk of birth defects and genetic and epigenetic abnormalities in the child, but it is not clear if the ICSI or the underlying infertility is responsible for the defects [124, 125]. The risk of birth defects was found significantly higher in children conceived through IVF/ICSI than those naturally conceived children [26, 125]. Moreover, de novo chromosomal abnormalities were significantly lower in naturally conceived children than those children conceived by ICSI [125]. These results may link spermatozoa DNA damage with the development of childhood diseases.

21.4.7 Oxidative Stress in Male Reproductive System and ART Outcomes

The quality of the spermatozoa used for ART is crucial to get successful fertilization and pregnancy [126]. Spermatozoa are vulnerable to oxidative stress because their membranes are rich in unsaturated fatty acids [126] and they have no mechanisms for repairing their damaged DNA [127]. This will result in decrease in membrane fluidity and reduce the activity of membrane and ion channels which will end with impaired spermatozoa fertilization capacity. Hammadeh et al. demonstrated that using DNA-damaged spermatozoa during IVF correlated with high rates of failed fertilization, defective embryo development, implantation failure, and early abortion [128].

Swim-up technique is not suitable for semen preparation of ejaculates that contain large numbers of leukocytes and immature and damaged spermatozoa because they will be a source of ROS which will destroy the functional spermatozoa [129].

The other technique used to prepare spermatozoa for ART is density-gradient centrifugation; this technique employs centrifugation to separate fractions of spermatozoa based on motility, size, and density [130]. This technique is used to isolate mature, leukocyte-free spermatozoa [129], but centrifugation process may activate the leukocytes in the semen samples which will generate high levels of ROS with subsequent adverse effect on the functional spermatozoa [62]. Thus, minimizing centrifugation time reduces the generation of ROS and may ensure the use of high-quality sperm in ART [62]. Therefore, using spermatozoa preparation techniques that reduce the generation of ROS will positively influence the results of IVF and ICSI procedures [128].

21.4.7.1 Varicocele

Infertile patients with varicocele present with higher levels of ROS and lower levels of seminal antioxidants leading to increased OS. Published data have shown that the varicocele grade has an impact on the levels of seminal ROS, which were proven to decline following the surgical correction of varicocele [131]. The relationship between varicocele and OS is not fully understood; however, some mechanisms have been proposed. Significant amount of nitric oxide (NO) production are released from the dilated pampiniform plexus of veins leading to OS [132]. Additionally, NO may also interact with superoxide anion resulting in the production peroxynitrate, which leads to further sperm damage [133].

Patients diagnosed with infertility and varicocele have significantly higher levels of OS and sperm DNA damage compared to healthy controls. This could indicate a potential pathway for sperm DNA damage resulting from OS in these patients with varicocele [134]. OS and its resulting sperm dysfunction was estimated to cause infertility in 15% of males with varicocele. Another potential cause of sperm DNA damage in varicocele patients is apoptosis. Levels of apoptosis are higher in ejaculated spermatozoa from varicocele patients than in spermatozoa from healthy men [135].

21.4.7.2 Malignant Conditions

Testicular and systemic malignant conditions have been proven to result in male infertility even before the initiation of cancer treatments. Patients with testicular and systemic malignancies present with higher levels of DNA fragmentation compared to fertile controls [136]. The levels appear to be more significant in patients diagnosed with Hodgkin's and non-Hodgkin's diseases [137]. In addition, management of malignant conditions including chemo and radiotherapy has a negative impact on male fertility. The different therapeutic agents used for cancer treatment result in sperm DNA damage, which may present with mutations, repeated miscarriages, and carcinogenesis in offspring [138].

21.5 Oxidative Stress and Female Fertility

ROS play crucial roles in the female reproduction; they may act as mediators in hormone signaling, oocyte maturation, ovarian steroidogenesis, ovulation, luteolysis, luteal maintenance in pregnancy, implantation, compaction, blastocyst development, germ cell function, and corpus luteum formation [135].

Oocytes and spermatozoa can also experience direct damage, which can lead to impaired fertilization due to an environment of OS in the peritoneal cavity.

Many events related to infertility may occur due to OS in the female reproductive tract, such as endometriosis, hydrosalpinx, polycystic ovarian disease, unexplained infertility, and recurrent pregnancy loss [129]. In addition, apoptosis may cause embryo fragmentation, implantation failure, abortion, or congenital abnormalities in offspring.

21.5.1 Oxidative Stress and Ovarian Function

Different levels of ROS were detected in the follicular fluid environment that may interfere with folliculogenesis and steroidogenesis [139, 140]. Folliculogenesis refers to the maturation of the ovarian follicle, a densely packed shell of somatic cells that contains an immature oocyte. The production of cytokines, kinins, prostaglandins, proteolytic enzymes, nitric oxide, and steroids before last stage of oocyte production enhances the flow of blood to the ovary [141]. Low intrafollicular oxygenation due to poor blood flow to follicles reduces the developmental potential of oocytes [142].

Physiologically, normal development of the oocyte (ovulation) and the subsequent embryo may require certain levels of ROS in follicular fluid [7]. Follicular fluid may contain leukocytes, macrophages, and cytokines, which can all produce ROS [7]. Oxidative stress found to affect the midluteal corpus luteum and steroidogenic capacity both in vitro and in vivo. The presence of ROS in the follicular fluid environment surrounding the oocytes may play a critical role in fertilization and embryo development, influencing IVF outcome parameters such as fertilization, embryo cleavage, and pregnancy rates [5]. Some studies reported a correlation between follicular growths and programmed follicular cell death with nitric oxide (NO) concentration; at a low level, may prevent apoptosis, while at high concentration it may promote cell death by peroxynitrite generation [143]. LPO seen in the preovulatory Graafian follicle suggests an important role of oxidative stress in ovarian function [144].

On the other hand, the presence of lower concentrations of ROS in follicular fluid than that in serum indicate to the presence of high concentrations of antioxidant systems in follicular fluid which help protect an oocyte from oxidative damage [144].

Superoxide dismutase (SOD), an antioxidant enzyme, was found in the ovary, particularly in the theca interna cells in the antral follicles protectes the cells from harmful free radicals of oxygen [145]. Therefore, theca interna cells may act as important protectors of the oocyte from OS during oocyte maturation. In another study, it was found that glutathione peroxidase may reduce levels of hydroperoxides inside follicle [146].

Other antioxidant enzymes such as catalase and other nonenzymatic antioxidants such as vitamin E, ascorbic acid, reduced glutathione, and the carotenoid lutein have been suggested to protect the oocyte and the embryo from OS by detoxifying and neutralizing ROS production [7].

21.5.2 Oxidative Stress and Embryo Development

Oxidative stress may cause retardation in embryo development and growth due to cell-membrane damage, DNA damage, and apoptosis. Sources of ROS production in oocytes and embryos include oxidative phosphorylation, NADPH oxidase, and xanthine oxidase. Poor quality of spermatozoa due to oxidative stress has been correlated with poor fertilization rates, impaired embryo development, and increased rates of pregnancy loss [147].

Burton et al. [46] reported that in vivo fertilization and embryo development require low oxygen tension. Using high levels of O_2 during in vitro cultures may enhance hydrogen peroxide (H_2O_2) production, increase DNA fragmentation, and reduce embryo development [46]. ROS such as H_2O_2 are responsible for apoptosis initiation and may cause the failure of blastocyst development and preimplantation embryo death. For this, low oxygen tension is used during culture to improve the implantation and pregnancy rate [45]. Dumoulin et al. [148] reported an improvement in preimplantation embryonic viability under low oxygen concentrations in patients undergoing IVF and ICSI.

Similarly, addition of antioxidants to the media used in ART improves the implantation and clinical pregnancy rates [45]. For example, addition of metal ion chelating agents like EDTA to the culture media may decrease the production of ROS and help in successful embryo development and pregnancy [45].

In a study designed to determine the level of H_2O_2 concentration within embryos and the morphological features of cell damage induced by H_2O_2, 31 fragmented embryos, 15 nonfragmented embryos, and 16 unfertilized oocytes were evaluated. Transmission electron microscopy and an in situ apoptosis detection kit were used to evaluate DNA fragmentation. The H_2O_2 concentration was significantly higher in the fragmented embryos than in the nonfragmented embryos and unfertilized oocytes. Interestingly, apoptosis was observed only in the fragmented embryos, as confirmed by electron microscopy. From this human study, it is clear that a direct relationship exists between increased ROS concentration and apoptosis. Further support of this finding comes from the observation that 5% O_2 decreases the relative concentration of H_2O_2 and results in improved embryo development in terms of quantity and quality in a mouse animal model. Therefore, a low O_2 concentration during in vitro culture of embryos decreases the H_2O_2 content and, therefore, reduces DNA fragmentation and improves developmental competence [149].

The oxidative status of the early embryo in IVF is associated with reduced chances of implantation. Assessment of the oxidative status of embryos in culture media before transfer may serve as an applicable tool for improving embryo selection in light of the legal limitations of the number of transferred embryos allowed [150].

21.5.3 Oxidative Stress and Implantation

Embryo implantation is a mutual interaction between blastocyst and uterus. Successful implantation is dependent on the cellular and molecular dialog between competent embryos and receptive uterus [151, 152]. Although many specific factors have been identified and characterized during embryo implantation, the molecular mechanism underlying embryo implantation still remains unknown. ROS have been implicated as a major cause of embryonic arrest, delayed and/or cell death [153].

A study compared gene expression profiles of prereceptive (2 days after the LH surge) vs. receptive (7 days after the LH surge) endometria obtained from the same proven fertile woman in the same menstrual cycle. Endometrial biopsies were analyzed using a DNA chip containing approximately 12,000 genes. Approximately 211 regulated genes were found. Validation of array data was accomplished by mRNA quantification by real-time quantitative fluorescent PCR (Q-PCR) of three upregulated genes (glutathione peroxidase 3 [GPx-3], claudin-4, and solute carrier family 1 member 1 [SLC1A1]). Human claudin-4 peaked specifically during the implantation window, whereas GPx-3 and SLC1A1 displayed the highest expression in the late secretory phase. In situ hybridization experiments demonstrated that GPx-3 and SLC1A1 expression was restricted to glandular and luminal epithelial cells during the midluteal and late luteal phases. This important experiment highlights the fact that GPx-3 may not be associated with the earliest stages of implantation, but it takes over later in the process [154].

21.5.4 Oxidative Stress and Pregnancy

Zorn et al. [119] showed that ROS levels were correlated negatively with fertilization and pregnancy rates after IVF. Besides that, in ICSI, ROS negatively correlated with embryo development to the blastocyst stage has been observed and also significant fewer ICSI-derived embryos reached the morula-blastocyst stage on day 4. In a study conducted by Hammadeh et al. [128], ROS concentration in seminal plasma and semen parameters in partners of pregnant and nonpregnant patients after IVF/ICSI were compared; they found that in IVF group, 42.3% of the cases were pregnant with a 47.8% implantation rate per embryo transferred. Besides, in ICSI patients, 18.2% were pregnant with a 40.7% implantation rate per embryo transferred [65].

21.5.5 Oxidative Stress and the Endometrium

It has been reported that both ROS and SOD play important roles in the regulation of endometrial function [58, 155–157]. SOD activity decreases in the late secretory phase, while ROS levels increase [156]. The levels of prostaglandin F2 increase towards the late secretory phase, whereas ROS triggers the release of prostaglandin F2 in vitro [158]. Stimulation of the cyclooxygenase enzyme is brought about by ROS via activation of the transcription factor NFKappa β, similar to menstruation mechanism [140]. Moreover, estrogen or progesterone withdrawal results in increased expression of cyclooxygenase-2 (COX-2) mRNA and increased prostaglandin F2α (PGF2α) synthesis in endometrial cells in vitro indicating the role of

ROS-mediated NFκB activation in the endometrium [140]. The oxidative stress was higher in the stroma of thin endometrium ′6 mm [159] as demonstrated by Azumaguchi et al. [159] who studied the relationship between implantation rate and endometrium thickness. In addition, expression of progesterone and estrogen receptors was higher, and TGF α expression was significantly lower in thin endometrium. Altered regulation of oxidative stress and levels of steroid receptors and TGF α appear to underlie the low implantation rate seen in patients with thin endometrium [156].

21.5.6 Oxidative Stress and Placenta/Abortions

Placental oxidative stress may affect the embryo implantation causing early pregnancy loss [160]. Oxidative stress also may have a role in patients with recurrent abortions with unknown reason [161]. The elevation of leukocytes numbers during pregnancy may result in ROS generation [162], which may modulate expression of cytokine receptors in the placenta, cytotrophoblasts, vascular endothelial cells, and smooth muscle cells [163]. Patients with recurrent abortions were associated with elevated levels of the T-helper cell 1(Th 1) response and glutathione, while GSH depletion leads to the inhibition of TH1-type cytokines [164].

21.5.7 Oxidative Stress in Female Reproductive System and ART Outcomes

Of the patients attending an infertility clinic, 40% have detectable levels of ROS formation in their semen and 25% have levels higher than the normal limits [165].

Men with high levels of ROS may have a lower fertility potential compared to those with low ROS [39, 110]. A reduction in ROS production could significantly improve ART outcome [166].

Oxidative stress can originate from the early steps of ART involving the oocyte, sperm, and embryo, as well as later on in the endometrial environment following embryo transfer.

Poorly vascularized follicles may produce oocytes with increased frequency of cytoplasmic defects, impaired cleavage, and abnormal chromosomal segregation resulted in impaired development [142]. Follicular fluids of the patients undergoing IVF found to contain oxidative stress markers [167]. High levels of ROS may increase embryo fragmentation and increase apoptosis ended with impaired development of the embryo [168]. ROS levels in follicular fluid can be used as potential marker for predicting success with IVF. Levels of OS in follicular fluid from women undergoing IVF were negatively associated with the outcome [169]. Low concentrations of ROS in follicular fluid are found to be critical for good IVF outcomes [93]. Moreover, Pasqualooto et al. [170] reported that low levels of ROS were required to achieve pregnancy. On the other hand, it was found that increased concentration of ROS in follicular fluid leads to reduction of the fertilization capability of the oocytes in ART cycles [171].

ROS and LPO could be used as markers for success with IVF. In addition, total antioxidants capacity (TAC) levels in day 1 culture media could be another marker reflecting the OS status during early embryonic growth. Pregnant women after treatment by IVF or ICSI had higher LPO levels and TAC in their follicular fluids. TAC levels were found to be higher in fluids yielded from follicles that produce fertilized oocytes than those with unfertilized oocytes [167, 170]. Bedaiwy et al. [172] reported significant correlation between day 1 TAC levels and clinical pregnancy rates in ICSI. Moreover, in another study, they showed that high day 1 ROS levels in culture media had no relationship with the fertilization rates in conventional IVF cycles, but were significantly related to higher fertilization rates and blastocyst development rates with ICSI cycles [173].

Other OS markers include homocysteine that found to have negative correlation with embryo quality in women with endometriosis [174]. Also, 8-Hydroxy-2-deoxyguanosine (8-OHdG) is an indicator of DNA damage induced by OS. It was reported that high concentrations of 8-OHdG were inversely correlated with fertilization rates and embryo quality [175]. In addition, thiobarbituric acid–reactive substances, conjugated dienes, and lipid hydroperoxides in preovulatory follicular fluid are oxidative stress markers; they found to have no correlation with IVF outcomes [144]. Smoking, a good oxidative stress inducer, has negative effect on ovarian function [176].

Embryos development and the effect of oxidative stress have been thoroughly studied both in animals and human systems. The energy demands of the developing embryo are high; the embryo derives energy by generating ATP through oxidative phosphorylation and glycolysis [177].

In addition, OS arising from spermatozoa induce peroxidative damage to the oocyte and its DNA, reducing the likelihood of successful fertilization [127].

Current reports implicate oxidative stress in the etiology of defective embryo development [19]. Differential growth patterns have been correlated to ROS levels on days 1–6 of embryo culture; the same study demonstrated increased embryonic fragmentation and low cleavage in ICSI cycles with increased ROS levels on day [173].

In addition, studies in which the sperm was exposed to artificially produced ROS resulted in a significant increase in DNA damage in the form of modification of all bases, production of base-free sites, deletions, frame shift, DNA cross-links, and chromosomal rearrangement [178].

Various studies suggest that DNA fragmentation in sperm is induced, for the most part, during sperm transport through the seminiferous tubules and the epididymis [179, 180].

In a recent study, Greco et al. reported that DNA fragmentation in ejaculated sperm, as measured by TUNEL in a selected group of oligozoospermic and normozoospermic males, was significantly higher than that found in testicular sperm from these same males (23.6% vs. 4.8%, $P<0.001$). Pregnancy rates obtained with testicular sperm were significantly higher than those obtained with ejaculated sperm (44.4% vs. 5.6%, $P=0.001$) [180].

In a similar study, Steele et al. found that the level of DNA fragmentation in epididymal sperm was significantly higher than that of testicular sperm obtained from the same patients. All these findings support the hypothesis that ROS induces DNA fragmentation during sperm transport through the seminiferous tubules and epididymis, and that this is one of the main mechanisms for DNA damage in sperm. These results also underscore the significance of DNA fragmentation in ART outcome [181].

Several studies have shown that sperm DNA quality had robust power to predict fertilization in vitro [182–185]. In a recent report, the only parameter that showed a significant difference between pregnant and nonpregnant groups by IVF was the percentage of sperm with DNA damage after preparation, as assessed by in situ nick translation. It was significantly higher in those patients who did not establish a pregnancy [186].

High SDF levels have been shown to influence fertilization rate [187, 188] and embryo quality [189], leading to repeated pregnancy loss [190] and low ART outcome [82, 191]. Failures may therefore be due to poor sperm DNA quality.

The sperm DNA fragmentation index (DFI), as measured by the sperm chromatin structure assay (SCSA), determines the level of sperm DNA integrity in a semen sample. Semen samples containing more than 30% sperm with fragmented DNA have been associated with reduced pregnancy rates [192]. Increased DFIs have been shown in infertile men with normal semen analyses [193]. The SCSA has been proposed as an adjunct in the infertility clinic to identify couples with poor fertility prospects [192, 194–196].

In two studies including 380 couples attempting natural conception, increasing DFI values were correlated with low frequency of or failure to achieve pregnancy [197, 198]. Using a DFI cut-off value of >30%, all couples in one study either failed to achieve or achieved pregnancy only after 4 months [197]. Payne et al. [199] reported that 9 of 19 couples with DFI >27% achieved clinical pregnancy with IVF/ICSI. On the contrary, other studies reported no pregnancy after in vitro ART procedures, both standard IVF and ICSI, when the DFI in raw semen was more than 27% [184, 200].

Previously published data suggested that an abnormal sperm DNA Fragmentation Assay (SDFA™) test predicted low success with IUI, IVF, or ICSI [134, 184, 200]. Although prediction of IUI failure appears to have been confirmed [134], several subsequent studies questioned this theory with respect to IVF and ICSI [123, 201].

High SDF values have been shown to reduce the efficacy of intrauterine insemination (IUI) from 16 to 4% [194], or lower [202]. In contrast, the same SDF values do not seem to affect the outcome of IVF or intracytoplasmic sperm injection (ICSI) techniques—the best results being obtained in ICSI [194, 203]. This is probably due to the fact that there is a selection process involved during these techniques, in which the best spermatozoon and fertilized embryo are chosen before implantation, thus reducing the impact of low sperm quality.

Recently, in the most definitive research to date, a prospective analysis of approximately 1,000 ART cycles (387:IUI, 388:IVF, 223: ICSI) from 637 patients demonstrated that an abnormal SDFA™ test is highly predictive of IUI failure [196]. Live IUI delivery rates for patients with normal and abnormal SDFA™ scores were dramatically different (19 and 1.5%, respectively). Live IVF and ICSI delivery rates were no different between SDFA™ abnormal or normal patients.

21.5.8 Oxidative Stress and the Endometriosis

Researches did not report a definitive conclusion about the association between OS and endometriosis. Endometriosis may result in infertility due to mechanical blockage of the sperm-egg union by endometriomata, or adhesions, or pelvic anatomy malformations.

Peritoneal fluid of women with endometriosis has been found to be increased in size and contains high concentration of activated macrophages, cytokines, and prostaglandins. Activated macrophage may be responsible for increased production of ROS. Alpay et al. [204] suggest that ROS may enhance growth and adhesion of endometrial cells in the peritoneal cavity, promoting endometriosis adhesions and infertility. In addition, Murphy et al. [205] reported a correlation between high level of ROS produced by peritoneal macrophages and LPO in endometriosis patients, whereas Jackson et al. [206] reported a weak association between LPO and endometriosis, and other researchers reported no association [175].

High levels of NO generated by peritoneal macrophages were detected. NO may change the composition of the peritoneal fluid which may affect the processes of ovulation, gamete transport, sperm-oocyte interaction, fertilization, and early embryonic development [207]. Many studies reported that peritoneal fluid has a reduced TAC and individual antioxidant enzymes such as SOD, which reflect a weak defense ability [208].

A significant increase of ROS in the oviduct fluid could result in oxidative damage to the oocyte and spermatozoa viability and the process of fertilization and embryo transport in the oviduct. This increase in ROS may affect the spermatozoa plasma membrane and the acrosome causing inability to bind and penetrate the oocyte, respectively [204].

21.5.9 Role of Oxidative Stress in Polycystic Ovary Syndrome

A recent study by Kusçu et al. [209] demonstrated that PCOS subjects have significantly elevated concentration of plasma MDA independent of obesity. The results showed that MDA level is significantly higher in young, nonobese PCOS patients, even in the absence of IR when compared with controls.

Protein oxidation status often is assessed with a colorimetric assay that measures protein carbonyl (PC) content, after extracting the serum with dinitrophenylhydrazine. Fenkci et al. [210] demonstrated that the PC level was significantly higher in PCOS patients with normal BMI compared with controls. This observation of higher protein oxidation suggested that free radicals damage proteins in PCOS patients [210]. Furthermore, protein carbonyls were shown to have a positive correlation with fasting insulin, suggesting a strong association between insulin resistance and protein oxidation in PCOS.

Measuring plasma concentration of NO_3 and NO_2 assesses NO concentration. The sum of NO_3 and NO_2 is assumed as the best index of total NO. Nácul et al. [211] reported that NO levels in PCOS patients were similar to that of age and BMI-matched controls. Moreover, a significantly negative correlation was observed between NO and fasting insulin levels. These data suggested that NO was related to the presence of insulin resistance in PCOS patients, although further studies are needed to clarify the role of NO in PCOS.

21.5.10 The Influence of ROS and Total Antioxidant Capacity in IVF and ICSI Cycles

Women who have undergone IVF have a tendency toward higher levels of ROS in their follicular fluid than those who conceive naturally [170]. Bedaiwy et al. [173] assessed day 1 culture media ROS effects on fertilization rates, cleavage rates, fragmentation, and blastocyst formation after prolonged culture. The authors compared conventional IVF with ICSI in order to determine which technique produced higher ROS levels. The results showed that slow development, high fragmentation, and reduced formation of morphologically normal blastocysts were associated with increased levels of day 1 ROS. This study did not show any significant difference in fertilization rates and blastocyst development rates in IVF cycles, but significant reduction in these outcomes was reported in ICSI cycles with higher day-1 ROS levels [173].

In more recent study, Aurrekoetxea et al. describe, for the first time, serum oxidizability and antioxidant status in women undergoing an IVF cycle. The author found that the serum is less protected from oxidation after the cycle, showing a lower resistance to in vitro oxidation, reduced total antioxidant activity (TAA), and decreased levels of hydrophilic antioxidants and vitamin E. These results strongly suggest increased ROS production and the presence of oxidative stress [212].

21.5.11 Generation of ROS During Assisted Reproduction Techniques

Literature has documented role of OS in the pathophysiology of infertility and assisted fertility [135, 161, 194]. ROS may originate directly from gametes and embryos (endogenous sources) as well as their surroundings (exogenous sources). Several exogenous factors identified in culture media enhance embryo production of ROS: oxygen concentration, the presence of metallic cations, and visible light exposure. Spermatozoa may also aid in the production of ROS during assisted reproduction techniques.

Elevated ROS levels generated exogenously or endogenously influence the gametes, gamete interaction, fertilization, and pregnancy rates with IVF/ICSI. Oxidative insult to embryos can lead to two-cell block, embryonic arrest, or even embryonic demise [213, 214]. Higher levels of ROS in the follicular fluid and semen are associated with poor fertility outcomes with assisted reproduction [171, 174]. In a meta-analysis from our group, ROS in semen have been reported to significantly effect the fertilization rate with IVF [161]. Measurement of ROS levels may help counsel patients on their adverse effects. The effects of oxidative stress in an ART setting may be amplified due to the lack of physiological defense mechanisms available and due to the number of potential sources of ROS at play [128].

21.6 Oxidative Stress and Gamete Cryopreservation

Cryopreservation enhances production of ROS that induce membrane LPO, DNA damage, and apoptosis in frozen spermatozoal [206]. Removing the seminal plasma leaves spermatozoa unprotected by antioxidants which will affect cryopreserved spermatozoa negatively. The addition of antioxidants to the thawing media improves spermatozoa function and in vitro fertilizing capabilities [215]. DNA damage was reported in cryopreserved oocytes due to oxidative stress [216]. Using the antioxidant vitamin C in the culture media reduces the stromal tissue apoptosis and damage [217].

21.7 Strategies to Overcome OS in Assisted Reproduction

ROS may originate from the male or female gamete or the embryo or indirectly from the surroundings, which includes the cumulus cells, leucocytes, and culture media. In human IVF/ICSI procedures, the clinical pregnancy rates have remained unchanged at 30–40% [218].

It is hypothesized that the altered redox state in in vitro conditions may play a role in poor ART outcomes, and controlling OS may improve ART outcomes. Fertilization and embryo development in vivo occur in an environment of low oxygen tension [46].

It has been noted that blastocyst development in vitro always lags behind blastocyst development in vivo as there is a variation in the ability of IVF media and its components to scavenge ROS and prevent DNA damage and apoptosis [219]. During ART procedures, it is important to emulate in vivo conditions by avoiding conditions that promote ROS generation. Accomplishing that has been shown to lead to a reduction in blastocyst degeneration, increased blastocyst development rates; increased hatching of blastocysts, and reduction in embryo apoptosis, and other degenerative pro-oxidant influence has been reported [19].

The available strategies include the following:

1. Ensuring in vitro culture under low oxygen tension conditions: During culture, low oxygen tension conditions improve the implantation and pregnancy rate better than high oxygen tension [45].
2. Metal ion culture media supplementation: It has been shown that metal ions may enhance the production of oxidants. As a result, it was suggested that it may be useful to add metal ion chelating agents to culture media to decrease the production of oxidants [45].
3. Enzymatic and nonenzymatic antioxidant culture media supplementation: Higher implantation and clinical pregnancy rates are reported when antioxidant-supplemented media is used rather than standard media without antioxidants. Various nonenzymatic antioxidants including beta-mercaptoethano [220], protein [219], vitamin E [149], vitamin C [221], cysteamine [222], cysteine [223], taurine and hypotaurine [224], and thiols are added to the culture media with the purpose of improving the developmental ability of the embryos by reducing the effects of ROS. Also, the addition of the enzymatic antioxidant, for example SOD, to the culture media prevented the deleterious effects of OS on sperm viability and on the embryo development both in vivo and in vitro. This was demonstrated by increased development of the two-cell stage embryos to the expanded blastocyst stage in the SOD-supplemented media. Mechanical removal of ROS in IVF/ET has been studied as a method to improve IVF outcome [225]. The rinsing of cumulus oophorus has been shown to overcome the deleterious effects of ROS in patients with ovarian endometriosis [225].
4. Control of sperm ROS production and sperm chromatin damage: Spermatozoa are particularly susceptible to ROS-induced damage because their plasma membranes contain large quantities of PUFA and their cytoplasm contains low concentrations of the scavenging enzymes [193].
5. Reducing sperm-oocyte coincubation time: Reports suggest that a prolonged sperm-oocyte coincubation time (16–20 h) increases the generation of ROS. Two prospective randomized controlled studies have advocated using a shorter sperm-oocyte coincubation time [64]. Coincubation times of 1–2 h resulted in better quality embryos and significantly improved fertilization and implantation rates [226].

21.8 Conclusions

The literatures provide some evidence of oxidative stress influencing the entire reproductive stages of a woman, even the menopausal years. OS plays a role in multiple physiological processes from oocyte maturation to fertilization and embryo development. OS can arise as result of excessive production of free radicals and/or impaired antioxidant defense mechanism.

An increasing number of published studies have pointed towards increased importance of the role of OS in female reproduction. Clearly, we have much to learn, but what we do know is that the role of OS in female reproduction cannot be underestimated. There is evidence that OS plays a role in conditions such as abortions, preeclampsia, hydatidiform mole, fetal embryopathies, preterm labor, and preeclampsia and gestational diabetes, which lead to an immense burden of maternal and fetal morbidity and mortality.

We emphasize that free radicals have important physiological functions in the female reproductive tract as well as excessive free radicals precipitate female reproductive tract pathologies. Reference values for ROS, minimum safe concentrations, or physiologically beneficial concentrations have yet not been defined. Patients should be assessed according to the etiological factors and analyzed separately.

Most of the published studies on oxidative stress are either observational or case control studies. Newer studies should be designed with more patient numbers, similar outcome parameters, and uniform study populations so that results can be more easily compared.

Measurement of OS in vivo is controversial. The sensitivity and specificity of various oxidative stress markers are not known. Measurement of biomarkers of OS is subject to interlaboratory variations and interobserver differences. A uniform method with comprehensive assessment of the OS biomarkers should be used so that the results can be compared across the studies.

Treatment strategies of antioxidant supplementation, directed toward reducing OS need to be investigated in randomized controlled trials. Antioxidants may be advised when specific etiology cannot be identified as in idiopathic infertility as there is no other evidence-based treatment for idiopathic infertility and reports indicate the presence of OS.

Strategies to overcome OS in vitro conditions and balancing between in vivo and in vitro environments can be utilized in ART to successfully treat infertility. Interventions for overcoming oxidative stress in conditions such as abortions, preeclampsia, preterm labor and gestational diabetes, and intrauterine growth retardation are still investigational with various randomized controlled trials in progress.

References

1. Sikka SC. Role of oxidative stress and antioxidants in andrology and assisted reproductive technology. J Androl. 2003;25(1):5–18.
2. Kenkel S, Rolf C, Nieschlag E. Occupational risks for male fertility: an analysis of patients attending a tertiary referral centre. Int J Androl. 2001;24(6):318–26.
3. Sharlip ID, Jarow JP, Belker AM, et al. Best practice policies for male infertility. Fertil Steril. 2002;77(5):873–82.
4. Agarwal A, Makker K, Sharma R. Clinical relevance of oxidative stress in male factor infertility: an update. Am J Reprod Immunol. 2008;59(1):2–11.

5. Agarwal A, Saleh RA, Bedaiwy MA. Role of reactive oxygen species in the pathophysiology of human reproduction. Fertil Steril. 2003;79(4):829–43.
6. Valko M, Morris H, Cronin MTD. Metals, toxicity and oxidative stress. Curr Med Chem. 2005;12(10):1161–208.
7. Attaran M, Pasqualotto E, Falcone T, et al. The effect of follicular fluid reactive oxygen species on the outcome of in vitro fertilization. Int J Fertil Womens Med. 2000;45(5):314–20.
8. Sakkas D, Urner F, Bizzaro D, et al. Sperm nuclear DNA damage and altered chromatin structure: effect on fertilization and embryo development. Hum Reprod. 1998;13 Suppl 4:11–9.
9. Desai NR, Mahfouz R, Sharma R, Gupta S, Agarwal A. Reactive oxygen species levels are independent of sperm concentration, motility, and abstinence in a normal, healthy, proven fertile man: a longitudinal study. Fertil Steril. 2010;94(4):1541–3.
10. Bucak MN, Sariozkan S, Tuncer PB, et al. The effect of antioxidants on post-thawed Angora goat (Capra hircus ancryrensis) sperm parameters, lipid peroxidation and antioxidant activities. Small Rumin Res. 2010;89(1):24–30.
11. Kefer JC, Agarwal A, Sabanegh E. Role of antioxidants in the treatment of male infertility. Int J Urol. 2009;16(5):449–57.
12. Sanocka D, Kurpisz M. Reactive oxygen species and sperm cells. Reprod Biol Endocrinol. 2004;2:12–26.
13. Ochsendorf FR. Infections in the male genital tract and reactive oxygen species. Hum Reprod Update. 1999;5(5):399–420.
14. Agarwal A, Nallella KP, Allamaneni S, Said T. Role of antioxidants in treatment of male infertility: an overview of the literature. Reprod Biomed Online. 2004;8(6):616–27.
15. Rhee SG. H_2O_2, a necessary evil for cell signaling. Science. 2006;312(5782):1882–3.
16. Du Plessis SS, Makker K, Desai NR, Agarwal A. Impact of oxidative stress on IVF. Expert Rev Obstet Gynecol. 2008;3(4):539–54.
17. Palmieri B, Sblendorio V. Oxidative stress detection: what for? Part I. Eur Rev Med Pharmacol Sci. 2006;10:291–317.
18. Agarwal A, Gupta S, Sharma RK. Role of oxidative stress in female reproduction. Reprod Biol Endocrinol. 2005;3:28.
19. Guerin P, El Mouatassim S, Menezo Y. Oxidative stress and protection against reactive oxygen species in the pre-implantation embryo and its surroundings. Hum Reprod Update. 2001;7:175–89.
20. Pardon OF, Brackett NL, Sharma RK, et al. Seminal reactive oxygen species, sperm motility and morphology in men with spinal cord injury. Fertil Steril. 1997;67:1115–20.
21. Dong M, Shi Y, Cheng Q, Hao M. Increased nitric oxide in peritoneal fluid from women with idiopathic infertility and endometriosis. J Reprod Med. 2001;46:887–91.
22. Pierce JD, Cackler AB, Arnett MG. Why should you care about free radicals? RN. 2004; 67:38–42.
23. Van Langendonckt A, Casanas-Roux F, Donnez J. Oxidative stress and peritoneal endometriosis. Fertil Steril. 2002;77:861–70.
24. Ohl J, Lefebvre-Maunoury G, Wittemer C, Nisand G, Laurent MC, Hoffmann P. Nitric oxide donors for patients undergoing IVF. A prospective, double-blind, randomized, placebo-controlled trial. Hum Reprod. 2002;17:2615–20.
25. Rosselli M, Keller PJ, Dubey RK. Role of nitric oxide in the biology, physiology and pathophysiology of reproduction. Hum Reprod Update. 1998;4:3–24.
26. Lane M, Gardner DK. Understanding cellular disruptions during early embryo development that perturb viability and fetal development. Reprod Fertil Dev. 2005;17:371–8.
27. Ottosen LDM, Hindkjar J, Ingerslev J. Light exposure of the ovum and preimplantation embryo during ART procedures. J Assist Reprod Genet. 2007;24:99–103.
28. Girotti AW. Photosensitized oxidation of membrane lipids: reaction pathways, cytotoxic effects, and cytoprotective mechanisms. J Photochem Photobiol B. 2001;63:103–13.
29. Godley BF, Shamsi FA, Liang F-Q, et al. Blue light induces mitochondrial DNA damage and free radical production in epithelial cells. J Biol Chem. 2005;280:21061–6.

30. Korhonen K, Sjoevall S, Viitanen J, Ketoja E, Makarevich A, Peippo J. Viability of bovine embryos following exposure to the green filtered or wider bandwidth light during in vitro embryo production. Hum Reprod. 2009;24:308–14.
31. Beehler BC, Przybyszewski J, Box HB. Formation of 8-hydoxydeoxyguanosine within DNA of mouse keratinocytes exposed in culture to UVB and H_2O_2. Carcinogenesis. 1992;13: 2003–7.
32. Takenaka M, Horiuchi T, Zanagimachi R. Effect of light on development of mammalian zygotes. Proc Natl Acad Sci USA. 2007;104(36):4289–93.
33. Nakayama T, Noda Y, Goto Y, Mori T. Effects of visible light and other environmental factors on the production of oxygen radicals by hamster embryos. Theriogenology. 1994;41: 499–510.
34. Finkelstein JN, Johnston CJ. Enhanced sensitivity of the postnatal lung to environmental insults and oxidant stress. Pediatrics. 2004;113(4):1092–6.
35. Senft AP, Dalton TP, Nebert DW, Genter MB, Puga A, Hutchinson RJ, et al. Mitochondrial reactive oxygen production is dependent on the aromatic hydrocarbon receptor. Free Radic Biol Med. 2002;33:1268–78.
36. Ercal N, Gurer-Orhan H, Aykin-Burns N. Toxic metals and oxidative stress part I: mechanisms involved in metal-induced oxidative damage. Curr Top Med Chem. 2001;1(6): 529–39.
37. Naughton DP, Knappit J, Fairburn K, Gaffney K, Blake D, Grootveld M. Detection and investigation of the molecular nature of low-molecular-mass copper(II) ions in isolated rheumatoid knee joint synovial fluid. FEBS Lett. 1995;361:167–72.
38. Liochey SI, Fridovich I. Superoxide and iron: partners in crime. IUBMB Life. 1999;48: 157–61.
39. Sharma RK, Agarwal A. Role of reactive oxygen species in male infertility. Urology. 1996;48:835–50.
40. Stone JR, Yang S. Hydrogen peroxide: a signaling messenger. Antioxid Redox Signal. 2006;8:243–70.
41. Passos JF, Saretzki G, Ahmed S, Nelson G, Richter T, Peters H, Wappler I, Birket MJ, Harold G, Schaeuble K, et al. Mitochondrial dysfunction accounts for the stochastic heterogeneity in telomere-dependent senescence. PLoS Biol. 2007;5:e110.
42. Andreyev AY, Kushnareva YE, Starkov AA. Mitochondrial metabolism of reactive oxygen species. Biochemistry (Mosc). 2005;70(2):200–14.
43. Lam M, Oleinick NL, Nieminen AL. Photodynamic therapy-induced apoptosis in epidermoid carcinoma cells. Reactive oxygen species and mitochondrial inner membrane permeabilization. J Biol Chem. 2001;276(50):47379–86.
44. Orsi NM, Leese HJ. Protection against reactive oxygen species during mouse preimplantation embryo development: role of EDTA, oxygen tension, catalase, superoxide dismutase and pyruvate. Mol Reprod Dev. 2001;59:44–53.
45. Catt JW, Henman M. Toxic effects of oxygen on human embryo development. Hum Reprod. 2000;15 Suppl 2:199–206.
46. Burton GJ, Hempstock J, Jauniaux E. Oxygen, early embryonic metabolism and free radical-mediated embryopathies. Reprod Biomed Online. 2003;6:84–96.
47. Maas DH, Storey BT, Mastroianni Jr L. Oxygen tension in the oviduct of the rhesus monkey (Macaca mulatta). Fertil Steril. 1976;27:1312–7.
48. Booth PJ, Holm P, Callesen H. The effect of oxygen tension on porcine embryonic development is dependent on embryo type. Theriogenology. 2005;63:2040–52.
49. Bavister B. Oxygen concentration and preimplantation development. Reprod Biomed Online. 2004;9:484–6.
50. García JI, Sepúlveda S, Noriega-Hoces L. Beneficial effect of reduced oxygen concentration with transfer of blastocysts in IVF patients older than 40 years old. Health. 2010;2:1010–7.
51. Waldenstrom U, Engstrom AB, Hellberg D, Nilsson S. Low oxygen compared with high oxygen atmosphere in blastocyst culture, a prospective randomized study. Fertil Steril. 2009;9:2461–5.

52. Kovačič B, Sajko M, Vlaisavljević V. A prospective randomized trial on the effect of atmospheric versus reduced oxygen concentration on the outcome on intracytoplasmic sperm injection cycles. Fertil Steril. 2010;94(2):511–9.
53. Favetta LA, St John EJ, King WA, Betts DH. High levels of p66shc and intracellular ROS in permanently arrested early embryos. Free Radic Biol Med. 2007;42:1201–10.
54. Harvey AJ, Kind KL, Thompson JG. Regulation of gene expression in bovine blastocysts in response to oxygen and the iron chelator desferrioxamine. Biol Reprod. 2007;77:93–101.
55. Islam KN, Mendelson CR. Permissive effects of oxygen on cyclic AMP and interleukin-1 stimulation of surfactant protein A gene expression are mediated by epigenetic mechanisms. Mol Cell Biol. 2006;26:2901–12.
56. Wang X, Sharma RK, Sikka SC, Thomas Jr AJ, Falcone T, Agarwal A. Oxidative stress is associated with increased apoptosis leading to spermatozoa DNA damage in patients with male factor infertility. Fertil Steril. 2003;80:531–5.
57. Bian Q, Xu LC, Wang SL, Xia YK, Tan LF, Chen JF, Song L, Chang HC, Wang XR. Study on the relation between occupational fenvalerate exposure and spermatozoa DNA damage of pesticide factory workers. Occup Environ Med. 2004;61:999–1005.
58. Hammadeh ME, Hamad MF, Montenarh M, Fischer-Hammadeh C. P2 ratio in human spermatozoa from smokers and non-smokers. Hum Reprod. 2010;25(11):2708–20.
59. Kim JG, Parthasarathy S. Oxidation and the spermatozoa. Semin Reprod Endocrinol. 1998;16:235–9.
60. Watson PF. The causes of the reduced fertility with cryopreserved semen. Anim Reprod Sci. 2000;60–61:481–92.
61. Agarwal A, Ikemoto I, Loughlin KR. Effect of sperm washing on levels of reactive oxygen species in semen. Arch Androl. 1994;33:157–62.
62. Shekarriz M, DeWire DM, Thomas Jr AJ, Agarwal A. A method of human semen centrifugation to minimize the iatrogenic sperm injuries caused by reactive oxygen species. Eur Urol. 1995;28:31–5.
63. Chi HJ, Kim JH, Ryu CS, Lee JY, Park JS, Chung DY, et al. Protective effect of antioxidant supplementation in sperm-preparation medium against oxidative stress in human spermatozoa. Hum Reprod. 2008;23:1023–8.
64. Kattera S, Chen C. Short coincubation of gametes in in vitro fertilization improves implantation and pregnancy rates: a prospective, randomized, controlled study. Fertil Steril. 2003;80(4):1017–21.
65. Dirnfeld M, Bider D, Koifman M, Calderon I, Abramovici H. Shortened exposure of oocytes to spermatozoa improves in vitro fertilization outcome: a prospective, randomized, controlled study. Hum Reprod. 1999;14:2562–4.
66. Quinn P, Lydic ML, Ho M, Bastuba M, Hendee F, Brody SA. Confirmation of the beneficial effects of brief co-incubation of gametes in human in vitro fertilization. Fertil Steril. 1998;69:399–402.
67. Agarwal A, Saleh RA. Role of oxidants in male infertility: rationale, significance, and treatment. Urol Clin North Am. 2002;29:817–27.
68. Aitken RJ. The human spermatozoon—a cell in crisis? The Amoroso Lecture. J Reprod Fertil. 1999;115:1–7.
69. Rivlin J, Mendel J, Rubinstein S, Etkovitz N, Breitbart H. Role of hydrogen peroxide in sperm capacitation and acrosome reaction. Biol Reprod. 2004;70:518–22.
70. Lewis B, Aitken RJ. A redox-regulated tyrosine phosphorylation cascade in rat spermatozoa. J Androl. 2001;22:611–22.
71. Gavella M, Lipovac V. NADH-dependent oxidoreductase (diaphorase) activity and isozyme pattern of sperm in infertile men. Arch Androl. 1992;28:135–41.
72. Vernet P, Fulton N, Wallace C, Aitken RJ. Analysis of reactive oxygen species generating systems in rat epididymal spermatozoa. Biol Reprod. 2001;65:1102–13.
73. Evenson DP, Darzynkiewicz Z, Melamed MR. Simultaneous measurement by flow cytometry of sperm cell viability and mitochondrial membrane potential related to cell motility. J Histochem Cytochem. 1982;30:279–80.

74. Gil-Guzman E, Ollero M, Lopez MC, Sharma RK, Alvarez JG, Thomas Jr AJ, et al. Differential production of reactive oxygen species by subsets of human spermatozoa at different stages of maturation. Hum Reprod. 2001;16:1922–30.
75. Aitken RJ, West KM, Buckingham D. Leukocytic infiltration into the human ejaculate and its association with semen quality, oxidative stress, and sperm function. J Androl. 1994;15: 343–52.
76. Huszar G, Vigue L. Correlation between the rate of lipid peroxidation and cellular maturity as measured by creatine kinase activity in human spermatozoa. J Androl. 1994;15:71–7.
77. Gomez E, Buckingham DW, Brindle J, Lanzafame F, Irvine DS, Aitken RJ. Development of an image analysis system to monitor the retention of residual cytoplasm by human spermatozoa: correlation with biochemical markers of the cytoplasmic space, oxidative stress, and sperm function. J Androl. 1996;17:276–87.
78. Aitken RJ, Krausz C. Oxidative stress, DNA damage and the Y chromosome. Reproduction. 2001;122:497–506.
79. Alvarez JG, Touchstone JC, Blasco L, Storey BT. Spontaneous lipid peroxidation and production of hydrogen peroxide and superoxide in human spermatozoa. Superoxide dismutase as major enzyme protectant against oxygen toxicity. J Androl. 1987;8:338–48.
80. Aitken RJ, Fisher HM, Fulton N, Gomez E, Knox W, Lewis B, et al. Reactive oxygen species generation by human spermatozoa is induced by exogenous NADPH and inhibited by the flavoprotein inhibitors diphenylene iodonium and quinacrine. Mol Reprod Dev. 1997;47:468–82.
81. Agnihotri S, Purohit SB, Laloraya M, Kumar GP. Regional heterogeneity in intracellular distribution of superoxide and hydrogen peroxide within the sperm and its relation to sperm development. Arch Androl. 1999;43:113–21.
82. Sikka SC. Role of oxidative stress and antioxidants in andrology and assisted reproductive technology. J Androl. 2004;25:5–28.
83. Garrido N, Meseguer M, Simon C, Pellicer A, Remohi J. Pro-oxidative and anti-oxidative imbalance in human semen and its relation with male fertility. Asian J Androl. 2004;6: 59–65.
84. Venkatesh S, Singh G, Gupta NP, Kumar R, Deecaraman M, Dada R. Correlation of sperm morphology and oxidative stress in infertile men. Iranian J Reprod Med. 2009;7(1):29–34.
85. El-Demiry MI, Young H, Elton RA, Hargreave TB, James K, Chisholm GD. Leucocytes in the ejaculate from fertile and infertile men. Br J Urol. 1986;58:715–20.
86. Kiessling AA, Lamparelli N, Yin HZ, Seibel MM, Eyre RC. Semen leukocytes: friends or foes? Fertil Steril. 1995;64:196–8.
87. Sharma RK, Pasqualotto AE, Nelson DR, Thomas Jr AJ, Agarwal A. Relationship between seminal white blood cell counts and oxidative stress in men treated at an infertility clinic. J Androl. 2001;22:575–83.
88. Plante M, de Lamirande E, Gagnon C. Reactive oxygen species released by activated neutrophils, but not by deficient spermatozoa, are sufficient to affect normal sperm motility. Fertil Steril. 1994;62:387–93.
89. Pasqualotto FF, Sharma RK, Nelson DR, Thomas AJ, Agarwal A. Relationship between oxidative stress, semen characteristics, and clinical diagnosis in men undergoing infertility investigation. Fertil Steril. 2000;73:459–64.
90. Shekarriz M, Thomas Jr AJ, Agarwal A. Effects of time and sperm concentration on reactive oxygen species formation in human semen. Arch Androl. 1995;34:69–75.
91. Alvarez JG, Sharma RK, Ollero M, Saleh RA, Lopez MC, Thomas Jr AJ, Evenson DP, Agarwal A. Increased DNA damage in sperm from leukocytospermic semen samples as determined by the sperm chromatin structure assay. Fertil Steril. 2002;78:319–29.
92. Whittington K, Ford WC. Relative contribution of leukocytes and of spermatozoa to reactive oxygen species production in human sperm suspensions. Int J Androl. 1999;22:229–35.
93. Wang X, Falcone T, Attaran M, Goldberg JM, Agarwal A, Sharma RK. Vitamin C and vitamin E supplementation reduce oxidative stress-induced embryo toxicity and improve the blastocyst development rate. Fertil Steril. 2002;78(6):1272–7.

94. MacLeod J. The role of oxygen in the metabolism and motility of human spermatozoa. Am J Physiol. 1943;138:512–8.
95. Jones R, Mann T, Sherins R. Peroxidative breakdown of phospholipids in human spermatozoa, spermicidal properties of fatty acid peroxides, and protective action of seminal plasma. Fertil Steril. 1979;31:531–7.
96. Agarwal A, Gupta S, Sharma RK. Role of oxidative stress in female reproduction. Reprod Biol Endocrinol. 2005;3:28.
97. Aitken RJ. The Founders Lecture. Human spermatozoa: fruits of creation and seeds of doubt. Reprod Fertil Dev. 2005;16:655–64.
98. Balercia G, Armeni T, Mantero F, Principato G, Regoli F. Total oxyradical scavenging capacity toward different reactive oxygen species in seminal plasma and sperm cells. Clin Chem Lab Med. 2003;41:13–9.
99. Zalata AA, Ahmed AH, Allamaneni SS, Comhaire FH, Agarwal A. Relationship between acrosin activity of human spermatozoa and oxidative stress. Asian J Androl. 2004;6:313–8.
100. Griveau JF, Dumont E, Renard P, Callegari JP, Le Lannou D. Reactive oxygen species, lipid peroxidation and enzymatic defence systems in human spermatozoa. J Reprod Fertil. 1995;103:17–26.
101. Hammadeh ME, Filippos A, Hamad MF. Reactive oxygen species and antioxidant in seminal plasma and their impact on male fertility. Int J Fertil Steril. 2009;3(3):87–110.
102. Saleh RA, Agarwal A. Oxidative stress and male infertility: from research bench to clinical practice. J Androl. 2002;23:737–52.
103. Hammadeh ME, Stieber M, Haidl G, Schmidt W. Association between sperm chromatin condensation, morphology based on strict criteria, and fertilization, cleavage and pregnancy rates in an IVF program. Andrologia. 1998;30:29–35.
104. Hammadeh ME, Al-Hasani S, Stieber M, et al. The effect of chromatin condensation (aniline blue staining) and morphology (strict criteria) of Human spermatozoa on fertilization, cleavage and pregnancy rates in an intracytoplasmic sperm injection programme. Hum Reprod. 1996;11:2468.
105. Hammadeh ME, Al Hasani S, Rosenbaum P, Schmidt W, Fischer Hammadeh C. Reactive oxygen species, total antioxidant concentration of seminal plasma and their effect on sperm parameters and outcome of IVF/ICSI patients. Arch Gynecol Obstet. 2008;277:515–26.
106. Kao SH, Chao HT, Chen HW, Hwang TI, Liao TL, Wei YH. Increase of oxidative stress in human sperm with lower motility. Fertil Steril. 2008;8:2178–84.
107. Zini A, de Lamirande E, Gagnon C. Lower levels of nitric oxide promote human sperm capacitation in vitro. J Androl. 1996;16:424–31.
108. Leclerc P, de Lamirande E, Gagnon C. Regulation of protein-tyrosine phosphorylation and human sperm capacitation by reactive oxygen derivatives. Free Radic Biol Med. 1997;22:643–56.
109. Aitken RJ, Harkiss D, Knox W, Paterson M, Irvine DS. A novel signal transduction cascade in capacitating human spermatozoa characterised by a redox-regulated, cAMP-mediated induction of tyrosine phosphorylation. J Cell Sci. 1998;111:645–56.
110. De Lamirande E, Gagnon C. Impact of reactive oxygen species on spermatozoa: a balancing act between beneficial and detrimental effects. Hum Reprod. 1995;10 Suppl 1:15–21.
111. Herrero MB, de Lamirande E, Gagnon C. Nitric oxide regulates human sperm capacitation and proteintyrosine phosphorylation in vitro. Biol Reprod. 1999;61:575–81.
112. Aitken RJ, Clarkson JS, Fishel S. Generation of reactive oxygen species, lipid peroxidation, and human sperm function. Biol Reprod. 1989;41:183–97.
113. Sengoku K, Tamate K, Yoshida T, Takaoka Y, Miyamoto T, Ishikawa M. Effects of low concentrations of nitric oxide on the zona pellucida binding ability of human spermatozoa. Fertil Steril. 1998;69:522–7.
114. De Lamirande E, Gagnon C. Human sperm hyperactivation and capacitation as parts of an oxidative process. Free Radic Biol Med. 1993;14:255–65.
115. Sukcharoen N, Keith J, Irvine DS, Aitken RJ. Predicting the fertilizing potential of human sperm suspensions in vitro: importance of sperm morphology and leukocyte contamination. Fertil Steril. 1995;63:1293–300.

116. Sukcharoen N, Keith J, Irvine DS, Aitken RJ. Prediction of the in-vitro fertilization (IVF) potential of human spermatozoa using sperm function tests: the effect of the delay between testing and IVF. Hum Reprod. 1996;11:1030–4.
117. Aitken RJ, West KM. Analysis of the relationship between reactive oxygen species production and leucocyte infiltration in fractions of human semen separated on Percoll gradients. Int J Androl. 1990;13:433–51.
118. Kodama H, Kuribayashi Y, Gagnon C. Effect of sperm lipid peroxidation on fertilization. J Androl. 1996;17:151–7.
119. Zorn B, Vidmar G, Meden-Vrtovec H. Seminal reactive oxygen species as predictors of fertilization, embryo quality and pregnancy rates after conventional in vitro fertilization and intracytoplasmic sperm injection. Int J Androl. 2003;26:279–85.
120. Zini A, Bielecki R, Phang D, et al. Correlations between two markers of sperm DNA integrity, DNA denaturation and DNA fragmentation, in fertile and infertile men. Fertil Steril. 2001;75:674–7.
121. Ellington J, Evenson DP, Wright Jr RW, et al. Higher quality human sperm in a sample selectively attach to oviduct (fallopian tube) epithelial cells in vitro. Fertil Steril. 1999;71:924–9.
122. Huang CC, Lin DP, Tsao HM, et al. Sperm DNA fragmentation negatively correlates with velocity and fertilization rates but might not affect pregnancy rates. Fertil Steril. 2005;84: 130–40.
123. Zini A, Meriano J, Kader K, et al. Potential adverse effect of sperm DNA damage on embryo quality after ICSI. Hum Reprod. 2005;20:3476–80.
124. Hansen M, Kurinczuk JJ, Bower C, et al. The risk of major birth defects after intracytoplasmic sperm injection and in vitro fertilization. N Engl J Med. 2002;346:725–30.
125. Bonduelle M, Aytoz A, Van Assche E, et al. Incidence of chromosomal aberrations in children born after assisted reproduction through intracytoplasmic sperm injection. Hum Reprod. 1998;13:781–92.
126. Tarozzi N, Bizzaro D, Flamigni C, Borini A. Clinical relevance of sperm DNA damage in assisted reproduction. Reprod Biomed Online. 2007;14(6):746–57.
127. Agarwal A, Gupta S, Sekhon L, Shah R. Redox considerations in female reproductive function and assisted reproduction: from molecular mechanisms to health implications. Antioxid Redox Signal. 2008;10(8):1375–403.
128. Hammadeh ME, Radwan M, Al-Hasani S, et al. Comparison of reactive oxygen species concentration in seminal plasma and semen parameters in partners of pregnant and non-pregnant patients after IVF/ICSI. Reprod Biomed Online. 2006;13(5):696–706.
129. Agarwal A, Said TM, Bedaiwy MA, Banerjee J, Alvarez JG. Oxidative stress in an assisted reproductive techniques setting. Fertil Steril. 2006;86(3):503–12.
130. Falcone T, Hurd W. Clinical reproductive medicine and surgery. Philadelphia: Mosby Elsevier; 2007. p. 233–50.
131. Barbieri E, Hidalgo M, Venegas A, Smith R, Lissi E. Varicocele-associated decrease in antioxidant defenses. J Androl. 1999;20:713–7.
132. Ozbek E, Turkoz Y, Gokdeniz R, Davarci M, Ozugurlu F. Increased nitric oxide production in the spermatic vein of patients with varicocele. Eur Urol. 2000;37:172–5.
133. Weinberg JB, Doty E, Bonaventura J, Haney AF. Nitric oxide inhibition of human sperm motility. Fertil Steril. 1995;64:408–13.
134. Saleh R, Agarwal A, Sharma R, Said T, Sikka S, Thomas Jr A. Evaluation of nuclear DNA damage in spermatozoa from infertile men with varicocele. Fertil Steril. 2003;80: 1431–6.
135. Agarwal A, Prabakaran SA. Mechanism, measurement, and prevention of oxidative stress in male reproductive physiology. Indian J Exp Biol. 2005;43(11):963–74.
136. Said TM, Tellez S, Evenson DP, Del Valle AP. Assessment of sperm quality, DNA integrity and cryopreservation protocols in men diagnosed with testicular and systemic malignancies. Andrologia. 2009;41:377–82.
137. Kobayashi H, Larson K, Sharma R, et al. DNA damage in cancer patients before treatment as measured by the sperm chromatin structure assay. Fertil Steril. 2001;75:469–75.

138. Aitken RJ, De Iuliis GN. Origins and consequences of DNA damage in male germ cells. Reprod Biomed Online. 2007;14:727–33.
139. Behrman HR, Kodaman PH, Preston SL, et al. Oxidative stress and the ovary. J Soc Gynecol Investig. 2001;8:S40–2.
140. Sugino N, Karube-Harada A, Taketani T, et al. Withdrawal of ovarian steroids stimulates prostaglandin F2alpha production through nuclear factor-kappaB activation via oxygen radicals in human endometrial stromal cells: potential relevance to menstruation. J Reprod Dev. 2004;50:215–25.
141. Du B, Takahashi K, Ishida GM, Nakahara K, Saito H, Kurachi H. Usefulness of intra-ovarian artery pulsatility and resistance indices measurement on the day of follicle aspiration for the assessment of oocyte quality. Fertil Steril. 2006;85(2):366–70.
142. Van Blerkom J, Antczak M, Schrader R. The developmental potential of the human oocyte is related to the dissolved oxygen content of follicular fluid: association with vascular endothelial growth factor levels and perifollicular blood flow characteristics. Hum Reprod. 1997;12(5):1047–455.
143. Lee TH, Wu MY, Chen MJ, Chao KH, Ho HN, Yang YS. Nitric oxide is associated with poor embryo quality and pregnancy outcome in in vitro fertilization cycles. Fertil Steril. 2004;82(1):126–31.
144. Jozwik M, Wolczynski S, Szamatowicz M. Oxidative stress markers in preovulatory follicular fluid in humans. Mol Hum Reprod. 1999;5:409–13.
145. Sugino N, Takiguchi S, Kashida S, Karube A, Nakamura Y, Kato H. Superoxide dismutase expression in the human corpus luteum during the menstrual cycle and in early pregnancy. Mol Hum Reprod. 2000;6(1):19–25.
146. Paszkowski T, Traub AI, Robinson SY, et al. Selenium dependent glutathione peroxidase activity in human follicular fluid. Clin Chim Acta. 1995;236:173–80.
147. Baker MA, Aitken RJ. Reactive oxygen species in spermatozoa: methods for monitoring and significance for the origins of genetic disease and infertility. Reprod Biol Endocrinol. 2005;3:67–75.
148. Dumoulin JC, Meijers CJ, Bras M, Coonen E, Geraedts JP, Evers JL. Effect of oxygen concentration on human in-vitro fertilization and embryo culture. Hum Reprod. 1999;14(2):465–9.
149. Kitagawa Y, Suzuki K, Yoneda A, Watanabe T. Effects of oxygen concentration and antioxidants on the in vitro developmental ability, production of reactive oxygen species (ROS), and DNA fragmentation in porcine embryos. Theriogenology. 2004;62(7):1186–97.
150. Wiener-Megnazi Z, Shiloh H, Avraham L. Oxidative parameters of embryo culture media may predict treatment outcome in in vitro fertilization: a novel applicable tool for improving embryo selection. Fertil Steril. 2011;95:979–84.
151. Dey SK, Lim H, Das SK, Reese J, Paria BC, Daikoku T, Wang H. Molecular cues to implantation. Endocr Rev. 2004;25:341–73.
152. Wang H, Dey SK. Roadmap to embryo implantation: clues from mouse models. Nat Rev Genet. 2006;7:185–99.
153. Ashok BT, David L, Chen YG, Garikapaty VP, Chander B, Kanduc D, Mittelman A, Tiwari RK. Peptide mimotopes of oncoproteins as therapeutic agents in breast cancer. Int J Mol Med. 2003;11:465–71.
154. Riesewijk A, Martin J, van Os R, Horcajadas JA, Polman J, Pellicer A, et al. Gene expression profiling of human endometrial receptivity on days LH+2 versus LH+7 by microarray technology. Mol Hum Reprod. 2003;9:253–64.
155. Narimoto K, Noda Y, Shiotani M, et al. Immunohistochemical assessment of superoxide dismutase expression in the human endometrium throughout the menstrual cycle. Acta Histochem Cytochem. 1990;23:487–98.
156. Sugino N, Kashida S, Takiguchi S, et al. Induction of superoxide dismutase by decidualization in human endometrial stromal cells. Mol Hum Reprod. 2000;6:178–84.
157. Sugino N, Nakata M, Kashida S, et al. Decreased superoxide dismutase expression and increased concentrations of lipid peroxide and prostaglandin $F_2\alpha$ in the decidua of failed pregnancy. Mol Hum Reprod. 2000;6:642–7.

158. Sugino N, Karube-Harada A, Kashida S, Takiguchi S, Kato H. Reactive oxygen species stimulate prostaglandin F2 alpha production in human endometrial stromal cells in vitro. Hum Reprod. 2001;16:1797–801.
159. Azumaguchi A, Henmi H, Kanazawa T, Saito M. Mechanism underlying the low implantation rate in patients with thin endometrium. J Mamm Ova Res. 2009;26:134–8.
160. Jauniaux E, Hempstock J, Greenwold N, et al. Trophoblastic oxidative stress in relation to temporal and regional differences in maternal placental blood flow in normal a and abnormal early pregnancies. Am J Pathol. 2003;162:115–25.
161. Agarwal A, Allamaneni SS, Nallella KP, George AT, Mascha E. Correlation of reactive oxygen species levels with the fertilization rate after in vitro fertilization: a qualified meta-analysis. Fertil Steril. 2005;84:228–31.
162. Fait V, Sela S, Ophir E, et al. Peripheral polymorphonuclear leukocyte priming contributes to oxidative stress in early pregnancy. J Soc Gynecol Investig. 2005;12:46–9.
163. Banerjee S, Smallwood A, Moorhead J, et al. Placental expression of IFN-{gamma} and its receptor IFN-{gamma}R2 fail to switch from early hypoxic to late normotensive development in preeclampsia. J Clin Endocrinol Metab. 2005;90:944–52.
164. Peterson JD, Herzenberg LA, Vasquez K, et al. Glutathione levels in antigen presenting cells modulate Th1 compared with Th2 response patterns. Proc Natl Acad Sci USA. 1998; 95:3071–6.
165. Zini A, de Lamirande E, Gagnon C. Reactive oxygen species in semen of infertile patients: levels of superoxide dismutase- and catalase-like activities in seminal plasma and spermatozoa. Int J Androl. 1993;16:183–8.
166. Machaty Z, Thompson JG, Abeydeera LR, Day BN, Prather RS. Inhibitors of mitochondrial ATP production at the time of compaction improve development of in vitro produced porcine embryos. Mol Reprod Dev. 2001;58:39–44.
167. Oyawoye O, Abdel Gadir A, Garner A, Constantinovici N, Perrett C, Hardiman P. Antioxidants and reactive oxygen species in follicular fluid of women undergoing IVF: relationship to outcome. Hum Reprod. 2003;18(11):2270–4.
168. Yang HW, Hwang KJ, Kwon HC, Kim HS, Choi KW, Oh KS. Detection of reactive oxygen species (ROS) and apoptosis in human fragmented embryos. Hum Reprod. 1998;13(4): 998–1002.
169. Wiener-Megnazi Z, Vardi L, Lissak A, et al. Oxidative stress indices in follicular fluid as measured by the thermochemiluminescence assay correlate with outcome parameters in in vitro fertilization. Fertil Steril. 2004;82 Suppl 3:1171–6.
170. Pasqualotto EB, Agarwal A, Sharma RK, et al. Effect of oxidative stress in follicular fluid on the outcome of assisted reproductive procedures. Fertil Steril. 2004;81(4):973–6.
171. Das S, Chattopadhyay R, Ghosh S, et al. Reactive oxygen species level in follicular fluid—embryo quality marker in IVF? Hum Reprod. 2006;21(9):2403–7.
172. Bedaiwy M, Agarwal A, Said TM, et al. Role of total antioxidant capacity in the differential growth of human embryos in vitro. Fertil Steril. 2006;86(2):304–9.
173. Bedaiwy MA, Falcone T, Mohamed MS, et al. Differential growth of human embryos in vitro: role of reactive oxygen species. Fertil Steril. 2004;82(3):593–600.
174. Ebisch IM, Peters WH, Thomas CM, Wetzels AM, Peer PG, Steegers-Theunissen RP. Homocysteine, glutathione and related thiols affect fertility parameters in the subfertile couple. Hum Reprod. 2006;21:1725–33.
175. Seino T, Saito H, Kaneko T, Takahashi T, Kawachiya S, Kurachi H. Eight-hydroxy-2′-deoxyguanosine in granulosa cells is correlated with the quality of oocytes and embryos in an in vitro fertilization-embryo transfer program. Fertil Steril. 2002;77(6):1184–90.
176. Shiverick KT, Salafia C. Cigarette smoking and pregnancy I: ovarian, uterine and placental effects. Placenta. 1999;20(4):265–72.
177. Thompson JG, McNaughton C, Gasparrini B, McGowan LT, Tervit HR. Effect of inhibitors and uncouplers of oxidative phosphorylation during compaction and blastulation of bovine embryos cultured in vitro. J Reprod Fertil. 2000;118:47–55.

178. Duru NK, Morshedi M, Schuffner A, Oehninger S. Semen treatment with progesterone and/or acetyl-L-carnitine does not improve sperm motility or membrane damage after cryopreservation-thawing. Fertil Steril. 2000;74:715–20.
179. Alvarez JG. Efficient treatment of infertility due to sperm DNA damage by ICSI with testicular sperm. Hum Reprod. 2005;20:2031–2; author reply 2–3.
180. Greco E, Scarselli F, Iacobelli M, et al. Efficient treatment of infertility due to sperm DNA damage by ICSI with testicular spermatozoa. Hum Reprod. 2005;20:226–30.
181. Steele EK, McClure N, Maxwell RJ, Lewis SE. A comparison of DNA damage in testicular and proximal epididymal spermatozoa in obstructive azoospermia. Mol Hum Reprod. 1999;5:831–5.
182. Sun JG, Jurisicova A, Casper RF. Detection of deoxyribonucleic acid fragmentation in human sperm: correlation with fertilization in vitro. Biol Reprod. 1997;56:602–7.
183. Duran EH, Gurgan T, Gunalp S, Enginsu ME, Yarali H, Ayhan A. A logistic regression model including DNA status and morphology of spermatozoa for prediction of fertilization in vitro. Hum Reprod. 1998;13:1235–9.
184. Larson KL, DeJonge CJ, Barnes AM, Jost LK, Evenson DP. Sperm chromatin structure assay parameters as predictors of failed pregnancy following assisted reproductive techniques. Hum Reprod. 2000;15:1717–22.
185. Chan PJ, Calinisan JH, Corselli JU, Patton WC, King A. Updating quality control assays in the assisted reproductive technologies laboratory with a cryopreserved hamster oocyte DNA cytogenotoxic assay. J Assist Reprod Genet. 2001;18:129–34.
186. Tomlinson MJ, Moffatt O, Manicardi GC, Biyyaro D, Afnanm M, Sakkas D. Interrelationship between seminal parameters and sperm nuclear DNA damage before and after density gradients centrifugation implication for assisted reproduction. Hum Reprod. 2001;16:2160–5.
187. Muriel LL, Garrido N, Fernandez JL, Remohi J, Pellicer A, et al. Value of the sperm deoxyribonucleic acid fragmentation level, as measured by the sperm chromatin dispersion test, in the outcome of in vitro fertilization and intracytoplasmic sperm injection. Fertil Steril. 2006;85:371–83.
188. Muriel LL, Mesguer M, Ferdinanz JL, Al Varez J, et al. Value of the sperm chromatin dispersion test in predicting pregnancy outcome in intrauterine insemination: a blind prospective study. Hum Reprod. 2006;21:738–44.
189. Velez De La Calle JF, Muller A, Walschaerts M, Clavere J, et al. Sperm deoxyribonucleic acid fragmentation as assessed by the sperm chromatin dispersion test in assisted reproductive technology programs: results of a large prospective multicenter study. Fertil Steril. 2008;90:1792–9.
190. Carrell DT, Liu L, Peterson CM, et al. Sperm DNA fragmentation is increased in couples with unexplained recurrent pregnancy loss. Arch Androl. 2003;49:49–55.
191. Henkle R, Hajimohammad M, Stalf T, Hoogendijk C, Mehnert C, et al. Influence of deoxyribonucleic acid damage on fertilization and pregnancy. Fertil Steril. 2004;81:965–72.
192. Evenson D, Wixon R. Meta-analysis of sperm DNA fragmentation using the sperm chromatin structure assay. Reprod Biomed Online. 2006;12:466–72.
193. Saleh RA, Agarwal A, Nelson DR, Nada EA, El-Tonsy MH, Alvarez JG, Thomas Jr AJ, Sharma RK. Increased sperm nuclear DNA damage in normozoospermic infertile men: a prospective study. Fertil Steril. 2002;78:313–8.
194. Bungum M, Humaidan P, Spano M, Jepson K, et al. The predictive value of sperm chromatin structure assay (SCSA) parameters for the outcome of intrauterine insemination, IVF and ICSI. Hum Reprod. 2004;19:1401–8.
195. Erenpreiss J, Bungum M, Spano M, Elzanaty S, Orbidans J, Giwercman A. Intra-individual variation in sperm chromatin structure assay parameters in men from infertile couples: clinical implications. Hum Reprod. 2006;21:2061–4.
196. Bungum M, Humaidan P, Axmon A, Spano M, Bungum L, Erenpreiss J, Giwercman A. Sperm DNA integrity assessment in prediction of assisted reproduction technology outcome. Hum Reprod. 2007;22:174–9.

197. Evenson DP, Jost LK, Marshall D, et al. Utility of the sperm chromatin structure assay as a diagnostic and prognostic tool in the human fertility clinic. Hum Reprod. 1999;14:1039–49.
198. Spano M, Bonde JP, Hjollund HI, Kolstad HA, Cordelli E, Leter G. Sperm chromatin damage impairs human fertility. The Danish First Pregnancy Planner Study Team. Fertil Steril. 2000;73:43–50.
199. Payne JF, Raburn DJ, Couchman GM, Price TM, Jamison MG, Walmer DK. Redefining the relationship between sperm deoxyribonucleic acid fragmentation as measured by the sperm chromatin structure assay and outcomes of assisted reproductive techniques. Fertil Steril. 2005;84:356–64.
200. Larson-Cook KL, Brannian JD, Hansen KA, Kasperson KM, Aamold ET, Evenson DP. Relationship between the outcomes of assisted reproductive techniques and sperm DNA fragmentation as measured by the sperm chromatin structure assay. Fertil Steril. 2003;80: 895–902.
201. Gandini L, Lombardo F, Paoli D, et al. Full-term pregnancies achieved with ICSI despite high levels of sperm chromatin damage. Hum Reprod. 2004;19:1409–17.
202. Duran EH, Morshedi M, Taylor S, Oehninger S. Sperm DNA quality predicts intrauterine insemination outcome: a prospective cohort study. Hum Reprod. 2002;17:3122–8.
203. Benchaib M, Lorange J, Mazoyer C, et al. Sperm deoxyribonucleic acid fragmentation as a prognostic indicator of assisted reproductive technology outcome. Fertil Steril. 2007;87: 93–100.
204. Alpay Z, Saed GM, Diamond MP. Female infertility and free radicals: potential role in adhesions and endometriosis. J Soc Gynecol Investig. 2006;13(6):390–8.
205. Murphy AA, Palinski W, Rankin S, Morales AJ, Parthasarathy S. Macrophage scavenger receptor(s) and oxidatively modified proteins in endometriosis. Fertil Steril. 1998;69(6): 1085–91.
206. Jackson LW, Schisterman EF, Dey-Rao R, Browne R, Armstrong D. Oxidative stress and endometriosis. Hum Reprod. 2005;20(7):2014–20.
207. Gupta S, Agarwal A, Krajcir N, Alvarez JG. Role of oxidative stress in endometriosis. Reprod Biomed Online. 2006;13:126–34.
208. Szczepanska M, Kozlik J, Skrzypczak J, Mikolajczyk M. Oxidative stress may be a piece in the endometriosis puzzle. Fertil Steril. 2003;79(6):1288–93.
209. Kusçu NK, Var A. Oxidative stress but not endothelial dysfunction exists in non-obese, young group of patients with polycystic ovary syndrome. Acta Obstet Gynecol Scand. 2009;88(5): 612–7.
210. Fenkci IV, Serteser M, Fenkci S, Kose S. Paraoxonase levels in women with polycystic ovary syndrome. J Reprod Med. 2007;52(10):879–83.
211. Nácul AP, Andrade CD, Schwarz P, de Bittencourt Jr PI, Spritzer PM. Nitric oxide and fibrinogen in polycystic ovary syndrome: associations with insulin resistance and obesity. Eur J Obstet Gynecol Reprod Biol. 2007;133(2):191–6.
212. Aurrekoetxea I, Ruiz-Sanz J, del Agua AR, et al. Serum oxidizability and antioxidant status in patients undergoing in vitro fertilization. Fertil Steril. 2010;94:1279–86.
213. Harvey AJ, Kind KL, Thompson JG. REDOX regulation of early embryo development. Reproduction. 2002;123:479–86.
214. Dennery PA. Role of redox in fetal development and neonatal diseases. Antioxid Redox Signal. 2004;6:147–53.
215. Gadea J, Gumbao D, Matas C, Romar R. Supplementation of the thawing media with reduced glutathione improves function and the in vitro fertilizing ability of boar spermatozoa after cryopreservation. J Androl. 2005;26(6):749–56.
216. Chan PJ, Calinisan JH, Corselli JU, Patton WC, King A. Updating quality control assays in the assisted reproductive technologies laboratory with a cryopreserved hamster oocyte DNA cytogenotoxic assay. J Assist Reprod Genet. 2001;18(3):129–34.
217. Kim SS, Yang HW, Kang HG, et al. Quantitative assessment of ischemic tissue damage in ovarian cortical tissue with or without antioxidant (ascorbic acid) treatment. Fertil Steril. 2004;82(3):679–85.

218. Speroff L, Glass RH, Kase NG. Assisted reproduction clinical gynecologic endocrinology and infertility. 6th ed. Philadelphia: Lippincott Williams and Wilkins; 1999. p. 643–724.
219. Esfandiari N, Falcone T, Agarwal A, et al. Protein supplementation and the incidence of apoptosis and oxidative stress in mouse embryos. Obstet Gynecol. 2005;105(3):653–60.
220. Feugang JM, de Roover R, Moens A, et al. Addition of beta-mercaptoethanol or Trolox at the morula/blastocyst stage improves the quality of bovine blastocysts and prevents induction of apoptosis and degeneration by prooxidant agents. Theriogenology. 2004;61(1):71–90.
221. Dalvit G, Llanes SP, Descalzo A, Insani M, Beconi M, Cetica P. Effect of alpha-tocopherol and ascorbic acid on bovine oocyte in vitro maturation. Reprod Domest Anim. 2005; 40(2):93–7.
222. Oyamada T, Fukui Y. Oxygen tension and medium supplements for in vitro maturation of bovine oocytes cultured individually in a chemically defined medium. J Reprod Dev. 2004;50(1):107–17.
223. Ali AA, Bilodeau JF, Sirard MA. Antioxidant requirements for bovine oocytes varies during in vitro maturation, fertilization and development. Theriogenology. 2003;59(3–4):939–49.
224. Guerin P, Guillaud J, Menezo Y. Hypotaurine in spermatozoa and genital secretions and its production by oviduct epithelial cells in vitro. Hum Reprod. 1995;10(4):866–72.
225. Lornage J. Biological aspects of endometriosis in vitro fertilization. J Gynecol Obstet Biol Reprod (Paris). 2003;32(8 Pt 2):S48–50.
226. Dirnfeld M, Shiloh H, Bider D, et al. A prospective randomized controlled study of the effect of short coincubation of gametes during insemination on zona pellucida thickness. Gynecol Endocrinol. 2003;17(5):397–403.

Chapter 22
Oxidative Stress and the Use of Antioxidants for Idiopathic OATs

Ashok Agarwal, Anthony H. Kashou, and Lucky H. Sekhon

Abstract *Aim*: To examine the effects of ROS and OS on male fertility and to evaluate the use of antioxidants as a means of treatment to improve fertilization rates in subfertile males suffering from idiopathic oligoasthenoteratozoospermia (iOAT).
Methods: Review of PubMed database.

Results: Current research notes ROS-associated male factor infertility to be the most common potential etiology of impaired sperm quality. The various effects of these oxidants may be neutralized by antioxidants. Although antioxidant therapy has shown to potentially treat iOAT by improving semen parameters, its success remains limited. Our review calls for a deeper look and understanding of the type(s), dosage, and duration of antioxidant treatment used in order to apply its use in a clinical setting.

Keywords Male infertility • Oxidative stress • Antioxidant treatment • Idiopathic oligoasthenoteratozoospermia • Antioxidants • Anti-inflammatory drugs

A. Agarwal, PhD(✉)
Center for Reproductive Medicine, Cleveland Clinic,
Lerner College of Medicine, 9500 Euclid Avenue, Cleveland, OH 44195, USA
e-mail: agarwaa@ccf.org

A.H. Kashou, BS
Center for Reproductive Medicine, Cleveland Clinic, 9500 Euclid Avenue, Cleveland, OH 44195, USA

L.H. Sekhon, MD, BSc
OBGYN, Mount Sinai School of Medicine, Medical Center, New York, NY, USA

A. Agarwal et al. (eds.), *Studies on Men's Health and Fertility*, Oxidative Stress in Applied Basic Research and Clinical Practice, DOI 10.1007/978-1-61779-776-7_22,

22.1 Introduction

Infertility affects eighty million people worldwide [1]. Male factor subfertility is now thought to account for up to 50% of these cases [2]. However, the majority of the causes remain unknown [3]. Oxidative stress (OS) is implicated in pathogenesis of 30–80% of male factor subfertility cases [4]. This state arises when there is an imbalance between reactive oxygen species (ROS) generation and the semen's natural antioxidant defense.

Although the pathophysiological role of ROS in human sperm dysfunction and male infertility has been extensively explored in recent years, the etiology of suboptimal sperm quality has yet to be completely understood. There is substantial evidence to suggest that low levels of ROS are necessary for spermatozoa to acquire fertilizing capabilities [5, 6]. However, in excessive amounts, ROS can disrupt sperm maturation pathways. Exogenous factors such as pollution, high temperature, and electromagnetic radiation, and endogenous factors including infections, chronic disease, and autoimmunity can act as major sources of oxidative stress [4, 7, 8].

Endogenous seminal antioxidant compounds act as free radical scavengers to neutralize ROS. Semen of subfertile men was found to have lowered antioxidant levels compared to those found in healthy, fertile controls [9]. Idiopathic oligoasthenoteratozoospermia (iOAT) is considered one of the most prevalent causes of male infertility. iOAT is a disorder related to defects in sperm concentration, motility, and morphology and has been attributed to free radical-induced OS.

Although many treatments have been developed to overcome male factor infertility, no reliable strategy has been established. Intracytoplasmic sperm injection (ICSI), a highly efficacious but costly assisted reproductive technique to circumvent deficits in sperm function, has displayed some success in fertilization. A new area of study has focused on pharmacotherapy targeting the oxidative stress in iOAT. Men with a high dietary intake of antioxidants were reported to demonstrate improved semen quality [10]. This chapter investigates the effects of OS on semen quality, as well as the use of antioxidants as a potential treatment strategy to improve semen parameters, such as concentration, motility, and morphology, of men with iOAT.

22.2 Oxidative Stress

While oxygen is essential to sustaining normal cell function, the breakdown of its products may generate free radicals that can act as beneficial cell-signaling molecules or induce irreversible cellular damage and death [11]. The chemically unstable ROS by-products from aerobic cellular metabolism react almost instantaneously with neighboring species within their vicinity. The interaction of these stability-seeking agents causes them to propagate a cascade of reactions, which may ultimately disrupt and damage living cells and tissues. A few ROS include superoxide, hydrogen peroxide, and the hydroxyl radical, as well as the common ROS subclass—reactive nitrogen species (RNS)—such as nitric oxide (NO).

Generally, free radical production is counterbalanced by several mechanisms that include both enzymatic and nonenzymatic antioxidants. However, in a period of imbalance between ROS and the body's antioxidants OS ensues. OS may be a consequence of excess ROS production and/or reduced antioxidant capacity. The inability of the human biological system to detoxify/reduce oxidants or to repair detrimental damage disrupts physiological homeostasis. OS has been implicated in the pathogenesis of many other human diseases including cancer, diabetes, AIDS and Parkinson disease; however, only recently have ROS been considered toxic to human spermatozoa [12].

22.3 Effects on Sperm

Numerous studies continue to explore the role of ROS in male infertility. Virtually, every human ejaculate is thought to be contaminated with potential sources of ROS [13]. Physiologically, ROS are vital for modulating gene and protein activities for sperm proliferation, differentiation, and fertilization [6]. The exact levels and duration of normal exposure as well as the pathways that induce the pathological and physiological effects on sperm remain unclear.

ROS-induced sperm damage has been suggested to contribute to at least 40% of all male factor infertility cases [14, 15]. Excessive levels of nitric oxide have been shown to inhibit both motility and sperm competence for zona binding [12]. OS may impair spermatogenesis, impede sperm migration, and even cause abnormal sperm morphology and improper sperm function. Therefore, controlled regulation of ROS levels is essential for proper reproductive function.

22.4 Idiopathic Oligoasthenoteratozoospermia

iOAT is a complex medical condition that affects approximately 30% of all infertile men and is the most common cause of male infertility [16, 17]. As a three-part disorder, low sperm concentration and motility and morphologically abnormal sperm characterize iOAT. Three classifications have been put into place to evaluate the severity of the disease: (1) isolated astheno ± teratozoospermia (no alteration in sperm concentration); (2) moderate (sperm concentrations between $<20\times10^6$/mL and $>5\times10^6$/mL); and (3) severe (sperm concentrations $<5\times10^6$/mL) [18].

Although iOAT is often thought to be related to defects in spermatogenesis, its actual cause remains unknown and cannot be diagnosed using common laboratory techniques. In addition, a majority of the infertile patients suffering from iOAT have normal physical examinations, normal hormonal profiles, and no discernable cause for their subfertile status. The age of patients may have an adverse effect on sperm motility and volume, which have been reported to continuously decrease from the age of 22–80 [19]. The lower metabolic rate frequently observed in older age is thought to contribute to this decrease in sperm motility.

Additional etiologies for iOAT have been suggested. An increase of ROS in the tubules and seminal plasma may cause apoptosis, consequently affecting sperm concentration, motility, and morphology [16]. ROS are also able to penetrate through the sperm membrane, thereby disrupting sperm motility. Leukocytospermia has been reported to have similar adverse effects; however, much controversy remains behind the exact mechanistic pathways and reasoning of the noted positive correlation [20].

Infection and the subsequent release of cytokines, including IL-6, IL-8, and tumor necrosis factor-alpha, have been associated with decreased sperm quality. Inflammation-mediated cytokines have been shown to induce ROS production, causing lipid peroxidation (LPO) of the sperm cell membrane [20]. A recent study conducted by Fraczek et al. revealed that cytokines released during the oxidative burst from inflammation intensify the degree of OS associated with leukocytes [21]. Furthermore, it has been reported ROS levels in whole ejaculates to be negatively correlated with sperm concentration ($r=-0.52$, $p=0.0003$), motility ($r=-0.41$, $p=0.006$), and morphology ($r=-0.34$, $p=0.02$), while also correlating positively to seminal leukocyte concentration ($r=0.65$, $p<0.0001$) [22]. Another study that compared 167 infertile patients with 19 controls found that sperm concentration, motility, and morphology were significantly reduced, while the mean ROS levels were significantly higher in infertile subjects than controls [23]. The positive correlation between ROS levels in whole ejaculates and seminal leukocyte concentrations provides a rationale for treatment of OS-positive infertile patients with antioxidant supplementation and other strategies to control excessive leukocyte infiltration.

Cytoplasmic droplets have also been implicated in the pathogenesis of iOAT. These residues have been found to develop from defects in spermatogenesis. They contain large amounts of glucose-6-phosphate dehydrogenase (G_6PD), which is known to generate NADPH. NADPH activation generates ROS production in spermatozoa via the action of NADPH oxidase, thereby promoting OS in the sperm membrane [24]. Interestingly, male infertility seems to have a familial relation among brothers and maternal uncles [25]. A genetic etiology is currently considered the most accepted theory for iOAT. Bujan et al. suggests an autosomal recessive mode of inheritance [26]. Thus, it is essential for future genomic and proteomic studies to focus on identifying possible genetic defects or mutations that could give rise to male infertility.

Oligozoospermia. Multiple mutations in mitochondrial DNA (mtDNA) have been correlated with oligozoospermic patient [27]. An in vitro study using a mouse model revealed that different proportions of mutated mtDNA had respiratory chain defects from meiotic arrest during spermatogenesis [28]. This indicated that respiration-deficient spermatocytes were incapable of completing meiosis and eliminated by apoptosis. The study confirmed that the mutated mtDNA that affected mitochondrial respiration function also resulted in not only oligozoospermia, but also asthenozoospermia and teratozoospermia. Moreover, since mtDNA is more susceptible than the nuclear genome, any impairment of the electron transport chain consequently enhances ROS generation in mitochondria from incomplete reduction of oxygen [29]. Other reports indicate that mitochondrial dysfunction is associated with decreased membrane potential and increased production of superoxide anions, hydroxide radicals, and hydrogen peroxide [30, 31]. An experimental study suggests that genetic alterations found in sperm mtDNA may have occurred during cell differentiation, spermatogenesis, or from random segregation of mtDNA during cell division [32].

Asthenozoospermia. Impairment in sperm motility has been attributed to increased ROS generation [33]. Impaired motility is thought to result from a cascade of events, involving lowered axonemal protein phosphorylation and sperm immobilization, both of which are associated with reduced membrane fluidity that is essential for sperm–oocyte fusion [34]. It has also been hypothesized that membrane-soluble hydrogen peroxide may potentially diffuse across membranes into cells and become oxidized even further, thereby inhibiting the activity of enzymes, particularly G_6PD. This cytosolic enzyme controls the rate of glucose flux through the hexose monophosphate shunt, regulating the intracellular availability of NADPH. Spermatozoa use this as a source of electrons to stimulate the generation of ROS by NADPH oxidase [35]. Any impairment or inhibition of G_6PD leads to a decrease in the availability of NADPH and a concomitant accumulation of oxidized glutathione (GSH). Subsequently, this can lower antioxidant defense of spermatozoa and result in LPO [36]. Since energy production is critical in order to facilitate movement of sperm, any impairment in the process of ATP production may have a detrimental effect on sperm motility.

Teratozoospermia. Morphologically abnormal spermatozoa have been established as a predominant source of high ROS generation in human ejaculate [37]. The irregular structure is thought to result from impairments during spermatogenesis, whereby cytoplasmic extrusion pathways become impaired. As a result, spermatozoa released from the germinal epithelium during spermiation carry excess residual cytoplasm, and under such conditions, are believed to be immature and functionally defective [38]. Studies have shown that levels of ROS production in semen were negatively correlated with the percentage of normal sperm forms [39, 40].

22.5 Antioxidants

Antioxidants are naturally occurring or synthetic biomolecules that prevent free radical-induced damage by averting the formation of radicals, scavenging them, or promoting their decomposition in the body. Their neutralizing capabilities reside in their ability to donate an electron to ward off the deleterious effects of the highly reactive radicals or by converting ROS into different, less harmful molecules.

There are three main routes of defense antioxidants utilized in combating detrimental ROS levels: prevention, interception, and repair. The seminal plasma transition metals are capable of preventing ROS from initiating chain reaction. These chelating metals are vital in controlling LPO damage to sperm. However, as these chelators become loosely bound to reduced oxygen products, they can generate secondary or even more reactive oxidants.

Certain antioxidants can intercept and impede free radical-induced chain reactions, forming nonradical end products. α-Tocopherol is a form of vitamin E antioxidant with chain-breaking/intercepting abilities [41]. It is capable of inhibiting LPO in membranes by scavenging ROS, such as the peroxyl and alkoxyl radicals. However, this system is highly regulated and α-tocopherol's ability to maintain a steady rate of radical reduction in the plasma membrane is dependent on its recycling by external reducing agents. Therefore, α-tocopherol is able to function even at low concentrations.

The last mechanistic process of defense employed by antioxidants is to simply repair any damage caused. In cases of minimal oxidative damage, normal physiological function may be restored. However, if OS-induced damage subsists, spermatozoa are incapable of repairing this level of injury. This is often due to the lack of an intricate cytoplasmic enzyme/antioxidant system in mature sperm that is required to treat this type of damage, which demonstrates the high susceptibility of spermatozoa to oxidative insults [42].

Antioxidants come in a variety of forms, ranging from those generated endogenously by the body to those acquired exogenously through the consumption of certain foods or dietary supplements. When the natural balance between oxidants and antioxidants within the body is perturbed via antioxidant deficiency or overwhelming ROS production, the subsequent adverse effects have been found to be diminished, and sometimes resolved, through bodily antioxidant defense and supplementation [43]. Since human spermatozoa membranes are rich in unsaturated fatty acids, they are sensitive to oxygen-induced damage mediated by LPO.

Normally, the seminal plasma is well endowed with an array of antioxidant defense mechanisms to quench ROS and protect against any possible damage to spermatozoa. The variety of pathways compensate for the deficiency in cytoplasmic enzymes in sperm. Potential anti-ROS enzymes and other low molecular weight nonenzymatic substances make up the total antioxidant capacity (TAC) of the seminal plasma. Fertile men were reported to have a higher, more effective TAC in comparison to infertile male subjects [44]. If the male reproductive system becomes compromised by infection or any impairment in spermatogenesis that elevates ROS levels, antioxidant defense mechanisms may be overwhelmed and depleted, and thus result in OS. Given that various studies implicate OS in poor sperm samples, medical treatments that reduce or protect against OS may be an appropriate strategy to alleviate infertility in men with iOAT.

22.6 Antioxidants: A Practical Approach for Treating iOAT

As a multifactor disorder, iOAT presents a clinical challenge in the diagnosis and medical treatment of infertile men. ICSI has enabled iOAT to be circumvented mechanically; however, it fails to address the fundamental factors contributing to male factor infertility. Oral supplementation of hormones and antioxidants has been the primary area of focus for pharmacotherapy treatment of iOAT. Significant improvements in sperm concentration, motility, and morphology have been reported [45]. Therefore, if it were possible to achieve a particular regimen of antioxidant(s) with the proper dosage and duration of intervention necessary to fully repair each aspect of the disorder, these men would have a greater probability of gaining normal sperm function.

In order to consider a drug as active, it must improve sperm parameters and pregnancy rates in at least one blind, prospective, placebo-controlled trial, as well as additional trials from independent groups [46]. The aim of the following sections is to evaluate evidence gathered from experimental trials with oral antioxidants that may serve in potentially treating impaired sperm quality. The mode of action of each antioxidant will be discussed and evaluated for their clinical efficacy. Table 22.1 is provided to summarize collected data, along with critical commentary for each antioxidant.

22.7 Carnitine

Carnitine is a water-soluble antioxidant derived mostly from the human diet. Some sources include fish, meat, poultry, and dairy products. It is formed from the biosynthesis of lysine and methionine. Carnitine is found in nearly all cells of the body and is essential in sperm energy production by transporting long-chain fatty acids from the cytosol into the mitochondria. In addition, carnitine is involved in the removal of cytotoxic compounds from the cell. Normally, the body produces a sufficient amount of carnitine to meet daily needs; however, in some cases individuals are incapable of producing enough. For instance, vegetarians were found to be significantly deficient in carnitine as opposed to an omnivorous control group [47].

Carnitine is also important with respect to many properties specific to sperm maturation. In a meta-analysis by Ross et al., oral supplementation with carnitine improved spermatozoa motility in three out of four randomized controlled studies [48]. Intra- and extracellular carnitine may be essential in sperm energy metabolism, and thereby provide the fuel for posttesticular processes involved in sperm motility. Motile spermatozoa have been found to have a greater degree of carnitine acetylation than immotile spermatozoa [49]. Moreover, free-L-carnitine is acetylated only in mature spermatozoa [50]. Studies have suggested that the initiation of acquiring sperm motility occurs parallel to an increase in carnitine concentration in the epididymal lumen and acetyl-L-carnitine in spermatozoa [49, 50]. Carnitine that accumulates in the epididymis, in both free and acetylated forms, is utilized by spermatozoa for mitochondrial β-oxidation of long-chain fatty acids, as well as for enhancing the cellular energetic in mitochondria. This is done by facilitating the entry and utilization of fatty acids and restoring the phospholipid composition of its membrane by decreasing fatty acid oxidation. The resulting increase in metabolic rate allows for a greater production of energy and increased sperm motility. Table 22.1 summarizes the most recent studies on carnitine supplementation for improving sperm quality.

Table 22.1 Quality assessment of oral carnitine supplementation

References	Study design	Cases	Dosage	Duration	Ages	Main outcome
Costa et al. [51]	Cross-sectional	100	LC (3 g/day)	4m	–	↑ Motility ($p \leq 0.001$) ↑ Rapid linear progression ($p \leq 0.001$) ↑ Mean velocity ($p \leq 0.001$) ↑ Linearity index ($p \leq 0.001$) ↑ Total number ($p \leq 0.001$)
Vitali et al. [52]	Case–control	47	LC (3 g/day)	3m	–	↑ Motility (80%)
Lenzi et al. [53]	Double-blind, placebo-controlled, crossover	86	LC (2 g/day)^{A}; Placebo^{B}	2-m wash-out + 2-m therapy/placebo + 2-m wash-out + 2-m placebo/therapy + 2-m wash-out	20–40	↑ Total motility^{A,B} ($p = 0.04$) ↑ Forward motility^{A,B} ($p = 0.05$) ↑ Concentration^{A,B} ($p = 0.01$) ↑ Linearity^{A,B} ($p = 0.03$)
Cavallini et al. [54]	Double-blind, randomized controlled	195 (39^{A}; 44^{B}; 47^{C})	LC (2 g/day) + LAC (1 g/day)^{A}; LC (2 g/day) + LAC (1 g/day) + Cinnoxicam (30 mg/4 day)^{B}; Placebo^{C}	6-m intervention + 3-m follow-up	27–40	↑ Motility^{A,B} ($p < 0.05$) ↑ Concentration^{A,B} ($p < 0.05$) ↑ Morphology^{A,B} ($p < 0.05$) ↑ Pregnancy rates^{A,B,a}

Lenzi et al. [55]	Double-blind, randomized controlled	56 (26^{A}; 30^{B})	LC (2 g/day) + LAC (1 g/day)^{A}; Placebo^{B}	2-m wash-out + 6-m intervention + 2-m follow-up	20–40	↑ Total motility^{A} ($p<0.05$) ↑ Forward motility^{A} ($p<0.05$) ↑ Concentration^{A,B} ($p>0.05$) ↑ Morphology^{A,B} ($p>0.05$)
Balercia et al. [56]	Double-blind, randomized, controlled	59 (15^{A}; 15^{B}; 14^{C}; 15^{D})	LC (3 g/day)^{A}; LAC (3 g/day)^{B}; LC (2 g/day) + LAC (1 g/day)^{C}; Placebo^{D}	1-m wash-out + 6-m intervention + 3-m follow-up	24–38	↑ Total motility^{A,B,C} ($p<0.05$) ↑ Forward motility^{A,B,C} ($p<0.05$) ↑ Morphology^{A} ($p<0.05$) ↑ TAC^{A,B,C} ($p<0.05$)
Sigman et al. [57]	Double-blind, randomized controlled	21 (12^{A}; 9^{B})	LC (2 g/day) + LAC (1 g/day)^{A}; Placebo^{B}	24-w intervention	18–65	↓ Motility^{A,B} ($p>0.05$) ↑ Total motility^{A,B} ($p>0.05$)

LC L-carnitine; *LAC* L-acetyl-carnitine; *m* month; *w* week; *TAC* total antioxidant capacity

[a]Group A had significantly higher pregnancy rate compared with group C ($p<0.01$), and group B had a significantly increased pregnancy rate compared with group A ($p<0.05$)

22.7.1 Critical Commentary

The findings of preliminary, uncontrolled studies suggest that oral carnitine supplementation improves sperm motility both in a quantitative and qualitative manner [51, 52]. This is in agreement with recent literature which reported that both L-carnitine (LC) and L-acetyl-carnitine (LAC) improved sperm motility parameters, especially in groups with lower baseline motility [54, 55]. A metabolic alteration has been postulated. Defective function in one of the intracellular carnitine-dependent systems may lead to a reduction in fatty-acid oxidation, thereby impairing energy-dependent sperm functions such as motility. Carnitine supplementation provides additional substrate for sperm energy metabolism and motility.

Carnitine's action to improve sperm parameters may differ according to baseline semen characteristics. Lenzi et al. were unable to demonstrate a change in seminal plasma levels in the carnitine arm [53], whereby baseline levels may have possibly not allowed, although unproven, detection of small but physiologically important increases in carnitine levels. Hence, the lack of improvement in semen parameters may in fact be due to no increase in seminal plasma or sperm carnitine levels.

LC is thought to work in a dose-dependent manner, as 2 g/day was reported to elevate sperm concentration, while a 3 g/day yielded no improvement. The lack of significant variation in sperm concentration following therapy excludes a direct positive effect on the spermatogenic process. The effects of carnitine may be post-testicular, as many reports do not account for any improvement in morphology [53, 56]. Carnitine supplementation in combination therapy with the antioxidant cinnoxicam was seen to improve sperm concentration, motility, and morphology in patients with iOAT [54].

Importantly, no side effects were reported in any subjects receiving LC/LAC treatment. In some cases, sperm parameters failed to improve. However, many of the studies reported elevated pregnancy rates as a result of the intervention. Therefore, the mechanism by which long-term combined carnitine treatment may improve fertilization capacity may not be directly evident from microscopic analysis [55]. Large-scale randomized controlled and dose-finding trials are necessary to confirm carnitine's effects on sperm characteristics. In vitro studies may help to further elucidate the compound's mechanism of action.

22.8 Lycopene

Lycopene is a naturally synthesized carotenoid-class molecule commonly found in fruits and vegetables. It is particularly useful in the human redox defense mechanisms against free radicals. Lycopene has been reported to have the highest quenching rate constant with singlet oxygen and to have a higher plasma level than that of beta-carotene [58]. Normally, it is found in high concentrations in the testes and seminal vesicles; however, decreased levels have been demonstrated in men suffering from infertility. Table 22.2 summarizes recent studies pertaining to lycopene intervention for improving sperm parameters.

Table 22.2 Quality assessment of oral lycopene supplementation

References	Study design	Cases	Dosage	Duration	Ages	Main outcome
Mohanty et al. [59]	Case–control	50	Lycopene (8 mg/day)	3-month intervention + 12-month follow-up	21–50	↑ Count (70%)[a] ↑ Concentration (60%)[a] ↑ Motility (54%)[a] ↑ SMI (46%)[a] ↑ Morphology (38%)[a] ↑ Pregnancy (36%)[a]
Gupta et al. [58]	Case–control	30	Lycopene (4 mg/day)	3-month intervention + follow-up	23–45	↑ Concentration (67%)[a] ↑ Motility (53%)[a] ↑ Morphology (40%)[a] ↑ Pregnancy (20%)[a]
Mendiola et al. [60]	Case–control	61 (30[A]; 31[B])	Dietary habits and nutrient consumption recorded[A,B]; higher intake of lycopene[B] ($p<0.05$)	–	29–38	Low intake of lycopene[A] is associated with ↓ Concentration ↓ Motility ↓ Morphology
Oborna et al. [61]	Double-blind, placebo-controlled, crossover	44 (18[A]; 26[B])	Lycopene (20 mg/day); Placebo	Lycopene (20 mg/day) followed by placebo[A]; Placebo followed by lycopene (20 mg/day)[B]	27–38	↓ sRAGE levels in seminal plasma ($p=0.012$)[A], ($p=0.008$)[B]

SMI sperm motility index; *sRAGE* soluble isoforms of receptor for advanced glycation end products

[a]Percent of patients

22.8.1 Critical Commentary

Lycopene is one of the most potent, highly lipophilic antioxidants. Low lycopene intake has been associated with a decrease in sperm concentration, motility, and morphology levels [60]. While its main action is antioxidative, it also has antiproliferative, immunomodulatory, and anti-inflammatory effects and influences on cell differentiation, communication, and signaling. The receptor for advanced glycation end products (RAGE) is a cell surface multiligand receptor that is activated by advanced glycation end products (AGEs). Activation of RAGE induces cellular response and results in OS [62]. Since OS is recognized as an important contributing factor to male infertility [63], the release of sRAGE may aid in the removal/neutralization of proinflammatory ligands. Lycopene has been shown to reduce seminal sRAGE levels [61], illustrating a potential role in treating ROS-associated male infertility. Lycopene's effects are most likely an antioxidative course of action or from an increased clearance of AGE ligand/sRAGE complexes.

Preliminary, controlled studies indicate a promising role for lycopene in treating ROS-associated male infertility. Lycopene supplementation has been shown to improve sperm count, concentration, motility, morphology, and pregnancy rates [59, 64]. Moreover, there were no reported side effects or complications from treatment. Future studies using larger sample sizes must be conducted to further investigate the role of lycopene supplementation in the treatment of iOAT.

22.9 *N*-Acetyl-Cysteine

N-acetyl-cysteine (NAC) is a metabolite of L-cysteine produced within the human body. It helps to synthesize GSH, one of the body's most important and powerful natural antioxidant and detoxification mechanisms. NAC increases GSH levels, subsequently alleviating OS by neutralizing free radicals. GSH is also known to aid in the transport of nutrients to lymphocytes and phagocytes, as well as protect cell membranes. NAC plays a vital role in germ cell survival in human seminiferous tubules. Several studies have indicated an optimistic role of NAC in improving semen parameters (Table 22.3).

22.9.1 Critical Commentary

NAC is an *N*-acetyl derivative of the naturally occurring amino acid L-cysteine [67]. Due to the antioxidant nature of NAC, it is thought that NAC supplementation may improve redox status in idiopathic male infertility and semen parameters.

Table 22.3 Quality assessment of oral *N*-acetyl-cysteine supplementation

References	Study design	Cases	Dosage	Duration	Ages	Main outcome
Paradiso Galatioto et al. [65]	Randomized controlled	42 (20^{A}; 22^{B})	NAC 600 mg/d + vitamin-minerals^{A}; Placebo^{B}	90-d intervention + 12-m follow-up	23–36	↑ Count $^{A}(p=0.009)$, $^{B}(p=0.1)$ ↑ Motile number $^{A}(p=0.752)$, $^{B}(p=0.976)$ ↑ Morphology $^{A}(p=0.926)$, $^{B}(p=0.833)$
Safarinejad and Safarinejad [45]	Double-blind, randomized controlled	468 (116^{A}; 118^{B}; 116^{C}; 118^{D})	Se (200 μg/d)^{A}; NAC (600 μg/d)^{B}; Se (200 μg/d) + NAC (600 μg/d)^{C}; Placebo^{D}	26-week intervention + 30-w treatment-free	25–48	↑ Count $^{A}(p=0.02)$, $^{B}(p=0.04)$, $^{C}(p=0.01)$, $^{D}(p=0.08)$ ↑ Concentration $^{A}(p=0.03)$, $^{B}(p=0.04)$, $^{C}(p=0.01)$, $^{D}(p=0.1)$ ↑ Motility $^{A}(p=0.03)$, $^{B}(p=0.07)$, $^{C}(p=0.02)$, $^{D}(p=0.1)$ ↑ Morphology $^{A}(p=0.03)$, $^{B}(p=0.03)$, $^{C}(p=0.03)$, $^{D}(p=0.1)$

(continued)

Table 22.3 (continued)

References	Study design	Cases	Dosage	Duration	Ages	Main outcome
Ciftci et al. [66]	Randomized controlled	120 (60^A; 60^B)	NAC 600 mg/d^A; Placebo^B	3-m intervention	33	↓ Concentration ^{A}(NS) ↑ Motility ^{A}($p<0.05$) ↑ Morphology ^{A}(NS) ↑ Volume ^{A}($p<0.05$) ↓ Viscosity ^{A}($p<0.001$) ↑ TAC ^{A}($p<0.001$) ↓ TP ^{A}($p<0.001$) ↓ OSI ^{A}($p<0.001$)

NAC *N*-acetyl-cysteine; *Se* selenium; *TAC* total antioxidant capacity; *TP* total peroxide; *OSI* oxidative stress index; *NS* not significant; *d* day; *m* month

Oxidative stress is thought to cause inflammation and fibrosis leading to the entanglement of spermatozoa. This is thought to prevent proper migration of sperm through to cervical tract fluids and the site of fertilization [68]. NAC's reducing nature is thought to lower these effects of ROS activity in semen and decrease semen viscosity. Human semen samples incubated with NAC (1.0 mg/mL) at room temperature exhibited improved total sperm motility and significantly reduced ROS [69]. Ciftci et al. demonstrated that sperm volume, viscosity, liquefaction time, and motility, as well as oxidative status in the plasma were significantly improved in the NAC-treated group compared to a control group [66]. Another study confirmed the effect of NAC administration to decrease seminal oxidative stress. However, no improvement in sperm motility was seen [70].

The safe and effective dosage of NAC has yet to be determined. An overdose of NAC has been to cause a minimal allergic reaction. No contraindications other than hypersensitivity have been demonstrated. In a study by Ciftci et al., no side effects were reported in any patients following 3 months of NAC (600 mg/day) treatment [66]. Taken together, the evidence presented suggests that NAC can act as an antioxidant to protect sperm from ROS damage by reducing the production of oxygen radicals. Nevertheless, its efficacy and clinical usefulness remain controversial. To fully understand its mechanism of action, additional studies should be performed on a molecular level. Randomized controlled trials to evaluate the safety and efficacy of NAC-combination therapy are essential to confirm its benefits and standardize therapeutic values.

22.10 Vitamins C and E

Vitamin C, also known as ascorbic acid (AA), is a water-soluble antioxidant necessary for normal tissue growth and repair in the body. Since AA is not manufactured endogenously, it must be incorporated in one's daily diet. AA is commonly found in a variety of fruits and vegetables. It is an essential antioxidant for blocking some of the oxidative damage inflicted by free radicals and is found at concentrations higher in seminal plasma than in serum [71, 72]. A positive correlation was noted between seminal plasma AA concentration and the percentage of morphologically normal spermatozoa [73]. Additionally, poor semen samples associated with OS have been found to contain significantly lowered AA concentrations [74]. These studies demonstrate the importance of AA in warding off the adverse effects of ROS.

On the other hand, vitamin E is a lipid-soluble vitamin. Its antioxidant properties are involved in protecting vitamin A and essential fatty acids from oxidation and preventing breakdown of tissues. Similarly to vitamin C, the body cannot generate vitamin E. Hence, it must be supplemented into a daily diet from corn, lentils, wheat, rice, or nuts. Vitamin E has family of eight isomers; α-tocopherol is the only form that is actively maintained in the human body, and thus, found in the largest quantities in the blood and tissues, specifically in cell membranes. It is believed to display defensive mechanisms by acting as a reducing agent to inhibit LPO, as well as elevating the activity of other antioxidants to aid in the scavenging of free radicals [75, 76].

Combination therapy of vitamins C and E has been hypothesized to be potentially effective in the treatment of idiopathic male infertility. Administration of hydrophilic and lipophilic antioxidants concurrently may allow for a synergistic effect in reducing the amount of peroxidative attack on spermatozoa. Recent studies assessing vitamins C and E on sperm parameters are summarized in Table 22.4.

22.10.1 Critical Commentary

Vitamin E is an essential antioxidant, located mainly in the cell membrane. It is thought to have a chief role in interrupting free radical cascade reactions, as well as acting to scavenge free radicals during univalent reduction of molecular oxygen during electron transport between complexes in the mitochondria. Sperm motility oxidation of the fatty acid bilayer by ROS will damage the mitochondrial sheath of spermatozoa, impairing their motility.

Studies show an increase in sperm motility and in vitro fertilizing potential in response to Vitamin E treatment [76, 77]. Although combined vitamin C and E therapy failed to improve conventional semen parameters during the treatment stage, prolonged abstinence time was found to significantly increase ejaculate volume, sperm count, sperm concentration, and the total number of motile spermatozoa [78]. These substantial improvements following a period of prolonged abstinence suggests that Vitamin C and E may have a synergistic effect, which enhances long-term reproductive success.

While in vitro studies have demonstrated that vitamin E has a protective effect on the sperm motility [79, 80], no significant improvement was seen in conventional semen analysis parameters in vivo [77]. Moreover, vitamin E supplementation had no effect on ROS levels in semen [77]. Conflicting findings among studies may be attributed to the fact that some authors used a chemiluminescent assay that measures both intracellular and extracellular ROS. Since vitamin E is more likely to be chain-breaking rather than a scavenging antioxidant, it would be expected to protect the membrane components without influencing ROS production. Despite the fact that no significant improvements in sperm parameters have been confirmed, Kessopoulou et al. believe the in vivo administration of vitamin E has the potential to act as a successful treatment in treating male infertility and warrants further evaluation [77]. Additional randomized controlled trials using patients with identified sperm abnormalities as well as a deeper understanding of the synergistic mode of action of vitamin C and E are essential to confirm these reports and ascertain a standardized intervention for effective treatment.

Table 22.4 Quality assessment of oral vitamins C and E supplementation

References	Study design	Cases	Dosage	Duration	Ages	Main outcome
Kessopoulou et al. [77]	Double-blind randomized crossover	30 (15^A; 15^B)	Vitamin E (600 mg/d) followed by placebo^A; Placebo followed by vitamin E (600 mg/d)^B	3-m intervention + 1-m wash-out + crossed-over to other treatment	32	↑ Zona binding test $(p<0.004)^{A,B}$
Suleiman et al. [76]	Double-blind randomized controlled	87 (52^A; 35^B)	Vitamin E (300 mg/d)^A; Placebo^B	6-m intervention	–	↑ Motility $^A(p<0.001)$, $^B(p<0.05)$ ↓ LPO $^A(p<0.001)$, $^B(p<0.001)$
Rolf et al. [78]	Double-blind randomized controlled	31 (15^A; 16^B)	Vitamin C (1,000 mg/d) + vitamin E (800 mg/d)^A; Placebo^B	56-d intervention	–	↑ Total count $^{A,B}(p<0.05)$ ↑ Concentration[a] $^{A,B}(p<0.05)$ ↑ Total motile sperm[a] $^{A,B}(p<0.05)$ ↑ Ejaculate volume[a] $^{A,B}(p<0.05)$

mg milligram; *m* month; *d* day; *h* hour; *LPO* lipid peroxidation

[a]While significant ($p<0.05$), results only appeared upon prolonged abstinence; treatment did not improve conventional semen parameters nor 24-h sperm survival rate

22.11 Selenium

Selenium (Se) is a trace mineral found in plants, as well as in some meat and seafood. Se is incorporated into proteins to form selenoproteins that are essential antioxidant enzymes. These antioxidant properties aid in preventing and diminishing cellular damage from free radicals. Studies have noted at least 25 selenoproteins in human body to help maintain normal sperm structure integrity [46].

Se has been suggested to be vital for proper testicular development, spermatogenesis, and spermatozoa motility and function [81]. Deficiency of Se has been linked to a loss of sperm motility, instability of the mitochondrial midpiece, and morphological abnormalities [46, 81]. The loss of motility may be a result of depletion in energy supply from mitochondrial instability.

Much controversy still remains concerning the exact mode by which Se eliminates oxidative damage to improve semen quality. It has been suggested that Se may be mediated by selenoenzymes, such as phospholipid hydroperoxide GSH-peroxidase (GSH-Px) or the sperm capsular selenoprotein GSH-Px, which are related to the production of functional spermatozoa [82]. The most recent studies are summarized in Table 22.5.

22.11.1 *Critical Commentary*

Selenium (Se) is an essential element that has a demonstrated role in normal testicular development, spermatogenesis, and spermatozoa function [81]. The incorporation of selenium into proteins has allowed for them to work in maintaining membrane integrity. Sperm capsular selenoproteins play a structural role in spermatozoa in the form of GSH-Px [86, 87], which is an effective hydroperoxide scavenger in the prevention of oxidative damage to spermatozoa [88, 89]. Iwanier et al. reported elevated GSH-Px activity in the plasma and red cells of selenium-supplemented patients [83]. However, the exact mechanism by which Se exerts its antioxidant effect on semen is unknown.

Several reports have indicated correlations between semen Se concentration and sperm parameters. Bleau et al. reported maximal sperm motility in semen samples with Se levels ranging between 50 and 69 ng/mL, while concentrations below and above this range resulted in a high incidence of asthenozoospermia [90]. Additionally, significantly lowered Se concentrations were found in tetrazoospermic than in normozoospermic men [91]. Hence, Se supplementation appears to be dose-dependent such that antioxidant GSH-Px activity would increase upon Se intake until the dose–response relationship achieved a plateau. The dosage required to achieving optimal Se levels to maximize antioxidant enzyme activity needs to be determined in order to use this treatment to reduce ROS levels and improve sperm quality and fertilization rates.

Table 22.5 Quality assessment of oral selenium supplementation

References	Study design	Cases	Dosage	Duration	Ages	Main outcome
Iwanier and Zachara [83]	Double-blind randomized controlled	33 (16^{A}; 17^{B})	Se-rich-yeast (200 μg/d)^{A}; sodium selenite (200 μg/d) mixed with baker's yeast^{B}	12w	19–38	↑ Whole blood, plasma and seminal fluid [Se] $^{A,B}(p<0.001)$
Scott et al. [84]	Double-blind randomized controlled	64 (16^{A}; 30^{B}; 18^{C})	Se (100 μg/d)^{A}; Se(100 μg/d) + vitamins A (1 mg/d), C (10 mg/d), and E(15 mg/d)^{B}; Placebo^{C}	6m	33	↑ Count $^{A,B,C}(p<0.243)$ ↑ Motility $^{A,B,C}(p<0.068)$ ↑ Plasma Se $^{A,B,C}(p<0.001)$
Hawkes and Turek [85]	Double-blind randomized controlled	12 (6^{A}; 6^{B})	Se (47 μg/d) for 21 days + Se (13 μg/d) ^{A} or Se (297 μg/d) ^{B} for 99 days	120d	–	↓ Plasma [Se] ^{A}(50%) ↑ Plasma [Se] ^{B}(40%)
Safarinejad and Safarinejad [45]	Double-blind randomized controlled	468 (116^{A}; 118^{B}; 116^{C}; 118^{D})	Se (200 μg/d)^{A}; NAC (600 μg/d)^{B}; Se (200 μg/d) + NAC (600 μg/d)^{C}; Placebo^{D}	26-w intervention + 30-w treatment-free	25–48	↑ Count $^{A,B,C}(p\leq 0.05)$ ↑ Motility $^{A,B,C}(p\leq 0.05)$ ↑ Morphology $^{A,C}(p\leq 0.05)$, $^{B}(p=0.07)$

Se selenium; *NAC* *N*-acetyl cysteine; *w* week; *m* month; *d* day

The mechanism by which Se induced its effects on sperm is still poorly understood and its effects on male fertility and semen quality in humans remain controversial. Some studies have reported no effect of Se supplementation at all [92, 93]. While supplementation was found to cause incremental increases in seminal fluid Se levels, spermatozoal quality characteristics showed no improvement [83]. However, combination therapy with Se, Vitamin E, and NAC has shown promising results [45, 84]. These improvements were most likely supplement-dependent as all parameters were seen to return to baseline values during the posttreatment period. Conflicting study results may be explained by differences in the baseline fertility status of control subjects, andrological history, methodological variations in study design, and semen analysis, as well as demographic characteristics. There is a need for larger studies to fully assess the potential side effects of Se supplementation. Additional studies are necessary to evaluate the route and proper state by which Se directs its neutralizing effects, its side effects, as well as to confirm the findings of improving sperm motility and fertility rates.

22.12 Nonsteroidal Anti-Inflammatory Drugs

Although most commonly known as treatment for pain relief and inflammation, nonsteroidal anti-inflammatory drugs (NSAIDs) have exhibited antioxidant effects in improving sperm quality and fertility in rabbits [94]. They function mainly through the reversible inhibition of cyclooxygenase, thereby inhibiting the production of prostaglandins and thromboxanes. Elevated prostaglandin levels have been found in men suffering from OAT [95]. Combination therapy of NSAIDs and carnitine, which are thought to both undergo similar mechanistic pathways, may facilitate the beneficial results of NSAIDs therapy by suppressing excess prostaglandin production. Table 22.6 reviews recent studies conducted with NSAIDs on sperm.

22.12.1 Critical Commentary

Although there is a paucity of information in the literature, preliminary reports suggest a promising role for NSAIDs in improving sperm quality. They are thought to stabilize lysosomal membranes, thereby partially preventing apoptosis [94, 97, 98]. Some evidence has shown that an increase in prostaglandin concentration in seminal plasma can inhibit spermatic function [98]. One study found success in treating men with iOAT and prostaglandin-F2 in seminal plasma by treating them with a NSAID called flubiprofen [97]. Cinnoxicam is a lipophilic NSAID that has been shown to enhance sperm concentration, motility, and morphology [96] and even lead to higher pregnancy rates when combined with LC and LAC [54]. Its fat-soluble nature facilitates lymphatic and prostatic absorption [97, 99]. No specific biochemical study has found a mechanism by which cinnoxicam elicits its actions, although

Table 22.6 Quality assessment of oral supplementation of nonsteroidal anti-inflammatory drugs (NSAIDs)

References	Study design	Cases	Dosage	Duration	Ages	Main outcome
Cavallini et al. [96]	Double-blind randomized controlled	156 (41^{A}; 61^{B}; 54^{C})	Surgery^{A}; Cinnoxicam (30 mg/4d)^{B}; Placebo (glycerine suppository 1×/4d)^{C}	12-m intervention	34–37	↑ Concentration^{B,a,b} ↑ Motility^{B,b} ↑ Morphology^{B,b}
Cavallini et al. [54]	Double-blind randomized controlled	130 (47^{A}; 39^{B}; 44^{C})	Placebo (starch tablet 2×500 mg/d) + (glycerine suppository 1×/4d)^{A}; LC (1×2 g/d) + LAC (500×2 mg/d) + glycerine suppository (1×/4d)^{B}; LC (1×2 g/d) + LAC (500×2 mg/d) + cinnoxicam suppository (1×30 mg/d)^{C}	6-m intervention + 3-m/6-m follow-up	28–40	↑ Concentration $^{B,C}(p<0.05)$ ↑ Motility $^{B,C}(p<0.05)$ ↑ Morphology $^{B,C}(p<0.05)$ ↑ Pregnancy $^{C}(p<0.01)$

LC L-carnitine; *LAC* L-acetyl-carnitine; *d* day; *mg* milligram; *m* months

[a]Significant differences in sperm concentrations ($p<0.001$), but not with replicates

[b]Significant increases with oligoasthenospermia associated with a grade III ($p<0.05$), but not in those with grade IV or V varicocele

it is known that NSAIDs inhibit ROS and protstaglandin synthesis. Preliminary results appear to be best following 4 months of intervention, while returning to baseline levels following therapy suspension [54, 96]. A balance is critical in regulating ROS concentration for proper sperm function; too high of concentrations inhibit sperm motility and modify sperm morphology; however, too low of concentrations downregulate sperm capacitation and acrosomal reactions [100]. For instance, one study revealed chronic treatment with NSAIDs at low doses to improve sperm quality and fertility [99], while in vitro model showed that high cinnoxicam concentrations in seminal plasma inhibit sperm motility by most likely lowering the ROS in seminal plasma to excessive low levels [97]. Additional studies are required to establish the proper dosage of cinnoxicam treatment.

Only a few minor side effects were reported—mild euphoria, mild epigastralgia, and nausea—but never resulted in therapy suspension [54]. Future randomized controlled trials should test the utility of other common NSAIDs such as aspirin and ibuprofen in the treatment of male infertility. LAC/LC + cinnoxicam suppositories reveal much potential as a reliable treatment. However, there is still not enough data to support the use of NSAIDs in the treatment of iOAT.

22.13 Pentoxifylline

Pentoxifylline is an oral supplement commonly used to improve blood flow in patients with circulation problems by decreasing the viscosity of blood. As a xanthine derivative, pentoxifylline acts as a competitive nonselective phosphodiesterase inhibitor, thereby raising intracellular cAMP. It has also shown an anti-inflammatory effect in neutralizing ROS by controlling the release of superoxide anions [101, 102]. This presumably reduces the amount of OS by downregulating the body's ability to initiate an inflammatory response. Table 22.7 summarizes the most recent studies conducted with pentoxifylline on improving sperm parameters.

22.13.1 Critical Commentary

Pentoxifylline is an oral supplement commonly used to improve blood flow in patients with circulation problems by decreasing the viscosity of blood. As a xanthine derivative, pentoxifylline acts as a competitive nonselective phosphodiesterase inhibitor, thereby raising intracellular cAMP. It has also shown an anti-inflammatory effect in neutralizing ROS by controlling the release of superoxide anions [101, 102]. This presumably reduces the amount of OS by downregulating the body's ability to initiate an inflammatory response.

Pentoxifylline's antioxidant nature has been studied for its role in improving sperm quality [101, 102]. It has been demonstrated to be effective in preserving sperm motility in vitro [103] and in improving semen parameters in vivo [105].

Table 22.7 Quality assessment of oral pentoxifylline supplementation

References	Study design	Cases	Dosage	Duration	Ages	Main outcome
Pang et al. [103]	Prospective	–	Pentoxifylline (3.6 mM)^{A};	1-h incubation + 24-h incubation	–	↑ VCL $^{A}(p<0.05)$ ↑ ALH $^{A}(p<0.05)$
Okada et al. [104]	Case–control	71 (15^{A*}; 35^{B*})	Pentoxifylline (300 mg/day); Pentoxifylline (1,200 mg/day)	4-m intervention + 4-m intervention (follow-up every 4w)	–	↑ Motility $^{A*,B*}(p<0.05)$ ↑ Mean curvilinear velocity $^{A*,B*}(p<0.05)$

m month; *w* week; *VCL* curve linear velocity; *ALH* amplitude of lateral sperm head displacement

^{A*}Sperm preparations generated detected ROS levels at steady state

^{B*}18/35 with asthenospermia whose sperm preparations failed to generate detectable ROS levels at steady state

In a study by Okada et al., asthenozoospermic patients with detectable steady state levels of ROS were seen to have lowered ROS production levels and preserved sperm motion parameters following pentoxifylline treatment [104]. A low dosage (300 mg/day) treatment was found to be relatively ineffective in comparison to a higher dosage (1,200 mg/day), which increased sperm motility and motion parameters. However, no improvement in pregnancy rate was seen to result from either treatment regimen. Another report found that pentoxifylline increased the curvilinear velocity, path velocity, and straight-line velocity in both normozoospermic and asthenozoospermic specimens, but did not modify the percentage of motile spermatozoa [106].

These studies may suggest that pentoxifylline may serve to enhance sperm motility, making it a potential therapeutic approach for treatment of idiopathic male infertility. Nevertheless, data remain inconclusive. Studies focusing on the pharmacokinetics of pentoxifylline are essential in understanding its mechanism of action. In addition, prospective, double-blind clinical trials with fecundity as the main outcome measure are necessary to validate the full effects of pentoxifylline.

22.14 Zinc

Zinc (Zn) is a ubiquitous trace element that protects proteins and enzymes against free radical attack. The Zn molecule in Zn-containing enzymes is thought to shield specific regions of the enzyme from being oxidized, and thus, preserving its stability and activity. Additionally, since Zn does not readily undergo redox reactions, it functions to prevent free radical formation by other highly reactive metals, such as copper and iron. Therefore, Zn therapy may be an effective treatment for ROS-associated iOAT. It may act to neutralize the effects of ROS, thereby protecting sperm function. Interestingly, fertile and subfertile men have shown significant differences in their seminal plasma Zn concentrations and sperm motility [107]. Other reports indicate similar results, as it was hypothesized for Zn to work through various mechanisms in preventing OS by virtue of its stabilizing effects as an antioxidant [108, 109]. The basis of these findings may suggest that Zn may contribute to fertility through its positive effect on spermatogenesis. Recent studies conducted with Zn sulfate and its efficacy in combinational treatment are summarized in Table 22.8.

22.14.1 Critical Commentary

Zn is a ubiquitous trace element that protects proteins and enzymes from free radical attack. Since Zn does not readily undergo oxidation-reduction reactions, it can function to prevent free radical formation by other highly reactive metals such as copper and iron. The Zn molecule in Zn-containing enzymes shields specific

Table 22.8 Quality assessment of oral zinc supplementation

References	Study design	Cases	Dosage	Duration	Ages	Main outcome
Wong et al. [110]	Double-blind randomized controlled	94 (22^{A}; 23^{B}; 24^{C}; 25^{D})	Folic acid (5 mg/day)^{A}; Zn sulfate (66 mg/day)^{B}; Zn sulfate (66 mg/day) + Folic acid (5 mg/day)^{C}; Placebo^{D}	26-week intervention	–	↑ Concentration $^{C}(p<0.05)$ ↑ Morphology $^{A,B}(p<0.05)$ ↑ Total normal count $^{C}(p<0.05)$
Omu et al. [107]	Case–control	45 (11^{A}; 12^{B}; 14^{C}; 8^{D})	Zn sulfate (400 mg/day)^{A}; Zn sulfate (400 mg/day) + vitamin E (20 mg/day)^{B}; Zn sulfate (400 mg/day) + vitamin E (20 mg/day) + vitamin C (10 mg/day)^{C}; nontherapy control^{D}	3-month intervention	35 ± 6^{A} 35 ± 1^{B} 34 ± 9^{C}	↑ Motility $^{A,B,C}(p<0.001)$ ↑ Fertilizing capacity $^{A,B,C}(p<0.05)$ ↑ Total antioxidant capacity $^{A,B,C}(p<0.001)$ ↑ Seminal Bcl-2 $^{A,B,C}(p<0.05)$ ↓ Bax $^{A,B,C}(p<0.01)$ ↓ MDA $^{A,B,C}(p<0.01)$ ↓ TNF-α $^{A,B,C}(p<0.001)$

Zn zinc; *MDA* malone dialdehyde

regions of the enzyme from being oxidized, thereby preserving enzymatic stability and activity.

For such reasons, Zn therapy is thought to be an effective treatment for ROS-associated male infertility. Interestingly enough, fertile and subfertile men have shown significant differences in their seminal plasma Zn concentrations and sperm motility [107, 111]. Zn is hypothesized to exert its effects via numerous pathways to prevent OS formation, apoptosis, and sperm DNA fragmentation by virtue of its stabilizing nature as an antioxidant. Chia et al. noted seminal plasma zinc concentration to be significantly correlated with sperm density, motility, and viability [111]. These findings suggest that Zn may promote male fertility via its positive effect on critical steps in spermatogenesis.

Wong et al. reported an increase in total normal sperm count in both fertile and subfertile men following combination treatment with Zn sulfate (66 mg/day) and folic acid (5 mg/day) [110]. Hence, Zn and folic acid may work in a synergistic manner to protect spermatozoa. There is a need for additional randomized, placebo-controlled trials with larger sample sizes as well as various dosages and intervention periods to confirm the efficacy and safety of Zn and folic acid combination therapy. Establishing the beneficial effects of Zn treatment on fertility may aid in a therapeutic approach for treating iOAT.

22.15 Conclusion

Regulated levels of ROS are essential in physiologically regulating normal sperm function. However, in an environment with uncontrolled, elevated ROS levels, OS ensues and sperm function and viability are endangered. OS resulting from excessive production of ROS, impaired antioxidant defense mechanisms, or both precipitates in a wide range of pathologies that are currently believed to adversely affect sperm quality. Despite the established role of OS in the pathogenesis of male infertility, there is a lack of consensus as to the clinical utility of seminal OS testing in an infertility clinic. A major reason for this disconnect is related to the weakly defined standard protocol for assessing seminal OS.

Nevertheless, antioxidant therapies have illustrated promising results in improving the semen parameters of subfertile men suffering from iOAT. They have become the most widespread utilized and studied novel therapy for treating male factor infertility. However, the proper dosage, type, and duration of antioxidant treatment for clinicians to administer have yet to be confirmed and standardized. Many studies point to improvements in just one or two of the three parameters of iOAT, failing to fully address the disease as a whole. Additionally, safety becomes the primary concern, as high dosages of antioxidant therapy are capable of resulting in adverse effects. It has not been established whether antioxidant therapy is the proper management in cases of elevated ROS production, because intracellular sperm antioxidant status, abstinence time, sperm count, as well as other confounding factors must be considered. Since there are no reliable, predictive, and inexpensive methods in determining the extent of ROS exposure and antioxidant capacity in patients, further advances in this area may prove valuable for assigning ROS values to serve as potential indicators of the correct antioxidant therapy to prescribe. Moreover, if ROS exposure values are established, this may help to establish levels in which antioxidant treatment may be administered. Since the liberation of transition-metal ions from metalloproteins during a prooxidative state is thought to serve as catalysts for free radical damage, especially in the reduced state, antioxidant treatment may even enhance oxidation damage. The body faces an "antioxidant paradox" in which the administration of a potent antioxidant to reduce a prooxidative state can worsen conditions. Therefore, establishing "cut-off" values of ROS in which an antioxidant can be used will help to control ROS damage and determine therapeutic values necessary for treating iOAT.

Men with iOAT are often administered a number of various therapies based on experimental studies, yet their supporting evidence in controlled human studies is sparse. Thus, in the absence of approved and effective treatment, medications prescribed are based solely on rationale. Assessment of OS status may also help in selecting the patient population that would most benefit from antioxidant supplementation. Further studies in this area are essential. Antioxidant treatment holds for a promising future as a conservative, inexpensive remedy in treating infertility worldwide.

References

1. Tournaye H. Evidence-based management of male subfertility. Curr Opin Obstet Gynecol. 2006;18(3):253–9.
2. Attia AM, Al-Inany HG, Farquar C, Proctor M. Gonadotrophins for idiopathic male factor male subfertility. Cochrane Database Syst Rev. 2007;(4):CD005071.
3. Sharlip ID, Jarow JP, Belker AM, Lipshultz LI, Sigman M, Thomas AJ, Schlegel PN, Howards SS, Nehra A, Damewood MD, Overstreet JW, Sadovsky R. Best practice policies for male infertility. Fertil Steril. 2002;77(5):873–82.
4. Tremellen K. Oxidative stress and male infertility—a clinical perspective. Hum Reprod Update. 2008;14(3):243–58.
5. Aitken RJ. Molecular mechanisms regulating human sperm function. Mol Hum Reprod. 1997;3(3):169–73.
6. Aitken RJ. The Amoroso lecture. The human spermatozoon—a cell in crisis? J Reprod Fertil. 1999;115(1):1–7.
7. Aitken RJ, De Iuliis GN. Origins and consequences of DNA damage in male germ cells. Reprod Biomed Online. 2007;14(6):727–33.
8. Alvarez JG. Nurture vs nature: how can we optimize sperm quality? J Androl. 2003; 24(5):640–8.
9. Bykova M, Athayde K, Sharma R, Jha R, Sabanegh E, Agarwal A. Defining the reference value of seminal reactive oxygen species in a population of infertile men and normal healthy volunteers. Fertil Steril. 2007;88 Suppl 1(P-597):305.
10. Eskenazi B, Kidd SA, Marks AR, Sloter E, Block G, Wyrobek AJ. Antioxidant intake is associated with semen quality in healthy men. Hum Reprod. 2004;20(4):1006–12.
11. Reddy SV, Suchitra MM, Reddy YM, Reddy PE. Beneficial and detrimental actions of free radicals: a review. J Global Pharma Technol. 2010;2(5):3–11.

12. Agarwal A, Prabakaran SA. Mechanism, measurement, and prevention of oxidative stress in male reproductive physiology. Indian J Exp Biol. 2005;43(11):963–74.
13. Aitken RJ. Free radicals, lipid peroxidation and sperm function. Reprod Fertil Dev. 1995;7(4):659–68.
14. Iwasaki A, Gagnon C. Formation of reactive oxygen species in spermatozoa of infertile patients. Fertil Steril. 1992;57(2):409–16.
15. Zini A, de Lamirande E, Gagnon C. Reactive oxygen species in semen of infertile patients: levels of superoxide dismutase- and catalase-like activities in seminal plasma and spermatozoa. Int J Androl. 1993;16(3):183–8.
16. Cavallini G. Male idiopathic oligoasthenoteratozoospermia. Asian J Androl. 2006; 8(2):143–57.
17. Hirsch A. ABC of subfertility: male subfertility. BMJ. 2003;327(7416):669–72.
18. Bonanomi N, Lucente G, Silvestrini B. Male fertility: core chemical structure in pharmacological research. Contraception. 2002;65(4):317–20.
19. Eskenazi B, Wyrobeck AJ, Sloter E, Kidd SA, Moore L, Young S, Moore D. The association of age and semen quality in healthy men. Hum Reprod. 2003;18(2):447–54.
20. Kefer JC, Agarwal A, Sabanegh E. Role of antioxidants in the treatment of male infertility. Int J Urol. 2009;16(5):449–57.
21. Fraczek M, Sanoka D, Kamieniczna M, Kurpisz M. Proinflammatory cytokines as an intermediate factor enhancing lipid sperm membrane peroxidation in in vitro conditions. J Androl. 2008;29(1):85–92.
22. Saleh RA, Agarwal A. Oxidative stress and male infertility: from research bench to clinical practice. J Androl. 2002;23(6):737–52.
23. Pasqualotto FF, Sharma RK, Nelson DK, Thomas AJ, Agarwal A. Relationship between oxidative stress, semen characteristics, and clinical diagnosis in men undergoing infertility investigation. Fertil Steril. 2000;73(3):459–64.
24. Fisher H, Aitken RJ. Comparative analysis of the ability of precursor germ cells and epididymal spermatozoa to generate reactive oxygen metabolites. J Exp Zool. 1997;277(5): 390–400.
25. Younglai E, Collins JA, Foster WG. Canadian semen quality: an analysis of sperm density among eleven academic fertility centers. Fertil Steril. 1998;70(1):76–80.
26. Bujan L, Mansat A, Pontonnier F, Mieusset R. Time series analysis of sperm concentration in fertile men in Toulouse (France) between 1977 and 1992. BMJ. 1996;312(7029):471–2.
27. St. John JC, Cooke ID, Barratt CLR. Detection of multiple deletions in the mitochondrial DNA of human testicular tissue from azoospermic and severe oligozoospermic patients. In: Barratt CLR, De Jonge C, Mortimer D, Parinaud J, editors. Genetics of male fertility. Paris: Inserm EDK; 1997. p. 333–47.
28. Nakada K, Sato A, Yoshida K, Morita T, Tanaka H, Inoue S, Yonekawa H, Hayashi J. Mitochondria-related male infertility. Proc Natl Acad Sci USA. 2006;103(41):15148–53.
29. Suzuki H, Kumagai T, Goto A, Sugiura T. Increase in intracellular hydrogen peroxide and upregulation of nuclear respiratory gene evoked by impairment of mitochondrial electron transfer in human cells. Biochem Biophys Res Commun. 1998;249(2):542–5.
30. Wang X, Sharma RK, Gupta A, George V, Thomas AJ, Falcone T, Agarwal A. Alterations in mitochondria membrane potential and oxidative stress in infertile men: a prospective observational study. Fertil Steril. 2003;80 Suppl 2:844–50.
31. Liu CY, Lee CF, Hong CH, Wei YH. Mitochondrial DNA mutation and depletion increase the susceptibility of human cells to apoptosis. Ann NY Acad Sci. 2004;1011:133–45.
32. Selvi Rani D, Vanniarajan A, Gupta N, Chakravarty B, Singh L, Thangaraj K. A novel missense mutation C11994T in the mitochondrial ND4 gene as a cause of low sperm motility in the Indian subcontinent. Fertil Steril. 2006;86(6):1783–5.
33. Armstrong JS, Rajasekaran M, Chamulitrat W, Gatti P, Hellstrom WJ, Sikka SC. Characterization of reactive oxygen species induced effects on human spermatozoa movement and energy metabolism. Free Radic Biol Med. 1999;26(7–8):869–80.

34. de Lamirande E, Gagnon C. Impact of reactive oxygen species on spermatozoa: a balancing act between beneficial and detrimental effects. Hum Reprod. 1995;10 Suppl 1:15–21.
35. Aitken RJ, Fisher H, Fulton N, Gomez E, Knox W, Lewis B, Irvine S. Reactive oxygen species generation by human spermatozoa is induced by exogenous NADPH and inhibited by flavoprotein inhibitors diphenylene iodinium and quinacrine. Mol Reprod Dev. 1997;47(4):468–82.
36. Griveau JF, Dumont E, Renard B, Callegari JP, Lannou DL. Reactive oxygen species, lipid peroxidation and enzymatic defense systems in human spermatozoa. J Reprod Fertil. 1995;103(1):17–26.
37. Aitken RJ, West KM. Analysis of the relationship between reactive oxygen species production and leukocyte infiltration in fractions of human semen separated on Percoll gradients. Int J Androl. 1990;13(6):433–51.
38. Huszar G, Sbracia M, Vigue L, Miller DJ, Shur BD. Sperm plasma membrane remodeling during spermiogenic maturation in men: relationship among plasma membrane beta 1,4-galactosyltransferase, cytoplasmic creatine phosphokinase and creatine phosphokinase isoform ratios. Biol Reprod. 1997;56(4):1020–4.
39. Ollero M, Gil-Guzman E, Lopez MC, Sharma RK, Agarwal A, Larson K, Evenson D, Thomas Jr AJ, Alvarez JG. Characterization of subsets of human spermatozoa at different stages of maturation: implications in the diagnosis and treatment of male infertility. Hum Reprod. 2001;16(9):1912–21.
40. Gil-Guzman E, Ollero M, Lopez MC, Sharma RK, Alvarez JG, Thomas Jr AJ, Agarwal A. Differential production of reactive oxygen species by subsets of human spermatozoa at different stages of maturation. Hum Reprod. 2001;16(9):1922–30.
41. Buettner GR. The pecking order of free radicals and antioxidants, lipid peroxidation, alpha-tocopherol and ascorbate. Arch Biochem Biophys. 1993;300(2):535–43.
42. Garrido N, Meseguer M, Simon C, Pellicer A, Remohi J. Pro-oxidative and anti-oxidative imbalance in human semen and its relation with male fertility. Asian J Androl. 2004; 6(1):59–65.
43. Young IS, Woodside JV. Antioxidants in health and disease. J Clin Pathol. 2001; 54(3):176–86.
44. Mahfouz R, Sharma R, Sharma D, Sabanegh E, Agarwal A. Diagnostic value of the total antioxidant capacity (TAC) in human seminal plasma. Fertil Steril. 2009;91(3):805–11.
45. Safarinejad MR, Safarinejad S. Efficacy of selenium and/or N-acetyl-cysteine for improving semen parameters in infertile men: a double-blind, placebo controlled, randomized study. J Urol. 2009;181(2):741–51.
46. Altman DG, Schultz KF, Moher D, Egger M, Davidoff F, Elbourne D, Gotzsche PC, Land T, CONSOT GROUP (Consolidated Standards of Reporting Trials). The revised CONSORT statement for reporting randomized trials: explanation and elaboration. Ann Intern Med. 2001;134(8):663–94.
47. Agarwal A, Said TM. Carnitines and male infertility. Reprod Biomed Online. 2004; 8(4):376–84.
48. Ross C, Morriss A, Khairy M, Khalaf Y, Braude P, Coomarasamy A, El-Toukhy T. A systematic review of the effect of oral antioxidants on male infertility. Reprod Biomed Online. 2010;20(6):711–23.
49. Golan R, Weissenberg R, Lewin LM. Carnitine and acetylcarnitine in motile and immotile human spermatozoa. Int J Androl. 1984;7(6):484–94.
50. Jeulin C, Lewin LM. Role of free L-carnitine and acetyl-L-carnitine in post-gonadal maturation of mammalian spermatozoa. Hum Reprod Update. 1996;2(2):87–102.
51. Costa M, Canale D, Filicori M, D'Iddio S, Lenzi A. L-carnitine in idiopathic asthenozoospermia: a multicenter study. Italian Study Group on Carnitine and Male Infertility. Andrologia. 1994;26(3):155–9.
52. Vitali G, Parente R, Melotti C. Carnitine supplementation in human idiopathic asthenospermia: clinical results. Drugs Exp Clin Res. 1995;21(4):157–9.

53. Lenzi A, Lombardo F, Sgro P, Salacone P, Caponecchia L, Dondero F, Gandini L. Use of carnitine therapy in selected cases of male factor infertility: a double-blind crossover trial. Fertil Steril. 2003;79(2):292–300.
54. Cavallini G, Ferraretti AP, Gianaroli L, Biagiotti G, Vitali G. Cinnoxicam and L-carnitine/acetyl-L-carnitine treatment for idiopathic and varicocele-associated oligoasthenospermia. J Androl. 2004;25(5):761–70.
55. Lenzi A, Sgrò P, Salacone P, Paoli D, Gilio B, Lombardo F, Santulli M, Agarwal A, Gandini L. A placebo-controlled double-blind randomized trial of the use of combined L-carnitine and L-acetyl-carnitine treatment in men with asthenozoospermia. Fertil Steril. 2004; 81(6):1578–84.
56. Balercia G, Regoli F, Armeni T, Koverech A, Manero F, Boscaro M. Placebo-controlled double-blind randomized trial on the use of L-carnitine, L-acetylcarnitine, or combined L-carnitine and L-acetylcarnitine in men with idiopathic asthenozoospermia. Fertil Steril. 2005;84(3):662–71.
57. Sigman M, Glass S, Campagnone J, Pyor JL. Carnitine for the treatment of idiopathic asthenospermia: a randomized, double-blind, placebo-controlled trial. Fertil Steril. 2006;85(5): 1409–14.
58. Gupta NP, Kumar R. Lycopene therapy in idiopathic male infertility- a preliminary report. Int Urol Nephrol. 2002;34(3):369–72.
59. Mohanty NK, Sujit K, Jha AK, Arora RP. Management of idiopathic oligoasthenospermia with lycopene. Indian J Urol. 2001;18(1):57–61.
60. Mendiola J, Torres-Cantero AM, Vioque J, Moreno-Grau JM, Ten J, Roca M, Moreno-Grau S, Bernabeu R. A low intake of antioxidant nutrients is associated with poor semen quality in patients attending fertility clinics. Fertil Steril. 2010;93(4):1128–33.
61. Oborna I, Malickova K, Fingerova H, Brezinova J, Horka P, Novotny J, Bryndova H, Filipcikova R, Svobodova M. A randomized controlled trial of lycopene treatment on soluble receptor for advanced glycation end products in seminal and blood plasma of normospermic men. Am J Reprod Immunol. 2011;66(3):179–84. doi:10.1111/j.1600-0897.2011.00984.x.
62. Wertz K, Siler U, Goralczyk R. Lycopene: modes of action to promote prostate health. Arch Biochem Biophys. 2004;430:127–34.
63. Goldin A, Beckman JA, Schmidt AM, Creager MA. Advanced glycation end products: sparking the development of diabetic vascular injury. Circulation. 2006;114(6):597–605.
64. Aziz N, Novotny J, Oborna I, Fingerova H, Brezinova J, Svobodova M. Comparison of chemiluminescence and flow cytometry in the estimation of reactive oxygen and nitrogen species in human semen. Fertil Steril. 2010;94(7):2604–8.
65. Paradiso Galatioto G, Gravina GL, Angelozzi G, Sacchetti A, Innominato PF, Pace G, Ranieri G, Vicentini C. May antioxidant therapy improve sperm parameters of men with persistent oligospermia after retrograde embolization for varicocele? World J Urol. 2008;26(1):97–102.
66. Ciftci H, Verit A, Savas M, Yeni E, Erel O. Effects of N-acetylcysteine on semen parameters and oxidative/antioxidant status. Urology. 2009;74(1):73–6.
67. Aitken RJ, Buckingham D, Harkiss D. Use of a xanthine oxidase oxidant generating system to investigate the cytotoxic effects of reactive oxygen species on human spermatozoa. J Reprod Fertil. 1993;97(2):441–50.
68. Elzanaty S, Malm J, Giwercman A. Visco-elasticity of seminal fluid in relation to the epididymal and accessory sex gland function and its impact on sperm motility. Int J Androl. 2004;27(2):94–100.
69. Oeda T, Henkel R, Ohmori H, Schill WB. Scavenging effect of N-acetyl-L-cysteine against reactive oxygen species in human semen: a possible therapeutic modality for male factor infertility? Andrologia. 1997;29(3):125–31.
70. Akiyama M. In vivo scavenging effect of ethylcysteine on reactive oxygen species in human semen. Nihon Hinyokika Gakkai Zasshi. 1999;90(3):421–8.
71. Dawson EB, Harris WA, Rankin WE, Charpentier LA, McGanity WJ. Effect of ascorbic acid on male fertility. Ann N Y Acad Sci. 1987;498:312–23.

72. Thiele JJ, Friesleben HJ, Fuchs J, Ochsendorf FR. Ascorbic acid and urate in human seminal plasma: determination and interrelationships with chemiluminescence in washed semen. Hum Reprod. 1995;10(1):110–5.
73. Lewis SE, Sterling ES, Young IS, Thompson W. Comparison of individual antioxidants of sperm and seminal plasma in fertile and infertile men. Fertil Steril. 1997;67(1):142–7.
74. Ehrenkranz RA. Vitamin E and the neonate. Am J Dis Child. 1980;134(12):1157–68.
75. Palamanda JR, Kehrer JP. Involvement of vitamin E and protein thiols in the inhibition of microsomal lipid peroxidation by glutathione. Lipids. 1993;28(5):427–31.
76. Suleiman SA, Ali ME, Zaki ZM, el-Malik EM, Nasr MA. Lipid peroxidation and human sperm motility: protective role of vitamin E. J Androl. 1996;17(5):530–7.
77. Kessopoulou E, Powers HJ, Sharma KK, Pearson MJ, Russell JM, Cooke ID, Barratt CL. A double-blind randomized placebo cross-over controlled trial using the antioxidant vitamin E to treat reactive oxygen species associated male infertility. Fertil Steril. 1995;64(4):825–31.
78. Rolf C, Cooper TG, Yeung CH, Nieschlag E. Antioxidant treatment of patients with asthenozoospermia or moderate oligoasthenozoospermia with high-dose vitamin C and vitamin E: a randomized, placebo-controlled, double-blind study. Hum Reprod. 1999;14(4):1028–33.
79. Aitken RJ, Clarkson JS. Significance of reactive oxygen species and antioxidants in defining the efficacy of sperm preparation techniques. J Androl. 1988;9(6):367–76.
80. de Lamirande E, Gagnon C. Reactive oxygen species and human spermatozoa. I. Effects on the motility of intact spermatozoa and on sperm axonemes. J Androl. 1992;13(5):368–78.
81. Ursini F, Heim S, Kiess M, Maiorino M, Roveri A, Wissing J, Flohe L. Dual function of the selenoprotein PHGPx during sperm maturation. Science. 1999;285(5432):1393–6.
82. Comhaire FH, Christophe AB, Zalata AA, Dhooge WS, Mahmoud AM, Depuydt CE. The effects of combined conventional treatment, oral antioxidants and essential fatty acids on sperm biology in subfertile men. Prostaglandins Leukot Essent Fatty Acids. 2000;63(3):159–65.
83. Iwanier K, Zachara BA. Selenium supplementation enhances the element concentration in blood and seminal fluid but does not change the spermatozoal quality characteristics in subfertile men. J Androl. 1995;16(5):441–7.
84. Scott R, Macpherson A, Yates RW, Hussain B, Dixon J. The effect of oral selenium supplementation on human sperm motility. Br J Urol. 1998;82(1):76–80.
85. Hawkes WC, Turek PJ. Effects of dietary selenium on sperm motility in healthy men. J Androl. 2001;22(5):764–72.
86. Surai PF, Blesbois E, Grasseau I, Chalah T, Brillard JP, Wishart GJ, Cerolini S, Sparks NH. Fatty acid composition, glutathione peroxidase and superoxide dismutase activity and total antioxidant activity of avian semen. Comp Biochem Physiol B Biochem Mol Biol. 1998;120(3):527–33.
87. Kleene KC. The mitochondrial capsule selenoprotein—a structural protein in the mitochondrial capsule of mammalian sperm. In: Burk RF, editor. Selenium in biology and human health. New York: Springer; 1994. p. 134–49.
88. Maddipati KR, Marnett LJ. Characterization of the major hydroperoxide-reducing activity of human plasma: purification and properties of a selenium-dependent glutathione peroxidase. J Biol Chem. 1987;262(36):17398–403.
89. Brown DG, Burk RF. Selenium retention in tissue and sperm of rats fed a Torula yeast diet. J Nutr. 1973;103(1):102–8.
90. Bleau G, Lemabre J, Faucher G, Roberts KD, Chapdelaine A. Semen selenium and human fertility. Fertil Steril. 1984;42(6):890–4.
91. Saaranen M, Suistomaa U, Vanha-Perttula T. Semen selenium content and sperm mitochondrial volume in human and some animal species. Hum Reprod. 1989;4(3):304–8.
92. Roy AC, Karunanithy R, Ratman SS. Lack of correlation of selenium level in human semen with sperm count/motility. Arch Androl. 1990;25(1):59–62.
93. Behne D, Gessner H, Wolters G, Brotherton J. Selenium, rubidium and zinc in human semen and semen fractions. Int J Androl. 1988;11(5):415–23.
94. Bendvold E, Gottlieb C, Svanborg K, Bygdeman M, Eneroth P. Concentration of prostaglandins in seminal fluid of fertile men. Int J Androl. 1987;10(2):463–9.

95. Mangano NG, Sabella P, Mangano A. In vitro effects of L-carnitine on the inhibition of sperm motility induced by nonsteroidal anti-inflammatory drug. Clin Ther. 2000;151(3):353–64.
96. Cavallini G, Biagiotti G, Ferraretti AP, Gianaroli L, Vitali G. Medical therapy of oligoasthenospermia associated to left varicocele: an option. BJU Int. 2003;91(6):513–8.
97. Ito H, Fuse H, Minagawa H, Kawamura K, Marukami M, Shimazaky J. Internal spermatic vein prostaglandins in varicocele patients. Fertil Steril. 1982;37(2):218–22.
98. Fuse H, Minagawa H, Ito H, Shimazaki J. The effects of prostaglandin synthetase inhibitor on male infertility. Hinyokika Kiyo. 1984;30(10):1439–45.
99. Loescher W, Littgenau H, Schlegel W, Kruger S. Pharmacokinetics of non-steroidal anti-inflammatory drugs in male rabbits after acute and chronic administration and effect of chronic treatment on seminal prostaglandins, sperm quality and fertility. J Reprod Fertil. 1988;82(1):353–64.
100. Ochsendorf FR. Infections in the male genital tract and reactive oxygen species. Hum Reprod Update. 1999;5(5):399–420.
101. Gavella M, Lipovac V, Marotti T. Effect of pentoxifylline on superoxide anion production by sperm. Int J Androl. 1991;14(5):320–7.
102. Gavella M, Lipovac V. Pentoxifylline-mediated reduction of superoxide anion production by human spermatozoa. Andrologia. 1992;24(1):37–9.
103. Pang SC, Chan PJ, Lu A. Effects of pentoxifylline on sperm motility and hyperactivation in normozoospermic and normokinetic semen. Fertil Steril. 1993;60(2):336–43.
104. Okada H, Tatsumi N, Kanzaki M, Fujisawa M, Arakawa S, Kamidono S. Formation of reactive oxygen species by spermatozoa from asthenozoospermic patients: response to treatment with pentoxifylline. J Urol. 1997;157(6):2140–6.
105. Yovich JM, Edirisinghe WR, Cummins JM, Yovich JL. Influence of pentoxifylline in severe male factor infertility. Fertil Steril. 1990;53(4):715–22.
106. Tesarik J, Thebault A, Testart J. Effect of pentoxifylline on sperm movement characteristics in normozoospermic and asthenozoospermic specimens. Hum Reprod. 1992;7(9):1257–63.
107. Omu AE, Al-Azemi MK, Kehinde EO, Anim JT, Oriowo MA, Mathew TC. Indications of the mechanisms involved in improved sperm parameters by zinc therapy. Med Princ Pract. 2008;17(2):108–16.
108. Omu AE, Dahti H, Al-Othman S. Treatment of asthenozoospermia with zinc sulphate: andrological, immunological and obstetric outcome. Eur J Obstet Gynecol Reprod Biol. 1998; 79(2):179–84.
109. Landau B, Singer R, Klein T, Segenreich E. Folic acid levels in blood and seminal plasma of normo- and oligospermic patients prior and following folic acid treatment. Experientia. 1978;34(10):1301–2.
110. Wong WY, Merkus HM, Thomas CM, Menkveld R, Zielhuis GA, Steegers-Theunissen RP. Effects of folic acid and zinc sulphate on male factor subfertility: a double-blinded, randomized, placebo-controlled trial. Fertil Steril. 2002;77(3):491–8.
111. Chia SE, Ong CN, Chua LH, Ho LM, Tay SK. Comparison of zinc concentrations in blood and seminal plasma and the various sperm parameters between fertile and infertile men. J Androl. 2000;21(1):53–7.

Chapter 23
Leukocytospermia and Oxidative Stress

Margot Flint, Ashok Agarwal, and Stefan S. du Plessis

Abstract The invasion of microorganisms and infective bacteria in the genito-urinary tract leads to the rapid increase in white blood cells, a condition referred to as leukocytospermia. This inflammatory response, aimed at killing the microorganisms via the production and release of reactive oxygen species (ROS), can result in pathologically high concentrations of ROS. When these concentrations greatly exceed the level required for normal physiological function, the natural defense system of scavenging antioxidants can be overwhelmed, resulting in oxidative stress (OS) thereby compromising the integrity of spermatozoa and functional parameters vital for successful fertilization. The complexity of OS is furthered when additional factors (e.g., smoking, varicocele) increase ROS levels in the male genito-urinary system. At present, the association between semen parameters and leukocyte concentrations is a focal point in the field of male reproductive science. This chapter aims at exploring the relationship between leukocytospermia, OS, the harmful effects on male reproductive potential, as well as possible treatment regimes.

Keywords Leukocytospermia • Oxidative stress • Male genitalia tract infections • Reactive oxygen species • Leukocytes • Antioxidants

M. Flint, MSc • S.S. du Plessis, PhD, MBA (✉)
Division of Medical Physiology, Stellenbosch University, Francie van Zijl Avenue, PO Box 19063, Tygerberg 7505, South Africa
e-mail: ssdp@sun.ac.za

A. Agarwal, PhD
Center for Reproductive Medicine, Cleveland Clinic, Lerner College of Medicine, 9500 Euclid Avenue, Cleveland, OH 44195, USA

A. Agarwal et al. (eds.), *Studies on Men's Health and Fertility*, Oxidative Stress in Applied Basic Research and Clinical Practice, DOI 10.1007/978-1-61779-776-7_23,

23.1 Introduction

Various infective pathological conditions can result in the defensive reaction of tissues to the attack of microorganisms. An infection leads to the activation of specific immune cells, in particular a form of white blood cells (WBC's) known as leukocytes. The result which arises from such defensive action is the generation of reactive oxygen species (ROS). These are chemically reactive molecules containing oxygen and are produced by the invading leukocytes. The defensive oxidative pathway that the ROS take in killing microbes can have a biopositive effect when ROS concentrations are maintained at a low level or are counterbalanced by the protective antioxidant scavenging system which maintains homeostasis [1] In certain conditions whereby homeostasis is disrupted, an imbalance arises between oxidants and antioxidants, in favor of the former, resulting in a condition known as oxidative stress (OS) [1, 2].

The harmful effect of this imbalance is OS-induced damage, which is a threat to all cellular elements, including: amino acids, carbohydrates, lipids, and nucleic acids [3]. OS is considered as the circumstance which underlies the etiology of various human conditions [4]. In context of this, a wide variety of diseases exist in which the pathophysiological role of ROS and OS has been implicated in the pathogenesis of the condition. Examples of these include: cancer, diabetes, and inflammatory bowel disease [3].

In the male reproductive system, the harmful effect that ROS can have on sperm and their parameters has been known for a relatively long period of time. In the mid-1980s, Professor John Aitken and his group pioneered studies into the activities of ROS in male reproductive biology [3], which has continued comprehensively over the following years, in particular the Cleveland Clinic Foundation in America which has extensively researched the effects of OS in male infertility.

Research into the negative influence that leukocytospermia can have on semen parameters introduced investigations into the role of leukocyte-produced ROS. ROS are spontaneously generated and required at a basal level for certain spermatozoal physiological functions [5]. It has been proven that samples considered being peroxidase-positive have higher concentrations of ROS [6] and these polymorphonucleated leukocytes (PMNL) release oxygen radicals, such as hydrogen peroxide and superoxide which are known toxic factors towards spermatozoa. Irrelevant of the concentration of leukocytes in semen, the presence of these WBC's have been shown to be associated with OS which can negatively effect semen parameters such as sperm concentration and morphology [7].

23.2 Male Genital Tract Infections

Specific seminal parameters and chemical components which contribute towards the ejaculate can serve as diagnostic tools in assessing if the accessory sex glands are normally functioning [8]. Diverse irregularities or deviations from the standard reference values of these physical characteristics indicate a possible underlying pathophysiological condition [9].

The response of the genito-urinary tract to the invasion of microorganisms and inflammation is an important component of the immune defense system [10]. The reaction is considered to be extremely similar to the reaction in other body sites [11]. Therefore, a semen analysis can serve as a valuable diagnostic tool in assessing possible disorders of the male genital tract and the secretory pattern of the male accessory sex glands [9]. The most common infective bacteria in mixed accessory gland infection (MAGI) are *Chlamydia trachomatis* (41.4%), a common sexually transmissible pathogens in sexually active young men, followed by *Ureaplasma urealyticum* (15.5%) and *Mycoplasma hominis* (10.3%) [12, 13] as well as *Neisseria gonorrhoeae*, an additional marker of seminal tract infection [8, 14]. In men experiencing infertility issues, the presence or colonization of *U. urealyticum* and *M. hominis* in semen is a common finding [14] and semen cultures of bacterial pathogens remain the most common diagnostic method for seminal tract infections [8]. This passive or active invasion of these bacterial strains induces a generalized or local reaction in the urogenital tract [1]. This inflammatory response which can continue for extended periods of time leads to a pathological condition resulting in the activation of seminal WBC's [15]. Inadequate treatment of an infection and eradication of bacterial pathogens can lead to a chronic bacterial infection of the male accessory sex glands [4].

23.3 Leukocytospermia

Human semen is a heterogeneous fluid which contains a variety of cellular elements beyond spermatozoa. Several immunologic factors are present in human semen, such as chemokines, immunoglobulins, and growth factors [16], as well as a subset of WBC's. The microscopic evaluation of almost all semen samples will display nonspermatozoal cells, in particular WBC's [17], which encompass various forms such as granulocytes, lymphocytes, and macrophages [13]. An abnormally high concentration of WBC's in the semen is a condition called leukocytospermia, also referred to as leukospermia, pyospermia, or pyosaemia [18]. An increased concentration of leukocytes is the basic molecular defense mechanism against the detection of foreign organisms. The World Health Organization (WHO) criteria defines leukocytospermia as the presence of $>1 \times 10^6$ WBC's/mL of semen [18]. This threshold is regarded as a possible indicator of an ongoing male genital tract infection [18]. Specifically, the detection of pathological concentrations of leukocytospermia with the exclusion of a bladder infection or urethritis has been suggested as a basic diagnostic tool in recognizing genital tract infection [19, 20].

The activation of PMNL, which constitute 50–80% of the total seminal WBC count [8, 21], results in releasing a protease by degranulation known as elastase [22]. The presence of this particular protease is considered a highly reliable and sensitive marker of an asymptomatic infection [23] and can be used in diagnosing a clinically silent illness [24]. At present, due to the fact that the condition is asymptomatic and various sites in the reproductive system can be affected, the exact location of the origin of excess leukocytes is unknown [25, 26]. Due to the lack of understanding as to where exactly leukocytes are produced from, their release may be initially prompted by an inflammatory response of the genital tract to a bacterial invasion and then continually produced in their absence by immunological activity [7].

23.4 Effects of Leukocytospermia

Among the increasing number of partners experiencing fertility challenges, it has been estimated that the male factor is solely responsible for 30% of the failed fertilization rates [26]. Symptomatic and asymptomatic urinary tract infections which produce leukocytes is a condition frequently observed in infertility clinics [27, 28]. This negative effect on the male's fertilizing potential can be a result of the direct correlation which is found between increased concentrations of leukocytes and chromatin alterations and morphological abnormalities [6]. Despite the controversy which has been created from studies, the general biological and clinical conclusion is that the presence of leukocytes in semen, regardless of the concentration [7], is associated with OS and can be of a negative influence on certain parameters which can impair fertility [29, 30].

23.5 Reactive Oxygen Species

The metabolism of oxygen results in the generation of highly reactive agents belonging to the class of free radicals and are termed ROS [31]. Free radicals are short-lived atoms or molecules that contain one or more electrons with unpaired spin [1, 32]. These chemical intermediates can include hydroxyl radicals, superoxide anions, hypochlorite ions, and peroxyl radicals and are natural by-products of normal physiological processes [3, 31, 33]. An important factor to consider when examining the role of ROS is the concentration at which they are found. ROS illicit a biopositive influence when maintained at low concentrations. However, excessive generation of these oxidizing agents, accompanied by a lack of inactivation, results in damage to biomolecules [31, 32].

In the light of this counterbalance, spermatozoa face what is known as the "oxygen paradox." This contradiction lies in the fact that spermatozoa require oxygen for survival; however, the metabolites it produces, for example ROS, can compromise cell survival [30, 32]. When this environment develops, the natural defense system of scavenging antioxidants can be overwhelmed and basic semen parameters are negatively affected [15]. The consistent production of ROS by cellular aerobic metabolism makes it challenging to prevent OS injury in spermatozoa. The result is the counterbalance and OS which can cause pathological effects [34]. The extent of the damage that OS can have on semen and spermatozoa parameters is dependent on the level of ROS that is produced by the activated leukocytes and the extent by which it can overwhelm the natural antioxidant defense system [35].

23.5.1 Positive Effects of ROS

Like all cells living in aerobic conditions, spermatozoa need a consistent supply of oxygen. Under controlled physiological conditions, the generation of a low level of reactive oxidants is essential for maintaining normal physiological processes such as sperm–oocyte fusion, capacitation, and hyperactivation [31, 32]. In order to maintain cellular stability, this small amount of ROS needs to be continuously inactivated [5, 31, 32].

23.5.2 Negative Effects of ROS

In contrast to a beneficial role these oxidizing agents can play in various cellular events, it must be strongly considered that all cellular components, including nucleic acids, lipids, and proteins, are potential OS targets as a result of supra-physiological concentrations of ROS [1, 32]. Due to the fact that free radicals predominantly attack the closest stable molecule, which subsequently turns that specific particle into a free radical, ROS can be involved in a cascade of reactions which can damage a wide variety of biomolecules [31, 36]. The action which occurs on a molecular level by the interaction of ROS is the removal of hydrogen molecules [37] which results in a loss of motility [38], as well as other spermatozoa functions such as capacitation and acrosomal reactions [36]. Various conditions can promote the production of ROS which include: an increase in cellular metabolism, loss of antioxidant capability, and presence of inflammatory cells, for example leukocytes [1]. Two approaches prevent possible cell damage from pathologically or physiologically produced ROS. The first preventative step involves antioxidants which are compounds that scavenge and suppress ROS. The action of antioxidants counterbalances the damaging effect of OS. The second action is the control of the reactive oxidants in a microenvironment which limits the possible damaging effect that the ROS can have [1].

23.6 Generation of ROS in the Male Reproductive System

In contrast to normozoospermic semen samples, teratospermic, oligospermic, and asthenospermic subjects present semen samples with higher concentrations of ROS [39]. These free radicals are produced in spermatozoa by two predominant systems [31, 32, 40, 41]. First, as the generation of ROS is a result of oxygen metabolism, spermatozoal mitochondria are evidently the main contributor of the free radical [36]. Through the act of cellular respiration under normal physiological conditions, superoxide anions are the main ROS produced [36]. Spermatozoa require a constant source of energy to maintain motility and ROS is generated at a mitochondrial level by the nicotinamide adenine dinucleotide-dependant oxido-reductase (NADH) pathway [42].

Second, the sperm plasma membrane can generate ROS through the nicotinamide adenine dinucleotide phosphate-dependant oxidase system (NADPH) [35]. The plasma membrane is incredibly sensitive to ROS due to the high content of polyunsaturated fatty acids (PUFA), primarily docosahexaenoic acid [1, 4, 43]. This form of fatty acid provides fluidity to the plasma membrane with their highly specific lipid composition [36]. However, an unstable high concentration of ROS in an ejaculate can result in the peroxidation of these lipids [44]. The negative resultant effect from the peroxidative damage caused by the oxygen free radicals is increased permeability of the sperm plasma membrane [29, 44–46]. This process is induced by hydrogen peroxide, the most toxic compound for spermatozoa [3, 35, 47]. An additional and third source of ROS in seminal plasma is xanthine oxidase, an enzyme crucial in the catabolism of purine [48].

23.6.1 Generation of ROS by Leukocytes

Despite the fact that spermatozoa are capable of producing ROS, even in small amounts, the most predominant source of ROS formation by phagocytosis is peroxidase-positive leukocytes, mainly polymorphonuclear (PMN) neutrophilic granulocytes, which are crucial in diagnosing MAGI as they are active in the inflammatory process [41, 49–51]. In comparison to the amount of ROS produced by spermatozoa at the point of capacitation, the production rate by leukocytes is 1,000 times greater [52]. Various biological arguments between research groups have been created in determining the exact level of leukocytes that produce concentrations of ROS which are harmful towards spermatozoal parameters. The general consensus is that irrelevant of the exact concentration, the mere presence of these WBC's has been shown to be associated with OS [3, 39]. Since the presence of ROS in a semen sample can be the result of either spermatozoa in a pathological condition or leukocyte infiltration following infection, it is important in a clinical sense to accurately establish the source of seminal ROS [34].

23.7 Sources of OS

23.7.1 Extracellular Sources

When considering the OS that cells experience, it must be considered that this state can result from exogenous or endogenous influences [4]. The male genito-urinary system can be exposed to damaging exogenous factors, which are suggested sources of ROS and promoters of infection. These include lifestyle behaviors such as smoking and external influences such as environmental pollutants [35], as well as clinical diagnoses such as varicocele [53, 54] and spinal cord injuries [31]. Assisted reproductive therapy (ART) is also an established source of ROS [39]. The removal of seminal plasma is a procedure carried out in preparation for ART. This subsequently also removes natural antioxidants in the semen [7]. In this context, an additional exogenous source of OS can be certain processing techniques that are used in preparing semen, such as centrifugation [37].

23.7.2 Intracellular Sources

In contrast, when considering endogenous influences in the context of the male reproductive system, the seminal fluid contains a broad variety of cellular cells which can be considered possible sources. These include epithelial and round cells, spermatozoa at varying stages of spermatogenesis, as well as leukocytes [3]. Taking this into account demonstrates that essentially all ejaculate cells are potential providers of ROS [34], with seminal leukocytes and abnormal or immature spermatozoa being the predominant contributors [29, 41]. Immature spermatozoa have been postulated as being responsible for oxidative damage to their mature counterparts during the seminal migration from the seminiferous tubules to the epididymis [6]. Following an increase in the production of ROS by these two sources, the natural antioxidative capabilities are overwhelmed with the induction of OS [44].

23.8 Activation of Leukocytes

Infection of the genital tract is often observed as an asymptomatic subclinical inflammation [55, 56] with up to 80% of leukocytospermic semen samples showing no visual detection of microbial infection [26]. Following infiltration of infectious agents into the genital tract, the initial immune reaction is the increase in seminal leukocytes [15]. This inflammatory process aimed at killing the microorganisms results in the increase in leukocyte-produced ROS from the activated WBC's [11]. The infiltration and activation of these excess PMNL can cause excessively high concentrations of ROS which greatly exceed the required level for normal physiological functions. This elevation results in OS being initiated which produces a damaging prooxidant load [36]. Irrelevant of the concentration of leukocytes in semen, the presence of these WBC have been shown to be associated with OS and impairment in the quality of semen and sperm parameters such as concentration and morphology [7].

Two pathways are found in response to the activation of seminal leukocytes during infection. The first is the increase of NADPH through the hexose monophosphate shunt [32, 57]. The second route is the generation of high concentrations of ROS produced by a respiratory burst which acts as a protective mechanism following the infection [1, 32]. The result is the antioxidants in the cell being consumed. This oxidative burst is from the activation of the myeloperoxidase system present in both macrophages and PMNL and is controlled by various cytokines [1, 58]. The damaging effect of the infection is shown by the fact that leukocytes in an activated state are able to produce a 100-fold increase in the concentration of ROS when compared to leukocytes that are not activated [59]. This rapid and high increase of OS is an example of how damaging the condition of leukocytospermia can be.

23.9 Identification of an Infection

Biochemical evidence has shown that in males with clinically silent genital tract infections, the prostate and seminal vesicles are the organs predominantly affected and targeted by inflammation [60]. The primary role of the seminal vesicles is to provide high concentrations of fructose to the seminal plasma and is vital to the functional integrity of spermatozoa as it is the major source of glycolytic energy in order to maintain motility [18]. The reference value for normal concentrations of fructose is 13 μmol (2.34 mg) or more per ejaculate [18]. Determination of the concentration of the monosaccharide is commonly employed in laboratories for a variety of purposes including the auxiliary diagnosis of retrograde ejaculation, obstructive and nonobstructive azoospermia [61], and as a marker to assess seminal vesicular function [18, 62]. Changes in seminal vesicles secretory patterns can modify the composition of products of the vesicular fluid and of the ejaculate, affecting sperm function [63]. Conditions such as abnormal concentrations of zinc and fructose as well as hyperspermia, an increase in semen volume (>6 mL), and hypospermia (<2 mL) are all symptomatic of a glandular dysfunction in the seminal vesicles [64]. Inflammation from infection can cause the secretory epithelium lining the seminal vesicles to be affected, resulting in atrophy and a subsequent decrease in the seminal fructose concentration [61]. Biochemical evidence has shown that in males with clinically silent genital tract infections, the prostate is the organ that is predominantly affected and targeted by inflammation [60]. The total output of citric acid has been stated as the strongest discriminating power in terms of biochemical markers in differentiating between semen of infected and noninfected infertile men [65]. The threshold level for normal citrate concentrations is >52 μmol (9 mg) per ejaculate [18].

Inadequate treatment of an infection and removal of bacterial pathogens can lead to a chronic bacterial infection of these two male accessory sex glands [53]. An important symptomatic effect accompanying an infection is the damaging effect that the prostate and the seminal vesicles suffer from [18, 64]. Also, the leukocytes, predominantly PMN-granulocytes, are produced from these genito-urinary glands [66] due to the damaged tissue attracting WBC's to the site of infection [64]. During the upregulation of cell-mediated immunity in response to infection, a coordinated response allows for PMNL to release a protease by degranulation known as elastase [16, 22].

The presence of this particular protease is considered to be a highly reliable and sensitive marker of an asymptomatic infection [23] and can be used in diagnosing a clinical infection [24]. It can also be used in clinical application as a marker of the efficiency of anti-inflammatory treatment, as well as an alternative means of monitoring the levels of seminal WBC's [22]. An enzyme-linked immunoabsorbant assay (ELISA) can be used for the quantitative detection of PMN-elastase. PMN-elastase determination has been stated as being a useful screening method to detect leukocytospermia [67]

23.10 Identification of Leukoctyospermia

A wide variety of detection methods are available to identify leukocytes in seminal plasma. Determining the quantity of seminal ROS generation in a sample has been suggested as an important test to be performed in the clinical andrology assessment [57]. The identification of the presence of ROS-producing cells in a semen sample can be an important step taken in establishing the underlying cause for acquired or inherent defects that may be undetected [57].

The tests carried out to establish the condition of leukocytospermia include immunocytology, peroxidase, polymorphonuclear-elastase, and cytology [7]. The traditional method and technique recommended by the WHO [18] for counting leukocytes in human semen is to use a histochemical procedure to identify the peroxidase enzyme found in the cytoplasm that characterizes PMN-granulocytes [37]. In the examination of a semen sample, the difference between immature germ cells and leukocytes cannot be morphologically distinguished from each other, therefore cytological identification is an inaccurate method to detect leukocytes [68]. PMN-granulocytes and leukocytes are considered "round cells" of nonspermatogenic origin [69]. Due to a lack of diligence in differentiating between different categories of round cells, particularly spermatogenic germ cells and leukocytes, an overestimation of the number of leukocytes in a semen sample can often result [70] which can increase the chance of misdiagnoses. There is a significant relationship between the presence of PMN granulocyte elastase and the concentration of peroxidase-positive cells in a semen sample [68]. This particular cytological staining technique is considered to be a reliable method in the detection of leukocytes as it has minimal chances of misdiagnoses [27].

However, it must be considered that during the inflammatory process, degranulation and deregulation occur, resulting in the extracellular liberation of the leukocyte's cellular contents. The quantification and identification of leukocytes based on this specific method may therefore be affected as the peroxidase compound may be undetectable [71]. In order to ensure the most accurate quantification of leukocytes, an obvious recommendation would be to use peroxidase staining, which detects intracellular enzymatic activity [67], as well as an alternative method such as the biochemical analysis of the concentration of PMN-elastase, which identifies extracellular enzymes.

23.11 Measuring ROS Production

While detecting the presence of ROS-producing cells is possible, the direct measurement of in vivo ROS concentrations is challenging due to the fact that ROS are very short-lived [1]. In the field of male reproductive studies, the measurement of OS is quantified only indirectly by the presence of ROS. The quantification of ROS has been promoted as a valuable test to be performed in the assessment of seminal plasma that is to be used for ART as it may be beneficial in predicting end points such as the fertilization rate [5]. The detection and quantification of ROS as a means of studying the role of OS is carried out in varying ways.

One of these methods is the luminol-dependant chemiluminescence assay, which indirectly measures ROS produced by spermatozoa by means of a sensitive probe, which can either be luminol or lucigen [31, 41]. The principle of this test is based on the concentration of released luminescence in a luminometer [1, 72]. This is the amount of light that is generated with the interaction between the ROS present in the sample and a specific chemiluminescent probe [41]. This particular assay is a reproducible test with high sensitivity and specificity and allows for both the extra- and intracellular ROS to be measured [5, 31]. In conjunction with quantifying and identifying ROS in the semen sample by means of this particular assay, the relationship between the presence of leukocytospermia and the excessive ROS concentrations can be investigated. This can be done by determining leukocytospermia by performing the Endtz test, which is a myeloperoxidase staining technique based on the peroxidase activity of PMN [41]. An Endtz positive semen sample has been proven as an indicator of positive chemiluminescence for ROS [41].

The second test that can be used to determine if ROS is generated is by the measurement of lipid peroxidation through the Lipid Peroxidation Assay (LPO) [1]. Lipid peroxidation in spermatozoa is initiated by the presence of ROS and results in the generation of lipid hydroperoxides [1] which are stable molecules under normal physiological conditions [44]. However, under conditions of elevated ROS and resultant OS, malondialdehyde is produced from the decomposition of lipid peroxides [44]. The concentrations of this particular compound can be measured in biochemical assays to assess the level of peroxidative damage [73]. An additional third test to assess the level of seminal OS is the ROS-TAC score. This particular method combines the concentrations of TAC and ROS in semen samples and is quantified by using a statistical formula that allows for the optimal quantification of seminal OS [44, 54]. Owing to cases where normozoospermic males suffer from infertility, which may be the result of seminal OS, the ROS-TAC score has been suggested as an advantageous means of assessing OS in comparison to ROS alone when distinguishing between fertile and infertile subjects [54]. Due to the clinical relevance of identifying OS due to its causative role in infertility, research has led to the development of an inexpensive assay: the photometric nitro blue tetrazolium (NBT) assay which allows for the measurement of seminal ROS production [74]. This finding serves as a hopeful indicator of what future research into OS may hold.

23.12 Reproductive Sytems Defense Mechanisms

In examining the negative effects of high concentrations of ROS and the resultant OS, it must be considered that a degree of protection is provided to the spermatozoa by antioxidant defense mechanisms [44, 75]. Due to the proven relationship existing between increased incidents of infertility and inept antioxidant functioning in the seminal plasma [48, 76], determining the concentration of antioxidants may allow for the assessment of the fertility capabilities [31]. The properties of antioxidants allow for varying degrees of protection in the seminal plasma by acting as scavengers and creating a defense system against OS by removing the extracellular free radicals [3, 32]. This is due to the aforementioned fact that all ejaculates contain forms of cellular elements which can protect the spermatozoa from free radical toxicity [49]. In assessing the antioxidative capabilities of a semen sample, the previously explained TAC (Total Antioxidant Capacity) score is utilized. In determining the TAC score of a semen sample, two measurement techniques can be performed: a colorimetric assay and enhanced chemiluminescence, which quantifies TAC levels according to a set equation [31].

23.12.1 Antioxidants

The line of defense employed against oxidative insult consists of antioxidants which are agents that reduce the level of OS by breaking oxidative chain reactions and can be classified according to two subsets: enzymatic or "natural" such as superoxide dimutase, catalase, and glutathione peroxidase, and nonenzymatic or "synthetic" chain-breaking antioxidants for example pyruvate, glutathione, and carnitine, all of which can offer protection to spermatozoa against hazardous OS produced by the leukocytes [1, 2, 10, 77, 78].

These defensive compounds against free radical-induced OS can be regarded in two categories: preventative and scavenging antioxidants [43] In human semen, examples of the three main antioxidants present are urate, thiols, and ascorbate [77]. Antioxidative action effective in offering protection to spermatozoa below a critical threshold of ROS can be considered to consist of two stages. The primary step may be viewed as preventative, whereby the ROS are scavenged directly by the antioxidative compounds present in the seminal plasma [1]. Following this initial response initiated by antioxidative reactions is the secondary stage which encapsulates the influence that antioxidants may have on the actions that the ROS have already initiated. This step is regarded as interception which includes interruptions of chain reactions, for example the process of lipid peroxidation [1] The cytoplasm of the spermatozoa also contain a low concentration of antioxidants such as catalase and superoxise dimutase, which are capable of offering additional protection against ROS [10, 79]. However, due to the small volume of cytoplasm in each spermatozoa, this defensive approach against extracellular free radicals is less effective compared to the antioxidants abundantly present in seminal plasma [80].

Despite the antioxidative system to provide protection against OS, this pathway can be easily overwhelmed. The rapid increase in the production of leukocyte-derived ROS can overpower the antioxidative capabilities of the seminal plasma and spermatozoa [57]. The effect of leukocytospermia is highly effective in causing potential DNA damage towards spermatozoa at varying stages of maturation which can decrease the chances of successful oocyte fertilization [6, 39]. Damaging effects on the sperm genome that can arise from OS include: chromatin cross-linkage, base modification, and strand breakages in spermatozoa [31]. These DNA damages work in a cascade-like manner [7] and abnormal spermatozoa from leukocytospermic samples have more than doubled levels of impaired DNA under the influence of OS when compared to similar forms of irregular spermatozoa in samples that are not leukocytospermic [81].

23.13 Future Research and Treatment Approaches

A predominant contributing factor that idiopathic male infertility has been shown to have is the imbalance between OS and the antioxidant-induced scavenging of ROS present in the semen [10, 82]. Numerous factors may negatively influence semen parameters, for example posttesticular damage in the epididymis, abnormal spermatogenesis, environmental variables, or infection [18]. However, the presence of high concentrations of ROS can add to a damaging effect on the sperm parameters, such as morphology and motility [40], which are essential for normal spermatozoa functioning. By maintaining low concentrations of the free radicals in semen, the chances of successful fertilization can be considerably enhanced for a number of couples facing fertility challenges [36].

Past studies have shown the negative influence that elevated concentrations of ROS may elicit on the fertilizing capabilities of male subjects through decreasing spermatozoa motility and DNA damage when it was demonstrated that 40–88% of these samples had markedly high levels of ROS [77].

23.13.1 Clinical Treatment

Various forms of drug therapies against the presence of ROS have been developed as a result of research showing the harmful effect of these biomolecules, the first approach being drugs which can inhibit the formation of ROS [1]. The second is the stimulation of endogenous defense mechanisms because ROS are part of the initial protection against inflammation [1, 11]. The use of antioxidants is regarded as the most logical approach in the clinical intervention to treat the free radical's harmful effects [31]. It has been put forward that semen from males presenting with high concentrations of leukocytes and therefore OS may benefit from supplementary treatment with selected oral antioxidants [10, 15], even if subjects have normal semen analysis [54]. The intake may improve the functional properties of spermatozoa such as sperm's capability of successful oocyte fusion by providing protection from peroxidative damage [31, 83]. At present, the Cochrane Review provides evidence which suggests that subfertile couples involved in ART have increased chances of pregnancy and live birth rates, if the male subject receives an oral supplementation of antioxidants [84].

The supplementation with antioxidants during the preparation processes performed for ART such as density gradient centrifugation has been proposed and evaluated in the treatment pathway for ROS generation [31]. In addition, sperm–oocyte fusion can be enhanced by halting the production of ROS by immature spermatozoa and leukocytes by the addition of antioxidants to the culture media [85].

In addition to antioxidants, the pharmacological treatment of leukocytospermia with therapeutic concentrations of specific antibiotics has been shown to decrease leukocyte-produced ROS. Example of such drugs includes tetracycline and erythromycin, which can inhibit the actions of PMN-granulocyte [1]. Pharmacological findings further open the avenue for the treatment of leukocytospermia with drugs to avoid a state of damaging OS. Diagnostic and prognostic tests which allow for the measurement of the OS levels could be a considerable contributing factor in the treatment of idiopathic infertility [38]. Dietary supplementation with foods containing high concentrations of antioxidants such as carotenoids, vitamin C, and the main chain-breaking antioxidant vitamin E offers an additional line of defense against OS-induced injury [82]. An additional avenue into the beneficial effect that antioxidants have is shown by the cyroprotective effect that they have over sperm when added as supplements in the cyropreservation process [86].

23.14 Conclusion

In order to maintain normal physiological functioning in the male reproductive tract, there is a delicate balance between the concentration of ROS and the seminal antioxidants which must be maintained. The predominant contributing factor to idiopathic male infertility has been shown to be this imbalance which is created during periods of genito-urinary infection. As covered in the chapter leukocytospermia, the inflammatory condition resulting from the rapid increase in leukocytes during infection results in a high elevation in the concentrations of ROS and OS, which puts a powerful negative effect on various semen and spermatozoa variables [10, 36, 43]. The complexity of OS is furthered when additional factors which can increase ROS levels in the male genito-urinary system are considered [57]. Extensive growth in knowledge has been provided by a broad number of research groups which have focused on the significance and effects of excessive leukocytes that have infiltrated into the semen. The harmful influence that ROS can have on sperm and their parameters has been known for a relatively long period of time in the field of male reproductive biology. At present, the association between semen parameters and leukocyte concentrations is still a focal point in the field of male reproductive science.

References

1. Ochsendorf FR. Infections in the male genital tract and reactive oxygen species. Hum Reprod Update. 1999;5(5):399–420.
2. Sikka SC, Rajasekaran M, Hellstrom WJ. Role of oxidative stress and antioxidants in male infertility. J Androl. 1995;16(6):464–8.
3. Makker K, Agarwal A, Sharma R. Oxidative stress & male infertility. Indian J Med Res. 2009;129(4):357–67.
4. Aydemir B, et al. The influence of oxidative damage on viscosity of seminal fluid in infertile men. J Androl. 2008;29(1):41–6.
5. Desai N, et al. Physiologic and pathologic levels of reactive oxygen species in neat semen of infertile men. Fertil Steril. 2009;92(5):1626–31.
6. Alvarez JG, et al. Increased DNA damage in sperm from leukocytospermic semen samples as determined by the sperm chromatin structure assay. Fertil Steril. 2002;78(2):319–29.
7. Sharma RK, et al. Relationship between seminal white blood cell counts and oxidative stress in men treated at an infertility clinic. J Androl. 2001;22(4):575–83.
8. Keck C, et al. Seminal tract infections: impact on male fertility and treatment options. Hum Reprod Update. 1998;4(6):891–903.
9. Lewis SE. Is sperm evaluation useful in predicting human fertility? Reproduction. 2007; 134(1):31–40.
10. Pasqualotto FF, et al. Seminal oxidative stress in patients with chronic prostatitis. Urology. 2000;55(6):881–5.
11. Zorn B, et al. Semen polymorphonuclear neutrophil leukocyte elastase as a diagnostic and prognostic marker of genital tract inflammation—a review. Clin Chem Lab Med. 2003;41(1): 2–12.
12. Gdoura R, et al. Screening for bacterial pathogens in semen samples from infertile men with and without leukocytospermia. Andrologia. 2008;40(4):209–18.
13. Lackner J, et al. Value of counting white blood cells (WBC) in semen samples to predict the presence of bacteria. Eur Urol. 2006;49(1):148–52; discussion 152–3.
14. Trum JW, et al. Value of detecting leukocytospermia in the diagnosis of genital tract infection in subfertile men. Fertil Steril. 1998;70(2):315–9.
15. Sanocka D, et al. Male genital tract inflammation: the role of selected interleukins in regulation of pro-oxidant and antioxidant enzymatic substances in seminal plasma. J Androl. 2003; 24(3):448–55.
16. Politch JA, et al. Concentrations and significance of cytokines and other immunologic factors in semen of healthy fertile men. Hum Reprod. 2007;22(11):2928–35.
17. el-Demiry MI, et al. Immunocompetent cells in human testis in health and disease. Fertil Steril. 1987;48(3):470–9.
18. World Health Organization. WHO laboratory manual for the examination and processing of human semen. 5th ed. Geneva: WHO Press; 2010. p. 271.
19. Kokab A, et al. Raised inflammatory markers in semen from men with asymptomatic chlamydial infection. J Androl. 2010;31(2):114–20.
20. Weidner W, Krause W, Ludwig M. Relevance of male accessory gland infection for subsequent fertility with special focus on prostatitis. Hum Reprod Update. 1999;5(5):421–32.
21. Vicari E. Seminal leukocyte concentration and related specific reactive oxygen species production in patients with male accessory gland infections. Hum Reprod. 1999;14(8):2025–30.
22. Eggert-Kruse W, et al. Clinical relevance of polymorphonuclear (PMN-) elastase determination in semen and serum during infertility investigation. Int J Androl. 2009;32(4):317–29.
23. Micic S, et al. Elastase as an indicator of silent genital tract infection in infertile men. Int J Androl. 1989;12(6):423–9.
24. Jochum M, Pabst W, Schill WB. Granulocyte elastase as a sensitive diagnostic parameter of silent male genital tract inflammation. Andrologia. 1986;18(4):413–9.

25. Aziz N, et al. Novel associations between specific sperm morphological defects and leukocytospermia. Fertil Steril. 2004;82(3):621–7.
26. Gambera L, et al. Sperm quality and pregnancy rate after COX-2 inhibitor therapy of infertile males with abacterial leukocytospermia. Hum Reprod. 2007;22(4):1047–51.
27. Kaleli S, et al. Does leukocytospermia associate with poor semen parameters and sperm functions in male infertility? The role of different seminal leukocyte concentrations. Eur J Obstet Gynecol Reprod Biol. 2000;89(2):185–91.
28. Arata de Bellabarba G, et al. Nonsperm cells in human semen and their relationship with semen parameters. Arch Androl. 2000;45(3):131–6.
29. Aitken RJ, et al. Differential contribution of leucocytes and spermatozoa to the generation of reactive oxygen species in the ejaculates of oligozoospermic patients and fertile donors. J Reprod Fertil. 1992;94(2):451–62.
30. de Lamirande E, Gagnon C. Capacitation-associated production of superoxide anion by human spermatozoa. Free Radic Biol Med. 1995;18(3):487–95.
31. Cocuzza M, et al. Clinical relevance of oxidative stress and sperm chromatin damage in male infertility: an evidence based analysis. Int Braz J Urol. 2007;33(5):603–21.
32. Saleh RA, et al. Leukocytospermia is associated with increased reactive oxygen species production by human spermatozoa. Fertil Steril. 2002;78(6):1215–24.
33. Aitken RJ, Buckingham D, Harkiss D. Use of a xanthine oxidase free radical generating system to investigate the cytotoxic effects of reactive oxygen species on human spermatozoa. J Reprod Fertil. 1993;97(2):441–50.
34. Aitken RJ, Baker HW. Seminal leukocytes: passengers, terrorists or good samaritans? Hum Reprod. 1995;10(7):1736–9.
35. Aitken RJ, et al. Reactive oxygen species generation by human spermatozoa is induced by exogenous NADPH and inhibited by the flavoprotein inhibitors diphenylene iodonium and quinacrine. Mol Reprod Dev. 1997;47(4):468–82.
36. Sanocka D, Kurpisz M. Reactive oxygen species and sperm cells. Reprod Biol Endocrinol. 2004;2:12.
37. Fariello RM, et al. Effect of leukocytospermia and processing by discontinuous density gradient on sperm nuclear DNA fragmentation and mitochondrial activity. J Assist Reprod Genet. 2009;26(2–3):151–7.
38. Athayde KS, et al. Development of normal reference values for seminal reactive oxygen species and their correlation with leukocytes and semen parameters in a fertile population. J Androl. 2007;28(4):613–20.
39. Sharma RK, Agarwal A. Role of reactive oxygen species in male infertility. Urology. 1996; 48(6):835–50.
40. Aitken RJ, West K, Buckingham D. Leukocytic infiltration into the human ejaculate and its association with semen quality, oxidative stress, and sperm function. J Androl. 1994;15(4): 343–52.
41. Shekarriz M, et al. Positive myeloperoxidase staining (Endtz test) as an indicator of excessive reactive oxygen species formation in semen. J Assist Reprod Genet. 1995;12(2):70–4.
42. Gavella M, Lipovac V. NADH-dependent oxidoreductase (diaphorase) activity and isozyme pattern of sperm in infertile men. Arch Androl. 1992;28(2):135–41.
43. Agarwal A, Prabakaran SA. Mechanism, measurement, and prevention of oxidative stress in male reproductive physiology. Indian J Exp Biol. 2005;43(11):963–74.
44. Saleh RA, Agarwal A. Oxidative stress and male infertility: from research bench to clinical practice. J Androl. 2002;23(6):737–52.
45. Alkan I, et al. Reactive oxygen species production by the spermatozoa of patients with idiopathic infertility: relationship to seminal plasma antioxidants. J Urol. 1997;157(1):140–3.
46. de Lamirande E, Gagnon C. Human sperm hyperactivation in whole semen and its association with low superoxide scavenging capacity in seminal plasma. Fertil Steril. 1993;59(6):1291–5.
47. Bansal AK, Biaspuri GS. Impacts of oxidative stress and antioxidants on semen functions. Vet Med Int. 2011;1(1):1–7.

48. Sanocka D, et al. Oxidative stress and male infertility. J Androl. 1996;17(4):449–54.
49. Krause W, et al. Cellular and biochemical markers in semen indicating male accessory gland inflammation. Andrologia. 2003;35(5):279–82.
50. Ford WC, Whittington K, Williams AC. Reactive oxygen species in human sperm suspensions: production by leukocytes and the generation of NADPH to protect sperm against their effects. Int J Androl. 1997;20 Suppl 3:44–9.
51. Whittington K, Ford WC. Relative contribution of leukocytes and of spermatozoa to reactive oxygen species production in human sperm suspensions. Int J Androl. 1999;22(4):229–35.
52. de Lamirande E, et al. Reactive oxygen species and sperm physiology. Rev Reprod. 1997;2(1): 48–54.
53. Acosta AA, Kruger TF, editors. Human spermatozoa in assisted reproduction. 2nd ed. Bath: Parthenon Publishing Group; 1996. p. 518.
54. Pasqualotto FF, et al. Oxidative stress in normospermic men undergoing infertility evaluation. J Androl. 2001;22(2):316–22.
55. Kokab A, et al. Raised inflammatory markers in semen from men with asymptomatic chlamydial infection. J Androl. 2010;31(2):114–20.
56. Bezold G, et al. Prevalence of sexually transmissible pathogens in semen from asymptomatic male infertility patients with and without leukocytospermia. Fertil Steril. 2007;87(5): 1087–97.
57. Esfandiari N, et al. Utility of the nitroblue tetrazolium reduction test for assessment of reactive oxygen species production by seminal leukocytes and spermatozoa. J Androl. 2003;24(6): 862–70.
58. Blake DR, Allen RE, Lunec J. Free radicals in biological systems—a review orientated to inflammatory processes. Br Med Bull. 1987;43(2):371–85.
59. Plante M, de Lamirande E, Gagnon C. Reactive oxygen species released by activated neutrophils, but not by deficient spermatozoa, are sufficient to affect normal sperm motility. Fertil Steril. 1994;62(2):387–93.
60. Wolff H, et al. Impact of clinically silent inflammation on male genital tract organs as reflected by biochemical markers in semen. J Androl. 1991;12(5):331–4.
61. Lu JC, et al. Standardization and quality control for determination of fructose in seminal plasma. J Androl. 2007;28(2):207–13.
62. Gonzales GF. Function of seminal vesicles and their role on male fertility. Asian J Androl. 2001;3(4):251–8.
63. Andrade-Rocha FT. Semen analysis in laboratory practice: an overview of routine tests. J Clin Lab Anal. 2003;17(6):247–58.
64. Comhaire FH, et al. Mechanisms and effects of male genital tract infection on sperm quality and fertilizing potential: the andrologist's viewpoint. Hum Reprod Update. 1999;5(5):393–8.
65. Comhaire FH, Vermeulen L, Pieters O. Study of the accuracy of physical and biochemical markers in semen to detect infectious dysfunction of the accessory sex glands. J Androl. 1989; 10(1):50–3.
66. Kessopoulou E, et al. Origin of reactive oxygen species in human semen: spermatozoa or leucocytes? J Reprod Fertil. 1992;94(2):463–70.
67. Ricci G, et al. Leukocyte detection in human semen using flow cytometry. Hum Reprod. 2000;15(6):1329–37.
68. Henkel R, et al. Urogenital inflammation: changes of leucocytes and ROS. Andrologia. 2003; 35(5):309–13.
69. Johanisson E, et al. Evaluation of 'round cells' in semen analysis: a comparative study. Hum Reprod Update. 2000;6(4):404–12.
70. Maegawa M, et al. A repertoire of cytokines in human seminal plasma. J Reprod Immunol. 2002;54(1–2):33–42.
71. Villegas J, et al. Indirect immunofluorescence using monoclonal antibodies for the detection of leukocytospermia: comparison with peroxidase staining. Andrologia. 2002;34(2):69–73.
72. Wang A, et al. Generation of reactive oxygen species by leukocytes and sperm following exposure to urogenital tract infection. Arch Androl. 1997;39(1):11–7.

73. Aitken RJ, Clarkson JS, Fishel S. Generation of reactive oxygen species, lipid peroxidation, and human sperm function. Biol Reprod. 1989;41(1):183–97.
74. Tunc O, Thompson J, Tremellen K. Development of the NBT assay as a marker of sperm oxidative stress. Int J Androl. 2010;33(1):13–21.
75. Sikka SC. Oxidative stress and role of antioxidants in normal and abnormal sperm function. Front Biosci. 1996;1:e78–86.
76. Miesel R, et al. Severe antioxidase deficiency in human semen samples with pathological spermiogram parameters. Andrologia. 1997;29(2):77–83.
77. Lewis SE, et al. Total antioxidant capacity of seminal plasma is different in fertile and infertile men. Fertil Steril. 1995;64(4):868–70.
78. Agarwal A, et al. Role of antioxidants in treatment of male infertility: an overview of the literature. Reprod Biomed Online. 2004;8(6):616–27.
79. Storey BT, Alvarez JG, Thompson KA. Human sperm glutathione reductase activity in situ reveals limitation in the glutathione antioxidant defense system due to supply of NADPH. Mol Reprod Dev. 1998;49(4):400–7.
80. Williams AC, Ford WC. Functional significance of the pentose phosphate pathway and glutathione reductase in the antioxidant defenses of human sperm. Biol Reprod. 2004;71(4): 1309–16.
81. Erenpreiss J, et al. Effect of leukocytospermia on sperm DNA integrity: a negative effect in abnormal semen samples. J Androl. 2002;23(5):717–23.
82. Agarwal A, Prabakaran SA, Said TM. Prevention of oxidative stress injury to sperm. J Androl. 2005;26(6):654–60.
83. Siciliano L, et al. Impaired seminal antioxidant capacity in human semen with hyperviscosity or oligoasthenozoospermia. J Androl. 2001;22(5):798–803.
84. Showell MG, et al. Antioxidants for male subfertility. Cochrane Database Syst Rev. 2011;(1):CD007411.
85. Irvine DS. Glutathione as a treatment for male infertility. Rev Reprod. 1996;1(1):6–12.
86. Bucak MN, et al. Effects of antioxidants on post-thawed bovine sperm and oxidative stress parameters: antioxidants protect DNA integrity against cryodamage. Cryobiology. 2010;61(3): 248–53.

Chapter 24
Clinical Consequences of Oxidative Stress in Male Infertility*

Tamer M. Said, Sheila R. Gokul, and Ashok Agarwal

Abstract Male infertility affects 40% of infertile couples in the USA and may be attributed to conditions such as varicocele, leukocytospermia, infection, and idiopathic infertility. Such conditions may be associated with elevated levels of reactive oxygen species (ROS), decreased antioxidants, and oxidative stress (OS). OS can lead to male infertility in both an in vitro and in vivo setting. The negative effects of ROS on male fertility present as sperm DNA damage, decreasing motility, apoptosis, and lipid peroxidation. ROS and antioxidant levels can be measured and quantified in order to detect OS in semen samples. Both oral antioxidant therapy and culture media supplementation have proven to be effective in reducing OS. Future research is still needed in order to better understand the mechanisms involved in oxidative damage in the context of male infertility and to improve the treatments available for patients with OS-mediated male factor infertility.

Keywords Male infertility • Oxidative stress • Reactive oxygen species • Varicocele • Infection • Leukocytospermia • Idiopathic infertility • DNA damage • Antioxidants

---

*This research was conducted at the Cleveland Clinic's Center for Reproductive Medicine, Cleveland, OH.

T.M. Said MD, PhD, HCLD (ABB)
Andrology Laboratory and Reproductive Tissue Bank, The Toronto Institute for Reproductive Medicine, 56 Aberfoyle Crescent, Toronto, ON, Canada M8X2W4

S.R. Gokul
Center for Reproductive Medicine, Cleveland Clinic, 9500 Euclid Avenue, Cleveland, OH 44195, USA

A. Agarwal, PhD (✉)
Center for Reproductive Medicine, Cleveland Clinic, Lerner College of Medicine, 9500 Euclid Avenue, Cleveland, OH 44195, USA
e-mail: agarwaa@ccf.org

A. Agarwal et al. (eds.), *Studies on Men's Health and Fertility*, Oxidative Stress in Applied Basic Research and Clinical Practice, DOI 10.1007/978-1-61779-776-7_24,

24.1 Introduction

Male factor infertility is a contributing factor in roughly 40% of infertile couples in the USA [1]. It is also thought to be the sole cause in about half of these cases [2]. Several conditions have been described as potential causes for male infertility including varicocele, leukocytospermia, genital infection, and idiopathic infertility, as well as exposure to environmental factors such as smoking and pollutants. These clinical conditions have been linked to oxidative stress (OS), which occurs when there is an imbalance between reactive oxygen species (ROS) and antioxidants in the body.

ROS are reactive molecules that contain oxygen, which includes oxidizing free radicals that are necessary for normal biological functions. When produced in excess, ROS can cause sperm damage and may lead to selective cell death (apoptosis) [3]. Spermatozoa are particularly susceptible to oxidative damage because of their lack of an antioxidant defense system outside of the cytoplasm, which is mostly lost during the maturation process [4]. When the sperm cytoplasm is retained as cytoplasmic droplets, higher OS is to be expected due to ROS-producing enzymes in the cytoplasm [5]. Another source of ROS is leukocytes in the seminal fluid. Oxidative damage can affect the lipids of the sperm plasma membrane, as well as the sperm DNA. This type of damage does not only affect sperm function but also lead to negative consequences with the developing embryo and pregnancy rates [6].

In this chapter, we review recent literature and research to assess the current knowledge regarding the impact of OS on male infertility. Also discussed are the mechanisms behind oxidative damage and its contribution to the pathogenesis of male infertility.

24.2 Manifestations of Oxidative Stress

OS can lead to sperm damage and infertility through several pathways. ROS can react with DNA, proteins, carbohydrates, and lipids to cause sperm dysfunction and cell death. These detrimental effects have consequences both in vivo and in vitro, and may ultimately lead to infertility in males. The extent of oxidative damage depends on the nature of the ROS as well as the amount of exposure to OS.

24.2.1 Lipid Peroxidation

Lipids are found in the sperm plasma membrane in form of polyunsaturated fatty acids (PUFA). These fatty acids contain double bonds that make them susceptible to attack by free radicals. The initial attack by the hydroxyl radical leads to a series of chemical reactions referred to as lipid peroxidation [7]. Lipid peroxidation results in damage of the axonemal structure, which has consequences related to sperm motility [8]. Lipid peroxidation is considered to be the first step to sperm dysfunction inflicted by ROS, as well as the most important pathologic process that results in the decrease of sperm parameters [9]. Oxidative damage to the sperm lipid membrane also leads to a loss of membrane integrity, and as a result DNA and proteins can become susceptible to damage.

24.2.2 Decreased Motility

There are several hypotheses showing how ROS and OS can lead to decreased sperm motility. One of these hypotheses postulates that hydrogen peroxide diffuses across the plasma membrane into the spermatozoa and affects enzymes such as glucose-6-phosphate dehydrogenase and NADPH oxidase that are vital to normal spermatozoa function [10]. Another hypothesis suggests that ROS can lead to a decrease in protein phosphorylation and mitochondrial activity, consequently leading to sperm immobilization by reducing membrane fluidity [11]. Damage to the mitochondrial membrane by ROS can further potentiate the effects of OS on motility by causing a loss of intracellular ATP, leading to axonemal damage [12]. Finally, it was also reported that oxidative stress impacts the sperm motility pattern and motion kinetics. While low levels of nitric oxide are needed for hyperactivation, excess levels of NO have been shown to inhibit motility by impairing sperm respiration [13].

24.2.3 DNA Damage

DNA is usually protected from oxidative damage by its condensed form and the antioxidants present in the cytoplasm. After lipid peroxidation, DNA in both the mitochondria and nucleus becomes more susceptible to react with ROS. Sperm DNA damage has been associated with poor sperm parameters, infertility, and poor in vitro fertilization (IVF) outcomes [14]. DNA damage may present as fragmentation, deletions, frameshifts, point mutation, and DNA cross-links, as well as other genetic modifications and rearrangements. OS can also lead to single-strand DNA and double-strand DNA breaks [15]. If the amount of oxidative DNA damage is considerable, then apoptosis and embryo fragmentation may also occur.

24.2.4 Apoptosis

Apoptosis is an ongoing physiological phenomenon that leads to the elimination of abnormal spermatozoa in order to limit the number of male germ cells that can be maintained by Sertoli cells in the testes. Apoptosis is initiated in spermatozoa when high levels of ROS damage the mitochondrial membranes, and cytochrome-c proteins are released [1]. This activates caspases 9 and 3, which play essential roles in apoptosis. High levels of cytochrome-c and caspases have been correlated with increased levels of sperm DNA damage such as single and double-stranded DNA strand breaks [16]. Caspases have been also implicated in the decrease of sperm motility [17].

24.2.5 *Sperm Morphology*

Caspase-mediated apoptosis and increased OS have a positive relationship with increased sperm damage and abnormal sperm morphology [3]. There is also a higher incidence of abnormal sperm morphology in conditions related to OS [8]. DNA damage and ROS production has also been found to correlate with abnormal head morphology and cytoplasmic retention in immature sperm, but not in mature sperm [4]. This may be a result of OS affecting the regulation of spermiogenesis, the final stage of spermatogenesis where immature spermatids develop into mature spermatozoa [4]. Morphologically abnormal and immature sperm can lead to even higher levels of ROS production during sperm migration, and consequently lead to OS-related damage in mature sperm [4]. Abnormal morphology related to OS is not limited to immature spermatids and can extend to mature spermatozoa [14].

24.3 Clinical Conditions Associated with OS

24.3.1 *Varicocele*

The incidence of varicoceles in the general male population is 20% [18], while varicoceles are thought to be present in 40% of infertile males [19]. In males with secondary infertility, varicoceles are seen in 70–80% of patients [20]. Although the exact mechanisms through which varicocele damages spermatogenesis and sperm quality are not well understood, varicoceles have been associated with increased NO production [21], intratesticular temperatures, and low antioxidant capacity [22]. NO is released by phagocytes and endothelial cells in the male reproductive tract [23]. In patients with varicocele, excessive NO is released by the dilated veins, which can lead to the dysfunction of spermatozoa [24]. Also, NO reacts with the superoxide anion and produce peroxynitrate, a strong oxidant that can negatively impact sperm function [25].

While the exact mechanisms that lead to infertility and lower sperm parameters in men with varicocele are yet to be confirmed, reactive oxygen and nitrogen species may play a role [26]. OS and other forms of damage associated with increased reactive oxygen and nitrogen species leads to infertility in 15% of males with varicocele [27]. Patients with varicocele have higher ROS levels regardless of whether or not they are infertile [28]. A study by Allamaneni et al. found a positive correlation between the severity of varicocele and higher ROS production [29]. Other studies have reported lower antioxidant concentrations along with OS in patients with varicocele [30, 31]. On the other hand, some studies report no association between levels of OS and varicocele grade or fertility status in males [28, 32]. In support of the association of OS and varicoceles, varicocelectomy and antioxidant treatments have been shown to improve male fertility in patients with varicocele [33].

24.3.2 Leukocytospermia

Leukocytes should only make up 5% of the round cell population in semen in normal males [34]. Leukocytospermia is defined as the presence of peroxidase-positive leukocytes in concentrations of greater than 1×10^6 per mL of semen. Increased leukocyte infiltration in semen is common in conditions that involve inflammation such as genitourinary tract infection. Leukocytes exhibit increased ROS production in an early defense mechanism to kill infecting bacteria [35]. Therefore, high amounts of seminal leukocytes may even be used to diagnose a genitourinary tract infection [36]. However, leukocytospermia may be seen in situations unrelated to infection, such as smoking or heavy alcohol consumption [37].

Although ROS production by leukocytes is necessary for normal sperm function, leukocytes are considered to be one of the main sources of excessive ROS and OS in the male reproductive tract. Activated leukocytes have been shown to produce up to 100 times the amount of ROS produced by nonactivated leukocytes [38]. Leukocytes may also lead to oxidative damage in levels below the 1×10^6 per mL threshold for diagnosis of leukocytospermia [39]. Increased seminal leukocytes result in high levels of lipid peroxidation and DNA damage through ROS production. Nonsteroidal anti-inflammatory drugs and antioxidants have been proven to be effective treatments for male patients with leukocytospermia [40]. Sperm preparation may also be used to decrease the effects of ROS and OS on mature spermatozoa by separating leukocytes and immature sperm from the mature fraction [39].

A decline in sperm parameters has been correlated with leukocytospermia; a decrease that may be dependent on the concentration of leukocytes in the seminal fluid [41]. Parameters that were shown to be impaired by leukocytospermia include decreased motility, acrosomal damage, abnormal morphology [42, 43], decreased sperm concentration [44], decreased hyperactivation, sperm DNA damage, and impaired oocyte penetration [45]. However, other contradictory studies reported no significant relationship between leukocytospermia and decreased sperm function or parameters [46, 47].

24.3.3 Genitourinary Tract Infection

Genitourinary tract infections may be caused by a number of bacteria, including *Escherichia coli, Klebsiella pneumonia, Enterococcus faecalis, Chlamydia trachomatis,* and *Ureaplasma urealyticum* [36]. Genitourinary tract infection may originate in the kidney, bladder, epididymis, prostate, or urethra, and includes diagnoses such as prostatitis, epididymitis, orchitis, pyelonephritis, bacterial cystitis, and urethritis. These types of infections are associated with inflammation and increased leukocytes in the seminal fluid, which may lead to increased levels of ROS and OS [8].

High levels of ROS production by leukocytes occurs as a defense mechanism designed to kill microbes [35]. ROS generation may continue due to stimulation mediated by cytokines such as interleukin-6 and -8, which are associated with infection and inflammation [48]. Leukocytospermia and DNA damage have been found in patients with genitourinary tract infections [49]. Chronic infections may lead to increased damage over time. Patients with both varicocele and some type of genitourinary tract infection have been found to have higher levels of ROS than patients with just one of the conditions [44]. The most common method of treatment is with antibiotics such as tetracyclines, although antioxidant treatments involving carnitines have been also proven to be effective [50].

In male accessory gland infection, there is a high prevalence of abnormal semen quality and infertility [51]. Infection of the epididymis can cause asthenozoospermia, while impairment of the seminal vesicles can lead to obstruction, decreased semen volumes and fructose levels [52]. Infection of the epididymis is also a source of obstructive azoospermia [2]. Infection of the prostate can cause semen hyperviscosity due to inefficient secretion of proteolytic enzymes and antigens [53], which can affect the sperm motility and progression through the female reproductive tract. Semen samples of men with an accessory gland infection have been shown to have lower concentrations of certain key antioxidants such as citric acid and zinc, which are involved in maintaining seminal pH levels as well as DNA condensation and chromatin stability [54–57].

24.3.4 Idiopathic Infertility

Infertile men are diagnosed with idiopathic infertility when normal sperm parameters and no other clinical condition that may result in infertility are seen. Oxidative damage may be a contributing factor to infertility in these normozoospermic males. While the exact cause of infertility may be unknown, idiopathic infertility has been correlated with higher levels of ROS and lower antioxidant levels than in fertile men [58]. A study found that men with idiopathic infertility had the second highest level of OS among the clinical diagnoses studied [44]. Therefore, it would be expected that infertile men diagnosed with idiopathic infertility would benefit from antioxidant treatments. Idiopathic infertility has also been associated with increased sperm

DNA damage. A study by Wang et al. found a relationship between increased sperm damage by ROS and higher levels of cytochrome c and caspases 9 and 3 [3]. This may also indicate elevated apoptosis in males with idiopathic infertility.

Studies have reported that anywhere from 25 to 40% of males with idiopathic infertility have high ROS levels [59]. Idiopathic infertility may be explained by OS in infertile men with normal sperm and semen parameters [39]. Although idiopathic infertility may only be a temporary diagnosis, an estimated 64% of diagnosed males will remain infertile after 1 year [3]. The sperm mitochondria have also been found to be damaged by ROS production in men with idiopathic infertility. This mitochondrial damage can lead to the release of proteins such as cytochrome c, and result in apoptosis and DNA damage [3]. However, apoptosis, DNA damage, and reduced sperm quality do not always present together [50].

24.3.5 Environmental and Lifestyle Factors

Saleh et al. found a correlation between smoking and increased seminal leukocyte levels, as well as increased ROS production and lower antioxidant levels [60, 61]. Similarly, another study by Close et al. found a correlation between increased seminal leukocyte infiltration in cigarette-smoking men [37]. Additionally, high levels of sperm DNA damage have been reported in male smokers [60, 62]. This may be due to the mutagens and carcinogens associated with cigarette smoke [62].Cigarettes contain nicotine, cotinine, hydroxycotinine, alkaloids, and nitrosamines, which are all sources of free radicals in the body [63]. Smoking has been associated with up to a 48% increase in leukocyte levels in semen, as well as up to a 107% increase in seminal ROS levels [60].

Environmental pollutants are a major source of ROS production that have been implicated in the pathogenesis of poor quality sperm [64]. A study by De Rosa et al. found that NO and lead in air pollution can negatively affect semen quality [65]. OS has been suggested to play a role in the development of these negative effects of pollution [66]. The accumulation of pollutants that may act as endocrine disruptors has been shown to have a negative effect on male reproductive function [67]. Air pollution from motor vehicles may have negative impact on male fertility through the release of NO [65]. Pollution may be also a contributing factor to the overall decline in male fecundity that has been seen in the past few decades [68, 69].

24.4 OS and In Vitro Infertility

Reactive oxygen and nitrogen species that lead to OS have a negative impact not only on male fertility in natural conception but also on assisted reproductive technologies (ART) in an in vitro setting. In any ART procedure, there is a risk of oxidative damage from sources such as exposure to ambient air [70]. Other sources of OS in an ART setting include the oocyte, embryo, cumulus cells, and immature sperm cells [6]. In addition, spermatozoa used in any ART procedure originate from an environment conducive to OS, which can lead to DNA damage [71].

Studies have shown that the ROS levels in the seminal fluid correlate with fertilizing potential and the IVF rate in procedures such as intrauterine insemination (IUI), in vitro fertilization (IVF), and intracytoplasmic sperm injection (ICSI) [71, 72]. Also, high ROS levels in day 1 embryo culture media have been related to decreased blastocyst and cleavage rate, low fertilization rate, and increased embryo fragmentation in ICSI cycles. Consistently, high day 1 ROS levels in culture media have been correlated with decreased pregnancy rates in ICSI and IVF [73].

ROS production and sperm DNA damage are associated with apoptosis [3], which has been shown to be associated with a decrease in fertilization rate [74]. It is of importance to note that sperm with DNA damage has the potential to lead to poor embryo development and carries the risk of birth defects [14]. Miscarriage rates were found to be higher in ICSI than in IVF, which may be explained by the fact that in the ICSI procedure, there is a greater chance of DNA damaged sperm being injected into the oocyte [75]. DNA damaged sperm is less likely to be used to fertilize an oocyte in IVF or IUI because of associated damage to the sperm plasma membrane, which is necessary for fertilization [1].

The general sources, mechanisms, and consequences of OS on male fertility are summarized in Fig. 24.1. Clinical conditions related to OS include idiopathic infertility, leukocytospermia, varicocele, genitourinary tract infection, environmental and lifestyle factors. OS acts through several mechanisms that lead to subfertility, such as lipid peroxidation, DNA damage, and apoptosis. OS can lead to several consequences related to male fertility, both in an in vivo and in vitro setting.

24.5 Management of Oxidative Stress

OS in a clinical setting can be diagnosed through a variety of tests that measure levels of ROS and antioxidant capacities in semen. OS and the conditions that lead to OS can be treated through various means, including oral and surgical treatments as well as laboratory techniques.

24.5.1 Diagnosis of OS and ROS Levels

The chemiluminescence assay is a direct method of quantifying extracellular ROS levels. This assay uses lucigenin and luminol to assess the generation of the superoxide anion and hydrogen peroxide [76]. Lucigenin detects superoxide anion, while luminol detects hydrogen peroxide. These agents are oxidizable substrates that react with certain oxygen species in order to measure levels of ROS. The chemiluminescence assay or calorimetric assay can also be used to measure the total antioxidant

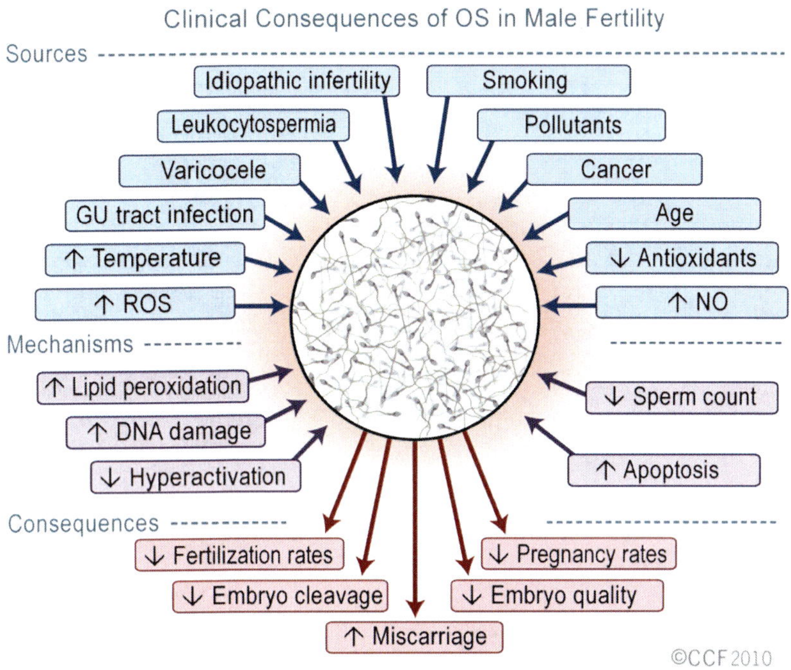

Fig. 24.1 The general sources, mechanisms, and consequences of oxidative stress (OS) on male fertility are summarized. Clinical conditions related to OS include idiopathic infertility, leukocytospermia, varicocele, genitourinary tract infection, environmental and lifestyle factors. OS acts through several mechanisms which lead to subfertility, such as lipid peroxidation, DNA damage, and apoptosis. OS can lead to several consequences related to male fertility, both in an in vivo and in vitro setting

capacity (TAC) of a semen sample. Subsequently, a ROS-TAC score is used to quantify if an imbalance exists between ROS and free radical scavengers. Low ROS-TAC scores are indicative of overall OS and have been identified in patients with idiopathic infertility, varicocele, and male accessory gland infection [77, 78]. These scores may be useful in predicting infertility when compared to using just ROS or TAC scores [78]. Patients with varicocele and other conditions associated with male factor infertility have low TAC scores [77], indicating an inability to scavenge ROS.

An indirect method of testing for ROS levels is through the examination of malondialdehyde (MDA) levels. MDA is a by-product of lipid peroxidation, which can be measured to detect the amount of lipid peroxidation in a semen sample. The MDA levels correlate with sperm motility and the potential for sperm–oocyte fusion [79]. This assay is useful in determining ROS levels before ART procedures and in diagnosing patients with subfertility [80]. Some studies have shown that high MDA levels correlate with other decreased sperm parameters such as concentration and motility [81, 82]. Higher MDA levels have been found in patients with varicocele when compared to controls, indicating increased peroxidative activity [31].

24.5.2 *In Vivo and In Vitro Antioxidant Treatments*

Oral antioxidant supplements may be used to counteract OS and treat male infertility. Studies have shown that antioxidants correlate with improvement in sperm parameters; however, in excess, these oral antioxidant supplements may have detrimental effects [83]. As an example, improvement in sperm motility and lower levels of ROS were attributed to vitamin E and selenium oral supplementation [84]. Other antioxidant supplementations that have been proven to be effective in reducing ROS levels include glutathione, L-carnitine, vitamin A, and *N*-acetyl cysteine [39, 50].

Antioxidant treatments can be also added to culture media in vitro during sperm preparation techniques in order to improve in vitro sperm fertilization ability. Vitamin C supplementation alone and in conjunction with vitamin E has been shown to be effective both orally and in culture media [85, 86]. Media supplementation with vitamin C and urate can lead to protection of spermatozoa from DNA damage [87].

24.5.3 *Sperm Preparation Techniques*

Sperm preparation techniques can be used to separate ROS producing agents such as leukocytes and immature spermatozoa from mature spermatozoa. The use of density gradients and centrifugation separates mature and immature sperm populations based on morphology and motility [88]. Swim up techniques may also be used to separate highly motile and morphologically normal sperm from the rest of the sperm population [89]. Percoll density gradient preparation is available, but is recommended for research purposes only.

24.5.4 *Specific Treatments*

Effective treatments have been developed for specific male conditions related to OS. One of the most effective treatments for patients with varicocele is varicocelectomy, the surgical means of varicocele repair [90]. Varicocelectomy is effective in decreasing oxidative stress, sperm DNA damage and increasing antioxidant capacity in seminal plasma of subfertile patients with varicocele [91]. Prevention and avoidance are also necessary in case of environmental factors induced oxidative stress such as smoking and pollutants that contribute to OS.

24.6 Conclusions

Conditions such as varicocele, genitourinary tract infection, leukocytospermia, and idiopathic infertility are associated with OS and have the potential of causing male factor infertility. Environmental factors such as smoking and pollution can also lead to elevated levels of ROS and subfertility in males. OS results in in vitro and in vivo clinical complications that may affect male fecundity. High ROS levels impair male fertility through lipid peroxidation, decreased motility, DNA damage, and apoptosis. OS can be detected and treated to lower ROS levels. Various approaches are currently available to prevent and treat OS in vivo and in vitro, which may lead to improvement of male fertility. There is still a need for further investigation into the mechanisms and pathways that lead to oxidative damage, as well as the need for improved fertility treatments specific to OS and the related conditions.

- High levels of ROS or low levels of antioxidants can lead to OS, affecting male fertility.
- OS can impair male fertility through lipid peroxidation, decreased motility, DNA damage, and apoptosis.
- Clinical conditions associated with OS include varicocele, leukocytospermia, genitourinary tract infection, and idiopathic infertility.
- ROS levels can be diagnosed by the chemiluminescence assay, and OS can be identified using the TAC levels, and the ROS-TAC levels.
- Oral antioxidants and antioxidant treatment of the media culture may be helpful in ART procedures to prevent OS and improve fertility.

References

1. Agarwal A, Makker K, Sharma R. Clinical relevance of oxidative stress in male factor infertility: an update. Am J Reprod Immunol. 2008;59(1):2–11.
2. Sharlip I, Jarow JP, Belker AM, Lipshultz LI, Sigman M, Thomas AJ, et al. Best practice policies for male infertility. Fertil Steril. 2002;77(5):873–82.
3. Wang X, Sharma RK, Sikka SC, Thomas Jr AJ, Falcone T, Agarwal A. Oxidative stress is associated with increased apoptosis leading to spermatozoa DNA damage in patients with male factor infertility. Fertil Steril. 2003;80(3):531–5.
4. Ollero M, Gil-Guzman E, Lopez MC, Sharma RK, Agarwal A, Larson K, Evenson D, et al. Characterization of subsets of human spermatozoa at different stages of maturation: implications in the diagnosis and treatment of male infertility. Hum Reprod. 2001;16(9):1912–21.
5. Gomez E, Buckingham DW, Brindle J, Lanzafame F, Irvine DS, Aitken RJ. Development of an image analysis system to monitor the retention of residual cytoplasm by human spermatozoa: correlation with biochemical markers of the cytoplasmic space, oxidative stress, and sperm function. J Androl. 1996;17(3):276–87.
6. Agarwal A, Said TM, Bedaiwy MA, Banerjee J, Alvarez JG. Oxidative stress in an assisted reproductive techniques setting. Fertil Steril. 2006;86(3):503–12.
7. Aitken J, Fisher H. Reactive oxygen species generation and human spermatozoa: the balance of benefit and risk. Bioessays. 1994;16(4):259–67.
8. Saleh R, Agarwal A, Kandirali E, Sharma RK, Thomas AJ, Nada EA, et al. Leukocytospermia is associated with increased reactive oxygen species production by human spermatozoa. Fertil Steril. 2002;78(6):1215–24.
9. Jones R, Mann T, Sherins R. Peroxidative breakdown of phospholipids in human spermatozoa, spermicidal properties of fatty acid peroxides, and protective action of seminal plasma. Fertil Steril. 1979;31(5):531–7.
10. Aitken R. Molecular mechanisms regulating human sperm function. Mol Hum Reprod. 1997;3(3):169–73.

11. de Lamirande E, Gagnon C. Impact of reactive oxygen species on spermatozoa: a balancing act between beneficial and detrimental effects. Hum Reprod. 1995;10 Suppl 1:15–21.
12. Sikka S. Relative impact of oxidative stress on male reproductive function. Curr Med Chem. 2001;8(7):851–62.
13. Herrero M, de Lamirande E, Gagnon C. Nitric oxide is a signaling molecule in spermatozoa. Curr Pharm Des. 2003;9(5):419–25.
14. Sakkas D, Mariethoz E, Manicardi G, Bizzaro D, Bianchi PG, Bianchi U. Origin of DNA damage in ejaculated human spermatozoa. Rev Reprod. 1999;4(1):31–7.
15. Kodama H, Yamaguchi R, Fukuda J, Kasai H, Tanaka T. Increased oxidative deoxyribonucleic acid damage in the spermatozoa of infertile male patients. Fertil Steril. 1997;68(3):519–24.
16. Said T, Paasch U, Glander HJ, Agarwal A. Role of caspases in male infertility. Hum Reprod Update. 2004;10(1):39–51.
17. Weng S, Taylor SL, Morshedi M, Schuffner A, Duran EH, Beebe S, Oehninger S. Caspase activity and apoptotic markers in ejaculated human sperm. Mol Hum Reprod. 2002;8(11): 984–91.
18. Pasqualotto F, Agarwal A. Varicocele and male infertility: an evidence based review. Arch Med Sci. 2009;5(1A):S20–7.
19. Madgar I, Weissenberg R, Lunenfeld B, Karasik A, Goldwasser B. Controlled trial of high spermatic vein ligation for varicocele in infertile men. Fertil Steril. 1995;63(1):120–4.
20. Naughton C, Nangia A, Agarwal A. Varicocele and male infertility: part II: pathophysiology of varicoceles in male infertility. Hum Reprod. 2001;7(5):473–81.
21. Romeo C, Ientile R, Santoro G, Impellizzeri P, Turiaco N, Impalà P, et al. Nitric oxide production is increased in the spermatic veins of adolescents with left idiophatic varicocele. J Pediatr Surg. 2001;36(2):389–93.
22. Agarwal A, Prabakaran S, Allamaneni SS. Relationship between oxidative stress, varicocele and infertility: a meta-analysis. Reprod Biomed Online. 2006;12(5):630–3.
23. Davies M, Fulton GJ, Hagen PO. Clinical biology of nitric oxide. Br J Surg. 1995;82(12): 1598–610.
24. Ozbeka E, Turkozc Y, Gokdenizb R, Davarcia M, Ozugurluc F. Increased nitric oxide production in the spermatic vein of patients with varicocele. Eur Urol. 2000;37(2):172–5.
25. Weinberg J, Doty E, Bonaventura J, Haney AF. Nitric oxide inhibition of human sperm motility. Fertil Steril. 1995;64(2):408–13.
26. Agarwal A, Sharma RK, Desai N, Prabakran S, Tavares A, Sabanaegh E. Role of oxidative stress in pathogenesis of varicocele and infertility. Urology. 2009;73(3):461–9.
27. Schoor R, Elhanbly SM, Niederberger C. The pathophysiology of varicocele-associated male infertility. Curr Urol Rep. 2001;2(6):432–6.
28. Hendin B, Kolettis PN, Sharma RK, Thomas Jr AJ, Agarwal A. Varicocele is associated with elevated spermatozoal reactive oxygen species production and diminished seminal plasma antioxidant capacity. J Urol. 1999;161(6):1831–4.
29. Allamaneni S, Naughton CK, Sharma RK, Thomas Jr AJ, Agarwal A. Increased seminal reactive oxygen species levels in patients with varicoceles correlate with varicocele grade but not with testis size. Fertil Steril. 2004;82(6):1684–6.
30. Künzle R, Mueller MD, Hänggi W, Birkhäuser MH, Drescher H, Bersinger NA. Semen quality of male smokers and nonsmokers in infertile couples. Fertil Steril. 2003;79(2):287–91.
31. Abd-Elmoaty M, Saleh R, Sharma R, Agarwal A. Increased levels of oxidants and reduced antioxidants in semen of infertile men with varicocele. Fertil Steril. 2010;94(4):1531–4.
32. Cocuzza M, Athayde KS, Agarwal A, Pagani R, Sikka SC, Lucon AM, et al. Impact of clinical varicocele and testis size on seminal reactive oxygen species levels in a fertile population: a prospective controlled study. Fertil Steril. 2008;90(4):1103–8.
33. Unal D, Yeni E, Verit A, Karatas OF. Clomiphene citrate versus varicocelectomy in treatment of subclinical varicocele: a prospective randomized study. Int J Urol. 2001;8(5):227–30.
34. Eggert-Kruse W, Bellmann A, Rohr G, Tilgen W, Runnebaum B. Differentiation of round cells in semen by means of monoclonal antibodies and relationship with male fertility. Fertil Steril. 1992;58(5):1046–55.

35. Saran M, Beck-Speier I, Fellerhoff B, Bauer G. Phagocytic killing of microorganisms by radical processes: consequences of the reaction of hydroxyl radicals with chloride yielding chlorine atoms. Free Radic Biol Med. 1999;26(3–4):482–90.
36. Munuce M, Bregni C, Carizza C, Mendeluk G. Semen culture, leukocytospermia, and the presence of sperm antibodies in seminal hyperviscosity. Arch Androl. 1999;42(1):21–8.
37. Close C, Roberts PL, Berger RE. Cigarettes, alcohol and marijuana are related to pyospermia in infertile men. J Urol. 1990;144(4):900–3.
38. Plante M, de Lamirande E, Gagnon C. Reactive oxygen species released by activated neutrophils, but not by deficient spermatozoa, are sufficient to affect normal sperm motility. Fertil Steril. 1994;62(2):387–93.
39. Agarwal A, Said TM. Oxidative stress, DNA damage and apoptosis in male infertility: a clinical approach. BJU Int. 2005;95(4):503–7.
40. Vicari E, La Vignera S, Calogero AE. Antioxidant treatment with carnitines is effective in infertile patients with prostatovesiculoepididymitis and elevated seminal leukocyte concentrations after treatment with nonsteroidal anti-inflammatory compounds. Fertil Steril. 2002; 78(6):1203–8.
41. Sharma R, Pasqualotto AE, Nelson DR, Thomas Jr AJ, Agarwal A. Relationship between seminal white blood cell counts and oxidative stress in men treated at an infertility clinic. J Androl. 2001;22(4):575–83.
42. Aziz N, Agarwal A, Lewis-Jones I, Sharma RK, Thomas Jr AJ. Novel associations between specific sperm morphological defects and leukocytospermia. Fertil Steril. 2004;82(3):621–7.
43. Lackner J, Agarwal A, Mahfouz R, du Plessis SS, Schatzl G. The association between leukocytes and sperm quality is concentration dependent. Reprod Biol Endocrinol. 2010;8:12.
44. Pasqualotto F, Sharma RK, Nelson DR, Thomas AJ, Agarwal A. Relationship between oxidative stress, semen characteristics, and clinical diagnosis in men undergoing infertility investigation. Fertil Steril. 2000;73(3):459–64.
45. Maruyama DJ, Hale RW, Rogers BJ. Effects of white blood cells on the in vitro penetration of zona-free hamster eggs by human spermatozoa. J Androl. 1985;6(2):127–35.
46. Tomlinson M, Barratt CL, Cooke ID. Prospective study of leukocytes and leukocyte subpopulations in semen suggests they are not a cause of male infertility. Fertil Steril. 1993;60(6): 1069–75.
47. Aitken J, Krausz C, Buckingham D. Relationships between biochemical markers for residual sperm cytoplasm, reactive oxygen species generation, and the presence of leukocytes and precursor germ cells in human sperm suspensions. Mol Reprod Dev. 1994;39(3):268–79.
48. Agarwal A, Prabakaran S, Allamaneni S. What an andrologist/urologist should know about free radicals and why. Urology. 2006;67(1):2–8.
49. Ochsendorf F. Infections in the male genital tract and reactive oxygen species. Hum Reprod Update. 1999;5(5):399–420.
50. Agarwal A, Nallella KP, Allamaneni SS, Said TM. Role of antioxidants in treatment of male infertility: an overview of the literature. Reprod Biomed Online. 2004;8(6):616–27.
51. Rowe P, Comhaire F, Hargreave T, Mahmoud A, Rowe P, Comhaire F, Hargreave T, Mahmoud A. WHO manual for the standardized investigation, diagnosis and management of the infertile male. 1st ed. Cambridge: Cambridge University Press; 2000.
52. Comhaire F, Mahmoud AMA, Depuydt CE, Zalata AA, Christophe AB. Mechanisms and effects of male genital tract infection on sperm quality and fertilizing potential: the andrologist's viewpoint. Hum Reprod. 1999;5(5):393–8.
53. Shibata K, Kajihara J, Kato K, Hirano K. Purification and characterization of prostate specific antigen from human urine. Biochim Biophys Acta. 1997;1336(3):425–33.
54. Kvist U, Björndahl L. Zinc preserves an inherent capacity for human sperm chromatin decondensation. Acta Physiol Scand. 1985;124(2):195–200.
55. Kundu T, Rao M. DNA condensation by the rat spermatidal protein TP2 shows GC-rich sequence preference and is zinc dependent. Biochemistry. 1995;34(15):5143–50.
56. Comhaire F, Vermeulen L, Pieters O. Study of the accuracy of physical and biochemical markers in semen to detect infectious dysfunction of the accessory sex glands. J Androl. 1989;10(1): 50–3.

57. Kjellberg S, Björndahl L, Kvist U. Sperm chromatin stability and zinc binding properties in semen from men in barren unions. Int J Androl. 1992;15(2):103–13.
58. Pasqualotto F, Sharma RK, Kobayashi H, Nelson DR, Thomas Jr AJ, Agarwal A. Oxidative stress in normospermic men undergoing infertility evaluation. J Androl. 2001;22(2):316–22.
59. Alkan I, Simşek F, Haklar G, Kervancioğlu E, Ozveri H, Yalçin S, Akdaş A. Reactive oxygen species production by the spermatozoa of patients with idiopathic infertility: relationship to seminal plasma antioxidants. J Urol. 1997;157(1):140–3.
60. Saleh R, Agarwal A, Sharma RK, Nelson DR, Thomas Jr AJ. Effect of cigarette smoking on levels of seminal oxidative stress in infertile men: a prospective study. Fertil Steril. 2002; 78(3):491–9.
61. Fraga C, Motchnik PA, Wyrobek AJ, Rempel DM, Ames BN. Smoking and low antioxidant levels increase oxidative damage to sperm DNA. Mutat Res. 1996;351(2):199–203.
62. Traber M, van der Vliet A, Reznick AZ, Cross CE. Tobacco-related diseases. Is there a role for antioxidant micronutrient supplementation? Clin Chest Med. 2000;21(1):173–87.
63. Pasqualotto F, Sobreiro BP, Hallak J, Pasqualotto EB, Lucon AM. Cigarette smoking is related to a decrease in semen volume in a population of fertile men. BJU Int. 2006;97(2):324–6.
64. Koizumi T, Li ZG. Role of oxidative stress in single-dose, cadmium-induced testicular cancer. J Toxicol Environ Health. 1992;37(1):25–36.
65. De Rosa M, Zarrilli S, Paesano L, Carbone U, Boggia B, Petretta M, et al. Traffic pollutants affect fertility in men. Hum Reprod. 2003;18(5):1055–61.
66. Fowler B, Whittaker MH, Lipsky M, Wang G, Chen XQ. Oxidative stress induced by lead, cadmium and arsenic mixtures: 30-day, 90-day, and 180-day drinking water studies in rats: an overview. Biometals. 2004;17(5):567–8.
67. Kumar S. Occupational exposure associated with reproductive dysfunction. J Occup Health. 2004;46(1):1–19.
68. Skakkebaek N, Jørgensen N, Main KM, Rajpert-De Meyts E, Leffers H, Andersson AM, et al. Is human fecundity declining? Int J Androl. 2006;29(1):2–11.
69. Hauser R. The environment and male fertility: recent research on emerging chemicals and semen quality. Semin Reprod Med. 2006;24(3):156–67.
70. Agarwal A, Ikemoto I, Loughlin KR. Effect of sperm washing on levels of reactive oxygen species in semen. Arch Androl. 1994;33(3):157–62.
71. Saleh R, Agarwal A. Oxidative stress and male infertility: from research bench to clinical practice. J Androl. 2002;22(6):737–52.
72. Zorn B, Vidmar G, Meden-Vrtovec H. Seminal reactive oxygen species as predictors of fertilization, embryo quality and pregnancy rates after conventional in vitro fertilization and intracytoplasmic sperm injection. Int J Androl. 2003;26(5):279–85.
73. Bedaiwy M, Falcone T, Mohamed MS, Aleem AA, Sharma RK, Worley SE, et al. Differential growth of human embryos in vitro: role of reactive oxygen species. Fertil Steril. 2004;82(3):593–600.
74. Høst E, Lindenberg S, Smidt-Jensen S. The role of DNA strand breaks in human spermatozoa used for IVF and ICSI. Acta Obstet Gynecol Scand. 2000;79(7):559–63.
75. Aitken R. The Amoroso lecture. The human spermatozoon—a cell in crisis? J Reprod Fertil. 1999;115(1):1–7.
76. Agarwal A, Allamaneni SS, Said TM. Chemiluminescence technique for measuring reactive oxygen species. Reprod Biomed Online. 2004;9(4):466–8.
77. Pasqualotto F, Sharma RK, Pasqualotto EB, Agarwal A. Poor semen quality and ROS-TAC scores in patients with idiopathic infertility. Urol Int. 2008;81(3):263–70.
78. Sharma R, Pasqualotto FF, Nelson DR, Thomas Jr AJ, Agarwal A. The reactive oxygen species-total antioxidant capacity score is a new measure of oxidative stress to predict male infertility. Hum Reprod. 1999;14(11):2801–7.
79. Aitken R, Harkiss D, Buckingham DW. Analysis of lipid peroxidation mechanisms in human spermatozoa. Mol Reprod Dev. 1993;35(3):302–15.
80. Oral O, Kutlu T, Aksoy E, Fiçicioğlu C, Uslu H, Tuğrul S. The effects of oxidative stress on outcomes of assisted reproductive techniques. J Assist Reprod Genet. 2006;23(2):81–5.

81. Tavilani H, Doosti M, Saeidi H. Malondialdehyde levels in sperm and seminal plasma of asthenozoospermic and its relationship with semen parameters. Clin Chim Acta. 2005; 356(1–2):199–203.
82. Hsieh Y, Chang CC, Lin CS. Seminal malondialdehyde concentration but not glutathione peroxidase activity is negatively correlated with seminal concentration and motility. Int J Biol Sci. 2006;2(1):23–9.
83. Suleiman S, Ali ME, Zaki ZM, el-Malik EM, Nasr MA. Lipid peroxidation and human sperm motility: protective role of vitamin E. J Androl. 1996;17(5):530–7.
84. Keskes-Ammar L, Feki-Chakroun N, Rebai T, Sahnoun Z, Ghozzi H, Hammami S, et al. Sperm oxidative stress and the effect of an oral vitamin E and selenium supplement on semen quality in infertile men. Arch Androl. 2003;49(2):83–94.
85. Greco E, Iacobelli M, Rienzi L, Ubaldi F, Ferrero S, Tesarik J. Reduction of the incidence of sperm DNA fragmentation by oral antioxidant treatment. J Androl. 2005;26(3):349–53.
86. Greco E, Romano S, Iacobelli M, Ferrero S, Baroni E, Minasi MG, et al. ICSI in cases of sperm DNA damage: beneficial effect of oral antioxidant treatment. Hum Reprod. 2005;20(9): 2590–4.
87. Thiele J, Friesleben HJ, Fuchs J, Ochsendorf FR. Ascorbic acid and urate in human seminal plasma: determination and interrelationships with chemiluminescence in washed semen. Hum Reprod. 1995;10(1):110–5.
88. Ricci G, Perticarari S, Boscolo R, Simeone R, Martinelli M, Fischer-Tamaro L, et al. Leukocytospermia and sperm preparation—a flow cytometric study. Reprod Biol Endocrinol. 2009;7:128.
89. Mortimer D. Sperm preparation methods. J Androl. 2000;21(3):357–66.
90. Agarwal A, Deepinder F, Cocuzza M, Agarwal R, Short RA, Sabanegh E, Marmar JL. Efficacy of varicocelectomy in improving semen parameters: new meta-analytical approach. Urology. 2007;70(3):532–8.
91. Chen SS, Huang WJ, Chang LS, Wei YH. Attenuation of oxidative stress after varicocelectomy in subfertile patients with varicocele. J Urol. 2008;179(2):639–42.

Chapter 25
Oxidative Stress and Infection

Enzo Vicari, Sandro La Vignera, and Aldo E. Calogero

Abstract Male accessory gland infections (MAGI) are included among the conventional diagnostic categories recognized to cause male infertility. They constitute a clinical model of oxidative stress for a number of considerations: (a) some uropathogens or etiological agents of sexually transmitted diseases (*Chlamydia trachomatis, Ureaplasma urealyticum*) by themselves, microbial products, and/or toxic metabolites may contribute to an overproduction of reactive oxygen species (ROS); (b) the canalicular spread of pathogens to one or more male accessory glands causes a further increase of ROS production, since they become the site of inflammation as shown by the presence of morphostructural abnormalities. The infecting pathogen triggers an inflammatory process which includes a series of multiple persistent components, such as kinetic of leukocyte subpopulations, pattern of cytokine production, and morphostructural abnormalities of the infected glands. This results in a final impairment of conventional and non-conventional sperm parameters. Therefore, MAGI-related oxidative stress is the sum of a microenvironmental and sperm-related damage. This includes several redox imbalance in the gland (ratio of gland inflamed areas to noninflamed areas), pattern of cytokine release (prooxidative/antioxidant ratio), and sperm microenvironment.

Keywords Chronic bacterial prostatitis • Oxidative stress • Infection • MAGI • Sperm parameters • Seminal leukocyte concentration • ROS production

25.1 General Background

Reactive oxygen species (ROS) (superoxide anion, hydrogen peroxide, and hydroxyl) are products of the normal cell oxidative metabolism, mainly mitochondrial, through enzymatic oxygen reduction reactions to produce energy [1]. Under physiological

E. Vicari, MD • S.L. Vignera, MD • A.E. Calogero, MD (✉)
Department of Internal Medicine and Systemic Diseases,
Policlinico "G. Rodolico", Catania, Italy
e-mail: acaloger@unict.it

A. Agarwal et al. (eds.), *Studies on Men's Health and Fertility*, Oxidative Stress in Applied Basic Research and Clinical Practice, DOI 10.1007/978-1-61779-776-7_25,

conditions, ROS production plays an important regulatory role, named "mediator of conception" [2], in various steps of the male reproductive process (sperm capacitation, acrosome reaction, sperm hyperactivation, binding to the zona pellucida; sperm–oocyte fusion) [3]. On the other hand, oxidative stress (OS) occurs when the generation of ROS exceeds the system's ability to neutralize or eliminate them. Hence, OS may result from an elevated ROS production and/or a reduced total antioxidant capacity.

Before ROS production is measured, confounding factors should be excluded, since both the same cells that generate low concentrations of ROS under physiological conditions [4] and cells (leukocytes, macrophages, spermiophages) triggered by the infection and by the dynamic of the chronic inflammatory process contribute to ROS overproduction. Confounding factors include certain lifestyles (smoking, alcohol consumption, drugs) and systemic and male reproductive tract diseases, such as primary testiculopaties cause of OS (cryptorchidism, varicocele, postorchitis testicular atrophy, occupational/environmental pollutants, testicular cancer) [5].

25.2 Infections of the Male Genital Tract and Infertility

25.2.1 Initial Nosological Gaps

The state of the art in the field of prostatitis and infertility suffers some gaps originating from classification and definitions given by two parallel research groups ending up with two independent consensus conferences.

The first, originating mainly from the urologic area (NIH Chronic Prostatitis Collaborative Research Network), was inspired by the clinical aspects of prostatitis (difficult to diagnose and to effectively treat) and includes only two categories (II and IV) associated with male infertility [6, 7]. According to the NIH classification, seminal bacterial infection, corresponding to chronic bacterial prostatitis (category II), has a marginal epidemiologic role, since bacterial prostatitis (acute or chronic) accounts for a low number of patients with prostate symptoms (5–10%) [8, 9]. The NIH prostatitis classification recognizes a main role to one gland (prostate) and to two factors: presence of microbial agents and evidence of inflammation (leukocytes) in the semen, expressed prostatic secretion (EPS), or the third voided urine specimen, the so-called VB3.

The second of uroandrologic area was represented by a WHO Task Force on the Diagnosis and Treatment of Infertility [10] and inspired by excretory, posttesticular causes of male infertility. It identifies a diagnostic category affecting male reproductive function and fertility, named male accessory gland infection (MAGI). The presence of MAGI is diagnosed in patients with oligo-, astheno-, and/or teratozoospermia (OAT) who fulfill the WHO [10] conventional criteria (Fig. 25.1).

The NIH prostatitis classification is more useful for the diagnosis in presence of prostatitis symptoms, whereas the WHO criteria respond mainly to male infertility

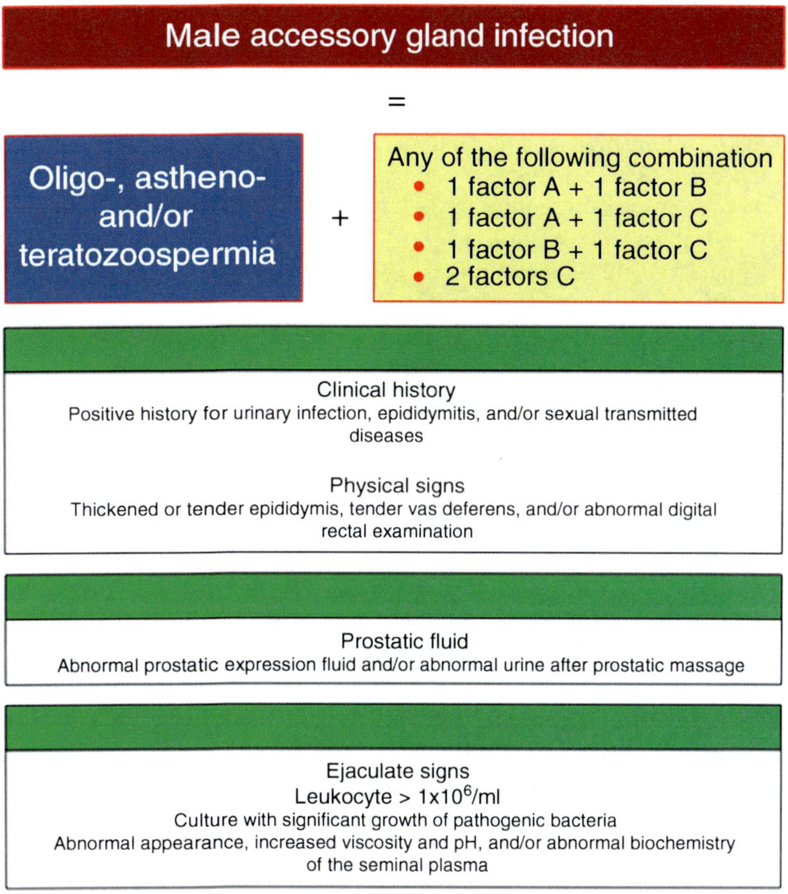

Fig. 25.1 Conventional WHO criteria to diagnose male accessory gland infections (MAGI) [10]

problem solving. The diagnostic workup linked to the original classifications is the corollary of these points of views. Therefore, both classifications should be maintained, though with a different field of application.

Among microbials, some Gram-negative bacteria such as Enterobacteriaceae (*Escherichia coli*, *Klebsiella* sp., *Proteus*, *Serratia*, *Pseudomonas* sp., etc.) have been recognized as known prostate pathogens (category II, NIH classification) because of their strong association with a clear positive clinical history (prior and/or recurrent urinary tract infection, sexually transmitted disease, congenital urogenital abnormalities) and some urogenital abnormalities at the physical examination. On the other hand, the only presence of some microorganisms is interpreted by a number of investigators as "probable" or "possible" prostate infection when, respectively, Gram-positive pathogens (*Enterococcus* sp.,

Streptococcus viridans, and *Streptococcus pyogens*, and *Staphylococcus aureus*) are present, or when coagulase-negative pathogens (*Staphylococcus epidermidis*), *Chlamydia trachomatis*, *Ureaplasma urealyticum*, anaerobes are present.

The major difficulty in interpreting microbiological findings is the presence of contaminating, indigenous microbial flora, inhibitory substances known to be in the prostatic secretions and/or previous antibiotic treatments. Thus, the diagnosis of bacterial prostatitis may be confirmed by quantitative bacteriological cultures in the semen (>10^3 pathogenic bacteria or >10^4 nonpathogenic bacteria in the seminal plasma diluted 1:2 with saline solution) [11] or segmented cultures, such as the four [12] or the two [13] glass test.

On the other hand, although theoretically the WHO definition of MAGI includes different diagnostic entities [prostatitis, prostatovesiculitis (PV), prostatovesiculoepididymitis (PVE)], practically it recognizes only the negative role of one gland (prostate) (identified by the factors of the group B) and one factor (infectious noxae: bacteria, *C. trachomatis*, *U. urealyticum*) on male reproductive function and fertility. Thus, the WHO conventional criteria, though not entirely coherent with the definition of MAGI which should include different diagnostic entities, represent a clinical aid to diagnose a prostate infection, but underestimate the important role of the chronic inflammatory response in terms of leukocytospermia and its products (ROS, cytokines) which, on the contrary, have been shown to play a relevant role.

Moreover, although acute or chronic bacterial prostatitis accounts for a low number of cases (5–10%) [7], MAGI has been reported to be present in a wide range (1.6–15%) of male infertile patients attending various infertility clinics [14–17]. Therefore, chronic, mainly symptomless MAGI may contribute to infertility to a various extent depending on the site of infection [9, 18–20] and/or on the host's inflammatory response in terms of leukocytospermia and leukocyte products: ROS [19–22] and/or cytokines [23–25]. Thus, the broad range of prevalence and the open debate with pros and cons on the role of MAGI in male infertility [26] relate to various factors:

1. Different consulting physician speciality, which has an impact on the workup of the infertile patient with MAGI
2. Incomplete and not well-defined diagnosis; for example, "sperm infection" is an improperly used laboratory definition, and it should be correctly replaced by the male diagnostic category defined conventionally as MAGI
3. Lack of appropriate morphostructural evaluation of the gland(s) involved in the inflammatory process
4. The highest estimated prevalence may be found in those studies which include patients who require specific andrological clinical counseling for persistent infection (even after antimicrobials) and/or previous assisted reproduction program failure

25.2.2 Negative Influences of the Initial Gaps

The lack of a precise diagnosis of site, based on a thorough clinical characterization of the accessory glands, has caused the following negative influences:

(a) The enrolment of heterogeneous cohorts of patients with infections. This has, in turn, produced contrasting and not comparable findings on the main conventional sperm parameters [9]
(b) The disproportionate number of items on Public Medline among putative clinical MAGI categories. According to a library search conducted up to February 2011, 5,250, 173, and 2,712 items were found for chronic microbial prostatitis, PV, and PVE, respectively
(c) Underutilization of ultrasound scanning in infertile patients with chronic MAGI
(d) Lack of synergy and of application of the recent basic research knowledge on the inflammatory response (leukocytospermia and ROS overproduction) and a clinical ultrasound-based approach [27]. This led us to hypothesize that for many years the above measurements of the inflammatory response have been preferentially evaluated in patients with idiopathic infertility rather than in patients with proven MAGI
(e) Finally, a careless diagnosis of MAGI does not allow making an early diagnosis and management of PV, which may be regarded as an intermediate condition of MAGI, before it becomes PVE

25.3 Call for a Review of This Topic

In the course of male genital tract infections, the invasion of microorganisms leads to a defense reaction of the accessory gland where they localize. This includes non-specific and specific immune reactions, with a transfer of oxidative damage from the infected gland to the spermatozoa present in the posttesticular sperm reservoirs. Prostate, seminal vesicles, and epididymis play various physiological actions on the posttesticular sperm reserve, and the time of interaction between the seminal plasma produced by these glands with spermatozoa is different. Therefore, the definition of male genital tract infection is improper and must be replaced with the appropriate definition of the site of infection on the male accessory glands. Thus, the NIH classification of prostatitis is reductive, and we prefer to use the definition of MAGI. Since the results on OS and infection have two very different nosological references represented respectively by chronic bacterial prostatitis (category II, NIH classification) and MAGI, this exciting topic is discussed in the next two sections.

25.4 Oxidative Stress and Chronic Bacterial Prostatitis

Bacteria responsible for prostate infection may originate from the urinary tract or can be sexually transmitted [28, 29]. The OS observed during chronic bacterial prostatitis is the result of elevated ROS production and/or reduced total antioxidant capacity [20, 30]. Indeed, infecting microorganisms trigger an inflammatory defense reaction in the prostate with a resulting OS due to ROS overproduction

[18, 25, 31, 32]. Current or past *C. trachomatis* infection has also been linked with an increased oxidative damage to spermatozoa [33].

Some bacteria (Gram-negative enteropathogens, *U. urealyticum*, *C. trachomatis*) may contribute to ROS overproduction by themselves, through products of their membrane (LPS from Gram negative or *C. trachomatis*) and/or through toxic metabolites (H_2O_2 and NH_3 produced by *U. urealyticum*) [32, 34, 35]. Only few studies have focused their attention on the correlation between the type of germ, OS, and sperm quality. In vitro incubation of spermatozoa from normozoospermic healthy men with various strains of bacteria resulted in a significant increase of malondialdehyde (MDA), an end-product of OS, after exposure to *Bacteroides ureolyticus*, *Staphylococcus hemolyticus,* or *E. coli* [34].

Shahed and Shoskes [36] showed that sperm OS in symptomatic patients with chronic bacterial prostatitis related to both ROS overproduction (especially with positive cultures) and reduced antioxidant capacity in men with category III prostatitis. Furthermore, the observed increased levels of OS markers and their decrease after treatment with antimicrobials (category II) or with the antioxidant dietary supplement (category III) suggested that Gram-positive bacteria in the EPS of some men with chronic pelvic pain syndrome may represent true pathogens on the basis of the clinical response to antibiotics. Zhou et al. [37] showed a significant increase of OS markers (plasma nitric oxide and erythrocyte MDA) in patients with chronic bacterial prostatitis compared to healthy normal volunteers, whereas plasma vitamin C, vitamin E, and β-carotene as well as erythrocyte superoxide dismutase, catalase, and glutathione peroxidase activities were significantly lower in patients with chronic bacterial prostatitis. Furthermore, OS and oxidative damage were closely related to the course of chronic bacterial prostatitis.

Although the demonstration of an increased OS is clear in these studies, the causal role of the infectious agents is limited by the number of studies on a specific strain of microorganisms and by the methodological approach which requires conventional peroxidase staining for the detection of seminal leukocytes which results in the identification of polymorphonuclear leukocytes and macrophages only.

Other studies about inflammatory prostatitis (category IIIA, NIH classification) cannot be included, since it represents another category of prostatitis, according to the NIH classification, improperly used to evaluate the involvement of leukocytes and/or proinflammatory cytokines (IL-1, IL-6, IL-8, TNF-α) during the infective process in the onset of the OS [18, 25, 34, 38].

25.5 Oxidative Stress and MAGI

MAGI are a consequence of microorganism canalicular spreading via urethra, prostate gland, seminal vesicles, deferent duct, epididymis, and testis. Hematogenous infections are rare.

In the diagnostic workup of MAGI, sperm analysis is viewed as a fundamental step. Indeed, MAGI is diagnosed in the presence of one or more sperm parameter abnormalities (OAT) associated with a combination of two or more of the following factors (Fig. 25.1):

- Factor A: history positive for urinary tract infections and/or sexually transmitted diseases and/or physical urogenital examination
- Factor B: expressed prostate signs of infection and/or inflammation
- Factor C: ejaculate signs of infection

The initiators of the infections responsible for MAGI are microorganisms originated from the urinary tract or sexually transmitted diseases [39]. A retrograde ascent of the pathogens is the usual route of infection. The main infectious agents are *Neisseria gonorrhoeae* and *C. trachomatis*, as well as Enterobacteria, although with a lower frequency [40]. Signs characteristic of MAGI are leukocytospermia and increased cytokines and ROS production. The following complications may also occur: epididymal and extraepididymal duct obstruction, spermatogenesis impairment in orchitis, sperm function impairment, and dysfunctions of the male accessory glands [40].

The initiating pathogens trigger an intermediate chronic inflammation, including a series of multiple persistent components, such as presence of specific leukocyte subpopulations, cytokine kinetic, morphostructural abnormalities of the infected accessory glands [19, 41–44], and a final sperm damage which impairs conventional and nonconventional sperm parameters. Thus, the negative impact on sperm quality during MAGI recognizes one or more of the following mechanisms: (1) secretory dysfunction of one or more accessory gland, (2) spermatogenesis deterioration, and (3) unilateral or bilateral obstruction or dysfunctional sperm distal urinary tracts (ampullar-vesiculo-ductal urosperm dyssynergia).

Chronic, mainly symptomless MAGI may contribute to infertility to a various extent depending on the site of inflammation [9, 18–20] and/or on the inflammatory response in terms of leukocytospermia and its products: ROS [16, 19, 22, 23] and/or cytokine production [17, 25, 45]. The glandular secretory dysfunction plays a relevant role, and it is expressed through a nonspecific chronic inflammatory reaction (leukocytospermia, increased seminal plasma proinflammatory cytokines, and ROS overproduction) and/or specific autoimmune response (production of antisperm antibodies) [19, 25, 41, 44]. These bioactive substances may persist even after an apparent postantibiotic microbial eradication, because of a constant OS, as antioxidant and scavengers are depleted gradually during the chronic inflammatory response [41, 44].

Gland structural abnormalities may, in addition, play a negative role. In fact, infertile patients with MAGI and elevated bacteriospermia ($\geq 10^5$ CFU/mL) or with *C. trachomatis* or *U. urealyticum* infection (at the urethral swabs after prostate massage) have the highest number of ultrasound abnormalities involving more glands (PVE) [19, 45]. These patients show an increased inflammatory response and an impaired semen quality directly related to the extension of MAGI (prostatitis < PV < PVE) [19] or a strong association with sperm abnormality [46].

Table 25.1 Ultrasound findings suggestive of prostatitis, vesiculitis, and/or epididymitis [19]

Prostatitis alone

Prostatitis alone is suspected in presence of two or more of the following ultrasound signs:

- Glandular asymmetry
- Hypoechogenicity associated with edema
- Hyperechogenicity associated with areas of calcification
- Dilation of the periprostatic venous plexus

Prostatovesiculitis

Prostatovesiculitis is suspected when in addition to the above-mentioned ultrasound features present in the prostate gland, two or more of the following ultrasound signs are present in the seminal vesicles:

- Enlargement and asymmetry
- Thickening and calcification of the glandular epithelium
- Polycyclic areas separated by hyperechogenic septa

Prostatovesiculoepididymitis

Prostatovesiculoepididymitis is suspected when in addition to the above-mentioned ultrasound features present in the prostate gland and in the seminal vesicles, two or more of the following ultrasound signs are present in the epididymis:

- Increased size of the head (craniocaudal diameter >12 mm) and/or tail (craniocaudal diameter >6 mm)
- Presence of multiple microcystic lesions confined to the epididymal head and/or tail
- Edematous hyperechoic epididymis
- Large hydrocele

25.6 Oxidative Stress According to the Type of the Infecting Microorganism and the Extension of MAGI

Between 1995 and 2002, we adopted a comprehensive approach in infertile patients which enabled us to identify an initial selected group of 1,127 patients with OAT without a concomitant female factor of infertility (ovarian, tubal, or cervical factor). According to the conventional WHO criteria [10], a diagnosis of excretory OAT due to MAGI was made in 547 patients (48.5%). It was microbial in 267 cases and amicrobial in the remaining 280 patients. On the basis of the presence of ultrasound abnormalities in the prostate, seminal vesicles, and epididymis (by scrotal and transrectal ultrasound scans) (Table 25.1), we could establish the site of the inflammatory process in 122 patients with microbial ($\geq 10^5$ CFU/mL) MAGI; in particular, 42.6% of them had prostatitis alone, 26.2% had PV, and the remaining 31.2% had PVE [16].

Continuing the same line of research, in a larger series of 382 patients out of 2,712 infertile couples (13.8%) who came to our observation in the last 11 years, the previous findings were confirmed. Particularly, 40.3% of them had prostatitis alone, 27.5% had PV, and the remaining 32.2% had PVE. Microbiological findings in these categories of MAGI showed a higher, but not statistically significant percentage of Gram-negative or Gram-positive microorganisms in patients with prostatitis alone or PV, respectively. Furthermore, *C. trachomatis*, detected at the urethral swab after prostate

Table 25.2 Microbiological findings in three categories of patients with MAGI exhibiting ultrasound abnormalities (ultrasound-positive)

Clinical categories	Prostatitis alone	Prostatovesiculitis	Prostatovesiculoepididymitis
Number of patients	154 (100%)	105 (100%)	123 (100%)
% of Clinical categories	40.3	27.5*	32.2*
Microorganisms $>10^5$ CFU/mL (sperm culture)			
Gram-negative	56 (36.4%)	40 (38.1%)	39 (31.7%)
• *Escherichia coli*	31	23	21
• *Proteus* spp.	13	9	12
• *Coliformi* spp.	6	4	3
• *Klebsiella*	6	4	3
Gram-positive	44 (28.6%)	38 (36.2%)	33 (26.8%)
• *Enterococcus*	25	21	20
• *Streptococcus* spp.	13	9	7
• *Staphylococcus* spp.	6	8	6
Microorganisms detected at urethral swab after prostate massage			
Chlamydia trachomatis	34 (22.1%)	14 (13.3%)	37 (30.1%)*
Ureaplasma urealyticum	20 (12.9%)	13 (12.4%)	14 (11.4%)

Values are expressed as number of the strains identified; Percentages are reported in *parentheses*.
*$p<0.01$ vs. prostatitis alone or prostatovesiculitis

massage, was significantly higher in patients with PVE, whereas *U. urealyticum* infection had a similar percentage of detection in the three categories of MAGI (Table 25.2). On the other hand, we found that patients with PVE had significantly worst sperm parameters (Table 25.3). No significant difference was observed in patients with prostatitis alone vs. patients with PV, MAGI ultrasound-negative patients vs. the fertile control group, or between patients with prostatitis alone or PV compared with ultrasound-negative MAGI patients and the fertile group.

Seminal leukocyte concentration, measured in all ultrasound-positive MAGI subgroups by immunocytochemical staining [47], using an antialkaline phosphatase monoclonal antibody CD45 (Dako Italia, Milan, Italy), was significantly higher than those found in MAGI ultrasound-negative patients and controls. Among ultrasound-positive MAGI patients, those with PVE had significantly higher values compared with patients with prostatitis alone, but not than those present in patients with PV (Table 25.3).

All ultrasound-positive MAGI patient groups showed significantly higher baseline and formyl-methionyl-leucyl-phenylalanine (fMLP)-stimulated ROS production than ultrasound-negative MAGI patients or controls (Table 25.3). Moreover, patients with complicated MAGI had a median baseline ROS production significantly higher than that found in patients with prostatitis alone; the highest fMLP-stimulated ROS production was observed in the patients with PVE. Furthermore, when we analyzed conventional sperm parameters, leukocyte concentration, and ROS production, these variables have been evaluated according to microbial pathogens (including *C. trachomatis* or *U. urealyticum*, detected at the urethral swab after prostate massage) found.

Table 25.3 Conventional sperm parameters, seminal leukocyte concentration, and basal and fMLP-stimulated radical oxygen species production from three categories of patients with microbial (≥10^5 CFU/mL or presence of *C. trachomatis* or *U. urealyticum*) MAGI exhibiting ultrasound abnormalities (altogether defined as ultrasound-positive MAGI), in patients with a low number of ultrasound abnormalities (altogether defined as ultrasound-negative MAGI) and in a control group of fertile men

	Ultrasound-positive MAGI ($n=382$)				
Parameters	Prostatitis alone ($n=154$)	Prostatovesiculitis ($n=105$)	Prostatovesiculoepididymitis ($n=123$)	Ultrasound-negative MAGI ($n=160$)	Fertile men ($n=34$)
Sperm parameters and seminal leukocytes					
Sperm concentration (mil/mL)	47.5 (11.7–132.6)	38 (8.4–150.6)	14.7*† (2.1–61.3)	49.5 (15–153.8)	56.5 (22.4–137.6)
Total sperm number (mil/ejaculate)	123.6 (19.5–459.6)	124.0 (29.9–615.7)	59.0*† (17–122)	134.5 (69–534.5)	155.0 (42.7–542)
Progressive motility (%)	21.5 (14–28)	19.5 (10–26)	11.5*† (2–18)	22.5 (18–31)	24.0 (20–32)
Normal forms (%)	20.5 (15–24)	17.0† (10–22)	11*† (3–19)	41.0 (28–47)	27 (20.2–32.5)
Leukocytes (mil/mL)	2† (0.8–4)	3.2† (1.1–9)	3.7† (1.9–9.5)	1.4 (0.8–2.0)	0.8 (0.4–1.2)
Radical oxygen species production (45% Percoll fraction)					
Baseline (<10^3 cpm)	185.5† (48–487.4)	303.6† (89.5–1,000)	496.0*† (148.6–1,000)	54.5 (18.4–73.2)	32.0 (12.7–44.6)
fMLP-stimulated (×10^3 cpm)	392.4† (168–879)	587.0† (304–1,000)	858.7*† (359–1,000)	72.4 (31.8–127.0)	58.7 (22.4–75.3)

Values were expressed as median and the 10th–90th percentiles in *parentheses*. $^{*}p<0.01$ vs. prostatitis alone or prostatovesiculitis; $^{\dagger}p<0.01$ vs. ultrasound-negative MAGI or fertile men

25.7 Sperm Parameters, Seminal Leukocyte Concentration, and ROS Production According to the Extension of MAGI and the Infecting Microorganism

25.7.1 MAGI Associated with Gram-Negative or Gram-Positive Microorganism Infection

Among ultrasound-positive MAGI categories, we found that patients with PVE have significantly lower sperm concentration, total sperm number, motile spermatozoa, and spermatozoa with normal morphology compared to patients with prostatitis alone or patients with PV, regardless of the presence of Gram-negative or Gram-positive microorganisms (Table 25.4).

Seminal leukocyte concentration, measured by immunocytochemical staining [47], was significantly higher in patients with PVE compared to patients with prostatitis alone, but similar to patients with PV (Table 25.4).

Basal and fMLP-stimulated ROS production had a pattern similar to that observed for seminal leukocytes.

25.7.2 MAGI Associated with C. trachomatis or U. urealyticum Infection

Among ultrasound-positive MAGI categories associated to *C. trachomatis* or *urealyticum* infection, sperm parameters, leukocyte concentration, and ROS production were similar to the findings described for MAGI caused by other microorganisms (Table 25.5). Nevertheless, it should be pointed out that in presence of *C. trachomatis* or *U. urealyticum* infection, patients with PVE have the worst parameters in absolute compared with the other two MAGI categories (prostatitis alone or PV) and microbiological etiology (Gram positive > Gram negative > *C. trachomatis* or *U. urealyticum*).

25.8 How to Correctly Diagnose MAGI?

Conventional WHO criteria for MAGI (Fig. 25.1) should be integrated with the ultrasound exploration of the male accessory glands and an accurate leukocyte assessment (by immunocytochemistry). Therefore, we propose an algorithm for the assessment of MAGI based on sequential levels (Fig. 25.2). Each level includes relative "processes" broken down into specific "actions" and "outcomes."

Table 25.4 Conventional sperm parameters, seminal leukocyte concentration, and basal and fMLP-stimulated radical oxygen species production from three categories of patients with ultrasound-positive MAGI according to the presence of Gram-negative or Gram-positive microorganisms (≥10^5 CFU/mL)

	Prostatitis alone ($n=100$)		Prostatovesiculitis ($n=78$)		Prostatovesiculoepididymitis ($n=72$)	
	Gram negative ($n=56$)	Gram positive ($n=44$)	Gram negative ($n=40$)	Gram positive ($n=38$)	Gram negative ($n=39$)	Gram positive ($n=33$)
Sperm parameters and seminal leukocytes						
Sperm concentration (mil/mL)	32 (10–102.6)	37 (13–162.7)	27 (8.0–100.2)	34 (15–118.5)	13.2* (2.0–56.4)	16.3* (2.0–64.5)
Total sperm number (mil/ejaculate)	101.4 (19.5–422.3)	115.6 (23.5–473.2)	89.0 (15.2–315.3)	104.0 (21.2–425.7)	57.5* (5–208)	62.0* (13–227)
Progressive motility (%)	17.0 (14–24)	18.0 (13–28)	16.0 (10–23)	20.0 (13–27)	10.0* (6–13)	13.0* (5–18)
Normal forms (%)	22.0 (15–24)	22.5 (15–23)	18.0 (10–22)	19.0 (12–22)	8.0* (5–19)	9.0* (7–19)
Seminal leukocytes (mil/mL)	2.2 (1.2–5.4)	1.9 (1.1–4.7)	3.5† (1.4–7.0)	3.0† (1.1–6.3)	3.8† (2.0–8.1)	3.4† (1.9–5.7)
Radical oxygen species production (45% Percoll fraction)						
Baseline (×10^3 cpm)	92.2 (58.6–178.4)	75.3 (48–143.5)	285.6† (89.5–489)	232.7† (71.9–462)	396.0† (152.2–1,000)	341.6† (128.4–1,000)
fMLP-stimulated (×10^3 cpm)	225.7 (168–583.3)	182.4 (120–879)	598.0† (304–1,000)	560.5† (243.8–1,000)	738.2* (337.7–1,000)	623.5* (286.3–1,000)

Values are expressed as median and the 10th–90th percentiles in *parentheses*. *$p<0.01$ vs. both prostatitis alone or prostatovesiculitis; †$p<0.01$ vs. prostatitis alone

Table 25.5 Conventional sperm parameters, seminal leukocyte concentration, and basal and fMLP-stimulated radical oxygen specie production from three categories of patients with ultrasound-positive MAGI according to the presence of *C. trachomatis* or *U. urealyticum*

	Prostatitis alone ($n=54$)		Prostatovesiculitis ($n=27$)		Prostatovesiculoepididymitis ($n=51$)	
Parameters	*C. trachomatis* ($n=34$)	*U. urealyticum* ($n=20$)	*C. trachomatis* ($n=14$)	*U. urealyticum* ($n=13$)	*C. trachomatis* ($n=37$)	*U. urealyticum* ($n=14$)
Sperm parameters and seminal leukocytes						
Sperm concentration (mil/mL)	23.0 (8.0–52.6)	29.0 (13.0–78.2)	21.9 (5.5–45.6)	25.2 (9.4–63.5)	10.0* (1.5–26.4)	13.8* (2.0–43.5)
Total sperm number (mil/ejaculate)	87.4 (14.5–178.3)	101.6 (18.2–373.2)	59.0 (12.0–146.7)	81.0 (21.2–325.7)	37.4* (4.5–78.0)	42.5* (6.3–87.3)
Progressive motility (%)	13.0 (6.0–18.0)	14.0 (8.0–20.0)	10.4 (2.4–15.0)	12.5 (4.0–18.0)	5.0* (0–10.3)	5.5* (0–12.2)
Normal forms (%)	20.0 (15.0–26.0)	22.4 (18.0–28.0)	14.0 (10.0–22.0)	16.5 (11.0–23.0)	6.5* (3.2–8.6)	8.5* (4.2–10.2)
Seminal leukocytes (mil/mL)	3.2 (2.6–6.2)	3.0 (1.8–5.2)	4.6† (3.0–7.4)	3.6† (1.1–6.3)	5.8* (3.8–11.3)	4.8* (2.4–7.7)
Radical oxygen species production (45% Percoll fraction)						
Baseline ($\times 10^3$ cpm)	273.6 (165.0–487.4)	197.3 (136.8–989.3)	463.6† (277.5–1,000)	339.7† (153.0–788.5)	570.7* (452.2–1,000)	443.6* (341.7–1,000)
fMLP-stimulated ($\times 10^3$ cpm)	525.7 (358.0–879.7)	397.2 (304.0–1,000)	775.0† (359–1,000)	577.5† (283.4–1,000)	789.6*† (683.2–1,000)	773.5*† (586.8–1,000)

Values are expressed as median and the 10th–90th percentiles in *parentheses*. *$p<0.01$ vs. both prostatitis alone or prostatovesiculitis; †$p<0.01$ vs. prostatitis alone

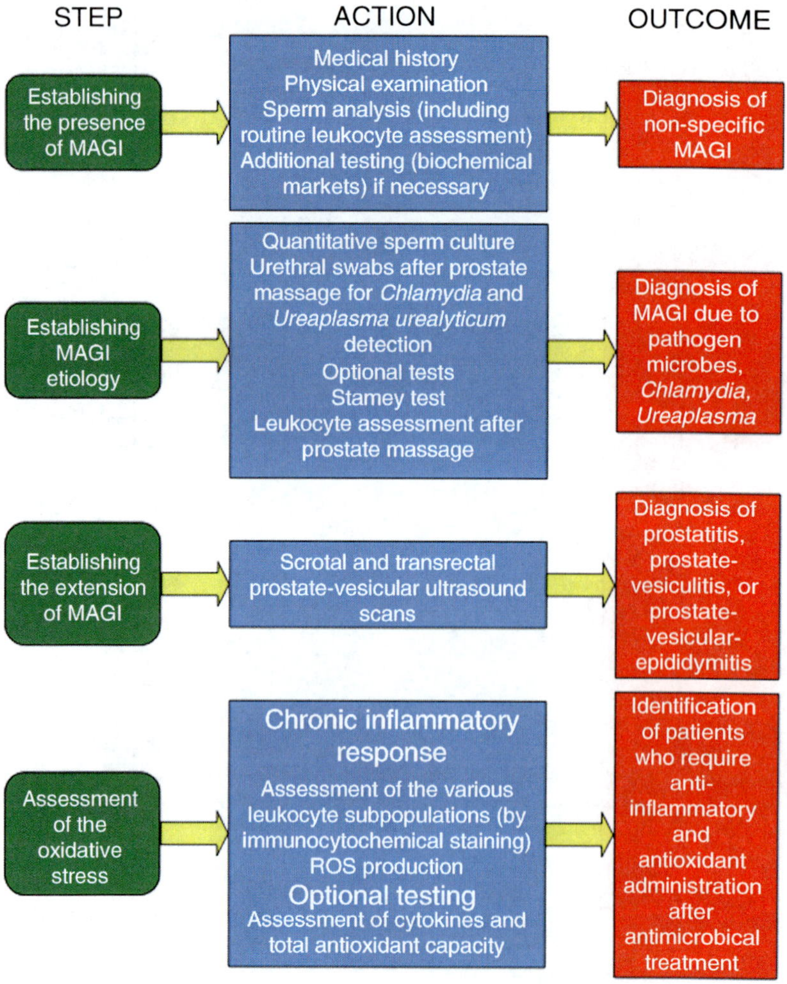

Fig. 25.2 Algorithm for the diagnosis of MAGI based on sequential levels

(a) To suspect/identify the presence of MAGI, only "first-line testing" is required. This includes collection of the clinical history, urogenital physical examination, and sperm analysis including seminal leukocyte measurement. The presence of OAT in combination with other factors allows to suspect/identify the presence of MAGI
(b) When a diagnosis of MAGI is suspected, "second-line testing" which included microbiological investigation (sperm culture and urethral swabs) allows identifying the etiological nature of MAGI
(c) "Second-line testing" includes also the attempt to localize the site(s) of the inflammatory process by ultrasound scans (didymoepididymal and transrectal

prostatovesicular ultrasonography). The localization of the inflammatory process is based on the presence of a significant number of ultrasound abnormalities, for each gland (Table 25.1). These abnormalities have been found strongly associated with elevated bacteriospermia (>10^5 CFU/mL) and ROS overproduction in studies on selected infertile infected patients [16, 19]

(d) In patients with complicated (microbial or amicrobial) MAGI (PV or PVE), "third-line testing" is needed to evaluate the host inflammatory response and to prescribe a rationale therapeutic strategy. This third-line testing includes the accurate assessment of leukocytospermia and the production of spermiotoxic leukocyte-related products (ROS and cytokines) [16, 18–20] which may impair sperm function by inducing DNA damage and/or apoptosis [20, 24, 48]

25.9 How to Treat MAGI?

MAGI should be treated with antibiotics when noxae influence sperm morphology and/or function, cause obstruction of the seminal tract, and express themselves with a proven mono- or multiple glandular involvement. On the other hand, the presence of bacteriospermia or leukocytospermia alone does not necessarily mean that there is a glandular infection, since temporary inflammatory episodes are likely present in the majority of sexually active men [27]. Furthermore, bacteriospermia may simply represent contamination or colonization.

A prolonged interaction between the inflammatory noxae and sperm and/or male accessory sex glands enhances an inflammatory response through the release of seminal ROS (and cytokines), which may persisted even following antibiotic treatment in patients with complicated MAGI (PV and PVE) [16]. This requires a multistep treatment. For example, patients with PVE, who have a significantly higher ROS production compared to patients with prostatitis alone or PV, should initially be treated with antibiotics (fluoroquinolones, macrolides, or doxycycline), subsequently with nonsteroidal anti-inflammatory compounds and, as a third-line therapy, with antioxidants. When this sequential pharmacological management is adopted, a significant amelioration of some sperm (forward motility and viability; leukocyte concentration and ROS production) and reproductive (pregnancy rate) parameters is achieved [22, 23].

25.10 Concluding Remarks

The role of infectious microorganisms as etiological agents of male infertility has been controversial. According to the clinical presentation of chronic bacterial prostatitis (NIH classification), the pathogenetic mechanisms that cause sperm abnormalities are simplistic and are not supported by a relationship with the putative causative factors (microorganisms and leukocytes). The research is stagnant unless it is extended to prostatitis category IIIA which causes urinary irritative symptoms, but may not be so frequent in patients with infertility. Greater interest has instead been laid on the research on seminal leukocyte subpopulations, cytokine and ROS production, and presence of morphostructural abnormalities of the accessory glands (evaluated by ultrasound scans) in patients with MAGI.

We found a significant role of the extension of the inflammatory process on sperm parameters and OS, in patients with various MAGI diagnostic categories and a significant microbiological load or in presence of *C. trachomatis* or *U. urealyticum.* Indeed, patients with PVE had the worst sperm parameters and the highest seminal leukocyte concentration (CD45-positive by immunocytochemistry), and the greatest basal and fMLP-stimulated ROS production in the 45% Percoll fraction (rich in leukocytes). All these measured end points were intermediate in patients with PV and less compromised in patients with prostatitis alone. According to the type of the microorganism responsible of the infection, patients with PVE due to *C. trachomatis* or *U. urealyticum* infection had higher seminal leukocyte concentration and ROS production than those found in patients with PVE due to Gram-negative or Gram-positive microorganisms.

The repertoire of defense mechanisms deployed by neutrophils includes production of ROS, proteolytic enzymes, and bactericidal peptides that can significantly contribute to damage the gland site of the infection, thus worsening the initial injury caused by the infectious microorganism. The critical role played by neutrophils in controlling infections and other inflammation can lead to unwanted collateral damage to adjacent cells and tissues. Thus, neutrophil activity and regulatory (juxtacrine, paracrine, and endocrine) mechanisms involved in the tissue damage need to be rigorously controlled. It may also be hypothesized that pathogen microorganisms involved in a more extended infectious process, though without a direct contact with spermatozoa, may trigger other harmful effects of infection and OS on the basis of their virulence or through the release of soluble factors of the bacterial metabolism [49, 50].

In conclusion, sperm parameters represent the "end-target" of many possible pathophysiological mechanisms which may contribute to the onset of infertility related to urogenital infections. Therefore, despite the fact that an open debate with pros and cons on the role of MAGI in male infertility is going on, the andrologist should consider MAGI as a risk factor of male infertility [46]. Thus, the presence MAGI should be thoroughly evaluated in the infertile male patient. The above-reported considerations suggest that the conventional diagnostic criteria to diagnose MAGI [10] (Fig. 25.1) should be revised and extended (Fig. 25.3). Accordingly, we propose that the diagnostic strategy should include scrotal and prostatovesicular ultrasound scans, among the factors of group B. Factors of the group C should include *C. trachomatis* and *U. urealyticum* detection by urethral swab following prostate massage and to establish a pathogenic bacterial load $>10^5$ UFC/mL by sperm culture. Finally, the measurement of the seminal leukocyte should be

Male accessory gland infection

=

Oligo-, astheno- and/or teratozoospermia

\+

Any of the following combination
- 1 factor A + 1 factor B
- 1 factor A + 1 factor C
- 1 factor B + 1 factor C
- 2 factors C

Clinical history

Positive history for urinary infection, epididymitis, and/or sexual transmitted diseases

Physical signs

Thickened or tender epididymis, tender vas deferens, and/or abnormal digital rectal examination

Prostate, seminal vesicles, and/or epididymis ultrasound abnormalities (by scrotal and transrectal ultrasound scans)

	Prostate	Seminal vesicles	Epididymis
Prostatitis alone	≥2	0	0
Prostate-vesiculitis	≥2	≥2	0
Prostate-vesiculo-epididymitis	≥2	≥2	≥2

In case of prostatitis alone, examine the prostatic fluid

Abnormal prostatic expression fluid and/or abnormal urine after prostatic massage

Ejaculate signs

Leukocytes > 1×10^6/ml (measured by immunocytochemistry)

Culture with significant growth of pathogenic bacteria (≥10^5 CFU/ml)

Chlamydia trachomatis or *Ureaplasma urealyticum* detected in the urethral swabs after prostate massage

Abnormal appearance, increased viscosity and pH, and/or abnormal biochemistry of the seminal plasma

Fig. 25.3 Proposed new criteria to diagnose MAGI integrating those proposed by the WHO [10]

performed by immunocytochemistry rather than peroxidase staining to evaluate all leukocyte subpopulations.

The new proposed guidelines for MAGI should be applied to infertile patients to gain more careful information about sex gland-related OS production and to establish a more appropriate therapeutic strategy.

References

1. Valko M, Leibfritz D, Moncol J, Cronin MT, Mazur M, Telser J. Free radicals and antioxidants in normal physiological functions and human disease. Int J Biochem Cell Biol. 2007;39: 44–84.
2. Ruder EH, Hartman TJ, Blumberg J, Goldman MB. Oxidative stress and antioxidants: exposure and impact on female fertility. Hum Reprod Update. 2008;14:345–57.
3. Tremellen K. Oxidative stress and male infertility—a clinical perspective. Hum Reprod Update. 2008;14:243–58.
4. Aitken RJ, Baker MA, De Iuliis GN, Nixon B. New insights into sperm physiology and pathology. Handb Exp Pharmacol. 2010;198:99–115.
5. Makker K, Agarwal A, Sharma R. Oxidative stress & male infertility. Indian J Med Res. 2009;129:357–67.
6. Krieger JN, Nyberg Jr L, Nickel JC. NIH consensus definition and classification of prostatitis. JAMA. 1999;282:236–7.
7. Nickel JC, Downey J, Hunter D, Clark J. Prevalence of prostatitis-like symptoms in a population based study using the National Institute of Health chronic prostatitis symptom index. J Urol. 2001;165:842–5.
8. Weidner W, Schiefer HG, Krauss H, Jantos Ch, Friederich HJ, Altmannsberger M. Chronic prostatitis: a thorough search for etiologically involved microorganisms in 1,461 patients. Infection. 1991;19 Suppl 3:119–25.
9. Weidner W, Krause W, Ludwig M. Relevance of male accessory gland infection for subsequent fertility with special focus on prostatitis. Hum Reprod Update. 1999;5:421–32.
10. Rowe P, Comhaire F, Hargreave TB, Mellows HJ. World health organization manual for the standardized investigation and diagnosis of the infertile couple. Cambridge: Cambridge University Press; 1993.
11. Comhaire F, Verschraegen G, Vermeulen L. Diagnosis of accessory gland infection and its possible role in male infertility. Int J Androl. 1980;3:32–45.
12. Meares EM, Stamey TA. Bacteriological localization patterns in bacterial prostatitis and urethritis. Invest Urol. 1968;5:492–518.
13. Nickel JC. The Pre and Post Massage Test (PPMT): a simple screen for prostatitis. Tech Urol. 1997;3:38–43.
14. Comhaire FH, Rowe PJ, Farley TMM. The effect of doxycycline in infertile couples with male accessory gland infection: a double blind prospective study. Int J Androl. 1986;9:91–8.
15. Andreessen R, Sudhoff F, Borgmann V, Nagel R. Results of ofloxacin therapy in andrologic patients suffering from therapy-requiring asymptomatic infections. Andrologia. 1993;25: 377–83.
16. Vicari E. Effectiveness and limits of antimicrobial treatment on seminal leukocyte concentration and related specific radical oxygen species production in patients with male accessory gland infection. Hum Reprod. 2000;15:2536–44.
17. Diemer T, Hales DB, Weidner W. Immune-endocrine interactions and Leydig cell function: the role of cytokines. Andrologia. 2003;35:55–63.
18. Ochsendorf FR. Infections in the male genital tract and reactive oxygen species. Hum Reprod Update. 1999;5:399–420.
19. Vicari E. Seminal leukocyte concentration and related specific radical oxygen species production in different categories of patients with male accessory gland infection. Hum Reprod. 1999;14:2025–30.
20. Agarwal A, Saleh RA, Bedaiwy MA. Role of reactive oxygen species in the pathophysiology of human reproduction. Fertil Steril. 2003;79:829–43.
21. Comhaire FH, Mahmoud AM, Depuydt CE, Zalata A, Cristophe AB. Mechanisms and effects of male genital tract infection on semen quality and fertilizing potential: the andrologist's viewpoint. Hum Reprod Update. 1999;5:393–8.

22. Vicari E, Calogero AE. Effect of treatment with carnitines in patients with prostato-vesciculo-epididymitis. Hum Reprod. 2001;16:2338–42.
23. Vicari E, La Vignera S, Calogero AE. Antioxidant treatment with carnitines is effective in infertile patients with prostato-vesciculo-epididymitis and elevated seminal leukocyte concentration after treatment with non-steroidal anti-inflammatory compounds. Fertil Steril. 2002;78:1203–8.
24. Sanocka D, Jedrzejczak P, Szumala-Kakol A, Fraczek M, Kurpisz M. Male genital tract inflammation: the role of selected interleukins in regulation of pro-oxidant and antioxidant enzymatic substances in seminal plasma. J Androl. 2003;24:448–55.
25. Garrido N, Meseguer M, Simon C, Pellicer A, Remohi J. Prooxidative and anti-oxidative imbalance in human semen and its relation with male fertility. Asian J Androl. 2004;6:59–65.
26. Weidner W, Colpi GM, Hargreave TB, Papp GK, Pomerol JM, Ghosh C; EAU Working Group on Male Infertility. EAU guidelines on male infertility. Eur Urol. 2002;42:313–22.
27. Purvis K, Christiansen E. Infection in the male reproductive tract. Impact, diagnosis and treatment in relation to male infertility. Int J Androl. 1993;16:1–13.
28. Fraczek M, Kurpisz M. Inflammatory mediators exert toxic effects of oxidative stress on human spermatozoa. J Androl. 2007;28:325–33.
29. Wagenlehner FM, Diemer T, Naber KG, Weidner W. Chronic bacterial prostatitis (NIH type II): diagnosis, therapy and influence on the fertility status. Andrologia. 2008;40:100–4.
30. Sharma RK, Pasqualotto FF, Nelson DR, Thomas Jr AJ, Agarwal A. The reactive oxygen species-total antioxidant capacity score is a new measure of oxidative stress to predict male infertility. Hum Reprod. 1999;14:2801–7.
31. Mazzilli F, Rossi T, Marchesini M, Ronconi C, Dondero F. Superoxide anion in human semen related to seminal parameters and clinical aspects. Fertil Steril. 1994;62(4):862–8.
32. Potts JM, Sharma R, Pasqualotto F, Nelson D, Hall G, Agarwal A. Association of *Ureaplasma urealyticum* with abnormal reactive oxygen species levels and absence of leukocytospermia. J Urol. 2000;163:1775–8.
33. Segnini A, Camejo MI, Proverbio F. *Chlamydia trachomatis* and sperm lipid peroxidation in infertile men. Asian J Androl. 2003;5(1):47–9.
34. Fraczek M, Szumala-Kakol A, Jedrzejczak P, Kamieniczna M, Kurpisz M. Bacteria trigger oxygen radical release and sperm lipid peroxidation in in vitro model of semen inflammation. Fertil Steril. 2007;88:1076–85.
35. Shang XJ, Huang YF, Xiong CL, Xu JP, Yin L, Wan CC. *Ureaplasma urealyticum* infection and apoptosis of spermatogenic cells. Asian J Androl. 1999;1:127–9.
36. Shahed AR, Shoskes DA. Oxidative stress in prostatic fluid of patients with chronic pelvic pain syndrome: correlation with gram positive bacterial growth and treatment response. J Androl. 2000;21:669–75.
37. Zhou JF, Xiao WQ, Zheng YC, Dong J, Zhang SM. Increased oxidative stress and oxidative damage associated with chronic bacterial prostatitis. Asian J Androl. 2006;8:317–23.
38. Sarkar O, Bahrainwala J, Chandrasekaran S, Kothari S, Mathur PP, Agarwal A. Impact of inflammation on male fertility. Front Biosci (Elite Ed). 2011;3:89–95.
39. Marconi M, Pilatz A, Wagenlehner F, Diemer T, Weidner W. Impact of infection on the secretory capacity of the male accessory glands. Int Braz J Urol. 2009;35:299–308; discussion 308–9.
40. Krause W. Male accessory gland infections. Andrologia. 2008;40:113–6.
41. Vicari E, La Vignera S, Arancio A, Calogero AE. Male accessory gland infections and infertility. In: Colpi GM, editor. Male infertility today. Fotolito e stampa grafiche Gelmini, Milan, Italy. 2004;4:139–51.
42. Ludwig M, Vidal A, Diemer T, Pabst W, Failing K, Weidner W. Seminal secretory capacity of the male accessory sex glands in chronic pelvic pain syndrome (CPPS)/chronic prostatitis with special focus on the new prostatitis classification. Eur Urol. 2002;42:24–8.
43. Lotti F, Corona G, Mancini M, Filimberti E, Degli Innocenti S, Colpi GM, Baldi E, Noci I, Forti G, Adorini L, Maggi M. Ultrasonographic and clinical correlates of seminal plasma

interleukin-8 levels in patients attending an andrology clinic for infertility. Int J Androl. 2011;34:600–13.

44. Vicari E, La Vignera S, Castiglione R, Calogero AE. Sperm parameters abnormalities, low seminal fructose and reactive oxygen species overproduction do not discriminate patients with unilateral or bilateral post-infectious inflammatory prostato-vesiculo-epididymitis. J Endocrinol Invest. 2006;29:18–26.
45. Henkel RR. Leukocytes and oxidative stress: dilemma for sperm function and male fertility. Asian J Androl. 2011;13:43–52.
46. Bayasgalan G, Naranbat D, Radnaabazar J, Lhagvasuren T, Rowe PJ. Male infertility: risk factors in Mongolian men. Asian J Androl. 2004;6:305–11.
47. World Health Organization. Laboratory manual for the examination of human semen and semen-cervical mucus interaction. 4th ed. Cambridge: Cambridge University Press; 1999.
48. Perdichizzi A, Nicoletti F, La Vignera S, Barone N, D'Agata R, Vicari R, Calogero AE. Effects of tumour necrosis factor-α on human sperm motility and apoptosis. J Clin Immunol. 2007;27:152–62.
49. Alteri CJ, Smith SN, Mobley HL. Fitness of *Escherichia coli* during urinary tract infection requires gluconeogenesis and the TCA cycle. PLoS Pathog. 2009;5:e1000448.
50. Schulz M, Sánchez R, Soto L, Risopatrón J, Villegas J. Effect of *Escherichia coli* and its soluble factors on mitochondrial membrane potential, phosphatidylserine translocation, viability, and motility of human spermatozoa. Fertil Steril. 2010;94:619–23.

Chapter 26
The Role of Obesity in ROS Generation and Male Infertility

Anthony H. Kashou, Stefan S. du Plessis, and Ashok Agarwal

Abstract

Aim: To discuss the relationship between obesity and male infertility, specifically exploring the role of reactive oxygen species (ROS) production in obesity and the subsequent generation of oxidative stress, as well as abnormal hypothalamus-pituitary-gonadal regulation associated with obese males.

Methods: Review of PubMed database.

Results: Both enhanced ROS generation and abnormal hormonal regulation due to obesity are strongly correlated to suboptimal semen quality and, thus, reduced male reproductive potential.

Conclusion: The continuing rise and prevalence of both obesity and declining male sperm count all over the world call for additional research and a greater awareness to obesity as a potential etiology of male infertility.

Keywords Male infertility • Obesity • Reactive oxygen species • Oxidative stress • Hormones

---

A.H. Kashou, BS
Center for Reproductive Medicine, Cleveland Clinic, 9500 Euclid Avenue, Cleveland, OH 44195, USA

S.S. du Plessis, PhD, MBA
Division of Medical Physiology, Stellenbosch University Francie van Zijl Avenue, PO Box 19063, Tygerberg 7505, South Africa

A. Agarwal, PhD (✉)
Center for Reproductive Medicine, Cleveland Clinic, Lerner College of Medicine, 9500 Euclid Avenue, Cleveland, OH 44195, USA
e-mail: agarwaa@ccf.org

A. Agarwal et al. (eds.), *Studies on Men's Health and Fertility*, Oxidative Stress in Applied Basic Research and Clinical Practice, DOI 10.1007/978-1-61779-776-7_26,

26.1 Introduction

With the onset of sedentary lifestyles, high-fat diets, and a general decline in physical activity, much evidence suggests that obesity is becoming a global pandemic [1, 2]. Obesity is reaching unprecedented levels in the Western world. One study revealed a prevalence of obesity in the USA of 19.8% [3], while another indicated a staggering 30% [4]. Reports have already shown that the world's overweight population has grown greater than its underweight population [4]. The World Health Organization (WHO) predicts that by 2015 approximately 2.3 billion adults will be overweight and that an additional 700 million will suffer from obesity [5].

Simultaneous to this alarming trend, there has been an apparent progressive increase in infertility rates over the past few decades. It has been indicated that 15% of all couples of reproductive age are infertile [6, 7], and up to 50% of all cases are believed to be due to the male factor alone [8]. Although still highly debated, the increased prevalence of overweight and obesity may account for the declining sperm counts over recent decades. According to a study conducted in 2000 by Swan et al., male sperm counts have been dropping by as much as 1.5% annually in the USA, as well as in other Western countries [9, 10]. Since these declines were not present in regions where obesity was less prevalent [9], this may further suggest a potential link between altered lifestyles, obesity, adverse health outcomes and, now, semen quality and male infertility [11].

As obesity trends continue to increase and expectation levels show no signs of decline, the interaction between obesity and fertility has received astonishing attention [3, 4]. Nevertheless, obesity and its effect on sperm count has just recently been documented [11, 12]. Studies have shown a negative correlation between obesity and various sperm parameters in the general population [11, 12]. Some have suggested a relationship between body mass index (BMI) and male infertility [13, 14]. Evidently, there has been a higher probability of abnormal spermatozoa and infertility found in obese men [1].

Furthermore, obesity has been shown to be associated with significant disturbances in the hormonal milieu, which can adversely affect the reproductive system [15, 16]. These reports have illustrated that fat tissue accumulation in men causes subsequent lowered serum levels of total and free testosterone while simultaneously elevating estrogen serum levels [15, 16]. A study demonstrated that such fat accumulation resulted in oxidative stress (OS) from a dysregulation of adipocytokine and reactive oxygen species (ROS) production [17]. Interestingly, data has been in agreement with recent studies suggesting that systemic OS correlates with BMI [18, 19]. The relationship of increasing levels of obesity and male-factor infertility calls for greater clinical awareness, as much evidence suggests that obesity may play a significant role with the rising subfertility rates.

The aim of this review is to discuss the relationships between obesity and male infertility. Moreover, it specifically explores the role of ROS production in obesity and its subsequent generation of OS, thereby directly and indirectly linking the effects of obesity on male reproductive potential.

26.2 Obesity

Obesity is a medical condition related to an excess accumulation of white adipose tissue in the body, which has the potential to inflict adverse health effects. The principal cause for this state is an energy imbalance between the body's energy intake and expenditure (Fig. 26.1). Excess energy is stored predominantly in the form of triglycerides and deposited in adipose tissue. Triglycerides play a vital role in metabolism serving as energy sources and for systemic transporters of dietary fat. Consequently, this fat accumulation and response to adipocyte hypertrophy has been suggested to compromise adipose tissue function and incite structural alterations to other organs [20]. The excessive fat accumulation in adipose tissue, liver, and other organs predisposes the onset of metabolic abnormalities that are often accompanied with hypertension, impaired glucose tolerance, insulin resistance leading to hyperinsulinemia, and dyslipidemia. Thus, obesity has been considered a chronic disease linked to a widespread range of physical, genetic and hormonal disorders.

Traditionally, obesity has been defined as a body weight of at least 20% above the weight corresponding to the lowest death rate for individuals of a specific height, gender and age, as well as additional specific requirements [21]. However, its present characterization has been more broadly classified as abnormal or excessive fat accumulation [10].

Several means of measuring obesity are currently utilized. The most accurate yet impractical measurements are to assess the weight of an individual underwater or to use an X-ray test called dual energy X-ray absorptiometry. Additional modes of evaluation include skinfold and waste to hip ratio (WHR) measurements, bioelectrical impedance analysis, and the risk factors associated with comorbidities. Two of the more common and simpler methods include the measurement of BMI and waist circumference.

BMI is a simple weight to height ratio that is defined as an individual's weight (kg) divided by the square of their height (m^2). There has been much controversy over the BMI ranges and cutoff points to deem an individual obese. The WHO has considered the normal range to lie between 18.5 and 24.9 kg/m^2, the risk of comorbidity to increase between 25 and 29.9 kg/m^2, and the onset of obesity to occur over 30 kg/m^2 [22]. In order to achieve optimum health, the median BMI for the adult population has been found to be in the range of 21–23 kg/m^2, while individuals should maintain a BMI in the range of 18.5–24.9 kg/m^2 [22]. These BMI values are age-independent and the same for both sexes [22]. Nevertheless, this may not represent the same degree of obesity in different populations due in part to different body proportions. A shortcoming of BMI as a measurement tool has been its overestimate in mesomorphic individuals. Hence, since calculations tend to diverge from accurate values, BMI values allow only for an estimate, at best.

The quantification of waist circumference has been believed to be more accurate and convenient marker of obesity. This measurement that is unrelated to height makes an approximate index of intra-abdominal fat mass and total body fat.

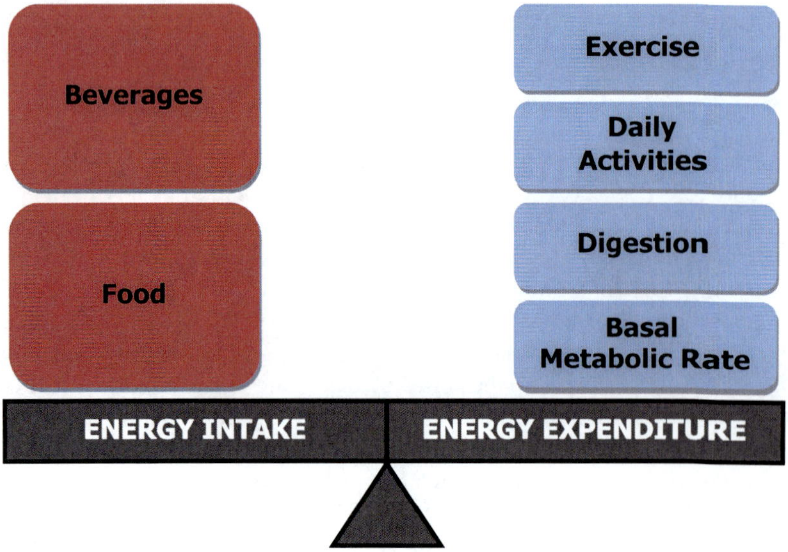

Fig. 26.1 Energy balance exists when energy intake is equal to energy expended. The principal cause of obesity is an imbalance between the body's energy intake and expenditure, where the energy consumed exceeds the energy expended

It has actually shown much of a correlation to BMI and risk factors of chronic diseases. A waist circumference ≥102 cm in males and ≥88 cm in females have been indicative of metabolic complications [22].

26.3 Obesity and Male Infertility

BMI has been found to be associated with altered sperm parameters in numerous reports. In a recent study investigating factors related to semen quality, the prevalence of infertility in obese men was found to be three times greater than in male partners of couples with idiopathic or female-factor infertility [12]. Moreover, sperm density and total count was shown to have a statistically significant negative correlation to increasing BMI [12]. Another study looked at normozoospermic partners in an infertile population and reported a reduction in sperm concentration among men with BMI greater than 30 kg/m^2 when compared to leaner members of the study group [23]. Kort et al. further examined the relationship between sperm parameters and BMI in a generally overweight selection of subjects [24]. After 520 semen samples were subjected to analysis, semen quality and the number of normal sperm per ejaculate exhibited declines with increasing BMI [24]. On the other hand, when obesity was expressed as a measurement of WHR, the similar trend between obesity and impaired sperm parameters was not seen [25]. This further illustrates

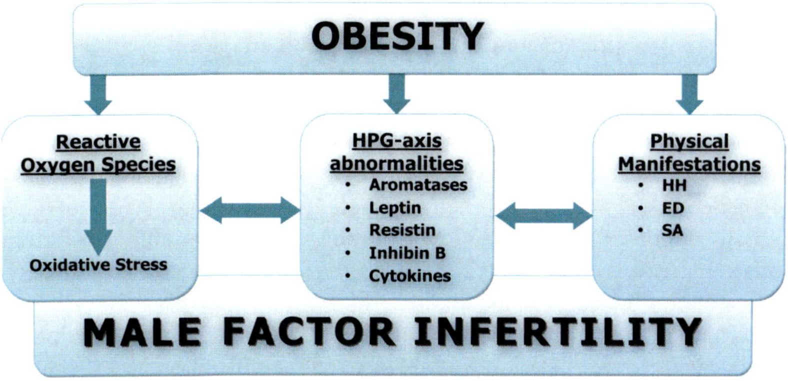

Fig. 26.2 Obesity can lead to male infertility via various mechanisms. These mechanisms manifest/act directly through or via interplay between reactive oxygen species (ROS) production, the hypothalamus-pituitary-gonadal (HPG) axis, as well as other physical factors (*HH* hypogonadotropic hypogonadism; *ED* erectile dysfunction; *SA* sleep apnea)

the inconsistency in obesity measurement techniques. Since an overwhelming evidence indicates that altered spermatogenesis and abnormal sperm parameters—reduced total sperm count and concentration—are correlated to the findings in obese males and that subfertility and infertility of couples are certainly related to such conditions [26], it may be postulated that obesity may induce semen abnormalities via the generation of ROS, dysregulation of the hypothalamic-pituitary-gonadal (HPG) axis, and/or physical manifestations (Fig. 26.2).

26.3.1 Reactive Oxygen Species

OS results from an impairment of a biological system's ability to reduce the formation of highly reactive species, repair detrimental damage, or reach a balance between ROS and antioxidants. Much research has focused on the etiology of OS, its link to male infertility, and its pathophysiological effects on male reproduction. Disturbances in reductases and functional redox state conditions disrupt cellular homeostasis. Numerous studies demonstrate a multitude of adverse effects induced by its causative factor—ROS. Environmental toxicants have been verified to impair a cell's reductive environment by decreasing its reduction potential, leading to subsequent reverse catalysis by oxioreductases, and possibly damaging cell membrane proteins, lipids and DNA [27]. Even moderate levels of OS may trigger molecules to initiate a cascade of reactions, thereby inducing programmed cell death (apoptosis) [28].

There are two general forms of free radical species: ROS and reactive nitrogen species (RNS). ROS are the more common free radicals with oxygen centers, whereas RNS are often considered to be a subclass thereof. Free radicals generated during oxygen reduction reactions of natural aerobic metabolic pathways typically form ROS. The three major forms of ROS are the superoxide anion ($O_2^{-\bullet}$), hydrogen peroxide (H_2O_2), and the extremely reactive hydroxyl radical ($OH^{-\bullet}$). These highly reactive pro-oxidant species can interact with antioxidants in order to maintain homeostasis. In the case of any imbalance between pro-oxidants and antioxidants, OS commences.

Of the inspired oxygen, 98% is reduced during lipolysis and ATP production, while the other 2% is incompletely reduced [29]. However, since many ROS are byproducts of aerobic cellular metabolism during oxidative phosphorylation in the mitochondria, any impairment due to OS may result in ATP depletion and potentially initiate cellular degradation [27]. Moreover, the mitochondria of spermatozoa have been found to be the primary source of ROS in infertile men [30]. The majority of ROS generated occurs at complexes I (NADH-Q dehydrogenase) and IV (conversion of ubiquinol to ubisemiquinone to ubiquinone) of the electron transport chain [31]. Since molecular oxygen is the final electron acceptor at complex IV in the formation of water, an extra electron may be captured during ATP generation becoming a major source of ROS - namely, superoxide anion [32]. This free radical can further propagate a series of reactions impairing cellular function and ultimately semen quality.

Many studies have pointed to ROS as independent biomarkers of semen quality due to their potential to cause suboptimal reproductive function [27, 33]. As highly reactive free radical oxygen molecules seek stability by attacking their neighboring stable species to obtain an electron, the targeted molecule itself becomes an unstable free radical, thereby generating a cascade of reactions. Consequently, structural and functional cellular damage may arise, both of which have been linked to irregular sperm function and motility via mitochondrial genome impairment [34, 35]. Extensive research supports the free radical-induced pathological effects on DNA damage, lipid peroxidation (LPO), and apoptosis in spermatozoa [28, 36].

Nevertheless, low levels of ROS are necessary in maintaining cellular homeostasis with their counteracting scavenging species, antioxidants, as well as in processes of the immune system, redox signaling and sperm maturation [33]. However, only a few studies have specifically discussed the physiological roles of free radicals in sperm function. The most well-documented ones have revealed their importance in controlling sperm maturation processes, capacitation, hyperactivation, acrosome reaction, and sperm–oocyte fusion; others have expressed ROS to be essential signal-transduction biomolecules and components of the complex cascade pathways in spermatozoa [36].

26.3.1.1 Sources of ROS

There are several sources of ROS, both endogenous and exogenous, found in the seminal plasma that can exert their effects on spermatozoa. Numerous studies report leukocytes and spermatozoa as the two main sources of free radicals found in semen [27, 34].

Leukocytes are the predominant source of endogenous ROS during sperm maturation, as well as one of the main mechanistic agents in combating pathogens [37, 38]. Moreover, leukocyte production is enhanced in obesity and participates in inflammatory pathways that are activated in adipose tissue of obese individuals [39]. Plante et al. demonstrated a positive correlation between levels of ROS and the degree of leukocyte contamination [30]. In times of infection or disease, peroxidase-positive leukocytes generate high levels of ROS through the nicotinamide adenine dinucleotide phosphate (NADPH) pathway [40, 41]. Subsequently, this elevated production of ROS during times of defense has been shown to have adverse effects on sperm function [42]. Furthermore, reports reveal signs of decreased motility and fertilization capacity from a lack of antioxidant-defense mechanisms in the testis and epididymis, rendering sperm extremely susceptible to infection [43–45].

Spermatozoa are an additional source of free radicals located in the semen. Normally, the cytoplasm is extruded during spermatogenesis. However, if spermatogenesis is impaired by any means, proper cytoplasm extrusion may not occur, and sperm are left arrested in an immature and functionally defective state. This defect ultimately results in increased ROS production through activation of the NADPH system, providing electrons for free radicals to initiate a series of events, eventually activating NADPH oxidase. Hence, there are two primary mechanisms by which spermatozoa may generate ROS: either (1) at the plasma membrane level through NADPH oxidase or (2) at the mitochondrial level through a NADH-dependent oxidoreductase system.

Exogenous sources of ROS also have a major influence on sperm quality and function through the production of ROS in pathological amounts. A few of these sources include industrial compounds, smoking, alcohol, spinal cord injury, and varicocele [36]. A majority of ROS comes from fat-soluble environmental toxins, allowing for a large accrual in white adipose tissue of obese males. Much evidence indicates that environmental pollutants increase ROS in the testes [34]. Accumulations of these highly reactive, unstable molecules cause the propagation of free radical reactions that have been proven disruptive to male reproductive function.

Furthermore, as lipophilic-toxin contaminants intensify in the scrotum of obese males, they may in turn cause direct effects on spermatogenesis and may be potentially linked to infertility. One such toxin, pthalate, a compound found in plastics and beauty products, was reported to induce sperm DNA damage and impair spermatogenesis [46, 47]. Studies also revealed elevated OS levels in male testes due to heavy metals (e.g. lead), pesticides, and sulfur dioxide (a common food preservative) [48–50]. Free radical generation and decreased antioxidant capacity have shown links to nicotine, a component of cigarettes, as well as cigarette smoke [51, 52]. Furthermore, evidence suggests that cigarette smoke, an environmental toxin, decreases sperm parameters—motility, morphology and concentration—all of which adversely affect male reproductive potential [52, 53]. A large amount of alcohol (ethanol) consumption plus a poor nutritional diet—often found in obese individuals—contribute to symptoms of elevated ROS with a simultaneous decrease in antioxidants [54]. The production of such reactive species further implies an additional pathway to stimulate substantial cellular damage to proteins, lipids and

DNA. Interestingly, it has been reported that over 90% of men with spinal cord injury are infertile due to possible elevated ROS levels resulting in poor semen motility and morphology [55, 56]. Other attributing factors include impairments in erectile and ejaculatory function. However, the more immotile sperm does not seem to be caused by lifestyle factors, such as elevated scrotal temperature, ejaculation frequency, and method of bladder management, and thus, may rather be related to factors within the seminal plasma [57]. Additionally, it has been reported that varicocele is a contributing factor of elevated levels of ROS [58]. Varicocele is a medical condition characterized by abnormal dilation and venous tortuosity in the pampiniform plexus around the male spermatic cord. Greater concentrations of nitric oxide (NO), a RNS, have been found in infertile men with varicocele [59]. Moreover, augmented xanthine oxidase activity, a source of superoxide, was observed [60]. This enhanced enzymatic activity and NO production appears to increase ROS, subsequently impairing sperm function [60]. Higher grades of varicocele in men have shown elevated levels of ROS in their semen, serving as biomarkers of OS from ROS-induced LPO and DNA damage, both of which decrease semen function and contribute to male infertility [61, 62].

However, the prevalence of varicocele does not appear to be proportional to BMI. In a clinical study, Handel et al. suggested this might be due to the increased adipose tissue in obese males preventing compression of the left renal vein or, simply, a decreased detection from the accumulation of adipose tissue in the spermatic cord [63]. Although studies have demonstrated that varicocele triggers ROS production, thereby affecting male reproductive potential, evidence remains limited to confirm a clear-cut relationship between obesity and varicocele.

Leukocytospermia is a medical condition associated with an elevated white blood cell count in semen and is often observed in obese men. It is usually seen in the process of warding off infection during the inflammatory response. Studies have linked sperm quality, sperm dysfunction and leukocytospermia to obesity; albeit, the evidence remains controversial [64–66]. Some have reported leukocytospermia to have an overall adverse effect on sperm function and quality, correlating to a decrease in sperm count, motility, morphology, hyperactivation and defective fertilization, while others show no effects [64–66]. Nevertheless, leukocytospermia can be regarded as a biological marker of systemic inflammation and potential sperm dysfunction.

26.3.1.2 Obesity and Oxidative Stress

An increase of ROS is triggered via high metabolic rates in order to maintain homeostasis in obese men. In localized areas near the testes, it can disturb spermatogenesis and result in a possible failure to discard the residual body into the Sertoli cell. Since testicular spermatozoa with proximal cytoplasmic retention lack adequate cytoplasmic reductive enzymes to control free radicals, a decline in antioxidant scavenging species would allow for a higher susceptibility to ROS impairment. The resulting imbalance of oxidant/antioxidant species (OS) is linked to suboptimal sperm function.

From the aforementioned evidence, it seems undoubtedly obvious to suggest that ROS-induced OS has a tremendous impact on male fertility, as well as on the detrimental obesity-implicated consequences. Since the absolute resting metabolic rates of obese inviduals are higher than that of non-obese individuals (these differences disappear when resting metabolic rate is adjusted for differences in body composition), it is plausible to suggest that an increased level of stress in the testicular environment may be due to an accumulation of white adipose tissue. This collection of lipocytes would lead to an augmented ROS production and increase the temperature in the testes environment. A study by Hjollund et al. deduced that a reduction in sperm concentration was associated with moderately elevated physiological temperatures of the scrotal skin [67]. Both ROS generation and increase temperature in the testes may denature enzymes involved in spermatogenesis, providing further evidence for a link between obesity and male infertility.

26.3.2 Abnormal HPG Regulation

Although several mechanisms that parallel obesity to infertility have been proposed, many remain ambiguous and relatively undefined. Studies indicate that the central factor linking the mechanisms associated with obesity and infertility is an abnormal regulation of the HPG axis, as well as the previously discussed OS. The HPG axis responds to fluctuations in hormones causing a range of widespread and local effects on the body and aspects of reproduction. Excess fat accumulation can impair the feedback regulation of the HPG axis and be a contributing factor to abnormal semen quality. Since sex steroids and glucocorticoids control the interaction between the hypothalamic-pituitary-adrenal (HPA) and the HPG axes, any imbalance may in turn affect spermatogenesis and male reproductive function. The abnormal endocrine changes observed in obese, infertile men are not similar to men with either obesity or infertility alone. Therefore, simultaneous irregular hormonal profile and adipose-derived hormone levels, such as with aromatase, leptin, resistin, inhibin B, cytokines, as well as many genetic factors and physical manifestations may further explain the connection between the escalating frequency of global obesity and subfertility.

26.3.2.1 Aromatase

White adipose tissue exhibits elevated aromatase activity and secretion of adipose-derived hormones in abdominal and visceral fat. Aromatase is an important cytochrome P450 enzyme involved in sexual development and is vital in the biosynthesis of estrogens from its precursor androgens, such as testosterone and dehydropiandrosterone. Ironically, obese men show signs of elevated estrogen levels as well as low levels of testosterone and follicle-stimulating hormone (FSH) [10]. Depleted levels of free and total testosterone are interrelated to aromatase overactivity in both intra-abdominal and subcutaneous fat. This condition of hypotestosteronemia—low levels of testosterone—and deregulated levels of sex hormones are related to a reduction in spermatogenesis and subsequent lowered sperm concentrations [68]. Therefore, both may potentially hinder additional aspects of male reproductive function causing suboptimal fertility in obese males.

In an in vitro study involving male mice, it was demonstrated that estrogen is required for fertility and that a mutation in the estrogen receptor gene leads to reduced mating frequency, lowered sperm numbers, and defective sperm function [69]. Nevertheless, since estrogen is more biologically active than testosterone, overproduction of estrogen from elevated expression levels of aromatase activity in obese men may elicit significant abnormal downstream effects in the testes. A report notes signs of both overexpressed levels and the absence of estrogen to elicit adverse effects on spermatogenesis, simultaneously affecting normal male reproductive potential [70].

The endocrine system, which is responsible for the regulation of metabolic activities, growth and development, as well as guiding reproduction, has estrogen receptors in the male hypothalamus involved in a negative feedback mechanism with gonadotropin-releasing hormone (GnRH), luteinizing hormone (LH), and FSH from the anterior pituitary gland. As estrogen agonist levels are elevated, an inhibitory effect on androgen biosynthesis is observed, pointing to a regulatory role of the HPG axis to cause detrimental effects on spermatogenesis and, in turn, further increasing the likelihood of subfertility in obese men.

26.3.2.2 Leptin

Leptin is an adipose-derived peptide hormone secreted from white adipocytes vital in the regulation of energy intake and expenditure. Human leptin is a protein made up of 167 amino acids. The level of secreted and circulating leptin is directly proportional to the total amount of body fat. It acts on hypothalamic neurons responsible for the secretion of GnRH. This tropic hormone stimulates the synthesis and secretion of gonadotropins, FSH and LH from the anterior pituitary.

Normally, elevated leptin levels are associated with an increase in weight gain and respond through a feedback mechanism in the hypothalamus to reduce food intake and to increase both energy expenditure and sympathetic activity. On the other hand, leptin deficiency from mutations in the *Ob(Lep)* gene located on chromosome 7 has also indicated relations to obesity [1]. A majority of obese males presented elevated serum concentrations of leptin with no mutation in their leptin receptors. This indicates that the development of an insensitivity and resistance to the action of endogenous leptin is one of the fundamental mechanisms of obesity [71].

In addition, the aromatase overactivity expressed in obese men that causes a higher conversion of testosterone to estrogen will induce a negative feedback signal on the hypothalamus and anterior pituitary to inhibit GnRH, FSH and LH secretion. The combination of effects from both the insensitivity to endogenous leptin and stimulation of the negative feedback pathway may have profound effects on male reproductive function via abnormal hormonal regulation. Furthermore, an excess secretion of leptin from adipose tissue in obese males illustrated deleterious effects on both spermatogenesis and the production of androgens from inhibitory receptors mediated by Leydig cells [72]. It has been speculated that the presence of leptin plasma membrane receptors in testicular tissue and semen samples may be the link between leptin and male reproductive function [73]. The findings reinforce a direct effect on sperm quality via abnormal HPG regulation and further suggest a plausible link between obesity and male infertility.

26.3.2.3 Resistin

Resistin is an endocrine secreted adipose-tissue specific factor. It is a cysteine-rich protein that serves in endocrine function and regulation. Resistin causes tissues, particularly the liver, to become insulin resistant. As a result, blood glucose levels rise from increased glycogenolysis and gluconeogenesis processes in the liver.

Glycogenolysis, glycolysis and the tricarboxylic acid cycle act to conserve energy as ATP from the catabolism of carbohydrates. If ATP supplies are sufficient, these pathways and cycles are allosterically inhibited. Although under conditions of excess ATP production, the liver will attempt to convert the excess mixture of molecules into glucose and/or glycogen. In general, during glycogenolysis, glycogen stored in the liver and muscle cells is converted to glucose-6-phosphate—the first byproduct of the glycolytic pathway.

In gluconeogenesis, glucose molecules are synthesized from non-carbohydrate sources, such as lactic acid, amino acids and glycerol. This process is constantly occurring in the liver in order to maintain glucose homeostasis. Gluconeogenesis proceeds in times of low acetyl CoA concentrations and high levels of ATP production.

Although both glycogenolysis and gluconeogenesis processes require ATP to take place, there is very little, if any, deficiency in ATP production reported that will hamper the motility of sperm. Although sperm motility is critical at the time of fertilization, as it allows the passage of sperm through the zona pellucida, a lack of ATP production from resistin and its subsequent impact on sperm motility cannot yet be confirmed as credible rationale for the increased incidence rate of male infertility.

Since resistin causes an increase in blood glucose levels from insensitivity to insulin, it is a primary factor associated with Non-Insulin-Dependent Diabetes Mellitus (NIDDM), or Type II diabetes. Over 80% of people with Type II diabetes suffer from obesity. Consequently, this resistance to insulin causes an increase in circulating insulin in the bloodstream—hyperinsulinemia—leading to an inhibitory effect on spermatogenesis and impacting male fertility potential. Interestingly, although diabetic men share normal semen parameters (concentration, morphology, and motility), the amount of impairment to nuclear and mitochondrial DNA was notably higher, again pointing to a reduction in reproductive capabilities and health.

Although resistin shows a strong association in humans with high levels of glucose, obesity and Type II diabetes, it is actually a major product of macrophages.

Macrophages are a type of white blood cells that ingest foreign material by means of phagocytosis. When the number of macrophages that reside in adipose tissue increases, this may result in elevated levels of ROS. Many studies have associated insulin resistance with elevated OS levels, inferring naturally produced ROS in obese males to maintain normal biological processes as a potential reason [74].

26.3.2.4 Inhibin B

Inhibin is a glycoprotein, growth-factor like hormone of gonadal origin. It is a dimer consisting of two covalently linked alpha and beta subunits. The beta subunit of inhibin exists in two forms, A and B. Although many studies have been conducted on inhibin, both in vivo and in vitro, they have failed to demonstrate and verify a systemic relationship between serum inhibin levels and spermatogenesis [75, 76].

The site of inhibin B production has been under much scrutiny as some studies indicate that germ cells and possibly Leydig cells can produce inhibin [77, 78]. However, the predominantly believed source of inhibin B originates from Sertoli cells, which play a supportive role in germ cell survival, in the testis into the seminal plasma [79, 80]. This hormone is involved in the HPG axis and displays a proportional decrease in obese males. The consequent decrease in germ cells demonstrates a decrease in sperm count and a reduced likelihood to fertilize.

The mechanistic pathway inhibin B follows to exert its biological effects remains unknown and is a subject of future study [81]. Normally, it acts to inhibit both FSH production and stimulation of testosterone release by Leydig cells. However, many studies have revealed that the expected compensatory increase in FSH levels in response to low levels of inhibin B were not observed in obese men. These low levels of inhibin B observed may have resulted from the suppressive effects of elevated estrogen levels from overly expressed aromatase in obese men.

Since inhibin B levels are directly related to sperm formation, low levels observed in obese males will result in abnormal spermatogenesis. As previously mentioned, the increased estrogen levels contribute to a negative feedback effect on the hypothalamus decreasing gonodoliberin and gonadotropin release, and subsequent lowered testosterone levels. As levels of testosterone fall, sperm function and quality become impaired, resulting in a reduction in male reproductive potential. Nevertheless, inhibin B seems to be an accurate biomarker of testicular damage and could become essential for future diagnosis of spermatogenic disorders in populations exposed to testicular toxicants.

Aside from both ROS and abnormal-induced HPG axis regulation, there continues to be a variety of other factors that have demonstrated evidence of recognized effects of obesity on male infertility. However, there still remain numerous cases of obese men with reproductive function and potential to fertilize. This present unexplainable link in some instances may be credited to unfavorable inherited genotypes. Although it has been well documented that obese-infertile men show significantly lower testosterone levels than obese-fertile men, genetic mutations may exist to clarify this discrepancy [10].

26.3.2.5 Cytokines

As excess fat storage accumulates in tissues other than adipose tissue, such as in the liver or striated tissue of the skeletal muscle, local insulin resistance may ensue and cause inflammation. Inflammation is the response to tissue injury and is often characterized by an increase in blood flow to the tissue, consequently increasing temperature in the localized area, as well as redness, swelling and pain.

Changes in morphology and composition of adipose tissue from obesity can cause alterations in protein production and secretion. Many of the secreted proteins may be proinflammatory mediators produced by macrophages residing in the adipose tissue. Proinflammatory cytokines originating from adipose tissue display elevated signs of insulin resistance during inflammatory response. Cytokines have demonstrated to directly interfere with insulin signaling pathways by tumor necrosis factor-alpha (TNF-α), inhibiting tyrosine phosphorylation of insulin receptor substrate-1 [82]. Recent studies initiated and conducted by Hotamisligil have illustrated a positive correlation between an increase in adipose tissue accumulation and proinflmmatory gene TNF-α expression [82]. It is indicated that the involvement of TNF-α and interleukin-6 cytokines results in a reduction in sperm motility during systemic inflammation response [83]. This decrease in motility may further result in the inability of the spermatozoa to progressively travel to the oocyte, thereby diminishing the likelihood of fertility.

Furthermore, an excess of white adipose tissue has shown to increase the secretion of adipocytokines, causing enhanced inflammation and a toxic effect on spermatozoa through the release of ROS [84]. This subsequent ROS release during periods of inflammation and its impact on sperm quality and function may be a causative aspect to male infertility. Therefore, it is reasonable to believe that excess fat buildup in obese men causing insulin resistance from elevated resistin levels, inflammation response, higher metabolic rates, release of ROS, and elevated temperatures may all be contributing factors to the previously noted nuclear and mitochondrial DNA damaged and, ultimately, decreased reproductive potential in Type II diabetic males.

26.3.3 Physical Manifestations

Obese males also encounter physical mechanisms that may enhance and, thus, further attribute to decreased fecundity and fertility. These problems include hypogonadotropic hypogonadism (HH), erectile dysfunction (ED) and sleeping disorders, such as sleep apnea.

26.3.3.1 Hypogonadotropic Hypogonadism

HH is a form of secondary hypogonadism in which a problem with the pituitary or hypothalamus gland causes the absence or a decreased function of the male testes.

This condition is elicited from a lack of gonadal stimulating hormones, including FSH and LH, which are essential in proper sexual function. Any disruption in the chain of events, from the hypothalamus in the brain secreting GnRH that stimulates the pituitary gland to release FSH and LH will cause a deficiency of sex hormones and prevent normal sexual maturation.

Many researchers have studied the relationship between obesity and HH, and their effects on male reproductive function. Strain et al. noted with significance that obese men had less than two-thirds the normal mean plasma levels of free testosterone, total testosterone and FSH; yet, 24-h LH levels appeared normal [85]. These findings represented a state of mild HH and appear to be characteristic of obese men [85]. It is speculated that the abnormality results from partial suppression of the pituitary by the elevated plasma estrogen levels [85, 86]. In addition, it has been noted that the subnormal levels of free and total testosterone and FSH are proportional to the degree of obesity [86].

Treatment with aromatase inhibitors or suppression of adrenocortical secretion of aromatase to stabilize estrogen levels in obese men has shown potential to normalize HH [86]. Additionally, the simple loss of weight has been shown to also normalize HH without any decrease in plasma estrogen levels [86]. It is suggested that weight loss in obese men results in diminished sensitivity of the GnRH-gonadotropin secretory mechanism to suppression by a given concentration of estrogen [86].

26.3.3.2 Erectile Dysfunction

ED is medical condition in which a male is incapable to get or keep an erection firm enough for sexual intercourse. Although obesity itself does not seem to be the underlying factor, it still does impose a risk to vasculogenic impotence through the development of chronic vascular disease [87]. A recent study by Corona et al. revealed that after adjustment for comorbidities, obese males with ED presented low androgen levels [88]. Moreover, lowered androgen levels have been associated with reduced plasma testosterone levels [89]. A decrease in testosterone levels in obese males with ED may further contribute to suboptimal semen quality, as testosterone is essential for the onset of sexual characteristics and the production and maturation of sperm in males.

Despite the well-documented studies that indicate the association of ED to lowered fertility rates and being more prevalent in obese men, evidence and information of the pathophysiological link between obesity and ED remains limited. It has been hypothesized that visceral obesity increases proinflammatory factors and, in doing so, promotes an inflammatory response and contributes to ED [90]. Since an erection depends on hemodynamics and vascular health, any factor that causes endothelial dysfunction or impairs endothelial NO release and the integrity of the vascular bed will contribute to ED. Nevertheless, a report illustrated that changes only in one's lifestyle improved sexual function in nearly one-third of obese men with ED [91].

26.3.3.3 Sleep Apnea

Sleep apnea is a common disorder in which an individual has one or more pauses in breathing or shallow breaths while sleeping. The chronic condition affects 4% of middle-aged men, but usually goes undiagnosed. Interestingly, it has been reported that about two-thirds of middle-aged men with obstructive sleep apnea suffer from obesity, particularly central obesity [92].

Patients suffering from sleep apnea often have a fragmented sleep course due to repetitive episodes of upper airway obstructions and hypoxia followed by arousal [93]. Furthermore, these patients demonstrate a blunted nocturnal rise of testosterone needed for normal spermatogenesis [94]. This obstructive condition has been linked to lowered mean testosterone and LH values compared to both young and middle-aged controls [94], as well as reduced morning testosterone levels [95]. Weight loss in patients with obstructive sleep apnea has shown to increase testosterone levels [96]. Therefore, sleep apnea is associated with decreased pituitary-gonadal function and, thus, may contribute to hypogonadism and further explain male-factor infertility and abnormal seminal parameters [95].

26.4 Discussion

With the advances in technology that revolutionize everyday lifestyles and industrialization all around the world, obesity has become a modern-day global pandemic. Moreover, as obesity is predicted to reach record numbers in the near future and fertility rates continue to plummet, scientists have began to link the two together. They have revealed that as much as half of all fertility problems come from male-factor defects.

Many researchers have pointed to the increased adipocyte accumulation to generate ROS that propagate systemic and detrimental effects on male reproductive health, while others have attributed the abnormal hormonal profile as the central factor. It appears to be a complex composition of both and, in fact, the increase in adipose-derived hormones and adipokine levels may better explain the association of BMI, altered sperm parameters, and infertility. Recent studies have began to examine genetic biomarkers, an excess of adipose-derived hormones, adipokine release as well as OS. It is suggested that the consistent decrease in hormonal levels and specific proteomic sperm mutations observed in obese males may adversely impact spermatogenesis. Hence, these markers would in turn hinder normal sperm production, maturation and quality, accounting for some of the male-factor defects related to obesity.

The inconsistency in the results from studies demonstrates the necessity for further investigation when examining the effects of obesity on semen parameters. The numerous signs of decreased testosterone levels from excess fat accumulation pose attention for additional focus and study in understanding the overall mechanistic pathway in order to make a definitive link to male infertility.

Simple lifestyle changes have shown to benefit hormonal levels, yet effective treatments, proper lifestyle changes, and surgical options should be further explored. Since antioxidants have been shown to reduce ROS levels, thereby minimizing damage via OS, and have become a hot topic in possibly treatming infertility, this natural remedy should be further explored as a potential treatment for obesity-related male infertility. Additionally, standard and accurate measurements to qualify an individual as obese should be established to confirm the links made to infertility and the health problems that accompany the condition. Nevertheless, the continuing rise and prevalence of both obesity and declining semen quality all over the world, both of which are associated to ROS, call for additional research and a greater awareness to obesity as a potential etiology of male infertility.

Acknowledgment The authors are grateful for the research support from the Center for Reproductive Medicine at Cleveland Clinic.

References

1. Du Plessis SS, et al. The effect of obesity on sperm disorders and male infertility. Nat Rev Urol. 2010;7(3):153–61.
2. World Health Organization. WHO technical report: diet, nutrition and the prevention of chronic disease. World Health Organization, Geneva. 2003. http://www.who/int/en. Accessed 6 Feb 2011.
3. Mokdad AH, Bowman BA, Ford ES, Vinicor F, Marks JS, Koplan JP. The continuing epidemics of obesity and diabetes in the United States. JAMA. 2001;286:1195–200.
4. Hedley AA, Ogden CL, Johnson CL, Carroll MD, Curtin LR, Flegal KM. Prevalence of overweight and obesity among US children, adolescents, and adults, 1999–2002. JAMA. 2004;291:2847–50.
5. World Health Organization. Obesity and overweight. 2006. http://www.who/int/en. Accessed 6 Feb 2011.
6. Stephen EH, Chandra A. Updated projections of infertility in the United States: 1995–2025. Fertil Steril. 1998;70:30–4.
7. Thonneau P, Marchand S, Tallec A, Ferial ML, Ducot B, Lansac J, Lopes P, Tabaste JM, Spira A. Incidence and main causes of infertility in a resident population (1,850,000) of three French regions (1988–1989). Hum Reprod. 1991;6:811–6.
8. Sharlip ID, Jarow JP, Belker AM, Lipshultz LI, Sigman M, Thomas AJ, Schlegel PN, Howards SS, Nehra A, Damewood MD, Overstreet JW, Sadovsky R. Best practice policies for male infertility. Fertil Steril. 2002;77:873–82.
9. Swan SH, Elkin EP, Fenster L. The question of declining sperm density revisited: an analysis of 101 studies published 1934–1996. Environ Health Perspect. 2000;108:961–6.
10. Cabler S, Agarwal A, et al. Obesity: modern man's fertility nemesis. Asian J Androl. 2010;12(4): 480–9.
11. Jensen TK, Andersson AM, Jorgensen N, Andersen AG, Carlsen E, Petersen JH, Skakkebaek NE. Body mass index in relation to semen quality and reproductive hormones among 1,558 Danish men. Fertil Steril. 2004;82:863–70.
12. Magnusdottir EV, Thorsteinsson T, Thorsteinsdottir S, Heimisdottir M, Olafsdottir K. Persistent organochlorines, sedentary occupation, obesity and human male subfertility. Hum Reprod. 2005;20:208–15.
13. Nguyen R, Wilcox A, Skjaerven R, Baird DD. Men's body mass index and infertility. Hum Reprod. 2007;22:2488–93.

14. Sallmen M, Sandler DP, Hoppin JA, Blair A, Baird DD. Reduced fertility among overweight and obese men. Epidemiology. 2006;17:520–3.
15. Giagulli VA, Kaufman JM, Vermeulen A. Pathogenesis of the decreased androgen levels in obese men. J Clin Endocrinol Metab. 1994;79:997–1000.
16. Schneider G, Kirschner MA, Berkowitz R, Ertel NH. Increased estrogen production in obese men. J Clin Endocrinol Metab. 1979;48:633–8.
17. Furukawa S, Fujita T, Shimabukuro M, Iwaki M, Yamada Y, Nakajima Y, Nakayama O, Makishima M, Matsuda M, Shimomura I. Increased oxidative stress in obesity and its impact on metabolic syndrome. J Clin Invest. 2004;114(12):1752.
18. Keaney Jr JF, Larson MG, Vasan RS, Wilson PWF, Lipinska I, Corey D, Massaro JM, Sutherland P, Vita JA, Benjamin EJ. Obesity and systemic oxidative stress. Arterioscler Thromb Vasc Biol. 2003;23(3):434–9.
19. Olusi SO. Obesity is an independent risk factor for plasma lipid peroxidation and depletion of erythrocyte cytoprotectic enzymes in humans. Int J Obes Relat Metab Disord. 2002; 26(9):1159–64.
20. Stienstra R, Duval C, Müller M, Kersten S. PPARs, obesity, and inflammation. PPAR Res. 2007;2007:95974.
21. Carlsen E, Giwercman A, Keiding N, Shakkebaek NE. Evidence for decreasing quality of semen during past 50 years. BMJ. 1992;305:609–13.
22. World Health Organization. Diet, nutrition and the prevention of chronic diseases. Report of a Joint WHO/FAO Expert Consultation. WHO Technical Report Series No. 916. World Health Organization, Geneva; 2003.
23. Koloszár S, Fejes I, Závaczki Z, Daru J, Szöllosi J, Pál A. Effect of body weight on sperm concentration in normozoospermic males. Arch Androl. 2005;51(4):299–304.
24. Kort HI, Massey JB, Elsner CW, Mitchell-Leef D, Shapiro DB, Witt MA, Roudebush WE. Impact of body mass index values on sperm quantity and quality. J Androl. 2006;27(3):450–2.
25. Fejes I, Koloszár S, Szöllosi J, Závaczki Z, Pál A. Is semen quality affected by male body fat distribution? Andrologia. 2005;37(5):155–9.
26. Slama R, Eustache F, Ducot B, Jensen TK, Jørgensen N, Horte A, Irvine S, Suominen J, Andersen AG, Auger J, Vierula M, Toppari J, Andersen AN, Keiding N, Skakkebæk NE, Spira A, Jouannet P. Time to pregnancy and semen parameters: a cross-sectional study among fertile couples from four European cities. Hum Reprod. 2002;17(2):503–15.
27. Kefer JC, Ashok A, Sabanegh E. Role of antioxidants in treatment of male infertility. Int J Urol. 2009;16:449–57.
28. Buttke TM, Sandstrom PA. Oxidative stress as a mediator of apoptosis. Immunol Today. 1994;15(1):7–10.
29. Agarwal A, Gupta S, Sekhon L, Shah R. Redox consideration in female reproductive function and assisted reproduction: from molecular mechanisms to health implications. Antioxid Redox Signal. 2008;10(8):1375–403.
30. Plante M, de Lamirande E, Gagnon C. Reactive oxygen species released by activated neutrophils, but not by deficient spermatozoa, are sufficient to affect normal sperm motility. Fertil Steril. 1994;62:387.
31. Inoue M, Sato EF, Nishikawa M, et al. Mitochondrial generation of reactive oxygen species and its role in aerobic life. Curr Med Chem. 2003;10(23):2495–505.
32. Cadenas E, Davies KJ. Mitochondrial free radical generation, oxidative stress, and aging. Free Radic Biol Med. 2000;29(3–4):222–30.
33. Agarwal A, Gupta S, Sikka S. The role of free radicals and antioxidants in reproduction. Curr Opin Obstet Gynecol. 2006;18(3):325–32.
34. Mathur PP. Impact of environmental toxicants on testicular function. Immun Endoc Metab Agents Med Chem. 2008;8:79–90.
35. Agarwal A, Cocuzza M, Abdelrazik H, Sharma RK. Oxidative stress management in patients with male or female factor infertility. In: Popov I, Lewin G, editors. Handbook of chemiluminescent methods in oxidative stress assessment. Kerala: Transworld Research Network; 2008. p. 195–218.

36. Kothari S, et al. Beneficial and detrimental role of free radicals. Indian J Exp Biol. 2010;48:425–35.
37. Whittington K, Ford WC. Relative contribution of leukocytes and of spermatozoa to reactive oxygen species production in human sperm suspensions. Int J Androl. 1999;22:229.
38. Tremellen K. Oxidative stress and male infertility—a clinical perspective. Hum Reprod Update. 2008;14:243.
39. Weisberg SP, McCann D, Desai M, Rosenbaum M, Leibel RL, Ferrante Jr AW. Obesity is associated with macrophage accumulation in adipose tissue. J Clin Invest. 2003;112(12):1796–808.
40. Blake DR, Allen RE, Lunec J. Free radicals in biological systems—a review orientated to inflammatory processes. Br Med Bull. 1987;43:371.
41. Aitken RJ, Baker MA. Oxidative stress and male reproductive biology. Reprod Fertil Dev. 2004;16:581.
42. Ochsendorf FR. Infection and reactive oxygen species. Andrologia. 1998;30 Suppl 1:81–6.
43. Cocuzza M, Sikka SC, Athayde KS, Agarwal A. Clinical relevance of oxidative stress and sperm chromatin damage in male infertility: an evidence based analysis. Int Braz J Urol. 2007;33:603.
44. Wolff H, Politch JA, Martinez A, et al. Leukocytospermia is associated with poor semen quality. Fertil Steril. 1990;53:528.
45. Maruyama Jr DK, Hale RW, Rogers BJ. Effects of white blood cells on the in vitro penetration of zona-free hamster eggs by human spermatozoa. J Androl. 1985;6:127.
46. Agarwal DK, Maronpot RR, Lamb JC, Kluwe WM. Adverse effects of butyl benzyl phthalate on the reproductive and hematopoietic systems of male rats. Toxicology. 1985;35:189.
47. Kasahara E, Sato EF, Miyoshi M, et al. Role of oxidative stress in germ cell apoptosis induced by di(2-ethylhexyl)phthalate. Biochem J. 2002;365:849.
48. Acharya UR, Acharya S, Mishra M. Lead acetate induced cytotoxicity in male germinal cells of Swiss mice. Ind Health. 2003;41:291.
49. Chitra KC, Sujatha R, Latchoumycandane C, Mathur PP. Effect of lindane on antioxidant enzymes in epididymis and epididymal sperm of adult rats. Asian J Androl. 2001;3:205.
50. Meng Z, Bai W. Oxidation damage of sulfur dioxide on testicles of mice. Environ Res. 2004;96:298.
51. Traber MG, van der Vliet A, Reznick AZ, Cross CE. Tobacco-related diseases. Is there a role for antioxidant micronutrient supplementation? Clin Chest Med. 2000;21:173.
52. Saleh RA, Agarwal A, Sharma RK, Nelson DR, Thomas Jr AJ. Effect of cigarette smoking on levels of seminal oxidative stress in infertile men: a prospective study. Fertil Steril. 2002; 78:491.
53. Vine MF. Smoking and male reproduction: a review. Int J Androl. 1996;19:323.
54. Koch OR, Pani G, Borrello S, et al. Oxidative stress and antioxidant defenses in ethanol-induced cell injury. Mol Aspects Med. 2004;25:191.
55. Padron OF, Brackett NL, Sharma RK, et al. Seminal reactive oxygen species and sperm motility and morphology in men with spinal cord injury. Fertil Steril. 1997;67:1115.
56. de Lamirande E, Leduc BE, Iwasaki A, Hassouna M, Gagnon C. Increased reactive oxygen species formation in semen of patients with spinal cord injury. Fertil Steril. 1995;63:637.
57. Brackett NL, Nash MS, Lynne CM. Male fertility following spinal cord injury: facts and fiction. Phys Ther. 1996;76(11):1221–31.
58. Hendin BN, Kolettis PN, Sharma RK, Thomas Jr AJ, Agarwal A. Varicocele is associated with elevated spermatozoal reactive oxygen species production and diminished seminal plasma antioxidant capacity. J Urol. 1999;161:1831.
59. Mehraban D, Ansari M, Keyhan H, et al. Comparison of nitric oxide concentration in seminal fluid between infertile patients with and without varicocele and normal fertile men. Urol J. 2005;2:106.
60. Mitropoulos D, Deliconstantinos G, Zervas A, et al. Nitricoxide synthase and xanthine oxidase activities in the spermatic vein of patients with varicocele: a potential role for nitric oxide and peroxynitrite in sperm dysfunction. J Urol. 1996;156:1952.
61. Agarwal A, Prabakaran S, Allamaneni SS. Relationship between oxidative stress, varicocele and infertility: a meta-analysis. Reprod Biomed Online. 2006;12:630.

62. Saleh RA, Agarwal A, Sharma RK, et al. Evaluation of nuclear DNA damage in spermatozoa from infertile men with varicocele. Fertil Steril. 2003;80:1431.
63. Handel L, Shettyl R, Sigman N. The relationship between varicoceles and obesity. J Urol. 2006;176(5):1912–3.
64. Aitken RJ, Baker HW. Seminal leukocytes: passengers, terrorists or good Samaritans? Hum Reprod. 1995;10:1736–9.
65. Tomlinson JM, Barratt CL, Cooke ID. Prospective study of leukocytes and leukocyte subpopulations in semen suggests they are not the cause of a male infertility. Fertil Steril. 1993; 60:1069–75.
66. Moskovtsev SI, Willis J, White J, Mullen BM. Leukocytospermia: relationship to sperm deoxyribonucleic acid integrity in patients evaluated for male factor infertility. Fertil Steril. 2009;88:737–40.
67. Hjollund NH, Bonde JP, Jensen TK, Olsen J. Diurnal scrotal skin temperature and semen quality. The Danish First Pregnancy Planner Study Team. Int J Androl. 2000;23(5):309–18.
68. Fejes I, Koloszár S, Závaczki Z, Daru J, Szöllösi J, Pál A. Effect of body weight on testosterone/estradiol ratio in oligozoospermic patients. Arch Androl. 2006;52(2):97–102.
69. Eddy EM, Washburn TF, Bunch DO, Goulding EH, Gladen BC, Lubahn DB, Korach KS. Targeted disruption of the estrogen receptor gene in male mice causes alteration of spermatogenesis and infertility. Endocrinology. 1996;137(11):4796–805.
70. O'Donnell L, Robertson KM, Jones ME, Simpson ER. Estrogen and spermatogenesis. Endocr Rev. 2001;22(3):289–318.
71. Rohner-Jeanrenaud F, Jeanrenaud B. Obesity, leptin, and the brain. N Engl J Med. 1996; 334:324–5.
73. Caprio M, Isidori AM, Carta AR, Moretti C, Dufau ML, Fabbri A. Expression of functional leptin receptors in rodent Leydig cells. Endocrinology. 1999;140(11):4939–47.
73. Jope T, Lammert A, Kratzsch J, Paasch U, Glander HJ. Leptin and leptin receptor in human seminal plasma and in human spermatozoa. Int J Androl. 2003;26(6):335–41.
74. Fraczek M, Kurpisz M. Inflammatory mediators exert toxic effects of oxidative stress on human spermatozoa. J Androl. 2007;28:325–33.
75. de Kretser DM, McLachlan RI, Robertson DM, Burger HG. Serum inhibin levels in normal men and men with testicular disorders. J Endocrinol. 1989;120:517–23.
76. de Kretser DM, McFarlane JR. Inhibin in the male. J Androl. 1996;17:179–82.
77. Risbridger GP, Clements J, Robertson DM, Drummond AE, Muir J, Burger HG, et al. Immuno- and bioactive inhibin and inhibin alpha-subunit expression in rat Leydig cell cultures. Mol Cell Endocrinol. 1989;66:119–22.
78. Andersson AM, Muller J, Skakkebaek NE. Different roles of prepubertal and postpubertal germ cells and Sertoli cells in the regulation of serum inhibin B levels. J Clin Endocrinol Metab. 1998;83:4451–8.
79. Grootenhuis AJ, Timmerman MA, Hordijk PL, de Jong FH. Inhibin in immature rat Sertoli cell conditioned medium: a 32 kDa alpha beta-B dimer. Mol Cell Endocrinol. 1990;70:109–16.
80. Handelsman DJ, Spaliviero JA, Phippard AF. Highly vectorial secretion of inhibin by primate Sertoli cells in vitro. J Clin Endocrinol Metab. 1990;71:1235–8.
81. Meachem SJ, Nieschlag E, Simoni M. Inhibin B in male reproduction: pathophysiology and clinical relevance. Eur J Endocrinol. 2001;145(5):561–71.
82. Hotamisligil GS. Inflammatory pathways and insulin action. Int J Obes Relat Metab Disord. 2003;27:S53–5.
83. Yoshida S, Harada T, Iwabe T, Taniguchi F, Mitsunari M, Yamauchi N, Deura I, Horie S, Terakawa N. A combination of interleukin-6 and its soluble receptor impairs sperm motility: implications in infertility associated with endometriosis. Hum Reprod. 2004;19(8):1821–5.
84. Fraczek M, Kurpisz M. Inflammatory mediators exert toxic effects of oxidative stress on human spermatozoa. J Androl. 2007;28:325–33.
85. Strain GW, Zumoff B, Kream J, Strain JJ, Deucher R, Rosenfeld RS, Levin J, Fukushima DK. Mild hypogonadotropic hypogonadism in obese men. Metabolism. 1982;31(9):871–5.

86. Zumoff B. Hormonal abnormalities in obesity. Acta Med Scand Suppl. 1988;723:153–60.
87. Chung WS, Sohn JH, Park YY. Is obesity an underlying factor in erectile dysfunction? Eur Urol. 1999;39(1):68–70.
88. Corona G, Mannucci E, Fischer AD, Lotti F, Petrone L, Balercia G, Baldini E, Forti G, Maggi M. Low levels of androgens in men with erectile dysfunction and obesity. J Sex Med. 2008;5(10):2454–63.
89. Feeley RJ, Traish AM. Obesity and erectile dysfunction: is androgen deficiency the common link? ScientificWorldJournal. 2009;9:676–84.
90. Traish AM, Feeley RJ, Guay A. Mechanisms of obesity and related pathologies: androgen deficiency and endothelial dysfunction may be the link between obesity and erectile dysfunction. FEBS J. 2009;276:5755–67.
91. Esposito K, Giugliano F, Di Palo C, Giugliano G, Marfella R, D'Andrea F, D'Armiento M, Giugliano D. Effect of lifestyle changes on erectile dysfunction in obese men: a randomized controlled trial. JAMA. 2004;291(24):2978–84.
92. Vgontzas AN, Papanicolaou DA, Bixler EO, Hopper K, Lotsikas A, et al. Sleep apnea and daytime sleepiness and fatigue: relation to visceral obesity, insulin resistance, and hypercytokinemia. J Clin Endocrinol Metab. 2000;85:1151–8.
93. Young T, Skatrud J, Peppard PE. Risk factors for obstructive sleep apnea in adults. JAMA. 2004;291:2013–6.
94. Luboshitzky R, Zabari Z, Shen-Orr Z, Herer P, Lavie P. Disruption of the nocturnal testosterone rhythm by sleep fragmentation in normal men. J Clin Endocrinol Metab. 2001; 86:1134–9.
95. Luboshitzky R, Lavie L, Shen-Orr Z, Herer P. Altered luteinizing hormone and testosterone secretion in middle-aged obese men with obstructive sleep apnea. Obes Res. 2005;13:780–6.
96. Semple PA, Graham A, Malcolm Y, Beastall GH, Watson WS. Hypoxia, depression of testosterone, and impotence in pickwickian syndrome reversed by weight reduction. Br Med J (Clin Res Ed). 1984;289(6448):801–2.

Chapter 27
Oxidative Stress in Benign Prostate Hyperplasia

Murat Savas

Abstract The greatest risk factor for developing benign prostatic hyperplasia (BPH) is advanced age. Potential molecular and physiologic contributors to the frequency of BPH occurrence in older individuals include the oxidative stress, chronic inflammation, and alterations in tissue microenvironment. As BPH and aberrant changes in reactive oxygen species become more common with aging, oxygen species signaling may play an important role in the development and progression of this disease. Increased oxidative stress is a result of either increased reactive oxygen species generation or a loss of antioxidant defense mechanisms. Oxidative stress is associated with several pathological conditions including inflammation and infection. Oxygen species are byproducts of normal cellular metabolism and play vital roles in stimulation of signaling pathways in response to changing intra and extracellular environmental conditions. This review is aimed to explore the mechanism of oxidative stress in prostate and the possibility of drug development against oxidative stress for prostatic disease prevention.

Keywords Oxidative stress • Benign prostate hyperplasia • Prostatic enlargement • Apoptosis • Steroid hormones • Prostate

27.1 Introduction

Historically, benign prostatic hyperplasia (BPH) has been a major focus of urologic practice and surgery. BPH assigns most of men after the age of 50 and represents the most common urologic disease among elderly males [1]. BPH is histologically

M. Savas, MD (✉)
Department of Urology, Harran University Medical School, Arastirma Hospital, Meterorology Caddesi, Sanliurfa 63200, Turkey
e-mail: mrtsvs@yahoo.com

A. Agarwal et al. (eds.), *Studies on Men's Health and Fertility*, Oxidative Stress in Applied Basic Research and Clinical Practice, DOI 10.1007/978-1-61779-776-7_27,

defined as an overgrowth of the epithelial and stromal cells from the transition zone and periurethral area. Incidence of histological BPH could be over 70% at 60 years and over 90% at 70 years [2]. However, histological BPH doesn't always lead to clinical manifestations. BPH symptoms can range over a wide scale from minimal bother to urinary retention and renal failure. To date, we still have no precise knowledge of the cellular and molecular processes underlying the pathogenesis of BPH and leading to a symptomatic disease [3]. Although the influence of androgens and estrogens has been demonstrated, hormonal factors alone may not fully explain BPH development.

Hydroxyl radicals, peroxides, and superoxides are reactive oxygen species (ROS) that are generated during everyday metabolic processes in a normal cell. ROS, generated either endogenously (mitochondria, metabolic process, inflammation, etc.) or from external sources [4], play a vital role in regulating several biologic phenomena. While increased ROS generation has traditionally been associated with tissue injury or DNA damage which are general manifestations of pathological conditions associated with infection, aging, mitochondrial DNA mutations, and cellular proliferation, new and exciting information points to an essential role for increased ROS generation in several cellular processes associated with neoplastic transformation and aberrant growth and proliferation [5, 6]. Processes associated with proliferation, apoptosis, and senescence may be a result of the activation of signaling pathways in response to intracellular changes in ROS levels [7]. Thus, the excessive production of ROS or inadequacy in a normal cell's antioxidant defense system (or both) can cause the cell to experience oxidative stress. This oxidative stress may play a broader role in cellular processes associated with initiation and development of BPH.

The role of inflammation in prostate diseases is suggested by the presence of inflammatory cells within the BPH and Prostate Cancer (PC). On histologic examinations from patients with BPH, inflammatory aspects are present in approximately 40% of cases. The men with inflammatory aspects inside the prostate have a significantly higher risk for BPH progression and acute urinary retention (AUR). Evidence shows that a cyclooxygenase-2 (COX-2) inhibitor can increase the apoptotic activity in human BPH tissue. In vitro studies showed an overexpression of inflammatory markers in BPH and PC compared with a normal tissue. There are significant inflammatory markers differences between BPH and PC, which is more severe inflammation process in PC. Another basic difference was a gene polymorphism in PC. Targeting the microenvironment may represent a promising therapeutic approach for prostate disease. Many epidemiological studies showed a beneficial effect of drug that influences inflammation such as nonsteroidal anti-inflammatory drugs (NSAIDs), antioxidant compound in food or supplements, and vitamin D receptor (VDR) agonists. These drugs need more investigation to prove their efficacy as chemopreventive agents for prostatic disease.

This chapter presents recent evidence that suggests a link between oxidative stress and BPH. This review is aimed to explore the mechanism of inflammation and oxidative stress in prostate and the possibility of drug development against inflammation for prostatic disease prevention.

27.2 The Etiology of Prostatic Enlargement

BPH is characterized histologically by a progressive enlargement of the prostate gland resulting from a nonmalignant proliferative process that includes both epithelial and stromal elements. Growth results from proliferation of fibroblasts/myofibroblasts and epithelial glandular elements near the urethra in the transition zone of the prostate gland [8–11]. The hyperplastic process is multifocal and exhibits a variegated histology with variable proportions of stromal nodules and glandular hyperplasia. The histology of BPH was carefully described by McNeal [12, 13], who described the BPH process in two phases. During the initial phase of BPH, small hyperplasic nodules appear in the periurethral area and gradually increase in number. A second phase of BPH, generally occurring in men older than 60 years, involves a dramatic and simultaneous increase in size of glandular nodules. McNeal [12] also noted that the histologic appearance of stromal tissue in BPH nodules resembled the histologic appearance of developmental mesenchyme. Thus, the author hypothesized that BPH is caused by "embryonic processes reawakened in a distorted form in adult life." Moreover, endocrine influences have been postulated to play an important role in BPH. Androgen stimulation is required for fetal prostate growth and development [14], but is considered to play only a permissive role in the pathogenesis of BPH. Androgen levels in the prostate are not significantly different in BPH and normal tissues, and currently no evidence shows an increase in BPH incidence for men undergoing androgen supplementation therapy [15–21]. However, estrogens or a changing ratio of androgens to estrogens in aging men have been speculated to play an important role in the pathogenesis of BPH. This hypothesis is based on two main observations. First, the ratio of testosterone to estradiol steadily declines in aging men [22]. Second, the experimental manipulation of estradiol levels in animal models can cause benign prostatic enlargement. Dogs and humans are believed to be the only mammals with a significant incidence of spontaneous BPH, and treatment of young dogs with androgen plus estrogen hormones leads to an earlier onset and greater extent of benign prostatic enlargement [23, 24]. Similarly, treatment of mice with androgen plus estrogen hormones leads to benign prostatic enlargement [25–28]. Recent studies implicating obesity and BPH could reflect an increased estrogen/testosterone ratio in obese men resulting from increased aromatization of testosterone in peripheral tissues.

On the other hand, prostatic inflammation is a common feature of the adult prostate and is associated with the development and progression of BPH. Acute and chronic prostatic inflammations are extremely common histological findings in the adult human [29–34]. McNeal [35] found inflammation in 44% of prostate tissue samples in an autopsy series in men without evidence of other prostate disease, whereas Bennett et al. [36] reported inflammation in 73% of prostates examined. The origin of inflammation in the prostate remains a subject of debate and is likely multifactorial. Evidence exists for urinary reflux into the prostatic ducts, and bacterial colonization/infection in surgical specimens of BPH seems to be common. Among patients who underwent transurethral resection of the prostate (TURP) and had preoperatively sterile urine, 38% of specimens grew bacteria when the tissues were morcellated and cultured [37]. Other possible causes of inflammation include noxious dietary constituents, autoimmune mechanisms, oxidative stress associated with androgen action, and systemic inflammation associated with the metabolic syndrome [38]. A retrospective study of 3,942 prostatic biopsies identified as consistent with BPH showed inflammation in 1,700 (43.1%) [32]. Furthermore, a study of specimens obtained from 80 men who had no symptoms of prostatitis but underwent TURP for treatment of BPH found inflammation to be uniformly present [39]. In another study that evaluated tissue removed with radical prostatectomy, inflammation was found in tissue samples of 35 of 40 patients who had BPH and that inflammation was associated with significantly greater prostatic weight than that observed in patients who had no inflammation [40]. In a prospective study of autopsy specimens obtained from 93 men who had histologic evidence of BPH, chronic inflammation was found (primarily in the transitional zone) in 75% of prostates examined compared with 55% of prostates not affected by BPH [29]. Prostate biopsy of 8,224 men enrolled in the Reduction by Dutasteride of Prostate Cancer Events (REDUCE) trial showed inflammation in more than three quarters of the biopsies. Chronic inflammation was more common than acute inflammation (78% vs. 15%, respectively). Inflammation also correlates with prostatic enlargement and symptomatic progression. Evidence of inflammation on baseline biopsy in the Medical Therapy of Prostatic Symptoms (MTOPS) trial correlated with prostate volume, suggesting a significant role in prostatic enlargement [41]. Inflammation also correlated with symptomatic progression, risk for urinary retention, and need for surgery [42]. In a recent analysis of the data from the REDUCE trial, Nickel and colleagues [43, 44] reported a weak but statistically significant association between chronic inflammation and symptoms severity. Several studies have identified associations suggesting metabolic risk factors for the development or progression of BPH. The Baltimore Longitudinal Study of Aging examined whether obesity, fasting plasma glucose, and diabetes were associated with prostatic enlargement [45, 46]. This analysis, authored by a collaborator in this proposal, showed a positive correlation of body mass index with prostate volume. The risk was increased for very obese men. The association of obesity with BPH has been supported by other studies. Hammersten and Hogstedt [47] observed that prostatic growth correlated with BMI, and Giovannucci et al. [48] found that obesity was associated with an increased risk for BPH surgery. One mechanism through which obesity has been postulated to promote hyperplasia is the increased peripheral aromatization of testosterone with a resulting increase in the estrogen/testosterone ratio. Another postulated mechanism involves the association of obesity with inflammation and oxidative stress; factors that have been associated with BPH. Other studies have shown that men diagnosed with BPH have a higher incidence of diabetes than the general population, and that diabetes is associated with more severe symptoms [49]. Part of the explanation may be that diabetes can be a primary cause of lower urinary tract symptoms (LUTS), but the studies cited earlier suggest that metabolic factors may influence the development and progression of LUTS indirectly by increasing the rate of prostatic enlargement.

27.3 Aging, Oxidative Stress, and Prostate

Aging is associated with many metabolic disorders and with increased incidence of various cancers [50, 51]. Prostate cancer is a major age-related malignancy. Many theories have been formulated to explain the molecular and biochemical aspect of aging, but Harman in 1956 proposed "free radical theory of aging" in which the author suggested that the accumulation of damage to biomolecules caused by free radicals plays a major role in human aging [52, 53]. It is also believed that cellular oxidative stress increases with age and the increase in mitochondrial mutations can lead to further increase in ROS generation due to defective oxidative phosphorylation and electron transport [54]. Thus, it is possible that the increase in ROS leads to a self-perpetuating cycle with an ever-increasing oxidative challenge placed on the cells.

Moreover, most of the cells in the prostate tumor express the androgen receptor and respond to androgens at an early stage, to facilitate their growth. Age-related changes in the levels of androgens and ratios of other androgenic hormones and changes in the balance between the auto/paracrine growth stimulatory factors [55] such as insulin growth factor (IGF), epidermal growth factor (EGF), and nerve growth factor (NGF), and the growth inhibitory factors such as transforming growth factors-β (TGF-β) and IGF-binding proteins (IGFBPs) are implicated for abnormal prostatic growth [56–58]. Interestingly, physiological stimulation of androgen receptor has been shown to increase ROS production [59, 60].

27.4 Apoptosis, Oxidative Stress, and BPH

An alternative mechanism for BPH may be related to metabolic disturbances. Obesity and elevated fasting glucose are components of the metabolic syndrome [61]. Both the obesity and the metabolic syndrome are associated with systemic inflammation and oxidative stress [62]. Inflammation has been previously implicated as a primary stimulus for prostate carcinogenesis [63], and in the same way, it is possible that BPH represents an alternate, nonmalignant pathway of unregulated prostate growth promoted by oxidative stress, inflammatory mediators, and IGFs [64]. Indeed, the analyses of surgical specimens have shown that BPH is usually associated with inflammation and that the extent and severity of the inflammation correspond to the amount of prostate enlargement [65].

BPH is a phenomenon characterized by an age-dependent increase in the volume of the prostate throughout the entire life of a man. The growth and involution of the prostate depend on the quantitative relationship between the rate of cell proliferation and cell death. Most previous studies have focused on proliferative and apoptotic rates in benign hyperplastic human prostates. Siegfried et al. [66] reported that there is an increase in the proliferation rate and decrease in the apoptosis rate in benign hyperplastic prostatic tissue. Therefore, the authors suggested that the growth of the aging prostate results from this disturbance in the balance between cell proliferation and apoptosis. Claus et al. [67] studied the apoptotic and proliferative rates in the epithelium and the stroma of BPH. They also demonstrated the apoptotic and proliferative rate in the epithelium of normal prostatic tissue. Interestingly, the results of this study indicate a decrease in the apoptotic rate in the stroma of BPH which may explain the enlargement of the prostatic tissue [68]. Apoptosis, also known as programmed cell death, plays an important role in all stages of an organism's development. While there are controversies in the literature regarding the role of apoptosis in aging, age-associated increases in apoptosis have been observed in several physiological systems, including the human immune system, human hair follicle, and rat skeletal muscle [69]. Apoptotic cell death is executed via two major signaling pathways, the intrinsic and extrinsic pathways, in either caspase-dependent or caspase-independent manners [70]. The intrinsic pathway involves the induction of various protein responses, such as posttranslational modifications, conformational changes, and interorganelle translocation of specific proteins. These responses can produce an alteration in mitochondrial membrane potential and the release of apoptogenic factors, such as cytochrome c and apoptosis-inducing factor, from the mitochondria to the cytoplasm. A cascade of downstream signals, including caspases, is then stimulated to orchestrate apoptotic responses. In contrast, the induction of apoptosis by extrinsic pathways requires binding of ligands to membrane receptors and recruitment of cytosolic adaptor proteins, which will, in turn, activate a series of initiator and effector caspases. It has been clearly established that ROS and ROS-modulated molecules participate in both intrinsic and extrinsic apoptotic pathways [71]. Some well-known exogenous ROS-generating stressors, such as radiation, proinflammatory cytokine treatment, growth factor withdrawal, and physiological challenges, such as heat stress, can stimulate apoptosis [72, 73]. One example of oxidative stress involvement in the extrinsic apoptotic signaling pathways is the redox activation of the Mitogen-activated protein kinases (MAPK) cascade upon sustained oxidative stress. A novel protein in the mitogen-activated protein kinase-kinase-kinase family, known as apoptosis signal-regulating kinase 1 (ASK1), has recently been identified as a critical redox sensor in the MAPK pathway [74]. Thioredoxin, a small enzyme that participates in redox reactions, can have a negative regulatory influence on ASK1 and, subsequently, apoptosis. Oxidative stress-induced apoptosis can also be reduced by a dominant-negative mutation of ASK1 [75]. Another example involving the extrinsic apoptotic pathway is the down-regulation of signaling pathways associated with growth factor receptor stimulation in response to oxidative stress [76]. In the intrinsic apoptotic pathway, it has been shown that proteins in the mitochondrial permeability transition pore complex, which controls mitochondrial membrane potential, are the direct targets of ROS. These proteins include the adenine nucleotide translocator in the inner membrane, the voltage-dependent anion channel in the outer membrane, and cyclophilin D at the matrix. Prooxidants capable of induction of mitochondrial permeability potential include not only chemicals, such as *tert*-butyl hydroperoxide and diamide, but also lipid peroxidation products such as 4-hydroxynonenal. Moreover, it has been increasingly recognized that oxidative damage to organelles, such as lysosomes and endoplasmic reticulum, stimulates crosstalk between these organelles and mitochondria and induction of apoptosis via the intrinsic signaling pathway [77].

More importantly, recent studies on p66Shc redox protein may provide a link between oxidative stress-mediated apoptosis and biological aging [78]. The p66Shc redox protein is the third isoform discovered in the Shc protein family. This group of proteins was initially identified as signal transduction adapters involved in mutagenic signaling through Ras, a small GTP-binding protein. Evidence has suggested that p66Shc is an atypical signal transducer that can be regulated by oxidative stress and also plays a role in H_2O_2 generation [79]. While mice lacking p66Shc (p66Shc$^{-/-}$) live 30% longer than the control animals, p66Shc$^{-/-}$ cells from knockout mice are resistant to ROS-induced apoptosis. Several lines of evidence indicate that p66Shc potentially acts at sites upstream of the mitochondrial permeability transition pore in oxidative stress-mediated apoptosis [80].

27.5 Steroid Hormones, Oxidative Stress, and Prostate

Prostate development, maturation, and normal function depends on the activity of the androgens testosterone and its derivative dihydrotestosterone (DHT). DHT, synthesized from testosterone in the prostate by 5α-reductase, has a more potent effect due to its higher affinity to the androgen receptor (AR) [81]. The AR in turn binds to androgen receptor elements (ARE) present in the promoter regions of many genes involved in cellular proliferation [82]. Traditionally, the initial stages of prostate cancer are controlled by androgen deprivation therapy; however, aberrant AR activity in prostate tumors finally leads to the development of a highly malignant state of disease unresponsive to androgen control [83].

Many studies have dwelt on the increased oxidative damage in cells due to ROS as a result of abnormal and increased androgen stimulation of androgen-sensitive prostate cancer cells [84, 85]. Though studies have not pointed out a potential mechanism for the increased levels of ROS after androgen stimulation, as discussed above, changes in the balance of prooxidant and antioxidant molecules in a cell may play an important role. Intracellular redox balance is largely the result of cyclic reduction and oxidation of Glutathione both in the cytoplasm and mitochondria of a cell [86]. Glutathione, synthesized in the cytosol and imported into the mitochondria, plays an important role in the protection of mitochondria from the deleterious effects of ROS generated as a result of electron transport [87]. Even though testosterone is the predominant hormone responsible for the regulation of prostate gland growth and functioning, recent discovery of estrogen receptors in the prostate has brought estrogen role in prostate cancer progression to prominence [88, 89]. Estradiol can be synthesized from testosterone in the prostate epithelial cells or taken up from general circulation. Certain isoflavonoids can also have weak estrogenic effects [90] and have been observed to cause significant infiltration of neutrophils and lymphocytes in the prostatic lobes of rats fed with dietary isoflavonoids. Chronic administration of DHT and estradiol to rats induces the expression of proinflammatory cytokines within the prostate [91]. These inflammatory infiltrates have been identified to be a major source of ROS production and the incidental oxidative injury to the prostate epithelium has been suggested to be the cause for the formation of proliferative inflammatory atrophy (PIA) [92]. These lesions generally form the basis for enhanced epithelial cell proliferation, regeneration, and give rise to prostatic intraepithelial neoplasia (PIN), and progressively to prostate cancer. It is generally believed that estrogen alone is not enough to cause malignancy, but abnormal estrogen receptor α (ERα) signaling in conjunction with elevated levels of testosterone have been shown to induce prostate hyperplasia and prostate cancer in mice [93, 94].

27.6 Role of Oxidative Stress and Inflammation in the Pathogenesis of BPH

Inflammation is a complex phenomenon consisting of humoral (cytokines) and cellular (leukocytes) components. Inflammation can influence the tissue microenvironment through the production of free radicals, COX activity, and NO synthesis, all linked to the deleterious oxidative effect of inflammation on prostatic tissue. These factors can alter protein structure and function, induce gene changes, cause post-translational modifications, including those involved in DNA repair and apoptotic processes, and provoke cellular proliferation.

All these aspects generate an important link between inflammatory processes and the induction of prostatic growth of preneoplastic and neoplastic lesions. Data show that chronic inflammation can induce proliferative events and posttranslational DNA modifications in prostatic tissue through oxidative stress [95]. In fact, repeated tissue damage and oxidative stress related to this event may provoke a compensatory cellular proliferation with the risk of hyperplastic growth or also of neoplastic modifications [96, 97]. It is well accepted that regions of prostatic inflammation can generate free radicals, such as nitric oxide (NO) and various species of oxygen. In particular, macrophages and neutrophil infiltrations provide a source of free radicals that can induce hyperplastic or precancerous transformations through the oxidative stress to the tissue and DNA [96]. A feature of these oxidative stress reactions is the production of arachidonic acid from membranes, a process associated with the generation of new reactive oxygen radicals [96]. It can also be converted by the COX enzymes to various eicosanoids, in particular, prostaglandins that have long been recognized as important factors in the regulation of prostatic cell proliferation [96]. Normally, prostate tissue is protected from oxidative stress reactions, free radicals, and highly ROS by the superoxide-dismutase and the glutathione-*S*-transferase (GST)–P1 enzyme systems, the body's natural protective mechanisms. It is important that estrogens, through the estrogen receptor β (ERβ), appear to influence the protective activity of GST on production of free radicals [98]. A modern context highlights that the transplacental transmission of an estrogen signal can promote cancer induction in later life. Estrogens can initiate molecular events, referred to as gene imprinting or gene silencing, that are related to the induction of an inflammatory response within the prostate and to the possibility that inflammation could induce preneoplastic lesions. Estrogens given to neonatal rodents result in a "developmental estrogenization" in which there are developmental defects, including a reduction in prostatic growth. This treatment also results in the development of lobe-specific inflammation, dysplasia, or PIN [97]. The spontaneous inflammatory response that is induced in animals by estrogens can be prevented by increasing soy intake or enhancing the levels of genistein [98]. Biochemically, phytoestrogens (isoflavones and lignans) are heterocyclic phenols structurally similar to the estrogenic steroids and thus possess estrogenic and anti-estrogenic activity. Because of their weak estrogenic activity, phytoestrogens may (a) act as antiestrogens by competing with the more potent, naturally occurring endogenous estrogens (e.g., 17β estradiol) for binding to the estrogen receptor; (b) inhibit the 5α reductase enzyme; (c) inhibit the aromatase enzyme; (d) inhibit tyrosine-specific protein kinases; and (e) inhibit angiogenesis. Additionally, certain phytoestrogens are antioxidants. Some phytoestrogens affect the topoisomerases, and many phytoestrogens inhibit the growth of experimental tumors [99].

In this pathogenetic hypothesis, NO and COX activity may both play an important role in determining the association between inflammation and prostate growth. In all the inflammatory cells that arrive in the prostate, the inducible nitric oxide synthase (iNOS) is the principal factor activating reactive nitrogens that can damage cells [100]. Gradini et al. [101] characterized NOS expression in human prostate tissue and, particularly for iNOS, they found an increased immunostaining in the epithelial cells in cases of BPH and more with high grade PIN (HGPIN) and PC, when compared with normal tissue. NO also enhances COX activity, the second factor. COX-2 activity has been detected in all inflammatory cells in the epithelium and interstitial spaces of human prostate tissue and it is increased in proliferative inflammatory lesions, generating proinflammatory prostaglandins [96, 97]. In human BPH tissue, Di Silverio et al. [102] showed that COX-2 inhibition can produce a significant increase in prostatic cell apoptotic activity. The mechanism used by inflammation to influence the development and progression of chronic prostatic diseases (BPH and PC) has been suggested and well supported by scientific evidence.

Three recent reviews on the pathogenesis of BPH have provided an evidence-based thesis that strongly suggests a role for inflammation in the propagation of histological BPH [3–104]. Kramer et al. [104] have recently outlined the current state of knowledge in regard to the influence of inflammation on the pathogenesis of BPH. Chronic inflammatory infiltrates, mainly composed of chronically activated T cells and macrophages, are frequently associated with BPH nodules [31–33]. These infiltrating cells are responsible for the production of cytokines (IL-2 and IFN-γ) which may support fibromuscular growth in BPH [104]. Immigration of T cells into the area is attracted by increased production of proinflammatory cytokines such as IL-6, -8, and -15 [3, 103, 105]. Surrounding cells become targets and are then killed by unknown mechanisms, leaving behind vacant spaces that are replaced by fibromuscular nodules with a specific pattern of a Th0/Th3 type of immune response [44].

In situ studies demonstrated elevated expression of proinflammatory cytokines in BPH. IL-6, -8, and -17 may perpetuate chronic immune response in BPH and induce fibromuscular growth by an autocrine or paracrine loop [44, 106] or via induction of COX-2 expression [107]. Immune reaction may be activated via Toll-like receptor (TLR) signaling and mediated by macrophages and T cells [106]. Conversely, anti-inflammatory factors such as macrophage inhibitory cytokine-1 (MIC-1) [108] may be decreased in symptomatic BPH tissues. Animal models provided evidence for the presence of unique T-cell subsets which may suppress autoimmunity in healthy Sprague–Dawley rats resistant to chronic nonbacterial prostatitis [109]. Based on the available scientific evidence, it is highly likely that age-dependent weakening of the immune system, coupled with modified hormonal secretion, leads to the deterioration of a postulated population of suppressor cells. This decline in T suppressor cells population, which actively suppresses the recognition of prostatic antigens, leads to gradual infiltration of the prostate by lymphocytes and subsequent cascade of events resulting in BPH [110].

27.7 Impact of Inflammation on Progression of Clinical BPH

An examination of baseline prostate biopsies in a subgroup of 1,197 patients in the MTOPS study found that there was a chronic inflammatory infiltrate in 43% of the men [111]. It was hypothesized that the presence of histological inflammation may be a predictor of BPH progression. There was a clinically significant difference in the progression rate based on the presence or absence of inflammation. Patients in all groups (placebo, finasteride, doxazosin, and combination finasteride and doxazosin) with inflammation were more likely to progress clinically in terms of symptoms, AUR- or BPH-related surgery.

For those with no inflammation, there was overall clinical progression in 13.2% of patients, while 3.9% had BPH-related surgery and none had AUR; corresponding values for those who had chronic inflammation were 21, 7.3 (not significant), and 5.6%. Chronic inflammation accounted for every AUR event in this subgroup of patients with prostate biopsies, while in patients with no inflammation there was no AUR. The observation that the presence of prostatic inflammation may be clinically relevant in terms of prediction of BPH-related progression is very important. This observation was also confirmed by the 4-year longitudinal follow-up of the 8,000 men enrolled in the REDUCE trial [112–114]. Many of the enrolled men would have had BPH at baseline (predicted by high baseline mean IPSS scores, elevated PSA, and negative initial biopsy). The baseline histological status of these men is documented and the progression data in terms of BPH symptom and event (surgery and AUR) progression are to be collected. A further important point is that inflammation may also have an important role in the pathogenesis of prostate cancer [114] and that particular association will become very clear when the REDUCE trial is completed.

27.8 Biomarkers for Inflammation in BPH

Inflammation has important association with the pathogenesis, symptoms, and progression of BPH. Then, discovery of a biomarker would be invaluable for monitoring the progress of the disease. There are a number of early candidates and many others are currently being assessed by international research groups. A small study [115] suggested that measurement of serum malondialdehyde (MDA), an index of inflammation and oxidative stress, may be a useful marker in BPH. Serum MDA levels were analyzed in 22 BPH patients and 22 healthy donors showing an increase in levels in the BPH patients and a positive correlation with PSA. To our knowledge, this association has not been replicated.

The association of serum C-reactive protein concentration, a nonspecific marker of inflammation, and LUTS, suggestive of BPH, was examined in 2,337 men who participated in the Third National Health and Nutrition Examination Survey between 1988 and 1994 [116]. They found that men with a C-reactive protein concentration above the limit of detection (>3.00 mg/L) were 1.47 times more likely to have three or more symptoms than men with a C-reactive protein concentration below the detection limit (not statistically significant).

Cytokines and chemokines, inflammatory mediators, are also believed to be important in the pathogenesis of prostatic inflammation. Increased expression of IL-8 is noted in BPH tissue culture [117], which by direct and indirect mechanisms could promote proliferation of no senescent epithelial and stromal cells. This enhanced proliferation contributes to the increased tissue growth seen in BPH. Such processes may lead to the discovery of potential biomarkers for prostate inflammation in BPH. Seminal plasma levels of 8 cytokines and 9 chemokines were evaluated in 83 men (20 healthy controls, 9 men with CP/CPPS IIIA [inflammatory], 31 with CP/CPPS IIIB [non-inflammatory], and 23 men with BPH) [118]. Prostate specimens from 13 BPH patients were analyzed to detect interleukin-8 (IL-8) producing cells and to characterize inflammatory infiltrates. IL-8 concentration in seminal plasma was positively correlated with symptom scores in both the CP/CPPS patients and BPH patients.

Although a number of potential markers (C-reactive protein, IL-8, markers of oxidative stress) have been evaluated, these markers are generally nonspecific for prostate or BPH. However, it opens the search for biomarkers that could be used to stratify patients as to risk of developing BPH or related BPH adverse outcomes, monitor symptoms, and response to medical therapy for BPH.

Inflammation is a complex phenomenon consisting of a humoral (cytokines) and cellular (leukocytes, monocytes, and macrophages) components [119, 120]. Cytokines that promote inflammation and act to make disease worse are called proinflammatory cytokines, whereas other cytokines that serve to reduce inflammation and promote healing are called anti-inflammatory cytokines [121]. Inflammation is usually a self-limited event, with initial proinflammatory cytokines and growth factor release and angiogenesis followed by anti-inflammatory cytokine-mediated resolution [122]. In normal tissues, anti-inflammatory cytokines are synchronically upregulated after the proinflammatory cytokines are produced, leading to inflammation resolution [123]. In chronic inflammation, mainly composed of chronically activated T cells and mononuclear phagocytes (monocytes and macrophages), there are persistence of promoters or a failure in the required mechanisms to resolve inflammation. This may release more progrowth cytokines as well as various growth factors and attract additional immune cells to the inflammation site which amplifies the inflammatory response [122, 124]. Prostate has a fully active immunologic response and involves a broad spectrum of immune responses against foreign antigens. Moreover, prostate contains scattered stromal and intraepithelial endogenous inflammatory cells such as T and B lymphocytes, macrophages, and mast cells [104, 125]. T cells increase with age, which correlates with the incidence of prostate inflammation during the aging process [126]. T cells are known to release factors that stimulate matrix formation and secrete potent epithelial and stromal mitogens, which could promote prostate stromal and epithelial proliferation/hyperplasia [127].

Stromal–epithelial prostate interactions play a pivotal regulatory role in the maintenance of homeostasis in health and development of disease [126, 128]. Prostate cell can induce inflammation reaction by expressing antigen presenting cells (APCs) and all of the TLRs. These expressions can produce proinflammatory cytokine and activate immune responses [104, 121, 129–131]. Konig et al. found a different expression pattern of the TLR in BPH and PC tissue. Most of the BPH tissues showed a strong expression for the TLR 4, 5, 7, and 9, whereas in PC increased expression was obtained for TLR 1, 2, and 3 [132]. But it is still unknown how it will influence the inflammatory process in both diseases.

T cells, prostatic stromal, and epithelial cells simultaneously secrete higher proinflammatory cytokines such as interleukins (IL-1, -1α, -2, -4, -6, -7, -8, and -17), the CXC-type chemokine, and their receptors in BPH and PC tissues compared to normal prostate tissue [121, 128, 130, 133]. These cytokines were thought to induce fibromuscular growth and proliferation of prostatic stromal or epithelial cell by an autocrine or paracrine loop or via induction of COX-2 expression [44, 127, 133]. IL-1α produced by epithelial cells induces fibroblast growth factor-7 (FGF-7) in prostate stromal cells that can induce benign growth of the prostate. IL-17 upregulates the secretion of other proinflammatory cytokines, such as IL-8 and -6 as well as of TGF-β. IL-8 and -6 are recognized as two potent growth factors for prostatic epithelial and stromal cells, with IL-8 playing a major role in stromal proliferation by the induction of FGF-2 [130, 134]. The expression of proinflammatory cytokines was different between BPH and PC. Mechergui et al. [121] and Konig et al. [132] found IL-6 and -8 were more overexpressed in PC tissue compared to BPH tissue. IL-6 regulates the growth of prostate carcinoma and activates the androgen receptor-dependant gene in prostatic cancer cells in the absence of androgen [121].

Chronic inflammation continuously produces COX-2 [123, 135, 136]. COX-2 increases production of prostaglandin (PG) E2 (an adhesion to the extracellular matrix) and concentrations of Bcl-2 (proapoptotic genes) and reduces the E-cadherin protein (with consequent loss of cell-to-cell adhesion). COX-2 also modulates production of angiogenic factors to induce angiogenesis.

Lastly, COX-2 increases the carcinogenic potential of cells through the oxidation of procarcinogens to carcinogens, increased cell growth, decreased apoptosis, as well as decreased immune response to abnormal or cancer cells matrix metalloproteinase overexpression with an associated increase in invasiveness [134, 137–139]. COX-2 is upregulated in a variety of malignancies including prostate cancer, throughout the tumorigenic process from early hyperplasia to metastatic disease [123, 140]. Many studies showed more overexpression of COX-2 in prostate cancer compared to that in BPH. COX-2 overexpression was also higher in PIN and poorly differentiated tumors [122, 132, 136, 140]. Chronic inflammation also produces a free radical substance/oxidative stress such as inducible nitric oxide (iNOS)/reactive nitric species (RNS) and various reactive oxygen species (ROS) [119, 123, 135, 141, 142]. These oxidative stresses can induce vascular tissue damage, protein structural and functional damage, and genomic damages and cause posttranslational modifications including those involved in DNA repair and apoptosis [119]. These can lead to oxidative DNA damage such as point mutations, deletions, or rearrangements and reduce DNA repair. These oxidative stresses also alter the stem cell population. Genomic alterations in cellular DNA result in the modulation of an imbalance between cell proliferation and cell death. A change in the normal regulation of programmed cell death leads to hyperplastic or precancerous transformation [122, 124, 143]. All of these active factors production also induce repetitive tissue damage and repair with the release cytokines, growth factors, and oncogenes, leading to increase of epithelial or stromal cell proliferation [143]. Normally, these highly oxidative stresses are removed by natural protective mechanism, the superoxide dismutase enzyme system, such as superoxide dismutase (SOD), glutathione peroxidase (GPX), and catalase (CAT) enzyme [119]. Human prostatic tissue is vulnerable to oxidative DNA damage due to more rapid cell turnover and fewer DNA repair enzymes [144]. Balance between oxidative stress and antioxidant component of the cells has a role in developing prostate disease [144]. There are increases in the oxidative stress and decrease in the antioxidant mechanisms in prostate disease. Sciarra et al. [119] characterized NOS expression in the human prostatic tissue and in particular for the iNOS-increased immunostaining in the epithelial cells of BPH and even more with HGPIN and PC than that of normal tissue. Khandrika et al. [144] also found increasing ROS generation for more aggressive phenotype in PC cell lines. Prostate cancer cell expresses lower levels of antioxidant enzymes or almost total inactivation of prooxidant scavenging enzyme than BPH [139]. Compared with normal prostate, the activity of antioxidant enzymes is decreased in BPH [139, 145–149].

Age was also one of contributing factors for changing the oxidant/antioxidant balance which is shifted towards oxidative stress [144]. Production of ROS and free radicals in the mitochondria is increased during aging [150]. There are also a down-regulation and/or underexpression of antioxidant with increasing age [151]. Therefore, oxidative stress can activate the transcription factor NF-κB (nuclear factor kappa-lightchain—enhancer of activated B cells) by TNF-α/AP-1 transduction pathway and NIK transduction pathway. NF-κB is known as a master inflammatory transcriptional regulator and is highly active in macrophages. Targets of NF-κB include genes regulating immune response, inflammation, cell proliferation, cell migration, and apoptosis. The nuclear translocation of NF-κB can activate target genes involved in carcinogenesis [137]. NF-κB potentially can lead to the amplification of the inflammatory response in the tumor environment. Dysregulation of the transcription factor NF-κB has been proposed as a putative molecular mechanism leading to chronic inflammation and cancer.

In a chronically inflamed tumor environment, it is difficult to distinguish whether the aberrant NF-κB activation originates from tumor cells or from immune infiltrates.

Wong et al. found that the exposure of prostatic epithelial cells to proinflammatory-soluble mediators directly activated NF-κB and induced local production of proinflammatory cytokines in the prostate epithelial cells. Narayanan et al. [120] and Wong et al. [124] found significant increased NF-κB in the prostatic specimen which has PIN and adenocarcinoma histopathology. The IL-1β-induced NF-κB pattern of intraprostatic chemo-attractive signals might have a capability for maintaining the chronic inflammation and PIA in the prostate, which are recognized as putative precursor lesions in the development of prostate cancer [152]. In normal prostate, the transduction pathway from NIK to NF-κB seems to be inactive in BPH. There is increasing TNF-α/AP-1 transduction pathway and also follows by increasing apoptotic pathway to inhibit uncontrolled cell proliferation [120]. In PC, the proapoptotic effect of TNF-α/AP-1 pathway was found decreased and also nuclear translocation of NF-κB increased which are an active form to stimulate prostate cancer cell proliferation [120, 124].

Induction of anti-inflammatory factors such as MIC-1 is an early response due to inflammation in the prostate [153]. MIC-1 was downregulated in BPH tissues compared to normal prostate tissue [44, 134, 154]. However, in PC, there are up-regulation and overexpression of MIC-1 in higher grade and more aggressive prostate cancers [153, 154]. Gene expression analysis between normal peripheral zones and transition zones of the specimens obtained from patients with prostate cancer revealed a preferential expression of MIC-1 in the peripheral zone (predominant site of tumor occurrence) compared with the transition zone (site of BPH) [154]. MIC-1 in tumor environment is assumed to reduce the tumor killing (functional) activity of macrophage [154].

Another factor that differentiates between BPH and PC was gene polymorphism. Polymorphisms in innate and adaptive immune genes may affect the nature and the extent of the immune response within the prostate, including the likelihood of persistent prostatic infection and chronic inflammation. There is much evidence that BPH has only rare genetic abnormalities [155]. Recently, multiple genes with regulatory roles on inflammatory pathways have been associated with prostate cancer risk, including Ribonuclease L (RNASEL), macrophage scavenger receptor 1 (MSR1), macrophage inhibitory cytokine-1 (MIC-1), interleukins (IL-8, -10), vascular endothelial growth factor (VEGF) and intercellular adhesion molecule (ICAM), ELAC2/HPC2, Machropaghe Scavenger Receptor (SR-A/MSR1), CHEK2, Breast Cancer gene 2 (BRCA2), Paraoxonase 1 (PON1), 8-oxoguanine glycosylase (OGG-1), TLRs, and COX-2 promoter. These genes are linked to cellular defenses against inflammation and oxidative stress, and defects in their function may lead to the inability to prevent tumor formation by this pathway [125, 136, 156–160].

27.9 Chemopreventation

The association of prostate diseases with inflammation and oxidative stress offers a framework to design therapeutic approaches. Targeting the microenvironment may represent a promising therapeutic approach for prostatic diseases. In the clinical practice, patients with chronic inflammatory aspects in the prostate gland could be stratified at higher risk of BPH progression or, in particular if associated focal atrophy, at higher risk of carcinogenic development. In both cases, the finding of chronic inflammation in the prostate may indicate the need of a preventive strategy.

27.10 Nonsteroidal Anti-inflammatory Drugs

The best evidence for the significance of inflammation during neoplastic progression comes from the study of cancer risk among long-term users of NSAIDs [119]. The ability of NSAIDs to inhibit COX-1 and -2 underlies their mechanism(s) of chemoprevention. COX-2 inhibition can produce a significant increase in prostate cell apoptosis through release of cytochrome c from mitochondria and subsequent activation of caspase-9 and -3 [123, 135, 139]. NSAIDs can also promote apoptosis by inhibiting NF-κB. Aspirin use has provided 10–39% risk reduction for prostate cancer [161, 162]. Aspirin is typically used on a regular basis for extended periods (e.g., when used in the prevention of coronary heart disease), whereas the other NSAIDs are typically prescribed to be used occasionally, as required (e.g., for pain relief) [162]. Selective COX-2 inhibitors, such as celecoxib, also reduce expression of COX-2, PGE2, and NF-κB and induction of apoptosis in prostate cancer cell [120, 163].

27.11 Antioxidant Therapy

It is generally accepted that prostatic inflammation frequently accompanies BPH [3]. The incidence of inflammation in surgically resected hyperplastic prostate tissue can amount up to 98.1% [31]. Although the cause of the inflammation is not well understood, the inflammatory microenvironment in the prostate is characterized by the accumulation of activated leucocytes and macrophages, which release proteases, angiogenic factors, chemokines, and superoxide anion (O_2^-). O_2^- is subsequently converted to other ROS, including hydrogen peroxide, singlet oxygen, and hydroxyl radicals (OH) [3]. Such a circumstance could aggravate the inflammation and lead to tissue destruction, contributing to the development of BPH [164].

In addition, bladder outlet obstruction due to BPH induces cyclical ischemia/reperfusion at the bladder wall, leading to perturbation of intracellular Ca^{2+} homeostasis and subsequent activation of a variety of Ca^{2+}-regulated enzymes and the generation of ROS such as O_2^- and OH [165]. The ROS-mediated oxidation of biomolecules probably causes partial denervation of the detrusor muscle with the subsequent development of supersensitivity, leading to bladder instability [166]. Therefore, ROS may be one of pathogenic factors in the inflammation of the dysfunctional detrusor and enlarged prostate, so that antioxidants may be useful in the treatment of BPH. The treatment goals for BPH are to relieve the irritable and obstructive symptoms. Options for treatment include surgical therapy and pharmaceutical intervention

[167–169]. Drugs for BPH commonly used in Japan are α1-adrenergic blockers, antiandrogenic agents, and phytotherapeutic agents.

Antioxidant compounds that scavenge oxidative stress may be useful in helping to overwhelm its mutagenic effects [170]. Dietary components/supplements that have antioxidant compound include carotenoids such as lycopenes (in tomato) and β-carotene (in yellow orange vegetables), vitamin A, retinol, vitamin C and E, and selenium.

Another additional food-derived antioxidant compound may also be beneficial in the prevention of prostate cancer including thioallyl components (garlic), sulphorophane (cruciferous vegetables), green tea polyphenol, and soy [149, 171–174]. Many in vitro, in vivo, epidemiological studies showed a significant effect of these antioxidant dietary components to prevent BPH and PC development [150, 173, 175–180]. But there are still controversies for the effectiveness of vitamin C and E. Latest study showed neither vitamin E nor C supplementation did reduce the risk of prostate cancer [181]. One last large study, SELECT (Selenium–Vitamin E Cancer Prevention Trial) study, has shown no differences in preventing prostate cancer in the generally healthy and heterogeneous population resulting in earlier discontinuation of the study. This result is probably due to lower efficacy of the given formulation in this study than in other study [149, 182].

VDR agonists, such as calcitriol, can promote innate immunity and regulate adaptive immune responses and exert anti-inflammatory and immunoregulatory properties potentially useful in the treatment of diseases characterized by chronic inflammation and cell proliferation. The prostate is a target organ of VDR agonists and represents an extrarenal synthesis site of 1,25-dihydroxyvitamin D3; there are marked inhibitory activity of the VDR agonist elocalcitol on basal and growth factor-induced proliferation of human prostate cells. Many mechanisms that could be proposed are (1) inhibition of the RhoA/ROCK pathway, a calcium-sensitizing pathway, to produce IL-8; (2) inhibition of the expression of COX-2; (3) decreasing PGE production and its receptors; and (4) prevention of NF-κB nuclear translocation.

The combination of calcitriol and NSAIDs results in a synergistic inhibition of BPH and PC cell growth and offers a potential therapeutic strategy by acting on a separate anti-inflammatory pathway [129, 130, 183].

Phytotherapy has long been used to alleviate the urodynamic symptoms of BPH. There is a clinical evidence that several prescriptions of herbal medications such as permixon® (from *Serenoa repens*) and tadenan® (from *Pygeum africanum*) improve lower urinary tract function, although the mechanisms of action are not fully understood [184, 185]. Previous work suggests that, like eviprostat, some of these drugs possess anti-inflammatory activity, which helps prevent the progression of obstruction-induced bladder dysfunction [186, 187]. Eviprostat® is the most widely used phytotherapeutic agent for LUTS in BPH. Eviprostat consists of five components and, though the precise mechanisms of action of eviprostat remain to be elucidated, the individual components are known to have diuretic, antiseptic, and anti-inflammatory effects [188]. Combination therapy with eviprostat and the α1-adrenergic blocker tamsulosin is even more effective, because of the combined effect of their different modes of action [189]. Anti-inflammatory activity is a characteristic of eviprostat, but it is still uncertain which components of eviprostat are responsible for this activity. Since inflammatory cell infiltration plays a role in the development of BPH, the ability of eviprostat to antagonize ROS production in human neutrophils may account, at least in part, for its beneficial effect in BPH. Eviprostate components (EVI-1, -2 and -4) have no effect on the rate of uric acid production in the xanthine/xanthine system, suggesting that these components act as O_2^- scavengers. On the other hand, it is unclear whether EVI-1, -2, -3, and -4 have inhibitory activity for OH formation or act as the OH scavenger because compounds of eviprostat, including several types of flavonoids, saponines, tannins, and so on, are known as both radical scavengers and iron chelators. To investigate the involvement of these anti-ROS activities of eviprostat in its anti-inflammatory effects, the individual components of eviprostat and a mixture of them were tested in the carrageenin-induced paw edema model in rats. This inflammation model is suitable for the in vivo assessment of the anti-inflammatory action of antioxidant or inhibitors of ROS generation, because SOD, a typical antioxidant enzyme, effectively reduces inflammation in this model [190, 191]. Although each component of eviprostat, when tested alone at an equivalent dosage, has no effect on the paw edema, both EVIs and EVI-1/2/4 ameliorate it and were significantly superior in its activity to EVI-5 alone, which is dominating in quantity ($p<0.05$, Dunnett's multiple comparison test), that is, percentages of inhibition in groups treated with EVI-1/2/4, EVIs, and EVI-5 were 35.1 ± 7.0, 41.5 ± 5.4, and $13.5 \pm 4.9\%$, respectively. Since several ROS are involved in inflammatory processes [3], the anti-ROS activities of EVI-1, -2, and -4 may contribute to the anti-inflammatory action of eviprostat. It is suggested that each extract has adequate activity on its own merits and the combination shows additive activity on the rat inflammatory model.

27.12 Conclusion

In all prostatic diseases, inflammatory processes and oxidative stress have a role in the pathogenesis as potential triggers of disease progression. Potential role for ROS in the regulation of cellular process controlling benign prostatic cell proliferation holds a lot of promise in understanding etiology and progression of BPH, as this may open doors for the development of novel therapeutics for BPH prevention and treatment.

References

1. Garraway WM, Collins GN, Lee RJ. High prevalence of benign prostatic hypertrophy in the community. Lancet. 1991;338:469.
2. Carter HB, Coffey DS. The prostate: an increasing medical problem. Prostate. 1990;16:39.
3. Lee KL, Peehl DM. Molecular and cellular pathogenesis of benign prostatic hyperplasia. J Urol. 2004;172:1784.

4. Barzilai A, Rotman G, Shiloh Y. ATM deficiency and oxidative stress: a new dimension of defective response to DNA damage. DNA Repair. 2002;22:3–25.
5. Naka K, Muraguchi T, Hoshii T, et al. Regulation of reactive oxygen species and genomic stability in hematopoietic stem cells. Antioxid Redox Signal. 2008;10:1883–94.
6. Lambeth JD. Nox enzymes, ROS, and chronic disease: an example of antagonistic pleiotropy. Free Radic Biol Med. 2007;43:332–47.
7. Sauer H, Wartenberg M, Hescheler J. Reactive oxygen species as intracellular messengers during cell growth and differentiation. Cell Physiol Biochem. 2001;11:173–86.
8. Bierhoff E, Vogel J, Benz M, et al. Stromal nodules in benign prostatic hyperplasia. Eur Urol. 1996;29:345–54.
9. Meigs JB, Mohr B, Barry MJ, et al. Risk factors for clinical benign prostatic enlargement in a community-based population of healthy aging men. J Clin Epidemiol. 2001;54:935–44.
10. Michel MC, Mehlburger L, Schumacher H, et al. Hyperinsulinaemia as a risk factor for developing benign prostatic hyperplasia. J Urol. 2001;163(6):1725–9.
11. Verhamme K, Dieleman J, Bleumink G, et al. Incidence and prevalence of lower urinary tract symptoms suggestive of benign prostatic enlargement in primary care—the Triumph project. Eur Urol. 2002;42:323–8.
12. McNeal J. Pathology of benign prostatic hyperplasia. Insight into etiology. Urol Clin North Am. 1990;17(3):477–86.
13. McNeal JE. Origin and evolution of benign prostatic enlargement. Invest Urol. 1978; 15(4):340–5.
14. Thomson AA, Marker PC. Branching morphogenesis in the prostate gland and seminal vesicles. Differentiation. 2006;74(7):382–92.
15. Gooren L. Androgen deficiency in the aging male: benefits and risks of androgen supplementation. J Steroid Biochem Mol Biol. 2003;85(2–5):349–55.
16. Green JS, Holden ST, Bose P, et al. An investigation into the relationship between prostate size, peak urinary flow rate and male erectile dysfunction. Int J Impot Res. 2001;13(6):322–5.
17. Kaufman JM. The effect of androgen supplementation therapy on the prostate. Aging Male. 2003;6(3):166–74.
18. Kirby RS, Lowe D, Bultitude MI. Intra-prostatic urinary reflux: an aetiological factor in abacterial prostatitis. Br J Urol. 1982;54(6):729–31.
19. Marks LS, Mostaghel EA, Nelson PS. Prostate tissue androgens: history and current clinical relevance. Urology. 2008;72(2):247–54.
20. Masumori N, Tsukamoto T, Kumamoto Y, et al. Japanese men have smaller prostate volumes but comparable urinary flow rates relative to American men: results of community based studies in 2 countries. J Urol. 1996;155(4):1324–7.
21. Mostaghel EA, Nelson PS. Intracrine androgen metabolism in prostate cancer progression: mechanisms of castration resistance and therapeutic implications. Best Pract Res Clin Endocrinol Metab. 2008;22(2):243–58.
22. Belanger A, Candas B, Dupont A, et al. Changes in serum concentrations of conjugated and unconjugated steroids in 40- to 80-year-old men. J Clin Endocrinol Metab. 1994;79(4):1086–90.
23. Coffey DS, Walsh PC. Clinical and experimental studies of benign prostatic hyperplasia. Urol Clin North Am. 1990;17(3):461–75.
24. Collins GN, Lee RJ, Russell EB, et al. Ultrasonically determined patterns of enlargement in benign prostatic hyperplasia. Br J Urol. 1993;71(4):451–6.
25. Fingerhut B, Veenema RJ. Histology and radioautography of induced benign enlargement of the mouse prostate. Invest Urol. 1966;4(2):112–24.
26. Ricke WA, McPherson SJ, Bianco JJ, et al. Prostatic hormonal carcinogenesis is mediated by in situ estrogen production and estrogen receptor alpha signaling. FASEB J. 2008;22(5):1512–20.
27. Ricke WA, Ishii K, Ricke EA, et al. Steroid hormones stimulate human prostate cancer progression and metastasis. Int J Cancer. 2006;118(9):2123–31.
28. Streng TK, Talo A, Andersson KE, et al. A dose dependent dual effect of oestrogen on voiding in the male mouse? BJU Int. 2005;96(7):1126–30.

29. Delongchamps NB, de la Roza G, Chandan V, et al. Evaluation of prostatitis in autopsied prostates—is chronic inflammation more associated with benign prostatic hyperplasia or cancer? J Urol. 2008;179:1736–40.
30. de la Rosette JJ, Witjes WP, Schafer W, et al. Relationships between lower urinary tract symptoms and bladder outlet obstruction: results from the ICS-"BPH" study. Neurourol Urodyn. 1998;17(2):99–108.
31. Kohnen PW, Drach GW. Patterns of inflammation in prostatic hyperplasia: a histologic and bacterial study. J Urol. 1979;121:755–60.
32. Kramer G, Marberger M. Could inflammation be a key component in the progression of benign prostatic hyperplasia? Curr Opin Urol. 2006;16:25–9.
33. Steiner G, Gessl A, Kramer G, et al. Phenotype and function of peripheral and prostatic lymphocytes in patients with benign prostatic hyperplasia. J Urol. 1994;151:480–4.
34. Theyer G, Kramer G, Assmann I. Phenotypic characterization of infiltrating leukocytes in benign prostatic hyperplasia. Lab Invest. 1992;66:96–107.
35. McNeal JE. Regional morphology and pathology of the prostate. Am J Clin Pathol. 1968;49(3):347–57.
36. Bennett BD, Culberson DE, Petty CS. Histopathology of prostatitis [abstract]. J Urol. 1990;143:265.
37. Weiss JP, Wein A, Jacobs J, et al. Use of nitrofurantoin macro-crystals after transurethral prostatectomy. J Urol. 1983;130:479.
38. De Marzo AM, Platz EA, Sutcliffe S, et al. Inflammation in prostate carcinogenesis. Nat Rev Cancer. 2007;7:256–69.
39. Nickel JC, Downey J, Young I, et al. Asymptomatic inflammation and/or infection in benign prostatic hyperplasia. BJU Int. 1999;84:976–81.
40. Gerstenbluth RE, Seftel AD, MacLennan GT, et al. Distribution of chronic prostatitis in radical prostatectomy specimens with up-regulation of bcl-2 in areas of inflammation. J Urol. 2002;167:2267–70.
41. Roehrborn CG. Benign prostatic hyperplasia: an overview. Rev Urol. 2005;7(9):S3–14.
42. Roehrborn CG. Definition of at-risk patients: baseline variables. BJU Int. 2006;97 Suppl 2:7–11.
43. Nickel JC, Roehrborn CG, O'Leary MP, et al. The relationship between prostate inflammation and lower urinary tract symptoms: examination of baseline data from the REDUCE trial. Eur Urol. 2008;54:1379–84.
44. Nickel JC. Inflammation and benign prostatic hyperplasia. Urol Clin North Am. 2007;35:109–15.
45. Parsons JK, Carter HB, Partin AW, et al. Metabolic factors associated with benign prostatic hyperplasia. J Clin Endocrinol Metab. 2006;91(7):2562–8.
46. Popp KA. Assignment of patients into the classification of cardiomyopathies. Circulation. 1992;86(5):1622–33.
47. Hammersten J, Hogstedt B. Clinical, anthropometric, metabolic and insulin profile of men with fast annual growth rates of benign prostatic hyperplasia. Blood Press. 1999;8(1):29–36.
48. Giovannucci E, Rimm EB, Chute CG, et al. Obesity and benign prostatic hyperplasia. Am J Epidemiol. 1994;140(11):989–1002.
49. Boon TA, Van Venrooij GE, Eckhardt MD. Effect of diabetes mellitus on lower urinary tract symptoms and dysfunction in patients with benign prostatic hyperplasia. Curr Urol Rep. 2001;2(4):297–301.
50. Wei YH, Lu CY, Lee HC, et al. Oxidative damage and mutation to mitochondrial DNA and age-dependent decline of mitochondrial respiratory function. Ann N Y Acad Sci. 1998; 854:155–70.
51. Finkel T, Holbrook NJ. Oxidants, oxidative stress and the biology of ageing. Nature. 2000;408:239–47.
52. Harman D. Aging: a theory based on free radical and radiation chemistry. J Gerontol. 1956; 11:298–300.

53. Harman D. The aging process. Proc Natl Acad Sci USA. 1981;78:7124–8.
54. Ishii T, Miyazawa M, Hartman PS, et al. Mitochondrial superoxide anion (O(2)(-)) inducible "mev-1" animal models for aging research. BMB Rep. 2011;44(5):298–305.
55. Untergasser G, Rumpold H, Hermann M, et al. Proliferative disorders of the aging human prostate: involvement of protein hormones and their receptors. Exp Gerontol. 1999;34: 275–87.
56. Rajah R, Valentinis B, Cohen P. Insulin-like growth factor (IGF)-binding protein-3 induces apoptosis and mediates the effects of transforming growth factor-beta1 on programmed cell death through a p53- and IGF-independent mechanism. J Biol Chem. 1997;272:12181–8.
57. Berry SJ, Coffey DS, Walsh PC, et al. The development of human benign prostatic hyperplasia with age. J Urol. 1984;132:474–9.
58. Juul A, Main K, Blum WF, et al. The ratio between serum levels of insulin-like growth factor (IGF)-1 and the IGF binding proteins (IGFBP-1, 2 and 3) decrease with age in healthy adults and is increased in acromegalic patients. Clin Endocrinol (Oxf). 1994;41:85–91.
59. Mehraein-Ghomi F, Lee E, Church DR, et al. JunD mediates androgen-induced oxidative stress in androgen dependent LNCaP human prostate cancer cells. Prostate. 2008;68:924–34.
60. Shigemura K, Sung SY, Kubo H, et al. Reactive oxygen species mediate androgen receptor- and serum starvation-elicited downstream signaling of ADAM9 expression in human prostate cancer cells. Prostate. 2007;67:722–31.
61. Haffner S, Taegtmeyer H. Epidemic obesity and the metabolic syndrome. Circulation. 2003;108:1541–5.
62. Furukawa S, Fujita T, Shimabukuro M, et al. Increased oxidative stress in obesity and its impact on metabolic syndrome. J Clin Invest. 2004;114:1752–61.
63. Nelson WG, De Marzo AM, Isaacs WB. Prostate cancer. N Engl J Med. 2003;349:366–81.
64. Wang W, Bergh A, Damber JE. Chronic inflammation in benign prostate hyperplasia is associated with focal upregulation of cyclooxygenase-2, Bcl-2, and cell proliferation in the glandular epithelium. Prostate. 2004;61:60–72.
65. Di Silverio F, Gentile V, De Matteis A, et al. Distribution of inflammation, pre-malignant lesions, incidental carcinoma in histologically confirmed benign prostatic hyperplasia: a retrospective analysis. Eur Urol. 2003;43:164–75.
66. Siegfried C, Wrenger M, Senge T, et al. Immunhistochemical determination of age related proliferation rates in normal and benign hyperplastic human prostates. Urol Res. 1993;21:305–8.
67. Claus S, Berges R, Senge T, et al. Cell kinetic in epithelium and stroma of benign prostatic hyperplasia. J Urol. 1997;158:217–21.
68. Berges RR, Vucanovic J, Epstein JI, et al. Implication of cell cycle kinetic changes during the progression of human prostate cancer. Clin Cancer Res. 1995;1:473–9.
69. Arck PC, Overall R, Spatz K, et al. Towards a "free radical theory of graying": melanocyte apoptosis in the aging human hair follicle is an indicator of oxidative stress induced tissue damage. FASEB J. 2006;20:1567–9.
70. Cho SG, Choi EJ. Apoptotic signaling pathways: caspases and stress-activated protein kinases. J Biochem Mol Biol. 2002;35:24–7.
71. Matsuzawa A, Ichijo H. Stress-responsive protein kinases in redox-regulated apoptosis signaling. Antioxid Redox Signal. 2005;7:472–81.
72. Assefa Z, Van Laethem A, Garmyn M, et al. Ultraviolet radiation-induced apoptosis in keratinocytes: on the role of cytosolic factors. Biochim Biophys Acta. 2005;1755:90–106.
73. Gupta S, Gollapudi S. Molecular mechanisms of TNF-alpha-induced apoptosis in aging human T cell subsets. Int J Biochem Cell Biol. 2005;37:1034–42.
74. Goswami A, Dikshit P, Mishra A, et al. Oxidative stress promotes mutant huntingtin aggregation and mutant huntingtin-dependent cell death by mimicking proteasomal malfunction. Biochem Biophys Res Commun. 2006;342:184–90.
75. Zhang R, Al-Lamki R, Bai L, et al. Thioredoxin-2 inhibits mitochondria-located ASK1-mediated apoptosis in a JNK-independent manner. Circ Res. 2004;94:1483–91.
76. Imoto K, Kukidome D, Nishikawa T, et al. Impact of mitochondrial reactive oxygen species and apoptosis signal-regulating kinase 1 on insulin signaling. Diabetes. 2006;55:1197–204.

77. Kevin CK, Hannah JZ. An integrated view of oxidative stress in aging: basic mechanisms, functional effects, and pathological considerations. Am J Physiol Regul Integr Comp Physiol. 2007;292:R18–36.
78. Migliaccio E, Giorgio M, Pelicci PG. Apoptosis and aging: role of p66Shc redox protein. Antioxid Redox Signal. 2006;8:600–8.
79. Migliaccio E, Giorgio M, Mele S, et al. The p66shc adaptor protein controls oxidative stress response and life span in mammals. Nature. 1999;402:309–13.
80. Pellegrini M, Finetti F, Petronilli V, et al. p66SHC Promotes T cell apoptosis by inducing mitochondrial dysfunction and impaired Ca^{2+} homeostasis. Cell Death Differ. 2007;14(2): 338–47.
81. Thompson IM, Coltman CA, Brawley OW, et al. Chemoprevention of prostate cancer. Semin Urol. 1995;13:122–9.
82. Wong YC, Wang YZ. Growth factors and epithelial-stromal interactions in prostate cancer development. Int Rev Cytol. 2000;199:65–116.
83. Asirvatham AJ, Schmidt M, Gao B, et al. Androgens regulate the immune/inflammatory response and cell survival pathways in rat ventral prostate epithelial cells. Endocrinology. 2006;147:257–71.
84. Pathak S, Singh R, Verschoyle RD, et al. Androgen manipulation alters oxidative DNA adduct levels in androgen-sensitive prostate cancer cells grown in vitro and in vivo. Cancer Lett. 2008;261:74–83.
85. Miyake H, Hara I, Kamidono S, et al. Oxidative DNA damage in patients with prostate cancer and its response to treatment. J Urol. 2004;171:1533–6.
86. Fernández-Checa JC, Kaplowitz N, García-Ruiz C, et al. GSH transport in mitochondria: defense against TNF-induced oxidative stress and alcohol induced defect. Am J Physiol. 1997;273:G7–17.
87. Armstrong JS, Steinauer KK, Hornung B, et al. Role of glutathione depletion and reactive oxygen species generation in apoptotic signaling in a human B lymphoma cell line. Cell Death Differ. 2002;9:252–63.
88. Tsurusaki T, Aoki D, Kanetake H, et al. Zone-dependent expression of estrogen receptors alpha and beta in human benign prostatic hyperplasia. J Clin Endocrinol Metab. 2003;88: 1333–40.
89. Linja MJ, Savinainen KJ, Tammela TL, et al. Expression of ERalpha and ERbeta in prostate cancer. Prostate. 2003;55:180–6.
90. Kwon SM, Kim SI, Chun DC, et al. Development of rat prostatitis model by oral administration of isoflavone and its characteristics. Yonsei Med J. 2001;42:395–404.
91. Harris MT, Feldberg RS, Lau KM, et al. Expression of proinflammatory genes during estrogen-induced inflammation of the rat prostate. Prostate. 2000;44:19–25.
92. Ho E, Boileau TW, Bray TM. Dietary influences on endocrine-inflammatory interactions in prostate cancer development. Arch Biochem Biophys. 2004;428:109–17.
93. Smith MR, Morton RA, Barnette KG, Sieber PR, et al. Toremifene to reduce fracture risk in men receiving androgen deprivation therapy for prostate cancer. J Urol. 2010;184(4): 1316–21.
94. Ricke WA, McPherson SJ, Bianco JJ, et al. Prostatic hormonal carcinogenesis is mediated by in situ estrogen production and estrogen receptor alpha signaling. FASEB J. 2008;22: 1512–20.
95. Naber K, Weidner W. Chronic prostatitis: an infectious disease? J Antimicrob Chemother. 2000;46:157–61.
96. Palapattu GS, Sutcliffe S, Bastain PJ, et al. Prostate carcinogenesis and inflammation: emerging insights. Carcinogenesis. 2004;26:1170–81.
97. Sugar LM. Inflammation and prostate cancer. Can J Urol. 2006;13 Suppl 1:46–7.
98. Coffey DS. Similarities of prostate and breast cancer: evolution, diet and estrogens. Urology. 2001;57(Suppl):31–8.
99. Brössner C, Petritsch K, Fink K, et al. Phytoestrogen tissue levels in benign prostatic hyperplasia and prostate cancer and their association with prostatic diseases. Urology. 2004;64(4):707–11.

100. Baltaci S, Orhan D, Cogus C, et al. Inducible nitric oxide synthase expression in BPH, low and high grade PIN and prostate carcinoma. BJU Int. 2001;88:100–3.
101. Gradini R, Realacci M, Petrangeli E, et al. Nitric oxide synthases in normal and benign hyperplastic human prostate: immunohistochemistry and molecular biology. J Pathol. 1999;189:224–9.
102. Di Silverio F, Bosman C, Salvatori M, et al. Combination therapy with rofecoxib and finasteride in the treatment of men with lower urinary tract symptoms (LUTS) and benign prostatic hyperplasia (BPH). Eur Urol. 2005;47:72–9.
103. Untergasser G, Madersbacher S, Berger P. Benign prostatic hyperplasia: age-related tissue remodeling. Exp Gerontol. 2005;40:121–8.
104. Kramer G, Steiner GE, Handisurya A, et al. Increased expression of lymphocyte- derived cytokines in benign hyperplastic prostate tissue, identification of the producing cell types, and effect of differentially expressed cytokines on stromal cell proliferation. Prostate. 2002;52:43–8.
105. Handisurya A, Steiner GE, Stix U, et al. Differential expression of interleukin-15, a pro-inflammatory cytokine and T-cell growth factor, and its receptor in human prostate. Prostate. 2001;49:251–62.
106. Konig JE, Senge T, Allhoff EP, et al. Analysis of the inflammatory network in benign prostate hyperplasia and prostate cancer. Prostate. 2004;58:121–9.
107. Wang W, Bergh A, Damber JE. Chronic inflammation in benign prostate hyperplasia is associated with focal upregulation of cyclooxygenase-2, Bcl-2, and cell proliferation in the glandular epithelium. Prostate. 2004;61:60–72.
108. Kakehi Y, Segawa T, Wu XX, et al. Down-regulation of macrophage inhibitory cytokine-1/prostate derived factor in benign prostatic hyperplasia. Prostate. 2004;59:351–6.
109. Vykhovanets EV, Resnick MI, Marengo SR. The healthy rat prostate contains high levels of natural killer-like cells and unique subsets of CD4+ helper inducer T cells: implications for prostatitis. J Urol. 2005;173:1004–10.
110. Kramer G, Mitteregger D, Marberger M. Is benign prostatic hyperplasia (BPH) an immune inflammatory disease? Eur Urol. 2007;51:1202–16.
111. Roehrborn CG. Definition of at-risk patients: baseline variables. BJU Int. 2006;97(S2):7–11.
112. Andriole G, Bostwick D, Brawley O, et al. Chemoprevention of prostate cancer in men at high risk: rationale and design of the reduction by dutasteride of prostate cancer events (REDUCE) trial. J Urol. 2004;172:1314–7.
113. Nickel JC, Roehrborn CG, O'Leary MP, et al. The relationship between prostate inflammation and lower urinary tract symptoms: examination of baseline data from the REDUCE trial. J Urol. 2007;177 Suppl 4:34–5.
114. Stock D, Groome PA, Siemens DR. Inflammation and prostate cancer: a future target for prevention and therapy? Urol Clin North Am. 2008;35(1):117–30; vii. Review.
115. Merendino RA, Salvo F, Saija A, et al. Malondialdehyde in benign prostatic hypertrophy: a useful marker? Mediators Inflamm. 2003;12:127–8.
116. Rohrmann S, De Marzo AM, Smit E, et al. Serum C-reactive protein concentration and lower urinary tract symptoms in older men in the Third National Health and Nutrition Examination Survey (NHANES III). Prostate. 2005;52:43–58.
117. Castro P, Xia C, Gomez L, Lamb DJ, Ittmann M. Interleukin-8 expression is increased in senescent prostatic epithelial cells and promotes the development of benign prostatic hyperplasia. Prostate. 2005;60:153–9.
118. Penna G, Mondaini N, Amuchastegui S, et al. Seminal plasma cytokines and chemokines in prostate inflammation: interleukin 8 as a predictive biomarker in chronic prostatitis/chronic pelvic pain syndrome and benign prostatic hyperplasia. Eur Urol. 2007;51:524–33.
119. Sciarra A, Mariotti G, Salciccia S, Gomez AA, et al. Prostate growth and inflammation. J Steroid Biochem Mol Biol. 2008;108:254–60.
120. Narayanan NK, Nargi D, Horton L, et al. Inflammatory processes of prostate tissue microenvironment drive rat prostate carcinogenesis: preventive effects of celecoxib. Prostate. 2009;69:133–41.
121. Mechergui YM, Jemaa AB, Mezigh C, et al. The profile of prostate epithelial cytokines and its impact on sera prostate specific antigen levels. Inflammation. 2009;32(3):202–10.

122. Sandhu JS. Prostate cancer and chronic prostatitis. Curr Urol Rep. 2008;9:328–32.
123. Caruso C, Balistreri CR, Candore G, et al. Polymorphisms of pro-inflammatory genes and prostate cancer risk: a pharmacogenomic approach. Cancer Immunol Immunother. 2009;58(12): 1919–33.
124. Wong CP, Bray TM, Ho E. Induction of proinflammatory response in prostate cancer epithelial cells by activated macrophages. Cancer Lett. 2009;276:38–46.
125. de Marzo AM, Platz EA, Sutcliffe S, et al. Inflammation in prostate carcinogenesis. Nature. 2007;7:256–69.
126. Harsch KM, Tasch JE, Heston WDW. Immunotherapies for prostate cancer. In: Chang C, editor. Prostate cancer: basic mechanisms and therapeutic approaches. Singapore: World Scientific Publishing; 2005. p. 33–54.
127. Penna G, Fibbi B, Amuchastegui S, Cossetti C, Aquilano F, Laverny G, et al. Human benign prostatic hyperplasia stromal cells as inducers and targets of chronic immunomediated inflammation. J Immunol. 2009;182:4056–64.
128. Kogan-Sakin I, Cohen M, Paland N, et al. Prostate stromal cells produce CXCL-1, CXCL-2, CXCL-3 and IL-8 in response to epithelia secreted IL-1. Carcinogenesis. 2009;30(4): 698–705.
129. Penna G, Fibbi B, Amuchastegui S, Corsiero E, et al. The vitamin D receptor agonist elocalcitol inhibits IL-8-dependent benign prostatic hyperplasia stromal cell proliferation and inflammatory response by targeting the RhoA/RhoKinase and NF-kB pathways. Prostate. 2009;69:480–93.
130. Lucia MS, Lambert JR. Growth factors in benign prostatic hyperplasia: basic science implications. Curr Urol Rep. 2008;9:272–8.
131. Begley LA, Kasina S, MacDonald J, et al. The inflammatory microenvironment of the aging prostate facilitates cellular proliferation and hypertrophy. Cytokine. 2008;43:194–9.
132. Konig JE, Senge T, Allhoff EP, et al. Analysis of the inflammatory network in benign prostate hyperplasia and prostate cancer. Prostate. 2004;58:121–9.
133. Steiner G, Stix U, Handisurya A, et al. Cytokine expression pattern in benign prostatic hyperplasia infiltrating T cells and impact of lymphocytic infiltration on cytokine mRNA profile in prostatic tissue. Lab Invest. 2003;83:1131–46.
134. Novara G, Galfano A, Berto RB, et al. Inflammation, apoptosis, and BPH: what is the evidence? Eur Urol Suppl. 2006;5:401–9.
135. Coussens LM, Werb Z. Inflammation and cancer. Nature. 2002;420:860–7.
136. Fernandez F, de Beer PM, van der Merwe L, Heyns CF. COX-2 promoter polymorphisms and the association with prostate cancer risk in South African men. Carcinogenesis. 2008;29(12):2347–50.
137. Sarkar FH, Adsule S, Li Y, Padhye S. Back to the future:COX-2 inhibitors for chemoprevention and cancer therapy. Med Chem. 2007;7:599–608.
138. Roberts RO, Jacobson DJ, Girman CJ, et al. A population-based study of daily nonsteroidal anti-inflammatory drug use and prostate cancer. Mayo Clin Proc. 2002;77:219–25.
139. Pathak SK, Sharma RA, Steward WP, et al. Oxidative stress and cyclooxygenase activity in prostate carcinogenesis: targets for chemopreventive strategies. Eur J Cancer. 2005;41:61–70.
140. Wang W, Bergh A, Damber JE. Chronic inflammation in benign prostate hyperplasia is associated with focal upregulation of cyclooxygenase-2, Bcl-2, and cell proliferation in the glandular epithelium. Prostate. 2004;61:60–72.
141. Untergasser G, Madersbacher S, Berger P. Benign prostatic hyperplasia: age-related tissue-remodelling. Exp Gerontol. 2005;40:121–8.
142. Rigas B, Sun Y. Induction of oxidative stress as a mechanism of action of chemopreventive agents against cancer. Br J Cancer. 2008;98:1157–60.
143. MacLennan GT, Eisenberg R, Fleshman RL, et al. The influence of chronic inflammation in prostatic carcinogenesis: a 5-year followup study. J Urol. 2006;176:1012–6.
144. Khandrika L, Kumar B, Koul S, et al. Oxidative stress in prostate cancer. Cancer Lett. 2009;282:125–36.

145. Goodwin AC, Jadallah S, Toubaji A, et al. Increased spermine oxidase expression in human prostate cancer and prostatic intraepithelial neoplasia tissues. Prostate. 2008;68:766–72.
146. Aydin A, Arsofa-Sarafinovska Z, Sayal A, et al. Oxidative stress and antioxidant in non-metastatic prostate cancer and benign prostatic hyperplasia. Clin Biochem. 2006;39:176–9.
147. Jung K, Seidel B, Rudolph B, et al. Antioxidant enzymes in malignant cell lines and in primary cultured prostatic cells. Free Radic Biol Med. 2007;23(1):127–33.
148. Yilmaz MI, Saglam K, Sonmez A, et al. Antioxidant system activation in prostate cancer. Biol Trace Elem Res. 2000;98:13–9.
149. Fleshner NE, Kucuk O. Antioxidant dietary supplements: rationale and current status as a chemopreventive agents for prostate cancer. Urology. 2001;57(Suppl 4A):90–4.
150. Sikka SC. Role of oxidative stress response elements and antioxidants in prostate cancer pathobiology and chemoprevention—a mechanistic approach. Curr Med Chem. 2003;10:2679–92.
151. Ripple MO, Henry WF, Rago RP, et al. Prooxidant–antioxidant shift induced by androgen treatment of human prostate carcinoma cells. J Natl Cancer Inst. 1997;89:40–8.
152. Vykhovanets EV, Shukla S, MacLennan GT, et al. Il-1b-Induced post-transition effect of NF-Kappa B provides time-dependent wave of signals for initial phase of intrapostatic inflammation. Prostate. 2009;69:633–43.
153. Karan D, Holzbeierlein J, Thrasher JB. Macrophage inhibitory cytokine-1: possible bridge molecule of inflammation and prostate cancer. Cancer Res. 2009;69(1):2–5.
154. Taoka R, Tsukuda F, Ishikawa M, et al. Association of prostatic inflammation with down-regulation of macrophage inhibitory cytokine-1 gene in symptomatic benign prostatic hyperplasia. J Urol. 2004;171:2330–5.
155. De Marzo AM, Coffey DS, Nelson WG. New concepts in tissue specificity for prostate cancer and benign prostatic hyperplasia. Urology. 1999;53(Suppl 3A):29–40.
156. Platz EA, De Marzo AM. Epidemiology of inflammation and prostate cancer. J Urol. 2004;171:S36–40.
157. Klein EA, Silverman R. Inflammation, infection and prostate cancer. Curr Opin Urol. 2008;18:315–9.
158. Kesarwani P, Ahirwar DK, Mandhani A, et al. IL-10-1082G>A: a risk for prostate cancer but may be protective against progression of prostate cancer in North Indian cohort. World J Urol. 2009;27(3):389–96.
159. Licastro F, Bertaccini A, Porcellini E, et al. Alpha 1 antichymotrypsin genotype is associated with increased risk of prostate carcinoma and PSA levels. Anticancer Res. 2008;28:395–400.
160. Sáenz-López P, Carretero R, Cózar JM, et al. Genetic polymorphisms of RANTES, IL1-A, MCP-1 and TNF-A genes in patients with prostate cancer. BMC Cancer. 2008;8:382.
161. Mahmud SM, Tanguay S, Begin LR, et al. Non-steroidal anti-inflammatory drug use and prostate cancer in a high-risk population. Eur J Cancer Prev. 2006;15:158–64.
162. Jacobs EJ, Rodriguez C, Mondul AM, et al. A large cohort study of aspirin and other non-steroidal anti-inflammatory drugs and prostate cancer incidence. J Natl Cancer Inst. 2005;97(13):975–80.
163. Hsu AL, Ching TT, Wang DS, et al. The cyclooxygenase-2 inhibitor celecoxib induces apoptosis by blocking Akt activation in human prostate cancer cells independently of Bcl-2. J Biol Chem. 2000;275(15):11397–403.
164. Handisurya A, Steiner GE, Stix U, Ecker RC, et al. Differential expression of interleukin-15, a pro-inflammatory cytokine and T-cell growth factor, and its receptor in human prostate. Prostate. 2001;49:251–62.
165. Azadzoi K, Pontari M, Vlachiotis J, et al. Canine bladder blood flow and oxygenation: changes induced by filling, contraction and outlet obstruction. J Urol. 1996;155:1459–65.
166. Levin RM, Levin SS, Zhao Y, et al. Cellular and molecular aspects of bladder hypertrophy. Eur Urol. 1997;32 Suppl 1:15–21.
167. Chapple CR. Pharmacological therapy of benign prostatic hyperplasia/lower urinary tract symptoms: an overview for the practising clinician. BJU Int. 2004;94:738–44.
168. Takeda M, Araki I, Kamiyama M, et al. Diagnosis and treatment of voiding symptoms. Urology. 2003;62:11–9.

169. Webber R. Benign prostatic hyperplasia. Clin Evid. 2004;11:1119–38.
170. Klein EA, Casey G, Silverman R. Genetic susceptibility and oxidative stress in prostate cancer: integrated model with implication for prevention. Urology. 2006;68:1145–51.
171. Kooiman GG, Martin FL, Williams JA, et al. The influence of dietary and environmental factors on prostate cancer risk. Prostate Cancer Prostatic Dis. 2000;3:256–8.
172. Pandey M, Gupta S. Green tea and prostate cancer: from bench to clinic. Front Biosci (Elite Ed). 2009;1:13–25.
173. Devrim E, Durak I. Is garlic a promising food for benign prostatic hyperplasia and prostate cancer? Mol Nutr Food Res. 2007;51:1319–23.
174. Chan R, Lok K, Woo J. Prostate cancer and vegetable consumption. Mol Nutr Food Res. 2009;53:201–16.
175. Rohrmann S, Giovannucci E, Willett WC, et al. Fruit and vegetable consumption, intake of micronutrients, and benign prostatic hyperplasia in US men. Am J Clin Nutr. 2007;85: 523–9.
176. Gann PH, Ma J, Giovannucci E, Willett W, et al. Lower prostate cancer risk in men with elevated plasma lycopene levels: results of a prospective analysis. Cancer Res. 1999;59: 1225–30.
177. Peters U, Takata T. Selenium and the prevention of prostate and colorectal cancer. Mol Nutr Food Res. 2008;52:1261–72.
178. Schwarz S, Obermuuller-Jevic OC, Hellmis E, et al. Lycopene inhibits disease progression in patients with benign prostate hyperplasia. J Nutr. 2008;138:49–53.
179. Wertz K, Siler U, Goralczyk R. Lycopene: modes of action to promote prostate health. Arch Biochem Biophys. 2004;430:127–34.
180. Bureyko T, Hurdle H, Metcalfe JB, et al. Reduced growth and integrin expression of prostate cells cultured with lycopene, vitamin E and fish oil in vitro. Br J Nutr. 2009; 101:990–7.
181. Gaziano JM, Glynn RJ, Christen WG, et al. Vitamins E and C in the prevention of prostate and total cancer in men: the physicians' health study II randomized controlled trial. JAMA. 2009;301(1):52–62.
182. Lippman SM, Klein EA, Goodman PJ, et al. Effect of selenium and vitamin E on risk of prostate cancer and other cancers: the selenium and vitamin E cancer prevention trial (SELECT). JAMA. 2009;301(1):39–51.
183. Krishnan AV, Moreno J, Nonn L, Swami S, et al. Calcitriol as a chemopreventive and therapeutic agent in prostate cancer: role of anti-inflammatory activity. J Bone Miner Res. 2007; 22(S2):V74–80.
184. Carraro JC, Raynaud JP, Koch G, Chisholm GD, et al. Comparison of phytotherapy (Permixon) with finasteride in the treatment of benign prostate hyperplasia: a randomized international study of 1098 patients. Prostate. 1996;29:231–40.
185. Levin RM, Das AK. A scientific basis for the therapeutic effects of *Pygeum africanum* and *Serenoa repens*. Urol Res. 2000;28:201–9.
186. Buck AC. Phytotherapy for the prostate. Br J Urol. 1996;78:325–36.
187. Buck AC. Is there a scientific basis for the therapeutic effects of *Serenoa repens* in benign prostatic hyperplasia? Mechanisms of action. J Urol. 2004;172:1792–9.
188. Ookita K, Shiraga T, Matsumura Y. Use of eviprostat for treatment of prostatic hypertrophy. Hinyokika Kiyo. 1966;12:511–7.
189. Kinoshita H, Kawamura N, Miyakita H, et al. The effects of combination therapy eviprostats and tamsulosin hydrochloride (Harnals) for benign prostatic hypertrophy. Hinyouki Geka. 1998;11:1297–305.
190. Boughton-Smith NK, Deakin AM, Follenfant RL, et al. Role of oxygen radicals and arachidonic acid metabolites in the reverse passive Arthus reaction and carrageenin paw oedema in the rat. Br J Pharmacol. 1993;110:896–902.
191. Iida M, Saito K. Role of endotoxin-like contaminants in the apparent anti-inflammatory activity of bovine superoxide dismutase. Inflamm Res. 1996;45:268–71.

Chapter 28
Role of Oxidative Stress in ED: Unraveling the Molecular Mechanism

Biljana Musicki and Arthur L. Burnett

Abstract Many advances in the understanding of erection physiology and pathophysiology have been made in recent years. These advances have revealed the importance of oxidative stress and a complex interaction between oxidative stress and regulatory pathways in the penis in the development and progression of erectile dysfunction (ED) associated with various disease states. In this chapter, we present current knowledge of the pathophysiology of ED pertaining to the mechanisms of reactive oxygen species (ROS) production, the interaction between ROS-generating sources and the main regulatory pathways in the penis, the status of the antioxidant systems that reduce ROS bioavailability, and cellular targets for ROS action in vasculogenic and neurogenic ED. We further discuss a therapeutic strategy to improve erectile function in disease states by targeting specific ROS mechanisms in the penis.

Keywords Oxidative stress • Erectile dysfunction • Molecular mechanism • Penis • Vasculogenic ED • Neurogenic ED • Superoxide • Peroxynitrite • Vasorelaxation • Vasoconstriction

28.1 Introduction

The pathophysiology of erectile dysfunction (ED) refers to the derangement of the normal erectile response, which involves neurological and vascular processes resulting in relaxation of the smooth muscles in the penis and penile erection. It is increasingly evident that oxidative stress, an imbalance between the production and

B. Musicki, PhD (✉) • A.L. Burnett, MD
Department of Urology, Johns Hopkins Hospital, 600 North Wolfe Street, Baltimore, MD 21287, USA
e-mail: bmusicki@jhmi.edu

A. Agarwal et al. (eds.), *Studies on Men's Health and Fertility*, Oxidative Stress in Applied Basic Research and Clinical Practice, DOI 10.1007/978-1-61779-776-7_28,

elimination of reactive oxygen species (ROS), plays an important role in the pathogenesis of vasculogenic and neurogenic ED associated with aging and diseases such as diabetes mellitus, hypercholesterolemia, and atherosclerosis. The objective of this chapter is to provide an understanding of the known mechanisms that link oxidative stress and ED.

28.2 Penile Erection

28.2.1 Physiology of Penile Erection

The penis contains two chambers (the corpora cavernosa), which run the length of the organ, and are surrounded by a membrane (the tunica albuginea). Erection or flaccidity of the penis results from relaxation or contraction, respectively, of smooth muscle cells in the corpora cavernosa. Penile erection is a complex neurovascular process involving relaxation of the corpus cavernosal smooth muscles and vasodilation of blood vessels supplying the corpora cavernosa. This is due to both increased arterial inflow into and restricted venous outflow from the penis. Increased blood flow causes physical expansion of sinusoidal spaces, creating a pressure in the corpora cavernosa and making the penis enlarge [1, 2]. Compression of the dilated blood vessels against the semielastic tunica albuginea restricts venous outflow, producing penile rigidity and sustaining erection. Detumescence occurs as a consequence of increased cavernosal smooth muscle tone and contraction of the sinusoids, reducing arterial inflow [3].

28.2.2 Mechanism of Penile Erection

The main mediator of penile erection is nitric oxide (NO). NO is synthesized by two enzymes: endothelial NO synthase (eNOS; NOS3) and neuronal NO synthase (nNOS; NOS1). nNOS-containing (nitrergic) nerve fibers course from the major pelvic ganglia (MPG) and terminate in the penis. eNOS localizes to the vascular and sinusoidal endothelium in the penis [4].

Sensory reflexogenic and psychogenic sexual stimulation activates nNOS, which initiates the erectile response. Sexual stimulation also neuronally releases acetylcholine, which stimulates eNOS in the endothelium. The resulting increase in blood flow activates eNOS and causes sustained endothelial NO release, accounting for the achievement and maintenance of full erection [1]. Upon its synthesis and release from nerve terminals and endothelial cells, NO diffuses to neighboring vascular and trabecular smooth muscle cells in the penis where it activates soluble guanylate cyclase to produce cyclic guanosine monophosphate (cGMP). NO/cGMP activates downstream targets such as ion channels, leading to the relaxation of the cavernosal

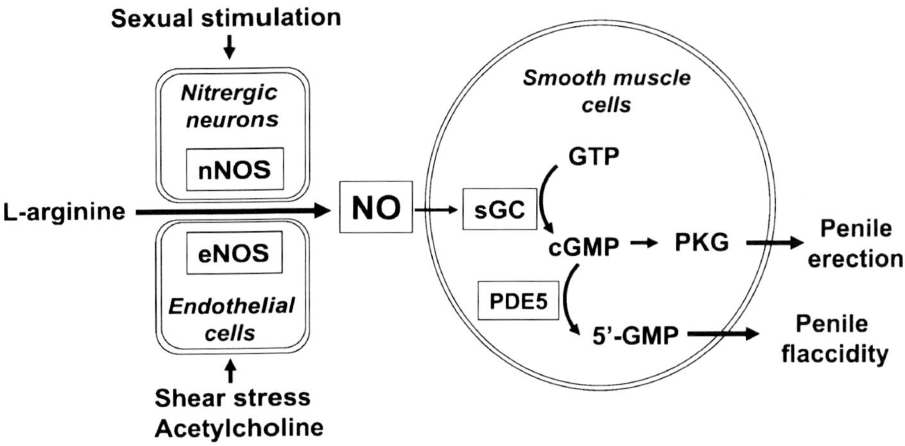

Fig. 28.1 NO-mediated penile erection. NO is the main mediator of penile erection. NO is synthesized from its precursor L-arginine by nNOS in nitrergic nerve terminals in response to a sexual stimulus, and by eNOS in endothelium in response to acetylcholine and shear stress elicited by increased blood flow in the corporeal sinusoids. NO diffuses to adjacent smooth muscle cells where it activates soluble guanylyl cyclase (sGC) and increases the production of 3′,5′-cyclic guanosine monophosphate (cGMP) from 5′-guanosine triphosphate (GTP). Subsequent activation of cGMP-specific protein kinase I (PKG) reduces contractile activity and promotes relaxation of smooth muscle cells and erection. cGMP in the penis is hydrolyzed primarily by type 5 phosphodiesterase (PDE5) to inactive 5′-GMP, which terminates NO signaling and returns the penis to the flaccid state

smooth muscle and vasodilation of blood vessels [5]. It also inhibits contractile regulatory pathways (see below).

cGMP is hydrolyzed by the phosphodiesterases, predominantly type 5 (PDE5), to inactive 5′-GMP, terminating penile erection [6]. PDE5 inhibitors such as sildenafil, vardenafil, and tadalafil inhibit PDE5, thereby augmenting cGMP levels and penile erection (see Fig. 28.1).

28.2.3 Contractile Pathways in the Penis

While activation of the NO pathway leads to penile erection, vasoconstrictive mechanisms maintain the penis in its flaccid state. Vasoconstriction is evoked by several mediators, such as norepinephrine, endothelins, angiotensins, and thromboxane A_2 [7]. Contraction of smooth muscles is mediated by calcium-dependent and calcium-independent pathways. The calcium-dependent pathway involves activation of receptors and/or opening of calcium channels on smooth muscle cells resulting in increase in intracellular levels of calcium. Once the cytosolic calcium returns to the basal levels, the calcium-sensitizing pathway takes over. This pathway causes contraction by increasing calcium sensitivity without changes in cytosolic calcium levels. This pathway involves RhoA, a small protein, which activates Rho-kinase (ROCK [8–10]). Tonic activity of RhoA/ROCK plays a key role in the control of erectile function and maintenance of penile flaccidity.

28.3 Erectile Dysfunction

ED is defined by the National Institutes of Health Consensus Conference on Impotence as the inability to achieve and maintain an erection sufficient to permit satisfactory sexual intercourse [11]. ED is a common worldwide clinical problem. It is estimated that 150 million men currently experience ED to a variable degree and by 2025 the prevalence will rise to over 300 million men [12]. ED occurs from multifaceted, complex mechanisms that can involve disruptions in neural, vascular, and hormonal signaling. The most common molecular features of ED are impaired formation and action of endothelial and neuronal NO and increased oxidative stress.

28.3.1 Vasculogenic ED

ED is predominantly a vascular disease. Vasculogenic ED is characterized by decreased production of vasorelaxant messengers, increased vasoconstriction, and reduced vasodilatory responses of smooth muscle cells [13]. The major mechanisms underlying vasculogenic ED are reduced endothelial NO bioavailability and increased oxidative stress. A variety of conditions that involve vascular abnormalities, such as diabetes, aging, hypercholesterolemia, hypertension, hyperhomocysteinemia, sedentary lifestyle, sickle cell disease (SCD), and cigarette smoking, are associated with the impairment of penile vascular function and vasculogenic ED in men and in a number of animal models ([13]; see Fig. 28.2). An early stage of vascular damage is represented by endothelial dysfunction, which can lead to increasingly severe changes such as atherosclerosis in the systemic vasculature and manifests clinically as coronary, renal, cerebral, and peripheral artery diseases. In fact, vasculogenic ED is a silent marker for cardiovascular and other systemic vascular diseases and carries an independent risk for future cardiovascular events [14].

28.3.2 Neurogenic ED

Neurogenic ED results from defects in neurotransmission of the smooth muscle response of the penis. It may arise due to trauma to nerves of the brain and spinal cord supplying the penis. Neurological conditions include, but are not limited to, Parkinson's disease, Alzheimer's disease, stroke, multiple sclerosis, spinal cord

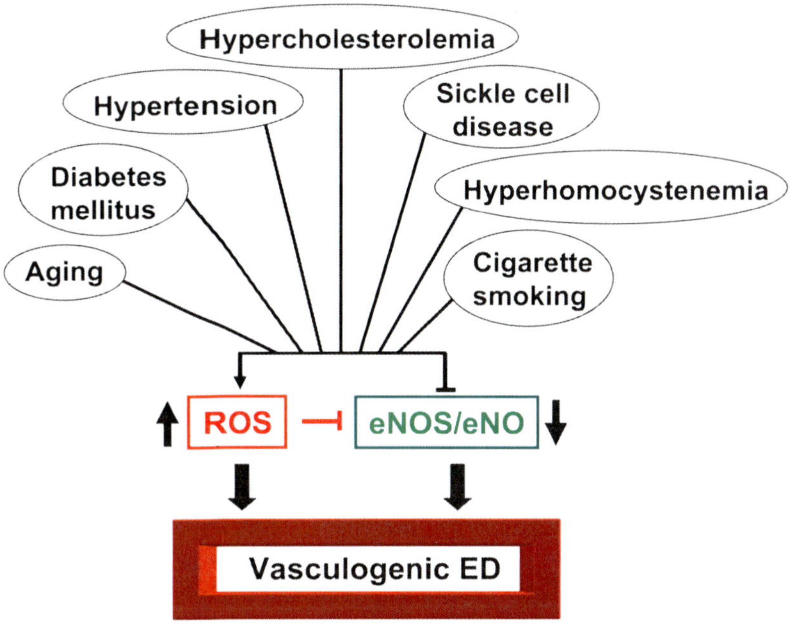

Fig. 28.2 ED is predominantly a vascular disease. A variety of chronic conditions that involve vascular disorders are associated with impairment of penile vascular function and characterize vasculogenic ED. The major mechanisms underlying vasculogenic ED are reduced endothelial NO bioavailability and increased oxidative stress (ROS)

injury, or injury of cavernous nerves as a consequence of surgeries for cancer of the prostate, bladder, and colon. More commonly, neurogenic ED results from the degeneration and loss of nerves associated with chronic diseases such as diabetes. Although the molecular mechanisms underlying neurogenic ED are not well understood, the principal theories include impaired nNOS function and reduced neuronal NO bioavailability, reduced blood supply to nerve tissue, deficiency of neurohormonal growth factors and neurotrophic support, and increased oxidative stress [15].

28.4 Oxidative Stress

ROS include a variety of oxygen containing substances with high reactivity with other biomolecules. ROS include both free radicals (containing one or more unpaired electrons), such as superoxide anion (O_2^-), hydroxyl ($^{\bullet}OH$), peroxyl ($RO_2^{\bullet}$), and hydroperoxyl ($HO_2^{\bullet}$) radicals, and nonradical species that are either easily converted into radicals or are oxidizing agents, such as hydrogen peroxide (H_2O_2) and other peroxides (ROOH). Under physiologic conditions, ROS are produced in a controlled manner and play important roles as secondary messengers in many intracellular signaling pathways [16, 17]. However, excessive production of ROS under pathophysiologic conditions, namely oxidative stress, results in tissue damage.

The principal ROS made by cells is superoxide anion, generated by a single electron donation to molecular oxygen. Superoxide exerts a range of pathophysiologic effects, such as induction of cell proliferation, apoptosis, and vasoconstriction. Superoxide has a relatively short half life and is cell-membrane impermeable; thus, it is unlikely to mediate effects distant from where it is produced. Many other ROS and reactive nitrogen species are formed secondary to reactions involving superoxide. Superoxide is rapidly dismutated to hydrogen peroxide. Hydrogen peroxide is lipid-soluble, has a longer biologic lifespan than superoxide, and crosses cellular membranes [17]. Hydroxyl radical could arise from superoxide and hydrogen peroxide. The hydroxyl radical is a highly reactive oxidant that has a very short half-life.

Oxidative stress is increasingly being recognized to operate as a main pathophysiologic mechanism underlying aging and a large number of diseases, including cancer, neurodegenerative diseases, vascular diseases, and ED.

28.5 Cellular Sources of ROS

The major cellular ROS-generating sources are NADPH oxidase, xanthine oxidase, mitochondrial electron transport, and uncoupled eNOS. ROS sources may interact with each other, and activation of one may augment the activity of others.

28.5.1 Nox Family of NADPH Oxidases

The NADPH oxidases are a family of enzymes that catalyze electron transfer from cytosolic NADPH to molecular oxygen to generate superoxide as its primary product. To date, seven NADPH oxidase isoforms have been identified in different mammalian cells and tissues [18]. The prototype $gp91^{phox}$-containing NADPH oxidase possesses cytosolic subunits ($p47^{phox}$, $p67^{phox}$, or homologues) and membrane-bound subunits ($gp91^{phox}$ and $p22^{phox}$), which form a functional enzyme complex upon activation [16].

NADPH oxidases are activated by diverse stimuli such as angiotensin II, proinflammatory cytokines, vasoconstrictors, hypoxia, growth factors, metabolic factors (hyperglycemia, hyperinsulinemia, free fatty acids, advanced glycation end products [AGEs]), mechanical stimuli, and superoxide itself. The NADPH oxidase-derived ROS have been implicated in aging and a variety of diseases, including hypertension, diabetes mellitus, hypercholesterolemia, hypertension, and SCD [19]. In recent years, emerging evidence suggests a role of NADPH oxidase in several ED states, as discussed later in this chapter.

28.5.2 Xanthine Oxidase

Xanthine oxidase catalyzes the oxidation of hypoxanthine to xanthine and uric acid in a reaction that involves a one-electron transfer to oxygen to yield superoxide. The source of xanthine oxidase is not completely clear, but increased cholesterol levels, liver damage, inflammation, and hypoxia may stimulate the release of the enzyme from the liver and other organs into the circulation. This circulating xanthine oxidase binds to endothelial cells to produce superoxide [20]. Endothelial cells themselves can express xanthine oxidase [21].

Increased xanthine oxidase expression and activity have been implicated in endothelial dysfunction in animal models of SCD, hypertension, diabetes mellitus, and hypercholesterolemia [16], but the role of xanthine oxidase in human cardiovascular diseases is controversial [22]. The role of xanthine oxidase-derived ROS in the penis is not well understood, and more studies are warranted to determine whether it plays a role in ED.

28.5.3 eNOS Uncoupling

Under physiologic conditions, NOS isoforms synthesize NO, which has a key role in vasodilation, antioxidation, and in mediation of vascular homeostasis. Under pathologic conditions, however, NOS isoforms can transform into prooxidants, generating superoxide rather than NO. This switch, which may involve constitutive eNOS and nNOS as well as the inducible NOS isoform (iNOS), is termed NOS uncoupling. NOS uncoupling refers to both the decreased NO-producing activity of the enzyme as well as its increased capacity to produce superoxide [23].

eNOS uncoupling is currently thought to play an important role in aging and cardiovascular diseases including diabetes mellitus, hypertension, hypercholesterolemia, atherosclerosis, cigarette smoking, and SCD [24]. Recent findings demonstrated that eNOS uncoupling in the penis is a significant contributor to oxidative stress and ED, as discussed later in this chapter.

28.5.4 Mitochondrial Electron Transport

Mitochondria generate ROS during normal oxidative phosphorylation and ATP synthesis through electron leak [25]. ROS production by the organelle, however, can be enhanced in response to numerous stimuli or alterations in the cellular environment, resulting in pathologic consequences.

Studies performed by Brownlee and coworker [26] suggest that hyperglycemia-induced oxidative stress from mitochondria is the earliest event in the development of diabetic vascular complications. However, in part because mitochondrial inhibitors are toxic and no experimental animal models are available, the role of excess ROS from mitochondria during in vivo pathologies is poorly understood [27]. The role of mitochondria-derived superoxide in ED has not been investigated.

28.6 Antioxidants

A series of antioxidant enzymes and numerous endogenous and dietary antioxidant compounds maintain defenses against oxidative stress by scavenging ROS. The susceptibility of cells to oxidative stress is the function of the overall balance between the degree of ROS formation and their antioxidant defense capability. Overproduction of ROS depletes enzymatic and nonenzymatic antioxidants leading to additional ROS accumulation and cellular damage [27]. The primary antioxidant enzymes include, but are not limited to, superoxide dismutase (SOD), catalase, and glutathione and thioredoxin peroxidases. The nonenzymatic antioxidants include, among others, vitamin C (ascorbic acid), vitamin E (α-tocopherol), β-carotene, and reduced glutathione [28].

Superoxide is converted spontaneously or by SOD to hydrogen peroxide. SOD exists in three isoforms: cytoplasmic copper-zinc (SOD1), mitochondrial manganese (SOD2), and extracellular copper-zinc (SOD3), the latter being the main vascular SOD. Catalase catalyzes the decomposition of hydrogen peroxide to water and molecular oxygen. Glutathione peroxidases (GPx) are selenium-containing enzymes which reduce hydrogen peroxide to water and lipid peroxides to their corresponding alcohols. Thioredoxin peroxidase (TrPx) is a member of the peroxiredoxin family, which reduces hydrogen peroxide [28].

Numerous nonspecific antioxidants scavenge ROS. Vitamin E is a lipid-soluble antioxidant, which protects LDL against oxidation. Vitamin C is the water-soluble antioxidant, which very effectively scavenges a wide array of ROS, and also prevents oxidation of BH_4, an essential NOS cofactor [22].

28.7 Oxidative Stress in ED

Oxidative stress has been implicated in vasculogenic and neurogenic ED, although the former has been evaluated in much more detail. Superoxide is generated from several vascular sources in the diseased penis, as discussed below. Furthermore, the reaction of superoxide and NO results in the formation of reactive nitrogen species such as the highly toxic molecule peroxynitrite, which causes oxidative damage to DNA, proteins, and lipids. ROS, produced in vascular smooth muscle cells and endothelial cells, may induce vasculogenic ED by scavenging NO or affecting eNOS expression and activity [17, 29], depleting NOS cofactors [30], generating vasoconstrictors, affecting smooth muscle cell integrity, inactivating antioxidants, and causing structural and functional changes within the vasculature [22, 27].

Increased oxidative stress may promote atherosclerotic disease through the oxidation of LDL, the major carrier of plasma cholesterol, with further potentiation of superoxide generation [31]. Several of these pathways have been described in the penis in association with ED, as detailed in further sections.

ROS also affect nitrergic neurotransmission via apoptosis of nitrergic nerves and decreased nNOS signaling, resulting in ED [32–35].

In the following sections, we describe the known sources of ROS and targets of ROS action in the penis in vasculogenic and neurogenic ED associated with aging and major disease states, detailing known molecular mechanisms of ROS production and action associated with each health condition.

28.7.1 Oxidative Stress in Aging-Associated ED

ED is highly associated with aging, with prevalence increasing from 6.5% in men aged 20–39 years to 77.5% in those 75 years and older, and it is projected to affect 322 million men worldwide by 2025 [12, 36].

Oxidative stress is an important factor contributing to ED associated with aging. Endothelial and smooth muscle cells of the aged rat penis produce ROS [37–40]. The sources and mechanisms responsible for ROS formation in the aged penis remain mostly unknown. It has recently been demonstrated that eNOS uncoupling is a source of ROS associated with age-related ED. Supplementation with sepiapterin, which prevents eNOS uncoupling, prevents age-associated ED and prevents increased oxidative stress [40]. Future studies are needed to determine in more detail the contribution and mechanism of eNOS uncoupling, as well as other possible sources of ROS in age-related ED.

Molecular mechanisms underlying decreased neurogenic-mediated corpus cavernosum relaxation associated with aging involve disturbances in the central and peripheral systems of neurotransmission. Central neuropathy involves increased apoptosis and excessive ROS production in the hypothalamic areas involved in the control of penile erection. Peripheral mechanisms have been attributed to a reduction in nitrergic nerve fibers in the penis and decreased nNOS expression and activity [33]. However, the sources of ROS and the mechanisms of ROS-induced impairment of nitrergic neurotransmission in age-associated ED are not known.

28.7.2 Oxidative Stress in Diabetes Mellitus-Associated ED

Diabetes mellitus is one of the major risk factors for ED. It has been estimated that 50–75% of diabetic men have ED [41].

Compelling data from molecular, cellular, and in vivo animal studies implicate a crucial role for oxidative stress in the development and progression of ED associated with diabetes. Both hyperglycemia and free fatty acids augment ROS production. Penile tissue and blood from diabetic men with ED [42–45] and type 1 diabetic animals [46–53] contain high levels of superoxide. Oxidative stress impairs neuronal and endothelial production of NO in the penis, increases cavernosal tissue apoptosis and fibrosis, and induces nerve damage via membrane lipid peroxidation [34, 54–59]. DNA damage and consequent activation of downstream signaling molecules leads to increased synthesis of proinflammatory molecules and inhibition of eNOS activity [52, 53, 60, 61]. All of these derangements contribute to diabetic ED.

The mechanisms for ROS production and the source of ROS in the diabetic penis are, however, only starting to be evaluated. In type 1 diabetic animals, increased protein expression of NADPH oxidase subunit $p47^{phox}$ implies the role of NADPH oxidase as a ROS-producing source [50, 62]. While the role of eNOS uncoupling in diabetic ED is not known, several studies in the diabetic penis [54, 63], and our unpublished studies (Musicki and Burnett, unpublished) indicate the role of eNOS uncoupling as another ROS source. Future studies are needed to establish the mechanism of NADPH upregulation and eNOS uncoupling in diabetes-associated ED.

Although the majority of diabetes related ED develops in men with type 2 diabetes [64], studies examining oxidative stress in animal models of type 2 diabetes are scant. One study demonstrated that type 2 diabetic rats and mice exhibit decreased antioxidant levels in the penis, indicating increased oxidative stress [65].

Diabetes is associated with progressive deterioration of nitrergic neurons in the penis. In type 1 diabetic rats, an early reversible decrease in nNOS content of nitrergic penile nerves is followed in more advanced diabetes by apoptosis of nitrergic nerve cell bodies in the MPG. The latter has been attributed to increased oxidative stress [34, 35].

These studies demonstrate that oxidative stress-mediated neural and vascular alterations play an integral role in diabetes-associated ED.

28.7.3 Oxidative Stress in Hypertension-Associated ED

Approximately 30% of male hypertensive patients have ED [66]. Despite many epidemiologic studies showing the link between hypertension and ED, scientific studies establishing the cellular and molecular mechanisms of hypertension-associated ED are sparse.

Angiotensin II is a potent vasoconstrictor implicated in the development and maintenance of hypertension. Within the vascular wall, angiotensin II acting through the angiotensin I (AT_1) receptor stimulates the production of ROS by activation of NADPH oxidase [67]. The corpus cavernosum of hypertensive rats exhibits increased lipid peroxidation [68–70]. Protein expressions of NADPH oxidase subunits $p47^{phox}$ [71] and $gp91^{phox}$ [70] are upregulated in hypertensive rat penis in parallel with increased oxidative stress and ED. Furthermore, apocynin, an inhibitor of NADPH oxidase, reduces oxidative stress and improves erectile function in hypertensive rats [71], implying a major role for NADPH oxidase in ROS production.

More studies are needed to delineate the mechanisms of ROS upregulation and ROS targets in the penis in hypertension-associated ED.

28.7.4 Oxidative Stress in Hypercholesterolemia-Associated ED

Hypercholesterolemia is a significant risk factor for ED attributed to both vasculogenic and neurogenic factors, although the molecular mechanisms of the latter are largely unknown.

Increased oxidative stress has been postulated to be major molecular factors contributing to hypercholesterolemia-induced vasculogenic ED [72–74]. Corpus cavernosal tissue of cholesterol-fed animals exhibits increased production of ROS [75–79]. In the mouse penis, hypercholesterolemia increases protein expressions of NADPH oxidase subunits $p67^{phox}$, $p47^{phox}$, and $gp91^{phox}$ [76], while inhibition of NADPH oxidase by diphenyleneiodonium chloride and apocynin inhibits ROS production and preserves erectile function [76, 77]. These findings indicate a crucial role for NADPH oxidase as a ROS-producing enzyme in ED associated with hypercholesterolemia. In addition to NADPH oxidase, eNOS uncoupling [75, 76], but not xanthine oxidase [77], also serves as a source of ROS in the penis of experimental hypercholesterolemic animals.

Oxidative modification of LDL has a major role in hypercholesterolemia-associated atherosclerosis development and ED. Oxidized LDL (oxLDL) is present in penile tissues of patients with vasculogenic ED [80] and animal models of hypercholesterolemia-induced ED [75]. In the general vasculature, oxLDL itself increases superoxide generation via the induction of NADPH oxidase, xanthine oxidase, mitochondrial electron transport chain, and uncoupled eNOS [81], but its effect in the penile vasculature is not known.

The effect of hypercholesterolemia on neurogenic neurotransmission remains largely unknown [73]. Future studies are needed to further understand the pathology of both neurogenic and vasculogenic ED associated with hypercholesterolemia and specifically the role of oxidative stress.

28.7.5 Oxidative Stress in ED Associated with Cigarette Smoking

Cigarette smoke is a complex mixture of chemical compounds containing a high concentration of ROS, NO, peroxynitrite, and free radicals of organic compounds. These constituents get into the bloodstream and can directly activate vascular ROS production. In the general vasculature, cigarette smoking (or products of cigarette smoke) increases superoxide generation by both endothelial and smooth muscle cells from NADPH oxidase and uncoupled eNOS, and upregulates proinflammatory cytokines and the RhoA/ROCK contractile pathway. This results in reduced NO bioavailability, increased vasoconstriction, and endothelial dysfunction [82, 83].

Cigarette smoking, both active and passive, is a risk factor for ED [84, 85]. Increasing evidence provided by basic science studies supports the concept that smoking-related ED is associated with reduced bioavailability of NO due to increased oxidative stress. Chronic cigarette smoke exposure impairs neuroregulatory control of penile erection and impairs NO bioavailability [86]. Many of the vascular effects of chronic smoking are attributed to nicotine. In rabbit cavernosal smooth muscle cells, nicotine increases superoxide formation apparently by inducing NADPH oxidase [87].

Epidemiologic evidence suggests that not only active, but passive, exposure to cigarette smoke also significantly predicts incident ED [88, 89]. Chronic passive cigarette smoking decreases NO bioavailability [90], decreases penile eNOS activity, and increases ROS production in penile arterial and corporeal vascular endothelium and smooth muscle cells [91]. These studies demonstrate that oxidative stress and endothelial dysfunction are the fundamental pathophysiological mechanisms linking both active and passive cigarette smoke exposure to ED. However, the exact pathogenic mechanisms underlying cigarette smoking-related ED, the sources of ROS, and cellular targets in the penis are not well understood and require further investigation.

28.7.6 Oxidative Stress in Hyperhomocysteinemia-Associated ED

Mutations in genes responsible for the metabolism of homocysteine, renal insufficiency, or nutritional deficiencies of B vitamins required for homocysteine metabolism, particularly folic acid and vitamins B6 and B12, can result in severe forms of hyperhomocysteinemia [92].

ED correlates with high levels of homocysteine in men [93–95] and experimental animals [96, 97]. Limited studies have been done that delineate the mechanisms of hyperhomocysteinemia-associated ED. In the rabbit corpus cavernosum, hyperhomocysteinemia impairs relaxation through a reduction of endothelial NO bioavailability, and increases cavernosal tissue superoxide production [96, 97]. However, the mechanisms of hyperhomocysteinemia-induced ROS production and sources of ROS are not known.

28.7.7 Oxidative Stress in Sickle Cell Disease-Associated Priapism

SCD is a hemoglobinopathy resulting from the expression of abnormal sickle hemoglobin (HbS), which leads to red blood cell rigidity. Abnormal red blood cells are associated with poor blood flow resulting in tissue hypoxia and ischemia [98]. In addition to hemoglobinopathy and red blood cell sickling, SCD features an independent spectrum of vascular dysfunction, involving abnormalities in NO bioavailability, enhanced responses to vasoconstrictors, and elevated oxidative stress [99, 100].

Priapism is a very common vasculopathy of SCD. This erection disorder consists of nonwillful, excessive, and often recurrent penile erection unrelated to sexual excitement. It afflicts about 40% of men with SCD [101, 102]. Ischemic priapism, the most common form of priapism in which blood flow in the corpora cavernosa is absent, is frequently associated with irreversible penile tissue necrosis, as well as permanent and irreversible ED [102–111].

The sources of ROS in the penis and the mechanism of ROS-mediated pathophysiology associated with priapism remain mostly unknown. One study recently reported increased lipid peroxidation and increased protein oxidation in corporal tissue of rats with priapism and in SCD mice [112]. Unpublished results obtained in our laboratory implicate eNOS uncoupling and NADPH oxidase as sources of ROS in the penis of SCD transgenic mice (Musicki and Burnett, unpublished). Future studies are warranted to examine the sources of ROS and the effect of oxidative stress on downstream signaling in the penis in SCD-associated priapism.

28.8 Endogenous Antioxidants in the Penis in ED

Endothelial and smooth muscle cells of the rat penis contain antioxidants such as SOD3 [37, 46, 113, 114], catalase [115], reduced glutathione [38, 65], and glutathione peroxidase [116]. Cavernosal tissue and blood from men contain vitamin C, albumin, bilirubin, uric acid, and glutathione [42, 117]. Controversial data were, however, obtained regarding the status of endogenous antioxidants in the diseased penis. Penis from aged animals exhibit decreased levels of reduced glutathione but increased cytoplasmic SOD expression [38], although extracellular SOD protein and RNA expressions and SOD activity remain unchanged [37]. The diabetic penis exhibits decreased mRNA expression for catalase [118], while glutathione levels were reportedly decreased [47] or increased [42]. Similarly, while mRNA expression of SOD2 is decreased in the diabetic rat penis [118], SOD activity remains unchanged [46]. The corpus cavernosum of hypertensive rats exhibits decreased SOD activity [68–70], while corpus cavernosal tissue of cholesterol-fed animals exhibits increased SOD activity without changes in the activities of catalase and glutathione peroxidase [119]. These studies demonstrate that information regarding antioxidant status in the penis in association with ED is scarce and inconclusive.

28.9 Therapeutic Strategies to Scavenge ROS in the Penis

The effect of antioxidants on ED was evaluated in several diseased animal models. Gene transfer of SOD in aged animals was found to reduce superoxide anion formation and normalize erectile function [37]. Several antioxidants, such as SOD, vitamin E, ascorbic acid, melatonin, alpha-lipoic acid and gamma-linolenic acid, and peroxynitrite decomposition catalyst, were found to partially or completely improve diabetic vasculopathy and autonomic neuropathy in the penis and erectile function

[46, 48, 49, 52, 118, 120–125]. SOD and catalase [96, 97] reverse hyperhomocysteinemia-induced superoxide production in the penis [96, 97]. More basic science and clinical studies are, however, needed to delineate the effects of antioxidant therapy on oxidative stress in the penis and ED in order to obtain scientific ground which could be exploited for therapeutic purposes. Currently, antioxidants are not considered for treatment or recommendation for therapy for ED.

It is, however, becoming increasingly recognized that removal of ROS in a nontargeted, nonspecific fashion by scavenging is not always efficacious in reversing downstream pathologies, such as cardiovascular pathologies and ED, possibly due to the low bioavailability of the antioxidants at the proper localization and at the optimal time point, and unwanted cessation of physiologic functions regulated by ROS [118, 120, 126]. Several randomized clinical trials demonstrate that long-term supplementation with antioxidants such as vitamin E [127] and vitamin C [128] may not prevent cardiovascular events and may even increase the risk of heart failure in patients with vascular diseases.

28.10 Therapeutic Strategies to Decrease ROS Production in the Penis

Continued research in the area of improving erectile function in disease states is pointing to the greater importance of targeting ROS formation, rather than ROS scavenging, for the development of an effective therapeutic strategy to reduce oxidative damage in the penis and ED. Several pharmacological agents have been suggested to inhibit ROS production and improve erectile function through mechanisms different and beyond their primary therapeutic actions, such as PDE5 inhibitors, angiotensin-converting enzyme (ACE) inhibitors and AT_1 receptor antagonists, statins, and BH_4. This type of therapeutic strategy falls within the first line of recommended treatment of ED. This line of intervention involves preventable lifestyle modifications (such as discontinuation of cigarette smoking, exercise and weight control, Mediterranean-style diet, and a reduction in caloric intake) and treatable noninvasive pharmacologic therapies.

The following sections summarize recent findings regarding strategies to decrease oxidative stress and improve erectile function by targeting specific ROS-generating sources in the penis.

28.10.1 Targeting NADPH Oxidase

28.10.1.1 PDE5 Inhibitors

PDE5 inhibitors, e.g., sildenafil citrate, tadalafil, vardenafil hydrochloride, promote penile erection in response to sexual stimulation by inhibiting PDE5-catalyzed degradation of cGMP. They are widely accepted as an efficacious option for the treatment of ED in most, but not all, cases [6]. Recently, interest has been generated in the additional effects of PDE5 inhibitors on preventing formation of ROS and inhibition of ROS-mediated PDE5 upregulation in the penis.

Several animal studies have demonstrated the beneficial effect of PDE5 inhibitors on reducing oxidative stress in the penis. Sildenafil citrate was found to decrease superoxide production in the penis of a mouse model of secondhand smoke-induced ED [91]. Sildenafil also decreases oxidative stress in rabbit penile vascular smooth muscle after exposure to several ROS-inducing agents such as nicotine [87], TNFα [87], combination of homocysteine and copper [129], and vasoconstrictor thromboxane A_2 mimetic [130]. This effect of sildenafil is apparently due to the inhibition of NADPH oxidase through the reduction in protein expression of its subunit $p47^{phox}$ [130]. Tadalafil also exerts a beneficial acute effect on the cardiovascular system by reducing serum levels of oxidative stress [131] in patients with ED. A vasorelaxant agent given in combination with PDE5 inhibitors proved more beneficial in reducing oxidative stress in the penis and improving penile erection compared to PDE5 inhibitors alone [77, 132]. In cavernosal tissue from hypercholesterolemic rabbits, sildenafil nitrate (NCX 911), an NO donating derivative of sildenafil, inhibits NADPH-dependent superoxide formation in a more efficient way than sildenafil citrate [77]. This effect may be due to antioxidant properties of exogenous NO [133–136]. In addition to NO donors, a hydrogen sulfide-donating derivative of sildenafil, ACS6, also inhibits superoxide formation in cavernosal tissue from hypercholesterolemic rabbits [132] by inhibition of the $p47^{phox}$ subunit of NADPH oxidase and activation of PKA and PKG pathways.

An additional beneficial effect of PDE5 inhibitors in the penis is to prevent upregulation of PDE5 by oxidative stress, which would further contribute to ED by depleting cGMP levels. Oxidative stress has been shown to increase PDE5 expression and activity in the vasculature [137], including that of the penis [87, 129, 138]. Animal studies demonstrated that, in response to increased oxidative stress, sildenafil prevents PDE5 upregulation by inhibition of NADPH oxidase [87, 129, 132]. It was therefore proposed that the therapeutic benefit of PDE5 inhibitors might be mediated, in part, through inhibition of NADPH oxidase-derived oxidative stress and inhibition of PDE5 upregulation, both of which impair normal erectile function.

28.10.1.2 Angiotensin-Converting Enzyme Inhibitors and AT_1 Receptor Antagonists

Angiotensin II promotes endothelial dysfunction by inducing oxidative stress through activation of NADPH oxidase via AT_1 receptor [67]. ACE inhibitors and AT_1 receptor antagonists decrease oxidative stress by preventing the activation of NADPH oxidase and enhancing clearance of ROS [139–141]. Indeed, some (but not all) clinical trials have demonstrated that blocking angiotensin II signaling improves endothelial function in patients with hypertension and metabolic syndrome, and reduces the incidence of death, myocardial infarction, and stroke [22, 142].

Several studies have demonstrated the beneficial effects of antihypertensive treatment with ACE inhibitors or AT_1 receptor antagonists on erectile function in men and animals with ED [143–149]. AT_1 receptor antagonists irbesartan and losartan decrease ROS production and increase NO production in the penis of hypercholesterolemic mice [78] and aged rats [147] independent from a blood pressure lowering effect. These limited data imply that an underlying mechanism for the beneficial effects of inhibition of the renin-angiotensin system on erectile function may be through prevention of ROS generation.

28.10.1.3 Statins (3-Hydroxy-3-Methylglutaryl Coenzyme A Reductase Inhibitors)

Beneficial pleiotropic effects of statins are due to the improvement of endothelial function at least in part by reducing oxidative stress and improving eNOS function independent from their cholesterol-lowering effect. Statins upregulate antioxidants SOD3 and catalase [150, 151] and inhibit endothelial superoxide formation. The latter effect is due to the inhibition of NADPH oxidase [152] and eNOS uncoupling [153].

Basic science studies demonstrate that several statins, such as rosuvastatin and atorvastatin, improve diabetes- [154, 155], metabolic syndrome- [156], and hypertension- [157] related ED and increase the responsiveness to sildenafil by inhibiting RhoA/ROCK signaling in the penis. A limited number of clinical studies which evaluated the effect of statin therapy in men with ED have produced mixed results. Some [158–162], but not all [163, 164], studies demonstrated a beneficial effect of atorvastatin on erectile function in men with and without concomitant sildenafil treatment. It remains to be determined whether this effect of statins is due to a reduction in cholesterol levels or non-lipid-lowering actions. The role of statins in preventing oxidative stress in the diseased penis has not been investigated. Future basic science and clinical studies are warranted to determine whether statins may improve erectile function by decreasing oxidative stress and improving endothelial function in the penis and define the underlying mechanism for this effect.

28.10.2 Targeting eNOS Uncoupling

BH_4, an essential NOS cofactor, and its precursor sepiapterin, are commonly used to prevent eNOS uncoupling and oxidative stress and to prevent the development or progression of cardiovascular diseases in animal models [24]. Several large clinical trials are testing the efficacy of the oral formulation of BH_4 in the treatment of systemic hypertension, peripheral arterial disease, coronary artery disease, pulmonary arterial hypertension, and SCD [23]. In addition, several agents are being tested for their ability to increase BH_4 bioavailability as a more rational therapeutic strategy to improve eNOS coupling than BH_4 supplementation per se. Therapeutic agents currently employed in cardiovascular medicine, such as statin-based drugs [165], erythropoietin [166], folic acid (or its active metabolite 5-methyltetrahydrofolate [167]), insulin [168], ascorbic acid [169], or inhibitors of angiotensin II signaling [170], recouple eNOS by stimulating the binding of BH_4 to NOS, promoting the biosynthesis of BH_4, or protecting BH_4 from oxidation.

Data supporting the improvement in penile erection by targeting eNOS uncoupling in the penis are very scarce. Treatment of diabetic rabbits with folic acid was found to decrease oxidative stress in the cavernosal tissue [50]. Recent basic science findings have demonstrated that pharmacologic supplementation with sepiapterin, a BH_4 precursor, prevented an increase in oxidative stress in the penis and preserved erectile function in aged rats [40]. Further studies are warranted to examine the therapeutic effect of controlling eNOS uncoupling on penile erection in disease states.

28.11 Commercial Antioxidants

Several beverages, such as pomegranate juice, red wine, blueberry juice, cranberry juice, orange juice, and green tea, have been touted to have roles in improving erectile function in disease states via their ROS scavenging capacities. In animal models of arteriogenic ED, pomegranate juice decreased oxidative stress and improved, although it did not normalize, erectile function [171, 172]. This beneficial effect of pomegranate juice may be due to its main active ingredients such as polyphenol antioxidants. The other natural polyphenol, resveratrol, found mostly in grapes and red wine, restores penile function in animal models of hypercholesterolemia [173]- and diabetes [174]-induced ED. Cardiovascular protective effects of resveratrol have been attributed to activation of eNOS and the improvement of endothelial function [175]. However, the exact role and mechanism of action of these commercial antioxidants on oxidative stress in the penis and erectile function await further basic science and clinical investigations.

28.12 Conclusion

Emerging work supports a strong relationship between ED and oxidative stress. Basic science studies show that NADPH oxidase and/or eNOS uncoupling are sources of ROS in the penis in several vascular disease states associated with ED and priapism (see Fig. 28.3) and define several cellular targets of oxidative stress in the penis. However, the sources of ROS, the precise mechanisms of ROS production, the interaction between different ROS generating sources, the interaction between ROS-generating sources and endothelial NO/RhoA/ROCK pathways, and precise cellular targets for ROS in the penis, are still not well characterized in vasculogenic ED. Furthermore, the status of the antioxidant systems that reduce ROS bioavailability in the diseased penis is controversial and needs further clarification.

	ROS sources in the penis	Antioxidants in the penis
Aging	eNOS uncoupling	-,↑ SOD ↓ GSH/GSSG
Diabetes mellitus	NADPH oxidase eNOS uncoupling	-, ↓ SOD ↓ Catalase ↓,↑GSH, - Gpx1
Hypertension	NADPH oxidase	↓ SOD
Hypercholesterolemia	NADPH oxidase eNOS uncoupling	↑ SOD - Catalase - GPx
Cigarette smoking	NADPH oxidase	?
Hyperhomocysteinemia	?	?
Sickle cell disease	NADPH oxidase eNOS uncoupling	?

Fig. 28.3 Summary of the known ROS-generating sources and antioxidants in the penis in ED states. - Indicates no change; ↑ indicates increase; ↓ indicates decrease

The molecular mechanisms of ROS production and their actions in nerves supplying the penis, which lead to neurogenic ED, are poorly defined to date. Continued research in the field should elucidate whether targeting ROS formation, rather than ROS removal, may be more effective as a therapeutic strategy to reduce oxidative damage in the penis and ED. More clinical studies designed to address the role of oxidative stress in ED are warranted to translate findings from basic science to clinical practice and to determine the clinical significance of decreasing oxidative stress in ED.

28.13 Key Points

- Increased oxidative stress has emerged as a major pathophysiologic mechanism underlying the development and progression of both vasculogenic and neurogenic ED.
- NADPH oxidase is a source of ROS in the penis in ED associated with diabetes mellitus, hypertension, hypercholesterolemia, cigarette smoking, and SCD.
- eNOS uncoupling is a source of ROS in the penis in ED associated with aging, diabetes mellitus, hypercholesterolemia, and SCD.

- In vasculogenic ED, the main consequences of increased oxidative stress are decreased endothelial function (NO availability) and increased vasoconstriction.
- In neurogenic ED, the main consequences of increased oxidative stress are dysfunctional neurotransmission and apoptosis of nitrergic nerves.
- Basic science data suggest potential effect of antioxidants on improving erectile function in disease states. However, there is no evidence basis to support recommendations for using antioxidants or agents to reduce ROS production in the treatment of ED.

References

1. Hurt KJ, Musicki B, Palese MA, et al. Akt-dependent phosphorylation of endothelial nitric oxide synthase mediated penile erection. Proc Natl Acad Sci USA. 2002;99:4061–6.
2. Burnett AL. Novel nitric oxide signaling mechanisms regulate the erectile response. Int J Impot Res. 2004;16 Suppl 1:S15–9.
3. Gratzke C, Angulo J, Chitaley K, et al. Anatomy, physiology, and pathophysiology of erectile dysfunction. J Sex Med. 2010;7(1 Pt 2):445–75.
4. Musicki B, Ross AE, Champion HC, et al. Post-translational modification of constitutive nitric oxide synthase in the penis. J Androl. 2009;30:352–62.
5. Burnett AL. The role of nitric oxide in erectile dysfunction: implications for medical therapy. J Clin Hypertens (Greenwich). 2006;8(12 Suppl 4):53–62.
6. Burnett AL. Molecular pharmacotherapeutic targeting of PDE5 for preservation of penile health. J Androl. 2008;29:3–14.
7. Sáenz de Tejada I, Angulo J, Cellek S, et al. Physiology of erectile function. J Sex Med. 2004;1:254–65.
8. Chitaley K, Wingard CJ, Webb RC, et al. Antagonism of Rho-kinase stimulates rat penile erection via a nitric oxide-independent pathway. Nat Med. 2001;7:119–22.
9. Wang H, Eto M, Steers WD, et al. RhoA-mediated Ca^{2+} sensitization in erectile function. J Biol Chem. 2002;277:30614–21.
10. Wingard CJ, Husain S, Williams J, et al. RhoA-Rho kinase mediates synergistic ET-1 and phenylephrine contraction of rat corpus cavernosum. Am J Physiol Regul Integr Comp Physiol. 2003;285:R1145–52.
11. National Institute of Health. Impotence. NIH Consens Statement. 1992;10(4):1–31.
12. Aytac IA, McKinlay JB, Krane RJ. The likely worldwide increase in erectile dysfunction between 1995 and 2025 and some possible policy consequences. BJU Int. 1999;84:50–6.
13. Musicki B, Burnett AL. eNOS function and dysfunction in the penis. Exp Biol Med (Maywood). 2006;231:154–65.
14. Gazzaruso C, Solerte SB, Pujia A, et al. Erectile dysfunction as a predictor of cardiovascular events and death in diabetic patients with angiographically proven asymptomatic coronary artery disease: a potential protective role for statins and 5-phosphodiesterase inhibitors. J Am Coll Cardiol. 2008;51:2040–4.
15. Musicki B, Burnett AL. Pathophysiology of erectile dysfunction: molecular basis. In: Carson C, Kirby R, Goldstein I, editors. Textbook of erectile dysfunction. 4th ed. New York: Informa Healthcare; 2009. p. 72–82.
16. Thomas SR, Witting PK, Drummond GR. Redox control of endothelial function and dysfunction: molecular mechanisms and therapeutic opportunities. Antioxid Redox Signal. 2008;10:1713–65.
17. Elahi MM, Kong YX, Matata BM. Oxidative stress as a mediator of cardiovascular disease. Oxid Med Cell Longev. 2009;2:259–69.

18. Leto TL, Morand S, Hurt D, et al. Targeting and regulation of reactive oxygen species generation by Nox family NADPH oxidases. Antioxid Redox Signal. 2009;11:2607–19.
19. Brandes RP, Weissmann N, Schröder K. NADPH oxidases in cardiovascular disease. Free Radic Biol Med. 2010;49:687–706.
20. Aslan M, Ryan TM, Adler B, et al. Oxygen radical inhibition of nitric oxide-dependent vascular function in sickle cell disease. Proc Natl Acad Sci USA. 2001;98:15215–20.
21. McNally JS, Davis ME, Giddens DP, et al. Role of xanthine oxidoreductase and NAD(P)H oxidase in endothelial superoxide production in response to oscillatory shear stress. Am J Physiol Heart Circ Physiol. 2003;285:H2290–7.
22. Fostermann U. Nitric oxide and oxidative stress in vascular disease. Pflugers Arch. 2010; 459:923–39.
23. Katusic ZS, d'Uscio LV, Nath KA. Vascular protection by tetrahydrobiopterin: progress and therapeutic prospects. Trends Pharmacol Sci. 2009;30:48–54.
24. Li H, Förstermann U. Prevention of atherosclerosis by interference with the vascular nitric oxide system. Curr Pharm Des. 2009;15:3133–45.
25. Stowe DF, Camara AK. Mitochondrial reactive oxygen species production in excitable cells: modulators of mitochondrial and cell function. Antioxid Redox Signal. 2009;11:1373–414.
26. Giacco F, Brownlee M. Oxidative stress and diabetic complications. Circ Res. 2010;107: 1058–70.
27. Cohen RA, Tong X. Vascular oxidative stress: the common link in hypertensive and diabetic vascular disease. J Cardiovasc Pharmacol. 2010;55:308–16.
28. Vassort G, Turan B. Protective role of antioxidants in diabetes-induced cardiac dysfunction. Cardiovasc Toxicol. 2010;10:73–86.
29. Dudzinski DM, Igarashi J, Greif D, et al. The regulation and pharmacology of endothelial nitric oxide synthase. Annu Rev Pharmacol Toxicol. 2006;46:235–76.
30. Cunnington C, Channon KM. Tetrahydrobiopterin: pleiotropic roles in cardiovascular pathophysiology. Heart. 2010;96:1872–7.
31. Pritchard KA, Ackerman AW, Ou J, et al. Native low-density lipoprotein induces endothelial nitric oxide synthase dysfunction: role of heat shock protein 90 and caveolin-1. Free Radic Biol Med. 2002;33:52–62.
32. Gardiner J, Barton D, Overall R, et al. Neurotrophic support and oxidative stress: converging effects in the normal and diseased nervous system. Neuroscientist. 2009;15:47–61.
33. Shi JP, Zhao YM, Song YT. Effect of aging on expression of nitric oxide synthase I and activity of nitric oxide synthase in rat penis. Asian J Androl. 2003;5:117–20.
34. Cellek S, Qu W, Schmidt AM, et al. Synergistic action of advanced glycation end products and endogenous nitric oxide leads to neuronal apoptosis in vitro: a new insight into selective nitrergic neuropathy in diabetes. Diabetologia. 2004;47:331–9.
35. Cellek NA, Foxwell NA, Moncada S. Two phases of nitrergic neuropathy in streptozotocin-induced diabetic rats. Diabetes. 2003;52:2353–62.
36. Jackson G, Rosen RC, Kloner RA, et al. The second Princeton consensus on sexual dysfunction and cardiac risk: new guidelines for sexual medicine. J Sex Med. 2006;3:26–36.
37. Bivalacqua TJ, Armstrong JS, Biggerstaff J, et al. Gene transfer of extracellular SOD to the penis reduces O2-* and improves erectile function in aged rats. Am J Physiol Heart Circ Physiol. 2003;284:H1408–21.
38. Ferrini MG, Davila H, Valente EG, et al. Aging-related induction of inducible nitric oxide synthase is vasculo-protective in the arterial media. Cardiovasc Res. 2004;61:796–805.
39. Ferrini MG, Magee TR, Vernet D, et al. Aging-related expression of inducible nitric oxide synthase and markers of tissue damage in the rat penis. Biol Reprod. 2001;64:974–82.
40. Johnson JM, Bivalacqua TJ, Lagoda GA, et al. eNOS-uncoupling in age-related erectile dysfunction. Int J Impot Res. 2011;23(2):43–8.
41. Musicki B, Burnett AL. Endothelial dysfunction in diabetic erectile dysfunction. Int J Impot Res. 2007;19:129–38.
42. Tuncayengin A, Biri H, Onaran M, et al. Cavernosal tissue nitrite, nitrate, malondialdehyde and glutathione levels in diabetic and non-diabetic erectile dysfunction. Int J Androl. 2003; 26:250–4.

43. Costa C, Soares R, Castela A, et al. Increased endothelial apoptotic cell density in human diabetic erectile tissue-comparison with clinical data. J Sex Med. 2008;6:826–35.
44. Esposito K, Ciotola M, Giugliano F, et al. Phenotypic assessment of endothelial microparticles in diabetic and nondiabetic men with erectile dysfunction. J Sex Med. 2008;5:1436–42.
45. Burnett AL, Strong T, Bivalacqua TJ, et al. Serum biomarker measurements of endothelial function and oxidative stress after daily dosing of sildenafil in type 2 diabetic men with erectile dysfunction. J Urol. 2009;181:245–51.
46. Bivalacqua TJ, Usta MF, Kendirci M, et al. Superoxide anion production in the rat penis impairs erectile function in diabetes: influence of in vivo extracellular superoxide dismutase gene therapy. J Sex Med. 2005;2:187–97.
47. Ryu JK, Kim DJ, Lee T, et al. The role of free radical in the pathogenesis of impotence in streptozotocin-induced diabetic rats. Yonsei Med J. 2003;44:236–41.
48. De Young L, Yu D, Bateman RM, et al. Oxidative stress and antioxidant therapy: their impact in diabetes-associated erectile dysfunction. J Androl. 2004;25:830–6.
49. Paskaloglu K, Sener G, Ayangolu-Dulger G. Melatonin treatment protects against diabetes-induced functional and biochemical changes in rat aorta and corpus cavernosum. Eur J Pharmacol. 2004;499:345–54.
50. Shukla N, Hotston M, Persad R, et al. The administration of folic acid improves erectile function and reduces intracavernosal oxidative stress in the diabetic rabbit. BJU Int. 2009;103: 98–103.
51. Jin HR, Kim WJ, Song JS, et al. Functional and morphologic characterizations of the diabetic mouse corpus cavernosum: comparison of a multiple low-dose and a single high-dose streptozotocin protocols. J Sex Med. 2009;6:3289–304.
52. Angulo J, Peiró C, Cuevas P, et al. The novel antioxidant, AC3056 (2,6-di-t-butyl-4-((dimethyl-4-methoxyphenylsilyl)methyloxy)phenol), reverses erectile dysfunction in diabetic rats and improves NO-mediated responses in penile tissue from diabetic men. J Sex Med. 2009;6:373–87.
53. Wan ZH, Li WZ, Li YZ, et al. Poly(ADP-ribose) polymerase inhibition improves erectile function in diabetic rats. J Sex Med. 2011;8(4):1002–14.
54. Musicki B, Kramer MF, Becker RE, et al. Inactivation of phosphorylated endothelial nitric oxide synthase (Ser-1177) by O-GlcNAc in diabetes-associated erectile dysfunction. Proc Natl Acad Sci USA. 2005;102:11870–5.
55. Angulo J, Cuevas P, Fernández A, et al. Enhanced thromboxane receptor-mediated responses and impaired endothelium-dependent relaxation in human corpus cavernosum from diabetic impotent men: role of protein kinase C activity. J Pharmacol Exp Ther. 2006;319:783–9.
56. Nangle MR, Cotter MA, Cameron NE. Protein kinase C beta inhibition and aorta and corpus cavernosum function in streptozotocin-diabetic mice. Eur J Pharmacol. 2003;475:99–106.
57. Fatehi-Hassanabad Z, Chan CB, Furman BL. Reactive oxygen species and endothelial function in diabetes. Eur J Pharmacol. 2010;636:8–17.
58. Picchi A, Capobianco S, Qiu T, et al. Coronary microvascular dysfunction in diabetes mellitus: a review. World J Cardiol. 2010;25:377–90.
59. Keegan A, Jack AM, Cotter MA, Cameron NE. Effects of aldose reductase inhibition on responses of the corpus cavernosum and mesenteric vascular bed of diabetic rats. J Cardiovasc Pharmacol. 2000;35:606–13.
60. Nangle MR, Cotter MA, Cameron NE. Poly(ADP-ribose) polymerase inhibition reverses nitrergic neurovascular dysfunctions in penile erectile tissue from streptozotocin-diabetic mice. J Sex Med. 2010;7:3396–403.
61. Nangle MR, Cotter MA, Cameron NE. IkappaB kinase 2 inhibition corrects defective nitrergic erectile mechanisms in diabetic mouse corpus cavernosum. Urology. 2006;68:214–8.
62. Jin HR, Kim WJ, Song JS, et al. Intracavernous delivery of a designed angiopoietin-1 variant rescues erectile function by enhancing endothelial regeneration in the streptozotocin-induced diabetic mouse. Diabetes. 2011;60:969–80.
63. Escrig A, Marin R, Abreu P, et al. Changes in mating behavior, erectile function, and nitric oxide levels in penile corpora cavernosa in streptozotocin-diabetic rats. Biol Reprod. 2002; 66:185–9.

64. Chitaley K, Kupelian V, Subak L, et al. Diabetes, obesity and erectile dysfunction: field overview and research priorities. J Urol. 2009;182:S45–50.
65. Kovanecz I, Ferrini MG, Vernet D, et al. Pioglitazone prevents corporal veno-occlusive dysfunction in a rat model of type 2 diabetes mellitus. BJU Int. 2006;98:116–24.
66. Kloner R. Erectile dysfunction and hypertension. Int J Impot Res. 2007;19:296–302.
67. Touyz RM. Intracellular mechanisms involved in vascular remodelling of resistance arteries in hypertension: role of angiotensin II. Exp Physiol. 2005;90:449–55.
68. Ushiyama M, Morita T, Kuramochi T, et al. Erectile dysfunction in hypertensive rats results from impairment of the relaxation evoked by neurogenic carbon monoxide and nitric oxide. Hypertens Res. 2004;27:253–61.
69. Ushiyama M, Kuramochi T, Yagi S, et al. Antioxidant treatment with alpha-tocopherol improves erectile function in hypertensive rats. Hypertens Res. 2008;31:1007–103.
70. Claudino MA, Franco-Penteado CF, Priviero FB, et al. Upregulation of gp91phox subunit of NAD(P)H oxidase contributes to erectile dysfunction caused by long-term nitric oxide inhibition in rats: reversion by regular physical training. Urology. 2010;75:961–7.
71. Jin L, Lagoda G, Leite R, et al. NADPH oxidase activation: a mechanism of hypertension-associated erectile dysfunction. J Sex Med. 2008;5:544–51.
72. Yesilli C, Yaman O, Anafarta K. Effect of experimental hypercholesterolemia on cavernosal structures. Urology. 2001;57:1184–8.
73. Gholami SS, Rogers R, Chang J, et al. The effect of vascular endothelial growth factor and adeno-associated virus mediated brain derived neurotrophic factor on neurogenic and vasculogenic erectile dysfunction induced by hyperlipidemia. J Urol. 2003;169:1577–81.
74. Park K, Kim SW, Rhu KS, et al. Chronic administration of an oral Rho kinase inhibitor prevents the development of vasculogenic erectile dysfunction in a rat model. J Sex Med. 2006;3:996–1003.
75. Musicki B, Liu T, Strong T, et al. Low-fat diet and exercise preserve eNOS regulation and endothelial function in the penis of early atherosclerotic pigs: a molecular analysis. J Sex Med. 2008;5:552–61.
76. Musicki B, Liu T, Lagoda GA, et al. Hypercholesterolemia-induced erectile dysfunction: endothelial nitric oxide synthase (eNOS) uncoupling in the mouse penis by NAD(P)H oxidase. J Sex Med. 2010;7:3023–32.
77. Shukla N, Jones R, Persad R, et al. Effect of sildenafil citrate and a nitric oxide donating sildenafil derivative, NCX 911, on cavernosal relaxation and superoxide formation in hypercholesterolaemic rabbits. Eur J Pharmacol. 2005;517:224–31.
78. Baumhäkel M, Custodis F, Schlimmer N, et al. Improvement of endothelial function of the corpus cavernosum in apolipoprotein E knockout mice treated with irbesartan. J Pharmacol Exp Ther. 2008;327:692–8.
79. Baumhäkel M, Schlimmer N, Büyükafsar K, et al. Nebivolol, but not metoprolol, improves endothelial function of the corpus cavernosum in apolipoprotein e-knockout mice. J Pharmacol Exp Ther. 2008;325:818–23.
80. Zouaoui Boudjeltia K, Roumeguere T, Delree P, et al. Presence of LDL modified by myeloperoxidase in the penis in patients with vascular erectile dysfunction: a preliminary study. Eur Urol. 2007;51:262–8.
81. Schulz E, Anter E, Keaney Jr JF. Oxidative stress, antioxidants, and endothelial function. Curr Med Chem. 2004;11:1093–104.
82. Barbieri SS, Zacchi E, Amadio P, et al. Cytokines present in smokers' serum interact with smoke components to enhance endothelial dysfunction. Cardiovasc Res. 2011;90(3): 475–83.
83. Lowe ER, Everett AC, Lee AJ, et al. Time-dependent inhibition and tetrahydrobiopterin depletion of endothelial nitric-oxide synthase caused by cigarettes. Drug Metab Dispos. 2005;33:131–8.
84. Rosen RC, Fisher WA, Eardley I, et al. The multinational Men's Attitudes to Life Events and Sexuality (MALES) study: I. Prevalence of erectile dysfunction and related health concerns in the general population. Curr Med Res Opin. 2004;20:607–17.

85. Tostes RC, Carneiro FS, Lee AJ, et al. Cigarette smoking and erectile dysfunction: focus on NO bioavailability and ROS generation. J Sex Med. 2008;5:1284–95.
86. Imamura M, Waseda Y, Marinova GV, et al. Alterations of NOS, arginase, and DDAH protein expression in rabbit cavernous tissue after administration of cigarette smoke extract. Am J Physiol Regul Integr Comp Physiol. 2007;293:R2081–9.
87. Hotston MR, Jeremy JY, Bloor J, et al. Sildenafil inhibits the up-regulation of phosphodiesterase type 5 elicited with nicotine and tumour necrosis factor-alpha in cavernosal vascular smooth muscle cells: mediation by superoxide. BJU Int. 2007;99:612–8.
88. Xie Y, Garban H, Ng C, et al. Effect of long-term passive smoking on erectile function and penile nitric oxide synthase in the rat. J Urol. 1997;157:1121–6.
89. Kupelian V, Link CL, McKinlay JB. Association between smoking, passive smoking, and erectile dysfunction: results from the Boston Area Community Health (BACH) survey. Eur Urol. 2007;52:416–22.
90. Göçmez S, Utkan T, Duman C, et al. Secondhand tobacco smoke impairs neurogenic and endothelium-dependent relaxation of rabbit corpus cavernosum smooth muscle: improvement with chronic oral administration of L-arginine. Int J Impot Res. 2005;17:437–44.
91. Bivalacqua TJ, Sussan TE, Gebska MA, et al. Sildenafil inhibits superoxide formation and prevents endothelial dysfunction in a mouse model of secondhand smoke induced erectile dysfunction. J Urol. 2009;181:899–906.
92. Austin RC, Lentz SR, Werstuck GH. Role of hyperhomocysteinemia in endothelial dysfunction and atherothrombotic disease. Cell Death Differ. 2004;11 Suppl 1:S56–64.
93. Demir T, Comlekçi A, Demir O, et al. Hyperhomocysteinemia: a novel risk factor for erectile dysfunction. Metabolism. 2006;55:1564–8.
94. Al-Hunayan A, Thalib L, Kehinde EO, et al. Hyperhomocysteinemia is a risk factor for erectile dysfunction in men with adult-onset diabetes mellitus. Urology. 2008;71:897–900.
95. Lombardo F, Tsamatropoulos P, Piroli E, et al. Treatment of erectile dysfunction due to C677T mutation of the MTHFR gene with vitamin B6 and folic acid in patients non responders to PDE5i. J Sex Med. 2010;7:216–23.
96. Jones RW, Jeremy JY, Koupparis A, et al. Cavernosal dysfunction in a rabbit model of hyperhomocysteinaemia. BJU Int. 2005;95:125–30.
97. Koupparis AJ, Jeremy J, Angelini G, et al. Penicillamine administration reverses the inhibitory effect of hyperhomocysteinaemia on endothelium-dependent relaxation in the corpus cavernosum in the rabbit. BJU Int. 2006;98:440–4.
98. Kato GJ, Hebbel RP, Steinberg MH, et al. Vasculopathy in sickle cell disease: biology, pathophysiology, genetics, translational medicine, and new research directions. Am J Hematol. 2009;84:618–25.
99. Wood KC, Granger DN. Sickle cell disease: role of reactive oxygen and nitrogen metabolites. Clin Exp Pharmacol Physiol. 2007;34:926–32.
100. Wood KC, Hsu LL, Gladwin MT. Sickle cell disease vasculopathy: a state of nitric oxide resistance. Free Radic Biol Med. 2008;44:1506–28.
101. Montague DK, Jarow J, Broderick GA, et al. Members of the erectile dysfunction guideline update panel, American Urological Association. American Urological Association Guideline on the management of priapism. J Urol. 2003;170:1318–24.
102. Burnett AL, Bivalacqua TJ. Priapism: current principles and practice. Urol Clin North Am. 2007;34:631–42.
103. Bivalacqua TJ, Musicki B, Hsu LL, Gladwin MT, Burnett AL, Champion HC. Establishment of a transgenic sickle-cell mouse model to study the pathophysiology of priapism. J Sex Med. 2009;6:2494–504.
104. Broderick GA, Kadioglu A, Bivalacqua TJ, et al. Priapism: pathogenesis, epidemiology, and management. J Sex Med. 2010;7:476–500.
105. Mi T, Abbasi S, Zhang H, et al. Excess adenosine in murine penile erectile tissues contributes to priapism via A2B adenosine receptor signaling. J Clin Invest. 2008;118:1491–501.
106. Wen J, Jiang X, Dai Y, et al. Increased adenosine contributes to penile fibrosis, a dangerous feature of priapism, via A2B adenosine receptor signaling. FASEB J. 2010;24:740–9.

107. Kanika ND, Tar M, Tong Y, et al. The mechanism of opiorphin-induced experimental priapism in rats involves activation of the polyamine synthetic pathway. Am J Physiol Cell Physiol. 2009;297:C916–27.
108. Champion HC, Bivalacqua TJ, Takimoto E, et al. Phosphodiesterase-5A dysregulation in penile erectile tissue is a mechanism of priapism. Proc Natl Acad Sci USA. 2005;102:1661–6.
109. Burnett AL, Bivalacqua TJ, Champion HC, et al. Long-term oral phosphodiesterase 5 inhibitor therapy alleviates recurrent priapism. Urology. 2006;67:1043–8.
110. Burnett AL, Bivalacqua TJ, Champion HC, et al. Feasibility of the use of phosphodiesterase type 5 inhibitors in a pharmacologic prevention program for recurrent priapism. J Sex Med. 2006;3:1077–84.
111. Musicki B, Champion HC, Hsu LL, et al. Posttranslational inactivation of endothelial nitric oxide synthase in the transgenic sickle cell mouse penis. J Sex Med. 2011;8:419–26.
112. Kanika ND, Melman A, Davies KP. Experimental priapism is associated with increased oxidative stress and activation of protein degradation pathways in corporal tissue. Int J Impot Res. 2010;22:363–73.
113. Zhang Q, Radisavljevic ZM, Siroky MB, et al. Dietary antioxidants improve arteriogenic erectile dysfunction. Int J Androl. 2011;34(3):225–35.
114. Azadzoi KM, Golabek T, Radisavljevic ZM. Oxidative stress and neurodegeneration in penile ischaemia. BJU Int. 2010;105:404–10.
115. Uluocak N, Atılgan D, Erdemir F, et al. An animal model of ischemic priapism and the effects of melatonin on antioxidant enzymes and oxidative injury parameters in rat penis. Int Urol Nephrol. 2010;42:889–95.
116. Lagoda G, Jin L, Lehrfeld TJ, et al. FK506 and sildenafil promote erectile function recovery after cavernous nerve injury through antioxidative mechanisms. J Sex Med. 2007;4(4 Pt 1):908–16.
117. Tagliabue M, Pinach S, Di Bisceglie C, et al. Glutathione levels in patients with erectile dysfunction, with or without diabetes mellitus. Int J Androl. 2005;28:156–62.
118. Kawakami T, Urakami S, Hirata H, et al. Superoxide dismutase analog (Tempol: 4-hydroxy-2, 2, 6, 6-tetramethylpiperidine 1-oxyl) treatment restores erectile function in diabetes-induced impotence. Int J Impot Res. 2009;21:348–55.
119. Kim SC, Kim IK, Seo KK, et al. Involvement of superoxide radical in the impaired endothelium-dependent relaxation of cavernous smooth muscle in hypercholesterolemic rabbits. Urol Res. 1997;25:341–6.
120. Goçmen C, Seçilmiş A, Kumcu EK, et al. Effects of vitamin E and sodium selenate on neurogenic and endothelial relaxation of corpus cavernosum in the diabetic mouse. Eur J Pharmacol. 2000;398:93–8.
121. Keegan A, Cotter MA, Cameron NE. Corpus cavernosum dysfunction in diabetic rats: effects of combined alpha-lipoic acid and gamma-linolenic acid treatment. Diabetes Metab Res Rev. 2001;17:380–6.
122. Khan MA, Thompson CS, Jeremy JY, et al. The effect of superoxide dismutase on nitric oxide-mediated and electrical field-stimulated diabetic rabbit cavernosal smooth muscle relaxation. BJU Int. 2001;87:98–103.
123. Nangle MR, Cotter MA, Cameron NE. Effects of the peroxynitrite decomposition catalyst, FeTMPyP, on function of corpus cavernosum from diabetic mice. Eur J Pharmacol. 2004;502:143–8.
124. Gur S, Karahan ST, Ozturk B, et al. Effect of ascorbic acid treatment on endothelium-dependent and neurogenic relaxation of corpus cavernosum from middle-aged non-insulin dependent diabetic rats. Int J Urol. 2005;12:821–8.
125. Hirata H, Kawamoto K, Kikuno N, et al. Restoring erectile function by antioxidant therapy in diabetic rats. J Urol. 2009;182:2518–25.
126. Koh KK, Oh PC, Quon MJ. Does reversal of oxidative stress and inflammation provide vascular protection? Cardiovasc Res. 2009;81:649–59.

127. Lonn E, Bosch J, Yusuf S, et al. HOPE HOPE-TOO Trial Investigators. Effects of long-term vitamin E supplementation on cardiovascular events and cancer: a randomized controlled trial. JAMA. 2005;293:1338–47.
128. Sesso HD, Buring JE, Christen WG, et al. Vitamins E and C in the prevention of cardiovascular disease in men: the Physicians' Health Study II randomized controlled trial. JAMA. 2008;300:2123–33.
129. Hotston M, Jeremy JY, Bloor J, et al. Homocysteine and copper interact to promote type 5 phosphodiesterase expression in cavernosal smooth muscle cells. Asian J Androl. 2008;10: 905–13.
130. Koupparis AJ, Jeremy JY, Muzaffar S, et al. Sildenafil inhibits the formation of superoxide and the expression of gp47 NAD[P]H oxidase induced by the thromboxane A2 mimetic, U46619, in corpus cavernosal smooth muscle cells. BJU Int. 2005;96:423–7.
131. Verit A, Savas M, Ciftci H, et al. Assessment of the acute effects of tadalafil on the cardiovascular system based on examination of serum oxidative status and paraoxonase activity in men with erectile dysfunction: a preliminary study. Int J Impot Res. 2010;22:115–9.
132. Shukla N, Rossoni G, Hotston M, et al. Effect of hydrogen sulphide-donating sildenafil (ACS6) on erectile function and oxidative stress in rabbit isolated corpus cavernosum and in hypertensive rats. BJU Int. 2009;103:1522–9.
133. Selemidis S, Dusting GJ, Peshavariya H, et al. Nitric oxide suppresses NADPH oxidase-dependent superoxide production by S-nitrosylation in human endothelial cells. Cardiovasc Res. 2007;75:349–58.
134. Duerrschmidt N, Stielow C, Muller G, et al. NO-mediated regulation of NAD(P)H oxidase by laminar shear stress in human endothelial cells. J Physiol. 2006;576:557–67.
135. Harrison CB, Drummond GR, Sobey CG, et al. Evidence that nitric oxide inhibits vascular inflammation and superoxide production via a p47phox-dependent mechanism in mice. Clin Exp Pharmacol Physiol. 2010;37:429–34.
136. Muzaffar S, Shukla N, Angelini GD, et al. Nitroaspirins and morpholinosydnonimine, but not aspirin, inhibit the formation of superoxide and the expression of $gp91^{phox}$ induced by endotoxin and cytokines in pig pulmonary artery vascular smooth muscle cells and endothelial cells. Circulation. 2004;110:1140–7.
137. Muzaffar S, Shukla N, Bond M, et al. Superoxide from NADPH oxidase upregulates type 5 phosphodiesterase in human vascular smooth muscle cells: inhibition with iloprost and NONOate. Br J Pharmacol. 2008;155(6):847–56.
138. Hotston M, Jeremy JY, Persad R, et al. 8-isoprostane F2α up-regulates the expression of type 5 phosphodiesterase in cavernosal vascular smooth muscle cells: inhibition with sildenafil, iloprost, nitric oxide and picotamide. BJU Int. 2010;106:1794–8.
139. Hornig B, Landmesser U, Kohler C, et al. Comparative effect of ace inhibition and angiotensin II type 1 receptor antagonism on bioavailability of nitric oxide in patients with coronary artery disease: role of superoxide dismutase. Circulation. 2001;103:799–805.
140. Kuster GM, Nietlispach F, Kiowski W, et al. Role of RAS inhibition in the regulation of Cu/Zn-SOD in the cardiac and peripheral arterial beds in humans. Clin Pharmacol Ther. 2010;87:686–92.
141. Landmesser U, Drexler H. Effect of angiotensin II type 1 receptor antagonism on endothelial function: role of bradykinin and nitric oxide. J Hypertens Suppl. 2006;24:S39–43.
142. Friedrich EB, Teo KK, Böhm M. ACE inhibition in secondary prevention: are the results controversial? Clin Res Cardiol. 2006;95:61–7.
143. Yang R, Yang B, Wen Y, et al. Losartan, an Angiotensin type I receptor, restores erectile function by downregulation of cavernous renin-angiotensin system in streptozocin-induced diabetic rats. J Sex Med. 2009;6:696–707.
144. Fogari R, Zoppi A, Poletti L, et al. Sexual activity in hypertensive men treated with valsartan or carvedilol: a crossover study. Am J Hypertens. 2001;14:27–31.
145. Llisterri JL, Lozano Vidal JV, Aznar Vicente J, et al. Sexual dysfunction in hypertensive patients treated with losartan. Am J Med Sci. 2001;321:336–41.

146. Ferrario CM, Levy P. Sexual dysfunction in patients with hypertension: implications for therapy. J Clin Hypertens (Greenwich). 2002;4:424–32.
147. Park K, Shin JW, Oh JK, et al. Restoration of erectile capacity in normotensive aged rats by modulation of angiotensin receptor type 1. J Androl. 2005;26:123–8.
148. Abdel Aziz MT, El Asmer MF, Mostafa T, et al. Effects of losartan, HO-1 inducers or HO-1 inhibitors on erectile signaling in diabetic rats. J Sex Med. 2009;6:3254–64.
149. Toblli JE, Stella I, Mazza ON, et al. Different effect of losartan and amlodipine on penile structures in male spontaneously hypertensive rats. Am J Nephrol. 2004;24:614–23.
150. Landmesser U, Bahlmann F, Mueller M, et al. Simvastatin versus ezetimibe: pleiotropic and lipid-lowering effects on endothelial function in humans. Circulation. 2005;111:2356–63.
151. Adam O, Laufs U. Antioxidative effects of statins. Arch Toxicol. 2008;82:885–92.
152. Wassmann S, Laufs U, Bäumer AT, et al. HMG-CoA reductase inhibitors improve endothelial dysfunction in normocholesterolemic hypertension via reduced production of reactive oxygen species. Hypertension. 2001;37:1450–7.
153. Wenzel P, Daiber A, Oelze M, et al. Mechanisms underlying recoupling of eNOS by HMG-CoA reductase inhibition in a rat model of streptozotocin-induced diabetes mellitus. Atherosclerosis. 2008;198:65–76.
154. Morelli A, Chavalmane AK, Filippi S, et al. Atorvastatin ameliorates sildenafil-induced penile erections in experimental diabetes by inhibiting diabetes-induced RhoA/Rho-kinase signaling hyperactivation. J Sex Med. 2009;6:91–106.
155. Nangle MR, Cotter MA, Cameron NE. Effects of rosuvastatin on nitric oxide-dependent function in aorta and corpus cavernosum of diabetic mice: relationship to cholesterol biosynthesis pathway inhibition and lipid lowering. Diabetes. 2003;52:2396–402.
156. Wingard CJ, Moukdar F, Prasad RY, et al. Reversal of voltage-dependent erectile responses in the Zucker obese-diabetic rat by rosuvastatin-altered RhoA/Rho-kinase signaling. J Sex Med. 2009;6 Suppl 3:269–78.
157. Fibbi B, Morelli A, Marini M, et al. Atorvastatin but not elocalcitol increases sildenafil responsiveness in spontaneously hypertensive rats by regulating the RhoA/ROCK pathway. J Androl. 2008;29:70–84.
158. Miner M, Billups KL. Erectile dysfunction and dyslipidemia: relevance and role of phosphodiesterase type-5 inhibitors and statins. J Sex Med. 2008;5:1066–78.
159. Herrmann HC, Levine LA, Macaluso Jr J, et al. Can atorvastatin improve the response to sildenafil in men with erectile dysfunction not initially responsive to sildenafil? Hypothesis and pilot trial results. J Sex Med. 2006;3:303–8.
160. Hong SK, Han BK, Jeong SJ, et al. Effect of statin therapy on early return of potency after nerve sparing radical retropubic prostatectomy. J Urol. 2007;178:613–6.
161. Dadkhah F, Safarinejad MR, Asgari MA, et al. Atorvastatin improves the response to sildenafil in hypercholesterolemic men with erectile dysfunction not initially responsive to sildenafil. Int J Impot Res. 2010;22:51–60.
162. Saltzman EA, Guay AT, Jacobson J. Improvement in erectile function in men with organic erectile dysfunction by correction of elevated cholesterol levels: a clinical observation. J Urol. 2004;172:255–8.
163. Rizvi K, Hapson JP, Harvey JN. Do lipid-lowering drugs cause erectile dysfunction? A systematic review. Fam Pract. 2002;567:95–8.
164. Solomon H, Samarasinghe YP, Feher MD, et al. Erectile dysfunction and statin treatment in high cardiovascular risk patients. Int J Clin Pract. 2006;60:141–5.
165. Shinozaki K, Nishio Y, Ayajiki K, et al. Pitavastatin restores vascular dysfunction in insulin-resistant state by inhibiting NAD(P)H oxidase activity and uncoupled endothelial nitric oxide synthase-dependent superoxide production. J Cardiovasc Pharmacol. 2007;49:122–30.
166. d'Uscio LV, Katusic ZS. Erythropoietin increases endothelial biosynthesis of tetrahydrobiopterin by activation of protein kinase B{alpha}/Akt1. Hypertension. 2008;52:93–9.
167. Antoniades C, Shirodaria C, Warrick N, et al. 5-Methyltetrahydrofolate rapidly improves endothelial function and decreases superoxide production in human vessels: effects on

vascular tetrahydrobiopterin availability and endothelial nitric oxide synthase coupling. Circulation. 2006;114:1193–201.
168. Ishii M, Shimizu S, Nagai T, et al. Stimulation of tetrahydrobiopterin synthesis induced by insulin: possible involvement of phosphatidylinositol 3-kinase. Int J Biochem Cell Biol. 2001;33:65–73.
169. Heller R, Unbehaun A, Schellenberg B, et al. L-ascorbic acid potentiates endothelial nitric oxide synthesis via a chemical stabilization of tetrahydrobiopterin. J Biol Chem. 2001;276: 40–7.
170. Oak JH, Cai H. Attenuation of angiotensin II signaling recouples eNOS and inhibits nonendothelial NOX activity in diabetic mice. Diabetes. 2007;56:118–26.
171. Azadzoi KM, Schulman RN, Aviram M, et al. Oxidative stress in arteriogenic erectile dysfunction: prophylactic role of antioxidants. J Urol. 2005;174:386–93.
172. Zhang Q, Radisavljevic ZM, Siroky MB, et al. Dietary antioxidants improve arteriogenic erectile dysfunction. Int J Androl. 2011;34:225–35.
173. Soner BC, Murat N, Demir O, et al. Evaluation of vascular smooth muscle and corpus cavernosum on hypercholesterolemia. Is resveratrol promising on erectile dysfunction? Int J Impot Res. 2010;22:227–33.
174. Fukuhara S, Tsujimura A, Okuda H, et al. Vardenafil and resveratrol synergistically enhance the nitric oxide/cyclic guanosine monophosphate pathway in corpus cavernosal smooth muscle cells and its therapeutic potential for erectile dysfunction in the streptozotocin-induced diabetic rat: preliminary findings. J Sex Med. 2011;8:1061–71.
175. Wallerath T, Deckert G, Ternes T, et al. Resveratrol, a polyphenolic phytoalexin present in red wine, enhances expression and activity of endothelial nitric oxide synthase. Circulation. 2002;106:1652–8.

About the Authors

Ashok Agarwal is the Director of the Clinical Andrology Laboratory and Reproductive Tissue Bank, and the Director of Research at the Center for Reproductive Medicine since 1993. He holds these positions at The Cleveland Clinic Foundation, where he is a Professor at the Lerner College of Medicine of Case Western Reserve University. Ashok received his PhD in 1983 and did his post-doctoral training in Reproductive Biology under a fellowship from The Rockefeller Foundation at Harvard Medical School, Boston. He was an Instructor in Surgery and then an Assistant Professor of Urology at Harvard Medical School from 1988 to 1993. Ashok has published over 500 scientific papers and reviews in peer-reviewed scientific journals. He is currently an editor of ten medical text books/ manuals related to male infertility, ART, fertility preservation, DNA damage, and antioxidants. Ashok is active in basic and clinical research, and his laboratory has

A. Agarwal et al. (eds.), *Studies on Men's Health and Fertility*, Oxidative Stress in Applied Basic Research and Clinical Practice, DOI 10.1007/978-1-61779-776-7,

trained over 150 basic scientists and clinical researchers from the United States and abroad. In addition, more than 150 medical, undergraduate, and high school students have worked in his laboratory. His current research interests include studies on molecular markers of oxidative stress, DNA integrity, and apoptosis in the pathophysiology of male and female reproduction, effect of radio frequency radiation on fertility and fertility preservation in patients with cancer.

R. John Aitken's research career began with a PhD in reproductive biology from the University of Cambridge. Following postdoctoral positions at the Institute of Animal Genetics, University of Edinburgh and the University of Bordeaux, he joined the World Health Organization in Geneva, where he managed two independent WHO task forces dealing with different approaches to fertility regulation. In 1977, he joined the Medical Research Council's Reproductive Biology Unit, University of Edinburgh, to establish a research group in gamete biology with clinical outreach into male infertility. In 1992, John was awarded an Honorary Professorship within the Faculty of Medicine, University of Edinburgh, and in 1995 was elected a Fellow of the Royal Society of Edinburgh. In 1998 he received an ScD degree from the University of Cambridge in recognition of his contribution to the field of gamete biology. In the same year, he moved to the University of Newcastle, NSW, as Chair of Biological Sciences and, later, Director of the ARC Centre of Excellence in Biotechnology and Development. In 2011 he was elected to the Australian Academy of Science and is currently Laureate Professor of Biological Sciences and Director of the Priority Research Centre in Reproductive Science at the University of Newcastle.

Professor Juan G. Alvarez received his Medical Doctor degree at the University of Santiago of Compostela in Spain, has a PhD in Physical Chemistry from the University of Villanova, Pennsylvania, and was Associate Professor of Obstetrics and Gynecology and Reproductive Biology at Harvard Medical School. During over 3 decades he has developed his scientific career in world leading research centers such as the Reproductive Biology Unit at the University of Pennsylvania and Harvard Medical School, where he was Director of Research at the Department of Obstetrics and Gynecology and Codirector of the Women's Health Research Institute at Beth Israel Deaconess Medical Center in Boston. He was also Scientific Director of Institut Marques in Barcelona, one of the leading IVF centers in Spain. He carried out basic and clinical research in the field of infertility with particular emphasis on oxygen radical chemistry and biochemistry in mammalian spermatozoa and free radical-induced sperm DNA damage in human spermatozoa. Currently, he is Director of Centro ANDROGEN, one of the leading male infertility clinics in Spain. He combines his clinical activities with research in the area of male infertility. He collaborates with the Cleveland Clinic Glickman Urological Institute, Harvard Medical School, Yale University, and the Massachusetts Institute of Technology (MIT).

Index

N

P

S

T

X

Printed by Publishers' Graphics LLC USA
SO20120323-052
2012